Smart Innovation, Systems and Technologies

Volume 33

About this Series

The Smart Innovation, Systems and Technologies book series encompasses the topics of knowledge, intelligence, innovation and sustainability. The aim of the series is to make available a platform for the publication of books on all aspects of single and multi-disciplinary research on these themes in order to make the latest results available in a readily-accessible form. Volumes on interdisciplinary research combining two or more of these areas is particularly sought.

The series covers systems and paradigms that employ knowledge and intelligence in a broad sense. Its scope is systems having embedded knowledge and intelligence, which may be applied to the solution of world problems in industry, the environment and the community. It also focusses on the knowledge-transfer methodologies and innovation strategies employed to make this happen effectively. The combination of intelligent systems tools and a broad range of applications introduces a need for a synergy of disciplines from science, technology, business and the humanities. The series will include conference proceedings, edited collections, monographs, handbooks, reference books, and other relevant types of book in areas of science and technology where smart systems and technologies can offer innovative solutions.

High quality content is an essential feature for all book proposals accepted for the series. It is expected that editors of all accepted volumes will ensure that contributions are subjected to an appropriate level of reviewing process and adhere to KES quality principles.

More information about this series at http://www.springer.com/series/8767

Lakhmi C. Jain · Himansu Sekhar Behera
Jyotsna Kumar Mandal
Durga Prasad Mohapatra
Editors

Computational Intelligence in Data Mining - Volume 3

Proceedings of the International Conference on CIDM, 20-21 December 2014

Editors
Lakhmi C. Jain
University of Canberra
Canberra
Australia

and

University of South Australia
Adelaide, SA
Australia

Himansu Sekhar Behera
Department of Computer Science and Engineering
Veer Surendra Sai University of Technology
Sambalpur, Odisha
India

Jyotsna Kumar Mandal
Department of Computer Science and Engineering
Kalyani University
Nadia, West Bengal
India

Durga Prasad Mohapatra
Department of Computer Science and Engineering
National Institute of Technology Rourkela
Rourkela
India

ISSN 2190-3018 ISSN 2190-3026 (electronic)
Smart Innovation, Systems and Technologies
ISBN 978-81-322-3568-2 ISBN 978-81-322-2202-6 (eBook)
DOI 10.1007/978-81-322-2202-6

Springer New Delhi Heidelberg New York Dordrecht London

Softcover reprint of the hardcover 1st edition 2015

Printed on acid-free paper

Springer (India) Pvt. Ltd. is part of Springer Science+Business Media (www.springer.com)

Preface

The First International Conference on "Computational Intelligence in Data Mining (ICCIDM-2014)" was hosted and organized jointly by the Department of Computer Science and Engineering, Information Technology and MCA, Veer Surendra Sai University of Technology, Burla, Sambalpur, Odisha, India between 20 and 21 December 2014. ICCIDM is an international interdisciplinary conference covering research and developments in the fields of Data Mining, Computational Intelligence, Soft Computing, Machine Learning, Fuzzy Logic, and a lot more. More than 550 prospective authors had submitted their research papers to the conference. ICCIDM selected 192 papers after a double blind peer review process by experienced subject expertise reviewers chosen from the country and abroad. The proceedings of ICCIDM is a nice collection of interdisciplinary papers concerned in various prolific research areas of Data Mining and Computational Intelligence. It has been an honor for us to have the chance to edit the proceedings. We have enjoyed considerably working in cooperation with the International Advisory, Program, and Technical Committees to call for papers, review papers, and finalize papers to be included in the proceedings.

This International Conference ICCIDM aims at encompassing a new breed of engineers, technologists making it a crest of global success. It will also educate the youth to move ahead for inventing something that will lead to great success. This year's program includes an exciting collection of contributions resulting from a successful call for papers. The selected papers have been divided into thematic areas including both review and research papers which highlight the current focus of Computational Intelligence Techniques in Data Mining. The conference aims at creating a forum for further discussion for an integrated information field incorporating a series of technical issues in the frontier analysis and design aspects of different alliances in the related field of Intelligent computing and others. Therefore the call for paper was on three major themes like Methods, Algorithms, and Models in Data mining and Machine learning, Advance Computing and Applications. Further, papers discussing the issues and applications related to the theme of the conference were also welcomed at ICCIDM.

The proceedings of ICCIDM have been released to mark this great day in ICCIDM which is a collection of ideas and perspectives on different issues and some new thoughts on various fields of Intelligent Computing. We hope the author's own research and opinions add value to it. First and foremost are the authors of papers, columns, and editorials whose works have made the conference a great success. We had a great time putting together this proceedings. The ICCIDM conference and proceedings are a credit to a large group of people and everyone should be there for the outcome. We extend our deep sense of gratitude to all for their warm encouragement, inspiration, and continuous support for making it possible.

Hope all of us will appreciate the good contributions made and justify our efforts.

Acknowledgments

The theme and relevance of ICCIDM has attracted more than 550 researchers/academicians around the globe, which enabled us to select good quality papers and serve to demonstrate the popularity of the ICCIDM conference for sharing ideas and research findings with truly national and international communities. Thanks to all who have contributed in producing such a comprehensive conference proceedings of ICCIDM.

The organizing committee believes and trusts that we have been true to the spirit of collegiality that members of ICCIDM value, even as maintaining an elevated standard as we have reviewed papers, provided feedback, and present a strong body of published work in this collection of proceedings. Thanks to all the members of the Organizing committee for their heartfelt support and cooperation.

It has been an honor for us to edit the proceedings. We have enjoyed considerably working in cooperation with the International Advisory, Program, and Technical Committees to call for papers, review papers, and finalize papers to be included in the proceedings.

We express our sincere thanks and obligations to the benign reviewers for sparing their valuable time and effort in reviewing the papers along with suggestions and appreciation in improvising the presentation, quality, and content of this proceedings. Without this commitment it would not be possible to have the important reviewer status assigned to papers in the proceedings. The eminence of these papers is an accolade to the authors and also to the reviewers who have guided for indispensable perfection.

We would like to gratefully acknowledge the enthusiastic guidance and continuous support of Prof. (Dr.) Lakhmi Jain, as and when it was needed as well as adjudicating on those difficult decisions in the preparation of the proceedings and impetus to our efforts to publish this proceeding.

Last but not the least, the editorial members of Springer Publishing deserve a special mention and our sincere thanks to them not only for making our dream come true in the shape of this proceedings, but also for its brilliant get-up and in-time publication in Smart, Innovation, System and Technologies, Springer.

I feel honored to express my deep sense of gratitude to all members of International Advisory Committee, Technical Committee, Program Committee, Organizing Committee, and Editorial Committee members of ICCIDM for their unconditional support and cooperation.

The ICCIDM conference and proceedings are a credit to a large group of people and everyone should be proud of the outcome.

Himansu Sekhar Behera

About the Conference

The International Conference on "Computational Intelligence in Data Mining" (ICCIDM-2014) has been established itself as one of the leading and prestigious conference which will facilitate cross-cooperation across the diverse regional research communities within India as well as with other International regional research programs and partners. Such an active dialogue and discussion among International and National research communities is required to address many new trends and challenges and applications of Computational Intelligence in the field of Science, Engineering and Technology. ICCIDM 2014 is endowed with an opportune forum and a vibrant platform for researchers, academicians, scientists, and practitioners to share their original research findings and practical development experiences on the new challenges and budding confronting issues.

The conference aims to:

- Provide an insight into current strength and weaknesses of current applications as well as research findings of both Computational Intelligence and Data Mining.
- Improve the exchange of ideas and coherence between the various Computational Intelligence Methods.
- Enhance the relevance and exploitation of data mining application areas for end-user as well as novice user application.
- Bridge research with practice that will lead to a fruitful platform for the development of Computational Intelligence in Data mining for researchers and practitioners.
- Promote novel high quality research findings and innovative solutions to the challenging problems in Intelligent Computing.
- Make a tangible contribution to some innovative findings in the field of data mining.
- Provide research recommendations for future assessment reports.

So, we hope the participants will gain new perspectives and views on current research topics from leading scientists, researchers, and academicians around the world, contribute their own ideas on important research topics like Data Mining and Computational Intelligence, as well as network and collaborate with their international counterparts.

Conference Committee

Patron
Prof. E. Saibaba Reddy
Vice Chancellor, VSSUT, Burla, Odisha, India

Convenor
Dr. H.S. Behera
Department of CSE and IT, VSSUT, Burla, Odisha, India

Co-Convenor
Dr. M.R. Kabat
Department of CSE and IT, VSSUT, Burla, Odisha, India

Organizing Secretary
Mr. Janmenjoy Nayak, DST INSPIRE Fellow, Government of India

Conference Chairs

Honorary General Chair
Prof. P K. Dash, Ph.D., D.Sc., FNAE, SMIEEE, Director
Multi Disciplinary Research Center, S 'O'A University, India
Prof. Lakhmi C. Jain, Ph.D., M.E., B.E.(Hons), Fellow (Engineers Australia), University of Canberra, Canberra, Australia and University of South Australia, Adelaide, SA, Australia

Honorary Advisory Chair
Prof. Shankar K. Pal, Distinguished Professor
Indian Statistical Institute, Kolkata, India

General Chair
Prof. Rajib Mall, Ph.D., Professor and Head
Department of Computer Science and Engineering, IIT Kharagpur, India

Program Chairs
Dr. Sukumar Mishra, Ph.D., Professor
Department of EE, IIT Delhi, India
Dr. R.P. Panda, Ph.D., Professor
Department of ETC, VSSUT, Burla, Odisha, India
Dr. J.K. Mandal, Ph.D., Professor
Department of CSE, University of Kalyani, Kolkata, India

Finance Chair
Dr. D. Dhupal, Coordinator, TEQIP, VSSUT, Burla

Volume Editors
Prof. Lakhmi C. Jain, Ph.D., M.E., B.E.(Hons), Fellow (Engineers Australia), University of Canberra, Canberra, Australia and University of South Australia, Adelaide, SA, Australia
Prof. H.S. Behera, Reader, Department of Computer Science Engineering and Information Technology, Veer Surendra Sai University of Technology, Burla, Odisha, India
Prof. J.K. Mandal, Professor, Department of Computer Science and Engineering, University of Kalyani, Kolkata, India
Prof. D.P. Mohapatra, Associate Professor, Department of Computer Science and Engineering, NIT, Rourkela, Odisa, India

International Advisory Committee

Prof. C.R. Tripathy (VC, Sambalpur University)
Prof. B.B. Pati (VSSUT, Burla)
Prof. A.N. Nayak (VSSUT, Burla)
Prof. S. Yordanova (STU, Bulgaria)
Prof. P. Mohapatra (University of California)
Prof. S. Naik (University of Waterloo, Canada)
Prof. S. Bhattacharjee (NIT, Surat)
Prof. G. Saniel (NIT, Durgapur)
Prof. K.K. Bharadwaj (JNU, New Delhi)
Prof. Richard Le (Latrob University, Australia)
Prof. K.K. Shukla (IIT, BHU)
Prof. G.K. Nayak (IIIT, BBSR)
Prof. S. Sakhya (TU, Nepal)
Prof. A.P. Mathur (SUTD, Singapore)
Prof. P. Sanyal (WBUT, Kolkata)
Prof. Yew-Soon Ong (NTU, Singapore)
Prof. S. Mahesan (Japfna University, Srilanka)
Prof. B. Satapathy (SU, SBP)

Prof. G. Chakraborty (IPU, Japan)
Prof. T.S. Teck (NU, Singapore)
Prof. P. Mitra (P.S. University, USA)
Prof. A. Konar (Jadavpur University)
Prof. S. Das (Galgotias University)
Prof. A. Ramanan (UJ, Srilanka)
Prof. Sudipta Mohapatra, (IIT, KGP)
Prof. P. Bhattacharya (NIT, Agaratala)
Prof. N. Chaki (University of Calcutta)
Dr. J.R. Mohanty, (Registrar, VSSUT, Burla)
Prof. M.N. Favorskaya (SibSAU)
Mr. D. Minz (COF, VSSUT, Burla)
Prof. L.M. Patnaik (DIAT, Pune)
Prof. G. Panda (IIT, BHU)
Prof. S.K. Jena (NIT, RKL)
Prof. V.E. Balas (University of Arad)
Prof. R. Kotagiri (University of Melbourne)
Prof. B.B. Biswal (NIT, RKL)
Prof. Amit Das (BESU, Kolkata)
Prof. P.K. Patra (CET, BBSR)
Prof. N.G.P.C Mahalik (California)
Prof. D.K. Pratihar (IIT, KGP)
Prof. A. Ghosh (ISI, Kolkata)
Prof. P. Mitra (IIT, KGP)
Prof. P.P. Das (IIT, KGP)
Prof. M.S. Rao (JNTU, HYD)
Prof. A. Damodaram (JNTU, HYD)
Prof. M. Dash (NTU, Singapore)
Prof. I.S. Dhillon (University of Texas)
Prof. S. Biswas (IIT, Bombay)
Prof. S. Pattnayak (S'O'A, BBSR)
Prof. M. Biswal (IIT, KGP)
Prof. Tony Clark (M.S.U, UK)
Prof. Sanjib ku. Panda (NUS)
Prof. G.C. Nandy (IIIT, Allahabad)
Prof. R.C. Hansdah (IISC, Bangalore)
Prof. S.K. Basu (BHU, India)
Prof. P.K. Jana (ISM, Dhanbad)
Prof. P.P. Choudhury (ISI, Kolkata)
Prof. H. Pattnayak (KIIT, BBSR)
Prof. P. Srinivasa Rao (AU, Andhra)

International Technical Committee

Dr. Ranjan Behera, Ph.D., Assistant Professor
Department of EE, IIT, Patna, India
Dr. A.K. Barisal, Ph.D., Reader
Department of EE, VSSUT, Burla, Odisha, India
Dr. R. Mohanty, Reader
Department of CSE, VSSUT, Burla

Conference Steering Committee

Publicity Chair
Prof. A. Rath, DRIEMS, Cuttack
Prof. B. Naik, VSSUT, Burla
Mr. Sambit Bakshi, NIT, RKL

Logistic Chair
Prof. S.P. Sahoo, VSSUT, Burla
Prof. S.K. Nayak, VSSUT, Burla
Prof. D.C. Rao, VSSUT, Burla
Prof. K.K. Sahu, VSSUT, Burla

Organizing Committee
Prof. D. Mishra, VSSUT, Burla
Prof. J. Rana, VSSUT, Burla
Prof. P.K. Pradhan, VSSUT, Burla
Prof. P.C. Swain, VSSUT, Burla
Prof. P.K. Modi, VSSUT, Burla
Prof. S.K. Swain, VSSUT, Burla
Prof. P.K. Das, VSSUT, Burla
Prof. P.R. Dash, VSSUT, Burla
Prof. P.K. Kar, VSSUT, Burla
Prof. U.R. Jena, VSSUT, Burla
Prof. S.S. Das, VSSUT, Burla
Prof. Sukalyan Dash, VSSUT, Burla
Prof. D. Mishra, VSSUT, Burla
Prof. S. Aggrawal, VSSUT, Burla
Prof. R.K. Sahu, VSSUT, Burla
Prof. M. Tripathy, VSSUT, Burla
Prof. K. Sethi, VSSUT, Burla
Prof. B.B. Mangaraj, VSSUT, Burla
Prof. M.R. Pradhan, VSSUT, Burla
Prof. S.K. Sarangi, VSSUT, Burla
Prof. N. Bhoi, VSSUT, Burla
Prof. J.R. Mohanty, VSSUT, Burla

Prof. Sumanta Panda, VSSUT, Burla
Prof. A.K. Pattnaik, VSSUT, Burla
Prof. S. Panigrahi, VSSUT, Burla
Prof. S. Behera, VSSUT, Burla
Prof. M.K. Jena, VSSUT, Burla
Prof. S. Acharya, VSSUT, Burla
Prof. S. Kissan, VSSUT, Burla
Prof. S. Sathua, VSSUT, Burla
Prof. E. Oram, VSSUT, Burla
Dr. M.K. Patel, VSSUT, Burla
Mr. N.K.S. Behera, M.Tech. Scholar
Mr. T. Das, M.Tech. Scholar
Mr. S.R. Sahu, M.Tech. Scholar
Mr. M.K. Sahoo, M.Tech. Scholar
Prof. J.V.R. Murthy, JNTU, Kakinada
Prof. G.M.V. Prasad, B.V.CIT, AP
Prof. S. Pradhan, UU, BBSR
Prof. P.M. Khilar, NIT, RKL
Prof. Murthy Sharma, BVC, AP
Prof. M. Patra, BU, Berhampur
Prof. M. Srivastava, GGU, Bilaspur
Prof. P.K. Behera, UU, BBSR
Prof. B.D. Sahu, NIT, RKL
Prof. S. Baboo, Sambalpur University
Prof. Ajit K. Nayak, S'O'A, BBSR
Prof. Debahuti Mishra, ITER, BBSR
Prof. S. Sethi, IGIT, Sarang
Prof. C.S. Panda, Sambalpur University
Prof. N. Kamila, CVRCE, BBSR
Prof. H.K. Tripathy, KIIT, BBSR
Prof. S.K. Sahana, BIT, Meshra
Prof. Lambodar Jena, GEC, BBSR
Prof. R.C. Balabantaray, IIIT, BBSR
Prof. D. Gountia, CET, BBSR
Prof. Mihir Singh, WBUT, Kolkata
Prof. A. Khaskalam, GGU, Bilaspur
Prof. Sashikala Mishra, ITER, BBSR
Prof. D.K. Behera, TAT, BBSR
Prof. Shruti Mishra, ITER, BBSR
Prof. H. Das, KIIT, BBSR
Mr. Sarat C. Nayak, Ph.D. Scholar
Mr. Pradipta K. Das, Ph.D. Scholar
Mr. G.T. Chandrasekhar, Ph.D. Scholar
Mr. P. Mohanty, Ph.D. Scholar
Mr. Sibarama Panigrahi, Ph.D. Scholar

Mr. A.K. Bhoi, Ph.D. Scholar
Mr. T.K. Samal, Ph.D. Scholar
Mr. Ch. Ashutosh Swain, MCA
Mr. Nrusingh P. Achraya, MCA
Mr. Devi P. Kanungo, M.Tech. Scholar
Mr. M.K. Sahu, M.Tech. Scholar

Contents

Editors' Biography

Prof. Lakhmi C. Jain is with the Faculty of Education, Science, Technology and Mathematics at the University of Canberra, Australia and University of South Australia, Australia. He is a Fellow of the Institution of Engineers, Australia. Professor Jain founded the Knowledge-Based Intelligent Engineering System (KES) International, a professional community for providing opportunities for publication, knowledge exchange, cooperation, and teaming. Involving around 5,000 researchers drawn from universities and companies worldwide, KES facilitates international cooperation and generates synergy in teaching and research. KES regularly provides networking opportunities for the professional community through one of the largest conferences of its kind in the area of KES. His interests focus on artificial intelligence paradigms and their applications in complex systems, security, e-education, e-healthcare, unmanned air vehicles, and intelligent agents.

Prof. Himansu Sekhar Behera is working as a Reader in the Department of Computer Science Engineering and Information Technology, Veer Surendra Sai University of Technology (VSSUT) (A Unitary Technical University, Established by Government of Odisha), Burla, Odisha. He has received M.Tech. in Computer Science and Engineering from N.I.T, Rourkela (formerly R.E.C., Rourkela) and Doctor of Philosophy in Engineering (Ph.D.) from Biju Pattnaik University of Technology (BPUT), Rourkela, Government of Odisha respectively. He has published more than 80 research papers in various international journals and conferences, edited 11 books and is acting as a member of the editorial/reviewer board of various international journals. He is proficient in the field of Computer Science Engineering and served in the capacity of program chair, tutorial chair, and acted as advisory member of committees of many national and international conferences. His research interest includes Data Mining and Intelligent Computing. He is associated with various educational and research societies like OITS, ISTE, IE, ISTD, CSI, OMS, AIAER, SMIAENG, SMCSTA, etc. He is currently guiding seven Ph.D. scholars.

Prof. Jyotsna Kumar Mandal is working as Professor in Computer Science and Engineering, University of Kalyani, India. Ex-Dean Faculty of Engineering, Technology and Management (two consecutive terms since 2008). He has 26 years of teaching and research experiences. He was Life Member of Computer Society of India since 1992 and life member of Cryptology Research Society of India, member of AIRCC, associate member of IEEE and ACM. His research interests include Network Security, Steganography, Remote Sensing and GIS Application, Image Processing, Wireless and Sensor Networks. Domain Expert of Uttar Banga Krishi Viswavidyalaya, Bidhan Chandra Krishi Viswavidyalaya for planning and integration of Public domain networks. He has been associated with national and international journals and conferences. The total number of publications to his credit is more than 320, including 110 publications in various international journals. Currently, he is working as Director, IQAC, Kalyani University.

Prof. Durga Prasad Mohapatra received his Ph.D. from Indian Institute of Technology Kharagpur and is presently serving as an Associate Professor in NIT Rourkela, Odisha. His research interests include software engineering, real-time systems, discrete mathematics, and distributed computing. He has published more than 30 research papers in these fields in various international Journals and conferences. He has received several project grants from DST and UGC, Government of India. He received the Young Scientist Award for the year 2006 from Orissa Bigyan Academy. He has also received the Prof. K. Arumugam National Award and the Maharashtra State National Award for outstanding research work in Software Engineering for the years 2009 and 2010, respectively, from the Indian Society for Technical Education (ISTE), New Delhi. He is nominated to receive the Bharat Shiksha Ratan Award for significant contribution in academics awarded by the Global Society for Health and Educational Growth, Delhi.

Conceptual Modeling of Non-Functional Requirements from Natural Language Text

S. Abirami, G. Shankari, S. Akshaya and M. Sithika

Abstract The conceptual model is an intermediate model that represents the concepts (entities), attributes and their relationships that aids in the visualization of requirements. In literature, explicit design elements such as concept, attribute, operations, relationships are extracted as functional requirements and they are represented in the conceptual model. The conceptual model cannot be complete unless Non-Functional Requirements are included. This problem has motivated us to consider both functional and Non Functional Requirements in this research for the development of a conceptual model. Therefore, this paper presents a framework for the development of conceptual model by designing a classifier which segregates the functional and Non Functional Requirements (NFR) from the requirements automatically. Later, these requirements are transformed to the conceptual model with explicit visualization of NFR using design rules. In addition, the results of this model are also validated against the standard models.

Keywords Conceptual modeling · Non-functional requirements · Natural language processing · NFR classifier

S. Abirami (✉) · G. Shankari · S. Akshaya · M. Sithika
Department of Information Science and Technology, College of Engineering,
Anna University, Chennai, India
e-mail: abirami_mr@yahoo.com

G. Shankari
e-mail: shankarigiri2@gmail.com

S. Akshaya
e-mail: akshaya.it2010@gmail.com

M. Sithika
e-mail: sithika1008@gmail.com

L.C. Jain et al. (eds.), *Computational Intelligence in Data Mining - Volume 3*,
Smart Innovation, Systems and Technologies 33, DOI 10.1007/978-81-322-2202-6_1

1 Introduction

Automated Requirement Analysis is a kind of Information Extraction (IE) task in the domain of Natural Language Processing (NLP). In fact, it is the most significant phase of Software Development Life Cycle (SDLC) [1]. Errors caused during this phase can be quite expensive to fix it later. Requirements Gathering and Analysis is the most critical phase of the SDLC and this could greatly benefit through their automation as said by VidhuBala et al. [2]. Automation of conceptual model not only saves time but also helps in visualizing the problem. Conceptual model contains only domain entities and attributes according to Larman [3].

All the researches in the literature contribute towards the extraction of function requirements from natural language text. In addition to functional requirements, Non-Functional Requirements (NFR) should also be considered while creating a conceptual model. This is because the non-satisfaction of NFRs has been considered as one of the main reason for the failure of various software projects according to Saleh [4]. Therefore, we have proposed a system in this research which extracts Non-Functional Requirements with functional requirements automatically to make the conceptual model a complete one. The novelty of this paper lies in the development of the conceptual model named CMFNR thereby extracting and visualizing Non-Functional Requirements too.

This paper is organized as follows: Sect. 2 discusses about the related works and Sect. 3 describes our system of CM-NFR whereas Sect. 4 discusses the performance obtained by our system and Sect. 5 concludes the paper.

2 Related Work

Here, literature survey has been done in two areas namely explicit functional requirements modeling and Non-Functional Requirements modeling. Text mining has been used to assist requirement analysis in [5] where in the responsibilities were first identified using POS tags and then grouped semantically using clustering algorithms. But this approach requires human interaction to identify the conceptual classes. Later, Use case diagram, Conceptual model, Collaboration diagram and Design class diagram were generated using a tool called UMGAR (UML Model Generator from Analysis of Requirements) in [6]. But, it requires both requirements specification and use case specification in active voice form to generate the UML diagrams. Requirements and use cases were processed to produce classes, attributes and functional modules using DAT (Design Assistant Tool) in [7].

Tools like NLARE have been developed in literature to evaluate the quality of requirements like atomicity, ambiguity and completeness in [8]. Significant work in automating the conceptual model creation has been done in [2]. But the above system assumes that the input is grammatically correct and handles only explicit requirements. Non-Functional Requirements (NFR) are not handled here.

Taking Use Case Description (UCD) written in natural language as input, a parsed UCD and a UML class model were developed in [9], using a tool called Class-Gen. Parsed Use Case Description were subjected to rules to identify a list of candidate classes and relations. But the classes extracted should be refined by a human expert since the rules were not complete. All the above models discussed in the literature concentrate on functional requirements and they lack on the extraction of Non-Functional Requirements (NFR).

A novel approach to automatically transform NL specification of software constraints to OCL constraints was done by Bajwa [10]. According to UML, a note is used to represent a NFR. But this is not sufficient. Non-Functional Requirements comes in the second conceptual model of the class. Name of the class corresponded to the NFR being modeled and the entire requirement was specified in a note using OCL (Object Constraint Language). Hence, a class diagram can be used as extension to model NFR [4] in an effective way. As a result, in this research, we intend to develop an extended class diagram which could visualize the Non-Functional Requirements extracted automatically.

3 Conceptual Model-Non functional Requirements (CM-NFR) Architecture

The CMNFR shown in Fig. 1 illustrates the proposed system. This provides an enhancement to the model developed in [2] where in NFRs were not considered.

3.1 Tokenization

The natural language requirements in a text document is tokenized into a list of sentences using the Punkt Sentence Tokenizer [11] that is available as a part of the NLTK package for python.

3.2 NFR Classification

NFR classifier uses a set of pre-defined knowledge base to classify the list of sentences into two lists namely functional and Non-Functional Requirements. The knowledge base used by the classifier is specified below.

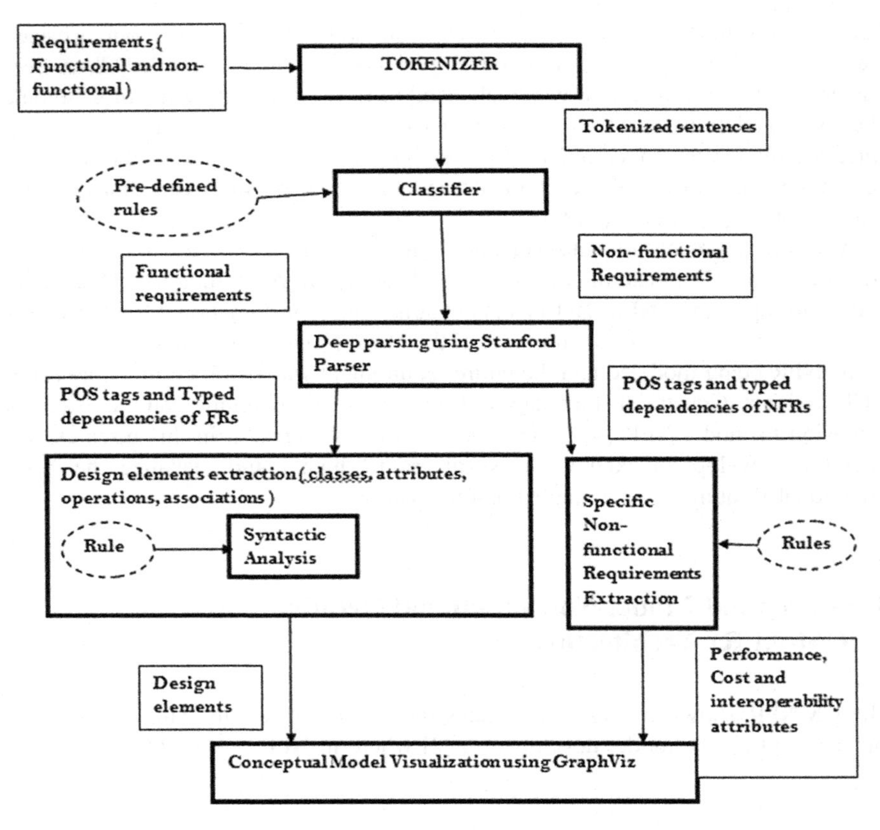

Fig. 1 Architecture of conceptual model generator

3.2.1 Extracting Specific Non-functional Requirements and Its Attributes

The classification of the NFR into a specific category namely performance, cost and interoperability is achieved by the following rules that are stated at a Bird's-eye view level. The novelty of our system lies in the construction of CM-NFR which extracts Non-Functional Requirements automatically from requirements text visualizes both functional and Non-Functional Requirements in a complete way. Syntactic rules used by the NFR Classifier to discriminate the Non-Functional Requirements from requirements has been listed below:

Classifier Rule 1: Whenever time is preceded by 'at least within', 'atmost within', 'within' clauses and the sentence contains the terms 'throughput' or 'response time', then it is a non-functional performance requirement.
Example: The system should retrieve the results within 3 s.
The system should have very less response time.

Classifier Rule 2: Whenever a sentence contains memory units and terms like 'memory', 'storage', 'space', then it is a non-functional memory requirement.
Example: The application should not consume a memory space of more than 2 GB.
Classifier Rule 3: Whenever a sentence contains the phrase '/second' or '/minute' or 'per second' etc. then it is a non-functional performance requirement that emphasizes speed.
Example: the system should handle 100 transactions/s.
Classifier Rule 4: Whenever a number is followed by rs, $, yen, pounds or any monetary unit then it is a non-functional cost requirement.
Classifier Rule 5: Whenever a sentence contains the stem 'interoperate' then it is a non-functional interoperability requirement.
Example: The application shall properly interoperate with the legacy Customer Oracle database.

The above rules were formulated after analyzing various requirements document done on number of projects in the development of software systems. Sentences which satisfy the classifier rules would be segregated either as functional or Non-Functional Requirements.

3.3 Deep Parsing

The output of the previous module (functional and Non-Functional Requirements) is first subjected to POS tagging using Stanford-POS tagger [12] to get POS tags (Part-Of-Speech) and then it is subjected to deep parsing using Stanford-parser to get typed dependencies and a parse tree.

3.4 Syntactic Analysis

The POS tagged functional requirements and their typed dependencies are used by the following rules to extract the required design elements as done in the conceptual modeling [2] later. The class rule, attribute and relation rules which were used in this research to extract design elements specified explicitly has been identified and used over the POS tagged sentences (both functional and non functional) as listed below. They are illustrated with an example for each in the following subsections:

3.4.1 Rules Used for Class Generation

Class Rule 1: *Any Noun that appears as a subject is always a class* [2].
Example: A *bank* owns an ATM.
Class Rule 2: *Nouns are always converted to their singular form.* This is done by obtaining the stemmed form of the word.

Class Rule 3: *Nouns occurring as objects participate in relations, but are not created as classes explicitly* [2].
Example: The user presses the button.
Class Rule 4: *Gerunds are created as classes* [9].
Gerund forms of verbs are treated as class names.
Examples: Borrowing is processed by the staff. Identified classes: Borrowing.

3.4.2 Rules Used for Attribute Generation

Attribute Rule 1: A noun phrase which follows the phrase 'identified by', 'recognized by' indicates presence of an attribute [9].
Example: An employee is identified by employee id. → Employee has attribute employee id.
Attribute Rule 2: An intransitive verb with an adverb may signify an attribute [9].
Example: the train arrives in the morning at 8 AM. → Train has an attribute, time.
Attribute Rule 3: A classifiable adjective, either in the predicate form or in attributive form signifies an attribute [2].
Example: The red button glows when pressed. → Button has an attribute, color.
Attribute Rule 4: possessive apostrophe signifies an attribute [9].
Example: employee's name is save → Name is an attribute of the employee.
Attribute Rule 5: The genitive case when referring to a relationship of possessor uses the 'of' construction [9].
Example: The name of the employee is printed → Name is an attribute of the employee.

3.4.3 Rules Used for Relation Generation

Relation rule 1: A transitive verb is a candidate for a relationship [9].
A transitive verb links a subject and an object. If the object translates to a class, the verb becomes a relation.
Example: The banker issues a cheque → Relation (banker, 1, cheque, 1, issues).
Relation rule 2: A verb with a preposition is a candidate for a relationship [2].
A verb that contains a prepositional object linked with a transitive verb along with a preposition, combines with the preposition to form a relationship.
Example: The cheque is sent to the bank. → sent _to is the relation.
Relation rule 3: A sentence of the form 'the r of a is b' [9].
Relation rule 4: A sentence of the form 'a is the r of b' [2].
Example: The bank of the customer is SBI → Relation (bank, 1, customer, 1, SBI).
Example: SBI is the bank of the customer → Relation (bank, 1, customer, 1, SBI).
In the above sentences, r is the relation that relates class a to class b. The above two rules are examples of the relation being in the form of a noun, rather than a verb.

3.4.4 Rules Used for Operation Generation

Operation rule 1: An intransitive verb is an operation [9].
An intransitive verb is a verb that does not need an object.
Example: The laptop hibernates. → hibernate is an operation of laptop.
Operation rule 2: A verb that relates an entity to a candidate class that is not created as an entity (fails Class Rule 3 after completion of class identification) becomes an operation [2].

3.4.5 Post Processing

Post processing is mainly done to remove unnecessary design elements. Post processing is done in three areas. Trivial relations are converted to operations, trivial associations are converted to attributes and trivial classes are removed. These are done based on the rules proposed in [2]. Since our system, mainly concentrates on the extraction of Non-Functional Requirements, design of elements has been adopted from [2, 9].

4 Results and Performance Analysis

4.1 Experimentation Setup and Metrics

The input to the system can be a either natural language text document or SRS (Software Requirements Specification). We have taken three standard input systems as datasets namely ATM, Online Course Registration System and Payroll System. This system has been implemented in Ubuntu environment by employing the languages Python and Shell Script for extraction of concepts, attributes and relationships [13] and dot script for visualizing the conceptual model generated [14]. Figure 2 depicts the conceptual model generated and visualized [14] with NFR for an ATM system requirements specification.

In previous model, performance of the functional requirements modeling has been measured based on three metrics namely Recall, Precision and Over-Specification. Equations 1, 2 and 3 depict the formulas for precision, recall and over-specification respectively.

$$\text{Precision} = N_{correct}/(N_{correct} + N_{incorrect}) \tag{1}$$

$$\text{Recall} = N_{correct}/\left(N_{correct} + N_{missing}\right) \tag{2}$$

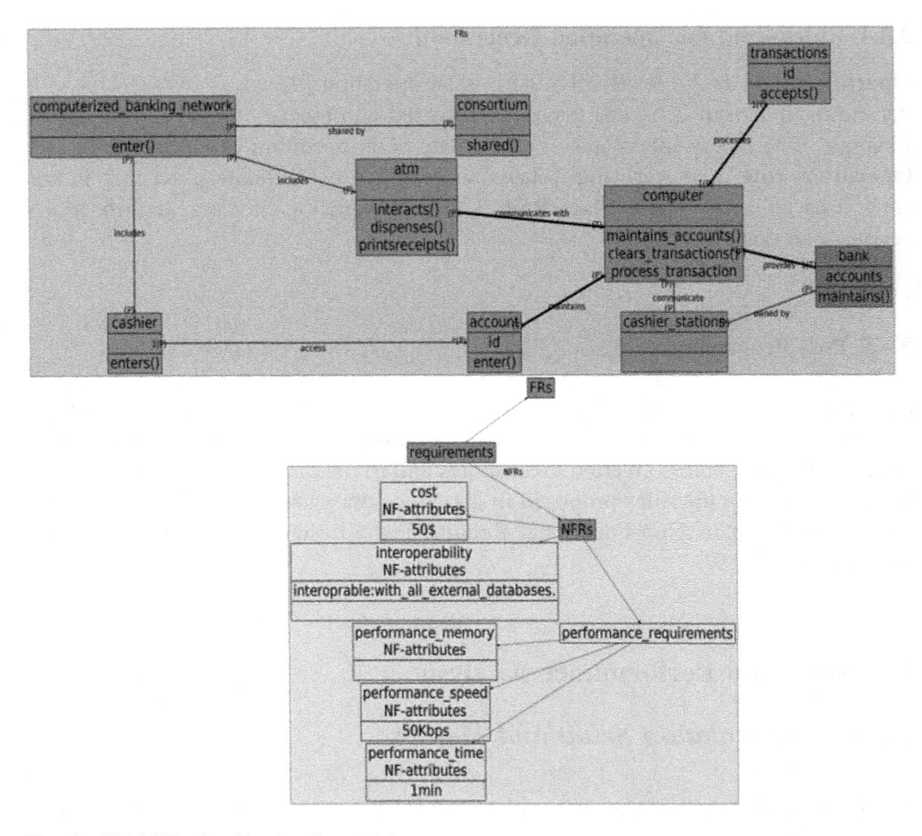

Fig. 2 CM-NFR visualization for ATM system

$$\text{Over-Specification} = N_{extra} / \left(N_{correct} + N_{missing}\right) \tag{3}$$

where,

$N_{correct}$ specifies the number of design elements correctly identified by the system and $N_{incorrect}$ indicates the number of wrong design elements that are identified as correct. $N_{missing}$ indicates the number of design elements that are in the answers but not identified by the system and N_{extra} shows the number of valid extra classes retrieved by the system.

4.2 Performance Evaluation

The three standard input documents taken to evaluate our system are ATM [15], Course Registration system [16] and Payroll System where payroll system is an user generated input. Initially, performance of the NFR classifier has been evaluated

Table 1 Performance evaluation of the classifier

	Hit rate (%)					Error rate (%)				
ATM	80	100	100	100	100	20	0	0	0	0
Payroll system	100	100	100	100	100	0	0	0	0	0
Course registration system	100	100	100	80	100	0	0	0	20	0

to assure whether the Non-Functional Requirements have been fully extracted by the system. Later, extraction rate of NFR attributes are evaluated using measures such as precious, recall and f-measures. In addition to that, a comparative analysis of the identification of NFR attributes between manual and automated process has been evaluated.

4.2.1 Evaluation of the Classifier

The performance of the classifier can be measured by calculating the number of requirements that are classified correctly versus the number of requirements classified incorrectly. The evaluation was done by taking the requirements documents of three domains and operating the classifier over it. The above process was repeated 5 times to ensure that the results are reliable. The results obtained are summarized in the Table 1 shown above.

4.2.2 Evaluation of NFR Extraction

The performance evaluation of the process of extracting specific Non Functional Requirements can be done by calculation precision, recall and F-score based on True Positive (TP), False Positive (FP), False Negative (FN) values. TP indicates correct classification. F-score gives a weighted average of the precision and recall values. The performance was evaluated again by running the process for three case studies and the results are summarized in the Table 2 and the Fig. 3 shown below.

Table 2 Performance evaluation of NFR extraction

	Precision	Recall	F-score
ATM	100	90	94.73
Payroll system	100	100	100
Course registration system	100	90	94.73

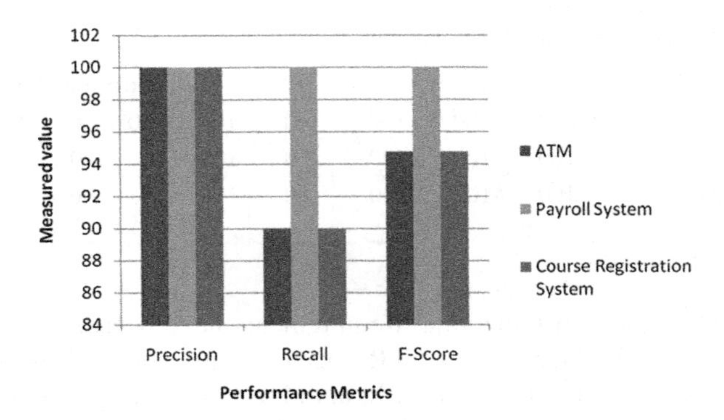

Fig. 3 Performance evaluation of NFR

Table 3 Time taken by individuals in identifying NFR attributes

NFR attributes	Requirements (s)	Automatically generated conceptual model (s)
Identification of NFR attributes	150	65
Specific NFR categories	85	50

4.2.3 Identifying NFR Attributes

The subjects were given time to analyze NFR attributes and to group them according to their category such as cost, interoperability and performance (memory, time, speed). The time taken by the individuals to identify NFR attributes from a Payroll system requirements and automatically generated conceptual model using the same payroll system were measured separately and the results are shown in Table 3.

5 Conclusion and Future Work

This research provides an efficient approach to extract design elements as well as their Non-Functional Requirements from requirements text document. The extraction and representation of the NFRs also known as soft goals is a novel aspect of our work through this NFR extraction and visualization in conceptual models, Non-Functional Requirements could be better grasped by the individuals rather from a natural language text. In future, this could be extended to more number of NFRs.

Acknowledgments This research is getting supported by UGC, New Delhi, India under Major Research project scheme of Engineering Sciences under the File – F.no. 42-129/2013(SR).

References

1. Meth, H., Brhel, M., Maedche, A.: The state of the art in automated requirements elicitation. J. Inf. Softw. Technol. **55**(10), 1695–1709 (2013) (Elsevier)
2. VidhuBala, R.V., Abirami, S.: Conceptual modeling of explicit natural language functional specifications. J. Syst. Softw. **88**(1), 25–41 (2013) (Elsevier)
3. Larman, C.: Applying UML and Patterns—An Introduction to Object-Oriented Analysis and Design and Iterative Development 3rd edn. Pearson Education, Ghaziabad (2005)
4. Saleh, K., Al-Zarouni, A.: Capturing non-functional software requirements using the user requirements notation. In: Proceedings of The International Research Conference on Innovations in Information Technology, India (2004)
5. Casamayor, A., Godoy, D., Campo, M.: Functional grouping of natural language requirements for assistance in architectural software design. J. Knowl. Based Syst. **30**(1), 78–86 (2012) (Elsevier)
6. Deeptimahanti, D.K., Sanyal, R.: Semi-automatic generation of UML models from natural language requirements. In: Proceedings of ISEC, ACM, India (2011)
7. Sarkar, S., Sharma, V.S., Agarwal, R.: Creating design from requirements and use cases: bridging the gap between requirement and detailed design. In: Proceedings of ISEC, ACM, India (2012)
8. Huertas, C., Reyes, J.R.: NLARE, A natural language processing tool for automatic requirements evaluation. In: Proceedings of CUBE, ACM, India (2012)
9. Elbendak, M., Vickers, P., Rossiter, N.: Parsed use case descriptions as a basis for object-oriented class model generation. J. Syst. Softw. **84**(7), 1209–1223 (2011) (Elsevier)
10. Bajwa, I.S., Lee, M., Bordbar, B.: Translating natural language constraints to OCL. J. King Saud Univ. Comput. Inf. Sci. **24**(2), 117–128 (2012) (Production and hosting by Elsevier)
11. http://www.nltk.org the official site for NLTK
12. http://nlp.stanford.edu/software/lex-parser.shtmt Parser and POS Tagger
13. Gowsikhaa, D., Abirami, S., Baskaran, R.: Construction of image ontology using low level for image retrivel. In: Proceedings of the International Conference on Computer Communication and Informatics, (ICCCI 2012), pp. 129–134 (2012)
14. VidhuBala, R.V., Mala, T., Abirami, S.: Effective visualization of conceptual class diagrams. In: Proceedings of International Conference on Recent Advances in Computing and Software Systems. pp. 1–6 (2012). doi:10.1109/RACSS.2012.6212688
15. Rumbaugh, J., Blaha, M., Premerlani, W., Eddy, F., Lorensen,W.: Object-Oriented Modeling and Design. Pearson Education, Ghaziabad (1991)
16. Section 1: Course Registration Requirements, IBM Corp, IBM Rational Softwares

Conceptual Modeling of Non-Functional Requirements.

References

1. [illegible] ... requirements classification ... IEEE Software Journal ... [illegible]
2. [illegible]
3. [illegible] ... Requirements Analysis and ... [illegible]
4. [illegible] ... requirements ... [illegible] ... Proceedings of ... Conference on ... [illegible]
5. [illegible] ... Journal of ... [illegible]
6. [illegible] ... models ... [illegible]
7. [illegible] ... requirements ... [illegible]
8. [illegible] ... requirements ... [illegible]
9. [illegible]
10. [illegible]
11. [illegible]
12. [illegible]
13. [illegible] ... and Applications ... [illegible]
14. [illegible] ... Systems ... [illegible]
15. [illegible] ... Modeling ... [illegible]
16. [illegible]

A Genetic Algorithm Approach for Multi-criteria Project Selection for Analogy-Based Software Cost Estimation

Sweta Kumari and Shashank Pushkar

Abstract This paper presents genetic algorithms as multi-criteria project selection for improving the Analogy Based Estimation (ABE) process, which is suitable to reuse past project experience to create estimation of the new projects. An attempt has also been made to create a multi-criteria project selection problem with and without allowing for interactive effects between projects based on criteria which are determined by the decision makers. Two categories of projects are also presented for comparison purposes with other traditional optimization methods and the experimented results show the capability of the proposed Genetic Algorithm based method in multi-criteria project selection problem and it can be used as an efficient solution to the problem that will enhance the ABE process. Here, Mean Absolute Relative Error (MARE) is used to evaluate the performance of ABE process and it has been found that interactive effects between projects may change the results.

Keywords Software cost estimation · Analogy based estimation · Genetic algorithm · Multi-criteria decision making · Nonlinear integer programming

1 Introduction

Success of software organizations rely on proper management activities such as planning, budgeting, scheduling, resource allocation and effort requirements for software projects. Software effort estimation is the process of making an approximate judgment of the costs of software. Inaccurate estimation of software cost lowers the proficiency of the project, wastes the company's budget and can result in failure of the

S. Kumari (✉) · S. Pushkar
Department of CSE, BIT, Mesra, Ranchi, India
e-mail: swetak44@gmail.com

S. Pushkar
e-mail: shashank_pushkar@yahoo.com

L.C. Jain et al. (eds.), *Computational Intelligence in Data Mining - Volume 3*,
Smart Innovation, Systems and Technologies 33, DOI 10.1007/978-81-322-2202-6_2

entire project. Broadly, there are two types of cost estimation methods: Algorithmic and Non-algorithmic. Algorithmic method calculates cost of the software by using formula or some experimental equations whereas Non-algorithmic methods use a historical data that is related to the previously completed similar projects for calculating the cost of the software [1]. Analogy based estimation is a Non-algorithmic method in that a critical role is played by the similarity measures between a pair of projects. Here, a distance is calculated between the software project being estimated and each of the historical software projects. It then finds the most similar project that is used to estimate the cost. Estimation by analogy is essentially a form of Case Based Reasoning [2]. However, as it is argued in [3] there are certain advantages in respect with rule based systems, such as the fact that users are keen to accept solutions from analogy based techniques, rather than solutions derived from uncomfortable chains of rules or neural nets. Naturally, there are some difficulties with analogy-based estimation such as lack of appropriate analogues and issues with selecting and using them. Choosing an appropriate set of projects participating in cost estimation process are very important for any organizations to achieve their goals. In this process, several reasons involve such as the number of investment projects, presence of multiple decision criteria such as takings maximization or risk minimization, business and operational rules such as budget limits and time windows for starting dates. Project selection is a difficult task, if the project interactions in terms of multiple selection criteria and information of preferences of decision-makers are taken into account, mainly in the presence of a huge number of projects.

2 Related Works

Various methods and mathematical models have been developed to deal with the problem of selecting and scheduling projects. Since the pioneering ranking method from [4] other methods have been proposed: Scoring [5], Analytical Hierarchy Process [6, 7] and Goal Programming [8] among others. However, these methods think that project interdependencies do not exist [9]. Other authors have proposed project selection models that deal with the existence of interdependencies. According to [8], these models, classified by the fundamental solution method are: Dynamic Programming [10] models reflect interdependency in special cases; Quadratic/Linear 0–1 programming models have a quadratic objective function and linear constraints, they limit interdependency only in the objective and between two projects. Quadratic/Quadratic 0–1 programming [11] with interdependency between two projects in the objective function and in the resource constraints; and Nonlinear 0–1 programming, with interdependency reflected in the objective function and the constraints among as many projects as necessary [9, 12] presents a linear 0–1 programming model for the selection and scheduling of projects, that includes technical interdependency. Medaglia et al. [13] also present some features of technical interdependence in linear project selection and scheduling models. According to the literature review, it has been found that estimation by analogies requires a significant amount of computations as well as lack of appropriate analogous and issues with selecting and using them. And it has been also observed that interactive effects between projects are significant for selecting a project because it may give unwanted result [14]. In this research, we propose Genetic Algorithms (GA) as multi-criteria project selection for improving the ABE system's performance and try to make a multi-criteria project selection problem allowing for project interactions that are based on multiple criteria and Decision makers preference information based on some significance. Here, the performance of ABE system is analyzed and compared in terms of MARE. Rest of the paper is divided in different sections as follows: a detail formulation of a multi-criteria project selection problem is described in Sect. 2. In Sect. 3, GA is used as the optimization technique for Project Selection problem. In Sect. 4, Numerical examples are also given for illustration purpose. Experiments and comparison results are described in Sect. 5 and in the end, the concluding remark is discussed.

3 Formulation of the Project Selection Problem

In project selection, a decision-maker is deals with the problem of selecting an appropriate subset of projects from an inappropriate set of projects based on a set of selection criteria. This process is known as the multi-criteria decision making (MCDM) process [15–17]. In general, there are two types of MCDM problems: multi-attribute decision making (MADM) problem and multi-objective decision making (MODM) problem. Based on this categorization, multi-criteria project selection problem is seen as a distinctive MADM problem in terms of the characteristics of project selection. In this paper N projects are considered for selection as well as evaluation, and decision variable x_n denotes whether the project is selected or not. If there are no interactive effects between projects, the project selection problem can be formulated in the following form:

$$\begin{aligned} Max\, E &= \sum_{n=1}^{N}\left(\sum_{j=1}^{J} p_j c_{nj}\right) x_n \\ \text{s.t.} \sum_{n=1}^{N} x_n &= R \\ x_n &= \{0,1\}, x_n = 1,2,\ldots,7 \end{aligned} \tag{1}$$

Therefore, the multi-criteria project selection problem with interactive effects between projects based on criteria can be formulated in the following form:

$$
\begin{aligned}
& Max\,E = \sum_{n=1}^{N}\left(\sum_{j=1}^{J} p_j c_{nj}\right) x_n + \sum_{j=1}^{J}\sum_{m=1}^{M}\left(p_j\left(b_j(c_m)\right)\left(\sum_{n=1}^{Q} c_{nj}\right)\right)\prod_{n=1}^{Q} x_n \\
& \text{s.t.} \sum_{n=1}^{N} x_n = R \\
& x_n = \{0, 1\}
\end{aligned}
\tag{2}
$$

Here, E is the total effects of selected projects, R is the number of the selected projects based on criteria, N is the number of projects to be evaluated and selected, P_j is preference degree of decision makers on criterion j, j = 1, 2 … J, C_{nj} is the value of projects n in criteria j, $b_j(c_m)$ is the value of interactive effects in a combination of m projects in j, j = 1, 2 … J, c_m is the Combination of m projects, m = 1, 2 … M and Q is the number of variables with interactive effects. It has been seen that, Eq. (2) becomes a 0–1 nonlinear integer programming problem. This kind of problem may be solved by branch and bound algorithm and dynamic programming method [18]. To solve this kind of problem, we incorporated the Genetic Algorithms for project selection.

4 GA-Based Optimization Approach

GA is optimization algorithms in evolutionary computing techniques, proposed in 1975 by a scientist Holland and extensively studied by De Jong, Goldberg [19, 20] and others [21]. It is inspired by natural biological evolution. GAs operates on a population of potential solutions applying the principle of survival of the fittest to produce better and better approximations to a solution. At each generation, a new set of approximations is created by the process of selecting individuals according to their level of fitness in the problem domain and breeding them together using operators borrowed from natural genetics. This process leads to the evolution of populations of individuals that are better suited to their environment than the individuals that they were created from, just as in natural adaptation. A detailed process of the Project Selection for ABE using GA is as follows:

1. Chromosome: Each individual chromosome consists of a number of digits. The value of each bit is set to be either 0 or 1: 0 means the related project in the historical data set is not selected and 1 means it is selected.
2. Population generation: After the encoding of the individual chromosome, the system then generates a population of chromosomes. It often represented in an array of chromosomes, while a chromosome is composed of many genes. In the mathematical optimization problem, a gene corresponds to a variable x_n, and a chromosome corresponds to a solution represented in a set of genes $x = (x_1, x_2 \ldots x_R)$ if R variables exist.

3. Fitness function: Each individual is evaluated by using the fitness function in GA. The GA is designed to maximize the fitness value. Generally the fitness function defines a score which gives each chromosome the probability to be chosen for reproduction or to survive.
4. Selection: Selection operator is used to create the population with higher fitness. Here, roulette wheel approach is used to select chromosomes from the current population with higher fitness.
5. Crossover: The main goal of crossover operator is to generate different offspring chromosomes to obtain a more optimal solution than their parents. Apply crossover techniques such as one-point, two-point and multi-point to initial chromosomes to produce new offspring chromosomes.
6. Mutation: Each bit of the chromosomes in the new population is chosen to change its value with a probability of 0.1, in such a way that a bit "1" is changed to "0" and a bit "0" is changed to "1". The main goal of the mutation operation is to prevent the GA from converging too quickly in a small area of the search space.
7. Stopping criteria: The new chromosome will be evaluated to verify whether it is a best solution or not. If it is satisfied, then the optimized results are obtained. If not satisfied, then it is repeated from Step 3 to Step 7 until a certain number of generations, a best fitness value or a convergence criterion of the population is reached.

5 Numerical Examples

The COCOMO data set [22] is chosen for the experiments and comparisons. This dataset consists of two independent variables-Size and EAF (Effort Adjustment Factor) and one dependent variable-Effort. Size is in KLOC (thousands of lines of codes) and effort is given in man-months. In this research our main task to reduce the whole project into an appropriate set of projects participating in cost estimation process that could save computing time and produce accurate results since it eliminates unwanted projects and contains only related projects. Here, we also compare the efficiency of GA-based optimization approach to the traditional approach. ILOG CPLEX barrier is a typical nonlinear optimizer which is used for computation and comparison purpose. This dataset is divided into two categories of projects. The first category of projects contains only seven variables and our objective is to select two best projects from the given projects and the second category of projects contains twenty-one variables in the project selection. Here, we want to select eleven best projects based on criteria which are decided by decision makers. Let the two criteria are j = {1: Magnitude of Relative Error (MRE),

2: Absolute Relative Error (ARE)}. The MRE and ARE can be calculated by the following equation:

$$MRE = \frac{|Actual - Estimated|}{Actual} \quad (3)$$

$$ARE = |Actual - Estimated| \quad (4)$$

Let P_j represents the preference degree which is determined by decision makers in terms of criteria *j*. In general, preference can be calculated by the following equation:

$$p_j' = \frac{p_j}{\sum_{j=1}^{J} p_j} \quad (5)$$

Table 1 shows the MRE and ARE value of seven different projects and Table 2 contains data which was obtained by using the Eq. (5).

Here, the interactive effects between these projects are not considered. But, it can be calculated by some statistical methods such as analysis of variance (ANOVA) [23]. For comparison purpose, interactive effects between the projects are also given in Table 3. Here, interactive effects between projects are assumed.

When the interactive effects between projects are not considered, the optimization problem in terms of Eq. (1) for project selection can be formulated as

$$\begin{aligned} \text{Max}\, E &= 20.793x_1 + 3.897x_2 + 8.378x_3 + 1.214x_4 \\ &\quad + 344.213x_5 + 99.798x_6 + 107.477x_7 \\ \text{s.t. } & x_1 + x_2 + x_3 + x_4 + x_5 + x_6 + x_7 = 2 \\ x_n &= \{0, 1\}, \quad n = 1, 2, \ldots 7 \end{aligned} \quad (6)$$

Table 1 Seven different projects of COCOMO dataset

Criteria (*j*)	Preference (P_j)	Projects (*n*)						
		*e*1	*e*2	*e*3	*e*4	*e*5	*e*6	*e*7
ARE	2	31.44	5.82	12.55	1.73	519.11	151.03	162.59
MRE	1	0.13	0.17	0.29	0.22	4.85	0.36	0.51

Table 2 Data which was obtained by using the Eq. (5) for seven different projects of COCOMO dataset

Criteria (*j*)	Preference (P_j)	Projects (*n*)						
		*e*1	*e*2	*e*3	*e*4	*e*5	*e*6	*e*7
ARE	0.66	31.44	5.82	12.55	1.73	519.11	151.03	162.59
MRE	0.33	0.13	0.17	0.29	0.22	4.85	0.36	0.51

Table 3 Interactive effects between projects based on criteria

Criteria (*j*)	Projects pairs (C_m)									
	e1e2	*e1e3*	*e1e4*	*e1e5*	*e1e6*	*e1e7*	*e2e3*	*e2e4*	*e2e5*	*e2e6*
ARE	0.55	0.60	0.30	0.20	0.50	0.20	0.45	0.40	0.65	0.45
MRE	−0.15	0	−0.20	−0.05	−0.25	0	−0.42	−0.15	0	−0.15

Criteria (*j*)	Projects pairs (C_m)										
	e2e7	*e3e4*	*e3e5*	*e3e6*	*e3e7*	*e4e5*	*e4e6*	*e4e7*	*e5e6*	*e5e7*	*e6e7*
ARE	0.35	0.55	0.40	0.60	0.55	0.40	0.65	0.45	0.55	0.60	0.30
MRE	−0.10	0	0	−0.80	−0.35	−0.15	−0.05	−0.03	0	−0.15	−0.22

Table 4 Comparison of objective value and computational time for GA-based method and CPLEX optimizer for seven variables

Equation no.	ILOG CPLEX barrier optimizer based objective value	GA based objective value	Computational time (s)	
			CPLEX	GA
6	451.691	728.6767	9	11
7	721.378	857.2201	13	15

If interactive effects between projects are considered, then the problem will become complex and from Eq. (2), the optimization problem allowing for project interactions can be formulated as

$$\begin{aligned}
\text{Max}\, E = {} & 20.793x_1 + 3.897x_2 + 8.378x_3 + 1.214x_4 + 344.213x_5 + 99.798x_6 \\
& + 107.477x_7 + 0.02x_1x_2 + 17.420x_1x_3 + 6.544x_1x_4 + 72.591x_1x_5 \\
& + 60.174x_1x_6 + 25.612x_1x_7 + 5.392x_2x_3 + 1.974x_2x_4 + 255.195x_2x_5 \\
& + 46.558x_2x_6 + 38.88x_2x_7 + 5.184x_3x_4 + 140.358x_3x_5 + 64.605x_3x_6 \\
& + 63.483x_3x_7 + 137.251x_4x_5 + 65.525x_4x_6 + 48.796x_4x_7 \\
& + 243.261x_5x_6 + 269.688x_5x_7 + 62.034x_6x_7 \\
\text{s.t. } & x_1 + x_2 + x_3 + x_4 + x_5 + x_6 + x_7 = 2 \\
x_n = {} & \{0, 1\}, \quad n = 1, 2, \ldots ..7
\end{aligned} \tag{7}$$

The above Eqs. (6) and (7) can be solved by a typical nonlinear optimizer, ILOG CPLEX barrier optimizer as well as GA-based method and the results are shown in Table 4.

Here, it is also noticed that from Eqs. (6) and (7), when interactive effects between projects are not considered, then the optimal solution x = (0, 0, 0, 0, 0, 1, 1) and the optimal solution x = (0, 0, 0, 0, 1, 0, 1) when it is considered. It indicates

Table 5 Comparison of objective value and computational time for GA-based approach and CPLEX optimizer for twenty-one variables

Equation no.	ILOG CPLEX barrier optimizer based objective value	GA based objective value	Computational time (s)	
			CPLEX	GA
8	1139.955	1579.1074	1.07	15
9	6699.812	8812.295	1.16	22

that interactive effects are very important for project selection and the computational time of CPLEX barrier optimizer is better than the GA-based optimization method because of small number of different projects and quadratic terms because GA-based optimization is a heuristic optimization method that may change the finale result, when a small size problem is encountered.

In this section, we want to check the efficiency of the proposed GA-based optimization method to the other nonlinear optimization method. Suppose there are twenty-one projects and more than two evaluations are presented and we want to select eleven best projects from the given selection problem. When the interactive effects between projects are not considered, then the objective function can be computed by using Eq. (1) as

$$\begin{aligned}\operatorname{Max} E = {} & 20.793x_1 + 3.897x_2 + 8.378x_3 + 1.214x_4 + 344.213x_5 + 99.798x_6 \\ & + 107.477x_7 + 27.166x_8 + 27.888x_9 + 1.647x_{10} + 9.016x_{11} + 8.326x_{12} \\ & + 2.996x_{13} + 1.512x_{14} + 133.983x_{15} + 85.022x_{16} + 157.615x_{17} \\ & + 112.854x_{18} + 12.851x_{19} + 23.146x_{20} + 0.128x_{21} \\ \text{s.t.} & \sum_{i=1}^{21} x_n = 11 \\ x_n = {} & \{0,1\}, \quad n = 1, 2, \ldots 21 \end{aligned} \tag{8}$$

Interactive effects between projects are considered and similarly, the objective function E can be calculated by using Eq. (2) and the results are shown in Table 5.

$$
\begin{aligned}
\text{Max } E = {} & 4.903x_1x_2 + 13.065x_1x_3 + 8.729x_1x_4 + 236.186x_1x_5 + 54.194x_1x_6 + 44.792x_1x_7 \\
& + 26.316x_1x_8 + 19.427x_1x_9 + 14.775x_1x_{10} + 16.338x_1x_{11} + 8.674x_1x_{12} + 5.902x_1x_{13} \\
& + 11.092x_1x_{14} + 108.212x_1x_{15} + 79.24x_1x_{16} + 35.618x_1x_{17} + 86.748x_1x_{18} + 18.428x_1x_{19} \\
& + 15.287x_1x_{20} + 6.244x_1x_{21} + 4.83x_2x_3 + 3.226x_2x_4 + 69.208x_2x_5 + 46.56x_2x_6 + 66.624x_2x_7 \\
& + 17.021x_2x_8 + 7.915x_2x_9 + 1.901x_2x_{10} + 1.903x_2x_{11} + 7.258x_2x_{12} + 3.388x_2x_{13} \\
& + 1.048x_2x_{14} + 41.311x_2x_{15} + 39.946x_2x_{16} + 104.826x_2x_{17} + 69.881x_2x_{18} + 13.257x_2x_{19} \\
& + 14.777x_2x_{20} + 1.378x_2x_{21} + 3.769x_3x_4 + 122.593x_3x_5 + 86.349x_3x_6 + 75.135x_3x_7 \\
& + 15.925x_3x_8 + 10.831x_3x_9 + 2.474x_3x_{10} + 3.433x_3x_{11} + 10.751x_3x_{12} + 6.744x_3x_{13} \\
& + 4.372x_3x_{14} + 35.503x_3x_{15} + 18.634x_3x_{16} + 107.735x_3x_{17} + 42.267x_3x_{18} + 6.281x_3x_{19} \\
& + 7.827x_3x_{20} + 3.781x_3x_{21} + 85.721x_4x_5 + 10.076_{x4x6} + 32.535_{x4x7} + 22.492_{x4x8} \\
& + 20.273_{x4x9} + 0.548_{x4x10} + 3.495_{x4x11} + 7.988_{x4x12} + 1.844_{x4x13} + 0.887x_4x_{14} \\
& + 26.992x_4x_{15} + 43.002x_4x_{16} + 87.233x_4x_{17} + 45.475x_4x_{18} + 8.301x_4x_{19} + 13.255x_4x_{20} \\
& + 0.483x_4x_{21} + 154.803x_5x_6 + 359.76x_5x_7 + 258.72x_5x_8 + 314.114x_5x_9 + 68.849x_5x_{10} \\
& + 105.471x_5x_{11} + 227.731x_5x_{12} + 189.702x_5x_{13} + 68.809x_5x_{14} + 95.295x_5x_{15} \\
& + 192.041x_5x_{16} + 199.68x_5x_{17} + 295.958x_5x_{18} + 159.925x_5x_{19} + 127.888x_5x_{20} \\
& + 188.341x_5x_{21} + 82.724x_6x_7 + 76.018x_6x_8 + 70.124x_6x_9 + 45.593x_6x_{10} + 32.582x_6x_{11} \\
& + 37.715x_6x_{12} + 46.183x_6x_{13} + 60.667x_6x_{14} + 93.416x_6x_{15} + 101.514x_6x_{16} + 128.518x_6x_{17} \\
& + 180.45x_6x_{18} + 84.343x_6x_{19} + 24.542x_6x_{20} + 34.91x_6x_{21} + 53.721x_7x_8 + 81.018x_7x_9 \\
& + 21.762x_7x_{10} + 52.319x_7x_{11} + 46.167x_7x_{12} + 71.673x_7x_{13} + 48.934x_7x_{14} + 84.368x_7x_{15} \\
& + 105.718x_7x_{16} + 105.908x_7x_{17} + 132.011x_7x_{18} + 66.048x_7x_{19} + 39.029x_7x_{20} + 26.83x_7x_{21} \\
& + 27.463x_8x_9 + 20.114x_8x_{10} + 27.031x_8x_{11} + 7.056x_8x_{12} + 19.541x_8x_{13} + 15.681x_8x_{14} \\
& + 56.299x_8x_{15} + 33.561x_8x_{16} + 83.056x_8x_{17} + 48.924x_8x_{18} + 29.913x_8x_{19} + 30.081x_8x_{20} \\
& + 10.881x_8x_{21} + 19.138x_9x_{10} + 7.349x_9x_{11} + 16.215x_9x_{12} + 18.449x_9x_{13} + 16.088x_9x_{14} \\
& + 40.42x_9x_{15} + 39.421x_9x_{16} + 55.554x_9x_{17} + 84.316x_9x_{18} + 20.282x_9x_{19} + 10.16x_9x_{20} \\
& + 8.382x_9x_{21} + 4.767x_{10}x_{11} + 6.423x_{10}x_{12} + 2.747x_{10}x_{13} + 2.396x_{10}x_{14} + 74.524x_{10}x_{15} \\
& + 30.285x_{10}x_{16} + 63.639x_{10}x_{17} + 40x_{10}x_{18} + 11.524x_{10}x_{19} + 16.032x_{10}x_{20} + 0.614x_{10}x_{21} \\
& + 3.424x_{11}x_{12} + 5.361x_{11}x_{13} + 4.12x_{11}x_{14} + 92.832x_{11}x_{15} + 42.248x_{11}x_{16} + 58.217x_{11}x_{17} \\
& + 66.904x_{11}x_{18} + 8.694x_{11}x_{19} + 19.191x_{11}x_{20} + 1.811x_{11}x_{21} + 5.044x_{12}x_{13} + 5.754x_{12}x_{14} \\
& + 56.819x_{12}x_{15} + 51.236x_{12}x_{16} + 82.815x_{12}x_{17} + 102.756x_{12}x_{18} + 15.775x_{12}x_{19} \\
& + 6.257x_{12}x_{20} + 5.00^{6}x_{12}x_{21} + 2.396x_{13}x_{14} + 82.091x_{13}x_{15} + 26.34x_{13}x_{16} + 32.084x_{13}x_{17} \\
& + 57.824x_{13}x_{18} + 3.147x_{13}x_{19} + 6.491x_{13}x_{20} + 1.538x_{13}x_{21} + 47.353x_{14}x_{15} + 38.82x_{14}x_{16} \\
& + 47.668x_{14}x_{17} + 51.363x_{14}x_{18} + 7.777x_{14}x_{19} + 4.873x_{14}x_{20} + 1.008x_{14}x_{21} + 164.093x_{15}x_{16} \\
& + 72.823x_{15}x_{17} + 110.862x_{15}x_{18} + 58.655x_{15}x_{19} + 70.592x_{15}x_{20} + 87.087x_{15}x_{21} \\
& + 48.454x_{16}x_{17} + 118.582x_{16}x_{18} + 34.189x_{16}x_{19} + 32.387x_{16}x_{20} + 42.488x_{16}x_{21} \\
& + 67.528x_{17}x_{18} + 127.681x_{17}x_{19} + 31.519x_{17}x_{20} + 94.544x_{17}x_{21} + 62.721x_{18}x_{19} \\
& + 88.175x_{18}x_{20} + 84.621x_{18}x_{21} + 12.494x_{19}x_{20} + 10.962x_{19}x_{21} + 8.101x_{20}x_{21} \\
\text{s.t. } & \textstyle\sum_{i=1}^{21} x_n = 11 \\
& x_n = \{0, 1\}, \quad n = 1, 2, \ldots 21
\end{aligned}
\tag{9}
$$

Similarly, from Eqs. (8) and (9), it has been found that when interactive effects between projects are not considered, then the optimal solution x = (1, 0, 0, 0, 1, 1, 1, 1, 1, 0, 0, 0, 0, 0, 1, 1, 1, 1, 0, 1, 0) and the optimal solution x = (1, 0, 0, 0, 1, 1, 1, 1, 1, 0, 0, 0, 0, 0, 1, 1, 1, 1, 1, 0, 0) when it is considered. Here, it has been found that

Table 6 Results and comparisons on COCOMO dataset

Project category	MARE of before project selection	MARE of after project selection with without consideration of project interactions	MARE of after project selection with consideration of project interactions
Seven variable	126.33	44.81	97.38
Twenty-one variable	85.62	82.05	81.31

the computational times are increased for both GA-based optimization method and CPLEX barrier optimizer when the number of deferent projects and quadratic terms are increased but the results shows that GA is better than CPLEX for solving this type of problem.

6 Experimental Results and Comparisons

This approach has been experimented on a COCOMO dataset. To evaluate the performance of proposed GA as multi-criteria project selection for improving the ABE, the Mean Absolute Relative Error (MARE) is used here. The MARE is defined in Eq. (10) as follows:

$$MARE = \frac{1}{n}\sum\nolimits_{i=1}^{n} |f_i - y_i| \tag{10}$$

where f_i is the Estimated and y_i is the Actual value respectively, n is the number of projects. Table 6 shows the comparison of results among various projects. It shows that MARE value of after project selection gives better results in comparison with MARE value of whole projects and also it has been found that the MARE value is changed with and without allowing for interactive effects between projects. These results suggest that the selection of an appropriate projects can not only reduce the information required to approximate software effort, but it can also result in obtaining better approximations and it also specify that GA-based optimization method can be used as an effective solution to other optimization method (i.e., CPLEX).

7 Concluding Remarks

In this study, a novel approach based on Genetic Algorithm has been proposed to solve the problem of multi-criteria project selection that will improve the performance of analogy based software cost estimation. It formulates a multi-criteria project selection problem with and without interactive effects between projects based on criteria which are determined by the decision makers. Detailed examples have also been illustrated that describe the effectiveness of the GA-based optimization method to enhance the process of Software Cost Estimation. The proposed approach has also been compared with other optimization methods used. The well-known COCOMO datasets are chosen for the experiments and comparisons. The results show that the GA based project selection approach has lowest MARE value, so that it will be able to provide good estimation capabilities for ABE system. It has also been observed that the proposed approach has a limitation. In some cases this GA method is not able to find the exact global optimum because there is no absolute assurance to find the best solution. However, this limitation is for any such meta-heuristic technique used for optimization. The future direction can be experimenting with some more methods for project selection that can help to overcome the above limitation and can further improve the process of software cost estimation.

References

1. Kumari, S., Pushkar, S.: Performance analysis of the software cost estimation methods: a review. Int. J. Adv. Res. Comput. Sci. Softw. Eng. **3**, 229–238 (2013)
2. Mukhopadhyay, T., Vicinanza, S.S., Prietula, M.J.: Examining the feasibility of a case-based reasoning model for software effort estimation. MIS Q. **16**, 155–171 (1992)
3. Shepperd, M.J., Schofield, C.: Estimating software project effort using analogies. IEEE Trans. Softw. Eng. **23**(12), 736–743 (1997)
4. Lorie, J.H., Savage, L.J.: Three problems in rationing capital. J. Bus. **28**(4), 229–239 (1955)
5. Rengarajan, S., Jagannathan, P.: Projects selection by scoring for a large R&D organization in a developing country. R&D Manage. **27,** 155–164 (1997)
6. Lockett, G., Hetherington, B., Yallup, P.: Modelling a research portfolio using AHP: a group decision process. R&D Manage **16**(2), 151–160 (1986)
7. Murahaldir, K., Santhanam, R., Wilson, R.L.: Using the analytical hierarchy process for information system project selection. Inf. Manage. **17**(1), 87–95 (1990)
8. Lee, J.W., Kim, S.H.: Using analytic network and goal programming for interdependent information systems project selection. Comput. Oper. Res. **19**, 367–382 (2000)
9. Santhanam, R., Kyparisis, G.J.: A decision model for interdependent information system project selection. Eur. J. Oper. Res. **89**, 380–399 (1996)
10. Nemhauser, G.L., Ullman, Z.: Discrete dynamic programming and capital allocation. Manage. Sci. **15**(9), 494–505 (1969)
11. Aaker, D.A., Tyebjee, T.T.: Model for the selection of interdependent R&D projects. IEEE Trans. Eng. Manage. **25**(2), 30–36 (1978)
12. Ghasemzadeh, F., Archer, N., Iyogun, P.: Zero-one model for project portfolio selection and scheduling. J. Oper. Res. Soc. **50**(7), 755 (1999)
13. Medaglia, A.L., Hueth, D., Mendieta, J.C., Sefair, J.A.: Multiobjective model for the selection and timing of public enterprise projects. Soc. Econ. Plann. Sci. (in press, 2007). http://dx.doi.org/10.1016/j.seps.2006.06.009
14. Carlsson, C., Fuller, R.: Multiple criteria decision making: the case for interdependence. Comput. Oper. Res. **22**, 251–260 (1995)
15. Korhonen, P., Moskowitz, H., Wallenius, J.: Multiple criteria decision support—a review. Eur. J. Oper. Res. **63**, 361–375 (1992)

16. Stewart, T.J.: A critical survey on the status of multiple criteria decision making theory and practice. Omega **20**, 569–586 (1992)
17. Ostermark, R.: Temporal interdependence in fuzzy MCDM problems. Fuzzy Sets Syst. **88**, 69–79 (1997)
18. Hammer, P., Rudeanu, S.: Boolean Methods in Operations Research and Related Areas. Springer, Berlin (1968)
19. De Jong, K.A.: Analysis of behavior of a class of genetic adaptive systems. Ph.D. thesis. University of Michigan, Ann Arbor, MI (1975)
20. Goldberg, D.: Genetic Algorithms in Search, Optimization and Machine Learning. Addison-Wesley, New York (1989)
21. Back, T., Schwefel, H.P.: An overview of evolutionary algorithms for parameter optimization. Evol. Comput. **17**(1), 87–95 (1993)
22. Boehm, B.W.: Software Engineering Economics. Prentice-Hall, Englewood Cliffs (1981)
23. Cox, D.R.: Interaction. Int. Stat. Rev. **52**, 1–25 (1984)

A Service-Oriented Architecture (SOA) Framework Component for Verification of Choreography

Prachet Bhuyan, Abhishek Ray and Durga Prasad Mohapatra

Abstract Service-Oriented Architecture (SOA) is a paradigm that encourages organization to understand, how their information technology infrastructure capabilities can be organized to achieve business goals? SOA promises a challenging generation of information systems application based on a new set of standards for enabling self describing, interoperable web services. The web service composition supports choreography in SOA. The web service composition is mainly managed through orchestration and choreography of services. In this paper, we propose a Framework for Service Choreography (FSC) to control the business processes in SOA to reduce complexity in web service composition. To address choreography security issues while passing messages between the services choreographed in SOA, we also propose a Verification Model (VM) using Security Assertions Markup Language (SAML 2.0) to provide authentication and authorization. Then, we have implemented the proposed choreography model.

Keywords SOA · Web services · Choreography · Orchestration · Framework for service choreography (FSC) · Verification model (VM)

1 Introduction

SOA is an architectural approach which supports service composition to ensure reusability and productivity. SOA is an approach for organizing by using Information Technology to match and combine needs with capabilities in order to

P. Bhuyan (✉) · A. Ray
School of Computer Engineering, KIIT University, Bhubaneswar, India
e-mail: pbhuyanfcs@gmail.com

A. Ray
e-mail: arayfcs@kiit.ac.in

D.P. Mohapatra
Department of Computer Science & Engineering, NIT, Rourkela, India
e-mail: durga@nitrkl.ac.in

L.C. Jain et al. (eds.), *Computational Intelligence in Data Mining - Volume 3*,
Smart Innovation, Systems and Technologies 33, DOI 10.1007/978-81-322-2202-6_3

support the overall mission of an enterprise. Generation ago, we had learned to abstract from hardware but currently we have learned to abstract from software in terms of SOA. So, SOA together with web service technology concentrates on the definition of a software system as a complex distributed business application.

Services are communicated through network and messages are transmitted through interfaces in a platform neutral, standardized format like Extensible Markup Language (XML).

Currently, web service compositions are considered to be the most widespread possibility of implementing SOA. Web service composition is created by combining pre-existing services. Web services are self-constrained and perform business activities. These business activities help the stakeholders to meet the desired design process.

Business Process Execution Language (BPEL) follows the orchestration paradigm and Web Service Choreography Description Language (WS-CDL) covers the choreography. Choreography is also covered by other standards such as WSCI. Orchestration refers to a composed business process which use both internal and external web services to fulfill its task. The Orchestration process is described at the message level that is in terms of message exchanges and execution order. Orchestration mainly focuses on the internal behavior of a business process (Process Model). Choreography addresses the interaction that implements the collaboration between services. Choreography focuses on the external perspective that is process interaction (Interaction Model). Choreography has altered much interact in the research field and researchers have worked on its model, analysis and implementation issues. In order to create business processes for an organization, effective message passing between clients is necessary.

To strengthen the business processes security is the major concerned for every organization's authentication and authorization.

We have proposed a Framework for Service Choreography (FSC) which is developed to control business processes in choreography scenario to reduce complexity. Further, in this paper we have proposed a Verification Model using Security Assertions Markup Language (SAML 2.0) for providing security to map SOA based choreography services which lack in choreography security of message passing.

The rest of the paper is organized as follows: Sect. 2 describes the related work. The proposed Choreography Framework is described with a case study in Sect. 3. The Verification Model (VM) for Choreography is described in Sect. 4. Section 5 addresses the implementation showing effective choreographed message passing scenario of OTBA. Section 6 presents the conclusion and future work.

2 Related Works

To the best of our knowledge many works [1–3, 5–12] concentrates on the web service composition and implementation of web service composition. In this section, we discuss various works in the area of choreography and their need of framework and verification of choreography.

Gómez-Gutiérrez and Rivera-Vega [5] defined that the web service composition is a process that usually requires advanced programming skills and vast knowledge about specific technologies. Gómez-Gutiérrez and Rivera-Vega proposed a framework for the smooth composition of web services.

Peltz [9] introduced the basic concept that a web service composition mainly depends on two aspects that is orchestration and choreography.

Braker et al. [1] proposed that web service choreography language demonstrates how service choreographies can be specified, verified and enacted with a simple process language by using Multi Agent Protocols (MAP). MAP provides open source framework for the enactment of distributed choreographies and the verification is done through model checking.

Zhao et al. [2] introduced the concept of Unified Modeling Language (UML) for the concept of modeling of web services and business processes. They also described the semantic web service functionalities. They generated web service and business processes by mapping patterns through transformation rules.

Rebai et al. [11] introduced an approach addressing the transition for choreography to orchestration, which is then accompanied with a concept of verification phase by using model checking methods.

Roglinger [10] discussed a correctness called as conformance to functional requirement which is a prerequisite for the quality of service of web service composition. Roglinger [10] also proposed a requirement framework for service-oriented modeling techniques that focuses on correctness properties shown by verification of web service compositions.

Kovac and Trcek [7] introduced formal methods that support the survey based on BPEL and WS-CDL languages which are mainly the core building blocks for business processes and pointed out that orchestration is more flexible than choreography. So this is the reason to concentrate on orchestration aspects using formal models like: error, event and compensation handling using extended version of pie calculus, process algebra and control flow. They concentrated on formal models which allow the analysis and verification of BPEL processes that focus on investigation on selected BPEL issues.

Pahl and Zhu [8] proposed ontology based framework for web service processes which provides techniques for web service composition, description and matching. They also discussed a description logic knowledge representation and reasoning framework which provides foundations. Their ontological framework was based on operational model for behavior of service process and composition.

Yoon and Jacobsen [12] proposed a novel distributed service choreography framework in order to overcome the problem of resolving semantic conflicts. This is challenging as services are loosely coupled and their interactions are not carefully governed. They deployed safety constraints in order to prevent conflicting behavior, enforce reliable and efficient service interactions. And to minimize runtime overhead via federated publish/subscribe messaging along with strategic placement of distributed choreography agents.

Fonseca et al. [3] presented a framework that allows the creation/development of SOA based application in mobile environment. The objective of the framework is to provide developers with tools for provision of services in mobile environment.

Kamatchi [6] provided a collaborative security framework for the implementation of SOA with web services.

3 Introduction of Proposed Choreography Framework

The choreography framework has been proposed in order to target the Research Queries (RQ) considering the Refs. [2, 5, 8, 9, 12]. The RQ are as follows:

RQ1 What do we choreograph and how?
RQ2 What is the message content exchanged between clients in an organization?
RQ3 What strategy each selected study uses to deal with choreography adaption?
RQ4 How issues related to choreographies are resolved or filtered?
RQ5 How the main structure of choreography is developed mainly the Target, Intervention degree and Necessity of model?

Target—specifies whether the adaptation of framework support functional or non functional requirement changes.
Intervention Degree—specifies whether the framework for adaptation of choreography automatically performed or is human intervention necessary.
Necessity of Model—specifies which choreography models representations or standards (WSCDL. WSCI, BPMN) are used in the strategy.

3.1 Proposed Framework for Service Choreography

Our FSC proposal is developed to solve the research queries (RQ). FSC is a set of assumptions, concepts, values and practice which is the way of viewing the current IT environment. FSC works on the idea for operation of components such as processes, policies, equipments and data for overall effectiveness of an organization. FSC helps to ensure that the design and configuration of services are synchronized throughout the enterprise. This helps to speed up the turnaround time for new and updated applications that evolve due to business demands.

Today's IT organization must establish strong governance to ensure that the services are properly designed, maintained and reused. So, framework can help minimize the complexities of integration by taking into account incompatibility and integration challenges. FSC provides robust guidelines to promote that services are reusable by focusing on governance to identify the best integration logic and

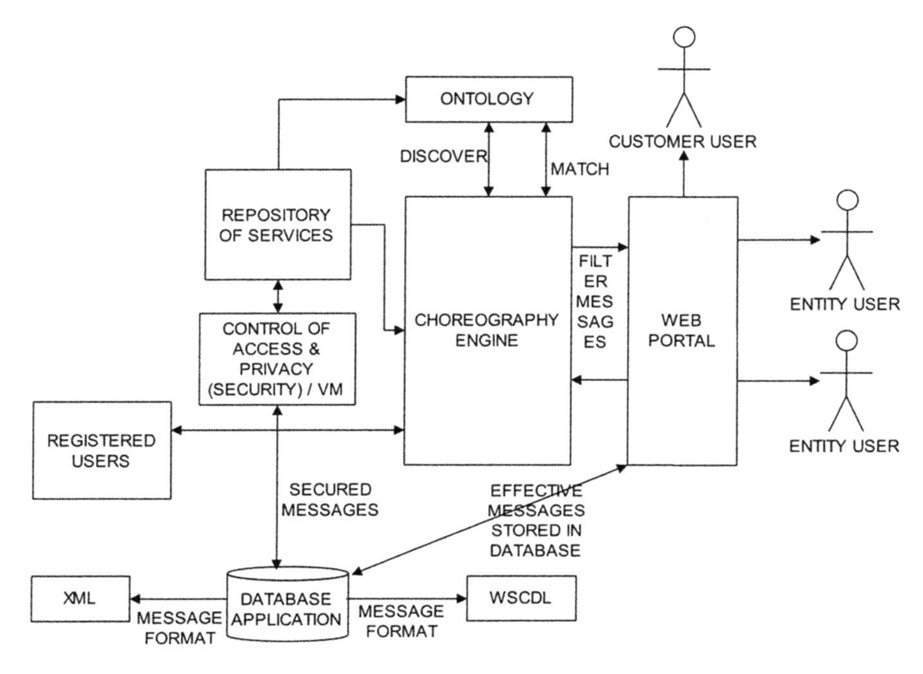

Fig. 1 Framework for service choreography (FSC)

streamlining the IT processes and maintenance. While building FSC framework we focused on three key areas: the management team, business process and integration platform. Our proposed FSC framework is shown in Fig. 1.

3.2 Description of FSC

The FSC components are shown in Fig. 1 and are described as follows:

Register of Users: It is the component of FSC in which all the users or clients can register for authentication through the access points in order to start the business process in an organization for effective message passing through choreography.
Control of Access and Privacy: It is responsible for ensuring that an incoming or outgoing message is expected at this point of time in order to transfer secured messages ultimately to the client. This has been supported by verification model (VM) for choreography in Sect. 4.
Ontology: Ontology is a classical approach for verification of choreography properties. Ontology definition depends on these factors such as concept-conceptual entity of the domain that capture the domain of message from repository of services. Attribute: discover the authenticate messages and sends it to choreography engine. Relation: maps the relationship between concepts and attribute properties of messages. Axiom: coherent description between concepts or properties or relations via

logical expressions and match the logical expressions with messages, then sends the messages to choreography engine for filtration.
Choreography Engine: In choreography engine requester and provider have their own instance of observable communication patterns. The first instance in choreography of the FSC is that the instances of message interaction are loaded for the requester and provider for their own descriptions. Choreography engine performs the basic activities.

- Choreography engine is used for evaluation of messages that means filter the available message by resolving issues or challenges.
- Choreography engine sends data of interactive messages to ontology format for discovering and matching to get the effective messages. The choreography engine component of FSC filters manages changes. It even replaces data for creation of effective message passing scenario to develop business processes in an organization.
- Choreography engine receives data from ontology format and forward it to the web portal component so that from this destination, data is finally sent or resulted output data is sent to the communication partner that is entity client or customer users who are waiting for their requested response.
  - *Web portal*: It is the gateway of the FSC where the entity user and customer can access to the organization for effective business process scenario in choreography.
  - *Database Application*: It is the component of FCC framework which stores information about entities and users.
  - *WSDL*: A Web Service Description Language (WSDL) document is an XML document that describes web services that are accessible over a network. WSDL describes the following in the FCC framework:

    Logically the message interaction performs with the location of the web service.
    Method to use to access the service of interacted message including the protocol that the web service uses for the messaging format.
    Through the use of WSDL format, a client that is entity user and customer involved in the FCC framework invoke the message. They receive the response along with input parameters that the client or customer must supply to the database. Hence in return database component acknowledges to the clients. The ontology concept and working of FSC is described through a case study.

3.3 Case Study: Online Ticket Booking of Airline

As per the FCS shown in Fig. 1 the service choreography defines communication which involves protocol for all partners. In this scenario, we choose the case study of Online Ticket Booking of Airline, which assumes two or more partners involved in the choreography scenario.

In case of OTBA, the different partners involved in choreography are: the Travel Agent, the Client or Traveler and the Airline Service. These partners are considered as registered users of FCS. Overall this scenario shows that the Travel Agent receive request quote from the Client or Passenger and answers back whether the requested reservation is available or not. If the flight is available, the Client or Passenger places an order to book the ticket; the Travel Agent processes the order and forwards the flight reservation details and payment details to the Airline Service. Then the travel agent sends the ticket to the passenger or client and informs the Airline Service about all the details of processing. Next the Travel Agent sends the bill to the passenger or client. When the passenger or client has received both, the bill and the ticket then make payment to the Travel Agent. The process ends when the receipt of the payment is done by the travel agent. All these activities are coordinated by the choreography engine in FCS.

In OTBA, ontology is used as specification to carry out the framework based analysis of verification of message interaction in choreography. This is also controlled by the choreography engine in FCS. Message interaction verification is done through the sequence diagram for each scenario. The Login scenario is used for Ticket check in status, Booking of Ticket and Payment. Logout is a process which describes an interaction process of an online travelling user starting with a login, and then repeatedly executing ticket enquiries, flight reservation enquiries, money transfer or payment before logging out. Each of these services implements a process internally called orchestration. The interactions resulting from the service invocations (e.g. Ticket Availability) and service provision (e.g. Login) are the result of service choreography.

For instance, Payment Process is a client of Travel agent and Travel agent is a client of Airline. Here, we formalize orchestration, choreography and develop FSC that defines, supports the entire framework's component composition activities. The FSC serves to capture the requirements of message exchange from an underlying layer for the ontology in FSC.

Ontology concept uses pre-conditions and post-conditions. Pre-conditions describe what a web service expects in order to provide in service. Post-conditions describe the result of FSC in relation to the input and conditions on it. We have developed the ontology in FSC framework in terms of description logic taking into account the example of OTBA. The ontology concept based on pre-condition and post condition is described in Fig. 2.

Here, two essential state components are pre and post, which denotes abstract pre and post states for effective message service process transitions in FSC. The above described is the entire procedure of work flow of FSC in order to resolve the error and exceptions so that effective choreograph message interaction can be sent as a result to the clients to make their business process effective in an organization.

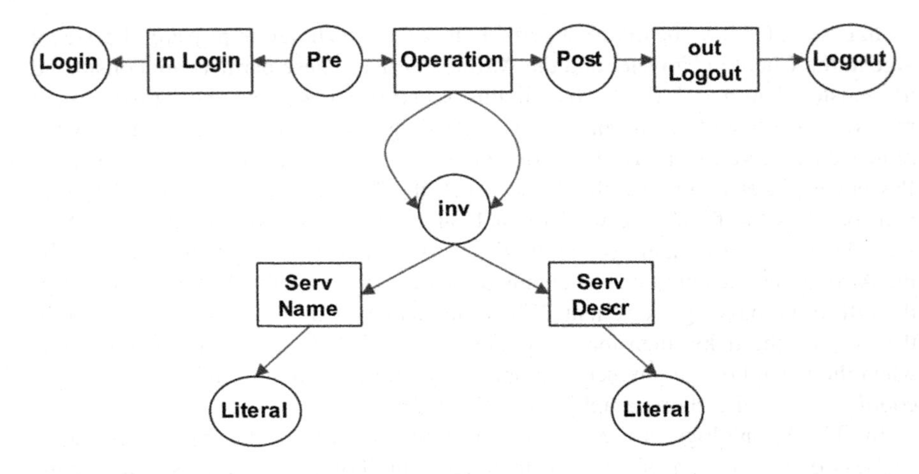

Fig. 2 Ontology based on pre-condition and post-condition for OTBA

4 Verification Model for Choreography

The proposed VM allows a message level security for choreography components to be implemented with SAML. The main aim of verification model is to provide security to message exchange between clients in choreography platform:

a. Ensuring adherence to all messages stored in service repository of FSC.
b. Ensuring interoperability among the clients participated in the organization.
c. Provide adaptability which is necessary for changes in message exchange in FSC.

The need for sophisticated message level security end-to end becomes a necessity for choreography platform in FSC. In order to consider the breaches of security, the Verification Model must contain at least the basic requirements as shown in Fig. 3.

Secured transmission of interacted messages in choreography is achieved by ensuring the confidentiality and integrity of the data, while the authentication and authorization will ensure that the service is accessed only by trusted requesting from the clients involved in the choreography platform.

The security requirements for Verification Model are

I. Authentication: It is the process of verifying the identity of choreographed messages and user or client.
II. Authorization: It is the permission to use a resource or data for effective result in choreography.
III. Trust: Ensure correct choreography of services.

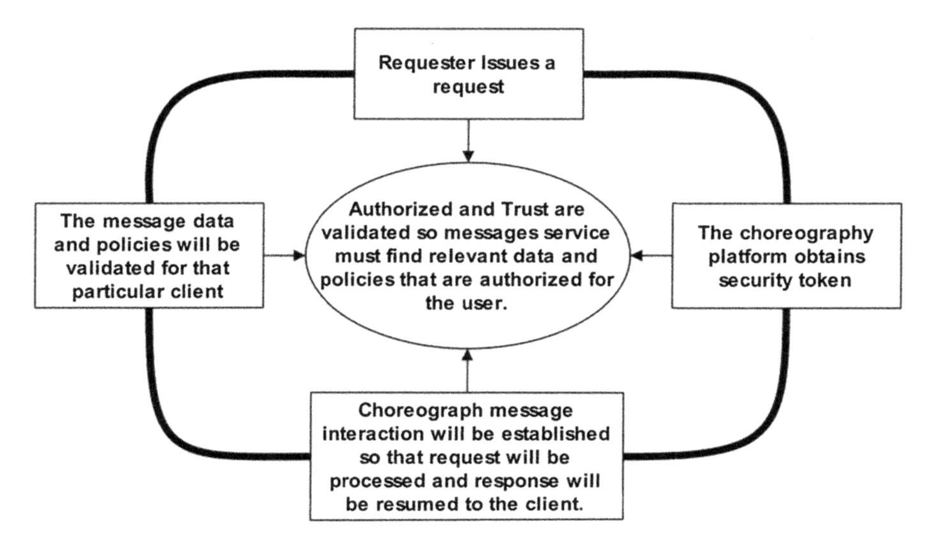

Fig. 3 Verification model for choreography

IV. Policy: Political trust sets a general policy statement for security while building relationships among interacted choreographed messages providing an appropriate standard of safety.

Authentication service component in Verification Model comprises of:

(a) *Message Source of Authority* (*MSA*): The top most root of Verification Model of trust, sometimes also referred as trust anchor for choreography.
(b) *Message Attribute Authority* (*MAA*): The issuer of an attribute certificate for client to access permission. The security of message token authorized in Verification Model for choreography is shown in Fig. 4.

Authorization is an aspect of security that comprises of secure conversation that is secure interaction, policy, trust, privacy and federation in VM. Privacy make sure no one can access other's data or effective message in choreography platform. If authorization were to be on a layer working with other services then it would work in conjunction with federation layer. In this paper our approach in Verification Model mainly depends on Security Assertions Markup Language (SAML2.0) as the industry standard for security.

SAML is the framework for exchanging information between testing parties which are security related to message interaction scenario in choreography. SAML exchanges information which are expressed in a XML format. SAML uses assertions made in the code which conveys information about message authentication function and authorization decisions. A SAML 2.0 assertion contains a packet of security information. SAML 2.0 is used in this Verification Model because it provides VM for authentication and authorization as follows:

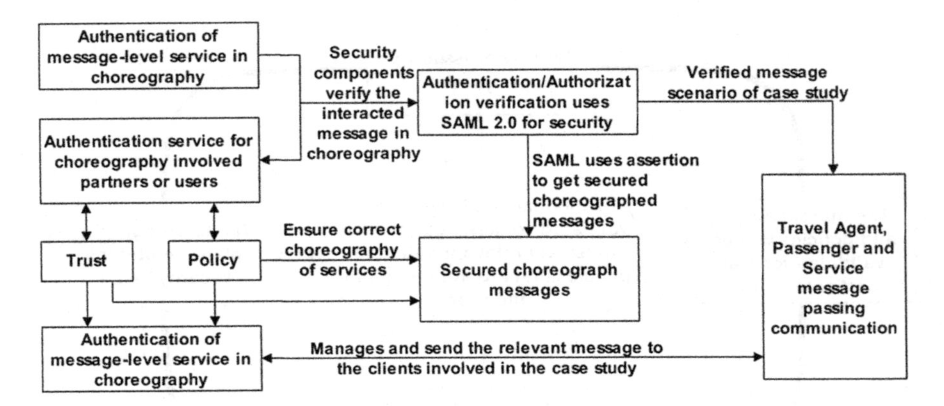

Fig. 4 Message token authorization in verification model for choreography

- ***Step 1***: It allows a user to authenticate once against a server that helps in validating the identity of choreographed messages.
- ***Step 2***: Once authenticated the server will issue an authentication to the user. The server can also create an authorization assertion which is a permission that grants privilege to the authenticate user involved in the choreography.
- ***Step 3***: The authenticate user can pass these assertions on to other application in order to verify the user without having any prior knowledge of the authenticate user of the choreography process.

This VM can be very useful for an enterprise perspective because enterprise rely on its partners to authenticate their own users. SAML has three components-assertion, protocol and binding. SAML 2.0 assertions are issued by SAML authorities, namely authentication authorities and attribute authorities. SAML binding 2.0 provides tremendous flexibility. Binding is a mapping of a SAML protocol message onto standard messaging format for effective communication in choreography. SAML 2.0 protocol describes the packaging of SAML elements within SAML request and response elements.

5 Implementation

This section deals with the implementation of our proposed work. Firstly, we have created the web services and choreographed them using NetBeans IDE 7.2. Next, we have created a database which deals with all the created web services and the mapping between them. The created database represents the choreographed messages for OTBA. Then, we have tested an effective choreographed message passing scenario (flight reserve successfully) of choreography for OTBA and the message "Confirm Flight Reservation" is effectively tracked by our proposed model.

6 Conclusion

In this paper, we proposed a Framework for Service Choreography (FSC). Our proposal mainly addresses the issues to overcome the research queries through ontology concept. We have also proposed a Verification Model (VM) with the concept of SAML 2.O for message level security at the time of choreography through authentication and authorization by trust, policy and federation. We have implemented our proposed model by considering a case study "Online Ticket Booking of Airline (OTBA)".

References

1. Braker, A., Walton, C.D., Roberston, D.D.: Choreographing web services. IEEE Trans. Serv. Comput. **2**, 152–166 (2009)
2. Zhao, C., Duan, Z., Zhang, M.: Model-driven approach for dynamic web service composition
3. Fonseca, J., Abdelouahab, Z., Lopes, D., Labidi, S.: A security framework for SOA applications in mobile environment. Int. J. Netw. Secur. Appl. (IJNSA) **1**(3), 90–107
4. Rao, P.J., Martin, M.: Logic-based web services composition: from service description to process model. In: International Conference on Web Services (ICWS), IEEE Press
5. Gómez-Gutiérrez, J.A., Rivera-Vega, P.I.: A framework for smooth composition of choreographies of web services. In: International Conference on Web Services (2007)
6. Kamatchi, R.: Security visualization collaborative security framework for service-oriented architecture. Int. J. Modell. Optim. **2**(4), 558–562
7. Kovac, D., Trcek, D.: A survey of web services orchestration and choreography with formal models
8. Pahl, C., Zhu, Y.: A semantical framework for the orchestration and choreography of web services. Electron. Notes Theoret. Comput. Sci. **151**, 3–18 (2006)
9. Peltz, C.: Web service orchestration and choreography. IEEE Comput. Soc. **36**(10), 46–52
10. Roglinger, M.: Verification of web service compositions: an operationalization of correctness and a requirements framework for service-oriented modeling techniques. Bus. Inf. Syst. Eng. **1**, 429–437
11. Rebai, S., Kacem, H.A., Kacem, H.H.: Position paper: an integration approach of service composition models: from choreography to orchestration. In: IEEE 21st International WETICE
12. Yoon, Y., Ye, C., Jacobsen, H.A.: A distributed framework for reliable and efficient service choreographies. Department of Electrical and Computer Engineering University of Toronto

Migration of Agents in Artificial Agent Societies: Framework and State-of-the-Art

Harjot Kaur, Karanjeet Singh Kahlon and Rajinder Singh Virk

Abstract Artificial agent societies are made up of heterogeneous intelligent software agents, which operate locally and cooperate/coordinate with each other in order to achieve their individual goals and the collective goals of a society. Also, these member agents can move from one location/society to another in order to achieve their own (individual) goals as well as to help achieve collective goals of the society. This movement of agents in/between societies is called agent migration. Agent migration is a multi-faceted, dynamic process with various types of dynamics associated to it. In order to facilitate and perpetuate the process of agent migration and study various types of dynamics existing in it, an artificial agent society should be equipped with some module or subsystem, which can handle the task of migration and its consequences in a structured manner. In this paper, we present an agent migration subsystem for an artificial agent society with all its constituents, which will be governing the process of agent migration and handling its consequences. In addition to that, we also discuss various possible typologies of agent migration process.

Keywords Agent migration · Agent migration subsystem · Migration norms · Migration protocols

---

H. Kaur (✉) · K.S. Kahlon · R.S. Virk
Department of Computer Science and Engineering, Guru Nanak Dev University, Amritsar, Punjab, India
e-mail: harjotkaursohal@gmail.com

K.S. Kahlon
e-mail: karanvkahlon@yahoo.com

R.S. Virk
e-mail: tovirk@yahoo.com

L.C. Jain et al. (eds.), *Computational Intelligence in Data Mining - Volume 3*,
Smart Innovation, Systems and Technologies 33, DOI 10.1007/978-81-322-2202-6_4

1 Introduction

An *Artificial Agent Society* [1] can be defined as a collection of intelligent software agents interacting with each other for some purpose and/or inhabiting a specific locality, possibly in accordance to some common norms/rules. These societies are analogous to human and ecological societies, and are an expanding and emerging field in research about social systems. Social networks, electronic markets and disaster management organizations can be viewed as such artificial (open) agent societies and can be best understood as computational societies. The members of such artificial agent societies are heterogeneous intelligent software agents, which are operating locally and cooperating and coordinating with each other in order to achieve goals of an agent society. Artificial Agent Societies can also be viewed as *normative systems*, as in them, the member agents residing have to obey certain rules/norms. These norms/rules are created and abolished from time-to-time in a society and this process is dynamic in nature. Also, the agents residing in the society, while obeying the norms of the society and in order to achieve their individual as well as societal goals move in/between them, and this movement can be termed as *agent migration*.

1.1 Migration

A **migration system** can be defined as a set of places linked by flows and counter-flows of people, goods, services and information [2], which tends to facilitate further exchange, including migration between the places. And, according to demographers, every act of migration involves [3] an origin, destination, and an intervening set of obstacles [4]. The dictionary meaning of word **Migration** according to Oxford online dictionary is

> Movement of people to a new area or country in order to find work or better living conditions.

Considering its *broader demographic perspective*, the term *migration* [3] can also be defined as temporary or permanent move of individuals or groups of people from one geographic location to another for various reasons ranging from better employment possibilities to persecution. Talking in terms of **agents** (intelligent/software), in artificial agent societies (open-agent societies, as only in their case, agents can leave or enter any time from/into the society [5]) as mentioned above, *agent migration* can be defined as movement of agents in and between societies to new locations for various reasons [6], ranging from societal, economic, social or personal. Although, this movement is selective, but it forms an integral part of the broader process known as *development* of societies, which itself is a linear, universal process consisting of successive stages [2]. The study of migration is not only confined to social sciences, but it has got with a vast and significant importance in the disciplines of political sciences, economics, anthropology, biology, psychology, history, law, media, demography, cultural studies, literature and finally in distributed artificial intelligence (computer sciences), because it embraces all dimensions of human experience.

In general, migration is considered as a complex dynamic process and a spatial phenomenon with multi-faceted nature [7] and has its own sophisticated theory and dynamics associated with it. The concept of agent migration has been derived from human or animal migration, because like human or animal migration, it has also got associated with it reasons, circumstances, patterns as well as consequences [3] of/for migration.

1.2 Dynamics in Artificial Agent Societies

Artificial agent societies like any other working system have some kind of *dynamics* [8] existing in them, which are in terms of dynamics of *Agent Migration, Role-Assignment, Norm-Emergence, Security, Agent Reproduction* and *Agent Interaction* [9]. These dynamics are basically changes that occur in the behavioral and emergent properties of the societies due to sensing of a stimulus from an environment as well as due to some external agent(s), who want(s) to migrate into the society. Also, all these types of dynamics are in one way or another related to each other, as all of them are either directly or indirectly affected by one another. In order to maintain the dynamic functionality of the society and manage various types of dynamics existing in it, various *subsystems* [9] can be created inside an artificial agent society, because if a society itself has to manage all these dynamics, then the *functioning* and *structure* of the society will be become inherently very complex.

As, an important issue in the development of artificial agent societies (*especially open*), in which agents can enter and leave any time, is the need to account for the various types of dynamics existing in them, at all phases of the development methodology. Therefore, *formalization of constructs and subsystems* for the above mentioned dynamics existing in artificial agent societies is very important from the point of view of *study* and *development* of artificial agent societies. The subsystems created for managing and governing various types of dynamics [8] existing in an artificial agent society are *Agent Migration Subsystem* (for managing agent migration dynamics), *Role-Assignment Subsystem* (for managing role assignment dynamics), *Norm-Emergence Subsystem* (for managing norm emergence dynamics), *Trust and Security Subsystem* (for managing trust and security dynamics), *Agent Reproduction Subsystem* (for managing crossing-over process of agents and its dynamics), *Agent Interaction Subsystem* (for managing dynamics of agent communication/interaction process).

Related to dynamics existing in the subsystems [9] of role-assignment, norm-emergence, security and agent interaction, a plenty of literature is available, but very less work has been performed in the direction of dynamics existing in agent migration and agent reproduction processes occurring in agent societies. Hence, in this paper, we have chosen to uncover various dynamics aspects of agent migration process [10] existing in artificial agent societies. In order to reveal the various dynamic aspects [7] of agent migration process, it is very important to formulate a *formal framework* [9] for *an agent migration subsystem* in artificial agent societies, and that is what, we have formulated and presented in our work.

This paper is organized has follows. In Sect. 2, we have discussed background work already performed in the area of agent migration. In Sect. 3, we have discussed the various typologies of existing agent migration process in artificial agent societies, and in Sect. 4, we have presented various formal notations and constructs for agent migration subsystem and its constituents, along with description of agent migration subsystem itself. In Sects. 5 and 6, conclusions and future work are covered respectively.

2 Previous Works

In this section, we have tried to summarize various works, which have been performed and are related to agent migration in artificial agent societies. Also, we have illustrated various possible extensions, which can be performed related to them.

The categorization of agent migration has also been performed by Costa et al. [6], as *Internal Migration* and *External Migration* where external migration occurs if agent moves between societies and internal migration occurs if agent moves within a society from one site (group) to another. Also, various consequences related to agent migration are discussed by Glaser and Morignot in [11] and Dignum et al. in [12], and they are in terms of *reorganization* of societies that takes place, i.e., society in which agent enters, has to modify (means modification of structure of its organization) itself in order to accommodate the agent in society's already existing roles, which are distributed amongst its member agents. Also, the agent which has joined the society needs to adapt itself so as to internalize in the organization of the society, it now lives in. The type of reorganization demonstrated by Glaser and Morginot is *static reorganization* and by Dignum et al. is *dynamic reorganization* as in the case of latter, modifications in structure and behavior of an artificial agent society after addition, removal or substitution of an agent are done to the society while it is executing (i.e., without bringing down the system). This particular type of reorganization is in the form of *dynamic adaptation*.

The works presented by above three authors are one of the basic and initial works done in the area of agent migration dynamics, but all these can be expanded also, by adding more aspects and factors to the agent migration dynamics. For instance, the only types of migration presented by Costa et al. [6] are internal and external migration, which are categorized taking into consideration only the factor of geographical distance spanned by an agent while moving from one place to another. But, there are many other factors, which can be considered (i.e., circumstances and adaptability) while agent is to move. Similarly, the only consequence of migration is listed by Dignum et al. [12] and Glaser and Morginot [11], as reorganization (static and dynamic) of society, which is one of the societal consequences, where as there are many other societal and individual consequences, which can be explored upon, while working with the process of agent migration.

The basic motivation behind the reorganization of society is either increase in the *agent's utility* or the *society's utility*. Utility [11, 13] of society means what society has gained out of reorganization after the new agent has joined the society and in order to evaluate the utility of convention of society, cost of reorganization and utility of existing structure once without the newly migrated agent and then with newly migrated agent is evaluated in the form of a utility function, where convention means distribution of roles between agents of the society. And if, the utility of the society increases with the above mentioned agent migration process. The reorganization and hence agent migration is considered beneficial for the society.

The utility of an agent depends on the roles it desires, the roles it has committed to and the confidence with which it is playing its roles. Normally, an agent chooses to integrate into that society, which increases its utility and in that case agent is beneficiary. And on the other hand, a society chooses to integrate with an agent that increases its utility. Initially, an agent joining an agent society will be consulting the institutional layer [13] of the society to commit itself to certain role, which it has to play in a society and in order to see, whether it satisfies the norms and rules enforced in the society by this layer.

The reorganization, an after effect of agent migration, (i.e., integration of a new agent with a society), demonstrated by Dignum et al. [12] is dynamic in nature. They have classified it into two types, i.e., *Behavioral* and *Structural. Behavioral Change,* which is a temporary change occurs when organizational structure remains the same but the agents enacting roles [14, 15], decide to use different protocols for the same abstract interaction described in the structure, in case an agent leaves the society and a new agent joins it. But in *Structural Change*, which is a permanent change [16], a decision is made concerning the modifications of one of the structural elements, i.e., societies adapt to environmental changes due to addition, deletion or modification of its structural elements, (i.e., agents, roles, norms, dependencies, ontologies and communication primitives).

Two different *issues* related to agent integration to an existing agent society are described by Eijk et al. [17] and they are based upon open-ended nature of agent societies that allows for the dynamic integration of new agents into an existing open system. These issues are distinguished as **Agent Introduction** and **Agent Creation**. *Agent Introduction* is addition of a new agent which is existing outside the society into a society and *Agent Creation* is similar to object creation, i.e., a newly integrated agent constitutes a previously non-existent entity and is created in a society for agents with limited resources who may face an overloading of tasks to be performed by them.

When agents migrate between societies, some patterns emerge for migration and they are illustrated by Hafizoglu and Sen [18] as *Conservative*, *Moderate* and *Eager* migration. These patterns emerge from opinions or choices of agents to migrate. For experimenting between choices and patterns they have used two-dimensional toroidal grid in which simulation proceeds in discrete time steps with agents having two types of opinions, i.e., binary or continuous.

For the cultural adaptation of migrant agents (i.e., those agents who migrate) into the destination society (i.e., a society to which agents migrate), Bordini [19, 20] has described an anthropological approach, which aims at allowing interactions among agents from different communities.

For experimenting social simulations based on agent-based models, Filho et al. [9, 10], have used agent migration in artificial agent societies for simulating human migration between societies, by using the concepts of social networks. They have a developed a multi-evolutionary model (MEAM) for social simulations, in which member agents of the society were comprised of multi-layered architecture, that made them to perceive and interpret external stimuli from different perspectives.

All the above mentioned works, which are related to agent migration in artificial agent societies, have no doubt considered agent migration, a vital dynamic process occurring in artificial agent societies, but how it is being initialized, happening and perpetuating in agent societies, and which sub-module or subsystem is responsible for agent migration in agent societies is described by none of them. Therefore, our objective in this paper to illustrate agent migration subsystem (described in next section) and its constituents, which are responsible for initializing the process of agent migration in agent societies [21], and for its perpetuation as well. In this article, we want to address the following issues:

1. Which concepts play a crucial role in each of the development stages (analysis, design and implementation) of artificial agent societies' methodologies for defining agent migration process occurring in artificial agent societies?
2. How can we in general specify concepts such as migration process, migration types, and migration dynamics?
3. What type of constructs can we use to formalize above stated concepts?
4. In addition to migration dynamics, how can various migration norms and protocols be identified and formalized?

Our work aims at developing open agent societies in which agent migration dynamics [22] is an important issue. This consideration of artificial agent societies, therefore, forms the main motivation of our work.

3 Types of Agent Migration

The typologies of agent migration process described by us are inspired from human migration process [23], because both can be considered as same kind of dynamic processes. The process of agent migration can be classified [22] according to three factors, i.e., according to geographical distance spanned [24] in migration, circumstances [23] under which migration takes place, and adaptability [24, 25] of an agent to cope up with new surroundings in the destination society, when it migrates from its society of origin. The society/site of origin from which, agent starts its migration can be termed as **migration source** or **sending society** and the destination society/site to which agent migrates is called **migration sink** or **receiving**

society. The agent who moves from source society to destination society by undertaking an agent migration process is called *migrant agent.*

The first set of migration categories, which are classified according to factor of *geographical distance* spanned by agent are *internal* and *external* migration [17], where

- **Internal Migration** can be defined as migration of agents between two sites in the same society, and
- **External Migration** can be defined as migration of agents between two different societies.

The second set of migration categories, which are classified according to *circumstances* [26] which lead to migration, they are *forced* or *involuntary*, *voluntary* or *imposed* migrations, where

- **Forced or Involuntary Migration** can be defined as migration in which agent is forced to migrate from source of migration to another society or site due to certain unfavorable circumstances at the migration source.
- **Voluntary Migration** can be defined migration in which agent voluntarily migrates to another society or migration sink, due to some favorable circumstances there.
- **Imposed Migration** can be defined as migration in which agent is not forced to migrate to another society but due to persistent unfavorable circumstances at the migration source, agent leaves source.

The third set of migration categories, which are classified according to *adaptability* [2] of an agent, with which it copes up with the new surroundings of the destination or sink society are *conservative* and *innovative* migrations, where,

- **Innovative Migration** can be defined as migration, in which migrating agent after migration into a new destination society, starts playing a new role allocated to it there, and also adapts to its (societies') norms.
- **Conservative Migration** can be defined as migration, in which migrating agent after migration into a new destination society, sticks to its old role, which was played by it in the source society, follows the same norms which it was following there and preserves its identity.

The categories of migration process described in this section are not disjoint categories, as internal and external migrations can be conservative, innovative, forced, voluntary or imposed migrations as well. Also, forced, voluntary and imposed migrations can be innovative and conservative in nature. In addition to this, one another type of classifications, which is the main backbone of our research work is classification of agent migration into **physical** and **organizational** agent migrations in artificial agent societies, where

- **Physical Migration** can be defined as physical movement of agents (mobile agents) in/between agent societies, which are hosted on different platforms, but on same or different computers, which can be geographically spanning different locations.

- **Organizational Migration** can be defined as logical movement of agents between two organizations (societies), which can be geographically spanning different locations. It can also be defined as detachment from organization of activities in one society and the movement of the total round of activities to another [27].

In the case of physical migration, integration of mobile agents [21] to a new society is adjusted by the infrastructure of the network itself, as they move from one site to another. This is because, they find standard environments at every site they visit, in addition to standard script interpreters for the execution of code and standard communication constructs. Whereas, in the case of organizational migration [19], it is the society infrastructure, which has to adjust and work out, in order to integrate the incoming agent, and also, agent has to adapt also to already existing destination society surroundings. Also, organizational migration can be internal, external, forced, voluntary, imposed, conservative or an innovative migration.

4 An Agent Migration Subsystem in Artificial Agent Societies

An agent migration process can be specified in terms of agent migration function, which a potential migrant agent executes to realize the same, the utility function [11] that migrant agent should use to calculate the cost and benefits of its migration, and the migration norms and protocols, which can be used to govern and supervise this process.

An agent migration subsystem [28, 29] can be viewed as an interdependent dynamic subsystem with its own dynamics [30], but is interlinked with other subsystems existing in the same and other societies, feedback and adjustments [28] coming from the agent migration process itself. All the terms that are described below are constituents of an agent migration subsystem, in an artificial agent society.

Definition 4.1 (*the agents world*) The Agents World [1] is a set of artificial agent societies existing in the world of agents, and is defined as a set $S = \{S_1, S_2, \ldots, S_n\}$, with a finite number of n societies.

Definition 4.2 (*Population*) A Population [1] is a set of member agents of any artificial agent society S, and is defined $P_i = \{a_{i1}, a_{i2}, \ldots, a_{im}\}$, with m being finite number of residing member agents in a society S_i.

Definition 4.3 (*Roles*) Roles [14] are place- holders assigned to various member agents of the society according to their capability and role-assignment/allocation protocols of the society. They are defined using a finite set $R = \{r_1, r_2, \ldots, r_k\}$, where k, is the number of roles in the role set R.

The tasks performed by member agents in artificial agent societies are completely determined by the roles they play. And, in open agent societies, in contrast to closed societies, roles have existence outside the agents in the implemented system. In these societies, agents are not completely defined by the roles they play and part of their behavior is determined by their own wishes and objectives as well, which are set and motivated from outside from an artificial agent society.

Definition 4.4 (*Agent*) An Agent a Є P_i, [22, 31] in a society S_i is a tuple $a = \{r_{ia}, u_a, H_{i,} B_a, C_a\}$, where

- r_{ia}—the role allocated to an agent a in society $S_{i,}$
- u_a—utility function of an agent a,
- H_i —the list of the neighboring societies of source society S_i, to which a can migrate,
- B_a—behavior of an agent a, according to which the agent migrates, and
- C_a—corresponds to the internal behavior.

The difference between B_a and C_a is that, the former one describes the behavior of an agent a from external point of view and is influenced by many factors as described next, whereas latter one describes agent from internal point of view, is based on deliberative agent reasoning process and helps agent in vital decision-making (including migration). Also, H_i is set of neighboring societies or one of the potential destination societies' for agent a, if it plans to migrate.

As, in our methodology for specifying the dynamics of agent migration process, we need to assume a certain agent architecture. Therefore, for this purpose, we assume the agent populations in our societies world, comprises of agents with *cognitive architecture*. In agents with cognitive architecture, agent behavior is determined by reasoning (deliberating) with their mental attitudes.

Definition 4.5 (*utility function*) The Utility Function [14] of an agent a Є P_i, associates a value with all roles, corresponding to the interest of the agent a in the considered particular role, i.e., u_a: $R^s \rightarrow \mathrm{R}$ ($s \le k$; a Є P_i,).

Here, in our work, we can assume utility functions to be normalized and additive functions with the following two properties:

- normalized $\leftrightarrow u_i(\phi) = 0$, i Є P_i
- additive $\leftrightarrow u_i(R_i) = \sum_{r \in R_i} u_i(r)$, i Є P_i.

Definition 4.6 (*agent behavior*) An Agent Behavior [28] B_a of an agent a Є P_i, can be defined as migration behavior of the agent, which is based on various push-pull factors [23] or reasons [22] present in source and destination societies, responsible for initialization and perpetuation of migration process.

The Agent Behavior B_a can be stated as

$$B_a = f(RI + SE \text{ factors } + \text{ Individual factors}) \quad (1)$$

where, RI = Relevant Information or input variables,
SE = Societal and Economic Factors, and
I = Individual Factors responsible for agent migration respectively.

Definition 4.7 (*agent migration subsystem*) An Agent Migration Subsystem [9, 32] existing in a society S_i with agent population $P_{i,}$ in agents world S can be defined as a hex-tuple $AMS_i = <P_{m,i,}\ F_i,\ S_i,\ S_j,\ N_{i,}\ MP_i\ M_{i,j}>$ and is an artificial agent society sub-module, responsible for the movement of agents amongst various societies or sites. This subsystem will be having its own migration protocols and mechanisms named as $MP_{i,}$ by which policy-making decisions for agent migration will be structured. The basic elements of this agent migration subsystem can be described as follows.

- $P_{m,i}$ —a set of migration agents.
- F_i —a set of facilitator agents who help in migration of potential migrant agents $P_{m,i}$.
- S_i —a sending or source society or society of origin, from which process of migration initiates.
- S_j —a receiving or sink society or society of destination, where process of migration terminates.
- N_i —a set of migration norms existing in an artificial agent society S_i.
- MP_i —a set of migration protocols existing in an artificial agent society S_j.
- $M_{i,j}$ —the migration function, which facilitates migration process for an agent $a \in P_i$, from source society S_i with population P_i to destination society S_j using facilitator agents $F_{i,}$ which will help agents decide whether to migrate or not.

Definition 4.8 (*migrant agents*) Migrant Agents [7] are denoted by $P_{m,i} \subset P_{i,}$ and is defined as a set of member agents who are potential migrants in society S_i with agent population P_i. These agents will be termed as migrant agents. This set will be comprised of only selective member agents who are interested or willing to migrate from source society S_i to some destination society or have migrated from some source society to destination society S_i. This set will be comprised of two types of agents, i.e.,

Definition 4.9 (*immigrant agents*) Immigrant Agents [30] are denoted by $P_{im,i} \subset P_{m,i}$, and are defined as a set of member agents who have migrated from source society to destination society S_i with agent population P_i.

Definition 4.10 (*emigrant agents*) Emigrant Agents [30] are denoted by $P_{em,i} \subset P_{m,i}$, and are defined as a set of member agents who want to migrate from source society S_i, with existing agent population P_i in it, to some other destination society.

Definition 4.11 (*facilitator agents*) Facilitator Agents [7] are denoted by $F_i \subset P_i$, and are defined as a set of member agents residing in source society S_i, who help potential migrant agents $P_{m,i}$ to move from their source society to destination society (S_j) by taking into consideration the behavior of agent towards migration. For any potential migrant agent a, its behavior will be B_a.

Definition 4.12 (*Neighborhood*) Neighborhood [4] is defined as list of societies $H_i \subset S_i$, which lie close or adjacent to the source society S_i, and can be said as one of the candidate destination societies, to which members agents of source society S_i may plan to migrate while realizing the process of migration.

And, in addition to migrant agents and facilitator agents in an agent migration subsystem, migration norms and protocols also play a very vital role in the process of agent migration. Although norms and protocols are either *rules or guidelines*, which are framed by a particular society and are to be followed by the members of the society, but still they are different in terms of how they are followed by various members of the society. For instance, protocols remain constant and there are no behavior expectations related to individuals by whom they are to be followed, when they are framed in a society. But on the other hand, norms are dynamic and variable in nature and they are created and changed frequently in the society from time to time. Also, related to norms is behavior expectation, i.e., an individual following a particular norm is expected to behave in certain specific manner, as specified by the norm. Therefore, in our agent migration subsystem, *migration norms and protocols* will be considered as two different elements with different specifications.

Definition 4.13 (*migration norms*) Migration Norms N_i are set of norms (rules or guidelines) followed by potential migrant agents $P_{m,i}$ in an artificial agent society S_i, when either they have to migrate from S_i to some other destination society, or when they have migrated from any other source society to destination society S_i.

For the managing migration norms and other types of norms existing in the society, i.e., the norm life-cycle, starting from the creation, identification, emergence and abolishment of various norms, there is a separate module or subsystem existing in an artificial agent society, known as *Norm emergence subsystem*. And, how this module works for emergence of migration norms perhaps can be explored in another research paper.

Although migration process is more identified by its lawlessness [26], but still there are laws or rules associated with this process, in order to govern this process and its internal decision-making in a structured manner. These rules or laws associated with the process of migration are called *migration protocols or policies*.

Definition 4.14 (*migration protocols*) Migration Protocols MP_i in a society S_i are meta- protocols (set of rules and policies), associated with its agent migration subsystem AMS_i for governing the process of agent migration of agents from society S_i to other societies, and from other societies to S_i.

The set of migration protocols, used in our agent migration subsystem are inspired from laws of migration used in human societies, which are commonly known as *Ravenstein's laws of migration* [33]. The Ravenstein's laws of migration (*7 in number*) can be listed as:

1. Most migrants mainly move over short distances.
2. Most migration is from agricultural to industrial areas.
3. Large towns grow more by migration than by natural increase.
4. Migration increases along with the development of industry, commerce and transport.
5. Each migration stream produces a counter-stream.
6. Females are more migratory than males.
7. The major causes of migration are economic.

From the above mentioned set of laws of migration presented by Ravenstein [33], except 6th law, rest are very relevant with respect to agent migration in artificial agent societies, and the migration policies or protocols framed for an agent migration subsystem in artificial agent societies can be based on them. The formulation of migration protocols of artificial agent societies is itself a complete a research direction in itself, so a complete research article in itself is required to work on them. Therefore, we leave this work as one on of the future research directions in the area of agent migration dynamics.

Finally, the *migration function* leading to the process of migration can be described as:

Definition 4.15 (*a migration function*) A Migration function $\mathrm{M}_{\mathrm{i},j}$ [22] for agent $a \in P_i$ in source society S_i for migrating to destination society S_j at any time $t \in T$, where P_i and P_j are populations of source and destination society respectively, and $D_{i,j}$, is distance between them, leading to the process of agent migration amongst them can be defined as:

$$\mathrm{M}_{\mathrm{i},j} : (a \times \mathrm{T}) \rightarrow S_j \tag{2}$$

Here, T is a time set comprised of discrete time steps, $\{t, t + 1 ,\ldots, t + n\}$.

Also, The Migration Function $\mathrm{M}_{\mathrm{i},j}$ can be written as:

$$\mathrm{M}_{\mathrm{i},j} = \frac{GP_iP_j}{D_{i,j}^2} \tag{3}$$

where, G is constant, and $D_{i,j}$ is the distance between the two societies S_i and S_j and it is measured in terms of number of hops agent has to take to move from one society S_i to S_j.

The above relationship states that *migration function* is directly related to the populations of source and destination societies' population sizes [34] and inversely related to square of distance between them, and this relationship is derived from *gravity model* of human migration [23], which is based on modified version of *Newton's Law of Gravitation*.

All the constituents of agent migration subsystem which are described above in the brief form, function together collectively in order to initiate, perpetuate and realize the agent migration process [35] in artificial agent societies. The details of these constituents can be studied in detail and can be formalized and implemented in any of the agent programming languages using tools such as MIMOSE, NetLogo or RePast for modeling artificial agent societies in social simulation [36, 37].

5 Conclusion

In this work, we have presented agent migration process in artificial agent societies from its very basic idea, and how it is inspired from human migration process. We have also presented various typologies related to agent migration process (again inspired from human migration process) according to various factors, i.e., geographical area spanned by migrant agents, circumstances in which migration takes place, and adaptability of the migrant agents in the destination society after the process of migration takes place. Also, an agent migration subsystem, is formulated for artificial agent societies with its constituents (migrant agents, facilitator agents, source and destination societies, neighborhood, migration norms, migration protocols and migration function itself), in order to facilitate the process of agent migration in/between agent societies. The constituents of agent migration subsystem, along with their notations, terminology and functionality are presented by us, in brief manner. This agent migration subsystem can be used to simulate artificial social evolution and various types of migration processes existing in human societies, in order to study and analyze various trends, patterns existing in it.

6 Future Research Directions

An agent migration subsystem proposed in this paper, along with various typologies of agent migration process can be very helpful in social simulation [36, 37], for performing experiments related to human migration in demographic societies, as such type of experiments are very difficult to conduct with real world human beings. In the agent migration subsystem, which is presented in our paper, migration protocols and migration norms are presented in brief manner, they can be elaborated in any of future works, pursued in the direction of agent migration dynamics in artificial agent societies. What will be role played by the utility function, how it will be calculated, when an emigrant agent in some source society, becomes an immigrant agent in another destination society, can be pursued as one of the future research directions. Also, how this agent migration subsystem can be linked with various other subsystems of its own societies and other societies can be one of the research directions. The net-migration, in-migration and out-migration of an artificial agent society can be studied upon by devising a proper set of migration metrics for artificial agent societies. Still, the area of patterns of agent migration has to be explored completely; and it can be also be pursued as one of the future research directions. In future, we also plan to implement the above described agent migration subsystem module in an artificial agent society with its constituents using any of artificial agent societies' modeling tools like RePast, MIMOSE or NetLogo, so that it can be used in social simulation.

References

1. Gilbert, N., Conte, R.: Artificial Societies: The Computer Simulation of Social Life. UCL Press, London (1995)
2. de Haas, H.: Migration and development: a theoretical perspective. Int. Migrat. Rev. **44**(1), 1–38 (2010)
3. Weeks, J.R.: The migration transition—chapter 9. In: Population: An Introduction to Concepts and Issues, 9th edn. Wadsworth Learning, USA (2005)
4. Lee, E.S.: The theory of migration. Demography **3**(1), 47–57 (1966)
5. Lekeas, G., Stathis, K.: Agents acquiring resources through social positions: an activity based approach. In: Proceedings of 1st International Workshop on Socio-Cognitive Grids: The Net as a Universal Human Resource, Santorini, Greece (2003)
6. da Costa, A.C., Hubener, T.F., Bordini, R.H.: On entering an open society. In: XI Brazilian Symposium on AI, Fortaleza, Brazalian Computing Society, pp. 535–546 (1994)
7. de Haas, H.: The internal dynamics of migration processes: a theoretical inquiry. J. Ethn. Migr. Stud. **36**(10), 1587–1617 (2010)
8. Brown, R.: Group Processes: Dynamics Within and Between Groups. Wiley, Chichester (1991)
9. Kaur, H., Kahlon, K.S., Virk, R.S.: A formal dynamic model of artificial agent societies. Int. J. Inf. Commun. Technol. **6**(1–2), 67–70 (2013)
10. Filho, H.S.B., de Lima Neto F.B., Fusco, W.: Migration and social networks: an explanatory multi-evolutionary agent-based model. In: 2011 IEEE Symposium on Intelligent Agents (IA), IEEE, Paris (2011)
11. Glasser, N., Morignot, P.: The reorganization of societies of autonomous agents. In: Proceedings of 8th European Workshop on Modeling Autonomous Agents in Multi Agent World, Ronneby, Sweden, LNCS 1237. Springer, Berlin (1997)
12. Dignum, V., Dignum, F., Sonen Berg L.: Towards dynamic reorganization of agent societies. In: Proceedings of CEAS: Workshop on Coordination in Emergent Agent Societies at ECAI2004, Valencia, Spain, 22–27 Sept 2004
13. Dignum, V., Dignum, F.: modeling agent societies: coordination frameworks and institutions. In: Proceedings of 10th Portuguese Conference on Progress in AI, Knowledge Extraction, MAS Logic Programming and Constraint Solving (2001)
14. Dastani, M., Dignum, V., Dignum, F.: Role-assignment in open agent societies. In: Proceedings of AAMAS'03: Second International Joint Conference on Autonomous Agents and Multi-Agent Systems, Melbourne, Australia, ACM (2003)
15. Dastani, M., van Riemsdijk, B.M., Hulstijn, J., Dignum, F., Meyer, J.J.C.: Enacting and deacting roles in agent programming. In: Odell J.J., Giorgini P., Muller J.P (eds.) Agent Oriented Software Engineering, LNCS, vol. 3382, pp. 189–204. Springer, Heidelberg (2004)
16. Dignum, V., Dignum, F., Furlado, V., Melo, A.: Towards a simulation tool for evaluating dynamic reorganization of agent societies. In: Proceedings of WS, on Socially Inspired Computing, AISB Convention (2005)

17. Eijk, R.M., de van Boer, F.S., van der Hoek, W., Meyer, C.J.J.: Open multi-agent systems: agent communication and integration. In: Proceedings of Intelligent Agents VI ATAL, pp. 218–222 (1999)
18. Hazoglu, F.M., Sen, S.: Patterns of migration and adoption of choices by agents in communities. In: Conitzer, V., Winiki, M., Padgham L., van der hock, W. (eds.) Proceedings of the 11th International Conference on Autonomous Agents and Multi Agent Systems (AAMAS 2012), International Foundation for Autonomous Agents and Multi Agent Systems, Valencia, Spain (2012)
19. Bordini, R.H.: Contributions to an Anthropological Approach to the Cultural Adaptation of Migrant Agents. Department of Computer Science, University College London, London (1999)
20. Bordini, R.H.: Linguistic support for agent migration. Masters Thesis, Universidade Federal do Rio Grands do Sul (UFRGS), Instituto de Informatica, Porto Alegre, RS–Brazil (1995)
21. Saber, M., Ferber, J.: MAGR: integrating mobility of agents with organizations. In: Proceedings of International Conference Intelligent Systems and Agents IADIS, Lisbonne, Portugal (2007)
22. Kaur, H., Kahlon, K.S., Virk, R.S.: Migration dynamics in artificial agent societies. Int. J. Adv. Res. Artif. Intell. **3**(2), 39–47 (2014)
23. King, R.: Theories and typologies of migration. International migration and ethnic relations. In: Willy Brandt series of working papers. 3/12. Malmo Institute for Studies of Migration, Diversity and Welfare (MIM), Malmo University, Sweden (2012)
24. Human Migration Guide.: Xepeditions, National Geographic Society, a tutorial
25. Human Migration.: Wikipedia, the free encyclopedia. http://wikipedia:org/wiki/Migration_(human)
26. Zanker, J.H.: Why do people migrate? A review of the theoretical literature. Working paper MGSoG/2008/WP002MPRA. Munich Personal RePEc Archive, Maastricht University, Maastricht Graduate School of Governance, Netherlands (2008)
27. Goldscheider, C.: Population, modernization and social structure. Little, Brown & Co., Boston (1971)
28. Andrew, D.Z.S., Zara, M.: Immigration behavior: towards social psychological model for research. In: Proceedings of ASBBS Annual Conference, ASBBS Annual Conference, vol. 20, no.1, Las Vegas (2013)
29. Bijwaard, E.G.: Modeling migration dynamics of immigrants. Tinbergen Institute discussion paper, no. 08-070/4 (2008)
30. Helmenstein, C., Yegorov, Y.: The dynamics of migration in presence of chains. Res. Memo. 334 (1993)
31. Nemiche, M., M'Hamdi, A., Chakraoui, M., Cavero, V., Pla Lopez, R.: A theoretical agent-based to simulate an artificial social evolution. Syst. Res. Behav. Sci. **30**(6), 693–702 (2013)
32. Filho, H.S., de Lima Neto, F.B., Fusco, W.: Migration, communication and social networks—an agent-based social simulation. In: Complex Networks, Studies in Computational Intelligence, vol. 424, pp. 67–74. Springer, Berlin (2013)
33. Ravenstein, E.G.: The laws of migration. J. Stat. Soc. London **48**(2), 167–235 (1885)
34. Greenwood, M.J.: Modeling migration. In: Encyclopedia of Social Measurement, vol. 2. Elsevier, New York (2005)
35. Biondo, A.E., Pluchino, A., Rapisanda, A.: Return migration after brain drain: a simulation approach. J. Artif. Soc. Soc. Simul. **16**(2), 11 (2013)
36. Gilbert, N.: Agent-based social simulation: dealing with complexity. Complex Syst. Netw. Excell. **9**(25), 1–14 (2005)
37. Bonabeau, E.: Agent-based modeling: methods and techniques for simulating human systems. Proc. Nat. Acad. Sci. U.S. Am **99**(3), 7280–7287 (2002)

Automated Colour Segmentation of Malaria Parasite with Fuzzy and Fractal Methods

M.L. Chayadevi and G.T. Raju

Abstract Malaria is an endemic, global and life threatening disease. Technically skilled person or an expert is needed to analyze the microscopic blood smears for long hours. This paper presents an efficient approach to segment the parasites of malaria. Three different colour space namely LAB, HSI and gray has been used effectively to pre-process and segment the parasite in digital images with noise, debris and stain. L and B plane of LAB, S plane of HSI of input image is extracted with convolution and DCT. Fuzzy based segmentation has been proposed to segment the malaria parasite. Colour features, fractal features are extracted and feature vectors are prepared as a result of segmentation. Adaptive Resonance Theory Neural Network (ARTNN), Back Propagation Network (BPN) and SVM classifiers are used with Fuzzy segmentation and fractal feature extraction methods. Automated segmentation with ARTNN has recorded an accuracy of 95 % compared to other classifiers.

Keywords Fuzzy segmentation · Fractal features · ARTNN · BPN · SVM

1 Introduction

Malaria is one of the top 10 most infectious and dangerous disease of the world which leads to maximum mortality and morbidity in most of the countries. More than 500 million people have died and 1 million deaths every year have been reported from malaria. It is caused by a unicellular micro-organism plasmodium which a parasitic protozoa and it is found in the blood as a carrier of malaria disease. Still, Malaria can be preventable and curable. The current state of the art involves manual counting by a laboratory technician who can distinguish staining

M.L. Chayadevi (✉)
JSSATE, Bangalore, India
e-mail: chayadevi1999@gmail.com

G.T. Raju
RNSIT, Bangalore, India
e-mail: gtraju1990@yahoo.com

L.C. Jain et al. (eds.), *Computational Intelligence in Data Mining - Volume 3*,
Smart Innovation, Systems and Technologies 33, DOI 10.1007/978-81-322-2202-6_5

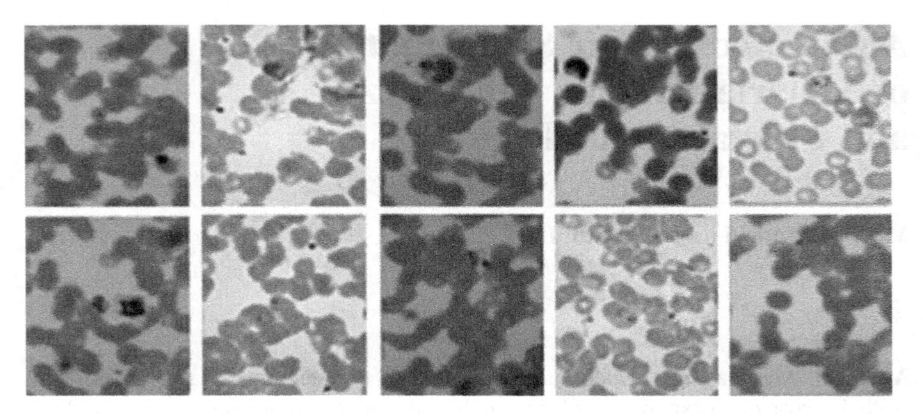

Fig. 1 Blood smear images of malaria parasite

artifacts from actual nuclei, white blood cells [1–5]. Although manual counting is relatively inexpensive to implement, adequate sensitivity requires proper training and supervision of technicians. In this paper an automated technique is proposed for malaria parasites segmentation from smear blood images. Conceivably, automation of this task could both facilitate laboratory efficiency as well as provide an alternative diagnostic tool in conjunction with microscopy in developing countries.

Clinical diagnosis of malaria starts with blood sample collection and analysis. Sample of blood is placed on the glass slide and then stained with Giemsa stain. This process involves lot of inherent technical limitations and human inconsistency problems. Figure 1 shows sample blood smear images of MP image dataset with lot of difference in staining colour, background, noise and debris.

In this paper, automated malaria parasite detection is proposed with fuzzy based colour segmentation with fractal feature extraction methodology. Input RGB images of blood smear were converted to gray, LAB and HSI colour models. L and B plane of the LAB and S plane of HSI images were used for colour segmentation. Fuzzy based image segmentation with fractal feature analysis is carried out for malaria parasite identification. Three prominent classifiers: Adaptive Resonance Theory Neural Network (ARTNN), Back Propagation Network with feed forward neural network (BPN) and Support Vector Machine (SVM) classifiers are used to classify the segmented objects as Malaria and Non-malaria parasites. Performance measure such as accuracy, precision, recall, sensitivity, specificity and f-measures are calculated for image dataset with each classifier. Proposed segmentation with ARTNN shows better results compared to other classifiers.

2 Automated Segmentation of Malaria Parasite

The input digital images have been captured using a CCD camera over bright field conventional microscope at JSS Medical College, Mysore. More than 500 digital images were obtained from suspected malaria patients at the hospital. The proposed

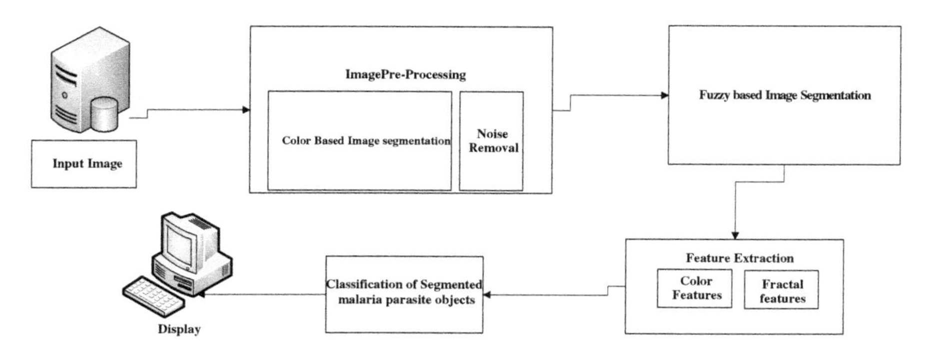

Fig. 2 Proposed architecture for malaria parasite segmentation

work has been shown in Fig. 2. There are 3 basic phases of the proposed work, pre-processing phase, segmentation phase and classification phase. In pre-processing phase, the input RGB image is converted to gray, YCbCr and HSI colour space. The noise and debris of the image is removed with the processed pre-processing methods. Fuzzy based image segmentation method is carried out to segment the stained malaria parasite (MP) objects from the input image. From the segmented objects, colour features and fractal features are extracted and further concatenated feature vectors are prepared as an input to classifiers.

2.1 Pre-processing of MP Images

Pre-processing is an essential step for any image analysis. It enhances the quality of image, which helps in extracting useful information from the image in later stages. The input to this stage is a medical image on which the malaria detection procedure has to be carried out. This stage includes colour conversion of the input image to gray scale, LAB and HSI. Figure 3 shows one of the test images. Conversion from RGB to gray also helps in data reduction. RGB to gray scale conversion is carried out using the equation:

$$\text{Gray} = (\text{Red} * 0.299 + \text{Green} * 0.587 + \text{Blue} * 0.114) \tag{1}$$

Lab is a colour opponent space with dimensions, L for lightness from bright to dark, a and b for colour opponent dimension. Figure 3 represents input RGB image of malaria blood smear and conversion into LAB colour space.

RGB to LAB colour conversion is given by the equation,

$$L = 0.2126\,(R) + 0.7152\,(G) + 0.0722\,(B) \tag{2}$$

$$A = 1.4749\,(0.2213\,(R) - 0.3390\,(G) + 0.1177\,(B)) + 128 \tag{3}$$

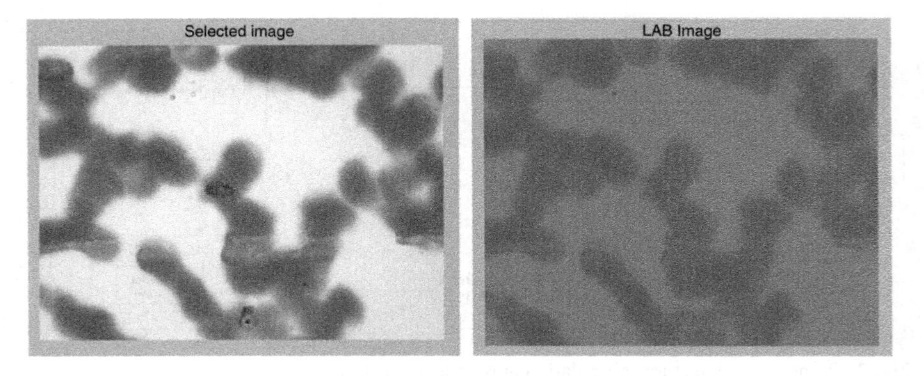

Fig. 3 Colour conversion: input RGB image, LAB converted input image

$$A = 0.6245\,(0.1949\,(R) + 0.6057\,(G) - 0.8006\,(B)) + 128 \tag{4}$$

HSI is a perceptual colour space model and device independent. It is a human perceptual colour model and very near to human vision system. HSI represent the colour model where H represents Hue, which varies from 0 to 360º, S represents saturation or purity, varies from 0 to 1 and I represents brightness value, ranges from 0 to1. To normalize RGB in [0,1], consider Max = max (R, G, B), Min = min (R, G, B). The HSV can be determined using

$$H = \begin{cases} 60 \cdot \left(0 + {G - B}/{(Max - Min)}\right) & if\ R = Max \\ 60 \cdot \left(2 + {B - R}/{(Max - Min)}\right) & if\ G = Max \\ 60 \cdot \left(4 + {R - G}/{(Max - Min)}\right) & if\ B = Max \end{cases} \tag{5}$$

$$H = 0 \quad \text{if} \quad Max = Min \tag{6}$$

$$\text{When } (H<0), \quad H = H + 360 \tag{7}$$

$$S = \begin{cases} 0 & if\ Max = 0 \\ 100, \frac{Max - Min}{Max} & otherwise \end{cases} \tag{8}$$

2.2 Automated MP Image Segmentation

Segmentation is the second phase of our proposed work and used for automated image interpretation. The goal of image segmentation is to simplify and segment the area of interest or object. Image segmentation used for object and boundary identification. Image segmentation is used to identify the malaria parasites from stained background of blood smears. The techniques available for segmentation of medical images are specific to application, imaging modality and type of body part to be studied. Fuzzy technique is considered one of the best methods for medical image segmentation. Fuzzy image segmentation is increasing in popularity because of rapid extension of fuzzy set theory, the development of various fuzzy set based mathematical modelling, synergistic combination of fuzzy, and its successful and practical application in image processing, pattern recognition and computer vision system. The stages involved in the fuzzy based segmentation [6] process are as follows.

1. Perform pixel classification of an image into desired number of regions with an appropriate clustering algorithm.
2. Find the key weight and threshold value, and the membership function for all pixel distributions and determine the centre of all the identified regions to define the membership function.
3. For all i = 1 to n pixels
   a. Select each unclassified pixel from the image and calculate membership function value in each region for that pixel.
   b. Apply fuzzy rule to classify the pixel and assign it to a particular region.

Gestalt, the eminent psychologist mentioned that the visual elements or objects or pixels can be perceptually grouped based on the proximity, good continuation, common fate, similarity, symmetry, relative size, closure and surround principles [7].

2.2.1 Membership Function for Spatial Relations

Membership function is based on fuzzy logic system and the capability of fuzzy rule to classify or segment the object. Proximity and good continuation are used to measure the particular membership function. A strong spatial relationship exists between neighbouring pixels of a particular region when pixels are close together and exhibit relatively smooth variations. Gray level pixel intensities are feature values which are used to construct the membership functions [8, 9]. A candidate pixel will be assigned to appropriate region based on neighbourhood relations. The neighbourhood radius r, o as candidate and # as neighbourhood pixels are shown in Fig. 4 for radius = 1, 2 and 4. In general, total number of neighbours is $(r + 1)^2$ when r = 1 and $(r + 1)^2 - 1$ when r is greater than 1.

```
(a)              (b)              (c) #   #   #   #   #
     #            #   #   #           #   #   #   #   #
#    O    #       #   O   #           #   #   O   #   #
     #            #   #   #           #   #   #   #   #
                                      #   #   #   #   #
```

Fig. 4 Neighbourhood representation. **a** r = 1, **b** r = 2, **c** r = 4

Total number of neighbourhood pixels and their distances from the candidate pixel must be considered to construct a membership function. The membership function μ should possess the following properties [7]:

1.

$$\mu \propto N \tag{9}$$

where N is number of neighbours

2.

$$\mu \propto \left(1/d\left(p_{x,y}, p_{s,t}\right)\right), \tag{10}$$

where $d\left(p_{x,y}, p_{s,t}\right)$ is the distance between pixels $p_{x,y}$ and $p_{s,t}$.

2.2.2 Determination of Weighting Factors and the Threshold

The following details the various stages of the algorithm to automatically determine this key weighting factor and its threshold [6].

1. Set the initial values for the three weighting factors as $W_1 = 1$, $W_2 = 1.8$, $W_3 = 1$
2. Define a set of all regions (*R*) and a set of centre pairs of all regions (*V*)

$$R = \{R_i | (1 \leq i \leq \Re)\}, \tag{11}$$

where R is the number of segmented image regions.

$$V = \left\{\left(C(R_i), C\left(R_j\right)\right) \middle| \forall i, jR_i, R_j \in R \wedge (i \neq j)\right\} \tag{12}$$

where $C(R_i)$ and $C\left(R_j\right)$ are centroids of region R_i and R_j respectively.

3. Compute the absolute sum of differences (*sofd*) between the elements of all pairs

$$sofd = \sum_{i=1}^{nc2} |V_i(1) - V_i(2)|, \tag{13}$$

where $nc2$ is the number of combination pairs of all regions.

4. Determine the approximate threshold T_a

$$T_a = \frac{sofd}{nc2 \times 4} \tag{14}$$

5. Calculate the average sum of differences (*arstd*) between the various standard deviations and approximate threshold

$$arstd = \frac{\sum_{i=1}^{\Re} rstd_i - T_a}{\Re}, \tag{15}$$

where $rstd_i$ is the standard deviation of the i-th region.
6. If the approximate threshold is overestimated, ($arstd < 0$), then the minimum of the standard deviation and T_a is taken as the final threshold value T, provided this value is neither too small (less than K % of T_a, where K is an arbitrary constant) nor zero. If this condition is invalid, then T_a becomes the final threshold.
7. Normalise the average sum of differences between the standard deviation and approximate threshold

$$narstd = arstd/\max(rstd_i, T_a) \tag{16}$$

8. Adjust the weight W_2 accordingly

$$W_2 = W_2 + narstd \tag{17}$$

Fuzzy based colour segmentation is carried out for the MP image dataset. Figure 5 shows segmented results of sample MP test image as binary segmented MP image and colour segmented MP Image.

3 Feature Extraction of MP Image

Smeared blood samples are stained with Giemsa stain to identify the cell nucleus from the image background. Colour segmentation alone cannot segment the bacterium; hence fractal features are also extracted for identification.

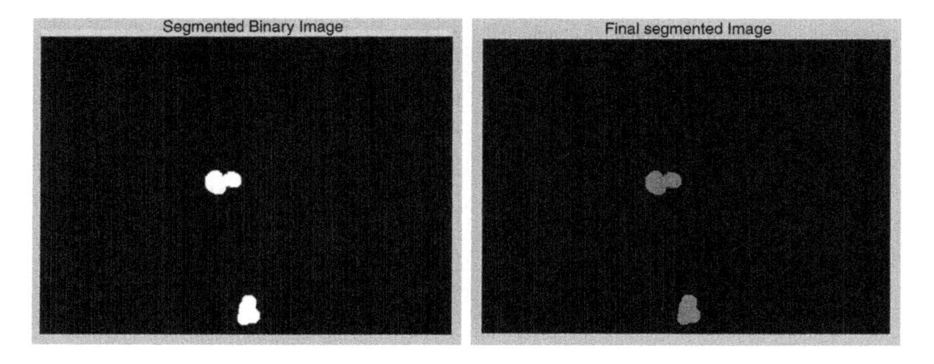

Fig. 5 MP segmentation: binary MP segmented Image, colour segmented MP image

3.1 Colour Feature Extraction

It is used to identify objects of interest, such as human faces, hand areas, for applications ranging from video conferencing, to perceptual interfaces and virtual environments. Use of colour histogram is the most common way for representing colour feature. One of disadvantage of the colour histogram is that it does not take the colour spatial distribution into consideration [10]. Colour feature extraction based on LAB and HSI has been proposed for MP extraction. This method generates 8 features for each plane and 32 features from 4 planes namely, L, B, S and gray image plane.

The steps followed in the extraction are

step 1. Extract the data from L plane and B plane of the LAB converted input image
step 2. Extract the data from S plane of HSI image and gray image
step 3. Perform convolution over 3 × 3 non overlapping matrices of each plane
step 4. Perform feature concatenation of each plane separately
step 5. Concatenate all the features and calculate the mean value of all planes.

Convolution is a mathematical operation to overlap one function over the other. It is used to blend one with other function. Two dimensional convolution equations in spatial form is represented by

$$\mathrm{c(n1,n2)} = \sum_{k_1=-\infty}^{\infty} \sum_{k_2=-\infty}^{\infty} a(k_1,k_2)b(n_1-k_1,n_2-k_2) \qquad (18)$$

where a and b are functions of two discrete variables, n_1 and n_2.

Table 1 represents the colour features of segmented malaria parasites and Table 2 represents segmented Non-malaria objects with fuzzy based segmentation.

Table 1 Colour features of malaria objects

L plane	B plane	S plane	Gray plane
2.0628 + 0.0043i	3.0308 − 0.0007i	0.0111 + 0.0000i	1.9075 + 0.0040i
0.6895 − 0.0006i	1.0099 + 0.0001i	0.0037 + 0.0000i	0.6376 − 0.0006i
1.1925 + 0.0036i	1.7206 − 0.0009i	0.0063 + 0.0000i	1.1022 + 0.0034i
1.1838 + 0.0000i	1.7224 + 0.0002i	0.0063 + 0.0000i	1.0949 − 0.0001i
−0.0015 + 0.0024i	0.0003 − 0.0006i	0.0000 + 0.0000i	−0.0013 + 0.0024i
0.7078 − 0.0000i	1.0051 − 0.0001i	0.0037 + 0.0000i	0.6542 − 0.0000i
0.2064 + 0.0007i	0.2947 − 0.0001i	0.0011 − 0.0000i	0.1908 + 0.0007i
−0.2062 + 0.0006i	−0.2947 − 0.0002i	−0.0011 + 0.0000i	−0.1906 + 0.0006i

Table 2 Colour features of non-malaria objects

L plane	B plane	S plane	Gray plane
2.0799 + 0.0164i	3.0297 − 0.0039i	0.0130 + 0.0001i	1.8991 + 0.0148i
0.6935 − 0.0014i	1.0100 + 0.0003i	0.0043 − 0.0000i	0.6330 − 0.0012i
1.1983 + 0.0079i	1.7195 − 0.0016i	0.0075 + 0.0000i	1.0933 + 0.0071i
1.1793 + 0.0083i	1.7266 − 0.0022i	0.0074 + 0.0000i	1.0759 + 0.0075i
−0.0033 − 0.0012i	0.0012 + 0.0007i	−0.0000 − 0.0000i	−0.0030 − 0.0012i
0.7112 + 0.0034i	1.0044 − 0.0008i	0.0044 + 0.0000i	0.6487 + 0.0031i
0.2078 − 0.0016i	0.2943 + 0.0005i	0.0013 − 0.0000i	0.1896 − 0.0015i
−0.2073 − 0.0004i	−0.2945 + 0.0002i	−0.0013 − 0.0000i	−0.1890 − 0.0004i

3.2 Fractal Feature Extraction

Medical image segmentation is the most challenging task due to the idiosyncrasies of the medical images. A medical image needs special attention for segmentation with unambiguous discrimination of the objects. Fractal features are the most prominent features in the segmentation as it carries geometric structure information [11]. In our research work, 3 planes L and B plane of LAB, S plane of HSI. Here each plane is divided in 4 × 4 matrices. Histogram peak is identified to find the local fractality of each plane [12]. From each plane 16 features are extracted, resulting 48 features from 3 planes. All the features are concatenated and mean value is calculated.

Steps followed in fractal feature extraction is given as follows

1. Extract the data from L, B plane LAB, subtract L and B plane from gray plane
2. Extract the data from S plane of HSI and gray, subtract S plane from gray plane
3. Apply DCT to the resulting planes of the image
4. Divide each plane into 4 × 4 matrices, for each matrix

    a. calculate the histogram features of 16 bins
    b. concatenate the results of each plane separately

5. Concatenate mean of all the plane features.

Table 3 represents Fractal features malaria and non-malaria objects with fuzzy based segmentation.

Table 4 represents comparision of performance measure with ARTNN, BPN and SVM classifier. Performance results shows that Fuzzy based segmentation with ARTNN classifier performs better compared to other classifiers. 70:30, 60:40, 50:50 are the three train:test ratios of image dataset.

Table 3 Fractal features of malaria and non-malaria objects

Malaria				Non-malaria			
L plane	B plane	S plane	Gray plane	L plane	B plane	S plane	Gray plane
−2.4102	1.3146	−1.6597	−9.9163	−3.8611	1.6735	−3.9398	−11.9521
−2.0998	1.6250	2.6589	−6.8936	−3.3999	2.1347	1.5637	−7.2402
−1.7894	1.9354	6.9774	−3.8709	−2.9386	2.5959	7.0672	−2.5283
−1.4790	2.2458	11.2959	−0.8481	−2.4774	3.0571	12.5707	2.1835
−1.1686	−36.2077	15.6144	2.1746	−2.0162	−47.9679	18.0742	6.8954
−0.8582	−31.8892	19.9329	5.1974	−1.5550	−42.4644	23.5777	11.6073
−0.5478	−27.5707	24.2514	8.2201	−1.0938	−36.9609	29.0812	16.3192
−0.2374	−23.2522	28.5699	11.2429	−0.6326	−31.4574	34.5847	21.0310
0.0730	−18.9337	−22.0073	14.2656	−0.1714	−25.9538	−30.7996	25.7429
0.3834	−14.6152	−18.9846	17.2883	0.2898	−20.4503	−26.0877	30.4548
0.6938	−10.2967	−15.9618	20.3111	0.7510	−14.9468	−21.3759	35.1667
1.0042	−5.9782	−12.9391	23.3338	1.2123	−9.4433	−16.6640	39.8786

Table 4 Classifier comparision with performance measures

Performance measures	ARTNN, BPN, SVM comparison for MP image analysis in %								
	ARTNN			BPN			SVM		
	70–30	60–40	50–50	70–30	60–40	50–50	70–30	60–40	50–50
Accuracy	94.93	92.43	90.03	81.10	80.41	76.16	85.71	81.78	79.72
Precision	96.76	94.71	92.70	85.08	78.23	75.14	81.67	76.31	75.38
Recall	94.32	93.14	91.30	78.26	79.56	73.50	71.01	69.36	66.67
Specificity	95.27	91.23	87.89	82.43	80.80	77.41	92.56	90.91	85.88
Sensitivity	94.32	93.14	91.30	78.26	79.56	73.50	71.01	62.36	66.67
F-measure	95.52	93.92	92.01	72.48	72.19	66.41	75.96	68.63	67.82

4 Conclusion

Development of automated segmentation techniques is still a field of interest. Malaria parasite segmentation over large image dataset with lot of noise, debris and colour variations were considered for image analysis. LAB and HSI colour planes with gray colour has been used for pre-processing malaria parasites of blood smear images. Fuzzy based colour segmentation with fractal feature extraction has been proposed for automated segmentation of malaria parasites. L and B planes of LAB images, S plane of HSI images and gray images were considered for feature extraction. Colour features and fractal features are extracted and concatenated feature vector is used further for classification. Our proposed segmentation method with ARTNN classification has shown better results with 95 % accuracy, can assist doctors to serve the mankind.

References

1. Makkapati, V.V., Rao, R.M.: Segmentation of malaria parasites in peripheral blood smear images. IEEE. 978-1-4244-2354-5/09 (2009)
2. Seman, N.A., Isa, N.A.M., Li, L.C., Mohamed, Z., Ngah, U.K., Zamli, K.Z.: Classification of malaria parasite species based on thin blood smears using multilayer perceptron network. Int. J. Comput. Internet Manag. **16**, 46–52 (2008)
3. Hirimutugoda, M.Y.M., Wijayarathna, G.: Image analysis system for detection of red cell disorders using artificial neural networks. J. Bio-Med. Inf. **1**, 35–42 (2010)
4. Savkare, S.S., Narote, S.P.: Automatic detection of malaria parasites for estimating parasitemia. Int. J. Comput. Sci. Secur. (IJCSS) **5**, 310 (2011)
5. Boray Tek, F., Dempster, A.G., Kale, I.: Malaria parasite detection in peripheral blood Images. In: Proceedings of Medical Imaging. Annual Conference, Manchester, UK (2006)
6. Karmakar, Gour C., Dooley, Laurence S.: A generic fuzzy rule based image segmentation algorithm. Pattern Recogn. Lett. **23**, 1215–1227 (2002)
7. Wertheimer, M.: Laws of organization in perceptual forms. Pshychol. Forsch. 6 (1923)
8. Kellogg, C.B., Zhao, F., Yip, K.: Spatial aggregation: language and applications. In: International Proceedings of AAAI (1996)
9. Yip, K., Zhao, F.: Spatial aggregation: theory and applications. J. Artif. Intell. Res. **5**, 1–26 (1996)
10. Alamdar, F., Keyvanpour, M.R.: A new colour feature extraction method based on dynamic colour distribution entropy of neighborhoods. IJCSI Int. J. Comput. Sci. Issues. **8**, 42 (2011)
11. Tang, Y.Y., Tao, Y., Ernest, C.M.: Lam: new method for feature extraction based on fractal behavior. Pattern Recogn. **35**, 1071–1081 (2002)
12. Dobrescu, R., Ionescu, F.: Fractal dimension based technique for database image retrieval. In: IONESCU (2003)
13. Jain, A.K., Mao, J., Mohiuddin, K.M.: IEEE (1996)

References

Syllable Based Concatenative Synthesis for Text to Speech Conversion

S. Ananthi and P. Dhanalakshmi

Abstract Speech Synthesis functions as a medium which converts text into speech. Speech Recognition and Speech Synthesis plays a vital role in Human-Machine Interaction. Synthesized speeches are extracted from concatenating the pieces of pre-recorded speech utterances from the database. The proposed work converts the written text into a syllables (syllable text representation) using rule based approach and subsequently it converts the syllable representation to modified syllable waveform clips that can be combined together to produce as sound. Syllabic transcription attempts to describe the individual variations that occur between speakers of a dialect or language. Syllable based concatenative synthesis aims to record the syllables that a speaker uses rather than the actual spoken variants of those syllables that are produced when a speaker converse a word. The Concatenative Speech Synthesis methods provide highly understandable speech utterance.

Keywords Concatenative speech synthesis · Concatenate wave segments · Syllable · Syllable transcription · Speech processing · Speech synthesis (SS) · Text normalization · Text to speech (TTS) conversion · Waveform concatenation

1 Introduction

Speech Synthesis refers to the capability of the computer to reproduce the written text into machine generated speech [1]. Systems may differ in their size of the stored database, which depend on its requirements or applications. Recent inclination is to collect huge database of fluent speech. The effectual method for Text to Speech conversion is concatenation method [2]. In this method, the system intends

S. Ananthi (✉) · P. Dhanalakshmi
Department of Computer Science and Engineering, Annamalai University, Chidambaram, Tamil Nadu, India
e-mail: ananthi68@gmail.com

P. Dhanalakshmi
e-mail: abidhana01@gmail.com

L.C. Jain et al. (eds.), *Computational Intelligence in Data Mining - Volume 3*, Smart Innovation, Systems and Technologies 33, DOI 10.1007/978-81-322-2202-6_6

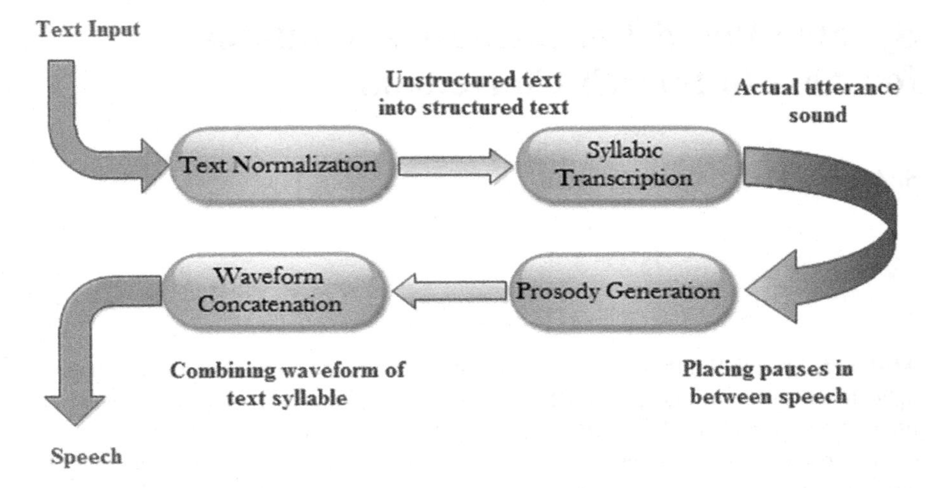

Fig. 1 Framework of proposed system

to select an optimal sequence of acoustic units at run time to synthesize a particular utterance. In this work, syllable utterances were considered because of its increasing usage. The framework of the proposed system is depicted in Fig. 1.

In SS, text document is given as input to the system which possibly consists of non-standard formats like digits/integers, numerical expressions, cardinal suffixes etc. Syllabic texts are extracted from the normalized form of the given input. Text normalization and Syllabic transcription are discussed in Sects. 2.1 and 2.3.

2 Concatenative Speech Synthesis

Text to Speech (TTS) Conversion system converts normal language of text into utterances [3]. Concatenation of speech synthesis can be created by concatenating pieces of recorded speech from the pre-recorded speech database, for the conversion of text into speech the system must generalize all the speech units uniquely. Any text may contain special elocution that should be stored in lexicon form. Concatenative Speech Synthesis converts the written text into syllabic representation and then it converts the generated syllabic representation into waveform. These waveforms are combined to produce as sound.

2.1 Text Normalization

Text to Speech Synthesizer mechanism works internally by synthesizing words. However, input text documents contains words with collection of written elements such as statistics, date, time, abbreviations, numbers, symbols etc., If the text consist of abbreviations and numbers, then the system must determine how these non-standards should be read out. Any text that has a special pronunciation should be stored in lexicon such as Abbreviation, Acronyms, Special symbols etc. All diverse elements must be first converted into general or actual utterances and then only the system can synthesized as speech [5]. Such conversion of diverse units into actual utterances which takes place internally within the synthesizer is expressed as Text Normalization.

In general, Text Normalization is the conversion of text that includes non-standard word such as statistics, abbreviations, misspelling into normal words. Each literal symbols are recognized as units. These symbols should be separated from adjacent text with white space. Speech unit can be either fixed size diphones or variable length units such as syllables and phones.

Open close brackets, equals, greater-than, dollar, percent, Pound sign, comma etc. are termed as Literal Symbols. Phrases in a context consist of literal symbols and numerical representation. The explanation for all the literals and numerical representation has to be declared initially for the conversion of non-structural to structural representation.

2.2 Syllable

A syllable is a basic unit of written language which consists of uninterrupted sound that can be used to form words. In other words, a syllable is the sound of a vowel that is created when pronouncing a word.

The word vowel comes from the Latin word *vocalis*, which means vocal [6]. According to phonetics, vowel is the pressure wave generated from the lungs and passed through vocal tract without any interruption. In English, A, E, I, O and U are the waves generated without any constriction by other vocal organ.

2.3 Syllabic Transcription

There are few rules for diving words into syllables. There are four ways to split up a word into its syllables:

Rule 1: Divide the word between two middle consonants

Rule 2: Usually divide before a single middle consonant

Rule 3: Divide before the consonant before an "-le" syllable

Rule 4: Divide off any compound words, prefixes, suffixes and roots which have vowel sounds

According to rule 1, split up words which have two middle consonants. For example consider the following words hap/pen, let/ter, din/ner. Happen consist of two consonants pp at the middle of the word, similarly for letter and dinner also it has two middle consonants.

The only exceptions are the consonant digraphs. Never split up consonant digraphs as they really represent only one sound. The exceptions are "th", "sh", "ph", "th", "ch", and "wh".

According to Rule 2, when there is only one syllable, it usually divides in front of it. "o/pen", "i/tem", "e/vil", and "re/port" are few examples for Rule2. The only exceptions in these are those times when the first syllable has an obvious short sound, as in "cab/in".

Based on Rule 3, a word that has the old-style spelling in which the "-le" sounds like "-el", divide before the consonant before the "-le". For example: "a/ble", "fum/ble", "rub/ble" "mum/ble" and "thi/stle". The only exceptions in these are "ckle" words like "tick/le".

According to Rule 4, the word is Split off into parts of compound words like "sports/car" and "house/boat". Divide off prefixes like "un/happy", "pre/paid", or "re/write". Also divide off suffixes as in the words "farm/er", "teach/er", "hope/less" and "care/ful". In the word "stop/ping", the suffix is actually "-ping" because this word follows the rule that when it is added with "-ing" to a word with one syllable, it doubles the last consonant and add the "-ing".

2.4 Prosody

Prosody is the pitch, volume and speed that words, sentences and phrases are spoken with. The system may also need to split the input into smaller chunks of output text to determine which words needs to be emphasized. The term prosody refers to both pitch and the placement of pauses in between speech for making synthetic natural speech sound. It is very easy for a human speaker to pause at any places in speech, but it's complicated for the machine to fabricate sounds. Prosody plays an important role in guiding listener for speaker attitude towards the message [7]. Prosody consists of systematic perception and recovery of speaker intentions based on Pauses, Pitch, Rate and Loudness [8]. Pauses are used to indicate phrases and separate the two words. Pitch refers the rate of vocal fold cycle as function of time. Rate denotes Syllable duration and time and Loudness represents the relative amplitude or volume.

Initially, the engine identifies the beginning and ending of sentences. The pitch will be likely to fall near the end of a statement and rise for a question. Similarly, the machines starts speaking the syllable and it fall on to the last word, and then pauses are placed in between the sentences or syllable for clear reading. So a prosodic system for a Text to Speech must provide suitable pauses for the speech and also adequate information to make the pitch sound realistic [9]. All the Text to Speech Engines have to convert the list of syllables and their volume, pitch and duration into digital audio. TTS Engine generates the digital audio by concatenating pre-recorded syllables which are stored in a database. The combined pre-recorded smallest unit of sound is given to the TTS Engines which speaks the sentence loud.

2.4.1 Duration

Time duration between the sentences or phones decides the clarity of the speech so that durational assignment plays a vital role in text to speech conversion. Each Phone is pronounced in various duration by the user. The duration of phone d is expressed as

$$d = d_{\min} + r\left(d' - d_{\min}\right) \tag{1}$$

where,

d Average duration of the phone
d_{min} Minimum duration of the phone
r correction

2.4.2 Pauses and Pitch

Pauses are mainly used in running text which is generated in the form of utterance output. In usual systems, the reliable location which is designate to insert pause is the pronunciation symbols [10]. For every Full stop or commas in a sentence the pause has to be placed for absolute reading of phrase in between those sentences. That is text may allow pausing for some duration while it found comma or Full stop in the sentence. So silence sound “Sil” is placed for few milliseconds then Connection “C” has to be made for the continuation for the phrases. Generally, Speech Synthesis engines need to express their usual pitch patterns within the broad limits specified by a Pitch markup [11].

2.4.3 Tone

The prosodic parameters in tones are used to generate the voice output. The tone is determined by calculating “TILT”. “TILT” is directly calculated from f0. From acoustic aspect, it is directly represented by the shape of fundamental frequency (F0) contour. The tone shape is represented by,

$$tilt = \frac{|A_{rise}| - |A_{fall}|}{|A_{rise}| + |A_{fall}|} \quad (2)$$

where,

A_{rise} = Amplitude of rise (in Hz)
A_{fall} = Amplitude of fall (in Hz)

3 Waveform Concatenation

A high quality speech synthesizer concatenates speech waveforms referenced by a large speech database. Speech quality is further improved by speech unit selection and concatenative smoothing [12, 13]. Synthesized speech can be produced by combining pieces of pre-recorded speech that are stored in a database. Based upon the similarity between the human voice and by its ability of understanding decides the quality of a speech synthesizer. The Intelligibility of the speech is very important quality for synthesized speech.

Concatenative synthesis is based on the concatenation or combining the progressive segment of recorded speech together. In general, concatenative synthesis produces the most natural-sounding synthesized speech. The waveform of the sound clips were combined together to form pleasant and audible speeches.

Concatenative TTS System produces very natural sounding speech. Since, they simply join pre-recorded segment or units to form sentences. Speech generated by this approach inherently possesses natural quality [14]. The system generates speech by searching for appropriate combinations of sound in a large database of human speech. The required syllabic fragments are found in database and hence joining or modifying certain speech unit can be avoided. The best combinations are found and their waveforms were concatenated. The following steps were followed by the Synthesis machine to convert text into speech.

Step 1: The text input is normalized and the sentences were converted into isolated or individual words

Step 2: Syllabic text representation is obtained from the normalized text based on syllabification rule

Step 3: Each syllabic unit present in the text is chosen form speech database by identifying its position from the database

Step 4: The duration and the pitch of syllable clips are changed according to their respective position

Step 5: The modified syllabic clips are concatenated using waveform concatenation method with respective pause within each word. To avoid discontinuities between syllables spectral smoothing is applied

Table 1 Text normalization for non-structure representation

Expression type	Normal text (unstructured)	Normalized text (structured)
Date	06/08/1988	Sixth august nineteen eighty eight
		6<digit>, th<Cardinal suffix>, August<word>, 1988<digit>
Tel number	9876543210	Tel{Nine, eight, seven, six, five, four, three, two, one, zero} individual numbers has to be read (int)
Vehicle number	TN 45	T<word_char>, N<word_char>fourty five<digit>
Time	Hour: minute: sec 12:53:49	Time{twelve hours fifty three minutes and fourty seconds}

4 Experimental Results

The quality of the synthetic voice is measured by voice generated by the system formal listening tests. In initial stage non-structural to structural conversion was performed using the concept of text normalization. Table 1 shows the sample conversion of normal structure to normalized structure.

Initially, the given input is normalized and sentences were broken up into the words. Further, syllables present in the words are analyzed and their appropriate positions are also analyzed by the system. It attempts to find the syllables present in the word. After finding the syllables, it searches for the syllable sound from the database by finding the position of the pronounced syllable. Finally, waveform of

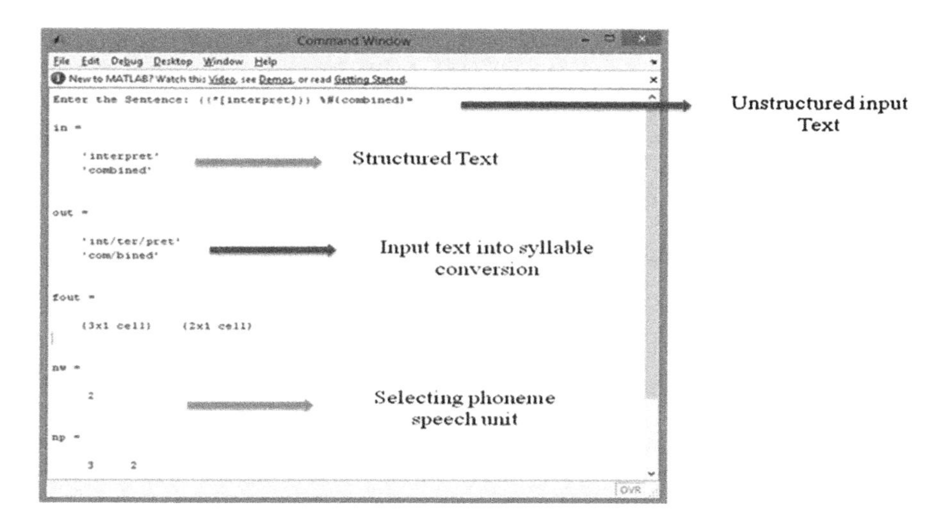

Fig. 2 An example for speech synthesis system

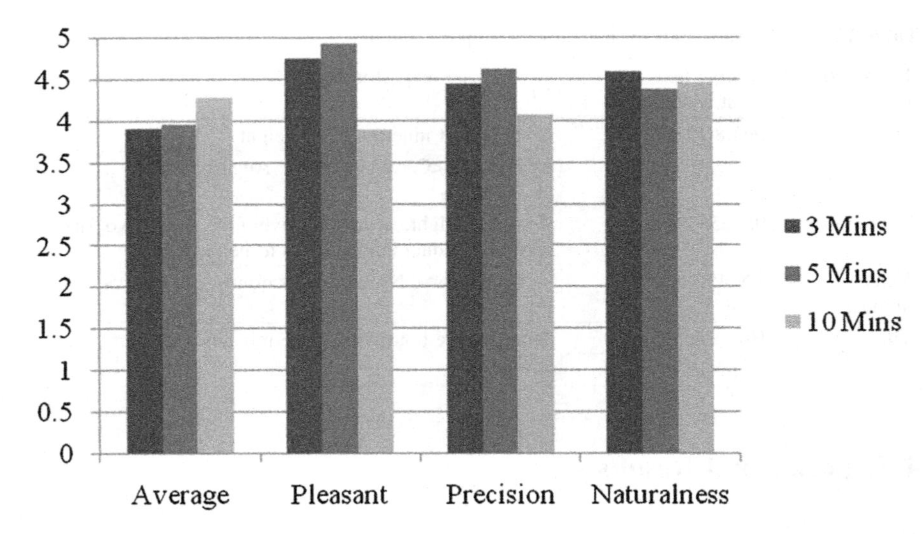

Fig. 3 Mean opinion score for synthesized speech

pronounced syllabic units were analyzed and their sound clips were combined together to form as natural speech. Figure 2 shows an example for speech synthesis system.

A set of 100 sentences were selected as a training set from the database. Training is given for each and every data used in the training stage. A testing data is chosen from the database which are not included or trained in the training set. A set of 75 sentences were selected from the speech database which is not in the training set. Rank of the voice quality decides the quality of the synthetic voice.

20 persons were chosen for testing the converted speech. The average scores given by these persons were considered for ranking this score and this score is termed as Mean Opinion Score (MOS) [15]. MOS for speech synthesis system using 4 point scale rating is shown in Fig. 3. Three categories of speeches were considered for measuring the quality of speech. Speech synthesis using 3, 5 and 10 min of speech were heard by different users. The quality of the speech is measured by the rank given by these 20 users. Rank is measured based on the score of 4 different scales.

The applications of Text to Speech conversions are

- Analysis by synthesis of pathological voices.
- Visual disability.
- Language teaching.

5 Conclusion

Concatenative Speech Synthesis method produces extremely widespread synthesis method. Huge collection of syllables were collected and stored in a database for further practice. Synthesized speech can be extracted by concatenating pieces of pre-recorded speech database. Speech Synthesis based on concatenative method generates the utterance of speech which is similar to human voice. Syllabic transBased upon the speed, accuracy and similarity between human voice the quality of the synthesized speech is analyzed. Mean opinion score using 4 point scale ratings were examined. Naturalness, precision and pleasantness were also analyzed, which are most commonly used criteria for high-quality speech.

References

1. Schuller, B., Zhang, Z., Weninger, F., Burkhardt, F.: Synthesized speech for model training in cross-corpus recognition of human emotion. Int. J. Speech Technol. **15**, 313–323 (2012)
2. Campbell, N.: Developments in corpus-based speech synthesis: approaching natural conversational speech. IEICE Trans. **87**, 497–500 (2004)
3. Campbell, N.: Conversational speech synthesis and the need for some laughter. IEEE Trans. Audio Speech Lang. Process. **17**(4), 1171–1179 (2006)
4. Sreenivasa Rao, K., Yegnanarayana, B.: Intonation modeling for Indian languages. Comput. Speech Lang. **23**, 240–256 (2009)
5. Vowel: Online etymology dictionary. http://www.etymonline.com/index.php?allowed_in_frame=0&search=vowel&searchmode=nl. Accessed 21 Nov 2013
6. Atal, B.S., Hanauer, S.L.: Speech analysis and synthesis by linear prediction of the speech wave. J. Acoust. Soc. Am. **50**, 637–655 (1971)
7. Badin, P., Fant, G.: Notes on vocal tract computation. Techical Report, STL-QPSR (1984)
8. Carlson, R., Sigvardson, T., Sjolander, A.: Data-driven formant synthesis. Technical Report, TMH-QPSR (2008)
9. Banks, G.F., Hoaglin, L. W.: An experimental study of duration characteristics of voice during the expression of emotion. Speech Monogr. **8**, 85–90 (1941)
10. Clark, R.A.J., Richmond, K., King, S.: Multisyn: opendomain unit selection for the festival speech synthesis system. Speech Commun. **49**, 317–330 (2007)
11. Courbon, J.L., Emerald, F.: A text to speech machine by synthesis from diphones. In: Proceeding of ICASSP. PTR, Upper Saddle River (2002)
12. Kim, J.K., Hahn, H.S., Bae, M.J.: On a speech multiple system implementation for speech synthesis. Wireless Pers. Commun. **49**, 533–543 (2009)
13. Saraswathi, S., Vishalakshy, R.: Design of multilingual speech synthesis system. Intell. Inform. Manage. **2**, 58–64 (2010)
14. Ahmed, M., Nisar, S.: Text-to-speech synthesis using phoneme concatenation. Int. J. Sci. Eng. Technol. **3**(2), 193–197 (2014)
15. Campell, N., Hamza, W., Hog, H., Tao, J.: Editorial special section on expressive speech synthesis. IEEE Trans. Audio Speech Lang. Process. **14**, 1097–1098 (2006)

Kannada Stemmer and Its Effect on Kannada Documents Classification

N. Deepamala and P. Ramakanth Kumar

Abstract Stemming is reducing a word to its root or stem form. Kannada is a morphologically rich language and words get inflected to different forms based on person, number, gender and tense. Stemming is an important pre-processing step in any Natural Language Processing application. In this paper, stemming is performed on Kannada words using unsupervised method using suffix arrays. An accuracy of 0.58 % was achieved with this method. The performance of the stemmer is further improved by using a stem-list dictionary in combination with the unsupervised method. A list of 18,804 stem words is created manually in Kannada Language as part of this work. A 10 % improvement in performance is observed. The effect of the proposed stemmer on text classification of Kannada documents using Naïve Bayes and Maximum Entropy methods are compared. It is shown in this paper, that stemming improves the performance of text classification.

Keywords Kannada Stemmer · Text classification · Unsupervised stemming · Naïve Bayes · Maximum entropy · Natural language processing

1 Introduction

Stemming is process of reducing a given word to its root form or stem. Stemming is an important pre-processing step in any Natural Language Processing application. It helps to reduce the number of features in Text classification application. There are

N. Deepamala (✉)
Department of Computer Science and Engineering,
R.V. College of Engineering, Bangalore, India
e-mail: deepamalan@rvce.edu.in

P. Ramakanth Kumar
Department of Information Science and Engineering,
R.V. College of Engineering, Bangalore 560059, India
e-mail: ramakanthkp@rvce.edu.in

L.C. Jain et al. (eds.), *Computational Intelligence in Data Mining - Volume 3*,
Smart Innovation, Systems and Technologies 33, DOI 10.1007/978-81-322-2202-6_7

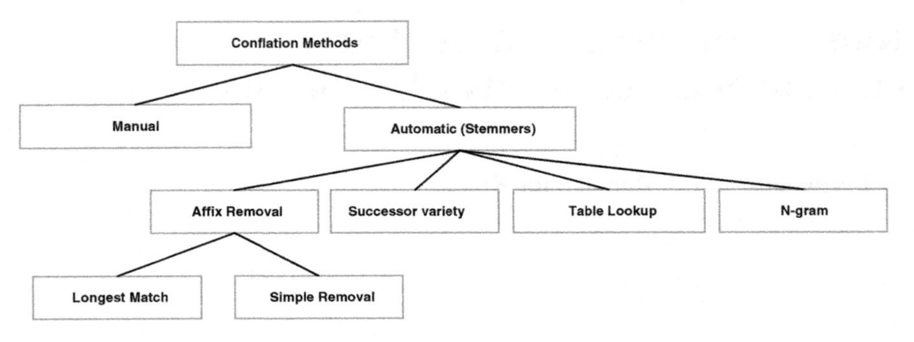

Fig. 1 Types of stemming [1]

many stemming algorithms both supervised and unsupervised available. According to [1], stemming methods or conflation methods can be classified as shown in Fig. 1.

Kannada being one of the 40 most spoken languages in the world requires tools for pre-processing to develop NLP applications. Stemming is an important pre-processing step in Natural Language Processing. Currently, lot of research is being carried out in morphological analysis of the Kannada language and also stemming.

In this paper, an effort is made to stem Kannada words using unsupervised method and to improve performance; a stem-list dictionary of 18,804 stems is added. Finally, the effect of stemming on text classification of Kannada Language documents is evaluated using Naïve Bayes and Maximum Entropy algorithms.

2 Literature Survey

Method of affix removal to reduce a word to its stem form is called stemming algorithms. There are different stemming algorithms developed in English and other languages. The following section explains some of the important stemming algorithms developed for English and other languages.

Lovins Stemmer proposed in [2] uses longest match principle for stemming. Considering it being an initially developed stemmer for English, it performed quite well. It had 297 word endings and was context sensitive. Each match had contextual restrictions to prevent wrong stemming. Paice in [3] stemmer is rule based iterative stemmer. It involves a rules file which specifies to remove or add characters to the endings. It is different from Lovins [2], that there is no recode phase. Porter stemmer [4] defines a suffix stripping algorithm where the suffixes are removed in linear steps.

Stemming of Indian Languages:

Indian languages are highly inflectional, and the word gets inflected to a different word based on gender, number, tense and person. Recently lot of work is being done in different Indian languages on different NLP applications and other related tasks. Stemming being an important pre-processing task, some important work carried out for Indian languages is listed below.

In [5], Ramanathan et al. developed a stemmer which removes the longest matching suffix from a word. For this they handcrafted a list of suffixes by considering the inflections for different parts of speech. With this method, an accuracy of 88 % was achieved with error of over-stemming of 13.84 % and under-stemming of 4.68 %. Islam et al. [6] proposed a similar suffix stripping stemmer for Bengali which could also detect mis spelled words and suggest corrections. With single error misspelling, an accuracy of 90.8 % was achieved and with multi-error misspellings an accuracy of 67 % was achieved. In Majumder et al. [7] stemmer using distance similarity measure, different string distance measures $\{D_1, D_2, D_3, D_4\}$ were defined for clustering of lexicon. For given two strings $X = x_0x_1x_2.....x_n$ and $Y = y_0y_1y_2.......yn$, a Boolean function p_i for penalty is given as

$$P_i = \begin{cases} 0 & \textit{if } x_i = y_i \; 0 \leq i \leq \min(n, n') \\ 1 & \textit{otherwise} \end{cases} \tag{1}$$

$$D_1(X,Y) = \sum_{i=0}^{n} \frac{1}{2^i} p_i \tag{2}$$

$$D_2(X,Y) = \frac{1}{m} \times \sum_{i=m}^{n} \frac{1}{2^{i-m}} \; \textit{if } m > 0, \; \infty \; \textit{otherwise} \tag{3}$$

$$D_3(X,Y) = \frac{n-m+1}{m} \times \sum_{i=m}^{n} \frac{1}{2^{i-m}} \; \textit{if } m > 0, \infty \; \textit{otherwise} \tag{4}$$

$$D_4(X,Y) = \frac{n-m+1}{n+1} \times \sum_{i=m}^{n} \frac{1}{2^{i-m}} \tag{5}$$

These distance measures were used to cluster the words into groups based on similarity. Each cluster represents a class which represent different morphological variant of a root word [7].

Pandey et al. [8] uses unsupervised method to develop a stemmer for Hindi. With heuristic approach to derive with the suffixes and prefix for Hindi and F-measure of 94.96 % was achieved. Dasgupta et al. [9] proposed unsupervised morphological parsing of Bengali. First a list of candidate affixes is created by following the method explained by Keshava et al. [10]. The spurious affixes in the list was filtered by counting number of distinct words to which each of those affixes attach and also considering the generative strength. The score of the affix m was calculated with two equations:

$$score(m) = length(m) \times (number\ of\ different\ words\ to\ which\ m\ attaches) \tag{6}$$

$$score(m) = length(m) \times \sum_{w} strength(w) \tag{7}$$

where w is the word to which m attaches.

Candidate roots were extracted using the affixes derived from above step. As an improvement, length dependent thresholds, identifying composite suffixes and improvement in root induction using word-similarity have been incorporated. They achieved an F-Score of 83 %. Majgaonkar et al. [11] has developed a rule based stemmer, an unsupervised stemmer using the n-gram splitting and a frequency based suffix stripping algorithm was also tested using iterative suffix stripping and statistical stripping methods. With statistical suffix stripping a maximum accuracy of 82.5 % was observed.

Suba et al. [12] proposed 2 stemmers for Gujarati. A lightweight inflectional stemmer based on hybrid approach and a heavy weight derivational stemmer based on rule based approach. Different modules like a list of stems, which gets updated with new stem words, POS tagged files and linguistics rules were developed using unsupervised method. The inflectional stemmer performed with accuracy of 90.7 % and for derivational stemmer, an accuracy of 70.7 % was achieved.

Vishal et al. [13] proposed Punjabi stemmer for nouns and proper names. A list of suffixes for nouns and proper names were generated. An algorithm involving different steps is used to remove the suffix. An accuracy of 87.3 % was achieved. Rana et al. [14] use brute force algorithm to achieve stemming. First the given word is matched in a list of words, if found, it is the root word, else suffix list is looked up and suffix is removed to get the root word. An accuracy of 80.73 % was achieved with this stemmer.

In Kannada Language, Padma et al. [15] have developed a morphological stemmer, analyzer, and generator for Kannada Language. The model is based on suffix-stripping, rule-based, and paradigm based. The model has been tested on random nouns. Suma Bhat [16] has tried stemming of Kannada words using three methods. Distance measure as discussed in Majgaonker et al. [11]. It is based on similarity between two strings, a distance measure is calculated and assigned. Smaller the value, higher is the similarity between two strings. The second method followed was clustering using agglomerative algorithm as in Majgaonker et al. [11]. Least distant clusters are merged until single cluster remains. Efficiency is improved by grouping words into pre-clusters by truncating them to first n characters where n = 1, …, 7. The same is used to compare stemming performance. The third method is language independent morpheme induction method followed in Dasgupta and Ng [9]. The algorithm requires the corpus to be converted to list of words with its frequency. A forward and backward tree is created in the algorithm to calculate the transitional probability of a letter given the context. The third step uses the morphemes identified to segment words. Evaluation is done based on Paice [3] method and found that distance measure that perform longest prefix matches is the best performing algorithm. An accuracy of 88.82 % has been achieved using the unsupervised morpheme segmentation approach.

3 Implementation of Unsupervised Kannada Stemmer

Dasgupta et al. [9] proposed unsupervised segmentation algorithm which split the word statistically into [prefix][stem][suffix]. They first extract candidate affixes. Next detect and remove composite suffixes and then extract list of candidate roots. The undivide++ program [17] provided as part of the implementation require lot of effort to detect the correct threshold configuration to obtain correct affix split. Multiple re-runs are required to obtain optimal results. As discussed in the previous section, [16] has used the same method as one of the method to stem the Kannada words. All the methods to stem Kannada words implemented till date are supervised or require many re-runs to find threshold. The following section explains the new method for stemming Kannada words using unsupervised method, where no re-runs are required and only a Kannada word list is sufficient.

Later, text classification using Naïve Bayes and Maximum Entropy method are tested with and without stemming to find the effect of stemming on classification.

3.1 Kannada Corpus

The method discussed below requires a list of suffixes. To automatically generate suffix list, kn_IN.dic dictionary file is used which is freely distributed with Linux spell checker and has 60,985 Kannada words. The kn_IN.dic dictionary is chosen instead of a word list from EMILLE corpus [18] because, all words listed are valid words and no filtering is required. The dictionary consists of combination of different words in root form or in inflected form as shown below.

E.g. ಅಂಕ

ಅಂಕಅಂಶಗಳನ್ನು

ಅಂಕಗಣಿತೀಯವಾಗಿ

ಅಂಕಗಳನ್ನು

ಅಂಕಗಳಲ್ಲಿ

ಅಂಕಗಳಿಕೆ

3.2 Suffix and Prefix Word List

The Steps followed in the proposed unsupervised Kannada Language stemmer is as follows:

Step 1 (**Generate Suffixes list**): Generate suffix array entries for each word in kn_IN.dic.

E.g. Suffix array for word ಅಂಕಗಳನ್ನು is

◌ಂಕಗಳನ್ನು

ಅಂಕಗಳನ್ನು

ಕಗಳನ್ನು

ಗಳನ್ನು

ನು

ನ್ನು

ಳನ್ನು

◌ು

ನು

Step 2: Merge all the suffixes and calculate the frequency of occurrence. 60895 words in kn_IN.dicon converting to suffix arrays and merging generated a list of 239,373 suffixes. The Table 1 shows output of this step which shows 10 most frequent suffixes with its frequency.

Table 1 Highest frequency suffixes

ಲಿ	2600
ದೆ	2808
ನ್ನು	3558
◌್ನು	3563
ನು	3752
ದ	4349
◌್	6601
◌ಿ	7001
◌ಿ	9853
◌ು	12450

Step 3: Remove all the entries in the above generated list with frequency equal to one. Suffixes with frequency one will be the word itself and cannot be a valid suffix. This reduces the suffix list to 35,207 entries.
Step 4 (**Prefix List**): Instead of using a Prefix array or Trie for the prefix frequency generation, word list from Emille corpus [18] is used with regular expressions. The Emille word list consist of 239,373 words.

3.3 Stem a Given Kannada Word

To stem a given Kannada word to its root form, following steps are followed:

- If the word length is ≤3, the word itself is root word. In Kannada, word less than 3 characters are not in inflected form.
- Else, Make a pointer point to the last character of the word. The pointer is made to move backward after each iteration. The pointer act as a split. The characters to the right of the pointer are treated as suffix and the characters to the left are treated as prefix. Frequency of the given prefix is found from the prefix list generated from the Emille Corpus in the previous step. Similarly the frequency of the given suffix is obtained from Suffix list generated in the previous step using suffix arrays.
- Calculate the sum of the prefix frequency and the suffix frequency for the given split P as shown below. The frequency of shorter suffixes is high, hence to compensate this disparity the frequency is multiplied with the length of the prefix and suffix as discussed in [12].

$$\begin{aligned} \mathrm{P} = & ((\text{string length of prefix word}) * \log(\text{frequency of the prefix})) \\ & + ((\text{string length ofsuffix word}) * \log(\text{frequency of the suffix})) \end{aligned} \tag{8}$$

Repeat the above steps for all splits till the length of prefix becomes <3 or the suffix is not found in the list.

- The highest value of P is the optimal split and the corresponding prefix is considered as the stem of the given word.

For example, for the word ಅಂಕಅಂಶಗಳನ್ನು

prefix = ಅಂಕಅಂಶಗಳನ್ನfreq = 1 suffix = ುfreq = 12450Value of P = 9

prefix = ಅಂಕಅಂಶಗಳನ್freq = 1 suffix = ನುfreq = 3752Value of P = 16

prefix = ಅಂಕಅಂಶಗಳನfreq = 1 suffix = ್ನುfreq = 3563Value of P = 24

prefix = ಅಂಕಅಂಶಗಳfreq = 1 suffix = ನ್ನುfreq = 3558Value of P = 32

prefix = ಅಂಕಅಂಶಗfreq = 1 suffix = ಳನ್ನುfreq = 1101Value of P = 35

prefix = ಅಂಕಅಂಶfreq = 1 suffix = ಗಳನ್ನುfreq = 1088Value of P = 41

prefix = ಅಂಕಅಂfreq = 1 suffix = ಶಗಳನ್ನುfreq = 28Value of P = 23

prefix = ಅಂಕಅfreq = 1 suffix = ಂಶಗಳನ್ನುfreq = 11Value of P = 19

prefix = ಅಂಕfreq = 64 suffix = ಅಂಶಗಳನ್ನುfreq = 4Value of P = 24

P=41 is the highest. Hence, Stem is ಅಂಕಅಂಶ

3.4 Kannada Stemmer with Dictionary Lookup

With only unsupervised stemming, an accuracy of 58 % was achieved. The testing word list was random words extracted from Emille Corpus [18] and had words already in root form, nouns, inflected words, numbers etc. No filtering was done on the words since; the application of the proposed stemming is for text classification where testing corpus will have random text. To improve the stemming performance, a stem list of 18,804 stems was created manually.

In the previously explained unsupervised stemming method, every prefix is incrementally matched with the stem-list dictionary. If found, the longest matching prefix is the stem for the given word, else, the split value P is calculated as given by Eq. 8 and continued. The highest P becomes the optimal split and corresponding prefix is the stem for the given word.

4 Text Classification of Kannada Documents

The effect of stemming on classification of Kannada documents is experimented as part of this paper. Stemming act as a pre-processing step in text classification for feature reduction. Machine learning algorithms like Naïve Bayes and Maximum Entropy are applied on Kannada documents. They consider the bag-of-words as features for classification. Naïve Bayes is defined as:

$$P(c|d) = \frac{P(d|c)P(c)}{P(d)}. \quad (9)$$

where d is the document and c is the class. Maximum Entropy in parametric form is defined as:

$$P(c|d) = \frac{1}{Z(d)} \exp(\textstyle\sum_i \lambda_i f_i \ (d, c)) \quad (10)$$

where each f_i (d, c) is a feature, λ_i is the parameter to be estimated and Z(d) is the normalizing factor [19].

For Naïve Bayes and Maximum Entropy classification task, Mallet toolkit [20] is used. The Kannada corpus for classification was collected manually from websites. After removing tags and other invalid characters, the documents were used as training documents. For testing, fivefold cross validation was applied by randomly splitting the documents as 70 % train and 30 % test documents. Totally 10 categories were tested with 100 documents in each category, i.e. a total of 1,000 documents. The list of categories is given in Table 2.

Table 2 List of categories for classification

ದುರ್ಘಟನೆ ,ಅಪಘಾತ (Disaster or Accident)
ಕಲೆ ,ಸಂಸ್ಕೃತಿ ,ಮನೋರಂಜನೆ) Arts, culture or Entertainment)
ಹಣಕಾಸು) Commerce)
ಅಶಾಂತಿ ,ಗೊಂದಲ ,ಯುದ್ಧ) Conflict or war)
ಶಿಕ್ಷಣ) Education)
ಆರೋಗ್ಯ (Health)
ರಾಜಕೀಯ) Politics)
ಧರ್ಮ) Religion)
ವಿಜ್ಞಾನ) Science)
ಕ್ರೀಡೆ (Sports)

5 Results and Evaluation

5.1 Stemmer Results

Randomly chosen 6,500 words from EMILLE corpus [18] is used for testing. The word list consisted of nouns, numbers etc. without any kind of filtering. The performance of the unsupervised stemming algorithm as proposed in this paper as compared to stemming with dictionary lookup + unsupervised stemming is shown in Table 3.

The stemming result as depicted in Fig. 2 shows that, with dictionary look-up, the performance of stemming increases by 10 %. In [16], using the unsupervised stemming for Kannada Language (morphind) Stemming Quality of 88 % was achieved. The current result is lesser may be due to unfiltered test set. The performance of text classification of Kannada documents into pre-determined categories is shown in Table 4 and graphically represented in Fig. 3 for Naïve Bayes and Maximum Entropy algorithms. For feature selection, Information gain is used. Information gain of term t is defined as:

$$G(T) = -\sum_{i=1}^{m} P_r(c_i)\log p_r(c_i) + P_r(t)\sum_{i=1}^{m} P_r(c_i|t)\log p_r(c_i|t) + \\ P_r(t)\sum_{i=1}^{m} P_r(c_i|t)\log p_r(c_i|t). \tag{11}$$

$\{c_i\}_{i=1}^{m}$ is the set of categories in the target space.

Table 3 Accuracies achieved with stemming algorithms

	Unsupervised stemming	Stem-list + unsupervised stemming
Stemming accuracy	0.58	0.68

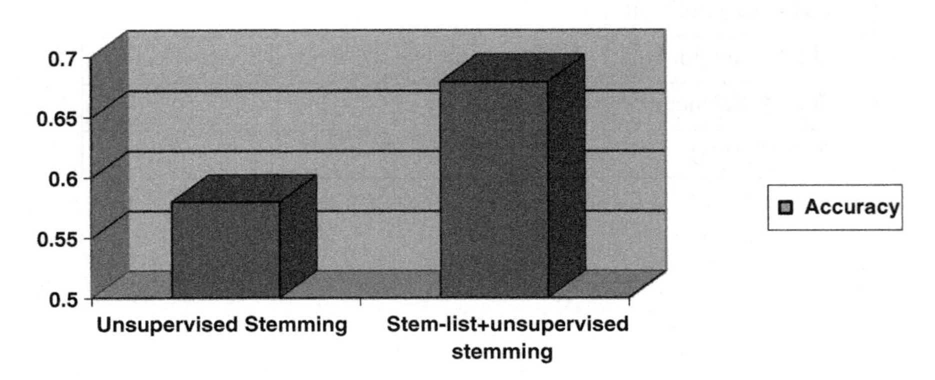

Fig. 2 Performance of unsupervised stemming and unsupervised + dictionary look-up

Table 4 Text classification with and without stemming

No. of features after infogain	Accuracy—Naïve Bayes (without stemming)	Accuracy—Naïve Bayes (with stemming)	Accuracy—maximum entropy (without stemming)	Accuracy—maximum entropy (with stemming)
100	0.59	0.64	0.62	0.66
200	0.63	0.71	0.69	0.71
500	0.70	0.75	0.72	0.77
1000	0.74	0.77	0.76	0.79
5000	0.78	0.80	0.80	0.82

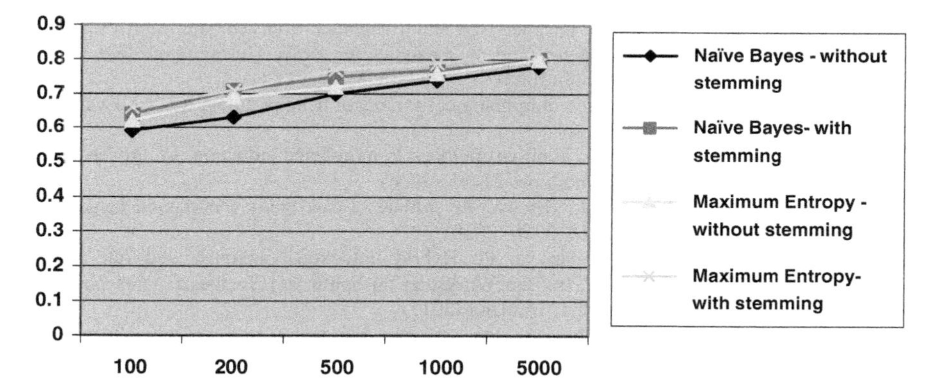

Fig. 3 Performance of text classification using Naïve Bayes and maximum entropy with and without stemming

6 Conclusion

An unsupervised stemming algorithm is proposed in this paper using suffix arrays. An accuracy of 58 % is achieved using unsupervised method and to improve the performance a stem-list dictionary look-up is added. An accuracy of 68 % is achieved using unsupervised method and stem-list dictionary. This proposed method does not require threshold calculation by multiple retries as done in previous implementations. When stemming is applied on text classification of Kannada Language text documents using Naïve Bayes and Maximum Entropy methods, the accuracy has increased compared to results without stemming.

References

1. Frakes, W.B., Baeza-Yates, R.: Information Retrieval: Data Structures and Algorithms. Prentice Hall, Englewood Cliffs (1992)
2. Lovins, J.: Development of a stemming algorithm. Mech. Transl. Comput. Linguist. **11**, 22–23 (1968)
3. Paice, C., Husk, G.: Another stemmer. ACM SIGIR Forum **24**(3), 566 (1990)
4. Porter, M.: An algorithm for suffix stripping. Program **14**(3), 130–137 (1980)
5. Ramanathan, A., Rao, D.D.: A lightweight stemmer for Hindi. In: Proceedings of EACL, ACL (2003)
6. Islam, Z., Uddin, N., Khan, M.: A light weight stemmer for bengali and its use in spelling checker. In: Proceedings of 1st International conference on Digital Communications and Computer Applications (DCCA 2007), Irbid, Jordan, pp. 87–93 (2007)
7. Majumder, P., Mitra, M., Parui, S.K., Kole, G., Mitra, P., Datta, K.: Yass: yet another suffix stripper. ACM Trans. Inf. Syst. **25**(4), 18 (2007)
8. Pandey, A.K., Siddiqui, T.J.: An unsupervised Hindi stemmer with heuristic improvements. In: Proceedings of the Second W.orkshop on Analytics for Noisy Unstructured Text Data, AND 2008, Singapore, pp. 99–105 (2008)
9. Dasgupta, S., Ng, V.: Unsupervised morphological parsing of bengali. Lang. Resour. Eval. **40**, 311–330 (2006)
10. Keshava, S., Pitler, E.: A simpler, intuitive approach to morpheme induction. In: Proceedings of 2nd Pascal Challenges Workshop, pp. 31–35 (2006)
11. Majgaonker, M.M., Siddiqui, T.J.: Discovering suffixes: a case study for Marathi language. Int. J. Comput. Sci. Eng. **04**, 2716–2720 (2010)
12. Suba, K., Jiandani, D., Bhattacharyya, P.: Hybrid inflectional stemmer and rule-based derivational stemmer for Gujrati. In: 2nd Workshop on South and Southeast Asian Natural Languages Processing, Chiang Mai, Thailand (2011)
13. Gupta, V., Lehal, G.S.: Punjabi language stemmer for nouns and proper names. In: Proceedings of the 2nd Workshop on South and Southeast Asian Natural Language Processing (WSSANLP) IJCNLP 2011, Chiang Mai, Thailand, pp. 35–39 (2011)
14. Kumar, D., Rana, P.: Design and development of a stemmer for Punjabi. Int. J. Comput. Appl. **11**(12), 0975–8887 (2010)
15. Padma, M.C., Prathibha, R.J.: Development of morphological stemmer, analyzer and generator for Kannada nouns. In: Proceedings of International Conference, ICERECT 2012, pp. 713–723 (2014)
16. Bhat, S.: Statistical stemming for Kannada. In: Proceedings The 4th Workshop on South and Southeast Asian NLP (WSSANLP), International Joint Conference on Natural Language Processing, Nagoya, Japan, pp. 25–33, 14–18 Oct 2013
17. http://www.hlt.utdallas.edu/~sajib/FinalDistribution.tar.gz. Accessed 24 July 2014
18. Emille corpus: http://www.emille.lancs.ac.uk (2003)
19. Nigam, K., Lafferty, J., McCallum, A.: Using maximum entropy for text classification. In: IJCAI 1999 Workshop on Machine Learning for Information Filtering, pp. 61–67 (1999)
20. McCallum, A.K.: MALLET: a machine learning for language toolkit (2002)

Boosting Formal Concept Analysis Based Definition Extraction via Named Entity Recognition

G.S. Mahalakshmi and A.L. Agasta Adline

Abstract E-learning involves learning materials of different standards which might be difficult for the e-learner to ponder ideas for fast reading. Extracting important definitions and phrases from the learning material will help in representation of knowledge in a more useful and attractive manner. This paper discusses a formal conceptualization based definition extraction approach for theoretical learning materials. The experiments have been conducted on Abraham Silberschatz, Peter B Galvin and Greg Gagne's Operating Systems Concepts—e-book and the results have outperformed the Named Entity based approach for Definition Extraction.

Keywords Definition extraction · Named entities · Formal concept analysis · Machine learning

1 Introduction

Definitions may be defined as statements that provide the meaning of a term. It gives exact information that is useful in many cases. In an eLearning environment, definitions could be used by a student to acquire quick knowledge of a topic. These definitions can be maintained as a collection of keywords for rapid reference. However, identifying definitions manually in a large text is a time consuming and tedious job. In texts with strong structuring (stylistic or otherwise), such as technical or medical texts, the automatic identification of definitions is possible through the use of the structure and possibly cue words. For instance, in most mathematical textbooks, definitions are explicitly marked in the text, and usually follow a regular

G.S. Mahalakshmi (✉)
Department of CSE, Anna University, Chennai, India
e-mail: gsmaha@annauniv.edu

A.L. Agasta Adline
Department of IT, Easwari Engineering College, Chennai, India
e-mail: agasta465@gmail.com

L.C. Jain et al. (eds.), *Computational Intelligence in Data Mining - Volume 3*,
Smart Innovation, Systems and Technologies 33, DOI 10.1007/978-81-322-2202-6_8

form. In less structured texts, such as programming tutorials, identifying the sentences which are definitions can be much more challenging, since they are typically expressed in a linguistically free form. In such cases, humans have to go through the entire text manually to tag definitional sentences.

One way of automating definition extraction is to identify linguistic forms definitions conform to, usually using either lexical patterns or through specific keywords or cue-phrases contained in the sentence [1]. Once such rules are identified, the sentences matching one or more of these forms can be identified. This approach has been shown to work with varying results. The main drawbacks of this method is that (i) the relative importance of the different linguistic forms is difficult to assess by human experts, and is thus usually ignored; and (ii) coming up with effective linguistic forms which tread the fine line between accepting most of the actual definitions, but not accepting non-definitions, requires time and expertise and can be extremely difficult. Since there typically is a numeric imbalance between definitions and non-definitions in a text, having a slightly over liberal rule can result in tens or hundreds of wrong positives (non-definitions proposed as definitions), which is clearly undesirable.

In this paper, we employ a special technique to automatically extract definitions from chapters of eBooks as a part of enhancing knowledge representation for eLearning. The candidate statements are trained by a machine learning algorithm which is assisted by both syntactic and semantic features. The system is further strengthened by the use of formal concept analysis which helps in discarding non-definition statements thus increasing the overall system accuracy.

2 Formal Concept Analysis

A formal context (L, D, I) consists of sets L, D and a binary relation $I \leq L \times D$. Formal concepts are particular clusters in $L X D$.

Consider an example where $x \,\varepsilon\, L$ and $y \,\varepsilon\, D$. The formal concept $\langle L_1, D_1 \rangle = \langle \{x_1, x_2, x_3, x_4\}, \{y_3, y_4\} \rangle$. Similarly other formal concepts are shown in shaded portions in Fig. 1.

3 Related Work

3.1 Definition Extraction

Researches in definition extraction started in early 90s where all the aspects of a definition right from philosophy to terminologies [2] where considered. In the recent periods, their contribution to dictionary building, question answering and information retrieval have gained popularity. In the TREC 2003 QA definition

Fig. 1 Formal concept table for ⟨L, D⟩

I	y_1	y_2	y_3	y_4
x_1	✓	✓	✓	✓
x_2	✓		✓	✓
x_3		✓	✓	✓
x_4		✓	✓	✓
x_5	✓			

I	y_1	y_2	y_3	y_4
x_1	✓	✓	✓	✓
x_2	✓		✓	✓
x_3		✓	✓	✓
x_4		✓	✓	✓
x_5	✓			

I	y_1	y_2	y_3	y_4
x_1	✓	✓	✓	✓
x_2	✓		✓	✓
x_3		✓	✓	✓
x_4		✓	✓	✓
x_5	✓			

I	y_1	y_2	y_3	y_4
x_1	✓	✓	✓	✓
x_2	✓		✓	✓
x_3		✓	✓	✓
x_4		✓	✓	✓
x_5	✓			

subtask evaluation, many different techniques for definition identification were introduced which used sophisticated linguistic tools [3]. Another method that uses definition patterns and terms that co-occur with the definiendum in on-line sources for both passage selection and definition extraction was suggested for question-answering [1].

The definition extraction in QA application concentrated mainly on the term to be defined. In our system that involves knowledge representation we focus on both the term as well the connectors between the statements. The initial works on definition extraction concentrated on pattern matching [4, 5]. The recent trends however have found this method to be ineffective and have combined the pattern based approach with machine learning models [6]. A machine learning based definition extraction for Dutch was carried out by [7] where structural and linguistic features were considered. The method showed that combining pattern matching with machine learning provides better results for definition extraction task.

Another useful method of definition extraction for glossary building was developed using evolutionary algorithms [8] where genetic programming concepts were combined with linguistic patterns. The algorithm was found to be scalable and could learn new rules coined by human linguistic experts.

3.2 Formal Concept Analysis in Information Retrieval

Knowledge representation and reasoning is inherent in FCA and this qualifies it for semantic processing of text documents in text mining [9]. Use of FCA in IR dates back to early 1968 [10] where document/term lattices were widely used. Later came the vector space model which has actually no lattice based approach but which dominated the text mining research space for decades. The next remarkable contribution of FCA to text mining was by Godin et al. [11]. This paper did not propose any visualization solution for the concept lattices and was merely text based. Following this, FCA was widely applied to software engineering in analyzing program codes as well [12]. Later FCA was applied to analyzing Google search results [13]. Here, Carpineto and Romano [13] argued that FCA shall find three varieties of applications in information retrieval: 1. Query refinement 2. Query and Navigation 3. Integrating with thesauri hierarchies

Concept lattices and logical concept analysis were used widely to support the task of browsing structured datasets [14–16]. Student interactions are captured with FCA for creation of e-learning semantic web [17]. Remarkably FCAIR 2013 workshop was conducted to explore the interdisciplinary applications of FCA and IR.

A more recent work on definition extraction suggested using a lattice called World Class Lattice (WCL) [18] in place of Lexico-Syntactic patterns. The words are represented in the form of a lattice which is a directed acyclic graph that is further clustered and classified. The system has greater accuracy rate over other linguistic patterns and was extended for hypernym extraction. Another recent work on DE using machine learning involved the use of Conditional Random Fields [19] along with statistical and linguistic features for classification and sequential labelling task. All these approaches were found to produce considerably low results and were manually annotated in several cases. The system discussed in this paper focuses on avoiding manual operations and attempts to rectify their shortcomings.

4 System Description

The definition extraction process proposed here consists of three main steps; (i) Identify definition candidate statements based on keywords, their pattern and rule, (ii) Extract position based, linguistic based and incidence based features, (iii) Train a machine learning model using the information obtained. The ML model is used to classify whether the statement is definitive or non definitive.

4.1 Candidate Statement Identification

A definition usually consists of three parts. First is the word to or group of words to be defined. Second is the definition or meaning of the word explained by one or more phrases. The last and the crucial part is the keywords that connect the head to its tail. Based on these keywords, the candidate sentences are sorted into three broad categories. They are the ‘is a’ type statements, ‘verb’ statements and ‘pronoun’ statements. The different types of candidate statements are shown along with few examples in Table 1.

Table 1 Type of sentences with example

Type of sentences	Meaning	Example
Is A	Sentences where the term and its definition are connected by a preposition+article term	A process is a program in execution
Verb	Sentences were the term and its definition are connected by a verb or verb phrase	A socket is defined as an endpoint for communication
Pronoun	Sentences where the definition and the term are in separate sentences and are joined together by a pronoun in the second sentence	Accounting information. This information includes the amount of CPU and real time used, time limits, account numbers, job or process numbers and so on

4.2 Feature Extraction

As the input data comprises of large amounts of words, it is essential to interpret them into specific set of features for further computation and machine understanding. When it comes to a sentence, there are a number of various features that can be considered ranging from the frequency of terms, their parts-of-speech, their senses and semantic values, syntactic patterns, their location both in a paragraph or entire text, supporting words, special characters and so on. In our system we extract three types of features.

- Structure based: Based on the position of the cue words, three types of sentences are identified.
  - First type of sentences is of the form "Word+Keyword+Definition". In this type of sentences, the word to be defined is followed by its definition which is connected by the cue words.
    E.g. *A Process IS A program in execution.*
  - The second type of sentences is denoted as "Definition+Keyword+Word". Here the definition is followed by the word to be defined. This also involves sentences starting with a pronoun while the previous statement contains the definition.

  E.g. Execution of a series of process without manual intervention is CALLED Batch processing.
  The sentences which do not fall into either category but contains the keywords are placed in the third type with smaller weight value.
  E.g. Each of these IS A[1]process waiting for control section.

[1] Relative position is calculated using the actual position and the total number of sentences.
RP = AP/N where RP—Relative Position, AP—Absolute Position, N—Total number of sentences.

- Position based: The location of statement in the text also plays a major role in determining its usage. The position can refer to the position a sentence in a paragraph as well as the entire text. Relative position[1] of the sentence is considered than actual position which produces more accurate information.
- Incidence based: It is a rather obvious observation that a definition statement contains mostly proper noun related to that term defined. Hence the count of noun phrases also contributes to the features list of definition statements. Thus numbers of proper noun are counted and their normalized score is calculated.
- Special features: Occurrence of special symbols (such as:, -, etc.) and other visually different characteristics (such as bold, italic) are also noted. Different weight scores are assigned to these characters and are considered as contributing factors.

4.3 Applying Formal Concept Analysis (FCA)

Prior to formal concept analysis, the learning material is preprocessed and *tf-idf* (term frequency—inverted document frequency) an inverted index is calculated. To reduce the dimensionality of processing huge *tf-idf*s, as a dimensionality reduction approach, formal concept analysis is explored. The terms are represented using a concept lattice representing the formal context. Concept lattice preserves the information contained in the formal context since the underlying principle is ordered lattice theory. In concept lattices, concepts are hierarchically ordered in a sub-concept and super-concept relations.

4.4 Machine Learning

We have used Naïve Bayes classifiers which are based on simple probability theory. They work with the principle of applying Bayes' theorem with strong independent assumptions called Naïve assumptions between features[2] derived for learning. Training is done by linear approximation theory. The Naïve Bayes model calculates the mean, standard deviation and the weighted sum for each attribute listed and finalizes the precision value for individual numeric feature or the total value for nominal features. The final classifier model considers these distribution values and is calculated as

$$Classify(f_1, \ldots, f_n) = \max p(C = c) \prod_{i=1}^{n} p(F_i = f_i | C = c) \quad (1)$$

where $p(F|C)$ is the probability model for small number of classes C over a set of n features $F_1 \ldots F_n$.

4.5 Definition Extraction and Named Entities

The machine learning approach used above shows satisfactory results when compared to other lexical pattern based methods and greedy approaches. These results can be boosted by the use of named entities. Named entity recognition is the process of identifying entities or atomic elements and placing them in their predetermined categories. These named entities indicate elements that are rigid designators and are mostly proper noun which in most cases stipulate a concept or definition. Hence these named entities when considered along with the linguistic patterns can eliminate the extraction of non-definitive statements thereby reducing the false positive rate and concurrently improving accuracy of the system.

For example, *The program counter (PC) is a processor register that indicates where a computer is in its program sequence.* In this statement, there are 4 proper nouns. When placed for entity recognition, 3 entities (program counter, processor register and program sequence) should be identified. Whereas in example, *The state of a process is defined in part by the current activity of that process* has 0 named entities even though it has the definitive linguistic pattern 'term+is defined+terms', thus dropping the sentence as a non-definitive statement. This approach can be applied for highly imbalanced dataset [20].

5 Experiments and Results

The experiment concentrates on extracting definition candidate statements based on certain set of rules from a non-annotated text data and building a definition identification model trained by machine learning algorithms and features of these extracted statements.

5.1 Experimental Setup

The dataset includes non annotated chapters from "Operating System concepts" by Abraham Silberschatz, Peter B Galvin and Greg Gagne. The input data has approximately 1,190 sentences that includes both definitive and non-definitive type of sentences. Certain rules such as presence of linguistic patterns and cue words and their order of occurrence were applied to extract candidate statements. This rule based approach yielded 102 candidate statements and further evaluation was carried out on them.

5.2 Case Study

Figure 2 shows a section from the book "Operating System concepts" by Abraham Silberschatz, Peter B Galvin and Greg Gagne. This section of data is given as input to the system. The system retrieves candidate statements based on the set of linguistic patterns given. For the above section the statements that are retrieved are shown in Fig. 3.

The above candidate statements are further processed to extract the structural, position and incidence based features. The values thus obtained are fed as attributes to the machine learning module (Fig. 4). The Naïve Bayes model identifies statement number 3 as a definition and tags the remaining statements as non definitive.

5.3 Evaluation

Table 2 deals with the precision of candidate statements extracted by rule based method. Separate rules were applied for three type of sentences (is a, verb, pronoun). Table 3 includes identifying three types of features and training a Naïve Bayes classification model. The features are ranked using *OneR* evaluation and

3.2.3 Context Switch

As mentioned in Section 1.2.1, interrupts cause the operating system to change a CPU from its current task and to run a kernel routine. Such operations happen frequently on general-purpose systems. When an interrupt occurs, the system needs to save the current *context* of the process running on the CPU so that it can restore that context when its processing is done, essentially suspending the process and then resuming it. The context is represented in the PCB of the process; it includes the value of the CPU registers, the process state (see Figure 3.2), and memory-management information. Generically, we perform a *state save* of the current state of the CPU, be it in kernel or user mode, and then a *state restore* to resume operations.

Switching the CPU to another process requires performing a state save of the current process and a state restore of a different process. This task is known as a *context switch*. When a context switch occurs, the kernel saves the context of the old process in its PCB and loads the saved context of the new process scheduled to run. Context-switch time is pure overhead, because the system does no useful work while switching. Its speed varies from machine to machine, depending on the memory speed, the number of registers that must be copied, and the existence of special instructions (such as a single instruction to load or store all registers). Typical speeds are a few milliseconds.

Fig. 2 Section from book (Operating System concepts)

```
1. The context is represented in the PCB of the process;
2. It includes the value of the CPU registers, the process state and
   memory-management information.
3. Switching the CPU to another process requires performing a state
   save of the current process and a state restore of a different
   process. This task is known as a context switch.
4. Typical speeds are a few milliseconds.
```

Fig. 3 Retrieved candidate statements

Type	Para	Line	Incidence	Structure
V	1	6	0.300	X
V	1	7	0.444	X
P	2	1	0.290	Y
I	2	6	0.333	Z

Fig. 4 Attributes for ML model

Table 2 Precision for rule based candidate sentence extraction

Type of sentence	Precision (%)
Is A	56
Verb	55
Pronoun	81

Table 3 Evaluation results for Naive Bayes model

Statement type	Precision	Recall	F-score	Accuracy
Is A	0.643	0.634	0.603	0.634
Verb	0.594	0.6	0.586	0.601
Pronoun	0.638	0.688	0.662	0.687
All	0.66	0.657	0.617	0.656

Ranker method. The training dataset are sent with the attributes in their increasing order of rank. The results of Naïve Bayes classification for individual statements are in Table 3.

5.4 Result and Discussion

The values in Table 3 show that the machine learning model gives fairly uniform results for the different types of statements. One main observation that has to be noted for the different types of statements is that their recall value was found to very high for the positive classifications than for the negative classification (Is a—87 %, Verb—76 %, Pronoun—84.6 %, Overall—90.7 %). This shows that the features were appropriate to identify definition statements (true positives). However the accuracy is brought down by the slight inefficiency to eliminate non-definitive statements.

There are few observed defects in the retrieval of candidate statements. Firstly, the use of keywords though effectively retrieves the definitive candidates; it also includes retrieval of non-definitive sentences. E.g. *Each of these is a process in waiting state* is not a definition even though it has the key word *is a*. Hence these types of statements require manual pruning.

A good solution is to consider the grammatical structure of elements before and after the keywords. For the above example sentence, the part of sentence before the keyword does not contain a proper noun. That is, it does not contain the definiendum (Latin for word to be defined) is not present. Hence this defect is rectified by the structural feature that eliminates sentences that are not of the accepted format.

Another common problem that was noted is that, a sentence retrieved contains a definitive clause but not entirely. For e.g. *Informally, as mentioned earlier, a process is a program in execution* is the complete sentence retrieved in which *a process is a program in execution* is the definitive phrase. Thus the retrieved sentences include certain amount of noise that lowers their accuracy. This can be minimized by taking into account the position of cue words and their surrounding clause. Weighted ranking the terms in a sentence based on their affinity to the cue words is one suggested remedy.

The final problem that needs to be addressed is definitions explained in more than one sentence. Our system considered definitions with a maximum of two sentences provided the second sentence starts with a pronoun and contains a verb-based keyword. For e.g. *The two-level model placed an intermediate data structure between the user and kernel threads. This data structure is known as a lightweight process*. Here the second sentence contains the keyword *known as* and it starts with a pronoun *this*. The problem with definitions having more than two sentences is yet to be addressed.

6 Conclusion

This paper discusses a highly efficient and direct method which is designed to extract candidate definitive statements from unstructured and imbalanced text and classify them into definitive and non definitive statements. Formal concept analysis is applied over terms in order to reduce the dimensionality. Naive Bayes classifier used was found to perform better than random forest classifier and entropy based classification because of two main reasons: (i) The classifier is feature independent, that is, it considers each feature individually and is not affected by the presence or absences of other features; (ii) It can be trained efficiently with lesser number of instances in the dataset when compared to the others. The work shall be enhanced using Named Entity Recognition which might help in boosting the true positive rate thereby improving the overall system efficiency. Thus in the field of eLearning where precise information are to be highlighted, the proposed approach for definition extraction shall be utilised for better knowledge representation. The system is further being studied for other linguistic patterns, informal definitions and special features that can help improve the accuracy of the system.

References

1. Saggion, H.: Identifying definitions in text collections for question answering. In: LREC (2004)
2. Wilks, Y., Slator, B.M., Guthrie, L.M.: Electric Words: Dictionaries, Computers, and Meanings. MIT Press, Cambridge (1996)
3. Wilks, Y., Slator, B.M., Guthrie, L.M.: Automatic extraction of legal concepts and Definitions. In: Legal knowledge and information systems: JURIX 2012, the twenty-fifth annual conference, vol 250, p. 157. IOS Press (2012)
4. Storrer, A., Wellinghoff, S.: Automated detection and annotation of term definitions in German text corpora. In: Proceedings of LREC, vol 2006 (2006)
5. Walter, S., Pinkal, M.: Automatic extraction of definitions from German court decisions. In: Proceedings of the Workshop on Information Extraction Beyond the Document, Association for Computational Linguistics, pp. 20–28 (2006)
6. Fahmi, I., Bouma, G.: Learning to identify definitions using syntactic features. In: Proceedings of the EACL 2006 Workshop on Learning Structured Information in Natural Language Applications, pp. 64–71 (2006)
7. Westerhout, E.: Definition extraction using linguistic and structural features. In: Proceedings of the 1st Workshop on Definition Extraction, Association for Computational Linguistics, pp. 61–67 (2009)
8. Borg, C., Rosner, M., Pace, G.: Evolutionary algorithms for definition extraction. In: Proceedings of the 1st Workshop on Definition Extraction, Association for Computational Linguistics, pp. 26–32 (2009)
9. Priss, U.: Formal concept analysis in information science. Ann. Rev. Info. Sci. Technol. **40**(1), 521–543 (2006)
10. Salton, G.: Automatic Information Organization and Retrieval. McGraw-Hill, New York (1968)
11. Godin, R., Gecsei, J., Pichet, C.: Design of browsing interface for information retrieval. In: Belkin, N.J., van Rijsbergen, C.J. (eds.) Proceedings of SIGIR '89, pp. 32–39 (1989)
12. Tilley, T., Cole, R., Becker, P., Eklund, P.: A survey of formal concept analysis support for software engineering activities. In: Ganter, B., Stumme, G., Wille, R. (eds.) Formal Concept Analysis, pp. 250–271. Springer, Berlin (2005)
13. Carpineto, C., Romano, G.: Exploiting the potential of concept lattices for information retrieval with CREDO. J. Univers. Comput. **10**(8), 985–1013 (2004)
14. Ducrou, J., Eklund, P., Wilson, T.: An intelligent user interface for browsing and searching MPEG-7 images using concept lattices. In: Yahia, S., Nguifo, E., Belohlavek, R. (eds.) Concept Lattices and Their Applications. Lecture Notes in Computer Science, vol. 4923, pp. 1–21. Springer, Berlin (2008)
15. Ferre, S.: Camelis: a logical information system to organize and browse a collection of documents. Int. J. Gen. Syst. **38**, 379–403 (2009)
16. d'Aquin, M., Motta, E.: Extracting relevant questions to an RDF dataset using formal concept analysis. In: Proceedings of the Sixth International Conference on Knowledge Capture (K-CAP '11), pp. 21–128. ACM, New York, NY, USA (2011)
17. Beydoun, G.: Formal concept analysis for an e-learning semantic web. Expert Syst. Appl. **36** (8), 10952–10961 (2009)

18. Navigli, R., Velardi, P.: Learning word-class lattices for definition and hypernym extraction. In: Proceedings of the 48th Annual Meeting of the Association for Computational Linguistics, Association for Computational Linguistics, pp. 1318–1327 (2010)
19. Anke, L.E.: Towards definition extraction using conditional random fields. In: RANLP, pp. 63–70 (2013)
20. Del Gaudio, R., Batista, G., Branco, A.: Coping with highly imbalanced datasets: a case study with definition extraction in a multilingual setting. Nat. Lang. Eng. **20**(03), 327–359 (2014)

Harmonic Reduction Using PV Fuzzy with Inverter

Subha Darsini Misra, Tapas Mohapatra and Asish K Nanda

Abstract In previous years fossil fuel acts as major resources for producing electricity. Although solar power is pollution free so it can be available at any place and anywhere in the world. A conventional photovoltaic system extracts solar energy with high boosting capacity. In conventional method buck boost converter was used and it has a drawback that input of the current has fewer amounts of ripples. In this topic single phase inverter is used by replacing R-load of boost converter. Advantages of boost converter are it has continuous input current that can help us to find the average current mode control techniques to force the line current to track changes in the line voltage. The dc to ac inverter has the capacity of bidirectional power supply. In the given method LC filters is used for filtering the harmonics. In this method fuzzy logic control technique is used and a program is written in mamdani fuzzy by the help of maxmin method and centroid method of fuzzy logic in the embedded mat lab block of MATLAB. Using the fuzzy control technique the harmonics is somehow reduced.

Keywords Array · Circuit · Equivalent · Model · Modelling · Photovoltaic · Maximum power · Fuzzy logic control

S.D. Misra (✉) · T. Mohapatra
Department of Electrical Engineering, SOA University, Bhubaneswar, Odisha, India
e-mail: tikinaiter@gmail.com

T. Mohapatra
e-mail: tapasmohapatra28@gmail.com

A.K. Nanda
Department of Electrical Engineering, VSSUT, Burla, Odisha, India
e-mail: asishoec@gmail.com

L.C. Jain et al. (eds.), *Computational Intelligence in Data Mining - Volume 3*,
Smart Innovation, Systems and Technologies 33, DOI 10.1007/978-81-322-2202-6_9

1 Introduction

Solar energy is the radiant energy from the sun. It is important to us because it provides the world directly or indirectly all most all its energy that sustains the world. Photovoltaic's (PVs) are arrays (combination of cells) that contain a solar voltaic material that converts the solar energy into electrical energy [1]. PV cell is the basic device for Photovoltaic Systems. Such systems include multiple components like mechanical and electrical connections and mountings and various modes of regulating and (if required) modifying the electrical output. The use of the new power control mechanisms called the Maximum Power Point Tracking (MPPT) algorithms has lead to the increasing efficiency of operation of the solar modules and thus is effective in the field of utilization of renewable energy sources [2]. Some of the conventional methods are used to track the maximum power point (MPP) are the perturb and observe method, the incremental conductance method, the constant voltage method and the short circuit method. Because the above methods are not able to track the MPP effectively under rapidly changing conditions the fuzzy logic controller is used [2, 3]. So the Fuzzy Logic Controller (FLC) is a robust method that improves MPP tracking as compared to the other method.

1.1 Objective

The basic objective would be to study MPPT and successful implementation the MPPT algorithms either in code form or using the Simulink models. Modelling the converter and the solar cell in Simulink and interfacing both with the MPPT algorithm to obtain the maximum power point operation would be of main importance.

(i) To reduce harmonics using PV with Fuzzy with boost converter.
(ii) In place of R-load of boost converter mathematical model of inverter is connected.
(iii) Input voltage is 30 V. And open circuit voltage is 32.9 V and short circuit current is 8.21 A.

2 PV Cell

Photovoltaic system directly converts the sunlight into Electricity. Basic device of the photovoltaic system is Photovoltaic cell shown in Fig. 1 is the most basic generation part of the PV system. Single diode mathematical model is applicable to simulate silicon photo voltaic cell. Which consist of a Photocurrent source, non linear diode and internal resistance. A solar cell is the building block of a solar

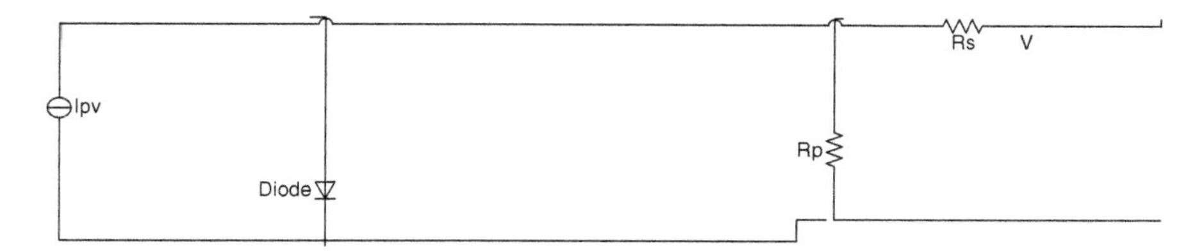

Fig. 1 PV cell

panel. A photovoltaic module is formed by connecting many solar cells in series and parallel. Considering only the single solar cell; it can be modeled by utilizing a current source, a diode and two resistors. This model is also known as a single diode model of solar cell. Two diode models are also available but only single diode model is considered here.

Where

R_p = Parallel Resistance.
R_s = Series Resistance.
I_{PV} = Light generated current.
I_D = Diode current.
I = Net current.

The characteristic equation for the photovoltaic cell is given by

$$I = I_{Lg} - I_{os*}\left[\exp\left\{q * V + \frac{IRs}{AKT}\right\} - 1\right] - V + IRs/Rsh \tag{1}$$

$$I_{os} = I_{Or*}\left(\frac{T}{T_r}\right)^3 * [\exp\left\{\frac{qE_g}{\frac{1}{Tr} - \frac{\frac{1}{T}}{AK}}\right\}] \tag{2}$$

where

I and V: Cell output current and voltage;
I_{os}: Cell reverse saturation current;
T: Cell temperature in Celsius;
K: Boltzmann's constant, 1.38×10^{-19} J/K;
Q: Electron charge, 1.6×10^{-23} C;
K_i: Short circuit current temperature coefficient at Iscr;
Lambda: Solar irradiation in W/m^2
I_{scr}: Short circuit current at 25 °C.

2.1 Modelling of Solar Cell

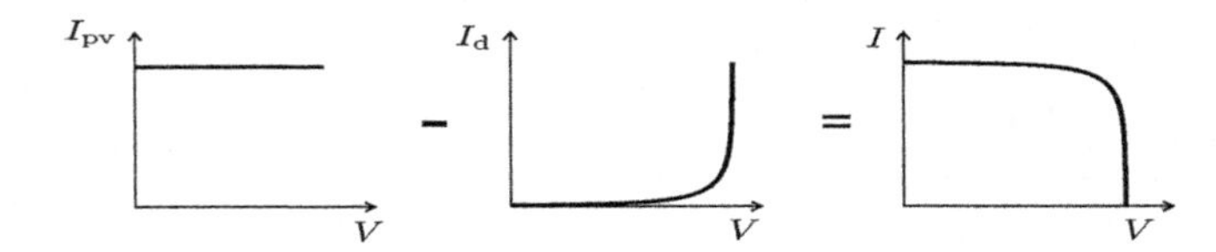

Fig. 2 Curve of solar cell

Where

I_{PV} = Light generated current.
I_D = Diode current.
I = Net current.

A solar cell is the building block of a solar panel. A photovoltaic module is formed by connecting many solar cells in series and parallel. Considering only the single solar cell; it can be modelled by utilizing a current source, a diode and two resistors. This model is also known as a single diode model of solar cell. Two diode models are also available but only single diode model is considered here. The characteristics of solar cell are shown in Fig. 2.

2.2 Effect of Variation of Solar Irradiation

The P–V and I–V curves of the solar cell shown in Fig. 3 are highly dependent on the solar irradiation values. The solar irradiation as a result of the environmental

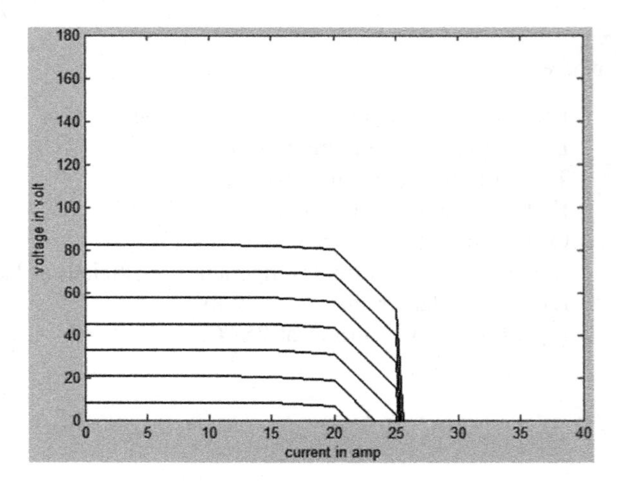

Fig. 3 Variation of V–I curve with solar irradiation

changes keeps on fluctuating, but control mechanisms are available that can track this change and can alter the working of the solar cell to meet the required load demands

3 Boost Converter

Boost converter which is shown in Fig. 4 steps up the input voltage magnitude to a required output voltage magnitude without the use of the transformer. The main components of the boost converter are an inductor, a diode and a high frequency switch. These in a co-ordinate manner supply power to the load at a voltage greater than the magnitude of input voltage [4].

3.1 Modes of Operation

The first mode is when the switch is closed; there are two modes of operation of a boost converter. Those are based on this is known as charging mode of operation. The second mode is when the switch is open; this is known as discharging mode of operation.

Charging Mode: In this mode of operation; the switch is closed and the inductor is charged by the source through the switch. The charging current is exponential in nature but for simplicity it is assumed to be linearly varying. The diode restricts the flow of current from the source to the load and the demand of the load is met by discharging of the capacitor.

Discharging Mode: In this mode of operation; the switch is open and the diode is forward biased. The inductor discharges and together with the source charges the capacitor and meets the load demand. The load current variation is very small and in many cases is assumed constant throughout the operation.

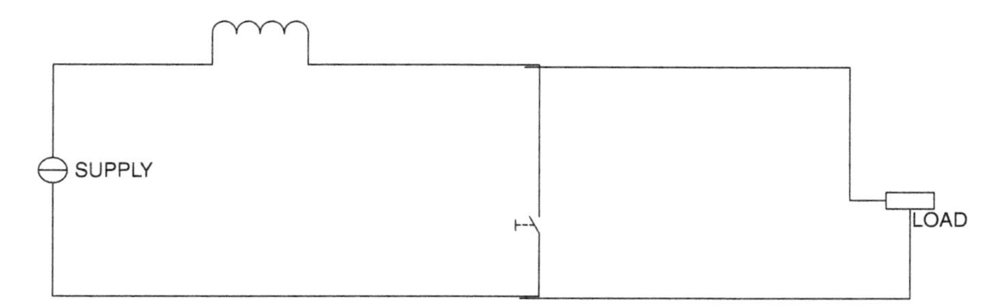

Fig. 4 Boost converter

4 Fuzzy Logic Controller Implementation in PV

The Fuzzy Logic Controller was used in this design because it is robust and does not require an accurate mathematical model while accommodating nonlinearities very well; however, it requires knowledge of the PV system by the designer [5, 6]. The inputs to this controller are the error, E and change in error, CE at a sample time k, which is defined by and, whiles the output is the duty cycle D.

E (k) = [P pvCk) − P pvCk − 1)]/[V pvCk) − V pvCk − 1]. The inputs to this controller are the error, E and change in error, CE at a sample time k, which is defined by and, whiles the output is the duty cycle D. CE (k) = E (k) − E (k − l) where P pv(k) and V pv(k) are the power and voltage of the PV module respectively. The input E(k) indicates if the load operating point at instant k is located on the left or right side of the PV characteristic curve, while CE(k) shows the direction of the moving point. During the process of Fuzzification, the input variables E and CE are translated into linguistic variables using membership functions as shown in Fig. 5. Triangular membership functions which are scaled between −1 and 1 are chosen to ensure that there is one dominant fuzzy set for any particular input. So the five fuzzy Subsets are NB (negative big), [7] NS (negative small), ZO (zero), PS (positive small) and PB (positive Big). The output is expressed as a linguistic variable and is determined by looking up a rule based table shown in Table 1.

In this process Mamdani method is used to control the output using maxmin method. Defuzzification the output is converted back to a numerical variable using the centroid method which provides an analog signal to control the converter [1, 6].

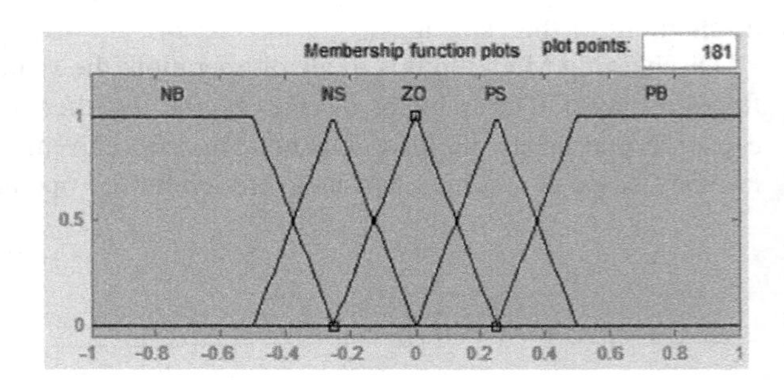

Fig. 5 Membership functions for inputs E, CE and D

Table 1 Fuzzy rule table

E/CE	NB	NS	Z0	PS	PB
NB	Z0	Z0	PB	PB	PB
NS	Z0	ZO	PS	PS	PS
ZO	PS	Z0	Z0	Z0	NS
PS	NS	NS	NS	Z0	Z0
PB	NB	NB	NB	ZO	ZO

5 Simulation Result

Following simulations has been done on the basis of following data

1. For PV cell: Refer Sect. 2 (i.e. PV cell.)
2. In P&O Method and incremental conductance method R = 40 Ω, C = 5 μF, L = 25 μH.

In fuzzy control method (Fig. 6) L, C is taken as 150 μH and 220, 5 μF. In place of R load of boost converter single phase inverter is connected in parallel with the LC filter. Where L = 2 mH, C = 30 μF.

In the fuzzy control method in place of DC source PV is taken then it is connected to boost converter. The output of the inverter in various voltage versus steady state time are shown in Figs. 7, 8 and 9. In the boost converter in place of R-load Single phase inverter is taken. And single phase inverter is connected to LC filter. And the MATLAB version 2010 is used in the fuzzy control method. We had

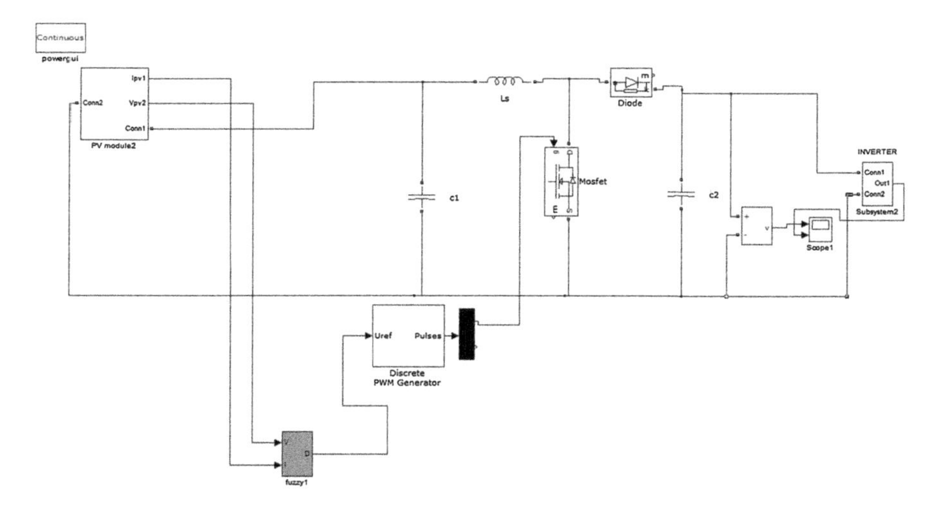

Fig. 6 Simulation of fuzzy control method

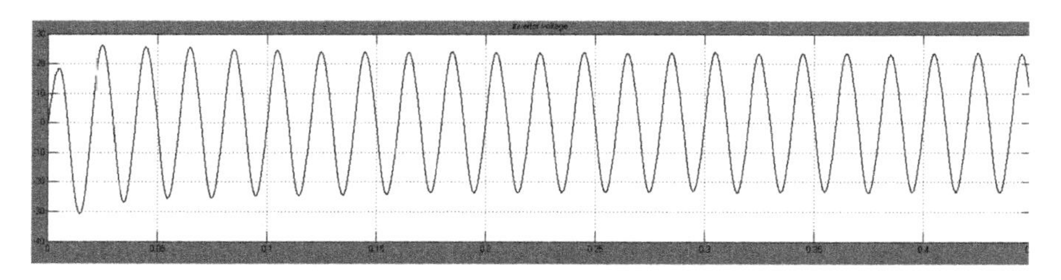

Fig. 7 Output of inverter voltage is 25 V and steady state time is 0.1 s using fuzz logic

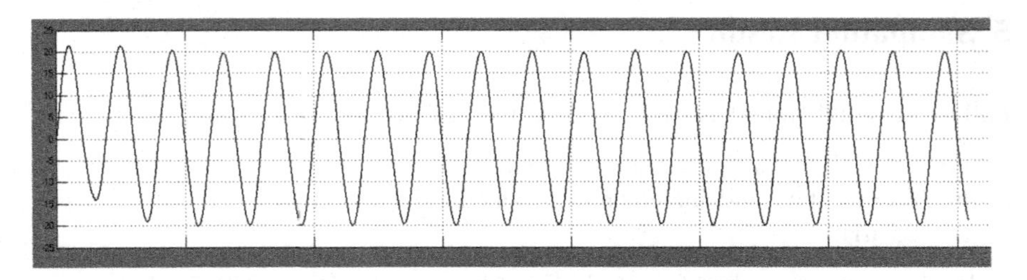

Fig. 8 Output of inverter voltage is 20 V and steady state time is 0.1 s using result of incremental conductance method

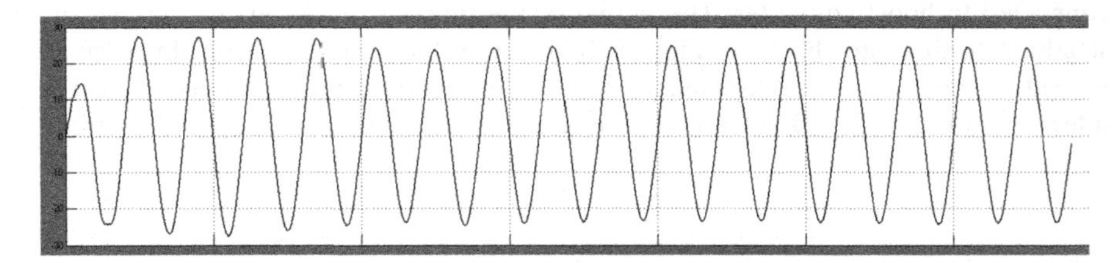

Fig. 9 Output of inverter voltage is 25 V and steady state time is 0.35 s using perturbation and observation method

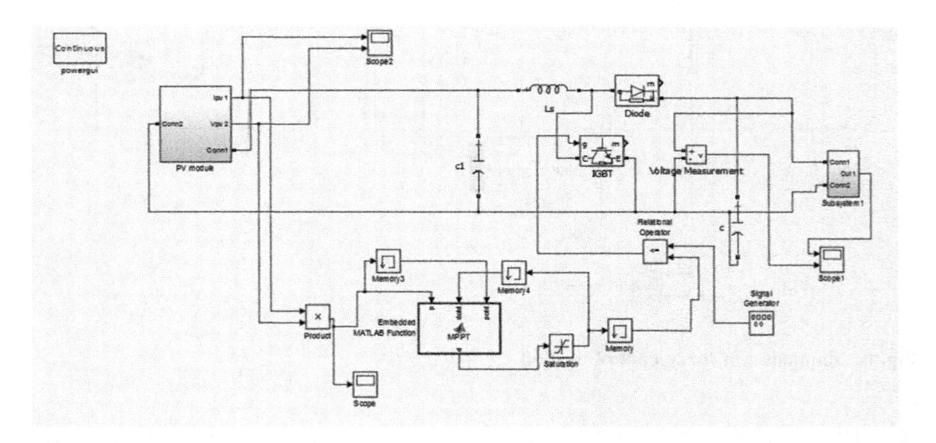

Fig. 10 Simulation of p&o method

given 30 V input and the obtained output is 25 V as some of the harmonics are reduced and 10 times boost has been done. And only 3 % harmonics left. In LC filter inductance and capacitance are taken as 2 mH and 30 μF respectively. In this paper Mathematical model of PV has been analyzed. Fuzzy control method is better than p&o and incremental because in p&o and incremental we get 220 V but in fuzzy control method 10 times boost has been done and we get the stable output. Simulation of p&o and incremental is same except the program. In p&o (Fig. 10), the program is written in comparison to power and change in power. Where as in incremental method, program is written in the form of change in voltage, change in current and duty cycle.

6 Conclusion

This paper gives an brief idea about a single-phase photovoltaic (PV) inverter topology with a fuzzy based MPPT control. The given paper has presented the fuzzy logic control to control the MPPT PV system. The proposed algorithm is fuzzy logic control which was simulated. The simulation results shows that this system is able accept fuzzy parameter for faster response, good performance in steady state. In addition the result of simulation and experiment has shown that MPPT controller by using fuzzy logic have provided more power than conventional. The system also provides energy to offer Improved output waveforms in steady state. In this paper an intelligent control techniques for the tracking of MPP were discovered in order to improve the efficiency of PV systems, under different temperature and irradiance conditions. The design and simulation of a fuzzy logic based MPPT controller was done by help of MATLAB. The given method has good response such as, fast responses with less fluctuation in the steady state, for rapid irradiance and temperature variations.

References

1. Amjad, A.M., Salam, Z.: A review of soft computing methods for harmonics elimination PWM for inverters in renewable energy conversion systems. Renew. Sustain. Energy Rev. **33**, 141–153 (2014). doi: 10.1016/j.rser.2014.01.080
2. DKM., Deepa, L.: Implementation of fuzzy logic MPPT control and modified H-bridge inverter in photovoltaic system . In: IEEE Conference Publications on 13–14 Feb 2014 doi:10.1109/ECS.2014.6892678(2014)
3. Altin, N., Ozdemir, S.: Three-phase three-level grid interactive inverter with fuzzy logic based maximum power point tracking controller. Energy Conv. Manage. **69**, 17–26 (2013). doi: 10.1016/j.enconman.2013.01.012
4. Subramanian, V., Murugesan, S.: An artificial intelligent controller for a three phase inverter based solar PV system using boost converter. In: 2012 Fourth International Conference on Advanced Computing (ICoAC), pp. 1–7, 13–15 Dec 2012. doi: 10.1109/ICoAC.2012.6416840 (2012)
5. Salam, A., Salim, A., Zaharah, S., Hamidon, F., Adam, I.: fuzzy logic controller for photovoltaic generation system in islanding mode operation. In: International Conference on Artificial Intelligence in Computer Science and ICT (AICS 2013), pp. 81–88. Langkawi, Malaysia, 25–26 Nov 2013. ISBN 978-967-11768-3-2

6. Sivagamasundari, M.S., Mary, P.M.: Fuzzy logic based cascaded H-bridge eleven level inverter for photovoltaic system using sinusoidal pulse width modulation technique. World Acad. Sci. Eng. Technol. Int. J. Electr. Robot. Electron. Commun. Eng. **8**(2), 443–447 (2014)
7. Sathy, S., karthikeyan, C.: Fuzzy Logic based Z-source inverter for hybrid energy resources. Int. J. Eng. Sci. Innov. Technol. **2**, 57–63 (2013). ISSN: 2319-5967

Experimental Validation of a Fuzzy Model for Damage Prediction of Composite Beam Structure

Deepak K. Agarwalla, Amiya K. Dash and Biswadeep Tripathy

Abstract Damage detection of beam structures have been in practice for last few decades. The methodologies adopted have been upgraded over the time depending upon the complexities of the damage or crack and the desired accuracy. The utilization of artificial intelligence (AI) techniques have also been considered by many researchers. In the current research, damage detection of a glass fiber reinforced cantilever beam has been done. A fuzzy based model using trapezoidal membership function has been developed to predict the damage characteristics i.e. damage position and damage severity. The inputs required for the fuzzy based model such as first three relative natural frequencies and first three mode shape differences have been determined by finite element analysis of the damaged cantilever beam subjected to the natural vibration. The results obtained from the proposed analysis have been experimentally validated.

Keywords Damage · Cantilever · Fuzzy · Finite element analysis · Trapezoidal membership function

1 Introduction

Damage detection of beam elements attracts various researchers in the past to carry out their research in the development of damage diagnostic techniques. A large number of damage detection tools are available for the beam members made of isotropic material. The AI techniques have also been used to diagnose the damage

D.K. Agarwalla (✉) · A.K. Dash · B. Tripathy
Department of Mechanical Engineering, Institute of Technical Education and Research, Siksha `O` Anusandhan University, Bhubaneswar, India
e-mail: deepak.cet08@gmail.com

A.K. Dash
e-mail: dash.amiya@gmail.com

B. Tripathy
e-mail: biswadeep143@gmail.com

L.C. Jain et al. (eds.), *Computational Intelligence in Data Mining - Volume 3*,
Smart Innovation, Systems and Technologies 33, DOI 10.1007/978-81-322-2202-6_10

effectively. Recently the composite materials are found to be engaged widely in various engineering applications. Therefore it is evident in the present scenario that a diagnostic tool for the damage diagnosis of composite beam members is highly required. The fuzzy system for damage diagnosis has been designed with six inputs (first three relative natural frequencies and first three relative mode shape differences) and two outputs (relative damage position, relative damage severity). A number of fuzzy linguistic terms and fuzzy membership functions (trapezoidal) have been used to develop the proposed damage identification technique. Chandrashekhar and Ganguli [1] have developed a fuzzy logic system (FLS) for damage detection. The probabilistic analysis has been performed using Monte Carlo simulation on a beam finite element (FE) model to calculate statistical properties. The frequencies were varied and the damage detection in the steel beam has been identified with 94 % accuracy. Moreover the FLS accurately classifies the undamaged condition reducing the possibilities of false alarms. Chandrashekhar and Ganguli [2] have come up with an idea that geometric and measurement uncertainty cause considerable problem in damage assessment which can be alleviated by using a fuzzy logic based approach for damage detection. Curvature damage factors (CDF) has been used as damage indicators and Monte Carlo simulation (MCS) has been used to study the changes in the damage indicator of a tapered cantilever beam. They have found that the method identifies both single and multiple damages in the structure. Silva et al. [3] have proposed an idea for structural health monitoring (SHM) using vibration data. The approach has been applied to data to a benchmark structure from LOS almost national Lab, USA. The two fuzzy clustering algorithms are compared and it shows that the Gustafson-Kessel algorithm marginally outperforms the fuzzy c-means but both are effective. Beena and Ganguli [4] have developed a new algorithmic approach for structural damage detection based on fuzzy cognitive map (FCM). A finite element model of a cantilever beam has been use to calculate the change in frequencies. The FCM works well for damage detection for ideal and noisy data and for further improvement in performance the hebbian learning have been used both FCM and hebbian learning give accurate results and represents a powerful tool for structural health monitoring. Radzienski et al. [5] have introduced a new method for structural damage detection. They have used damage indicators based on measured modal parameters in an aluminums cantilever beam. The special signal processing technique has been proposed to improve the effectiveness; however it has not been satisfactory. Hence new techniques have been proposed and were suitable for damage localization in beam like structure. Yan et al. [6] have reviewed about the development of vibration based on structural damage detection. The evaluation of the structural damage detection has been done and its diagnosis, application and development in the recent trends have been put forward. Reda Taha and Lucero [7] have come up with an idea of intelligent structural health monitoring with fuzzy sets to improve pattern recognition and damage detection. They have examined a pre stressed concrete bridge using data simulated from finite element analysis with new techniques. KO and NI [8] have investigated on recent technology development in the field of structural health monitoring. The aim was to evaluate structural health of the bridge which has been helpful for rehabilitation, maintenance and emergency repair. Erdogan and Bakir [9] have presented a fuzzy finite element model updating (FFEMU) method for damage detection problem. They have performed a probabilistic analysis using Monte Carlo simulation (MCS) and results are compared with the present method. Both the methods are accurate but the present method is not as computational expensive than MCS method. Radriguez and Arkkio [10] have come up with an idea for the detection of stator winding faults. They have used fuzzy logic to determine the stator motor condition. The finite element method (FEM) has been used and implemented in MATLAB. The proposed method has the ability to work with high accuracy which identifies the condition of motor. Kemru [11] have proposed a fuzzy logic method so as to improve the visual inspection and assessment of visual quality of industrial products. It is a unique application of vitreous china ceramic sanitary wares. Asnaashari and sinha [12] have proposed a probability distribution function which performs all the data processing in time domain for crack detection in beam. They have illustrated the numerical and experimental examples and their results are presented. Hasanzadeh et al. [13] have proposed an aligning method by a fuzzy recursive least square algorithm. The method of the adaptation for different crack shapes provides sufficient information for the training system which reduces the need for a large crack databases. Reza et al. [14] have proposed the fuzzy clustering method and logic rules which classified the defects. The detection and error has been compared with the bayes classifier. They have investigated the experimental images from eddy current, ultrasonic and radiography techniques which reduces the noise and drift leading to a better detection of defects. Saravanan et al. [15] have presented the use of decision tree that discriminates the fault conditions of the gear box from the vibration signals extracted. The vibration signal from a piezo-electric transducer captures different conditions for various loading and lubricating conditions. The statistical features have been extracted using decision tree which discriminate the fault conditions of the gear box.

From the literatures cited above, it has been observed that damage detection of different beam structures of various isotropic materials have been done in the past by various researchers. Various AI techniques have also been used by researchers to identify the damage position and damage severity for isotropic materials such as aluminum, steel etc. Damage identification of composite beam using AI techniques is still a potential area of research.

2 Finite Element Analyses

Finite element analysis (FEA) is a numerical method for solving a differential or integral equation. FEA is realized, first by dividing the structure into a number of small parts which are known as finite elements and the procedure adopted to attain these small elements is known as discretization. The solution for each finite element brought together to attain the global mass and stiffness matrix describing the dynamic response of the whole structure.

In the present research for damage identification in structural beam members, finite element based model is adopted to characterize the damage with respect to its severity and position. The dynamic behavior of the structure is altered with the presence of damage. The results obtained from finite element analysis of the damaged and undamaged structural beam members of Glass fiber reinforced composite is used as the inputs for the development of the fuzzy based model for damage diagnosis.

2.1 Finite Element Analysis of Damaged Cantilever Beam

In the current section, FEA is used for vibration analysis of a composite cantilever damaged beam (Fig. 1). The relationship between the displacement and the forces can be expressed as;

$$\begin{Bmatrix} u_j - u_i \\ \theta_j - \theta_i \end{Bmatrix} = C_{ovl} \begin{Bmatrix} U_j \\ \emptyset_j \end{Bmatrix} \tag{1}$$

where overall flexibility matrix C_{ovl} can be expressed as;

$$C_{ovl} = \begin{pmatrix} R_{11} & -R_{12} \\ -R_{21} & R_{22} \end{pmatrix}$$

The displacement vector in Eq. (1) is due to the damage.

The forces acting on the beam element for finite element analysis are shown in Fig. 1.

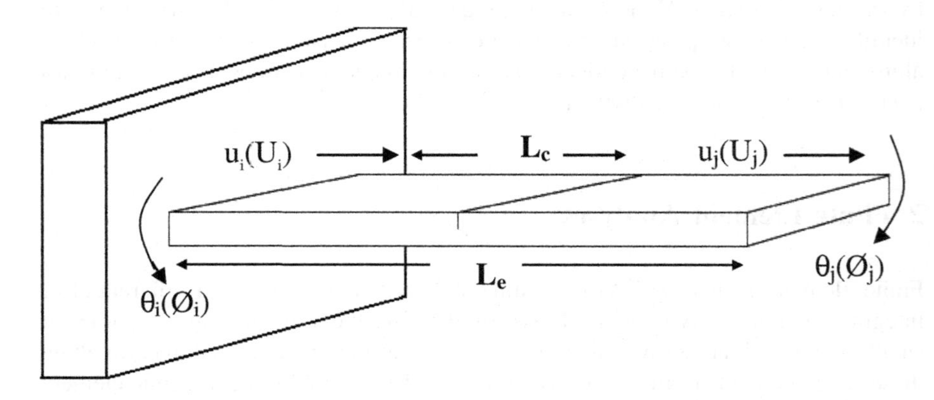

Fig. 1 Damaged beam element subjected to axial and bending forces

Where,

R_{11}: Deflection in direction 1 due to load in direction 1
$R_{12} = R_{21}$: Deflection in direction 1 due to load in direction 2
R_{22}: Deflection in direction 2 due to load in direction 2.

Under this system, the flexibility matrix C_{intact} of the intact beam element can be expressed as;

$$\begin{Bmatrix} u_j - u_i \\ \theta_j - \theta_i \end{Bmatrix} = C_{intact} \begin{Bmatrix} U_j \\ \emptyset_j \end{Bmatrix} \tag{2}$$

where,

$$C_{intact} = \begin{pmatrix} Le/EA & 0 \\ 0 & Le/EI \end{pmatrix}$$

The displacement vector in Eq. (2) is for the intact beam.

The total flexibility matrix C_{tot} of the damaged beam element can now be obtained by

$$C_{tot} = C_{intact} + C_{ovl} = \begin{bmatrix} Le/EA + R_{11} & -R_{12} \\ -R_{21} & Le/EI + R_{22} \end{bmatrix} \tag{3}$$

Through the equilibrium conditions, the stiffness matrix K_c of a damaged beam element can be obtained as

$$K_c = DC_{tot}^{-1}D^T \tag{4}$$

where D is the transformation matrix and expressed as;

$$D = \begin{pmatrix} -1 & 0 \\ 0 & -1 \\ 1 & 0 \\ 0 & 1 \end{pmatrix}$$

By solving the stiffness matrix Kc, the natural frequencies and mode shapes of the damaged cantilever beam can be obtained. This mathematical approach has been conceived by Computerized Analysis Software to estimate the natural frequencies and mode shapes of beam structures. In the current analysis, Computerized Analysis Software has been used to determine the vibration responses of damaged and undamaged cantilever of Glass fiber reinforced composite. The results of the finite element analysis for the first three modes of the damaged beam are used for the development of fuzzy model.

3 Fuzzy Based Model for Damage Diagnosis

The fuzzy system for damage diagnosis has been designed with six inputs (first three relative natural frequencies and first three relative mode shape differences) and two outputs (relative damage position, relative damage severity). A number of fuzzy linguistic terms and fuzzy membership functions (trapezoidal) have been used to develop the proposed damage identification technique. The trapezoidal membership function as shown in Fig. 2 has two base points (0.2, 0.5) and two shoulder points (0.3, 0.4). A mathematical expression for the trapezoidal membership function is presented below. The modal parameters obtained from the finite element analyses have been used to establish the rule base for designing of the fuzzy system. The performance of the proposed fuzzy based system for damage diagnosis has been compared with the results obtained from experimental analysis and presented in Table 3. A fuzzy logic system (FLS) essentially takes a decision by nonlinear mapping of the input data into a scalar output, using fuzzy rules. The FLS maps crisp inputs into crisp outputs. It can be seen from the figure that the FIS contains four components: the fuzzifier, inference engine, rule base, and defuzzifier.

3.1 Fuzzy Based Model Using Fuzzy Rule

$$\mu_F(X, 0.2, 0.3, 0.4, 0.5) = \begin{cases} 0 \text{ when } x \leq 0.2 \\ (x - 0.2)/(0.3 - 0.2) \\ \text{When } 0.2 \leq x \leq 0.3 \\ 1 \text{ when } 0.3 \leq x \leq 0.0.4 \\ (0.5 - x)/(0.5 - 0.4) \\ \text{When } 0.4 \leq x \leq 0.5 \end{cases}$$

Fuzzy inference system posses the approximation features by the help of fuzzy membership functions and fuzzy IF-THEN rules. This membership functions are designed by using the suitable fuzzy linguistic terms and fuzzy rule base. The fuzzy

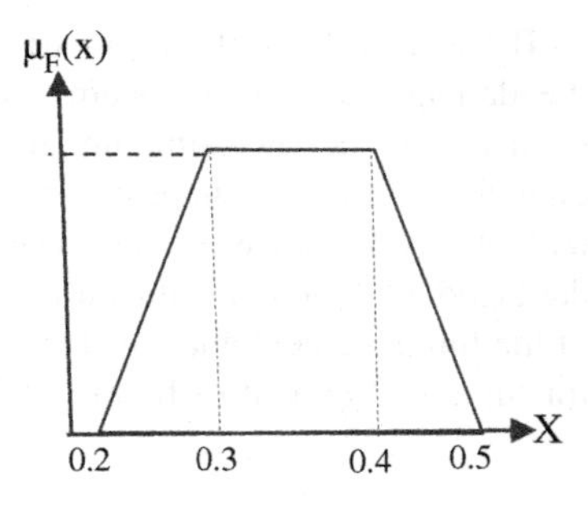

Fig. 2 Trapezoidal membership function

rule base or the conditional statements are used for fuzzification of the input parameters and defuzzification of the output parameters. Hence, the fuzzy model takes the input parameters from the application at a certain state of condition and using the rules it will provide a controlled action as desired by the system. A general model of a fuzzy inference system (FIS) is shown in Fig. 3.

The inputs to the fuzzy model for damage identification in the current analysis comprises,

Relative first natural frequency = FNF;
Relative second natural frequency = SNF;
Relative third natural frequency = TNF;
Relative first mode shape difference = FMD;
Relative second mode shape difference = SMD;
Relative third mode shape difference = TMD.
The linguistic term used for the outputs are as follows;
Relative damage position = RDP;
Relative damage severity = RDS.

The relationship between the fuzzy output set (F), defuzzifier and crisp output (K_0) can be established in the following equation;

$$K_0 = \text{defuzzifier (F)};$$

There are several defuzzification methods used for development of fuzzy system. Some of them are listed below;

- Centroid of the area
- Mean of maximum
- Weighted average method

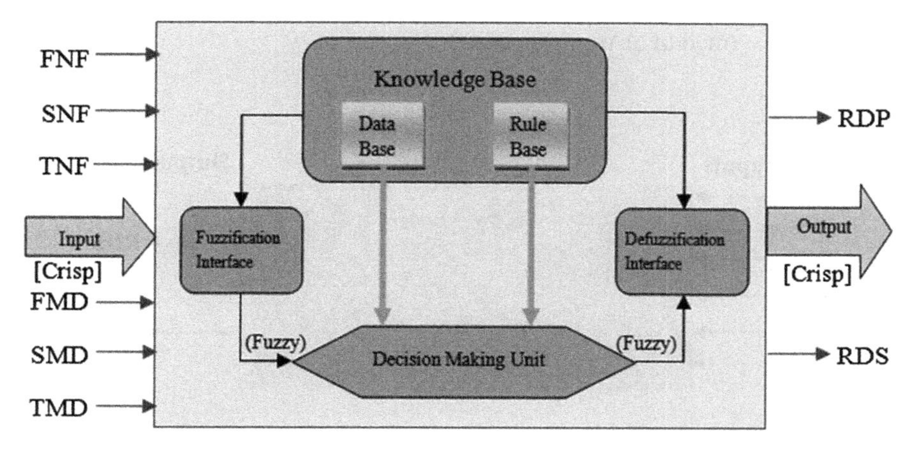

Fig. 3 Fuzzy controller for current analysis

3.2 Fuzzy Model Analysis

The pictorial view of the trapezoidal membership fuzzy models is shown in Fig. 4. Nine membership functions have been used for each input parameters to the fuzzy model. In designing the output membership functions for the output parameters such as relative damage position (RDP) and relative damage severity (RDS), twelve membership functions are considered. The fuzzy linguistic terms have been described in Table 1.

Based on the above fuzzy subsets, the fuzzy control rules are defined in a general form as follows:

If (FNF is FNF_i and SNF is SNF_j and TNF is TNF_k and FMD is FMD_l and SMD is SMD_m and TMD is TMD_n) then RDP is RDP_{ijklmn} and RDS is RDS_{ijklmn}

Where i = 1 to 9, j = 1 to 9, k = 1 to 9, l = 1 to 9, m = 1 to 9, n = 1 to 9

As "FNF", "SNF", "TNF", "FMD", "SMD", "TMD" have ten membership functions each. From Eq. (4) , two set of rules can be written as,

If (FNF is FNF_i and SNF is SNF_j and TNF is TNF_k and FMD is FMD_l and SMD is SMD_m and TMD is TMD_n) then RDS is RDS_{ijklmn}

If (FNF is FNF_i and SNF is SNF_j and TNF is TNF_k and FMD is FMD_l and SMD is SMD_m and TMD is TMD_n) then RDP is RDP_{ijklmn}

According to the usual fuzzy logic control method [16, 17], a factor W_{ijklmn} is defined for the rules as follows:

$$W_{ijklmn} = \mu_{fnf_i}(freq_i) \Lambda \mu_{snf_j}(freq_j) \Lambda \mu_{tnf_k}(freq_k) \Lambda \mu_{fmd_l}(moddif_l) \Lambda \mu_{smd_m}(moddif_m) \Lambda \mu_{tmd_n}(moddif_n)$$

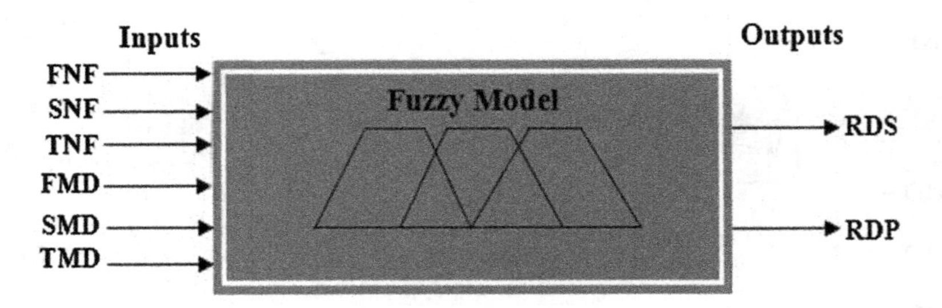

Fig. 4 Trapezoidal fuzzy model

Table 1 Description of fuzzy linguistic terms

Membership Functions Name	Linguistic Terms	Definition of the Linguistic terms
L1F1,L1F2,L1F3	$FNF_{1 \ldots 3}$	Low ranges of relative natural frequency for first mode of vibration
M1F1,M1F2, M1F3	$FNF_{4 \ldots 6}$	Medium ranges of relative natural frequency for first mode of vibration
H1F1,H1F2,H1F3	$FNF_{7 \ldots 9}$	Higher ranges of relative natural frequency for first mode of vibration
L2F1,L2F2,L2F3	$SNF_{1 \ldots 3}$	Low ranges of relative natural frequency for second mode of vibration
M2F1,M2F2,M2F3	$SNF_{4 \ldots 6}$	Medium ranges of relative natural frequency for second mode of vibration
H2F1,H2F2,H2F3	$SNF_{7 \ldots 9}$	Higher ranges of relative natural frequencies for second mode of vibration
L3F1,L3F2,L3F3	$TNF_{1 \ldots 3}$	Low ranges of relative natural frequencies for third mode of vibration
M3F1,M3F2,M3F3	$TNF_{4 \ldots 6}$	Medium ranges of relative natural frequencies for third mode of vibration
H3F1,H3F2,H3F3	$TNF_{7 \ldots 9}$	Higher ranges of relative natural frequencies for third mode of vibration
S1M1,S1M2,S1M3	$FMD_{1 \ldots 3}$	Small ranges of first relative mode shape difference
M1M1,M1M2,M1M3	$FMD_{4 \ldots 6}$	medium ranges of first relative mode shape difference
H1M1,H1M2,H1M3	$FMD_{7 \ldots 10}$	Higher ranges of first relative mode shape difference
S2M1,S2M2,S2M3	$SMD_{1 \ldots 3}$	Small ranges of second relative mode shape difference
M2M1,M2M2,M2M3	$SMD_{4 \ldots 6}$	medium ranges of second relative mode shape difference
H2M1,H2M2,H2M3	$SMD_{7 \ldots 10}$	Higher ranges of second relative mode shape difference
S3M1,S3M2,S3M3	$TMD_{1 \ldots 3}$	Small ranges of third relative mode shape difference
M3M1,M3M2,M3M3	$TMD_{4 \ldots 6}$	medium ranges of third relative mode shape difference
H3M1,H3M2,H3M3	$TMD_{7 \ldots 10}$	Higher ranges of third relative mode shape difference
SP1,SP2,SP3,SP4	$RDP_{1 \ldots 4}$	Small ranges of relative damage position
MP1,MP2,MP3,MP4	$RDP_{5 \ldots 8}$	Medium ranges of relative damage position
HP1,HP2, HP3,HP4	$RDP_{9 \ldots 12}$	Higher ranges of relative damage position
LS1,LS2,LS3,LS4	$RDS_{1 \ldots 4}$	Lower ranges of relative damage severity
MS1,MS2,MS3,MS4	$RDS_{5 \ldots 8}$	Medium ranges of relative damage severity
US1,US2,US3,US4	$RDS_{9 \ldots 12}$	Upper ranges of relative damage severity

Where $freq_i$, $freq_j$ and $freq_k$ are the first, second and third relative natural frequencies of the cantilever beam with damage respectively; $moddif_l$, $moddif_m$ and $moddif_n$ are the average first, second and third relative mode shape differences of the cantilever beam with damage respectively. By applying the composition rule of inference [16, 17], the membership values of the relative damage position and relative damage severity, $(position)_{RDP}$ and $(severity)_{RDS}$ can be computed as;

Table 2 Examples of twenty fuzzy rules to be implemented in fuzzy model

Sl. No.	Examples of some rules used in the fuzzy model
1	If FNF is H1F1, SNF is M2F2, TNF is M3F1, FMD is H1M2, SMD is H2M4, TMD is H3M3, then RDS is LS4 and RDP is SP3.
2	If FNF is L1F4, SNF is L2F4, TNF is L3F4, FMD is H1M1, SMD is H2M1, TMD is H3M2, then RDS is LS2 and RDP is SP4.
3	If FNF is L1F3, SNF is L2F4, TNF is L3F4, FMD is M1M2, SMD is H2M2, TMD is H3M3, then RDS is LS3 and RDP is SP2.
4	If FNF is H1F2, SNF is H2F1, TNF is H3F1, FMD is H1M3, SMD is H2M4, TMD is H3M4, then RDS is MS2 and RDP is SP3.
5	If FNF is M1F1, SNF is L2F2, TNF is L3F3, FMD is H1M1, SMD is H2M1, TMD is H3M2, then RDS is MS1 and RDP is SP2.
6	If FNF is L1F1, SNF is L2F2, TNF is L3F3, FMD is H1M3, SMD is M2M1, TMD is H3M4, then RDS is MS2 and RDP is MP1.
7	If FNF is L1F4, SNF is L2F4, TNF is L3F4, FMD is M1M2, SMD is H2M1, TMD is H3M1, then RDS is MS1 and RDP is SP1.
8	If FNF is H1F1, SNF is M2F2, TNF is M3F1, FMD is H1M2, SMD is H2M2, TMD is H3M2, then RDS is MS2 and RDP is SP2.
9	If FNF is L1F1, SNF is L2F4, TNF is L3F4, FMD is M1M1, SMD is M2M1, TMD is M3M2, then RDS is LS1 and RDP is MP2.
10	If FNF is M1F1, SNF is L2F2, TNF is L3F1, FMD is M1M2, SMD is M2M2, TMD is H3M1, then RDS is LS1 and RDP is SP2.
11	If FNF is M1F1, SNF is M2F1, TNF is M3F1, FMD is H1M3, SMD is H2M3, TMD is H3M4, then RDS is MS1 and RDP is SP4.
12	If FNF is M1F1, SNF is L2F1, TNF is L3F1, FMD is H1M3, SMD is H2M2, TMD is H3M3, then RDS is LS2 and RDP is MP1.
13	If FNF is M1F2, SNF is M2F1, TNF is M3F1, FMD is M1M1, SMD is H2M1, TMD is H3M2, then RDS is MS2 and RDP is MP2.
14	If FNF is H1F2, SNF is H2F1, TNF is H3F1, FMD is H1M4, SMD is H2M1, TMD is H3M1, then RDS is MS1 and RDP is SP4.
15	If FNF is M1F1, SNF is L2F1, TNF is L3F2, FMD is S1M1, SMD is S2M2, TMD is H3M1, then RDS is LS2 and RDP is MP1.
16	If FNF is L1F4, SNF is L2F4, TNF is L3F4, FMD is H1M2, SMD is S2M1, TMD is H3M2, then RDS is LS1 and RDP is MP3.
17	If FNF is M1F1, SNF is L2F3, TNF is L3F1, FMD is S1M2, SMD is M2M1, TMD is S3M1, then RDS is LS2 and RDP is MP3.
18	If FNF is L1F1, SNF is L2F1, TNF is L3F1, FMD is H1M2, SMD is H2M2, TMD is H3M2, then RDS is LS3 and RDP is MP2.
19	If FNF is H1F2, SNF is H2F1, TNF is H3F1, FMD is S1M2, SMD is H2M3, TMD is H3M1, then RDS is LS4 and RDP is MP1.
20	If FNF is L1F3, SNF is L2F4, TNF is L3F4, FMD is S1M3, SMD is S2M2, TMD is S3M3, then RDS is LS3 and RDP is SP3.

$$\mu_{\mathrm{RDP}_{\mathrm{ijklmn}}}(\text{position}) = \mathrm{W}_{\mathrm{ijklmn}} \wedge \mu_{RDP_{ijklmn}}(\text{position})$$
$$\mu_{RDS_{ijklmn}}(severity) = \mathrm{W}_{\mathrm{ijklmn}} \wedge \mu_{RDS_{ijklmn}}(severity)$$

The overall conclusion by combining the outputs of all the fuzzy rules can be written as follows:

$$\mu_{\mathrm{RDP}}(\text{position}) = \mu_{\mathrm{RDP}_{111111}}(\text{position}) \vee \cdots \vee \mu_{\mathrm{RDP}_{ijklmn}}(\text{position}) \vee \cdots \vee \mu_{\mathrm{RDP}_{101010101010}}(\text{position})$$
$$\mu_{\mathrm{RDS}}(\text{severity}) = \mu_{\mathrm{RDS}_{111111}}(\text{severity}) \vee \cdots \vee \mu_{\mathrm{RDS}_{ijklmn}}(\text{severity}) \vee \cdots \vee \mu_{\mathrm{RDS}_{101010101010}}(\text{severity})$$

The crisp values of relative damage position and relative damage severity are evaluated using the Centre of gravity method as:

$$\mathrm{Rel}ative damage position = RDP = \frac{\int (position \cdot \mu_{RDP}(position) \cdot d(position))}{\int \mu_{RDP}(position) \cdot d(position)}$$
$$\mathrm{Rel}ative damage severity = RDS = \frac{\int (severity) \cdot \mu_{RDS}(severity) \cdot d(severity)}{\int \mu_{RDS}(severity) \cdot d(severity)}$$

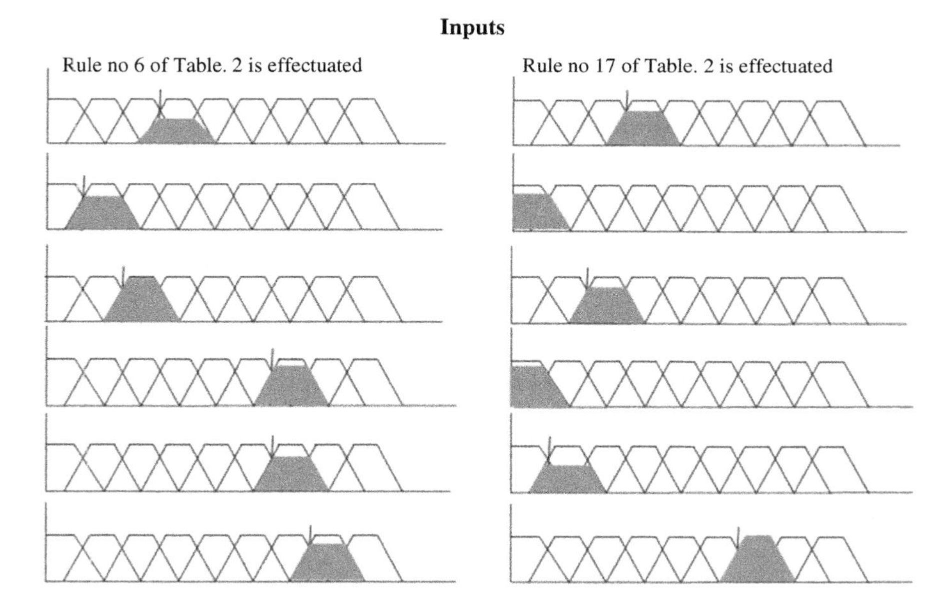

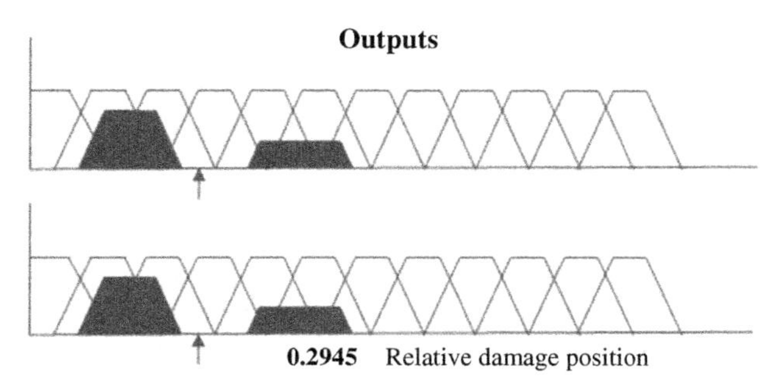

Fig. 5 Resultant values of relative damage severity and relative damage position from trapezoidal fuzzy model when Rules 6 and 17 of Table 2 are effectuated

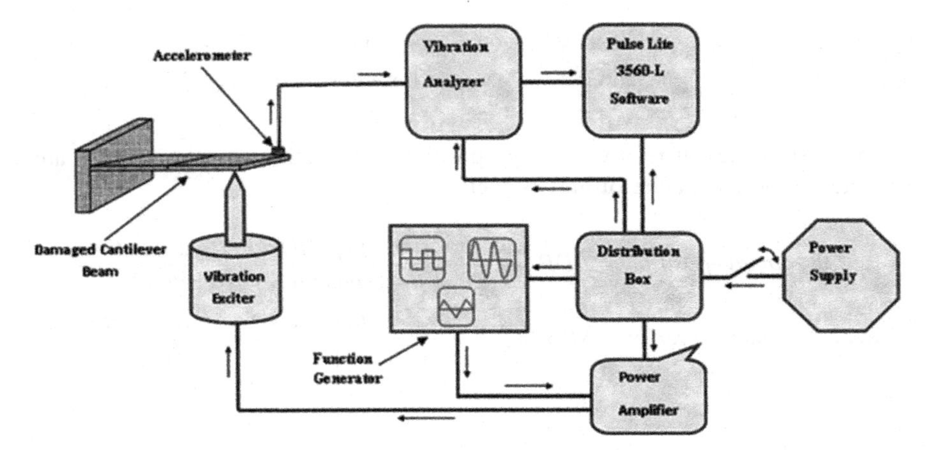

Fig. 6 Schematic diagram of experimental set up with cantilever beam

4 Experimental Validation

Experiments have been performed using the developed experimental set up (Fig. 6) for measuring the vibration responses (natural frequencies and amplitude of vibration) of the cantilever beams composite with dimension 1000 mm × 50 mm × 8 mm. During the experiment, the damaged and undamaged beams have been subjected to vibration at their 1st, 2nd and 3rd mode of vibration by using an exciter and a function generator. The vibrations characteristics of the beams correspond to 1st, 2nd and 3rd mode of vibration have been recorded by placing the accelerometer along the length of the beams. The signals from the accelerometer which contains the vibration parameters such as natural frequencies and mode shapes are analyzed and shown on the vibration indicator.

5 Results and Discussion

The results obtained from the fuzzy based model and experimental analyses have been presented in Table 3. The first three columns of Table 3 represent the first three relative natural frequencies and the next three columns represent the first three mode shape differences of the composite cantilever beam. Figure 5 depicts the relative damage severity and relative damage position of the beam member using the fuzzy based model with the trapezoidal membership function. The results obtained from the fuzzy based model are found to be in very good agreement with the experimental results for damage detection. The percentages of error for relative damage severity and relative damage position are found to be 2.85 and 3.1 % respectively.

Table 3 Comparison of modal parameters and damage characteristics of composite cantilever beam obtained from Fuzzy logic (Trapezoidal) based model and experimental analysis

FNF	SNF	TXF	FMD	SXD	TXD	Fuzzy trapezoidal		Experimental	
						RDS	RDP	RDS	RDP
0.9940	0.9999	0.9918	0.0043	0.0091	0.0066	0.3731	0.481	0.374	0.501
0.9872	0.9998	0.9828	0.0080	0.0123	0.0237	0.4982	0.490	0.500	0.501
0.9755	0.9997	0.9676	0.0119	0.0087	0.0411	0.6232	0.490	0.624	0.503
0.9999	0.9944	0.9937	0.1237	0.0265	0.0761	0.3733	0.273	0376	0.255
0.9998	0.9883	0.9870	0.3741	0.0465	0.0380	0.5028	0.243	0.501	0.254
0.9995	0.9768	0.9751	0.3699	0.0531	0.0534	0.6221	0.261	0.624	0.253

6 Conclusions

The fuzzy modeling implemented in the current section has been analyzed to get the following conclusions. The presence of damage in composite structural member has remarkable impact on the modal parameters of the dynamic structure. The first three relative natural frequencies and first three relative mode shape differences are engaged as inputs to the fuzzy model and relative damage positions and relative damage severity are the output parameters. The reliability of the proposed model has been established by comparing the results from the fuzzy models (trapezoidal) with that of the experimental analysis. The results are found to be well in agreement. Hence, the proposed trapezoidal fuzzy model can be effectively used as damage identification tools in vibrating composite structures.

References

1. Chandrashekhar, M., Ganguli, R.: Uncertainty handling in structural damage detection using fuzzy logic and probabilistic simulation. Mech. Syst. Signal Process. **23**, 384–404 (2009)
2. Chandrashekhar, M., Ganguli, R.: Damage assessment of structures with uncertainty by using mode-shape curvatures and fuzzy logic. J. Sound Vib. **326**, 939–957 (2009)
3. Silva, S.D., Junior, M.D., Junior, V.L., Brennan, M.J.: Structural damage detection by fuzzy clustering. Mech. Syst. Signal Process. **22**, 1636–1649 (2008)
4. Beena, P., Ganguli, R.: Structural damage detection using fuzzy cognitive maps and Hebbian learning. Appl. Soft Comput. **11**, 1014–1020 (2011)
5. Radzienski, M., Krawczuk, M., Palacz, M.: Improvement of damage detection methods based on experimental modal parameters. Mech. Syst. Signal Process. **25**, 2169–2190 (2011)
6. Yan, Y.J., Cheng, L., Wu, Z.Y., Yam, L.H.: Development in vibration-based structural damage detection technique. Mech. Syst. Signal Process. **21**, 2198–2211 (2007)
7. Reda Taha, M.M., Lucero, J.: Damage identification for structural health monitoring using fuzzy pattern recognition. Eng. Struct. **27**, 1774–1783 (2005)
8. Ko, J.M., Ni, Y.Q.: Technology developments in structural health monitoring of large-scale bridges. Eng. Struct. **27**, 1715–1725 (2005)
9. Erdogan, Y.S., Bakir, P.G.: Inverse propagation of uncertainties in finite element model updating through use of fuzzy arithmetic. Eng. Appl. Artif. Intell. **26**, 357–367 (2013)

10. Rodriguez, P.V.J., Arkkio, A.: Detection of stator winding fault in induction motor using fuzzy logic. Appl. Soft Comput. **8**, 1112–1120 (2008)
11. Kemru, M.: Assessing the visual quality of sanitary ware by fuzzy logic. Appl Soft Comput. **13**, 3646–3656 (2013)
12. Asnaashari, E., Sinha, J.K.: Crack detection in structures using deviation from normal distribution of measured vibration responses. J. Sound Vib. **333**, 4139–4151 (2014)
13. Hasanzadeh, R.P.R., Sadeghi, S.H.H., Ravan, M., Moghaddamjoo, A.R., Moini, R.: A fuzzy alignment approach to sizing surface cracks by the AC field measurement technique. NDT & E Int. **44**, 75–83 (2011)
14. Reza, H.P.R., Rezaie, A.H., Sadeghi, S.H.H., Moradi, M.H., Ahmadi, M.: A density-based fuzzy clustering technique for non-destructive detection of defects in materials. NDT & E Int. **40**, 337–346 (2007)
15. Saravanan, N., Cholairajan, S., Ramachandran, K.I.: Vibration-based fault diagnosis of spur bevel gear box using fuzzy technique. Expert Syst. Appl. **36**, 3119–3135 (2009)
16. Dash, A.K., Parhi, D.R.K.: Development of an inverse methodology for crack diagnosis using AI technique. Int. J. Comput. Mater. Sci. Surf. Eng. **4,** 143–167 (2011)
17. Parhi, D.R.: Navigation of mobile robots using a fuzzy logic controller. J. Intell. Robot. Syst. Theory Appl. **42**, 253–273 (2005)

A Comparative View of AoA Estimation in WSN Positioning

Sharmilla Mohapatra, Sasmita Behera and C.R. Tripathy

Abstract The location tracking becomes very useful aspect in security and data collection and communication based applications. The wireless sensors have got the preferable characteristics of low memory, low weight, and low power and low cost with its increasing speed of application. GPS location tracking system found to be inappropriate due to its line-of-sight and cost constraints. Hence techniques like AoA, ToA & RSS are in high demand for Position tracking of sensor nodes. But at the time of measurement of these above parameters there is a high chance that they get corrupted with the atmospheric impairments and multipath propagation errors resulting an erroneous location tracking. Here it's a part of the survey of different non-GPS location estimation systems of wireless sensor nodes. This paper is mainly focusing the estimation of angle of arrival (AoA) in different scenarios using different methods for location tracking.

Keywords WSN · LBS · AoA · SINR

1 Introduction

A wireless sensor network (WSN) consists of spatially distributed autonomous sensors to cooperatively monitor physical or environmental conditions, such as temperature, sound, vibration, pressure, motion or pollutants. The developments of wireless sensor networks was motivated by military applications such as battlefield

S. Mohapatra (✉) · S. Behera · C.R. Tripathy
Department of Computer Science and Engineering, Veer Surendra Sai University of Technology, Burla 768018, Odisha, India
e-mail: sharmilla.cs@gmail.com

S. Behera
e-mail: sasmita.mam@gmail.com

C.R. Tripathy
e-mail: crt.vssut@yahoo.com

L.C. Jain et al. (eds.), *Computational Intelligence in Data Mining - Volume 3*,
Smart Innovation, Systems and Technologies 33, DOI 10.1007/978-81-322-2202-6_11

surveillance and are now used in many industrial and civilian application areas, including industrial process monitoring and control, machine health monitoring, environment and habitat monitoring, healthcare applications, home automation, and traffic control [1]. In a Wireless Sensor Network the sensor nodes can be imagined as small computers, extremely basic in terms of their interfaces and their components. It usually consist of a *processing unit* with limited computational power and limited memory, *sensors* (including specific conditioning circuitry), a *communication device* (usually radio transceivers or alternatively optical), and a power source (usually in the form of a battery). The base stations are the one or more distinguished components of the WSN with much more computational energy and communication resources. A base station acts as a gateway between the sensor nodes and the end user. The unique characteristics of a WSN are limited power they can harvest or store, ability to withstand harsh environmental conditions, ability to cope with node failures, mobility of nodes, dynamic network topology, communication failures, heterogeneity of nodes, large scale of deployment, unattended operation.

Localization is a well studied problem in several areas including robotics, mobile ad hoc and vehicular networks and wireless sensor networks. The aim of localization is to assign geographic coordinates to each node in the sensor node. In Wireless Sensor Network, localization of sensors is a key enabling technology because sensors need to know their location in order to detect and record the events so that their data is meaningful. There are a couple of ways the sensors measure AoA. There are different techniques for position localization like Received Signal Strength (RSS), Time Of Arrival (TOA), Time Difference Of Arrival (TDOA) and Angle Of Arrival (AOA) [21]. The RSSI technique does not find out the exact location it helps in finding out a range. The TOA and TDOA technique requires clock synchronization between the base station and the mobile station whereas Angle of Arrival helps in estimating the exact location of a sensor node without any requirement of the clock synchronization.

2 AoA Measurement Techniques

In wireless Sensor Networks, localization of sensors is a key enabling technology because the sensors nodes need to know their locations to detect and record the events so that their data is meaningful. The use of GPS receiver with each sensor node is cost prohibitive in terms of both hardware and power refinements. Since GPS requires line of sight between the receiver and the satellite, it may not work well in buildings blocking the direct view to GPS satellite. So, manual assignment of node coordinates though impractical can be used in many cases so that we can estimate the position of the nodes. There is usually a sparse set of nodes called anchor–nodes, which have their coordinates either from GPS or manual configuration. Hence the main objective of this research is localization of sensor nodes through manual representation with the help of anchor nodes so that cost can be minimized. Through the technique of manual placement of the sensor nodes we can measure the angle and also the distance between the anchor nodes and the mobile node.

2.1 Angle of Arrival Estimation Using Array of Antenna

An antenna array [2] often called a 'phased array' is a set of 2 or more antennas. The signals from the antennas are combined or processed in order to achieve improved performance over that of a single antenna. The antenna array can be used to increase the overall gain, to cancel out interference from a particular set of directions, to determine the direction of arrival of the incoming signals, to maximize the SINR.

The general form of an antenna array can be illustrated as in Fig. 1 [3]. An origin and coordinate system are selected, and then the *N* elements are positioned, each at location given by $d_n = [x_n\ y_n\ z_n]$

The output of the antenna array can be written as $Y = \sum_{n=1}^{N} W_n X_n$.

To understand the benefits of antenna arrays, consider a set of 3-antennas located along the z-axis, receiving a signal (plane wave or the desired information) arriving from an angle θ relative to the z-axis as in Fig. 2. The antennas in the phased array are spaced one-half wavelength apart (centered at z = 0). From the antenna theory

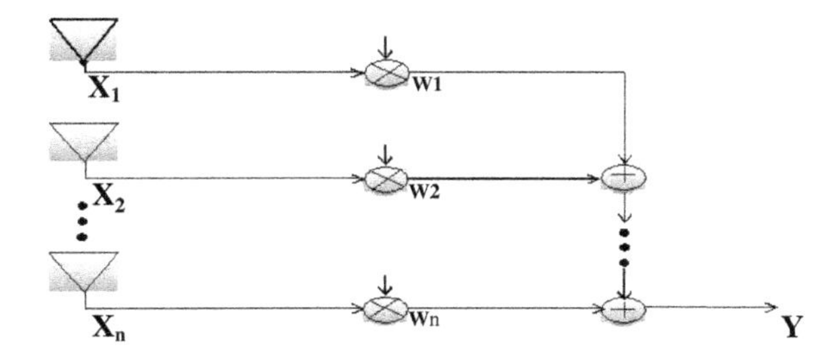

Fig. 1 Geometry of an arbitrary *N* element antenna array, weighting and summing of signals from the antennas to form the output in a phased array

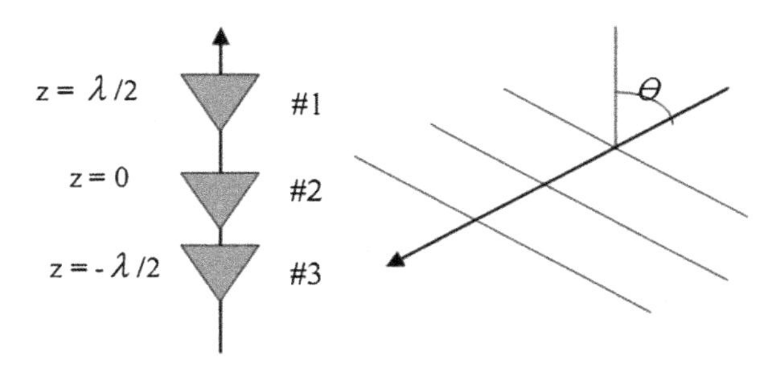

Fig. 2 3-element antenna array receiving a plane wave

[4], as the gain of the antenna is a function of angle of arrival (θ), θ can be measured at a particular receiver if the gain is known. And thus can be helpful in estimating the location of the source.

2.2 *Digital Compass*

A digital compass may be either a magnetometer such as a MEMS magnetometer in hand-held devices or a fiber optic gyrocompass as in a ship's navigation system. Cell phones and many other mobile devices now come packed with sensors capable of tracking them as they move. The digital compasses, gyroscopes, and accelerometers embedded in such devices have spawned a wide range of location-based services, as well as novel ways of controlling mobile gadgets—for instance, with a shake or a flick. Now a new way of making these sensors could make such technology smaller and cheaper. Conventional digital compasses are made using what's called complementary metal-oxide-semiconductor manufacturing, the most common method for making microchips and electronic control circuitry.

2.2.1 Fiber Optic Gyrocompass

A fiber optic is a compass and instrument of navigation. It is often part of a ships set of compasses, which also include a conventional gyrocompass and a magnetic compass. The fiber optic gyrocompass [2] is a complete unit, which unlike a conventional compass, has no rotating or other moving parts. It uses a series of fiber optic gyroscope sensors and computers to locate north. It has very high reliability and requires little maintenance during its service life. The entire system usually includes a sensor unit, a control and display unit, and an interface and power supply unit. It is often linked with the ship's other navigational devices including GPS [3].

2.2.2 MEMS Magnetometer

A magnetometer [4] is a measuring instrument used to measure the strength and perhaps the direction of magnetic fields. Magnetometers can be divided into *scalar* devices which only measure the intensity of the field and *vector* devices which also measure the direction of the field. Magnetometers [5] are widely used for measuring the Earth's magnetic fields and in geophysical surveys to detect magnetic anomalies of various types. They are also used militarily to detect submarines. Magnetometers can be used as metal detectors: they can detect only magnetic (ferrous) metals, but can detect such metals at a much larger depth than conventional metal detectors; they are capable of detecting large objects, such as cars, at tens of meters, while a metal detector's range is rarely more than 2 m. In recent years, magnetometers have been miniaturized to the extent that they can be incorporated in integrated circuits at very low cost and are finding increasing use as compasses in consumer devices such as mobile phones.

2.3 AoA Through Orientation Forwarding

Ad hoc network are mostly studied in the context of high mobility, high power nodes and moderate network sizes. The main features of ad hoc networks are large number of unattended nodes, lack of deploying supporting infrastructure and high cost of human supervised maintenance. This positioning and orientation algorithm is appropriate for indoor location aware applications, where the problem is not the unpredictable highly mobile topology rather the ad hoc deployment [6, 7, 8]. The orientation and positioning problems have been intensively studied in the context of mobile robot navigation [9]. The methods make extensive use of recognizable "landmarks". This method works under the assumption that a small fraction of network has only the positioning and orientation capability. This method is different from the other methods based on the capability of the nodes to sense the direction from which the signal is received which is known as the angle of arrival (AoA). This requires either an antenna array or several ultrasound receivers but along with positioning it also provides the orientation capability.

The network is a collection of ad hoc deployed nodes where any node can only communicate directly only with its immediate neighboring nodes within radio range. In the ideal case, when radio coverage of a node is circular, these networks are modeled as a fixed radius random graph.

But here ach node in this network is assumed to have one main axis against which all angles are reported and the capacity to estimate approximately accurate direction from which a neighbor is sending data. It assumes that after the deployment, the axis of the node has an arbitrary, unknown heading. Some of the nodes, called landmarks have additional knowledge about their position, from some external source, such as a GPS receiver, or manual input. The term bearing refers to an angle measurement with respect to another object [1, 10]. Radial is a reverse bearing, or the angle under which an object is seen from another point. The use of the term heading with the meaning of bearing to north is, the absolute orientation of the main axis of each node. In Fig. 3, for node B, bearing to A is $\widehat{ba}$ and radial from A is $\widehat{ab}$ and heading is $\widehat{b}$. So from the trigonometric methods, the headings, bearings and radials can be calculated of a node if the above information of the rest of the nodes is available. Thus heading of A can be calculated if heading of one of the neighbors, say B, is known, then it will be: $2\pi - (\widehat{ba} + \pi - \widehat{ab}) + \widehat{b}$.

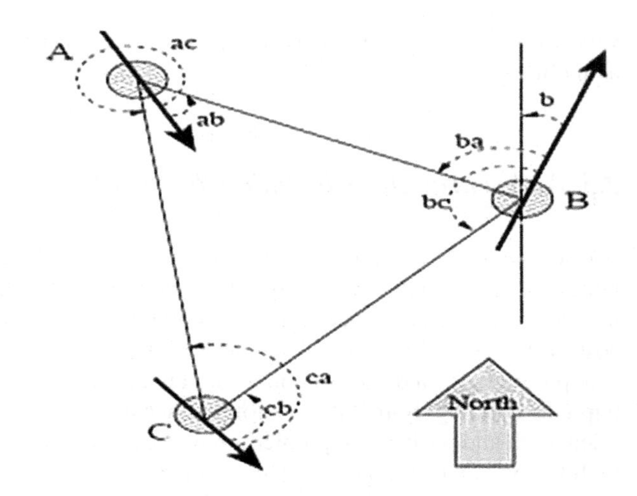

Fig. 3 Node with AoA capability

2.4 Cricket Compass for Context Aware Mobile Applications

By analogy to a traditional compass, knowledge of orientation through cricket compass attached to a mobile device enhances various applications including way-finding and navigation, directional service discovery. Context aware applications which adapt their behavior to environmental context such as physical location are an important class of applications in emerging pervasive computing environment [11]. The first motivating application is the *Way finder.* This application is designed to run in a handheld computer and help people navigate toward a destination in an unfamiliar setup. The second motivating application is called the *Viewfinder.* The user can point it any direction and specify a sweep angle and maximum distance. Using an active map integrated with a resource discovery system, the viewfinder then retrieves and displays a representation of the set of devices and services lying inside a sector of interest specified by the user and allow the user to interact with these devices through the representation on the map. Figure 4 shows a beacon B and a compass with two ultrasonic receivers R_1 and R_2 which are located at a distance L apart from each other. The angle of rotation of the compass θ with respect to the beacon B is related to the differential distance d_1 and d_2, where d_1 and d_2 are the distances of the receivers R_1 and R_2 from B.

Figure 4b shows the beacon B from Fig. 4a projected onto the horizontal plane along which the compass is aligned. In this figure, x_1 and x_2 are the projections of distances d_1 and d_2 on to the horizontal plane. It assumes the compass is held parallel to horizontal plane.

$$x_1^2 = d_1^2 - z^2 \tag{1}$$

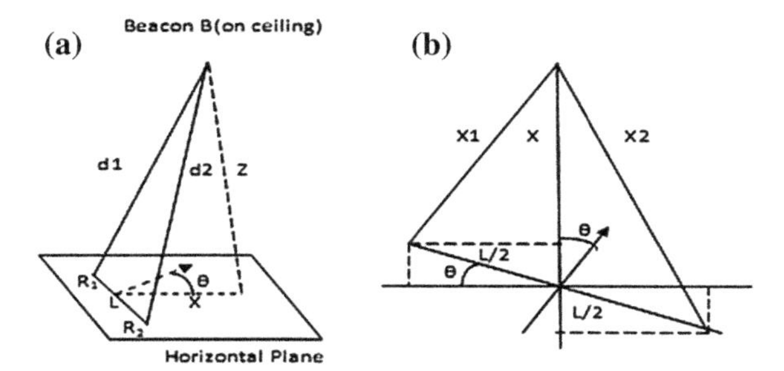

Fig. 4 **a** Determining the angle of orientation θ along the Horizontal plane. **b** A rotated compass leading to difference in distance between the beacon and the receivers

$$x_2^2 = d_2^2 - z^2 \tag{2}$$

$$x = (\bar{d}^2 - z^2)^{1/2}$$

where $\bar{d} \approx (d_1 + d_2)/2$ when $d_1, d_2 \gg L$.

From Fig. 4b

$$x_1^2 = (\frac{L}{2}\cos\theta)^2 + (x - \frac{L}{2}\mathrm{Sin}\theta)^2 \text{ and } x_2^2 = (\frac{L}{2}\cos\theta)^2 + (x + \frac{L}{2}\sin\theta)^2$$

$$\Rightarrow x_2^2 - x_1^2 = 2Lx\sin\theta$$

Substituting for x_1^2 and x_2^2 from Eqs. (1) and (2),

We get;

$$\sin\theta = (d_2 + d_1)(d_2 - d_1)/2Lx \tag{3}$$

So

$$\sin\theta = (d_2 - d_1)/\, L\sqrt{1 - (\frac{z}{d})^2} \tag{4}$$

Equation (4) implies that it suffices to estimate two quantities in order to determine the orientation of the compass with respect to a beacon: (i) $(d_2 - d_1)$, the difference in the distances of the two receivers from the beacon, and (ii) z/d, the ratio of the height of the beacon from the horizontal plane on which the compass is placed to the distance of the beacon from the center of the compass. The main goal of this method is to estimate each of these quantities with high precision so as to produce an accurate estimate of θ. The solution to this problem is to track the phase difference between the ultrasonic signals at the two different receivers and process the information.

The second quantity, z/d, is estimated by determining the (x, y, z) coordinates of the compass with respect to the plane formed by the beacons. This is done by placing multiple beacons in a room and estimating the time it takes for the ultrasonic signal to propagate between them and the compass. The differential distance can be measured as $\delta d = d_1 - d_2$. Let d_1 and d_2 be the distances to receivers R_1 and R_2 from the beacon B. Let w_1 and w_2 be the ultrasonic waveforms at the two receivers. ϕ can be expressed as: $\mathbf{\Phi} = \frac{\partial d}{\lambda} * 2\pi$.

2.5 AoA Through Maximum Likelihood Approach

Instead of using phase difference of arrival like: $\Phi 1 - \Phi 2 = 2\pi d \sin\theta/\lambda$, maximum likelihood function can be used [12], where an expected phase difference at each location can be considered and ten can be compared with the measured phase difference and likelihood can be evaluated. The location with the maximum likelihood will be the best estimate. So the target location of x is thus written as: $L(x) = p_x{*}\ (\Phi 1 - \Phi 2)$ where p_x is the probability of observing the phase difference at the antenna array when target is at x.

2.6 Angle of Arrival Estimation for Localization and Communication in Wireless Networks Using PSO-ML Paired Parameter

Wireless sensor network have drawn increasing attention due to potential utilization in a variety of civilian and military domains. Location derivation using bearing only [1, 13] and bearing hybrid measurements [14, 15] has been an active research area. This technique describes an way for accurate narrowband AOA estimation in unknown noise fields and harsh WSN scenarios [9]. By modeling the noise covariance as a linear combination of known weighting matrices, a maximum likelihood criterion is derived with respect to AOA and unknown noise parameters. ML criteria may yield superior statistical performance, but the cost function is multimodal, non linear and high dimensional. To tackle it accurately and efficiently, particle swarm optimization (PSO), a robust and fast global search tool is used. PSO is a recent addition to evolutionary algorithms first introduced by Eberhart and Kennedy [9]. Most of the applications demonstrated that PSO could give competitive or even better results in a much faster and cheaper way, compared to other heuristic methods such as genetic algorithms (GA).

Consider an array of M elements arranged in an arbitrary geometry on a sensor platform and N narrowband far-field sources at unknown locations. The complex M-vector of array outputs is modeled by the standard equation:

$$y(t) = A(\theta)s(t) + n(t) \tag{5}$$

where $\theta = [\theta_1 \ldots \theta_N]^T$ is the source AOA vector, $s\ (t)$ is composed of the emitter signals, and n(t) models the additive noise.

Particle swarm optimization is a stochastic optimization paradigm, which mimics animal social behaviors such as flocking of birds and the methods by which they find roosting places or food sources. The algorithm starts by initializing a population of particles in the normalized search space with random positions **x** and random velocities **v**, which are constrained between zero and one in each dimension. The position vector of the *i*th particle takes the form $x_i = [\theta_1, \ldots \theta_N, \eta_1, \ldots, \eta_j]$.

A particle position vector is converted to a candidate solution vector in the problem space through a mapping. The score of the mapped vector evaluated by the likelihood function $I_1\ (\theta, \eta)$, is regarded as the fitness of the corresponding particle. The *i*th particle's velocity is updated according to Eq. (6).

$$v_i^{k+1} = \omega^k v_i^k + c_1 r_1^k \odot (p_i^k - x_i^k) + c_2 r_2^k \odot (p_g^k - x_i^k) \tag{6}$$

where Pi is the best previous position of the *i*th particle, *Pg* is the best position found by any particle in the swarm, $\odot$ denotes element-wise product, $k = 1, 2,\ldots,$ indicates the iterations, ω is a parameter called the inertia weight, c_1 and c_2 are positive constants referred to as cognitive and social parameters respectively, $\mathbf{r}_1$ and $\mathbf{r}_2$ are independent random vectors.

3 Comparison

AoA measurement is an important technology of great practical interest in WSN. For long range transmissions and high accuracy, Angle of Arrival measurement, is the most preferred way than RSS and TOA based techniques. According to the behavior of the above techniques for AoA estimation, the different aspects of each methodology are compared and enlisted in the Table 1 in the harsh outdoor atmospheric wireless environment. The price is for the making and maintenance.

Table 1 Comparison of angle of arrival estimation techniques

Different techniques	Price	Hardware complexity	Effect of noise	Clock synchronization	Accuracy
Antenna array	High	High	Low	Not required	High
AoA through orientation and forwarding	Low	Low	Medium	Required	Medium
Cricket compass	Medium	Medium	High	Not required	High
Digital compass	Low	Low	Medium	Not required	Low
AoA trough ML	Low	Low	Medium	Required	Low
AoA through PSO-ML	Low	Low	Medium	Not required	Medium

4 Conclusion

Though the angle of arrival measurement becomes highly accurate, overcoming different interferences to achieve accurate output, use of a group of antenna and complex calculations increases the cost of the technology with the increase in connectivity, time and space complexity as well. In spite of all the above drawbacks, it still doesn't guarantee of the exact angle of arrival calculation. The size and the portability issues restrict the system from the use in sensors. The modern high resolution methods based on the concept of subspace such as MUSIC, ROOT MUSIC and ESPRIT are among the most efficient for estimating the directions of arrival (DOA) of signals using array antennas. Some other algorithms like spatial differencing method for DOA estimation prove to be more effective under the coexistence of both uncorrelated and coherent signals [16].

Determination of AoA using COTS antenna array [12] is rather accurate, cost effective and easily deployable than all other above technologies. But a little care has to be taken in the sense to reduce the size and increase robustness to be used in WSNs.

Such compasses as include, structures such as magnetic field concentrators that need to be added after the chip is made, adds complexity and cost. This is possible because the compass exploits a phenomenon called the Lorentz force. Most commercial digital compasses rely on a different phenomenon, called the Hall Effect, which works by running a current through a conductor and measuring changes in voltage caused by the Earth's magnetic field. But still the fiber optic Gyrocompass provides reliability at the cost of increasing cost of the system as it requires the help of the GPS system.

If no compass is available in any node, but each node knows its position, heading can still be found because the orientations for the sides of the triangle can be found from positions of its vertices [10] in Adhoc positioning.

But in a situation where the noise density is very less and the transmission range is very small like inside a closed room or in a closed boundary, if the sub-centimeter accuracy has to be achieved the devices like Cricket Compass plays a major role for Angle of Arrival and the Position Estimation. But in sensor networks, those which are meant for some specific purposes and placed in open areas and distributed in a long range of transmission the sub-centimeter accuracy is a very hard task to be achieved and the effect of the atmospheric noise is also very high which can affect the calculation of the angle of arrival and hence the position of a specific node. But the soft computing techniques like ML and PSO-ML helps improve accuracy with APS, where the environment is too noisy and not robust. The above descriptions inspire us to work further on an accurate and cost effective AoA calculation may be merged with soft computing techniques.

References

1. Niculescu, D., Nath, B.: Ad hoc positioning system (APS) using AOA. In: Proceedings of IEEE INFOCOM, pp. 1734–1743. San Francisco (Mar. 2003)
2. Lenz, J., Edelstein, A.S.: Magnetic sensors and their applications. IEEE Sens. J. **6**, 631–649 (2006)
3. House, D.J., Heinemann, B.: Seamanship Techniques: Shipboard and Marine Operations, p. 341. Butterworth-Heinemann, UK (2004)
4. Balanis, C.A.: Antenna. Wiley, New York (2005)
5. Lenz, J., Edelstein, A.S.: Magnetic sensors and their applications. IEEE Sens. J. **6**, 631–649 (2006)
6. Priyantha, N.B., Chakraborty, A., Balakrishnan, H.: The cricket location-support system. In: Proceedings of the 6th ACM MOBICOM, Boston, MA (August 2000)
7. Bulusu, N., Heidemann, J., Estrin, D.: GPS-less low cost outdoor localization for very small devices. In: IEEE Personal Communications Magazine, Special Issue on Smart Spaces and Environments (October 2000)
8. Oxygen home page. http://oxygen.lcs.mit.edu/
9. Li, M., Lu, Y.: Intelligent Systems Centre, Nanyang Technological University 50, Nanyang Drive, Angle of Arrival Estimation for localization and communication in wireless networks, Singapore. In: 16th European Signal Processing Conference (EUSIPCO 2008), pp. 25–29. Lausanne, Switzerland (August 2008)
10. Priyantha, N.B., Miu, A., Balakrishnan, H., Teller, S.: The cricket compass for context-aware mobile applications. In: Proceedings of the 7th ACM MOBICOM Conference, Rome, Italy (July 2001)
11. Yip, L., Comanor, K., Chen, J., Hudson, R.,Yao, K.,Vandenberghe, L.: Array processing for target DOA, localization and classification based on AML and SVM algorithms in sensor networks. In: Proceedings of IPSN'03, pp. 269–284. California (Apr 2003)
12. Patwari, N., Joshua, N., Kyperountas, S., Hero, A.O., Moses, R.L., Correal, N.S.: Locating the nodes. IEEE Signal Process. Mag. **22**, 54–69 (July 2005)
13. Ash, J., Potter, L.: Sensor network localization via received signal strength measurements with directional antennas. In: Proceedings of the 42nd Annual Allenton Conference on Communication, Control, and Computing, pp. 1861–1870. Champaign-Urbana, IL (Sept 2004)
14. Gu, Z., Gunawan, E.: Radiolocation in CDMA cellular system based on joint angle and delay estimation. Wireless. Pers. Commun. **23,** 297–309 (Mar 2002)
15. Liu, F., Wang, J., Sun, C., Du, R.: Spatial differencing method for DOA estimation under the coexistence of both uncorrelated and coherent signals. IEEE. Trans. Antennas. Propag, **60**(4), 2052–2062 (April 2012)
16. Chen, H.C., Lin, T.H., Kung, H.T., Lin, C.K., Gwon, Y.: Determining RF angle of arrival using cots antenna arrays. In: A Field Evaluation, Military Communication Conference Organized by IEEE (2012)
17. Niculescu, D., Nath, B.: Ad hoc positioning system (APS). In: Proceedings of GLOBECOM, San Antonio (November 2001)
18. Eberhart, R.C., Kennedy, J.: A new optimizer using particle swarm theory. In: Proceedings of the 6th Symposium on Micro Machine and Human Science, pp. 39–43. Nagoya, Japan (1995)
19. Bahl, P., Padmanabhan, V.N.: RADAR: An in-building RF-based user location and tracking system. In: Proceedings of IEEE INFO-COM. Tel-Aviv, Israel (March 2000)

Multi-strategy Based Matching Technique for Ontology Integration

S. Kumar and V. Singh

Abstract In modern era, ontology integration is most prominent and challenging problem in various domains of data mining. It helps greatly on defining interoperability in information processing systems. Ontology integration institutes interoperability by deriving semantic similarity and correspondences between the entities of ontologies. Ontology matching plays an integral role in whole integration process by identifying the similarity measure between source and target ontologies using various matching techniques, e.g. Element-level techniques and Structure level techniques. The scope of this paper is to discuss existing ontology matching techniques and propose a multi-strategy ontology matching approach. In Proposed algorithm is a multi-strategy matching approach, where multiple similarity measure are combined and finally an ontology tree based on the binary tree is created. Ontology mapping is used for creating ontology binary tree. The hybrid approach of Ontology integration performs significantly better than its predecessor with single similarity measures.

Keywords Ontology · String-based technique · Structure-based technique · Levenshtein distance · Matching · Similarity · Ontology integration

1 Introduction

In modern era, entire aspect of data integration is motivated by the information technology and social media, data cannot be stored on a single computer or with single semantics. "Semantic" means the meaning of data, the data stored at various geographical locations is linked based on their semantic similarity. The semantic

S. Kumar (✉) · V. Singh
Computer Engineering Department, NIT Kurukshetra, Kurukshetra, Haryana, India
e-mail: er.satishkumar149@gmail.com

V. Singh
e-mail: viks@nitkkr.ac.in

L.C. Jain et al. (eds.), *Computational Intelligence in Data Mining - Volume 3*,
Smart Innovation, Systems and Technologies 33, DOI 10.1007/978-81-322-2202-6_12

values of data can be a good source to define the relation with user query. The researchers presented the technology to help manage these data called semantic web. Semantic web has been used in various fields such as Information Systems, Search Engine etc. [1, 2]. Very large database systems to handle this with advent of data mining, because the large database analyze by finding patterns or relationships of data is an advantage of data mining. Research in area of semantic web integration is not yet widely explored, since there is a management tool for data mining of semantic web is less and data from the semantic web is stored in a format that cannot be used directly in data mining [3]. Semantic integration creates lots of interest of research in various disciplines, such as databases, information-integration, and ontologies. In the area of knowledge sharing and reuse, ontology integration is an important area of open research in domain of ontology engineering/management. Last few years, ontology integration catches the eyes of researcher in the field of ontology based systems development. There are few identified meanings of ontology integration. First, create ontology in a domain by reusing other existing ontologies in the same domain. Second, ontology is build by merging other existing ontologies into single ontology. Finally, when an application has built by using one or more ontologies, ontology integration is achieved through ontology matching and ontology alignment. Ontology matching concept is used to find contact between ontologies by finding the semantic similarities between the ontologies. Semantic similarity measures play a consequential role in text cognate area and application in ontology matching. Semantic similarities of data are being used in various fields e.g. Information retrieval, query answering etc. [4]. Identification of related attribute between the ontologies is the primary stage in ontology matching. In traditional applications ontology matching is a paramount operation e.g., Ontology evolution, Ontology integration, Data integration and Data warehouses [5]. An Ontology Integration algorithm is proposed, which begin with ontology identification followed by ontology matching and finally ontology mapping. Matching process is prime focus of this paper. The matching is base on some similarity measure. The similarity on words can be two ways, lexically and semantically. If the sequence of character is similar then they are lexically similar and if the two words are opposite of each other having similar meaning then they are semantically similar [6]. There are various types of similarity matching techniques to measure the similarity, the element-level techniques between two ontologies. An element-Level technique discusses the string-predicated techniques, Language-predicated techniques and Constraint-Based techniques etc. On the other hand Structure-Level techniques are used to measure the similarity based on the structure. Structure-Level techniques e.g. Graph-predicated technique, Taxonomy-predicated technique and Constraint-predicated techniques are discussed briefly in next section in paper. Traditional approach uses single technique for measuring similarity, between source and target ontology. The proposed algorithm for multi-strategy based matching based on the multiple similarity measures, the similarities between the ontologies and then ontology alignment by combining the string- predicated similarity and structure-predicated similarity. In string based similarity measure, primarily levenshtein distance is used to measure the similarity between the ontologies; while for structure based similarity measure WordNet based semantic similarity is used [3]. In algorithm similarity measure is calculated by combining the similarity measure of both previous approaches. Experimentation performed for the result verification and validation of algorithm, which shows that multi-strategy algorithm perform significant better than some of single strategy based algorithm.

2 Related Work

While many multi-strategy approaches has been proposed for ontology integration over the span of 10 years, among these approaches, recent approach of multi-strategy ontology integration using stack by Xia et al. [7]. Another similar effort made by Gang, L, Kunlun Wang and D. Liu, they proposed a multi-strategy ontology mapping. In algorithm characteristics of ontology structure and the instance of mapping on the ontology similarity are used for the ontology integration. Most recent work on the ontology integration is done by Fuqiang and Yongfu [8], as in proposed multi-strategy ontology integration approach, in which concept similarity between the ontologies is computed from the concept name, concept attribute and concept relationship.

3 Ontology Integration

Ontology provides the vocabulary of concepts that describe the domain of interest and a specific meaning of terms used in the vocabulary [5]. Ontology also provides the shared specification of conceptualization. Ontology reuse is one most adequate research issue in the ontology integration field. Ontology reuse, consist of two important processes, ontology merging and ontology integration. In ontology merging, ontology created in one domain by merging ontologies from two different domain ontologies. On the other hand in ontology integration, ontology is created by combining, assembling ontologies [9]. Therefore, ontology integration is major challenge and research issue. Main focus of this paper is on ontology integration by

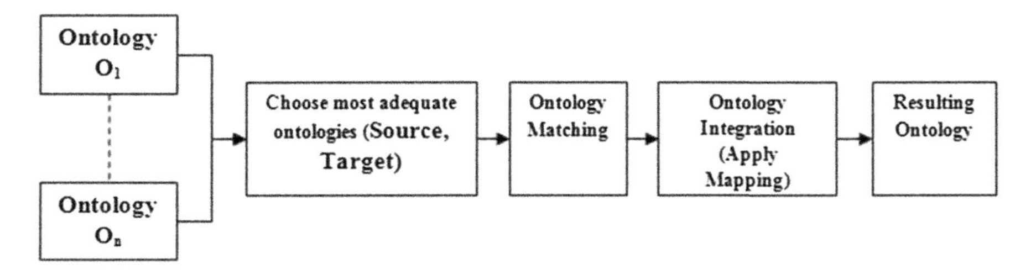

Fig. 1 Ontology integration process

using ontology matching combined with ontology mapping because it is a crucial part in ontology alignment. Ontology integration's typically involves four activities, shown in Fig. 1.

4 Ontology Matching

The ontology matching process is utilized to measures the homogeneity between set of ontologies [9]. The matching process determines an alignment A′, for two ontologies O_1 and O_2, for this some parameters of the matching process are used like,

(i) The utilization of an input alignment A
(ii) Matching parameter like weights, threshold
(iii) External resources utilized by the matching process like erudition-base and domain-categorical area.

Definition 1 (*Matching Process*) The matching process considered as function f which emanates from the ontologies to match O_1 and O_2, an input alignment 'A', a set of parameters p and a set of oracles and resources r, back to an alignment 'A' between these ontologies [4, 9].

$$A' = f\ (O_1, O_2, A, p, r)$$

Typical structure for ontology matching is shown in above Fig. 2. In the Ontology integration matching plays a crucial role and decider, how effectively ontologies are integrated. In ontology engineering, there are two ways of defining matching; first the matching is derived base on the element level e.g. String, Language, Constraint. Another ways matching is based on the structure used to store ontology i.e. Graph, Taxonomy, Instances etc. [9]. The classification is shown in below Fig. 3.

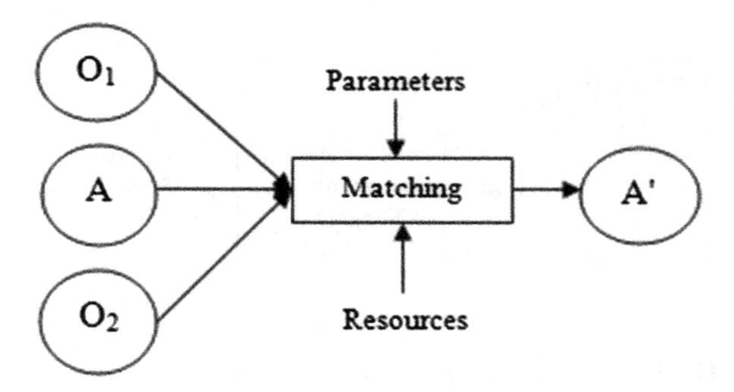

Fig. 2 Matching process

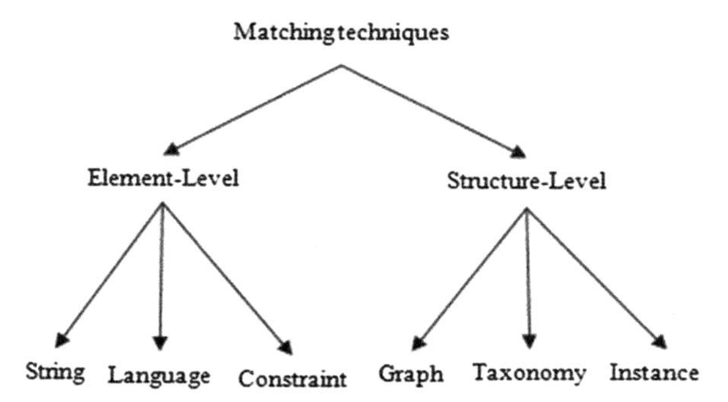

Fig. 3 Classification of ontology matching techniques

4.1 Element-Level Technique

An element-Level matching technique, in the measure of homogeneous attribute on ontologies and their instances is isolation from their relation with the attribute and their instances [9]. This category of ontology matching techniques, consist of string-predicated, language-predicated and constraint-predicated methods. A string-predicated technique, is employ to define the match description of ontology entities, this technique draws matching between input ontologies by considering ontology as a sequence of letter in an alphabet [2, 9], it also consider various distance measures methods for assertion the distance between strings, some such are popular like Edit-distance, Hamming-distance etc. Alternative approach, which is based on string parity method, in which is string parity defines the homogeneous attribute between the strings. If the strings are identical, string parity returns 1, otherwise returns 0, which means, strings are not identical.

(S is set of strings) (s, t are Input strings)
$|s|$ = length of the string s, $|t|$ = length of the string t
s$[i]$ for i $\in [1, |s|]]$ latter at position i of s.

String-predicated technique, which is contributory to quantify the similar attribute between the input strings, if we employ very similar string to represent the same concepts [10]. If we utilized the synonyms with dissimilar structure, then technique may be open-handed to the low kind attribute. The result of this technique is more supplementary, if we utilized the similar strings [11]. Language-predicated technique is also used to find the similarity between the words [9, 12]. This technique takes names as words/tokens, English language-predicated technique considered the inherent techniques. Language-predicated techniques utilized the various methods, which considered strings as sequence of characters. This method takes string as an input and fragment these string in words and these words may be identified sequence of words. Constraint-predicated technique [9, 12] is based on the internal structure of the ontology entities rather than comparing the designations or terms. This technique compares internal structure of the ontology entities. Later, these structure are called relational structure, because in this approach comparing the ontology entities with the other entities which are related.

4.2 Structure-Level Technique

Contrary to Element-level techniques, structure level measures the homogeneous attribute between the ontology and their instances to compare their relation with each other entities or their instances [9, 10]. In first among structure level techniques, Graph-predicated technique measures the homogeneous attribute between the pair of nodes of mapped graph of ontologies, as the ontology are predicated onto the graph with a specific positions in the graph [13]. Input for Graph-predicated technique is two ontologies labelled graphs. In the labelled graph, if two nodes are similar then their neighbours are additionally somehow homogeneous. Taxonomy-predicated technique is utilizing the graph-predicated approach but in this technique only the specialized relation are considered for the matching [9]. This type of technique is utilized as a comparison resource for matching classes. The perception after the taxonomy technique is that, specialization connect ontology/tokens, those are already homogeneous (as a super set or subset of each other), therefore their neighbour are additionally somehow homogeneous. Instance-predicated technique is utilized for the comparison of sets of instances of classes [9]. On the substructure of the comparison it is decided that classes are match or not. This technique is relies on the set-theoretic reasoning and statistical techniques. The approach proposed in the paper, for the ontology matching, combine both string based similarity and structure based similarity to be used to define the ontology similarity measures. In String-based similarity, levenshtein distance technique which is a number of edit operation such as, insertion; deletion subtractions that are required to convert a string into other string, is used for measuring the similarity between entities of ontologies [10]. WordNet in case of Structured-based similarity, which is based on semantic similarity, is used to measure the similarity between ontologies [3].

5 Proposed Approach and Example

A hybrid ontology matching algorithm is proposed (Fig. 4) in which, matching ontologies of input documents or domains are mapped into a binary tree/Ontology tree. In traditional techniques, ontology matching technique prefer to base their matching approach on single similarity measure, while including more similarity measures to identify the measure is helpful to achieve good set of integrated ontology. Multiple similarity measures are combined for matching process in proposed approach. First, similarity measure, is based on string (grammar) based similarity measure and second is structure based similarity measure. In previous section basic details are discussed. For string based similarity measure, primarily levenshtein distance is used to measure the similarity between the ontologies. In Levenshtein distance method [9], the number of edit operation insertion, deletion and subtraction of characters to convert the one string into another. Another

Proposed Algorithm:

Input (D_1, D_2)

Step 1: Ontology Identification/ creation from Documents.

Step 2: Store Ontologies into Document vectors.

Step 3: Begin

Apply ontology matching technique (Set of Ontologies) using binary search tree

/* String based similarity measure method using Levenshtein distance */

End

Step 4: Begin

Apply another ontology technique (Set of Ontologies) using binary search tree

/* Structure based similarity measure methodusing WordNet */

End

Step 5: Combine the results of above two methods by adding the similarity matrices.

Output (Global Ontology)

Fig. 4 Proposed algorithm

technique is structure based similarity measure, where similarity measures between entities using WordNet based semantic similarity. WordNet is a lexical database of English. In proposed algorithm, both matching techniques have been used for measuring the similarities between entities of concepts using binary tree. Proposed algorithm iteratively constructs matching searches for the entities in both Source ontology S and target ontology T (see illustration in example as given in the next section). Matrix will be evaluated using matching techniques (string and structure based technique) and finally the hybrid approach based calculated matrix shows the better evaluation result to the user. Flowchart of proposed algorithm is shown in Fig. 6. Designed algorithm is useful as it provides a better ontology matching approach to match the entities of ontologies.

As shown in Fig. 5 two ontologies are given source ontology S and target ontology T with different entities (S_1, S_2, S_3) and (T_1, T_2, T_3) are concepts. Other entities are relations, properties etc.

As shown above, ontologies/concepts have been compared from source ontology S with target ontology T, and based on the similarity other entities in S such as Name and instances can be matched with the corresponding entities such as Name in target ontology T. In proposed algorithm (Fig. 6), two similarity measures have been used to measure the similarity between source ontology S and target ontology T.

First, Similarity is string based similarity using (levenshtein distance), following Similarity matrix shows the similarity values of ontology matching,

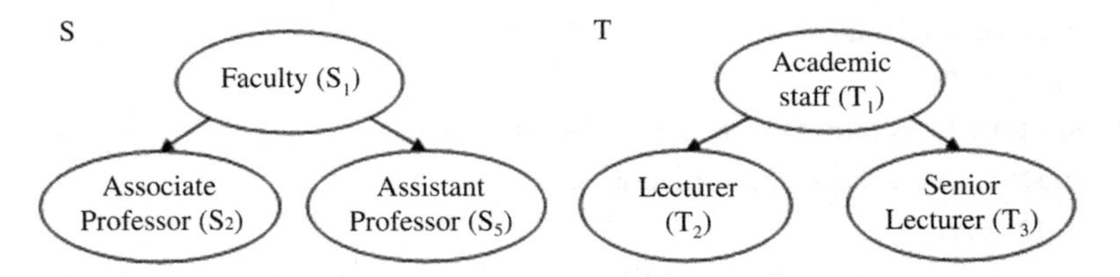

Fig. 5 Source ontology S and target ontology T

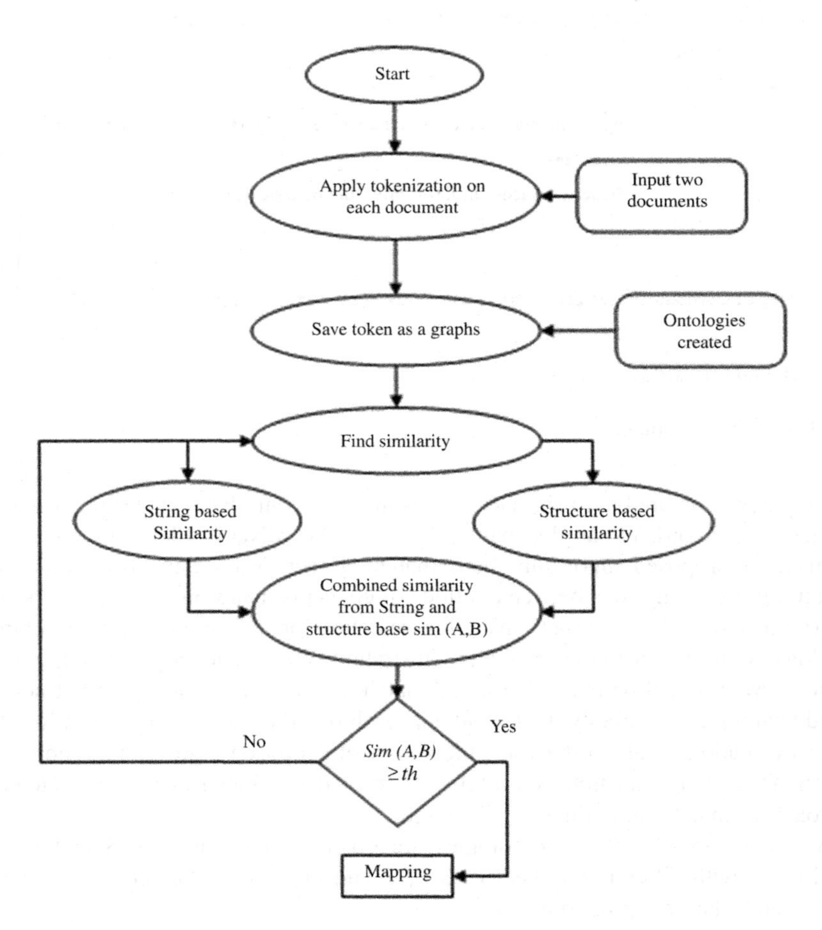

Fig. 6 Flowchart of proposed algorithm

$$\mathrm{Mat}_{(\text{String-Based})} = \begin{bmatrix} S_1rT_1 & S_1rT_2 & S_1rT_3 & S_1rT_4 \\ S_2rT_1 & S_2rT_2 & S_2rT_3 & S_2rT_4 \\ S_3rT_1 & S_3rT_2 & S_3rT_3 & S_3rT_4 \\ S_4rT_1 & S_4rT_2 & S_4rT_3 & S_4rT_4 \end{bmatrix} = \begin{bmatrix} 0.21 & 0.25 & 0.13 & 0.14 \\ 0.15 & 0.21 & 0.15 & 0.10 \\ 0.21 & 0.26 & 0.21 & 0.10 \\ 0.14 & 0.12 & 0.13 & 1.0 \end{bmatrix}$$

Similarity matrix for Structure based similarity using WordNet is shown below,

$$\mathrm{Mat}_{(\text{Structured-Based})} = \begin{bmatrix} S_1rT_1 & S_1rT_2 & S_1rT_3 & S_1rT_4 \\ S_2rT_1 & S_2rT_2 & S_2rT_3 & S_2rT_4 \\ S_3rT_1 & S_3rT_2 & S_3rT_3 & S_3rT_4 \\ S_4rT_1 & S_4rT_2 & S_4rT_3 & S_4rT_4 \end{bmatrix} = \begin{bmatrix} 0.0 & 0.0 & 0.0 & 0.11 \\ 0.0 & 0.0 & 0.0 & 0.0 \\ 0.0 & 0.0 & 0.0 & 0.0 \\ 0.0 & 0.0 & 0.0 & 1.0 \end{bmatrix}$$

$\mathrm{Mat}_{(\text{String-Based})}$ shows, the calculated string similarity matrix based on the levenshtein distance and $\mathrm{Mat}_{(\text{Structured-Based})}$ shows the calculated structured similarity matrix based on WordNet based semantic similarity. The matrix has been calculated on the basis of formal matching techniques. In proposed technique, for the similarity calculation, addition of both previous similarity matrixes is done, as shown below

$$\mathrm{Mat}_{(\text{Hybrid-Approach})} = \left(\mathrm{Mat}_{(\text{String-Based})} + \mathrm{Mat}_{(\text{Structured-Based})}\right)/2.0$$

$$\mathrm{Mat}_{(\text{Hybrid-Approach})} = \begin{bmatrix} S_1rT_1 & S_1rT_2 & S_1rT_3 & S_1rT_4 \\ S_2rT_1 & S_2rT_2 & S_2rT_3 & S_2rT_4 \\ S_3rT_1 & S_3rT_2 & S_3rT_3 & S_3rT_4 \\ S_4rT_1 & S_4rT_2 & S_4rT_3 & S_4rT_4 \end{bmatrix} = \begin{bmatrix} 0.10 & 0.12 & 0.06 & 0.12 \\ 0.07 & 0.10 & 0.07 & 0.05 \\ 0.10 & 0.13 & 0.10 & 0.05 \\ 0.07 & 0.06 & 0.06 & 1.0 \end{bmatrix}$$

In traditional techniques, prefer only either string based similarity method or structured based similarity. The performance of single similarity measure in matching is not significant as the approach with multiple similarity measure. In next section, hybrid approach of multiple similarity measures based approach is explained.

6 Experiment and Quality of Measures

In order to evaluate this method, we used two pairs of ontologies as shown in Table 1. Source ontology S and target ontology T. Source ontology contains the entities Faculty, Associate-Professor, Assistant-Professor and Name (S1, S2, S3, and S4). Target ontology contains the concepts Academic-Staff, Lecturer, Senior-Lecturer and Name (T1, T2, T3 and T4). First step, Sources and target ontologies are created.

Second step, apply the string based similarity measure using levenshtein distance. After apply this method a similarity matrix Mat (String-Based) between the entities is calculated on the basis of levenshtein distance.

$$\text{Matrix}_{(\text{String}-\text{Based})} = \begin{bmatrix} S_1rT_1 & S_1rT_2 & S_1rT_3 & S_1rT_4 \\ S_2rT_1 & S_2rT_2 & S_2rT_3 & S_2rT_4 \\ S_3rT_1 & S_3rT_2 & S_3rT_3 & S_3rT_4 \\ S_4rT_1 & S_4rT_2 & S_4rT_3 & S_4rT_4 \end{bmatrix} = \begin{bmatrix} 0.21 & 0.25 & 0.13 & 0.14 \\ 0.15 & 0.21 & 0.15 & 0.10 \\ 0.21 & 0.26 & 0.21 & 0.10 \\ 0.14 & 0.12 & 0.06 & 1.0 \end{bmatrix}$$

Table 1 Source and target ontology

Ontology name	Concepts
Ontology 1	Faculty, assistant-professor, associate-professor, name
Ontology 2	Academic-staff, lecturer, senior-lecturer, name

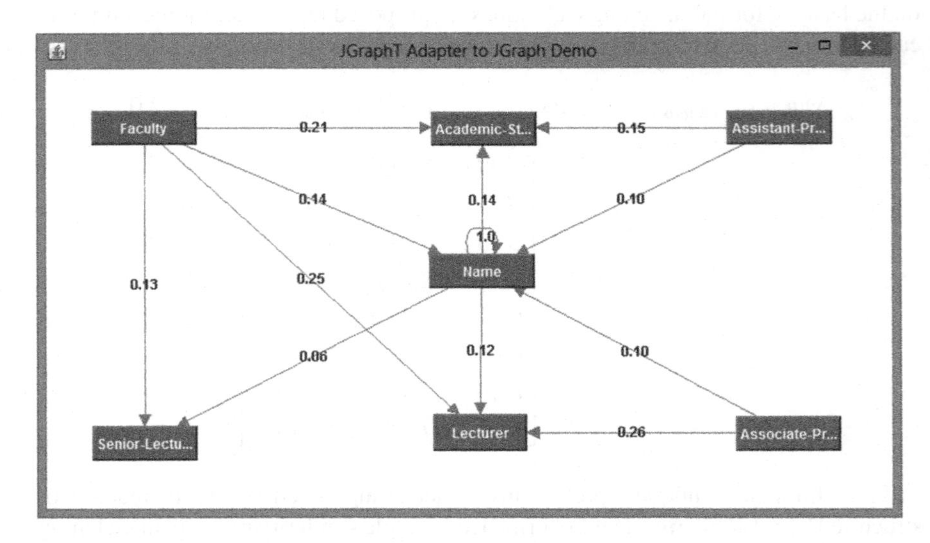

Fig. 7 Graph of string based similarity

As shown in Fig. 7 a tree or graph is constructed on the basis of calculated string-based matrix i.e. $\text{Matrix}_{(\text{String-Based})}$. Third step, structure based similarity measure method has been used for calculating the similarity matrix Mat (Structured-Based) between concepts based on their structure. Structure based similarity measure used the WordNet based semantic similarity measure method. WordNet is a lexical database of English.

$$\text{Mat}_{(\text{Structured-Based})} = \begin{bmatrix} S_1rT_1 & S_1rT_2 & S_1rT_3 & S_1rT_4 \\ S_2rT_1 & S_2rT_2 & S_2rT_3 & S_2rT_4 \\ S_3rT_1 & S_3rT_2 & S_3rT_3 & S_3rT_4 \\ S_4rT_1 & S_4rT_2 & S_4rT_3 & S_4rT_4 \end{bmatrix} = \begin{bmatrix} 0.0 & 0.0 & 0.0 & 0.11 \\ 0.0 & 0.0 & 0.0 & 0.0 \\ 0.0 & 0.0 & 0.0 & 0.0 \\ 0.0 & 0.0 & 0.0 & 1.0 \end{bmatrix}$$

After the matrix calculation a tree or graph is constructed on the basis of similarity matrix as shown in Fig. 8. Fourth step, the similarity matrix of both methods is combined. Similarity matrixes are added to evaluate the purpose.

$$\text{Matrix}_{(\text{Hybrid-Approach})} = \left(\text{Mat}_{(\text{String-Based})} + \text{Mat}_{(\text{Structured-Based})}\right)/2.0$$

$$\text{Matrix}_{(\text{Hybrid-Approach})} = \begin{bmatrix} S_1rT_1 & S_1rT_2 & S_1rT_3 & S_1rT_4 \\ S_2rT_1 & S_2rT_2 & S_2rT_3 & S_2rT_4 \\ S_3rT_1 & S_3rT_2 & S_3rT_3 & S_3rT_4 \\ S_4rT_1 & S_4rT_2 & S_4rT_3 & S_4rT_4 \end{bmatrix} = \begin{bmatrix} 0.10 & 0.12 & 0.06 & 0.12 \\ 0.07 & 0.10 & 0.07 & 0.05 \\ 0.10 & 0.13 & 0.10 & 0.05 \\ 0.07 & 0.06 & 0.06 & 1.0 \end{bmatrix}$$

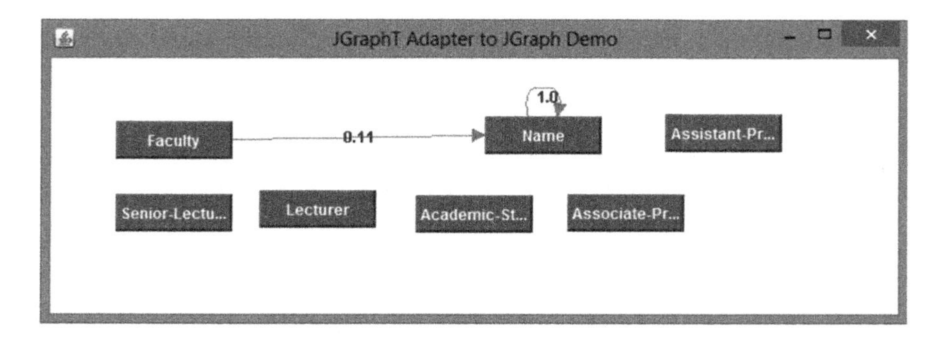

Fig. 8 Graph of structure based similarity

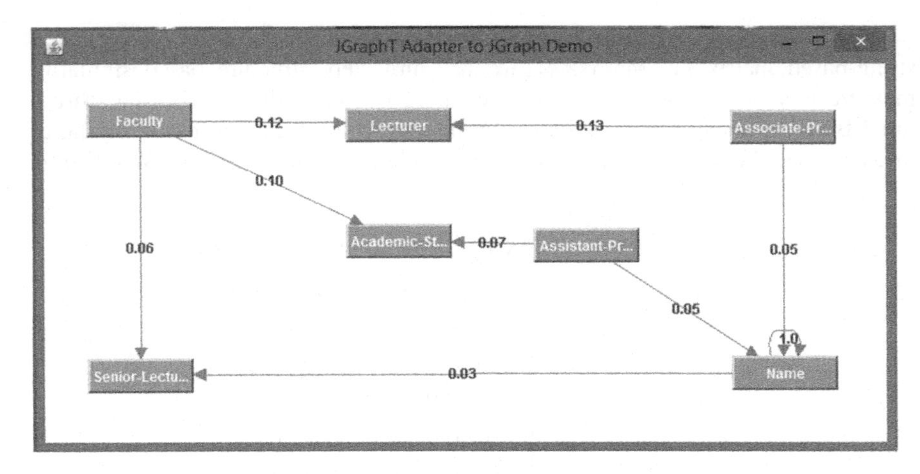

Fig. 9 Graph of Hybrid Based Similarity

After calculating similarity matrix measure for the individual, Matrix (Hybrid-Approach), based on the resulting matrix from methods, a tree or graph is constructed based on the combined matrix, and the resultant ontology is shown in Fig. 9. The given two ontologies are considered source ontology S and target ontology T, in example S contains 4 entities and T contain 4 concepts. Following mentioned parameters are used for comparative analysis of proposed algorithm in which quality of similarity measure use to compare the effectivity of matching techniques. These parameter are used in some traditional ontology matching techniques to justify the expediency of techniques also,

(1) **Precision** is a value defined in the range between 0 and 1; the higher the value, the fewer wrong mapping computed.

$$\text{Precision} = \frac{\text{number of correct found alignments}}{\text{number of found alignments}}$$

(2) **Recall**, is a value defined in the range between 0 and 1; the higher this value, the smaller the set of correct mappings which are not found.

$$\text{Recall} = \frac{\text{number of correct found alignments}}{\text{number of found alignments}}$$

(3) **F-measure** is a value defined in the range between 0 and 1, F-measure value also known as global measure of matching quality. F-Measure used the mean of precision and recall [3].

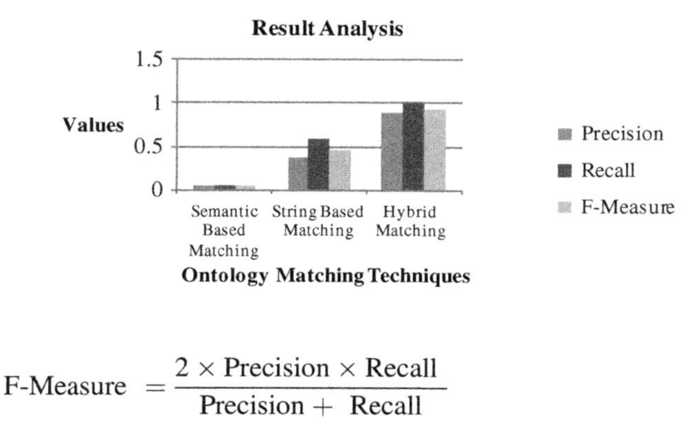

Fig. 10 Result graph

$$\text{F-Measure} = \frac{2 \times \text{Precision} \times \text{Recall}}{\text{Precision} + \text{Recall}}$$

On the basis of Precision, Recall and F-measure parameter values, as shown in above Fig. 10, various conclusions are pinched on the propped algorithm. The values used for comparison are various matching techniques e.g. Semantic based Techniques, String based matching, and Hybrid Based matching, Other hand values of three factors (Precision, Recall, F-measure) with respect to each of the techniques (Semantic based Techniques, String based matching, Hybrid Based matching). The Semantic based techniques are the poor among all techniques, as precision and recall is quite low. The String based matching techniques are better than the semantic based techniques, as the matching performed on the data structure on which the ontologies are stored/kept with some logical relation among them, due to this logical relation between the ontology the probability of correctness increased which result into improve values of Precision, Recall and F-measure. The proposed algorithm combine the similarity values of Semantic based techniques and String based techniques to derived the similarity measure and as shown in the above graph (Fig. 10). This algorithm outperformed the previous techniques in all three values, precision, recall, F-Measure.

7 Conclusion

Ontology matching between the entities plays an important role in ontology integration. Ontology integration is an important process of ontology reuse, other approach is ontology merging. Ontology integration is strongly based on similarity measure between input ontology (source, target). Traditionally, two types of techniques are used to evaluate the matching among the ontology and each approach have certain degree of limitations. The proposed algorithm for ontology matching is based on the multi-strategy approach for the ontology matching. In the approach similarity measures of the two traditional approaches are combined and this combined (hybrid) value is used for defining mapping between the set of ontology. The levenshtein distance similarity measures in string based techniques and WordNet similarity measure in Semantic based techniques are used. The experimentation are performed on large amount of datasets (documents, ontologies) to validate the authenticity of proposed algorithm, which shows that the proposed algorithm outperformed the traditional algorithm in all three aspect (Precision, Recall, F-Measure), which are considered for the comparative study.

References

1. Berners-Lee, T., Handler, J., Lassila, O.: The Semantic Web. Scientific American (2001)
2. Huiping, J.: Information retrieval and the Semantic Web. In: IEEE International Conference on Educational and Information Technology (ICEIT), vol. 3, pp. 461–463. Chongqing, China (2010)
3. Perez-Lopez, A., Blace, R. Fisher, M., Hebeler, J.: Semantic Web Programming. Wiley Publishing, Inc., New York (2009)
4. Huang, L., Hu, G., Yang, X.: Review of ontology mapping. In: 2nd International Conference on Consumer Electronics, Communications and Networks (CECNet), pp. 537–540. Yichang (2012)
5. Shvaiko, P., Euzenat, J.: Ontology matching: state of the art and future challenges. IEEE Trans. Knowl. Data. Eng. **25**(1), 158–176 (2013)
6. Ye, B., Chai, H., He, W., Wang, X., Song, G.: Semantic similarity calculation method in ontology mapping. In: 2nd International Conference on Cloud Computing and Intelligent Systems (CCIS), vol. 3, pp. 1259–1262. Hangzhou (2012)
7. Xia, H., Zheng, X., Hu, X., Tian, Y.: Multi-strategy ontology mapping based on stacking method. In: Sixth International Conference on Fuzzy Systems and Knowledge Discovery (2009)
8. Fuqiang, L.,Yongfu, X.: The method of multi-strategy ontology mapping. In: ICCIS (2011)
9. Euzenat, E., Shvaiko, P.: Ontology Matching. Springer, Heidelberg (2013)
10. Rodrfguez, A., Egenhofer, M.: Determining semantic similarity among entity classes from diferent ontologies. IEEE Trans. Knowl. Data. Eng. **15**(2), 442–456 (2003)
11. Li, C.M., Bo, S.J.: The research of ontology mapping method based on computing similarity. Sci. Technol. Inf. **1**, 552–554 (2010)
12. Zhe, Y.: Semantic similarity match of ontology concept based on heuristic rules. Comput. Appl. **12** (2007)
13. Doan, A.H., Jayant, M., Pedro, D. et al.: Learning to map between ontologies on the semantic web. In: Eleventh International World Wide Web Conference, Honolulu Hawaii, USA (2005)

Impact of Fuzzy Logic Based UPFC Controller on Voltage Stability of a Stand-Alone Hybrid System

Asit Mohanty and Meera Viswavandya

Abstract The main focus of the paper is to compensate the Reactive Power generated in the isolated wind–diesel hybrid system and enhancement of the Voltage stability by the use of Fuzzy PI based UPFC Controller. Linearised small signal model of the wind-Diesel System with different loading conditions is taken with IEEE Excitation system. In this paper the compensation has been carried out with UPFC Controller for different loading conditions. A self tunned Fuzzy PI controller is designed to tune the parameters of PI Controller Kp and Ki. Simulation result shows that the parameters of the system attend steady state value with less time due to the proposed controller.

Keywords Stand-alone hybrid system · UPFC · Fuzzy logic · Voltage stability

1 Introduction

Though renewables like wind, solar fuel cell are plentily available in the nature, they are intermittent and non-predictable in nature. Therefore Researchers have gone for combining one or more renewable energy sources to make a hybrid system so that any shortage due to one source can be compensated by the other. Small Hybrid systems not only provide power to the remote places, they are many times connected to the main grid to supply excess generated power.

One or more renewable sources are combined to form a Hybrid system where the shortage due to one

source is compensated by the other [1, 2]. Wind Diesel system is the most commonly used hybrid system in which a wind turbine combines with a diesel

A. Mohanty (✉) · M. Viswavandya
Department of Electrical Engineering, C.E.T, Bhubaneswar, India
e-mail: asithimansu@gmail.com

M. Viswavandya
e-mail: mviswavandya@rediffmail.com

L.C. Jain et al. (eds.), *Computational Intelligence in Data Mining - Volume 3*,
Smart Innovation, Systems and Technologies 33, DOI 10.1007/978-81-322-2202-6_13

generator to provide power in remote places. Normally a Synchronous generator is used as Diesel Generator and Induction Generator is used in Wind Turbine for an improved performance [3, 5]. Though Induction generators are advantageous in comparison to Synchronous generator due to its rugged characteristics, they need reactive power for their operation. In a hybrid power system having both synchronous generator and Induction generator, the Induction generator needs the reactive power and it is provided by the Synchrous generator. But the reactive power supplied by the synchronous generator is not sufficient and therefore a big gap is created between the demand and supply of reactive power. Because of these gap problems like voltage fluctuation and instability occur in the hybrid power system.

In the literature survey [6] authors have emphasized the use of capacitor banks to improve voltage stability and to compensate the reactive power in the system. [6–8]. But due to the uncertain nature of wind and wide variation of load the fixed capacitors fail to deliver the required reactive power to the system. [7]. To meet the challenge of these power quality issues like voltage instability and reactive power compensation researchers have advocated the use of FACTS (Flexible AC Transmission System) devices [9].

UPFC [10] is one of the important member of FACTS devices which is used like SVC and STACOM to compensate the reactive power of the power system. This device is used for voltage and angle stability studies of power system. In the absence of reactive power, the system goes through a wide voltage variations and unnecessary fluctuations. Papers have projected UPFC as a reactive power compensating device with PI controllers. This work proposes a fuzzified UPFC Controller for reactive power control in a wind diesel hybrid system. The proposed controller tunes the gains of the PI Controller to enhance the transient stability and reactive power compensating capability of the system.

Here in this paper a SIMULINK based Fuzzified wind diesel model (Fig. 2) with IEEE exciter 1 (Fig. 3) is taken to analyze the transient stability and reactive power compensation issue with the application of UPFC for 1 % change in load disturbance. In Section no II the complete description of the system and the mathematical modeling is depicted. The detailed work of self tuned fuzzy logic for tuning the gains of PI controller is mentioned in Sect. 3. The model simulation output results with description are represented in Sect. 4. The conclusion part is clearly represented in Sect. 5.

2 System Configuration and Its Mathematical Modeling

This hybrid system consists of one Induction generator (IG), a Synchronous generator (SG), electrical loads and Reactive power control in the form of UPFC supported by its control strategy. The block diagram is clearly depicted in Fig. 1. The hybrid system parameters are mentioned in Table 3.

The reactive power balance equation of the system as mentioned in Fig. 2 having (SG, UPFC, IG, and LOAD) is $\Delta Q_{SG} + \Delta Q_{UPFC} = \Delta Q_L + \Delta Q_{IG}$.

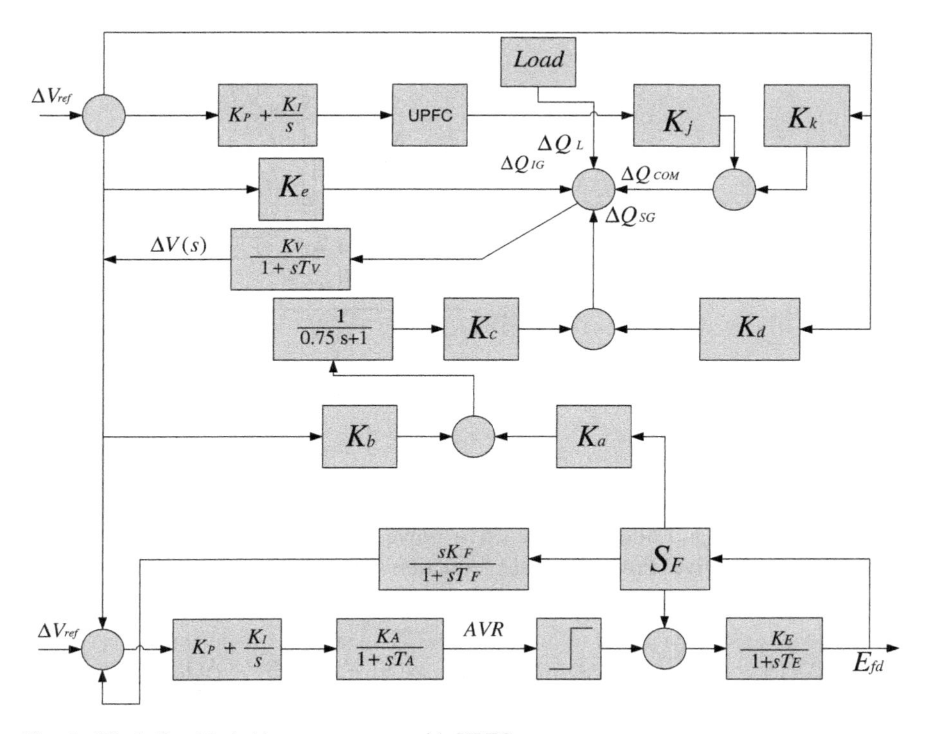

Fig. 1 Wind-diesel hybrid power system with UPFC

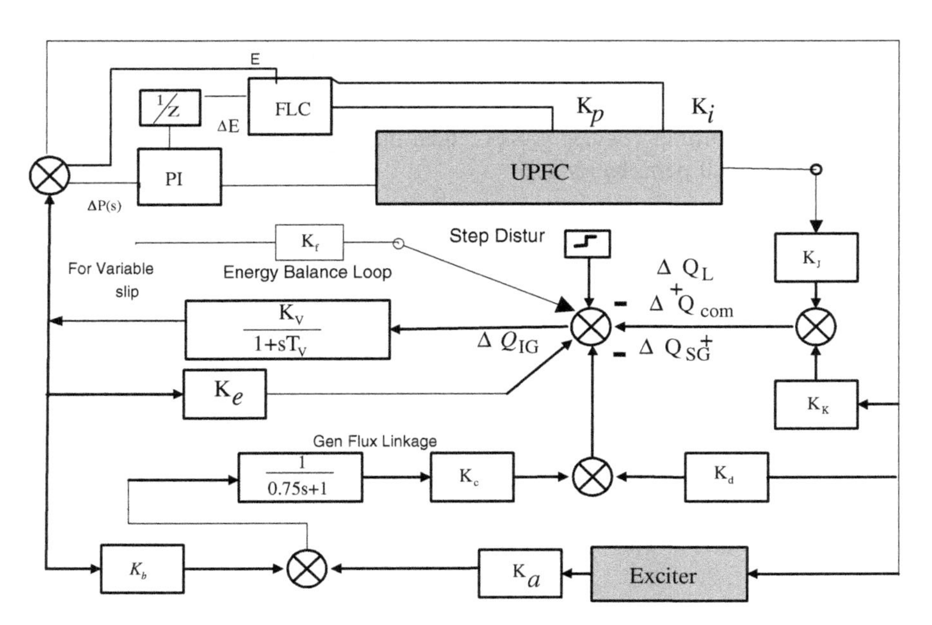

Fig. 2 Transfer function model with fuzzy logic based UPFC controller

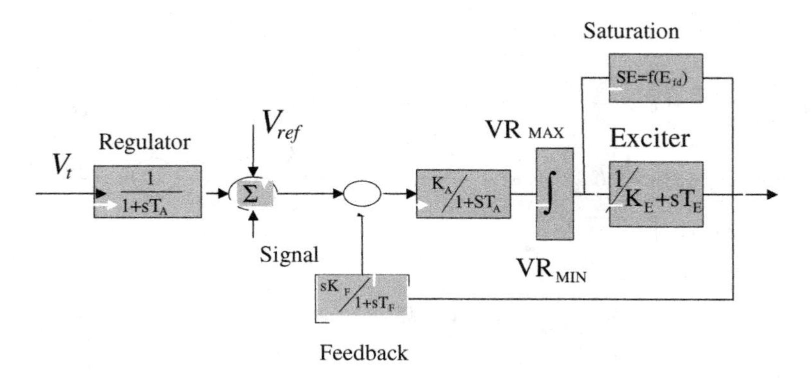

Fig. 3 Excitation system 1

Due to abrupt change in reactive power load ΔQ_L, the system terminal voltage varies accordingly and affects the reactive power balance equation of the whole system. Final reactive Power balance equation of the said system is $\Delta Q_{SG} + \Delta Q_{UPFC} - \Delta Q_L - \Delta Q_{IG}$ and this value deviates the system output voltage.

$$\Delta V(S) = \frac{K_V}{1 + ST_V}[\Delta Q_{SG}(S) + \Delta Q_{COM}(S) - \Delta Q_L(S) - \Delta Q_{IG}(S)] \tag{1}$$

$$\Delta Q_{SG} = \frac{V\cos\delta}{X'd\Delta E'q} + \frac{E'q\cos\delta - 2V}{X'd\Delta V} \text{ For small change} \tag{2}$$

$$\Delta Q_{SG}(S) = K_a \Delta E'q(s) + K_b \Delta V(s) \tag{3}$$

In case the controlling device is SVC, then the Laplace Transform of the said equation for a small perturbation [16]

$$\Delta Q_{SVC}(s) = K_c \Delta V(s) + K_d \Delta B_{SVC}(s) \tag{4}$$

2.1 UPFC Controller

The Structure of UPFC helps to control Reactive power and Active Power by injecting AC Voltage in series with amplitude and a phase angle to the Transmission Line. The role of first Inverter is to provide or absorb real power while the second Inverter produces and absorbs reactive power and there by provides shunt compensation. The injected powers depend on the injected voltages and bus voltages also. Buses *i* and *j* are taken as load buses in the load flow analysis as clearly shown in Figs.4 and 5. The injected powers are given as follow:

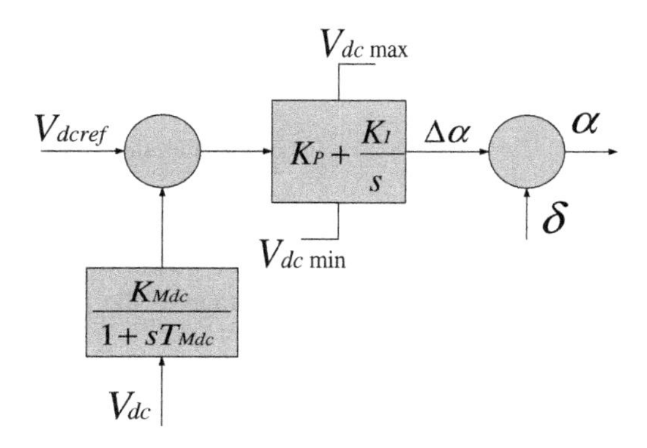

Fig. 4 Shunt controller of UPFC

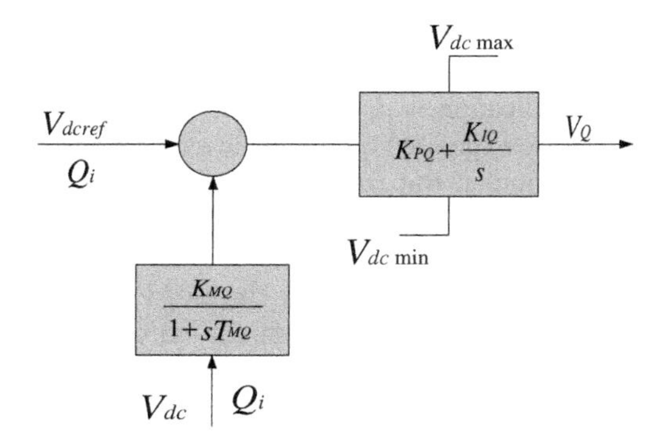

Fig. 5 Series controller of UPFC

$$P_{sh} = \frac{V_i V_{sh}}{x_{sh}} \sin(\delta_i - \delta_{sh}) \text{ and } Q_{sh} = \frac{V_i^2}{x_{sh}} - \frac{V_i V_{sh}}{x_{sh}} \cos(\delta_i - \delta_{sh}) \tag{5}$$

$$P_i = \frac{-V_i V_{pq}}{x_{ij}} \sin(\delta_i - \delta_{pq}) \text{ and } Q_i = \frac{V_i V_{pq}}{x_{ij}} \cos(\delta_i - \delta_{pq}) \tag{6}$$

$$P_j = \frac{V_J V_{pq}}{x_{ij}} \sin(\delta_j - \delta_{pq}) \text{ and } Q_j = \frac{-V_j V_{pq}}{x_{ij}} \cos(\delta_j - \delta_{pq}) \tag{7}$$

Multiplying V_j to the above two equations we can get

$$P_j = \frac{V_i V_j}{x_{ij}} \sin\delta - \frac{V_j V_{m2p}}{x_{ij}} \text{ and } Q_j = \frac{V_i V_j}{x_{ij}} \cos\delta - \frac{V_j V_{m2q}}{x_{ij}} - \frac{V_j^2}{x_{ij}} \tag{8}$$

$V_{m2p} = V_j\gamma(t)$ and $V_{m2p} = V_j\beta(t)$ where $\gamma(t)$ and $\beta(t)$ are control variables

After partial derivative $$\frac{dP_j}{dt} = \frac{dP_j}{d\delta}\frac{d\delta}{dt} + \frac{dP_j}{dV_{m2p}}\frac{dV_{m2p}}{dt} \quad (9)$$

$$\frac{dQ_j}{dt} = \frac{dQ_j}{d\delta}\frac{d\delta}{dt} + \frac{dQ_j}{dV_{m2p}}\frac{dV_{m2p}}{dt} \quad (10)$$

It can be said that the reactive power injected by UPFC depends upon V_{m2p} and angle δ,proportional to the voltage at the point of connection of UPFC.

3 Fuzzy Logic Controller (FLC)

Fuzzy control is a control mechanism based on Fuzzy Logic. Fuzzy Logic can be defined as a logic that are based on If-Then rules. A Fuzzy control uses rules that are used in operator controlled plants. A Fuzzy controller can work in three stages that are fuzzification, rule based and defuzzification.

Fuzzification is defined as the conversion of Crisp values to Fuzzified values. These are expressed by linguistic expressions and consists of membership functions like triangular, trapezoidal and rectangular etc.

Rule Base or Inference-This consists of all fuzzy IF-THEN rules in relation to the fuzzy sets.

Defuzzification- The last stage of Fuzzy controller is defuzzification where the fuzzy sets are again defuzzified and converted into crisp sets. There are several methods for defuzzification which includes Centre of gravity, Centroid method, bisector of area etc.

3.1 Self-Tuning Fuzzy PI Controller

The proposed Auto tuned Fuzzy logic PI controller is designed based on the concept of PI controller.

$$U(s) = KpE(s) + Ki\int E(s) \text{ and } E*Ki*\int E = U$$

The system performance can be improved by Changing the value of parameters Kp and Ki of the PI controller. The PI parameters (Kp and Ki) are slowly changed using fuzzy inference engine. This provides a mapping which is not linear with having inputs as error and change in error and ouputs Kp and Ki. Figure 6 shows the block having Fuzzy PI Controller.

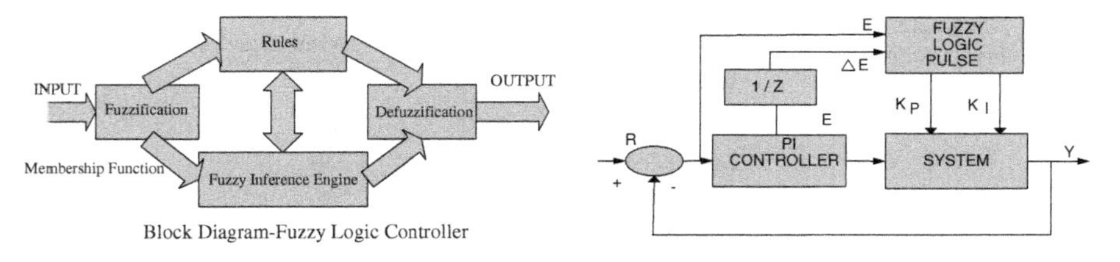

Fig. 6 Fuzzy PI controller block

$$\mu Pi = \min(\mu(E), \mu(\Delta E)), \mu Ii = \min(\mu(E), \mu(\Delta E))$$

Two input variables, error and change in error and two output variables Kp and Ki of the PI controller with seven linguistic variables of Triangular membership function is used for the proposed self tuning Fuzzy logic PI controller. The input membership functions are error(E) and change in error (ΔE) and output membership functions are Kp and Ki and are shown in Table 1 and 2.

Table 1 Fuzzy rule for Kp

E/ΔE	NL	NM	NS	Z	PS	PM	PL
NL	VL	VL	VB	VB	MB	M	M
NM	VL	VL	VB	MB	MB	M	MS
NS	VB	VB	VB	MB	M	M	MS
Z	VB	VB	MB	M	MS	VS	VS
PS	MB	MB	M	MS	MS	VS	VS
PM	VB	MB	M	MS	VS	VS	Z
PL	M	MS	VS	VS	VS	Z	Z

Table 2 Fuzzy rule for Ki

E/ΔE	NL	NM	NS	Z	PS	PM	PL
NL	Z	Z	VS	VS	MS	M	M
NM	Z	Z	VS	MS	MS	M	M
NS	Z	VS	MS	MS	M	MB	MB
Z	VS	VS	MS	M	MB	VB	VB
PS	VS	MS	M	MB	MB	VB	VL
PM	M	M	MB	MB	VB	VL	VL
PL	M	M	MB	VB	VB	VL	VL

Table 3 Parameters of wind-diesel hybrid system

System parameter	Wind diesel system
Wind capacity (KW)	150
Diesel capacity	150
Load capacity	250
Base power (KVA)	250
Synchronous generator	
$P_{SG,}$ (KW)	0.4
Q_{SG} (KW)	0.2
E_q (pu)	1.113
$E_{q,}$ (pu)	0.96
V (pu)	1.0
$X_{d,}$ (pu)	1.0
$T'_{do,}S$	5
Induction generator	0.6
$P_{IG,}pu$ (KW)	
$Q_{IG,}pu$ (Kvar)	0.189
$P_{IN,}pu$ (KW)	0.75
r1 = r2(pu)	0.19
X1 = X2(pu)	0.56
Load	1.0
P_L(pu) (KW)	
Q_L(pu) (Kvar)	0.75
α (Rad)	2.44
UPFC	
T_α(S)	0.05
T'_{do}	5.044
X'_d	0.3

4 Simulation Results

The complete Simulation was carried away taking the Fuzzy Logic PI Controller (Table 1 and Table 2)in the wind diesel hybrid power system in MATLAB/Simulink environment. Application of Fuzzy controller for reactive power and voltage control of the isolated hybrid system was taken with a step load change of 1 %. The variation of all the system parameters such as ΔQ_{SG}, ΔQ_{IG}, ΔQ_{UPFC}, $\Delta Q\alpha$, ΔV, ΔE_{fd}, ΔE_q and $\Delta E_q'$ etc., as shown in Fig. 7(a–h) are studied for the above disturbance using traditional PI Controller and the advanced Fuzzy PI Controller. In the study, The settling time and peak overshoot in the case of Fuzzy based UPFC controller is observed to be less as comparison to traditional PI Controller. The (FLC) shows better result by increasing the diesel power generation. Simulation results clearly show the output of Fuzzy PI Controller is better than the traditional PI Controller.

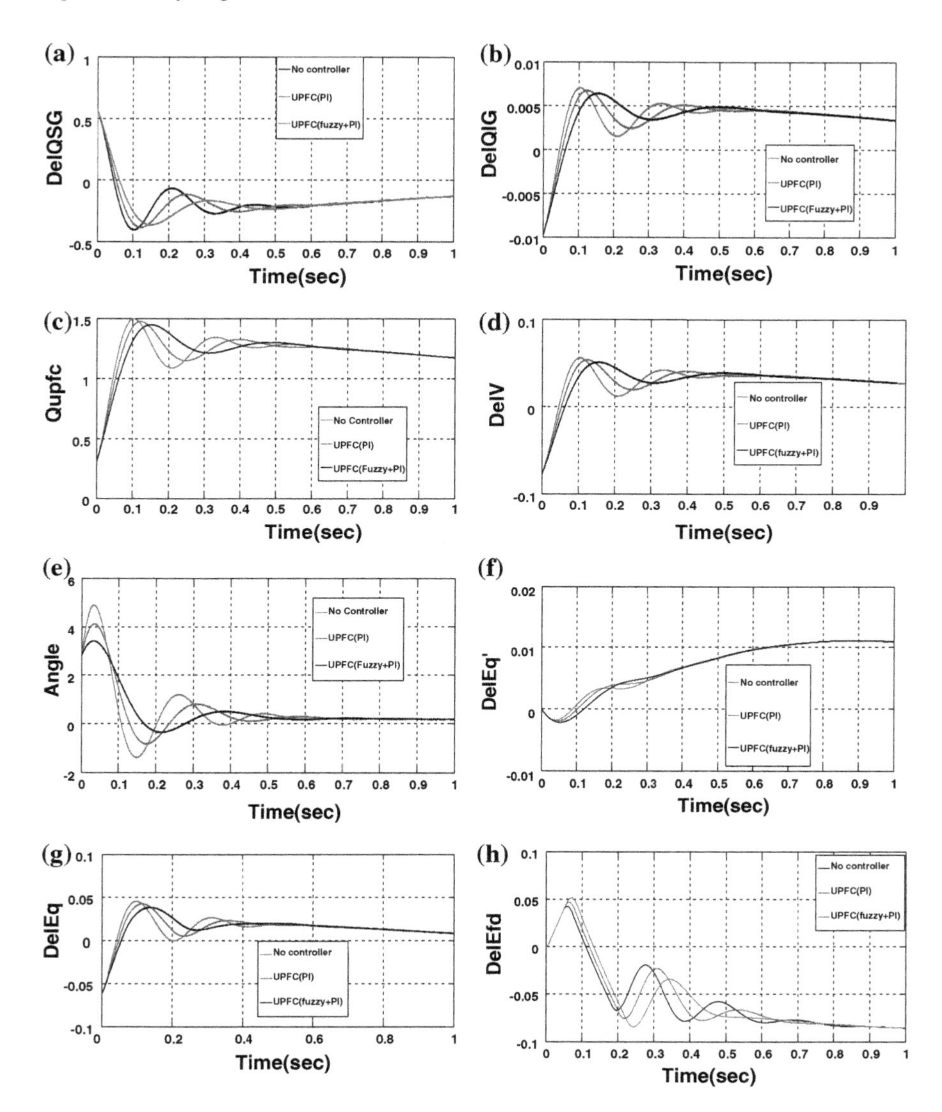

Fig. 7 Output of wind diesel hybrid system using UPFC for 1 % load change during transient condition (comparison with fuzzy controller).**a** Change in reactive power of DG, **b** Change in reactive power of IG, **c** Change in Reactive Power of UPFC,**d** Change in terminal voltage of WECS, **e** Thyristor change in angle ($\Delta\alpha$), **f** Change in internal emf, **g** Change in internal armature emf, **h** Change in voltage of Exciter.

5 Conclusion

Here in this paper the transient stability analysis has been done in the hybrid system of wind diesel with the incorporation of Fuzzy PI controller Reactive Power compensation of the model has been done and after comparing the performances of the proposed paper it can be said that the compensation in case of Fuzzified system is better than the conventional system having simple PI system. The proposed Fuzzified system shows better results in settling time and overshoot. The Fuzzified Controller shows effective improvement and changes in the output by tunning the values of Kp and Ki than the conventional PI controller. The Fuzzified system with UPFC seems Robust and effective in its approach.

References

1. Hingorani, N.G., Gyugyi, L.: Understanding FACTS, Concepts and Technology of Flexible AC Transmission System. IEEE Power Engineering Society, New York, pp. 63–65 (2000)
2. Kaldellis, J., et al.: Autonomous energy system for remote island based on renewable energy sources, In: Proceeding of EWEC 99, Nice
3. Murthy, S.S., Malik, O.P., Tondon, A.K.: Analysis of self excited induction generator. In: IEE Proceeding, vol. 129 (1982)
4. Padiyar, K.R.: FACTS Controlling in Power Transmission System and Distribution. New Age International Publishers (2008)
5. Tondon, A.K., Murthy, S.S., Berg, G.J.: Steady state analysis of capacitors excited induction generators. IEEE Transaction on Power Apparatus and Systems, March (1984)
6. Padiyar, K.R., Verma, R.K.: Damping torque analysis of static VAR system controller, IEEE Trans. Power Syst. **6**(2), 458–465 (1991)
7. Bansal, R.C.: Three phase self Excited Induction Generators, an over view, IEEE Transaction on Energy Conversion. (2005)
8. Saha, T.K., Kasta, D.: Design optimisation and Dynamic performance analysis of Electrical Power Generation System. IEEE Transaction on Energy Conversion Device (2010)
9. Ian, E., Gould, B.: Wind diesel power systems basics and examples, National Renewable Energy Laboratory. United States Department of Energy
10. Fukami, T., Nakagawa, K., Kanamaru,Y., Miyamoto, T.: Performance analysis of permanent magnet induction generator under unbalanced grid voltages, Translated from Denki Gakki Ronbunsi **126**(2), 1126–1133 (2006)

A Novel Modified Apriori Approach for Web Document Clustering

Rajendra Kumar Roul, Saransh Varshneya, Ashu Kalra and Sanjay Kumar Sahay

Abstract The Traditional apriori algorithm can be used for clustering the web documents based on the association technique of data mining. But this algorithm has several limitations due to repeated database scans and its weak association rule analysis. In modern world of large databases, efficiency of traditional apriori algorithm would reduce manifolds. In this paper, we proposed a new modified apriori approach by cutting down the repeated database scans and improving association analysis of traditional apriori algorithm to cluster the web documents. Further we improve those clusters by applying Fuzzy C-Means (FCM), K-Means and Vector Space Model (VSM) techniques separately. We use Classic3 and Classic4 datasets of Cornell University having more than 10,000 documents and run both traditional apriori and our modified apriori approach on it. Experimental results show that our approach outperforms the traditional apriori algorithm in terms of database scan and improvement on association of analysis.

Keywords Apriori · Association · K-means · Fuzzy C-means · VSM

1 Introduction

World Wide Web is the most important place for Information Retrieval (IR). Today volumes of data on web server are increasing exponentially which is almost doubling in every 6 months [1]. So we need to have efficient search engines to

R.K. Roul (✉) · S. Varshneya · A. Kalra · S.K. Sahay
BITS Pilani K. K. Birla Goa Campus, Zuarinagar 403726, Goa, India
e-mail: rkroul@goa.bits-pilani.ac.in

S. Varshneya
e-mail: saranshvarshneya@gmail.com

A. Kalra
e-mail: ashukalra7.17.27@gmail.com

S.K. Sahay
e-mail: ssahay@goa.bits-pilani.ac.in

L.C. Jain et al. (eds.), *Computational Intelligence in Data Mining - Volume 3*,
Smart Innovation, Systems and Technologies 33, DOI 10.1007/978-81-322-2202-6_14

handle large datasets. The query entered by users to any search engine will bring millions of documents from the web in which many of them are not of user choice. This problem can be further narrow down by making clusters on retrieved documents where each cluster will having similar kind of documents. To identify such similar kinds of documents from a corpus of huge size, association rule of data mining play a vital role. This rule discovers frequent itemset in a large dataset [2]. Association rules states that the knowledge of frequent itemset can be used to find out how an itemset is influenced by the presence of another itemset in the corpus. An itemset is frequent, if it is present in at least x % of the total transactions z in the database D, where x is the support threshold. When the numbers of items included in the database transaction are high and we are finding itemset with small support, the number of frequent itemset found are quite large, and it makes the problem very expensive to solve, both in time and space. Hence the minimum support count affects the computational cost of the higher (say kth) iteration of apriori algorithm. Thus, one can say that the cardinality of C^k and the size of D affect the overall computational cost. In fact, the traditional apriori algorithm generated and tested the candidate itemset in a level wise manner using iterative database scan which makes the computational cost very high [3].

Our approach considered documents as itemset and keywords as transaction so that we end up with clusters having a minimum frequency support threshold. By making keywords as transaction and document as itemset, we combined the support threshold idea of association rule mining algorithm with the output like that of a clustering algorithm. The *salient features* of our approach in comparison with traditional apriori algorithm can be described as follows:

- only one database scan by bringing the database into memory.
- reduces the time for accessing the transactions.
- avoid repeated database scan.
- nullifies the transactions which are no longer in use.
- decrease the number of candidate itemset during the candidate generation step.

We applied this modified apriori technique on a corpus of web documents to get some initial clusters. Next the FCM [2], K-Means [2] and VSM [2] techniques are applied separately to these initial clusters for further strengthening them as it missed out many documents because of the strong association rule of apriori with high minimum support. The experiment has been carried out on Classic3 and Classic4 dataset of Cornell University and the results demonstrate the applicability, efficiency and effectiveness of our approach.

The outline of the paper is organized as follows: Sect. 2 covers the related work based on different association techniques used for clustering. Background of the proposed work is described in Sect. 3. In Sect. 4, we describe the proposed approach adopted to form the clusters. Section 5 shows the experimental work and finally we concluded the present work with future enhancement in Sect. 6.

2 Related Work

Association rule of data mining is used to find out a set of frequent itemset from a list of itemset depending on the minimum support of user choice. It plays a vital role in clustering technique. Wang et al. [4] proposed a technique to find out frequent itemset in a large database and to mine association rules from frequent itemset. Though they improved the mining association rule algorithm based on support and confidence but it creates useless rules and lost useful rules out of which the useless rules are discarded to discover more reasonable association rules. Bodon [5] in his work, tried to store data which also stores frequent itemset with candidate itemset. Cao [6] includes grain bit binary extraction to mine frequent itemset. Wang et al. [7] in their approach tried to use normative database of users to get itemset. Li et al. [8] have used a cross-linker structure instead to substitute array based representation of transaction database. Tomanová and Kupka [9] applied properties of fuzzy confirmation measures in association rule mining process. Li et al. [10] uses a two-layer web clustering approach to cluster a number of web access patterns from web logs. At the first layer, Learning Vector Quantization (LVQ) is used and at the second layer rough K-Means clustering algorithm has been used. The experimental results show that the effect is close to monolayer clustering algorithm than the rough K-Means, and the efficiency is better than the rough K-Means. A Hierarchical Representation Model with Multi-granularity (HRMM), which consists of five-layer representation of data and a two-phase clustering process, is proposed by Huang et al. [11] based on granular computing and article structure theory. Roul et al. [12] in their work used Tf-Idf based apriori approach to cluster and rank the web documents. An effective clustering algorithm to boost up the accuracy of K-Means and spectral clustering algorithms is proposed in [13]. Their algorithm performs both bisection and merges steps based on normalized cuts of the similarity graph '*G*' to correctly cluster web documents. The proposed modified apriori approach scans the database only once and hence reduced the repeated database scan up to a maximum extent for making the initial clusters as compared to traditional apriori approach. The performances of the final clusters generated by FCM, VSM and K-Means techniques have been compared and the results show that FCM outruns VSM and K-Means.

3 Background

3.1 Traditional Apriori Algorithm

Apriori employs an iterative approach for finding frequent itemset. It starts by finding one frequent itemset and then progressively moves on to find (k + 1) itemset from k-frequent itemset. From this resulting (k + 1) itemset a set is formed which is the set of frequent (k + 1) itemset. This process continues until no more frequent itemset are found. Apriori uses *breadth first search* algorithm for association. This can be used to group the itemset based on their association which satisfied the minimum support.

3.2 Fuzzy C-Means Algorithm

FCM is a soft clustering technique which allows the feature vectors to belong to more than one cluster with different membership degree or belongingness of data. The value of the membership degree lies in the range from [0, 1]. It is simply an iterative fuzzy based algorithm that returns the centroid of the clusters. It will terminates when satisfy an objective function mentioned below.

$$\mathrm{T_m} = \sum_{i=1}^{V}\sum_{j=1}^{C} u_{ij}^{m}\left\|d_i - c_j\right\|^2 \qquad (1)$$

where,

m [1, ∞]	fuzzy coefficient
u_{ij}	membership degree of x_j w.r.t to c_j; range [0, 1]
c_j	centroid (vector) of cluster j
C	number of clusters
V	number of document vectors
d_i	document vector

1. Initialize the membership matrix U along with a fixed number of clusters. Number of rows and columns of U depend on the number of clusters and documents. $U^{(0)} = [u_{ij}]$, where 0 is 0th or first iteration.
2. Calculate the cluster center vector as $C_j = \frac{\sum_{i=1}^{V} u_{ij}^{m}.d_i}{\sum_{i=1}^{V} u_{ij}^{m}}$
   where, all symbols have same meaning as defined in Eq. 1. The membership values are calculated with respect to the new centers. Belongingness of the document to the cluster is calculated using Euclidian distance between the center and the document vector.
3. Update the $U^{(k)}$ matrix such that $u_{ij} = \frac{1}{\sum_{k=1}^{c}\left(\frac{d_i - c_j}{d_i - c_k}\right)^{2/(m-1)}}$
   where, all symbols having same meaning as in Eq. 1 and 'k' is the iteration step.
4. If $\|U^{(k+1)} - U^{(k)}\| < \varepsilon$, where, $\varepsilon < 1$ is the termination criterion, then stop, else goto step 2. These will coverage to a local minimum of the function T_m.

3.3 K-Means Algorithm

1. Collect the documents $d_1, d_2, \ldots, d_k$ which need to be clustered.
2. Pre-determine the number of clusters (*K*).
3. Initialize the means of those pre-determined *K* clusters.
4. Determine the Euclidean distance of each document from those means.
5. Group the documents having minimum distance to the corresponding mean.
6. Find the *new mean* of the each group formed in step 5.
7. If *new mean* = *previous mean* then stop else go to step 4.

3.4 Vector Space Model

Vector Space Model is an algebraic model which represents a set of documents as vectors in a common vector space. Thus, a document in its vector form can be look as, $D_i = [w_{1i}, w_{2i}, w_{3i}, \ldots, w_{ni}]$ where, w_{ji} is the weight of term j in document i.

If $x = \{x_1, x_2, \ldots, x_n\}$ and $y = \{y_1, y_2, \ldots, y_n\}$ are two documents in Euclidean n-space, then the distance between x to y, or from y to x is:

$$Euclidean\,Distance(x, y) = \sqrt{\sum_{i=1}^{n} (x_i - y_i)^2} \tag{2}$$

3.4.1 TF-IDF

TF the Term Frequency, which measures how often a term appears in a document. Tf(t) = (Number of times the term t appears in a document D)/(Total number of terms in the document D).

IDF the Inverse Document Frequency, which measures how important a term t is. Idf(t) = log(Total Number of documents)/(Number of documents in which the term t appears). If one combine Tf-Idf [14] then it can use to measure the relevance and importance of the term to a document. $Tf - Idf = Tf * Idf$.

3.4.2 Cosine_Similarity Measure

In Information Retrieval (IR) to measure the similarity between any two documents, a technique called Cosine_Similarity [14] has been used extensively. If A and B are two documents then cosine similarity between them can be represent as follows:

$$cosine\,Similarity(A, B) = \frac{(A \cdot B)}{(|A||B|)} \tag{3}$$

If Θ = 0°, then the two documents are similar. As Θ changes from 0° to 90°, the similarity between the two documents decreases.

3.5 F-Measure

It measures the system performance by combining the precision and recall of the system [2]. Precision is the fraction of the retrieved documents that are relevant and recall is the fraction of the relevant documents that are retrieved.

$$\text{F-measure} = 2 * \frac{(precision * recall)}{(precision + recall)}$$

4 Proposed Approach

4.1 Optimization Open Traditional Apriori

- Apriori generates frequent candidate itemset by generating all possible candidate itemset and then checking which itemset cross the minimum support count. Whereas our algorithm uses the following rule: If an itemset occurs (k − 1) times in the set of (k − 1) frequent itemset then only it is considered for kth frequent candidate itemset (since only then it has a chance to come in a 'k' sized frequent itemset) [15]. So, our algorithm rejects more number of unwanted candidate itemset than normal apriori (which are generally not going to be frequent in the next iteration) before the counting step.
- Apriori counts the occurrence of an itemset even if it does not appear in any of the frequent candidates but our algorithm removes that itemset from the array by setting 0 across it so that less number of checks is made for that itemset.
- Apriori does not take into account the unnecessary computations made if the size of the transaction is lesser than the size of the candidate itemset being generated. So to improve upon this, we ignore the corresponding transaction by putting a null across it [16].

All of the above statements basically remove the information which is no longer required from the 2D-array created initially and this reduces the unnecessary comparisons in the subsequent steps.

4.2 Finding Initial Cluster Centers

The proposed modified apriori algorithm described has been used to find the number of clusters and initial cluster centers, which latter used for all the three techniques (FCM, VSM and K-Means) for generating final clusters described in algorithm 3, 4 and 5 respectively. The frequent itemset generated are of frequency greater than the minimum support supplied by the user. In the generated frequent itemset of documents, the keywords are considered to be the transactions and the documents are the itemset. In this way the frequent itemset generated are the ones which have particular set of keywords in common and hence are closely related. This helps by deciding the number of clusters and also the centers of these clusters which is simply the centroid of the respective frequent itemset.

Algorithm 1: Modified Apriori Approach

Input: The dataset (D) and minimum support (min_sup)
Output: The maximum frequent itemset

1. Read the dataset into a 2D array and store the information of the database in binary form in the array with transactions as rows and itemset as columns.
2. $k \leftarrow 1$.
3. Find frequent itemset, L_k from C_k, the set of all candidate itemset.
4. Form C_{k+1} from L_k.
5. Prune the frequent candidates by removing itemset from C_k whose elements do not come atleast k-1 times in L_k.
6. Modify the entry in the 2D-array in memory to be zero for the itemset which are not occurring in any of the candidates in L_k.
7. Check the Size of Transaction (ST) attribute and remove transaction from 2D-array where ST<=k.
8. $k \leftarrow k+1$.
9. Repeat 5-8 until C_k is empty or transaction database is empty.

Step 5 is called the frequent itemset generation step. *Step 6* is called as the candidate itemset generation step and *step 7-9* are prune steps. Details of first two steps are described below.

Frequent itemset generation: Scan D and count each itemset in C_k, if the count is greater than min_sup, then add that itemset to L_k.

Candidate itemset generation: For k = 1, C_1 = (all itemset of length = 1).

For k > 1, generate C_k from L_{k-1} as follows:

The join step:

C_k = k-2 way join of L_{k-1} with itself.

If both $\{a_1,...,a_{k-2}, a_{k-1}\}$ & $\{a_1,..., a_{k-2}, a_k\}$ are in L_{k-1}, then add $\{a_1,...,a_{k-2}, a_{k-1}, a_k\}$ to C_k.

Algorithm 2: Finding Initial Clusters

Input: Documents set to be clustered, the user query
Output: Initial Clusters (C_i)

1. *Web page extraction and preprocessing:* Submit the query to a search engine and extract top 'n'web pages. Remove the stop and unwanted words. Select *noun* as the keywords. Stemming the keywords and stored the pre-processed 'n' pages as document, D_i, where i = 1, 2,.....,n. Consider each keyword as a transaction and D_i as transaction elements(itemset).
2. *Creation of Term-Document matrix(T) and finding Tf-Idf of D_i:* Term document matrix, T, is created by counting the number of occurrences of each term (i.e keyword) in each document D_i. Each row t_i of T shows a term's occurrence in each document D_i. Finding out the Tf-Idf of D_i from T (i.e D_{inew}←Tf-Idf(D_i)).
3. *Extraction of maximum frequent itemset:* Algorithm 1 is used to extract the maximum frequent itemset(i.e documents) from T and each maximum frequent itemset is treated as one cluster which gives the initial clusters(C_i).

Algorithm 3: Final Clusters using FCM

Input: Initial clusters C_i, Value of fuzziness parameter m, document vector D_{inew} generated in Algorithm 2
Output: Final clusters FC_i

1. *Centroid calculation:* *//find cluster centroids*

   for each frequent set $f_i \in C_i$ **do**
   c_i← Centroid;
   k← length(f_i);

   for each document $d_j \in f_i$ **do**
   c_i←c_i + d_j;
   end
   c_i←c_i/ k
   end

2. *Assignment documents to their respective clusters:* Final clusters FC_i can be obtained by applying FCM algorithm (discussed in section 3.2) on the set of documents D_{inew}, using initial cluster C_i, the centroid c_i and the fuzziness parameter m.

Algorithm 4: Final Clusters using VSM

Input: Initial clusters C_i, document vector D_{inew} generated in Algorithm 2
Output: Final clusters FC_i

1. *Centroid calculation:* *//find cluster centroids*

   for each frequent set $f_i \in C_i$ **do**
   $c_i \leftarrow$ Centroid;
   $k \leftarrow$ length(f_i);

   for each document $d_j \in f_i$ **do**
   $c_i \leftarrow c_i + d_j$;
   end
   $c_i \leftarrow c_i / k$
   end
2. *Assign the documents to their respective clusters:*

   for each $d_i \in D_{inew}$ **do**

   for each $c_j \in C_j$ (i.e centroid) **do**
   Similarity[i][j] = *cosineSimilarity*(d_i, c_j); // Equation 3
   end

   // find maximum similarity cluster(C_k, $k \in j$) among the clusters from 0
   //to j-1.
   $C_k \leftarrow$max(Similarity[i][0]….Similarity[i][j-1]); C_k.append(d_i); //initial clusters get updated .
   end
3. *Repeat step1* and *2 until no cluster changes and it gives* FC_i.

Algorithm 5: Final Clusters using K-Means

Input: Initial clusters C_i, document vector D_{inew} generated in Algorithm 2
Output: Final clusters FC_i

1. *Centroid calculation:* //*find cluster centroids:*

   for each frequent set $f_i \in C_i$ **do**
   $c_i \leftarrow$ Centroid;
   $k \leftarrow$ length(f_i);

   for each document $d_j \in f_i$ **do**
   $c_i \leftarrow c_i + d_j$;
   end
   $c_i \leftarrow c_i$ / k
   end
2. *Assign the documents to their respective clusters:*

   for each $d_i \in D_{inew}$ **do**

   for each $c_j \in C_j$ (i.e centroid) **do**
   Similarity[i][j] = *euclideanDistance*(d_i, c_j); //Equation 2
   end

   // find maximum similarity cluster(C_k, k∈j) among the clusters from 0
   // to j-1.
   $C_k \leftarrow$max(Similarity[i][0]....Similarity[i][j-1]);C_k.append(d_i); //initial clusters get updated.
   end
3. *Repeat step1 and 2 until no cluster changes and it gives* FC_i.

5 Experimental Results

For experimental purpose, to demonstrate the effectiveness of our approach, Classic3 and Classic4 dataset of Cornell University [17] have been used. The dataset contains four categories (CISI, CRAN, MED and CASM) set of documents. Each category set contains over 1,000 documents, but for our experimental analysis, the traditional and modified apriori algorithm were run over CISI, CASM and MED document sets by using 200, 400, 600, 800 and 1,000 documents from each set. The graphs depicted in Figs. 1, 2 and 3 suggests that the proposed algorithm has a better running time than the traditional apriori algorithm. The support count used for CASM data set is 5, for CISI is 10 and for MED is 25. It is clear from Fig. 4 that even on varying the support count for the dataset (600 documents of MED) the modified algorithm still outperforms the tradition apriori algorithm. The modified apriori algorithm focuses on limitations of the traditional apriori and as can be seen

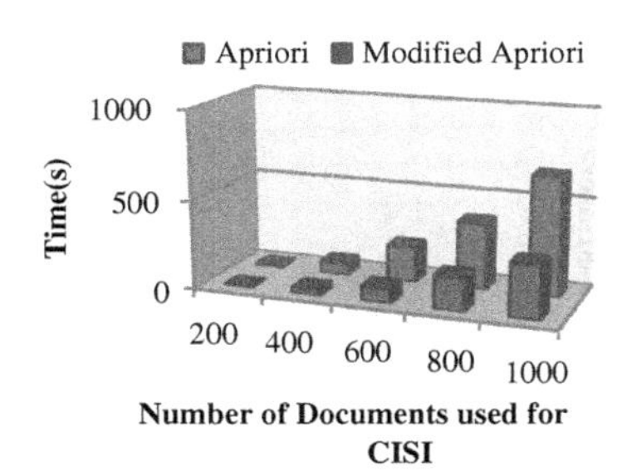

Fig. 1 Apriori versus modified apriori (CISI)

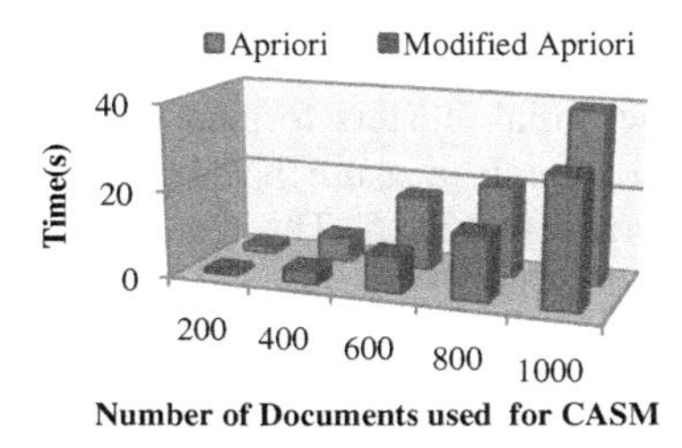

Fig. 2 Apriori versus modified apriori (CASM)

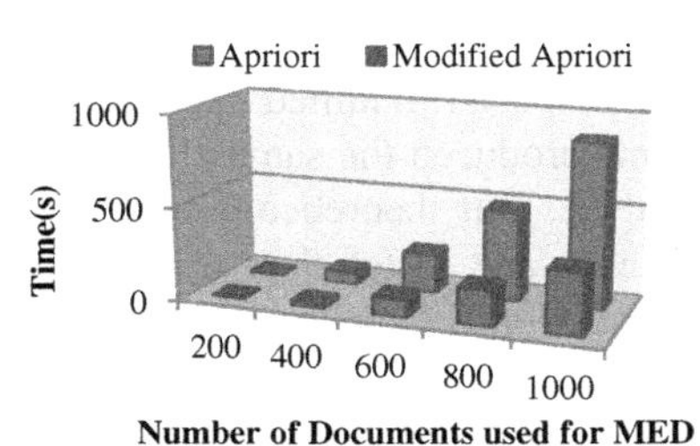

Fig. 3 Apriori versus modified apriori (MED)

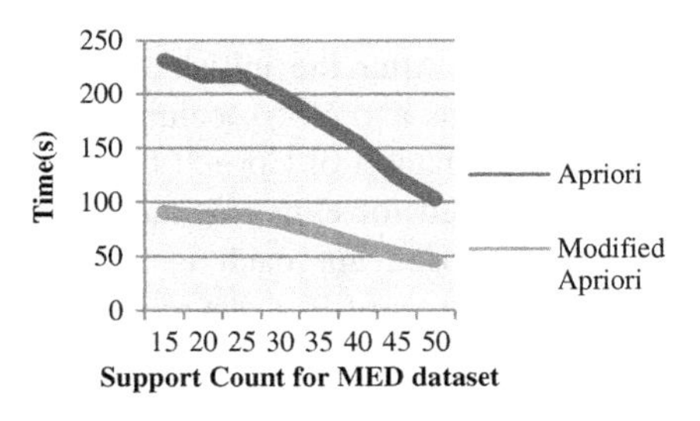

Fig. 4 Support count graph (MED)

from the graphs that it outruns the traditional apriori algorithm. It can also be observed from these graphs that as the number of documents increases the difference in running time between the traditional and modified apriori algorithm becomes significant. Finally FCM, K-Means and VSM techniques are applied on

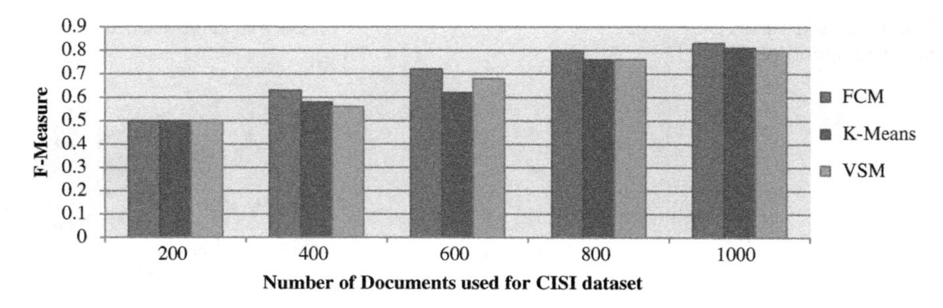

Fig. 5 Comparison of F-measure of different techniques (CISI)

the initial clusters formed by the modified apriori approach on CISI document set and the F-measure has been calculated separately for clusters of different sizes shown in Fig. 5. The results show that the performance measurement of FCM is better than VSM and K-Means for 400, 600, 800 and 1,000 documents of clusters.

6 Conclusion and Future Work

The proposed modified apriori approach when applied to a corpus of web documents produced the same clusters which is what a traditional apriori approach can produce. But theoretically and experimentally it has been shown that it is more efficient and faster than traditional apriori. This is so because at each step the information (i.e. documents) from the transactions has been removed which is no longer required and in turn it reduces the unnecessary comparisons. Also the additional pruning of frequent itemset reduces the number of candidate itemset for the counting step and the number of database scans is reduced to *one,* since once the vertical data layout in the form of a 2D-array is made then no more database scans are required and all above manipulations are done on that 2D-array. Hence it saves a lot of time. After the initial clusters formed by the modified apriori approach, FCM, K-Means and VSM techniques have been applied on it separately. Classic3 and Classic4 dataset of Cornell University with document set of different size have been considered for experimental purpose. From the results it has been found out that the proposed approach is better than traditional apriori. Performance measurement in terms of F-measure has been carried out for final clusters produced by each of the three techniques. We found that FCM can able to give better clusters compared to K-Means and VSM. This work can be extended by labeling each cluster and ranking the documents for every cluster so that user can restrict their search to some top documents instead of wasting their time to search the entire cluster. Further text summarization can also be done on each cluster to get information about the content of that cluster.

References

1. Barsagade, N.: Web usage mining and pattern discovery: asurvey paper. In: CSE8331, Dec 2003
2. Kumar, T.S.: Introduction to Data Mining. Pearson Education, Upper Saddle River (2006)
3. Agrawal, R., Mannila, H., Srikant, R., Toivonen, H., InkeriVerkamo, A.: Fast discovery of association rules in large databases. In: Fayyad, U.M., Piatetsky-Shapiro, G., Smyth, P. (eds.) Advances in Knowledge Discovery and Data Mining, pp. 307–328. AAAI press, Menlo Park (1996)
4. Wang, P., Shi, L., Bai, J., Zhao, Y.: Mining association rules based on apriori algorithm and application. In: International Forum on Computer Science-Technology and Applications, IFCSTA'09, vol. 1 (2009)
5. Bodon, F.: A trie-based APRIORI implementation for mining frequent itemset. In: ACM 1-59593-210-0/05/08 (2005)
6. Cao, X.: An algorithm of mining association rules based on granular computing. In: International Conference on Medical Physics and Biomedical Engineering (2012)
7. Wang, Y., Jin, Y., Li, Y., Geng, K.: Data mining based on improved apriori algorithm. Commun. Comput. Inf. Sci. **392**, 354–363 (2013)
8. Li, X., Shang, J.: A novel apriori algorithm based on cross linker. In: Proceedings of the International Conference on Information Engineering and Applications (IEA) (2012)
9. Tomanová, I., Kupka, J.: Implementation of background knowledge and properties induced by fuzzy confirmation measures in apriori algorithm. In: International Joint Conference CISIS'12-ICEUTE'12-SOCO', vol. 189, pp. 533–542 (2013)
10. Li, Y., Xing, J., Wu, R., Zheng, F.: Web clustering using a two-layer approach. LNCS, vol. 6988, pp. 211–218. Springer, Heidelberg (2011)
11. Huang, F., Zhang, S., He, M., Wu, X.: Clustering web documents using hierarchical representation with multi-granularity. World Wide Web **17**, 105–126 (2014). doi:10.1007/s11280-012-0197-x
12. Roul, R.K., Devanand O.R., Sahay S.K.: Web document clustering and ranking using Tf-Idf based Apriori Approach. In: IJCA Proceedings on International Conference on Advances in Computer Engineering and Applications, pp. 34–39 (2014)
13. Lee, I., On, B.W.: An effective web document clustering algorithm based on bisection and merge. Artif. Intell. **36**, 69–85 (2011). doi:10.1007/s10462-011-9203-4
14. http://www.nlp.fi.muni.cz/projekty/gensim/intro.html
15. Orlando, S., Palmerini, P., Perego, R.: Enhancing the apriori algorithm for frequent set counting. LNCS, vol. 2114, pp. 71–82. Springer, Heidelberg (2001)
16. Singh, J., Ram, H., Sodhi, J.S.: Improving efficiency of apriori algorithm using transaction reduction. Int. J. Sci. Res. Publ. **3**(1), 1–4 (2013)
17. http://www.dataminingresearch.com/index.php/2010/09/classic3-classic4-datasets

Sliding-Window Based Method to Discover High Utility Patterns from Data Streams

Chiranjeevi Manike and Hari Om

Abstract High utility pattern mining is one of the emerging researches in data mining. Mining these patterns from the evolving data streams is a big challenge, due the characteristics of data streams like high arrival rate, unbounded and gigantic in size, etc. Commonly there are three window models (landmark window, sliding window, time fading window) used in data streams. However, in most applications, users are interested in recent happenings. Hence, sliding window model has attracted high interest among three. Many approaches have been proposed based on the sliding window model. However, most of the approaches are based on level-wise candidate generation and text approach. In view of this, we propose an efficient one pass, tree based approach for mining high utility patterns over data streams. Experimental results show that the performance of our approach is better than the level-wise approach.

Keywords Data streams · Sliding window · High utility pattern · Frequent pattern mining · Data mining

1 Introduction

Utility pattern mining was introduced in 2003 to discover patterns based on the quantitative databases [1] Traditional frequent pattern mining was based on the support measure, where support is defined over the binary domain 0, 1, which is based on the occurrence frequency of items [2]. Frequent pattern mining considers purchased quantity of every item as 1. But in real time scenario a customer may purchase multiple items of different quantity. Let us consider a query of a sales

C. Manike (✉) · H. Om
Indian School of Mines, Dhanbad 826004, Jharkhand, India
e-mail: chiru.research@gmail.com

H. Om
e-mail: hariom.cse@ismdhanbad.ac.in

L.C. Jain et al. (eds.), *Computational Intelligence in Data Mining - Volume 3*,
Smart Innovation, Systems and Technologies 33, DOI 10.1007/978-81-322-2202-6_15

manager in a supermarket, "set of itemsets contributing more to the total profit?". To give response to such queries frequency of an itemset is not adequate measure [3]; we need to consider the purchased quantities of itemsets and their corresponding unit profits. Thus, high utility patterns are more significant in practical applications than frequent patterns. Utility pattern mining consider other useful measures of an itemset, they may be its purchased quantity, profit, cost, etc. [2]. Patterns having utility more than the user specified minimum utility threshold are called as high utility patterns. High utility pattern mining is the process of discovering patterns where the actual utility of the pattern is more than the user specified minimum utility threshold. Unlike frequent patterns, high utility patterns do not follow downward closure property [4], and it needs several utility calculations. Hence, high utility pattern mining is more complex than traditional frequent pattern mining. Liu et al. [5, 6] has observed an efficient pruning strategy among the transaction weighted utilities of itemsets, which is named as transaction weighted utility (TWU) downward closure property. Many algorithms have been developed based on the above efficient pruning strategy.

On the other hand pattern mining over data stream is a big challenge due to the evolving nature and requirements of data streams [7], as the data arriving continuously, the utilities of an itemsets may vary irregularly from one instance to other [8–10]. There are mainly three window models used in mining data streams, those are: landmark window, sliding window, and time fading window. Selecting particular window model is based on the type of knowledge to be discovered and the application domain [7, 11, 12]. Landmark window holds the information of itemsets from landmark time to current time, sliding window holds the recent information of itemsets like last 3 h, last 1,000 instances etc. Time fading window holds the information same as landmark window, in which the importance of an itemset decreases with time. Among these three models sliding window model gives the information of most recent occurred events and also many users are interested in this [13]. Especially in credit card fraud analysis, network intrusion detection, wireless sensor networks [7] where the event might last for a few milliseconds, or a few hours, or even a few days.

Unlike traditional databases, data streams do not allow algorithms to visit the data more than once. Hence, one pass algorithms are more suitable in these environments. Especially in sliding window model capturing incoming data as well as removing old data within window is main issue. In this paper, we have proposed an efficient approach, based on sliding window model, to effectively update the information of utility patterns and to efficiently discover high utility patterns.

2 Problem Definition

Let $I = \{i_1, i_2, \ldots, i_n\}$ be a finite set of n distinct items. A data stream $DS = \{T_1, T_2, \ldots, T_m\}$ be a stream of transactions, where each transaction Ti is a subset of I. An itemset X is a finite set of items $X = \{i_1, i_2, \ldots, i_k\} \in I$, where $k (1 \leq k \leq n)$

is the length of the itemset. An itemset with length k is also called as k-itemset. Each item or an itemset in a transaction associated with a value (For example, purchased quantity) called internal utility of an itemset, denoted as $iu(i_j)$. Each item $i_j \in I$ associated with a value (profit, price, etc.) outside the transaction data stream called external utility and it is denoted as $eu(i_j)$. Definitions in this section are adapted from the previous works [2, 5].

Definition 1 Utility of an item i_j in a transaction T_i is the product of its internal utility and external utility, denoted as $u(i_j, T_i) = iu(i_j, T_i) \times eu(i_j)$. For example, utility of an item in T_1 is 12 (i.e., 2×6), in Tables 1 and 2.

Definition 2 Utility of an itemset X in a transaction T_i is the sum of the individual utilities of items belongs to the itemset X, $u(X, T_i) = \sum_{i_j \in X \in T_i} u(i_j, T_j)$. For example, $u(abc, T_1) = 2 \times 6 + 3 \times 4 + 2 \times 10 = 44$, in Tables 1 and 2.

Definition 3 Transaction utility of a transaction T_i is the sum of all its items utilities, denoted as $tu(T_i)$. For example, $tu(T_2) = u(a) + u(b) + u(c) + u(e) = 52$, in Tables 1 and 2.

Definition 4 Transaction weighted utility of an itemset X is defined as the sum of all transaction utilities of transactions in which itemset X exists. For example, $twu(cd) = tu(T_1) + tu(T_5) + tu(T_6) = 195$, in Tables 1 and 2.

Definition 5 Minimum utility threshold is defined as the percentage of total transaction utility values of the data stream, defined as, $minUtil = \delta \times \sum_{T_i \in DS} tu(T_i)$.

Table 1 Example transaction data stream

Item/Tid	a	b	c	d	e	f	g
T_1	2	3	3	1	0	0	0
T_2	1	4	1	0	1	0	0
T_3	0	2	2	0	0	1	0
T_4	3	4	2	0	0	0	0
T_5	2	3	1	2	0	0	0
T_6	4	2	1	2	0	0	0
T_7	2	0	0	0	0	0	1
T_8	2	5	3	0	0	0	0

Table 2 Utility table

Item	a	b	c	d	e	f	g
Profit($)	6	4	10	15	20	30	40

Definition 6 High utility pattern mining can be defined as the process of finding set of patterns, where utilities of those patterns are more than the *minUtil*.

3 Related Works

Chan et al. [14], introduced high utility pattern mining in the year 2003. Theoretical model and basic definitions are given in [2] MEU (Mining with Expected Utility), two efficient strategies are also been introduced based on the utility and support values of itemsets. However, these strategies reduced candidate patterns effectively, it depends on the level-wise candidate generation-and-test problem and. The utility bound set by the property overestimating many patterns again filtering these huge set of patterns becomes complex task. Hence, same authors introduced two new algorithms namely UMining and UMining_H by incorporating above strategies [3]. Though, MEU, UMining, and UMining_H algorithms reduced candidate patterns, still depends on level-wise candidate generation and-test approach. Above algorithms are tried to reduce the search space using support and utility values of itemsets those are not efficient as Apriori anti-monotone property.

In 2005, Liu et al. [5, 6] has discovered an efficient heuristic pruning strategy called transaction weighted utility downward closure property. An efficient two pass algorithm was proposed by incorporating above property, which is called Two-Phase. The property says that TWU of an itemset is high then its subsets TWU values are also high. Even though, this property does not exist among the actual utilities of itemsets, it effectively used in finding potential patterns. The main drawback in this approach is this property allows some false itemsets, where the actual utilities of these itemsets are not high. Again filtering exact patterns from these potential set of patterns becomes main barrier in most of the cases. Afterwards, many approaches were proposed based on TWU downward closure property. But in data stream environment it may not possible to apply such an anti-monotone properties because the requirements of stream processing different from traditional databases. Moreover, low utility patterns may become high utility patterns when a new transaction arrives.

First approach that was proposed to mine the high utility patterns over data stream is THUI-Mine [15], called THUI-Mine. THUI-Mine is mainly integrates the tasks of Two-Phase [5, 6] algorithm for processing items in pre-processing procedure, and SWF [13] to use the concepts of filtering threshold in incremental procedure. THUI-Mine decomposed the problem of mining high utility patterns into two procedures: (1) Pre-processing procedure, (2) Incremental procedure. First, pre-processing procedure calculates the TWU of each item and discards the items whose TWU is less than the filtering threshold. Next generate the candidate 2-itemsets from the remaining items, and record their appeared partition number and

TWU value respectively. After processing each partition, candidate 2-itemsets with TWU value above filtering threshold value will be considered in next partition. Even THUI-Mine find complete set of high utility patterns, there is no pruning strategy incorporated in it to filter overestimated patterns.

To overcome the limitations in THU-Mine, two efficient algorithms were proposed in [16], called MHUI-TID and MHUI-BIT. These algorithms represent transaction information in two formats that is TIDlist and Bitvector respectively. For example, consider Table 1, in which item a encountered in transactions T_1, T_2, T_4, T5, T_6, T_7 and T_8; this can be represented in the form of Bitvector, that is ⟨11011111⟩ (assume window size is set to 8). Above information is also represented in the form of TIDlist that is {1, 2, 4, 5, 6, 7, 8}. This is a three-phase method, in first phase it builds information of items using either TIDlist or Bitvector representation. This algorithm processes the items as it was done by pre-processing procedure of THUI-Mine. In next phase it uses level-wise approach to generate the candidate itemsets of length more than 2. In second phase it builds a lexicographic tree structure to maintain the information of potential itemsets (k—itemsets, $k < 3$) and updates the information of newly identified itemsets when window slides. Third, high utility itemsets generation phase in which high utility patterns are generated based on the user queries. MHUI algorithms effectively reduced the number of potential candidates and optimal utilization of memory with these two representations. Hence, it achieved better performance over the THUI-Mine, but still need more than one database scans. MHUI keeps the information of itemsets with length 2, so remaining itemsets of length more than 2 will be processed in level-wise manner.

From the above discussion on related work of high utility pattern mining from data streams, we have identified two motivating factors to design a new algorithm which need at most one database scan and better runtime efficiency. In this paper we have designed a new one pass algorithm based on the prefix-tree structure and we did not consider the TWU values of an itemsets to disqualify items before starting the process.

4 Proposed Work

In this Section we present our proposed algorithm, in addition, in this section we briefly describe our method with example. Our method basically consists of three steps: (1) transactions processing: capturing information and storing in tree, (2) incremental: updating the information in tree when window is sliding, and (3) mining: tracing tree to produce output to the user query.

First step comprises of two subtasks, when the transactions arrives continuously in the data stream each transaction is processed one by one. After arriving transaction T_1, set of all possible patterns will be generated and utilities corresponding to those patterns are also being calculated in parallel. Second subtask is updating all generated patterns in sliding window tree called $RHUI_{SW}$-Tree by calling a procedure namely $RHUI_{SW}$-Tree-Update. For each transaction there will be $2^n - 1$ (n is the number of items) number of patterns generated. While updating all these patterns procedure $RHUI_{SW}$-Tree-Update follows lexicographical order, because order of processing significantly affects the execution time. Patterns which are sharing prefix will be updated first that is a, ab, ac, ad, abc, acd, abcd will appear along the same path of $RHUI_{SW}$-Tree, so redundant moves for each node from root node is avoided.

While updating patterns in tree the utility value of pattern is added to total utility and also enqueue to pattern utility queue. Figure 1, represents the $RHUI_{SW}$-Tree after updating transaction T_1. Same procedure will be followed for next and subsequent transactions, until we reach the window size limit, once we exceed the limit then sliding window phase will enter. Let us assume sliding window size is set to 4, so window accommodate first four transactions T_1, T_2, T_3, and T_4. During the above process updated patterns of all transactions will also be stored in a table called PTable (Pattern Table) that is Table 3, this table will be used as lookup table while deleting the information of obsolete transactions from the sliding window. Whenever the total count of processed transactions exceeds the window size limit, $RHUI_{SW}$-Tree-IncUpdate procedure will be invoked. $RHUI_{SW}$-Tree-IncUpdate updates the patterns and corresponding utilities of T_1 from the $RHUI_{SW}$-Tree with the help of PTable. PTable helps to remove all patterns corresponding to the transaction T_1 without revisiting all nodes. Next $RHUI_{SW}$-Tree-IncUpdate invokes the $RHUI_{SW}$-Tree-Update procedure to process first transaction of next window W_2 that is T_5. For instance, in incremental phase to delete all patterns of a transaction T_1 we need to find patterns and utilities of transaction T_1, and have to visit corresponding nodes in tree to delete. The cost of first step is reduced by maintaining those patterns in PTable and cost of traversing completed tree is also minimized by maintaining node links.

Last step is tracing $RHUI_{SW}$-Tree, when user queries the system algorithm trace the tree to produce high utility patterns. To visit all nodes generally we follow either top-down or bottom-up approach, in this case both approaches are unsuitable because no need to visit all nodes. To improve the query response time at this stage, we used another table called, HTable (Table 4) to keep information of nodes whose pattern total utility is more than the minimum utility threshold. While updating patterns if pattern total utility crosses utility threshold that corresponding node like immediately stored in HTable, Table 4 shows the status of HTable after processing each transaction from T_1 to T_5.

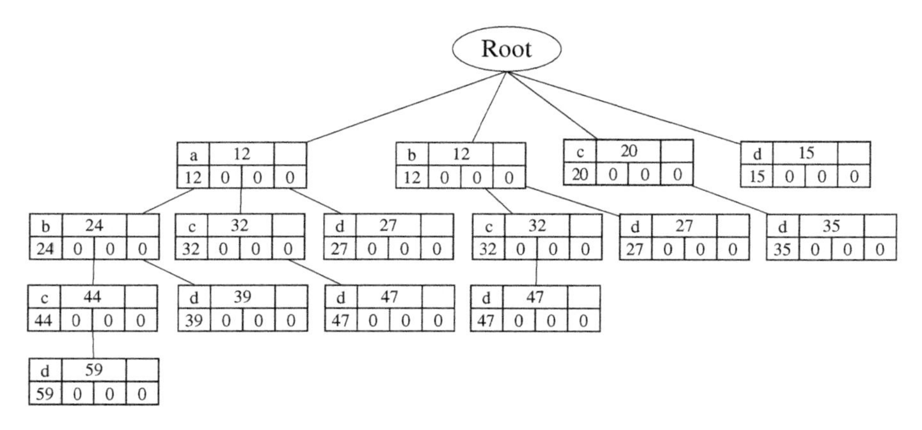

Fig. 1 $RHUI_{SW}$-Tree after updating transaction T_1

Table 3 PTable

T_1	T_2	T_3	T_4
a	a	b	a
b	b	c	b
c	c	f	c
d	e	bc	ab
…	…	…	…

Table 4 HTable

T_1	T_2	T_3	T_4	T_5
abcd	abcd	abcd	abcd	abcd
	bc	bc	bc	bc
	abc	abc	abc	abc
	abce	abce	abce	abce
		c	c	c
		…	…	…

```
Algorithm 1: RHUIMSW for mining recent high utility
             patterns
Input: A data stream DS, a pre-defined utility table, a
       user defined minimum utility threshold minUtil and
       a user specified sliding window size m
Output: Set of all high utility patterns
1 RHUI_SW-Tree ← Φ
2 Initialize sliding window size w ← m
3 PTable ← Φ
4 while a new transaction T_i arrives into DS do
5     for each transaction T_i with items n do
6           generate 2^n - 1 itemsets
7           calculate utility u of each itemset
8           while j ← 1 to 2^n - 1 do
9                 for each itemset X do
10                      call RHUI_SW-Tree-Update(X, u)
11                      store all updated nodes
                        information in PTable
12    call RHUI_SW-Tree-IncUpdate(RHUI_SW-Tree)
13    call RHUI_SW-Tree-Update(X, u)
14    store all updated nodes information in PTable
15 if user-request = true then
16    call RHUI_SW-Tree-Tracing (RHUI_SW-Tree)
17 return RHUI

Procedure 1: RHUI_SW-Tree-Update(X, u, RHUI_SW-Tree)
1 n ← size(X); X = {x_1, x_2, … x_n};
2 while l ← 1 to n do
3 search for x_i in children list of current node of
  HUI_SW-Tree;
4 if search = true then
5     case 1 search true and loop terminated, means
      patten X already exist so add utility to total
      utility and enqueue utility;
6     case 2 search true but not terminated means prefix
      itemset is already exist so continue create new
      pattern along that path;
7 else
8     make new node by creating a path from root node and
      set its utility;


Procedure 2: RHUI_SW-Tree-IncUpdate (PTable, RHUI_SW-Tree)
1 for i ← 1 to PTable.size()(i.e.,number of columns) do
2     for j ← 1 to PTable(i).size()(i.e., number of
                entries in ith column) do
3           nodelink ← PTable (i,j) delete corresponding
            pattern utility node which pointing nodelink
4 call RHUI_SW-Tree-Update(X, u, RHUI_SW-Tree)


Procedure 3: RHUI_SW-Tree-Tracing (HTable, RHUI_SW-Tree)
1 for index   1 to HTable.size() do
2     nodeline   HTable.get(index);
3     visit corresponding node of nodelink;
4     return pattern with utility;
```

5 Experimental Results

The experiments were performed on a PC with processor Intel(R) Core^{TM} i7 2600 CPU @ 3.40 GHZ, 2 GB Memory and the operating systems is Microsoft Windows 7 32-bit. All algorithms are implemented in Java. All testing data was generated by the synthetic data generator provided by Agrawal et al. in [4]. However, the IBM generator generates only the quantity of 0 or 1 for each item in a transaction. In order to adapt the dataset into the scenario of high utility pattern mining, the quantity of each item is generated randomly ranging from 1 to 5, and profit of each item is also generated randomly ranging from 1 to 20. To fit in the real time scenario we have generated the profits by following lognormal distribution (Fig. 2). In simulation of results we have fixed the sliding window size to 1,000 for all the experiments. Different data sets are used to compare the performance of our proposed algorithm against THUI-Mine, MHUI-TID.

5.1 Performance of $RHUIM_{SW}$ with Varying Minimum Utility Threshold

In this section, we show the performance of our proposed algorithm $RHUIM_{SW}$ over the dataset D50KT5N1000, parameters D, T and N represents the number of transactions, average number of items per transaction and total number of distinct items respectively. We have considered minimum utility threshold range from 0.5 to 1.5 %. From Fig. 3, we can observe that the performance of all algorithms increasing with minimum utility thresholds, because when the utility threshold is high then the number of high utility patterns is very low. From Fig. 3, we can

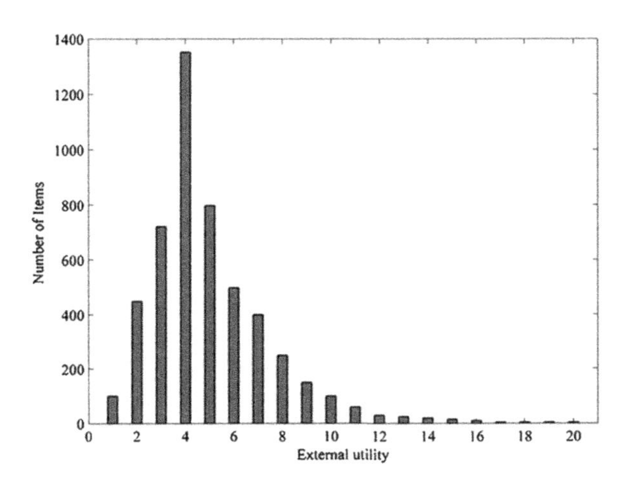

Fig. 2 Performance with varying minimum utility threshold

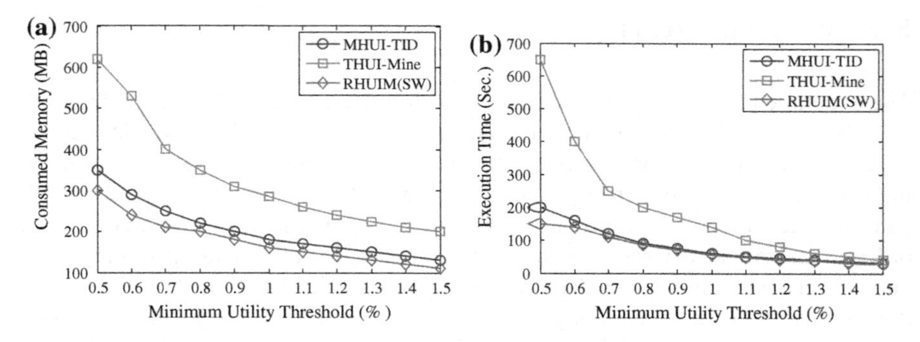

Fig. 3 Performance with varying minimum utility threshold. **a** Consumed memory. **b** Execution time

observe that proposed algorithm consumed less amount of memory than the remaining algorithms. There is some correlation in the amount of memory consumed by the MHUI-TID and proposed algorithm that we can observe in Fig. 3. The reason behind that is both the algorithms are using tree structure to maintain all the potential patterns. Even though, the execution time is very less among the three algorithms; our proposed algorithm does not give much better performance over MHUI-TID.

5.2 Performance of RHUIM$_{SW}$ with Varying Number of Transactions

In this section, we compare the performances of algorithms over the dataset DxKT5N1000, where x takes the values from 50 to 600. For all experiments we have set the minimum utility threshold to 0.9 %. In this experiment we have shown the scalability performance of algorithms with varying number of transactions. From Fig. 4, we can observe that the scalability of MHUI-TID and our proposed far better than the THUI-Mine algorithm.

5.3 Performance of RHUIM$_{SW}$ with Varying Parameters

In this section, the performance of our proposed algorithm is compared against MHUI-TID and THUI-Mine algorithms with varying average number of items per transaction and number of items. From Fig. 5, we can observe that execution time exponentially increasing with parameter especially from 6 and above values. From these results, we have understood that when the data set becomes dense the

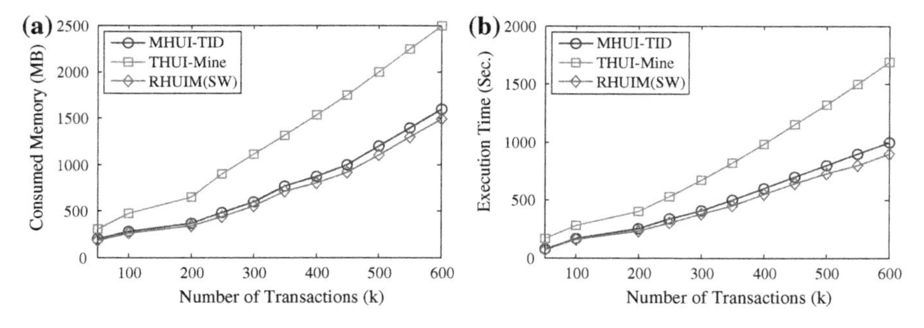

Fig. 4 Performance with varying number of transactions. **a** Consumed memory. **b** Execution time

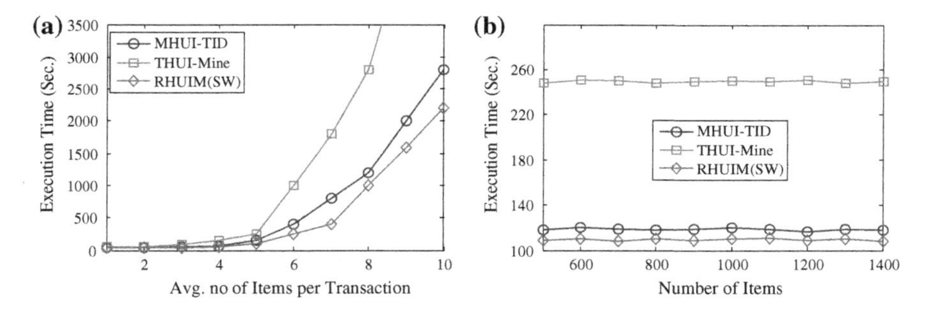

Fig. 5 Performance with varying parameters. **a** Avg. no of items per transactions. **b** No of items

performance completely worst. In this case, MHUI-TID and our algorithm performance is somewhat better than the THUI-Mine algorithm. This is achieved due to patterns prefix sharing in used tree structure. From Fig. 5, we can observe that the parameter, number of distinct items, will not make any significant effect on all algorithms.

6 Conclusions

In this paper, we have proposed an algorithm based on sliding window called, $RHUIM_{SW}$ for mining recent high utility itemsets over data streams. We have build PTable and HTable including the $RHUI_{SW}$-Tree to keep the information captured from sliding window and to make incremental update and tracing process faster. We have compared the performance against MHUI-TID and THUI-Mine. Experimental results shown that our approach is achieved good performance over THUI-Mine and MHUI-TID in terms of execution time and memory consumption.

References

1. Shen, Y.D., Zhang, Z., Yang, Q.: Objective-oriented utility-based association mining. In: Proceedings of 2002 IEEE International Conference on Data Mining, 2002, IEEE. ICDM 2002, pp. 426–433 (2002).
2. Yao, H., Hamilton, H.J., Butz, C.J.: A foundational approach to mining itemset utilities from databases. In: The 4th SIAM International Conference on Data Mining, pp. 482–486 (2004)
3. Yao, H., Hamilton, H.J.: Mining itemset utilities from transaction databases. Data Knowl. Eng. **59**(3), 603–626 (2006)
4. Agrawal, R., Srikant, R., et al.: Fast algorithms for mining association rules. In: Proceedings of 20th International Conference on Very Large Data Bases, VLDB, vol. 1215, pp. 487–499 (1994)
5. Liu, Y., Liao, W.K., Choudhary, A.: A two-phase algorithm for fast discovery of high utility itemsets. In: Ho, T., Cheung, D., Liu, H. (eds.) Advances in Knowledge Discovery and Data Mining, pp. 689–695. Springer, New York (2005)
6. Liu, Y., Liao, W.K., Choudhary, A.: A fast high utility itemsets mining algorithm. In: Proceedings of the 1st International Workshop on Utility-Based Data Mining, ACM, pp. 90–99 (2005)
7. Jiang, N., Gruenwald, L.: Research issues in data stream association rule mining. ACM Sigmod Rec. **35**(1), 14–19 (2006)
8. Cheung, D.W., Han, J., Ng, V.T., Wong, C.: Maintenance of discovered association rules in large databases: an incremental updating technique. In: Proceedings of the Twelfth International Conference on Data Engineering, IEEE, 1996, pp. 106–114 (1996)
9. Deypir, M., Sadreddini, M.H., Hashemi, S.: Towards a variable size sliding window model for frequent itemset mining over data streams. Comput. Ind. Eng. **63**(1), 161–172 (2012)
10. Chi, Y., Wang, H., Yu, P.S., Muntz, R.R.: Moment: maintaining closed frequent itemsets over a stream sliding window. In: Fourth IEEE International Conference on Data Mining, 2004, IEEE. ICDM'04, pp. 59–66 (2004)
11. Golab, L., Ozsu, M.T.: Issues in data stream management. ACM Sigmod Rec. **32**(2), 5–14 (2003)
12. Gaber, M.M., Zaslavsky, A., Krishnaswamy, S.: Mining data streams: a review. ACM Sigmod Rec **34**(2), 18–26 (2005)
13. Lee, C.H., Lin, C.R., Chen, M.S.: Sliding-window filtering: an efficient algorithm for incremental mining. In: Proceedings of the Tenth International Conference on Information and Knowledge Management, ACM, pp. 263–270 (2001)
14. Chan, R., Yang, Q., Shen, Y.D.: Mining high utility itemsets. In: Third IEEE International Conference on Data Mining, 2003, IEEE. ICDM 2003, pp. 19–26 (2003)
15. Chu, C.J., Tseng, V.S., Liang, T.: An efficient algorithm for mining temporal high utility itemsets from data streams. J. Syst. Softw. **81**(7), 1105–1117 (2008)
16. Li, H.F., Huang, H.Y., Chen, Y.C., Liu, Y.J., Lee, S.Y.: Fast and memory efficient mining of high utility itemsets in data streams. In: Eighth IEEE International Conference on Data Mining, 2008, IEEE. ICDM'08, pp. 881–886 (2008)

Important Author Analysis in Research Professionals' Relationship Network Based on Social Network Analysis Metrics

Manoj Kumar Pandia and Anand Bihari

Abstract Important author analysis is one of the key issues in the research professionals' relationship network. Research professionals' relationship network is a type of social network which is constitute of research professionals and there co-author relationship with other professionals. So many social network analysis metrics are available to analyze the important or prominent actor in the network. Centrality in social network analysis represents prestige or importance of a node with respect to other nodes in the network and also represents the importance of relationship between nodes. In this paper, we studied social network theory to understand how the collaboration of research professionals has impact in research world and performance of individual researcher. For this analysis, we use social network analysis metrics like normalize degree centrality, closeness centrality, betweenness centrality and eigenvector centrality.

Keywords Social network · Researcher relation · Centrality

1 Introduction

In research community, research professionals are involved in many research activities like publishing articles in Journal, conference and transaction. They are also involved in activities like publishing or editing books or chapters. These activities are accomplished individually by a single author or collaboratively by a

M.K. Pandia (✉)
Department of MCA, Silicon Institute of Technology, Bhubaneswar, Odisha, India
e-mail: manoj.pandia@gmail.com

A. Bihari
Department of CSE, Silicon Institute of Technology, Bhubaneswar, Odisha, India
e-mail: csanandk@gmail.com

L.C. Jain et al. (eds.), *Computational Intelligence in Data Mining - Volume 3*,
Smart Innovation, Systems and Technologies 33, DOI 10.1007/978-81-322-2202-6_16

group of authors. In a group of researcher, researchers may be from the same subject area or may be working in different subject areas. Research articles are helpful for enhancing technical or traditional education and also useful for new research. In research professionals' relationship network, one of the key analysis is to find the most influencing person in the research world, and also who is the most influencing person in a particular subject area. In this paper we discuss different type of social network analysis metrics for finding most influencing or prominent author in the professional network.

1.1 Social Network Analysis (Methodology)

A social network is a social structure made up of individuals (or organizations) called "nodes", which are tied (connected) by one or more specific types of interdependency, such as friendship, kinship, common interest, financial exchange, dislike, sexual relationships, or relationships of beliefs, knowledge or prestige [1–3].

Social network analysis views social relationships in terms of network theory consisting of nodes and ties (also called edges, links, or connections). Nodes are the individual actors within the network, and ties are the relationships between the actors. This representation allows researchers to apply graph theory to the analysis of social network for extracting some meaningful value [3]. There are two types of graphs commonly used in social network analysis: simple graphs and directed graphs. Simple graph connects nodes only if they have relationship with each other meanwhile directed graph uses arrows to indicate the direction if the relationship occurs from one to another [4, 5].

Social networks operate on many levels, from families up to the level of nations. They play a critical role in determining the way problems are solved, organizations are run, markets evolve, and the degree to which individuals succeed in achieving their goals. Social networks have been analyzed to identify areas of strengths and weaknesses within and among research organizations, businesses, and nations as well as to direct scientific development and funding policies [6–8].

In general, the benefit of analyzing social networks is that it can help people to understand how to share professional knowledge in an efficient way and to evaluate the performance of individuals, groups, or the entire social network [2].

1.2 Social Network Analysis Metrics for Important Author Analysis

Important or prominent actors are those that are linked or involved with other actors extensively. In the context of an organization, a person with extensive contacts (links) or communications with many other people in the organization is considered more important than a person with relatively fewer contacts. Thus, several types of social network analysis metrics are defined to find important or prominent actors in the social network. Here, we discuss degree centrality, closeness centrality betweenness centrality and eigenvector centrality.

Degree Centrality: Degree centrality of a node refers to the number of edges that are adjacent to this node [1, 9, 10]. Degree centrality represents the simplest instantiation of the notion of centrality since it measures only how many connection tie authors to their immediate neighbors in the network [3, 11]. Therefore important nodes usually have high degree. The normalized degree centrality is defined as the number of links of an actor divided by the maximal possible number [12–14]. The normalized degree centrality d_i of node 'i' is given as:

$$d_i = \frac{\mathrm{d(i)}}{(\mathrm{n}-1)} \quad (1)$$

Application: When a network refers to friendship network means the node represents a person and edge represents the relationship between the persons. Then the degree centrality is useful for finding more influence or important or prominent person in the network means the highest no of friends.

Closeness Centrality: Closeness measures how close an individual is to all other individual in a network, directly or indirectly [1, 2]. This metric can only be used in the connected network which means every node has at least one path to the other [15, 16]. The closeness centrality $C_c(n)$ of a node n is defined as the reciprocal of the average shortest path length and is computed as follows:

$$\mathrm{C_c(n)} = 1/\mathrm{avg(L(n,\ m))} \quad (2)$$

where L(n, m) is the length of the shortest path between two nodes n and m. The closeness centrality of each node is a number between 0 and 1 [17, 18]. Closeness centrality is a measure of how fast information spreads from a given node to other reachable node in the network [8, 19].

Application: A network refers to information flow network means information exchange between the sections. The closeness centrality is useful to find more influence node in the network which is nearest to average no of node.

Betweenness Centrality: Betweenness centrality is an indicator of an actor's potential control of communication within the networks [1, 3, 9]. Betweenness centrality is defined as the ratio of the number of shortest paths (between all pairs of nodes) that pass through a given node divided by the total number of shortest paths. The betweenness centrality $C_b(n)$ of a node n is computed as follows:

$$C_b(n) = \sum_{s \neq n \neq t} \left(\frac{\sigma_{st}(n)}{\sigma_{st}} \right) \quad (3)$$

where s and t are nodes in the network different from n, σ_{st} denotes the number of shortest paths from s to t, and $\sigma_{st}(n)$ is the number of shortest paths from s to t that n lies on. The betweenness value for each node n is normalized by dividing by the number of node pair excluding n: (N−1)(N−2)/2, where N is the total number of nodes in the connected component that n belongs to. Thus, the betweenness centrality of each node is a number between 0 and 1 [7, 20–22].

Eigenvector Centrality: Eigenvector centrality is a measure of the influence of a node in a network [23]. Eigenvector centrality is extension of degree centrality. In degree centrality, the degree centrality of a node is simply total no of node connected to that node but in Eigenvector centrality, the eigenvector centrality of a node is dependent on eigenvector centrality of neighbors' node [16]. In Eigenvector centrality not only the neighbors are important but also the score (eigenvector centrality) of the neighbors. A node has high eigenvector centrality it has many neighbors with high eigenvector centrality value than equal neighbors with low eigenvector centrality. Eigenvector centrality is simply dominant of eigenvector of the matrix. Philip Bonacich proposed eigenvector centrality in 1987 and Google's PageRank is a variant of it [24–26].

Eigenvector centrality *based on adjacency matrix*: For a given graph G = (V, E) with |V| number of vertices. Let $A = (a_{v,t})$ be the adjacency matrix of a graph G is the square matrix with rows and columns labeled by the vertices and entries [25, 27].

$$\text{i.e. } a_{v,t} = \begin{cases} 1, & \textit{if vertex v is linked to vertex t} \\ 0, & \textit{otherwise} \end{cases} \tag{4}$$

If we denote the centrality score of vertex v by X_v, then we can allow for this effect by making X_v proportional to the average of the centralities of v's network neighbors:

$$X_v = \frac{1}{\lambda} \sum_{t \in M(v)} X_t \quad = \frac{1}{\lambda} \sum_{t \in G} a_{v,t} X_t \tag{5}$$

where M(v) is a set of the neighbors of v and λ is a constant. With a small rearrangement this can be written in vector notation as the Eigenvector equation

$$AX = \lambda X \tag{6}$$

Hence we see that X is an eigenvector of the adjacency matrix of A with eigenvalue λ and λ must be the largest eigenvalue (using the Perron–Frobenius theorem) of adjacency matrix A and X is the corresponding eigenvector [25, 27].

The Eigenvector centrality defines in this way accords each node centrality depends on both numbers of connection and quality of its connection. The Eigenvector centrality of a node is high either it has large number of connections or it has less number of connections with high scoring node [22, 28].

The gist of the eigenvector centrality is to compute the centrality of node as a function of the centralities of its neighbors [25].

Eigenvector and Eigenvalue: An eigenvector of a square matrix A is a non-zero vector v that when the matrix is multiplied by v, yields a constant multiple of v, the multiplier being commonly denoted by λ. That is

$$Av = \lambda v. \tag{7}$$

The number λ is called eigenvalue of A corresponding to v.

2 Data Collection and Cleansing

For this study, we collect research articles details from IEEE which is published during year Jan. 2000–Jan. 2014 in different conferences and journals. Firstly, we search for articles in IEEE Explore, topic wise and then export search result. IEEE provide the search result in CSV format which contains following information "Document Title, Authors, Author Affiliations, Publication Title, Publication Date, Publication Year, Volume, Issue, Start Page, End Page, Abstract, ISSN, ISBN, EISBN, DOI, PDF Link, Author Keywords, IEEE Terms, INSPEC Controlled Terms, INSPEC Non-Controlled Terms, DOE Terms, PACS Terms, MeSH Terms, Article Citation Count, Patent Citation Count, Reference Count, Copyright Year, Online Date, Date Added To Xplore, Meeting Date, Publisher, Sponsors, Document Identifier".

We select only Document Title, Authors, Publication Title, Paper citation count and Document Identifier. After that we extract authors of papers. A paper has one or more author which is stored in authors' column and each name is separated by semicolon. So, here we split the Author's name based on the delimiter ";" and store each author name in separate column. We found that a paper has minimum 1 author and maximum of 59 authors.

Data cleaning has become necessary in order to allow processing of the extracted publication data. Most of the cleaning was due to the different types of language are used in author's naming in the publication. After cleaning of publication data, 26,802 publication and 61,546 authors were finally available for our analysis which is under all 16 subject category. After that we extract list of all journals and conference from IEEE [29]. We found 331 journals and magazines and 17,487 conferences which are under all 16 subject category. Finally, we import all publication, Journal and magazines and conference details into the MySQL database and do the database operation for extracting information.

3 Our Work

Our objective is to find out important author in the "Research professionals' relationship network". Research professionals' relationship network is a type of social network. In social network there are different network analysis metrics are available for finding important or prominent actors in the network. So, here we use social network analysis metrics to analyze the important researcher and academic groups in "Research professionals' relationship network".

In this work, firstly we draw the research professionals' relationships network by using python and network [30]. After that we use the social network analysis metrics to analyze the network of research professionals and find out important or prominent actors in the network.

3.1 Research Professionals' Relationship Network

Research professionals are involved in many research activities like publishing articles in Journal and conference, attend different conference, publish books, edit books etc. Earlier most of the articles focus in one subject area, but now days many inter disciplinary research articles are published as authors from different subject area are working collaboratively.

Research professionals' relationship network is one of the most tangible and well documented forms of research collaboration. It is represented as a weighted undirected graph. In this graph a node represent the author or researcher and edge between two nodes represent the co-authorship between the authors and the weight of the edge is the total no of citation by the authors together. Relationship between two nodes is recognized by strong or weak by the weight of edge.

3.2 Generating the Research Professionals' Relationship Network

Research professionals' relationship network is drawn based on co-author; means an author have connection with those authors who have published articles in journal and conference together, publish and edit books together, publish articles in transaction together, which can indicate individuals' status and the scientific collaboration. Based on the available publication data of researcher, we can build a network matrix which is representing the relationship between researchers according to following technique: Let there are three papers: P1, P2 and P3. P1 and P2 are conference paper and P3 is journal. P1 has two authors' a1 and a2 and citation is 12, P2 has three authors' b1, a3 and b2 and citation is 2. P3 has five authors' c1, a3, b2, c2 and b1 and citation is 11. It shows like this: P1-{a1, a2} P2-{b1, a3, b2} P3-{c1, a3, b2, c2, b1} [3, 31–34].

After that we extract author and its co-author which are involved in their research work and calculate the relationship weight according to their total citation count of all papers which is published together. We link the author {a1–a2}, {b1–a3, b1–b2, a3–b2}, {c1–a3, c1–b2, c1–c2, c1–b1, a3–b2, a3–c2, a3–b1, b2–c2, b2–b1, c2–b1}.

After that we generate the network by using python and network [30]. The network is like in Fig. 1. After generating a network, we can start the analysis.

3.3 Important Author Analysis and Result

In this part our aim is to find out important author in the research professionals' relationship network. We generate researchers' relationship network based on available publication data and calculated the degree centrality, closeness centrality, betweenness centrality and eigenvector centrality. Author with high degree centrality are important since they have more relationship with the others. Authors who have a good closeness centrality are also significant, means an author is nearest to average no. of author. Authors who have a good betweenness centrality are also prominent and an author who have a good eigenvector centrality is also prominent in the network [35, 36] means an author have a connection with a high score eigenvector centrality authors. Now, we calculate the normalized degree centrality, closeness centrality, betweenness centrality and eigenvector centrality for all researchers by using python and network [30]. The top 20 results based on eigenvector centrality, for the Communication Networking and Broadcasting topic are shown in Table 1.

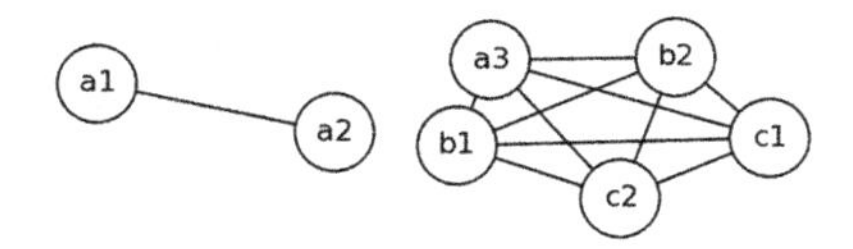

Fig. 1 Research professionals' relationship network an example

Table 1 Centrality report of top 20 research professionals

Sl	Name of authors	Degree centrality	Closeness centrality	Betweenness centrality	Eigenvector centrality
1	Candes, E.J.	0.0001074884	0.0001527466	0.0000000205	0.5773236046
2	Romberg, J.	0.0001433178	0.0002072990	0.0000000462	0.5773225899
3	Tao, T.	0.0000716589	0.0001451093	0.0000000000	0.5773225899
4	Loeliger, H.-A.	0.0000716589	0.0000716589	0.0000000000	0.0049250525
5	Frey, B.J.	0.0000716589	0.0000716589	0.0000000000	0.0049250525
6	Kschischang, F.R.	0.0000716589	0.0000716589	0.0000000000	0.0049250525
7	Plan, Y.	0.0000358295	0.0001074884	0.0000000000	0.0021034284
8	Falconer, D.	0.0002149767	0.0002149767	0.0000000231	0.0014050401
9	Ariyavisitakul, S.L.	0.0001074884	0.0001433178	0.0000000000	0.0014050284
10	Eidson, B.	0.0001074884	0.0001433178	0.0000000000	0.0014050284
11	Benyamin-Seeyar, A.	0.0001074884	0.0001433178	0.0000000000	0.0014050284
12	Rhea, S.C.	0.0001791473	0.0001791473	0.0000000000	0.0011241574
13	Stribling, J.	0.0001791473	0.0001791473	0.0000000000	0.0011241574
14	Joseph, A.D.	0.0001791473	0.0001791473	0.0000000000	0.0011241574
15	Zhao, B.Y.	0.0001791473	0.0001791473	0.0000000000	0.0011241574
16	Ling Huang	0.0001791473	0.0001791473	0.0000000000	0.0011241574
17	Kubiatowicz, J.D.	0.0001791473	0.0001791473	0.0000000000	0.0011241574
18	Rodriguez, J.	0.0006449301	0.0158868470	0.0001727501	0.0004759009
19	Bin Wu	0.0010032247	0.0171199200	0.0012234019	0.0004752933
20	Franquelo, L.G.	0.0003582945	0.0158851567	0.0000314376	0.0004748420

4 Conclusion

In this paper we have investigated the research professionals' relationship network and find out the prominent actor in the network based on different social network analysis metrics like normalized degree centrality, normalized closeness centrality, normalized betweenness centrality and eigenvector centrality. We found that different social network analysis metrics gives different value for each researcher. It means a researcher who may be most prominent in network based on degree centrality; he/she does not guarantee to be prominent author based on another centrality.

If we compare the different centrality, we can say that eigenvector centrality is more useful for finding important author in research professionals' relationship network, because it is based on direct linked researcher eigenvector centrality. This technique assigns a centrality value to an author, which depends not only on the number of links (degree) but also the quality of its connections. So an author may have a less degree but can have a probability of high centrality because of it prestige. In the research community an author may have less number of publications, but if the author has published articles with many other important authors, then that will be considers as more important. So eigenvector centrality is more suited for analyzing the relationship network in the research community.

References

1. Liu, B.: Web Data Mining. Springer International Edition, New York (2006)
2. Alireza Abbasi, J.A.: On the correlation between research performance and social network analysis measure applied to scholars collaboration networks, In: Proceedings HICSS (2011)
3. Liu, X., Bollen, J., Nelson, L.M., Sompel, H.: Co-authorship networks in the digital library research community, information processing and management. Int. Process. Manag. **41**, 1462–1480 (2005)
4. Jie Tang, D.Z., Yao, L.: Social network extraction of academic researchers. In: IEEE International Conference on Data Mining (ICDM) (2007)
5. Said, E.J.W.Y.H., Sharabati, W.K.: Retracted: social networks of author coauthor relationships. Comput. Stat. Data Anal. **52,** 2177–2184 (2005)
6. Carlos, D., Correa, T.C., Ma, K.-L.: Visual reasoning about social networks using centrality, sensitivity. IEEE Trans. Vis. Comput. Graph. **18**(1) (2012)
7. Landherr, A., Friedl, B., Heidemann, J.: A critical review of centrality measures in social networks. Bus. Inf. Syst. Eng. **2**(6), 371–385 (2010)
8. Wasserman, S., Faust, K.: Social network analysis: methods and applications. Cambridge University Press, New York (1994)
9. Miray Kas, L.R.C., Carley, K.M.: Monitoring social centrality for peer-to-peer network protection. IEEE Commun. Mag. **51**, 151–161 (2013)
10. Friedl, B., Landherr, A., Heidemann, J.: A critical review of centrality measures in social networks. BISE—STATE OF THE ART 371–385 (2009)
11. Shams-ul Arfeen, J.M., Kazi, A., Hyder, S.: Automatic co-authorship network extraction and discovery of central authors. Int. J. Comput. Appl. (0975 8887) **74**(4), 1–6 (2013)
12. Deng, Q., Wang, Z.: Degree centrality in scientific collaboration supernetwork. In: International Conference on Information Science and Technology, pp. 259–262. 26–28 March 2011
13. Erkan, G., Radev, D.R.: LexRank: graph-based lexical centrality as salience in text summarization. J. Artif. Intell. Res. **22,** 457–479 (2004)
14. Borgatti, S.P.: Centrality and AIDS. Connections **18**(1), 112–114
15. Jin, J., Liu, Y., Xu, K., Xiong, N., Li, G.: Multi-index evaluation algorithm based on principal component analysis for node importance in complex networks. Inst. Eng. Technol. **1**(3), 108–115 (2012)
16. Carlos, K.-L.M., Correa, D., Crnovrsanin, T., Keeton, K.: The derivatives of centrality and their applications in visualizing social networks
17. Borgatti, S.P., Everett, M.G.: A graph-theoretic perspective on centrality. Elsevier Soc. Netw. 1–19 (2005)
18. Daniel, G., Figueira, J.R., Eusébio, A.: Modeling centrality measures in social network analysis using bi-criteria network flow optimization problems. Eur. J. Oper. Res. **226,** 354–365 (2013)
19. Borgatti, S.P.: Centrality and network flow. Soc. Netw. **27**(1), 55–71 (2005)
20. Estrada, E., Rodríguez-Velázquez, J.A.: Subgraph centrality in complex networks, 1–29
21. Wang, G., Shen, Y., Luan, E.: A measure of centrality based on modularity matrix. Prog. Nat. Sci. **18**, 104–1043 (2008) (Elsevier)
22. Lohmann, G., Margulies, D.S., Horstmann, A., Pleger, B., Lepsien, J., Goldhahn, D., Schloegl, H., Stumvoll, M., Villringer, A., Turner, R.: Eigenvector centrality mapping for

analyzing connectivity patterns in fMRI data of the human brain. PLoS ONE **5**(4), 1–8 e10232 (2010). doi:10.1371/journal.pone.0010232
23. Spizzirri, L.: Justification and application of eigenvector centrality. Algebra in Geography: Eigenvectors of Network (2011)
24. Phillip Bonacich, P.L.: Eigenvector-like measures of centrality for asymmetric relations. Soc. Netw. **23,** 191–201 (2001) (Elsevier Press)
25. Newman, M.E.J.: The mathematics of networks
26. Ding, D., He, X.: Application of eigenvector centrality in metabolic networks. In: 2nd International Conference on Computer Engineering and Technology, vol. 1 (2010)
27. Straffin, P.D. Jr.: Linear algebra in geography: eigenvectors of networks. JSTOR Math. Mag. **53,** 269–276 (1980)
28. Kato, Y., Ono, F.: Node centrality on disjoint multipath routing. IEEE (2011)
29. http://ieeexplore.ieee.org/xpl/opac.jsp
30. Aric, D.A.S., Hagberg, A., Swart, P.J.: Exploring network structure, dynamics, and function using network. In: Proceedings of the 7th Python in Science Conference (scipy 2008) (2008)
31. Wang, B., Yang, J.: To form a smaller world in the research realm of hierarchical decision models. In: Proceedings PICMET'11 (2011)
32. Wang, B., Yao, X.: To form a smaller world in the research realm of hierarchical decision models. In: Proceedings of the 2011 IEEE IEEM', pp. 1784–1788 (2011)
33. Yulan, C.H.S.H.: Mining a web citation database for author co-citation analysis. Inf. Process. Manage. **38**, 491–508 (2002)
34. Newman, M.E.J.: Coauthorship networks and patterns of scientific collaboration. In: Proceedings of the National Academy of the United States of America (2004)
35. Said, Y.H., Wegman, E.J., Sharabati, W.K.: RETRACTED: social networks of author-coauthor relationships. Comput. Stat. Data Anal. **52**(4), 2177–2184 (2008)
36. Yulan, H., Siu, C. H.: Mining a web citation database for author co-citation analysis. Inf. Process. Manage. **38**(4), 491–508 2002

Predicting Software Development Effort Using Tuned Artificial Neural Network

H.S. Hota, Ragini Shukla and S. Singhai

Abstract Software development effort prediction using fixed mathematical formulae is inadequate due to impreciseness and nonlinearity exist in the software project data and leads to high prediction error rate, on the other hand Artificial Neural Network (ANN) techniques are very popular for prediction of software development effort due to its capability to map non linear input with output. This paper explores Error Back Propagation Network (EBPN) for software development effort prediction, by tuning some algorithm specific parameters like learning rate and momentum. EBPN is trained with two benchmark data sets: China and Maxwell. Results are analyzed in terms of various measures and found to be satisfactory.

Keywords Software effort prediction · Error back propagation network (EBPN)

1 Introduction

Software effort estimation is an important need of software development companies for delivery of software in due time, man power management and planning which refers to the prediction of possible amount of time in person/month, require to develop software. This topic has attracted the researchers during the last decade.

H.S. Hota (✉)
Guru Ghasidas Vishwavidyalaya Bilaspur (C.G.), Bilaspur, India
e-mail: proffhota@gmail.com

R. Shukla
Dr. C.V. Raman University Kota Bilaspur (C.G.), Bilaspur, India
e-mail: raginishukla008@gmail.com

S. Singhai
Government Engineering College Bilaspur (C.G.), Bilaspur, India
e-mail: singhai_sanjay@yahoo.com

L.C. Jain et al. (eds.), *Computational Intelligence in Data Mining - Volume 3*,
Smart Innovation, Systems and Technologies 33, DOI 10.1007/978-81-322-2202-6_17

In the software development process, the major ability is to accurately estimate the effort and costs in the initial stage. Inaccurate estimation of efforts wastes the company's budget and can result in failure of the entire project. Software effort estimation model [1] can be divided into two categories: algorithmic and non-algorithmic. The most popular algorithmic estimation models include Constructive cost model (COCOMO) [2] developed by Bohem and Software life cycle management (SLIM) model developed by Putnam, these models are based on fixed mathematical formula and both [3] take number of line of code as major input to their model. Line of code is difficult to obtain in early stage of a software development, hence these techniques are not so efficient also algorithmic model [3] are not able to provide suitable solution because they are often unable to capture non linear relationship in between input and output, on the other hand non-algorithm models are becoming popular. Soft computing based approach like ANN, Fuzzy Logic (FL) and its combination are widely used for software development effort prediction.

Many studies shows that soft computing technique is one of the reliable and efficient technique as compare to other traditional techniques. Kaur and Verma [4] have described the efficiency of neural networks based cost estimation model along with many traditional models like halsted model, Bailey-Basili model and Doty model. Reddy and KVSVN [5] have concerned with constructing software effort estimation model based on artificial neural networks. The model is designed accordingly to improve the performance of the network that suits to the COCOMO model. The network is trained with back propagation learning algorithm and resilient back propagation algorithm (RPROP) by iteratively processing a set of training samples and comparing the network's prediction with the actual effort. COCOMO dataset is used to train and to test the network and it was observed that proposed neural network model improves the estimation accuracy. Bhatnagar et al. [6] have investigated a novel approach towards the effort estimations using three different neural network models: Feed forward back propagation neural network (FFBPNN), cascade forward back propagation neural network (CFBPNN) and layer recurrent neural network (LRNN), results are compared based on the standard evaluation criteria such as Mean Magnitude Relative error (MMRE), Magnitude Relative error(MRE), and Balance Relative error (BRE). Heejun and Baek [7] have proposed and evaluated a neural network based software development estimation model which uses Function point and six input variables to be superior in the comparative evaluation of model performances. They showed that a neural network is useful in estimating effort. Sandhu et al. [8] have worked on Neuro-Fuzzy technique for software effort estimation using NASA software project data and the developed models are compared with the Halstead, Walston-Felix, Bailey-Basili and Doty models. The performance of the Neuro-Fuzzy based effort estimation Model and the other existing models are compared, the results show that the Neuro-Fuzzy system has the lowest MMRE and RMSSE values. Kaur et al. [9] have performed comparative analysis in between existing mathematical models such as Halstead model, Walston-Felix model, Bailey-Basili model, Doty model and Neural network based model. It is observed that NN has outperformed the other models.

Prabhakar and Dutt [10] have developed model to estimate software effort using various machine learning techniques like ANN, Adaptive Neuro-Fuzzy inference system (ANFIS) and decision tree (DT). These techniques have been used to analyze the results using China dataset. Tronto et al. [11], have compared a neural network based method with the regression approach to accomplish software effort estimation. ANN and multiple regressions were applied to COCOMO data set and results estimated by ANN are found better than others.

This paper utilizes EBPN technique to predict software development effort by tuning two algorithm specific parameters: Learning rate and momentum. Although EBPN has many features but may face problem of local minima and network paralysis. A higher or lower value of learning rate may create problem, hence another parameter momentum can be used to overcome these problems. In order to find out optimum values of these parameters, trial and error method is applied. EBPN is tested with two benchmark data sets: China and Maxwell.

To deal with the subject in detail, this paper is organized in the following order:

Sections 2 and 3 contains a description of the benchmark data sets and EBPN as prediction model while Sect. 4 discusses the experimental results obtained and at last, conclusion are drawn.

2 Benchmark Data Sets

There are many benchmark data sets available in the UCI repository site related to software effort. Two data sets used in this study are downloaded [12] from PROMISE (Predictor models in software engineering) data repository site popularly known as China and Maxwell data sets. These data sets have many features related to software development effort of numeric type. Detail of the data sets used for empirical experiment are shown in Table 1. As described in table, china data set has 499 instances with 19 features while Maxwell data set has 62 instances with 27 features. Maxwell data set has less number of instances as compare to China data set but with higher number of features as compare to China data set. These unbalanced feature and instances may influence ANN based model, since ANN based models are highly depends upon sample size. In order to feed these data sets to EBPN, it is divided into two parts as training and testing.

Table 1 Detail of data sets used for software effort estimation

Data set	No. of instance	No. of feature	Training data percentage	Testing data percentage
China	499	19	80	20
Maxwell	62	27	80	20

3 Error Back Propagation Network

EBPN is a kind of NN popularly used as predictor due to its capability of mapping high dimensional data. Error back propagation algorithm is used with EBPN which is one of the most important developments in neural networks. This network has re-awakened the scientific and engineering community to the modeling and processing of numerous quantitative phenomena. This learning algorithm is applied to multi-layer feed-forward networks consisting of processing element with continuous differentiable activation functions. For a given set of training input-output pair, this algorithm provides a procedure for changing the weights in neural network to predict the given input pattern correctly. The basic concept for this weight update algorithm is simply the gradient descent method.This is a method where the error is propagated back to the hidden unit. While designing ANN [13] for a specific problem,we have to decide number of neurons in hidden layer and number of hidden layer also, some other parameters like learning rate and momentum are to be optimized. EBPN faces problem of local minima and network paralysis, in order to overcome these a momentum factor can be added. Since there is no thumb rule to decide number of neurons in hidden layer hence models were tested by including neurons in hidden layer randomly. Effect of other parameters considered in the study are as below:

Learning rate—The learning rate, which controls how much the weights are adjusted at each update.

Momentum—A momentum term used in updating the weights during training which tends to keep the weight changes moving in a consistent direction and specify a value between 0 and 1.

An ANN designed to predict software effort using china data with 18 neurons in input layer 15 neurons in hidden layer and one neuron in output layer is shown in Fig. 1. Similarly an ANN for predicting software effort using Maxwell data is shown in Fig. 2.

4 Experimental Work

4.1 Performance Evaluation

Several measures used for performance evaluation of EBPN model are described below:

a. Mean Absolute Error (MAE): Is an average of the absolute errors which measures of how far the estimates are from actual values. It could be applied to any two pairs of numbers, where one set is “actual” and the other is an estimate prediction.

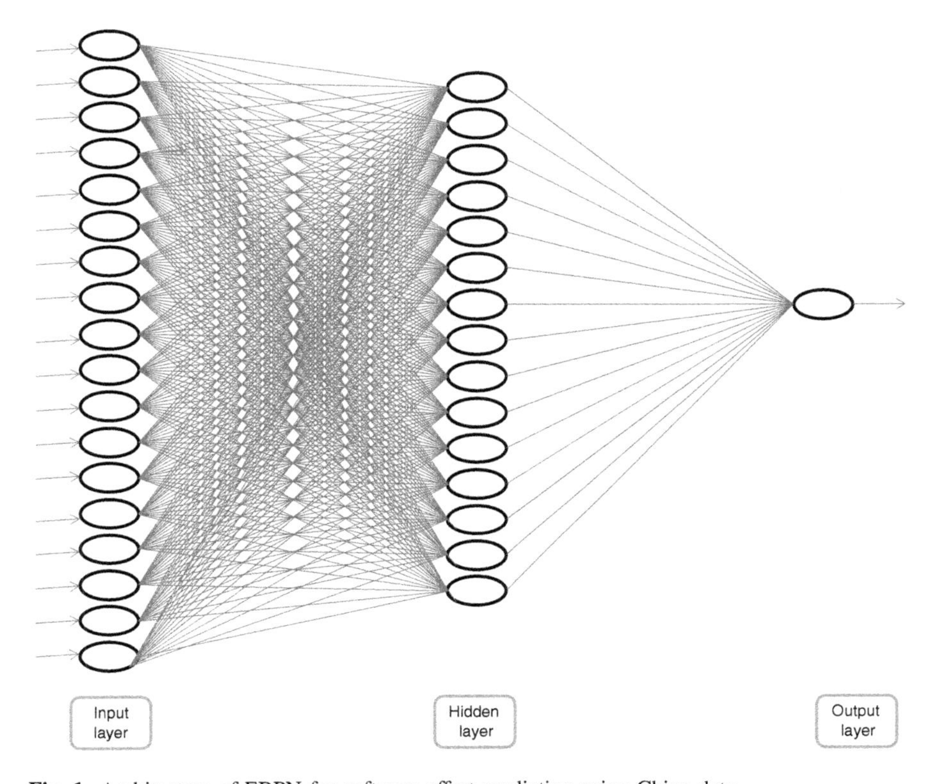

Fig. 1 Architecture of EBPN for software effort prediction using China data

$$\mathrm{MAE} = \frac{1}{n}\sum\nolimits_{i=1}^{n} |Pi - Ai| \tag{1}$$

b. Mean Absolute Percentage Error (MAPE): Used for accuracy estimation based on following formulae.

$$\mathrm{MAPE} = \frac{100}{n}\sum_{i=1}^{n} \frac{|Pi - Ai|}{Ai} \tag{2}$$

c. Mean Square Error (MSE): Gives high importance to large errors because the errors are squared before they are averaged. Thus MSE is used when large errors are undesirable.

$$\mathrm{MSE} = \left[\frac{1}{n}\sum_{i=1}^{n} (Pi - Ai)^2\right] \tag{3}$$

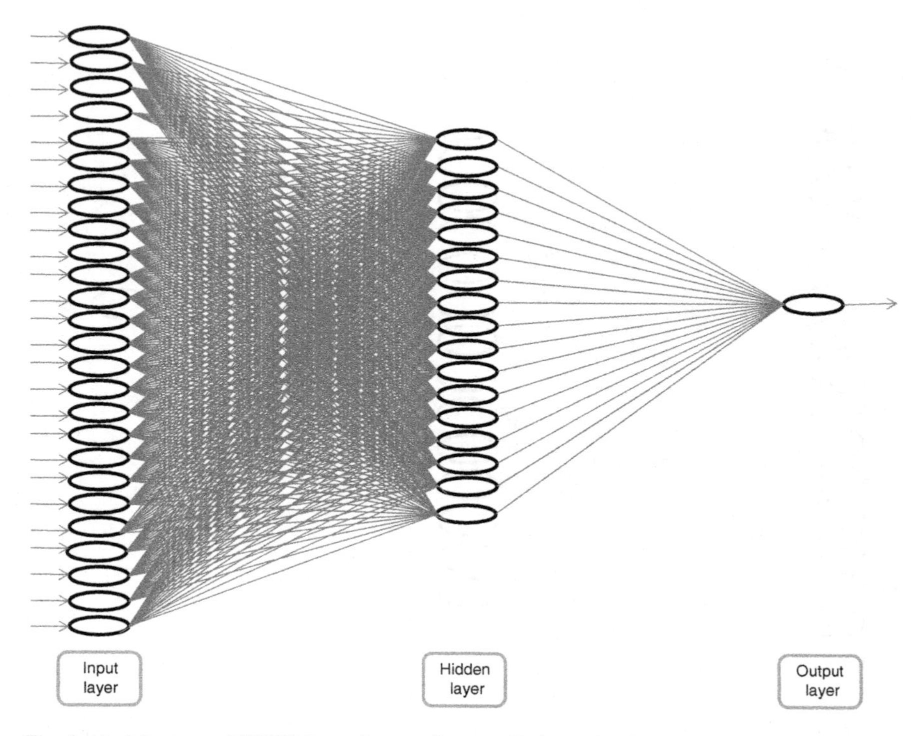

Fig. 2 Architecture of EBPN for software effort prediction using Maxwell data

d. Root Mean Square Error (RMSE): Is a frequently used measure of differences between values predicted by a model or estimator and the values actually observed from the thing being modeled or estimated. It is just the square root of the mean square error as shown in equation given below:

$$\text{RMSE} = \left[\sqrt{\frac{1}{n}\sum_{i=1}^{n}(Pi - Ai)^2}\right] \tag{4}$$

where Pi = Predicted value for data point i, Ai = Actual value for data point i, n = Total number of data points.

4.2 Result Discussion

Experimental work is carried out with WEKA open source data mining software which provides interactive way to develop and evaluate model. EBPN were trained and tested with two benchmark data sets. Numeric value of data were normalized in

Table 2 A comparative results obtained in case of China data set

	Neurons in hidden layer	Learning rate	Momentum	MAE	MAPE	MSE	RMSE
Training data	20	0.9	0.3	0.014	52.758	0.000	0.022
	17	0.9	0.2	0.015	50.955	0.000	0.029
	15	**0.9**	**0.1**	**0.011**	**39.455**	**0.000**	**0.020**
	13	0.5	0.3	0.038	150.39	0.004	0.070
	11	0.5	0.2	0.043	193.43	0.005	0.074
Testing data	20	0.9	0.3	0.041	184.87	0.006	0.082
	17	0.9	0.2	0.037	221.10	0.003	0.061
	15	**0.9**	**0.1**	**0.012**	**42.142**	**0.001**	**0.036**
	13	0.5	0.3	0.032	176.40	0.002	0.049
	11	0.5	0.2	0.042	281.34	0.004	0.064

between 0 and 1, so that there will be equal influence of each attribute during training of EBPN. A fixed ratio of training and testing samples were chosen randomly, training data are used to construct prediction model while testing data are unseen data and used to check the efficiency of model. In both the cases predicted values are used to calculate MAE, MAPE, MSE and RMSE using Eqs. (1), (2), (3) and (4) respectively and presented in Table 2 and 3 respectively for china and Maxwell data. One can see that results are consistent at training and testing stages for both the data sets. A lowest MAPE (As highlighted) obtained are 39.45 and 42.12 respectively at training and testing stages in case of china data set while these are 43.76 and 47.15 respectively at training and testing stages in case of Maxwell data set. It can also be observed that learning rate, momentum and number of neurons in hidden layer have influenced the efficiency of model.

Table 3 A comparative results obtained in case of Maxwell data set

Data partition	Neurons in hidden layer	Learning rate	Momentum	MAE	MAPE	MSE	RMSE
Training data	20	0.9	0.3	0.045	70.564	0.0049	0.070
	17	**0.9**	**0.2**	**0.031**	**43.761**	**0.001**	**0.043**
	15	0.9	0.1	0.034	51.186	0.0019	0.137
	13	0.5	0.3	0.041	54.369	0.0052	0.072
	11	0.5	0.2	0.045	72.574	0.0048	0.069
Testing data	20	0.9	0.3	0.054	60.062	0.015	0.123
	17	**0.9**	**0.2**	**0.028**	**47.159**	**0.001**	**0.042**
	15	0.9	0.1	0.053	142.13	0.003	0.061
	13	0.5	0.3	0.080	188.22	0.017	0.133
	11	0.5	0.2	0.029	60.559	0.001	0.044

5 Conclusion

Prediction of software effort is an important issue for software development company for proper planning and also to deliver software product in due time to the clients. Due to drawbacks of algorithmic model for prediction of software development effort non-algorithmic model based on soft computing techniques are widely accepted.

In this paper we have demonstrated the ability of ANN in form of EBPN for prediction of software development effort after tuning two algorithm specific parameters: Learning rate and Momentum. Number of hidden layer neurons are also optimized manually. Two benchmark data sets obtained from PROMISE data repository site were used to train and test the EBPN. EBPN with 0.9 and 0.1 learning rate and momentum respectively with 15 neurons in hidden layer has lowest value of MAE, MAPE, MSE and RMSE for china data set, similarly results are promising with learning rate = 0.9 and momentum = 0.2 with 17 neurons in hidden layer for Maxwell data set.

In future, for tuning various parameters of EBPN and for feature selection of benchmark data sets, Genetic algorithm (GA), Particle swarm optimization (PSO) and other optimization techniques can be used to enhance the results.

References

1. Pressman, R.S.: Software Engineering. McGraw Hill Higher Education, New York (2005)
2. Jalote, P.: An Integrated Approach to Software Engineering. Narosa Publishing House, New Delhi (2005)
3. Moataz, A.A., Saliu, M.O., AlGhamdi, J.: Adaptive fuzzy logic-based framework for software development effort prediction. Inf. Softw. Technol. **47**, 31–48 (2005)
4. Kaur, M., Verma, M.: Computing MMRE and RMSSE of software efforts using neural network based approaches and comparing with traditional models of estimation. IJCST **3**(1), 316–318 (2012)
5. Reddy, Ch.S., KVSVN, R.: An optimal neural network model for software effort estimation. Int. J. Softw. Eng. IJSE **3**(1), 63–78 (2010)
6. Bhatnagar, R., Ghose, M.K., Bhattacharjee, V.: A novel approach to the early stage software development effort estimations using neural network models: a case study. In: IJCA Special Issue on Artificial Intelligence Techniques—Novel Approaches and practical Applications AIT, pp. 27–30 (2011)
7. Heejun, P., Baek, S.: An empirical validation of a neural network model for software effort estimation. Sci. Direct. Expert Syst. Appl. **35**, 929–937 (2008)
8. Sandhu, P.S., Bassi, P., Brar, A.S.: Software effort estimation using soft computing techniques. World Acad. Sci. Eng. Technol. **46**, 488–491 (2008)
9. Kaur, J., Singh, S., Kahlon, K.S.: Comparative Analysis of the Software Effort Estimation Models. World Acad. Sci. Eng. Technol. **46**, 485–487 (2008)
10. Prabhakar, Dutta, M.: Application of machine learning techniques for predicting software effort. Elixir Compu. Sci. Eng. **56**, 13677–13862 (2013)

11. Tronto, I.F.de.B., Silva, J.D.S.da, Anna, N.S.: An investigation of artificial neural networks based prediction systems in software project management. Sci. Direct. J. Syst. Softw. **81**, 356–367 (2008)
12. Menzies, T., Caglayan, B., Kocaguneli, E., Krall, J., Peters, F., Turhan, B.: The PROMISE repository of empirical software engineering data, West Virginia University, Department of Computer Science. http://promisedata.goolecode.com (2012)
13. Zurada, J.M.: Introduction to Artificial Neural Systems. Jaico Publishing House, Mumbai (2006)

Computer Vision Based Classification of Indian Gujarat-17 Rice Using Geometrical Features and Cart

Chetna V. Maheshwari, Niky K. Jain and Samrat Khanna

Abstract Agricultural production has to be kept up with an ever-increasing population demand which is an important issue in recent years. The paper presents a solution for quality evaluation, grading and classification of INDIAN Gujarat-17 Rice using Computer Vision and image processing. In this paper basic problem of rice industry for quality assessment is defined and Computer Vision provides one alternative for an automated, non-destructive and cost-effective technique. In this paper we quantify the quality of ORYZA SATIVA SSP Indica (Gujarat-17) based on features which affect the quality of the rice. Based on these features CART (Classification and Regression Tree Analysis) is being proposed to evaluate the rice quality.

Keywords Machine vision · Combined parameters · Oryza sativa SSP indica · Mining techniques, CART

1 Introduction

The past few years were marked by the development of researches that contribute to reach an automatic classification technique of rice varieties grown in various Asian sub continents. After hours of working the operator may loose concentration which in turn will affect the evaluation process. Computer Vision system proved to be more efficient at the level of precision and rapidity. But, the natural diversity in appearance

C.V. Maheshwari (✉)
G.H. Patel College of Engineering and Technology, V.V. Nagar, India
e-mail: cmaheshwari23@gmail.com

N.K. Jain · S. Khanna
ISTAR, V.V. Nagar 388120, India
e-mail: nikykhabya@yahoo.com

S. Khanna
e-mail: sonukhanna@yahoo.com

L.C. Jain et al. (eds.), *Computational Intelligence in Data Mining - Volume 3*,
Smart Innovation, Systems and Technologies 33, DOI 10.1007/978-81-322-2202-6_18

of various rice varieties makes classification by Computer Vision a complex work to achieve. Many researches carried out to various techniques for classification of grains. Characterization models were based on morphological features, color features [1, 2] or textural features [3]. Other researchers [4–7] have tried to combine these features for the sake of improving the efficiency of classification. Recently, wavelet technique was integrated in grains characterization [8, 9].

In this hi-tech uprising, an agricultural industry has become more intellectual and automatic machinery has replaced the human efforts [10]. At present, data mining technology is used only when classical mining algorithm is applied in agricultural data processing, not to mention a set of systematic research and development methods. To solve the above problems, this paper presents a framework for agricultural data mining platform, including various soft computing techniques used worldwide in the classification process of rice in the Asian subcontinent.

In 2006, Rani et al. examined Historical significance, grain quality features and precision breeding for enhancement of export quality Basmati varieties in India like Badshahbhog, Kalanamak, Ambemohar 159, Jeeraga Samba, Chttimutyalu etc. [9]. But the work was limited to individual seeds and occlusion affected problem. In 2010, Bhupinder Verma determined various techniques for Grading and Categorization of three different Indian rice varieties. For this purpose, he developed semi-automatic FBS procedure for classification and grading of these rice. But the results achieved were not accurate as the system lacked proper illumination [11]. In 2010, Shantaiya et al. developed digital image analysis algorithm based on color, morphological and textural features to identify the six varieties rice seeds in Chhattisgarh region [2].

In 2012, we developed digital image analysis algorithm based on geometrical features to identify long, small and normal rice seed varieties in a given sample of Basmati rice widely grown in Gujarat region [7]. With the help of this paper we propose a data mining technique using Computer Vision perception for the purpose of classification of rice quality.

This paper comprises of six sections. Section 2 refers to problem being faced by rice industry working in non-automated mode i.e. the milling process and further quality evaluation module is still missing in most of the industries. Section 3 summarizes the need to propose a framework for both consumer and distributor to evaluate the quality of grain being supplied or consumed and the results so achieved by the framework are being shown in Sect. 4. Section 5 concludes the paper by showing a classification technique accepted by the industrialist for the purpose of quality. Last section acknowledges the rice mills of the near-by regions to provide adequate samples.

2 Methodology

2.1 Imaging Setup

A schematic diagram of the proposed system is in Fig. 1. In our proposed system there is a camera, which is mounted on the top of the box at point 1 in Fig. 1. The camera is having 12-mega pixels quality with 8x optical zoom. To evade problem of luminance and for good quality of image, we used two lights at point 2 and 3 as in the Fig. 1. In our system box contains opening which can be seen at point 4. At point 5 which is a tray in which rice seeds will be inserted for capturing an image [7].

Gujarat-17 rice (Oryza sativa L) seed contains foreign elements in terms of long as well as small seed as shown in Fig. 2. The blue highlighted circles are long, red colored ones are too small and green highlighted circles are normal seed. These seeds help in quantifying the quality. Proper removal of this seed is necessary if it is not so then it creates degradation in quality of rice seed. This paper proposes a new method for counting the number of Gujarat-17 rice (Oryza sativa L) seeds with these foreign elements as shown in Fig. 3 using Computer Vision.

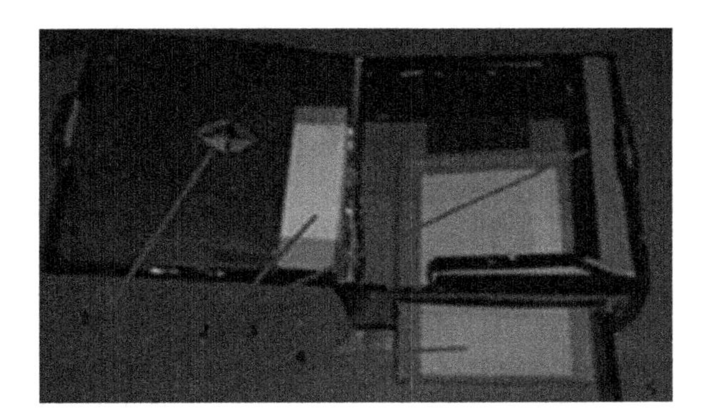

Fig. 1 Proposed computer vision system for analysis of seed [7]

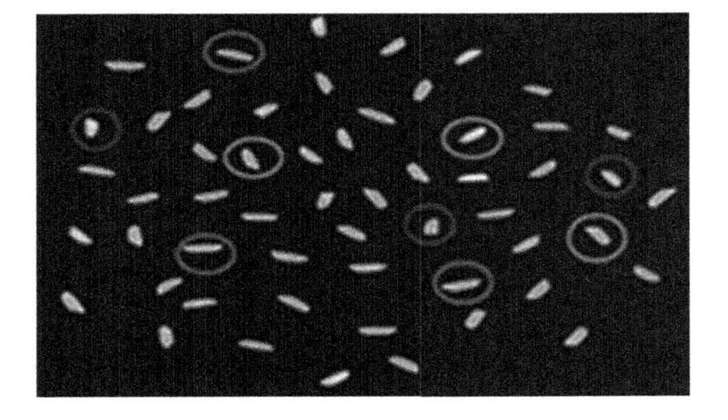

Fig. 2 Rice seed with and without foreign elements

Fig. 3 Normal, small and long rice seeds

2.1.1 Computer Vision

Computer Vision is a technology that merges mechanics, optical instrumentation, electromagnetic sensing, and digital input-output devices. It combines optical, infrared, ultra-violet, or X-ray sensing, with digital video technology and image processing. It is a technology for recognizing objects and extracting quantitative information from digital images to provide non-destructive quality based evaluation on image processing and analysis. Computer Vision systems are programmed to perform narrowly defined tasks such as counting objects on a conveyor, reading serial numbers, and searching for surface defects. Manufacturer's favors Computer Vision systems for visual inspections that require high-speed, high-magnification, 24-h operation, and/or repeatability of measurements. Computer Vision technology aims to emulate the function of human vision by electronically perceiving and evaluating an image [12]. Computer Vision is a immediate, financially viable, steady and objective inspection and evaluation technique which makes available one alternative for an automated, non-destructive and cost effective techniques for grain identification [13].

2.1.2 Cart

Classification identifies objects by classifying them into one of the finite sets of classes, which involves comparing the measured features of a new object with those of a known object or other known criteria and determining whether the new object belongs to a particular category of objects. A decision tree is a decision-modeling tool that graphically displays the classification process of a given input for given output class labels. Key components of a tree construction are the selection of splitting criteria, stopping criteria when a tree grows and the assignment of terminal nodes to a class. A general decision tree consists of one root node, a number of internal and leaf nodes, and branches. Leaf nodes indicate the class to be assigned to a sample. Each internal node of a tree corresponds to a feature, and branches represent conjunctions of features that lead to those classifications.

Classification and regression tree (CART) are ideally suited for the analysis of complex ecological data. As it is for discriminate analysis, the aim of CART is to classify a group of observations or a single observation into a subset of known classes. Comparing to classical parametric discriminate analysis CART offers a number of particular benefits like a high degree of results interpretability, high precision and fast computation. CART is a non-parametric tool of discriminate analysis, which is designed to represent decision rules in a form of so-called binary trees. Thus, CART is a special method of creating decision rules to distinguish between clusters of observations and determine the class of new observations. We use classification and CARTs to analyze survey data.

3 Result and Discussion

3.1 Proposed Algorithm to Detect Rice Seeds with Long and Small Seeds

According to our proposed algorithm as suggested in Table 1 first capture image of sample spread on the black or butter paper using camera as shown in Fig. 4.

This image is color image so we convert it into gray scale image as the color information is not of importance (Fig. 5). The identification of objects within an image is done by using an optimal edge detector, ISEF, for extracting edges of gray scale image. This phase identifies individual object boundaries and marks the center of each object for further processing. Thresholding is used to convert the segmented image to a binary image. The output binary image has values of 1 (White areas) for all pixels and 0 (black) for all other pixels as shown in Fig. 6.

Table 1 Proposed algorithm

1	Select the region of interest of the rice seeds
2	Convert the RGB image to gray images
3	Apply the edge detection operation and calculate the geometric parameters of the rice seeds
4	Based on parameters classify rice seed into three parts namely normal, long and small rice seeds
5	Display the count of normal, long and small rice seeds on screen using CART

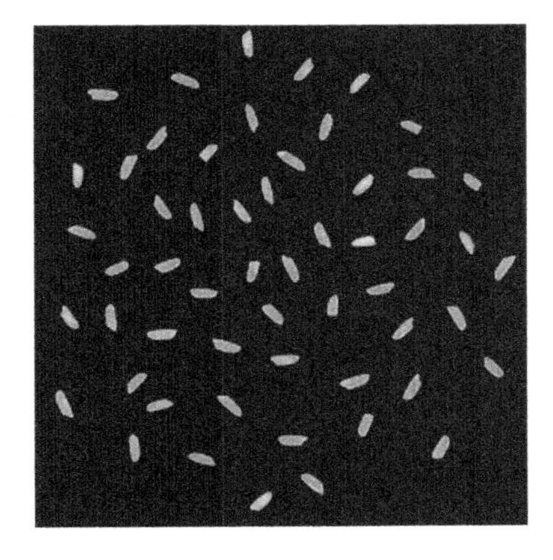

Fig. 4 RGB, gray-scale and edge detected image of one of the sample of Gujarat-17 Rice seed [7]

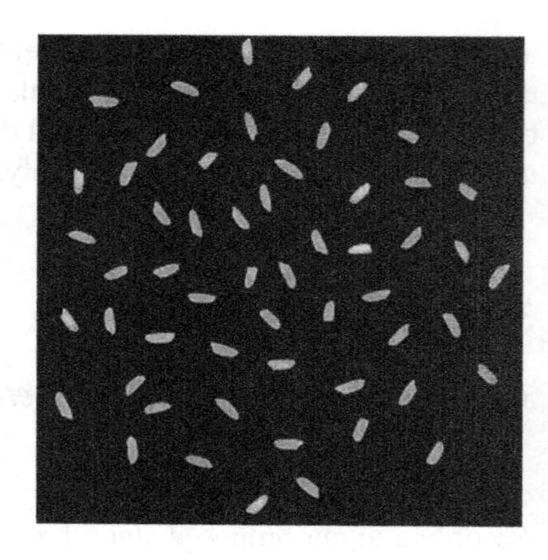

Fig. 5 RGB, gray-scale and edge detected image of one of the sample of Gujarat-17 Rice seed [7]

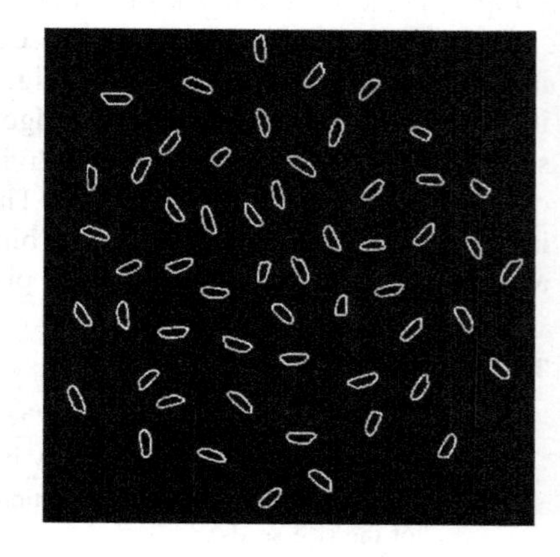

Fig. 6 RGB, gray-scale and edge detected image of one of the sample of Gujarat-17 Rice seed [7]

3.2 Parameter Calculation

Here we are extracting four parameters area, major axis, minor axis length and eccentricity for differentiating normal rice seed from long seed as well as small seed after in depth study and discussion with rice mill personnel.

The area **'A'** of any object in an image is defined by the total number of pixels enclosed by the boundary of the object. The major axis length **'N'** of an image is defined as the length (in pixels) of the major axis of the ellipse that has the same normalized second central moments as the region. The minor axis length **'M'** of an image is defined as the length (in pixels) of the minor axis of the ellipse that has the same normalized second central moments as the region. The eccentricity **'E'** is the ratio of the distance between the foci of the ellipse and its major axis length. The value is between 0 and 1.

Diagram for histogram for area, Major axis length, Minor axis length and Eccentricity calculation computed from sample is shown in Figs. 7, 8, 9 and 10.

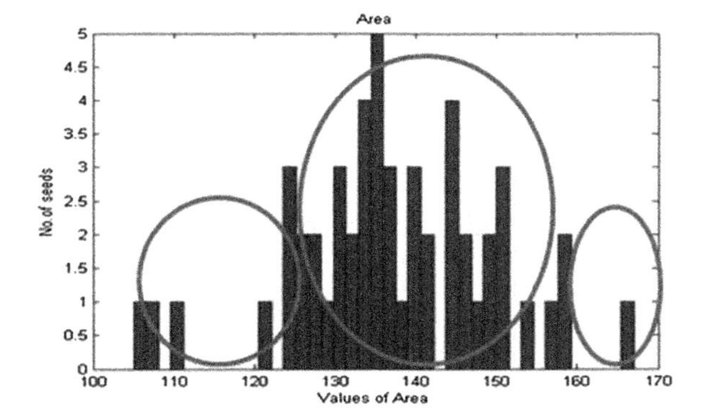

Fig. 7 Histogram showing area, major axis, minor axis and eccentricity of Gujarat-17 rice seeds [7]

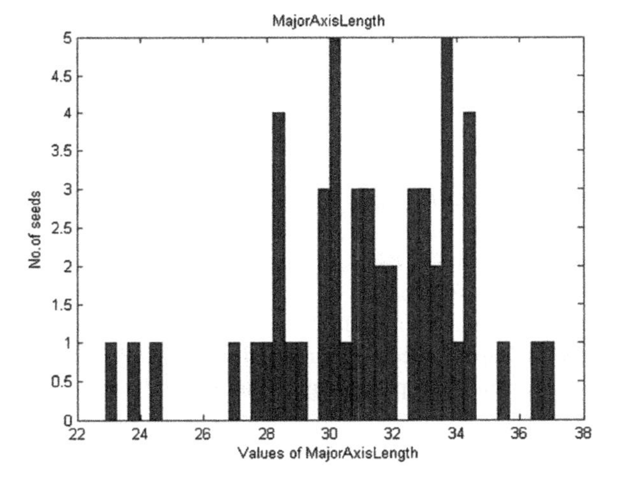

Fig. 8 Histogram showing area, major axis, minor axis and eccentricity of Gujarat-17 rice seeds [7]

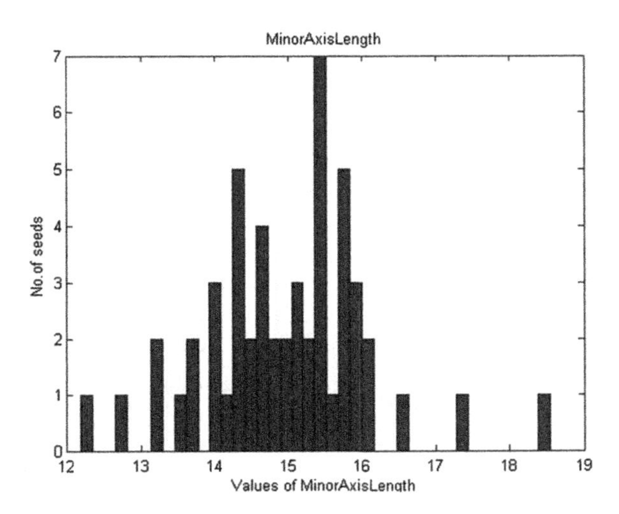

Fig. 9 Histogram showing area, major axis, minor axis and eccentricity of Gujarat-17 rice seeds [7]

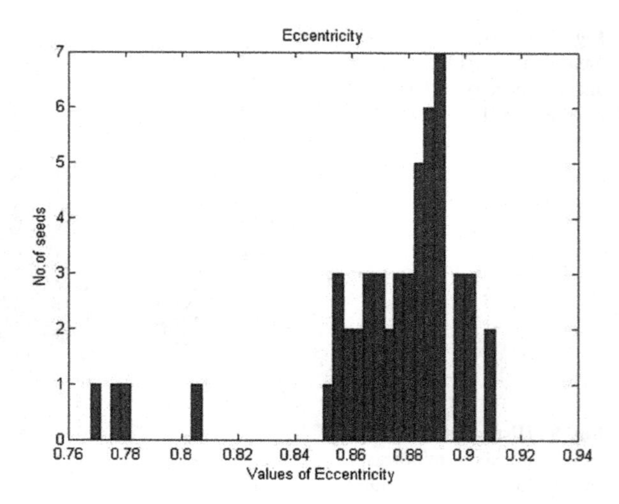

Fig. 10 Histogram showing area, major axis, minor axis and eccentricity of Gujarat-17 rice seeds [7]

From the histogram of Area we can clearly specify that the range of area for small seed can be easily distinguished as circle 1, long seed as circle 3 and normal seed as indicated by circle 2 (Fig. 7). Likewise, from the other histograms we can distinguish three classes of rice seeds. Use of Vernier caliper for quality evaluation by human inspector can be replaced by these geometric calculations. Table 2 shows intended parameters value based on histogram for normal seeds (NS), long seeds (LS) and small seeds (SS) respectively. Table 3 infers to results summarized based on algorithm and percentage values.

Table 2 Analysis for several seed available in one sample

Sr no.	A	N	M	E	Sr no.	A	N	M	E
1	136	39.364	12.994	0.8116	11	142	34.094	13.908	0.8970
2	162	31.154	13.897	0.8141	12	146	30.744	15.501	0.8645
3	185	36.729	13.481	0.8645	13	149	35.508	14.658	0.8959
4	198	37.456	15.169	0.8477	14	150	31.950	15.985	0.8659
5	105	28.739	11.614	0.8033	15	151	33.916	14.420	0.8915
6	114	31.568	13.289	0.8920	16	160	32.965	15.036	0.8812
7	116	24.978	12.341	0.8186	17	166	33.335	16.169	0.8763
8	127	30.993	12.117	0.9110	18	167	32.319	16.501	0.8677
9	134	31.820	15.669	0.8686	19	170	38.057	14.316	0.9080
10	140	31.826	15.782	0.8674	20	175	36.801	15.629	0.8839

Table 3 Result analysis of various samples based on algorithm and percentage values

S. no.	NS	LS	SS	Total seed	NS (%)	LS (%)	SS (%)
1	45	3	3	51	94	2	4
2	44	6	4	54	85	9	6
3	47	3	4	54	89	2	9
4	46	2	3	51	90	2	8
5	2	1	2	55	91	4	5
6	46	4	2	52	87	8	5
7	54	1	2	57	91	2	7
8	47	4	3	54	92	2	6
9	51	3	4	58	90	3	7
10	47	3	5	55	95	2	3
11	49	4	5	58	88	5	7
12	46	4	4	54	88	6	6
13	49	2	2	53	92	4	4
14	50	3	3	56	90	4	6
15	45	4	3	52	89	6	5
Average					92	3	5

Table 4 Computed threshold values

Parameters	Small rice	Normal rice	Long rice
Area	100–120	120–170	170–200
Major axis length	20–25	25–40	40–55
Minor axis length	10–15	15–20	20–25
Eccentricity	0.7–0.8	0.8–0.91	0.91–0.96

3.3 Classification of Rice Seed

For finding out the number of normal rice seeds, long rice seeds and small rice seeds we compute thresholds values using the histograms of Figs. 7, 8, 9 and 10 for area, minor axis length, major-axis length and eccentricity as mentioned in Table 4.

4 Cart

For the classification purpose, we have used these features and applied CART. For this technique first of all based on the histograms of calculated features the threshold value was calculated and then we have classified rice seed in three classes like Small Seed, Normal Seed and Long Seed. CART for one of the sample of Indian Gujarat-17 rice is shown in below Fig. 11, where x1 represents Area, x2 recognizes Major Axis Length, x3 represents Minor Axis Length and x4 indicates Eccentricity parameter.

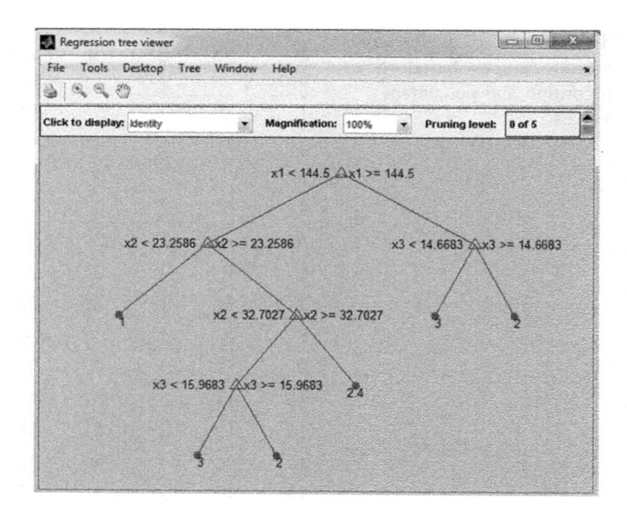

Fig. 11 CART for one of the sample of Gujarat-17 rice and CART for Gujarat-17 rice (15 samples)

Likewise we have tried on all the samples of Gujarat-17 rice individually. For improving the accuracy of classification we have combined all the samples and made one large database for the Gujarat-17 rice. The CART for the whole combined samples of Gujarat-17 rice is shown in Fig. 12.

There are many steps for decision tree. It starts by verifying calculated feature's threshold value and based on that it take decision. In Fig. 12 at node 1, decision taken from the x3 parameter which indicates Minor Axis Length. If the value of x3 is less than 14.9998 then node 2 is decided and if the value of x3 is greater or equal to 14.9998 then node 3 is decided for further derivation. At node 2 the decision about next node is based on the x4 parameter which is Eccentricity. If the value of x4 is less than 0.80137 then it is the terminated node for class 1 for small rice seed

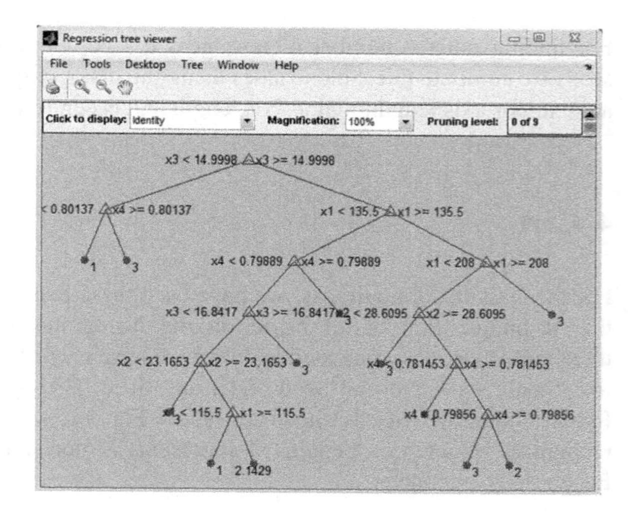

Fig. 12 CART for one of the sample of Gujarat-17 rice and CART for Gujarat-17 rice (15 samples)

but if the value of x4 is greater or equal to 0.80137 then it is the terminal node which gives class 3 means long rice seed. At node 3 the decision about next node is based on the x1 parameter which is Area. If the value of x1 is less than 135 then node 3 is decided but if the value of x1 is greater or equal to 135 then node 4 is decided. While At node 3 the decision about next node is based on the x4 parameter which is Eccentricity. If the value of x4 is less than 0.7989 then node 5 and if the value of x2 is greater or equal to 07989 then node 6 is decided for further tree description. At node 5 the decision about next node is based on the x2 parameter which is Major axis length. Now this steps are carried out based on values of different parameters up to the classification of seeds in small rice, long rice and normal rice as in Fig. 12. so from above result we can see all the features are verified one by one for detecting the high accuracy tree. In this tree we get average error of 1.1671 with accuracy of 98.83 %.

5 Conclusion

This paper presents a quality analysis of Gujarat-17 rice seeds via image analysis. Combined measurement techniques for counting normal seed and foreign element in terms of long as well as small seed for a given sample is being done with accuracy of 98.83 %. This paper illustrates new method which is non-destructive for quality analysis. The time taken to obtain such results is 35 s which is very less and clearly depicts its importance in the world of automation. Traditionally quality evaluation and assessment is done by human sensory panel which is time consuming, may be variation in results and costly. The results so achieved were testified by the rice mill owners and were satisfactory in nature.

Acknowledgments The author expresses their gratitude to Mr. H.N. Shah the owner of Shri Krishna Rice and Pulse Mill, Borsad for the system model as per the requirements desired by the image processing group.

References

1. Abdullah, M.Z., Fathinul-Syahir, A.S., Mohd-Azemi, B.M.N.: Automated inspection system for color and shape grading of star fruit (Averrhoa carambola L.) using Computer Vision sensor. Trans. Inst. Meas. Control. **27**(2), 65–87 (2005)
2. Shantaiya, S., Ansari, U.: Identification of food grains and its quality using pattern classification. In: International Conference [ICCT], 3rd–5th December. IJCCT, vol. 2 issue 2, 3, 4, pp. 70–74 (2010)
3. Amadasun, M., King, R.: Textural features corresponding to textural properties. IEEE Trans. Syst. Man Cybern. **19**(5), 1264–1274 (1989)
4. Sundaram, G., Ding, K.X.: Computer vision technology for food quality assurance. Trends Food Sci. Technol. **7**, 245–256 (1996)

5. Guzman, J.D., Peralta, E.K.: Classification of Philippine rice grains using computer vision and artificial neural networks. In: Iaald Afita WCCA World conference on Agricultural information and IT (2008)
6. Jain, K., Modi, Pithadiya, K.: Non Destructive quality evaluation in spice industry with specific reference to Cuminum Cyminum L (Cumin) seeds. In: International Conference on Innovations and Industrial Applications Malaysia, IEEE (2009)
7. Maheswari, C.V., Jain, K., Chintan, Modi, K.: Non destructive quality evaluation of Indian basmati Oryza sativa SSP Indica rice using image processing. In: CSNT-Rajkot, IEEE, pp. 189–193 (2012)
8. Salem, M., Alfatni, M., Rashid, A., Shariff, M., Osama, M., Saaed, B.: The design and process of the external inspection system in agricultural engineering. Map Asia and ISG (2010)
9. Rani, N.S., Pandey, M.K., Prasad, G.S.V., Sudharshan, I.: Historical significance, grain quality features and precision breeding for improvement of export quality Gujarat-17 varieties in India. Indian J. Crop Sci. **1**(1–2), 29–41 (2006)
10. Abdullah, M.Z., Aziz, A.S., Dos-Mohamed, A.M.: Quality inspection of bakery products using color-based computer vision system. J. Food Qual. **23**, 39–50 (2000)
11. Verma, B.: Image processing techniques for grading and classification of rice. In: International Conference on Computer and Communication Technology [ICCCT] (2010)
12. Ballard, D.A., Brown, C.M.: Computer Vision. Prentice-Hall, Englewood Cliffs (1982)
13. Blasco, J., Aleixos N., Molt, E: Computer vision system for automatic quality grading of fruit. Biosyst. Eng. **85**(4) (2003)
14. Guo, A., Yang, O., Gao, R.J., Liu, Yan de, Sun, X.D., Pan, Y., Dong, X.L.: An automatic method for identifying different variety of rice seeds using computer vision technology. In: Sixth International Conference on Natural Computation ICNC (2010)
15. Gumuş, B., Balaban, M.O., Unlusayın, M.: Computer vision applications to aquatic foods: a review. Turk. J. Fish. Aquat. Sci. **11**, 171–181 (2011)
16. Batchelor, B.G.: Lighting and viewing techniques In: Bachelor, B.G., Hill, D.A., Hodgson, D. C. (eds.) Automated Visual Inspection, pp. 103–179. IFS Publication Ltd, Bedford (1985)
17. Birewar, B.R., Kanjilal, S.C.: Hand operated batch type grain cleaner. J. Agric. Eng. Today **6** (6), 3–5 (1982)
18. Yadav, B.K., Jindal, V.K.: Modeling changes in milled rice (Oryza sativa L.) kernel dimensions during soaking by image analysis. Food Engineering and Bioprocess Technology. Asian Institute of Technology
19. Liu, C.C., Shaw, J.T., Poong, K. Y., Hong, M.C., Shen, M.L.: Classifying paddy rice by morphological and color features using computer vision. In: AACC International /CC-82-0649 (2005)
20. Du, C.J., Sun, W.: Learning techniques used in computer vision for food quality evaluation: a review. J. Food Eng. **72**(1), 39–55 (2006)
21. Openg, D.X., Yong, L.: Research on the rice chalkiness measurement based on the image processing technique. In: IEEE 978-1-61284-840-2/11/$26.00 (2011)
22. Parmar, R.R., Jain, K.R., Modi, K.: Unified approach in food quality evaluation using Computer Vision. In: ACC 2011, Part III, CCIS, vol. 192, pp. 239–248 (2011)

A Novel Feature Extraction and Classification Technique for Machine Learning Using Time Series and Statistical Approach

R.C. Barik and B. Naik

Abstract Curse of dimensionality is a major challenge for any arena of scientific research like data mining, machine learning, optimization, clustering etc. An optimized feature extraction or dimension reduction gives rise to better prediction and classification, which can be applied to various research areas such as Bioinformatics, Geographical Information System, Speech Recognition, Image processing, Biometric, Biomedical imaging, letter or character recognition etc. Extracting the informative feature as well as removing the redundant and unwanted feature by reducing the data dimension is a remarkable issue for many scientific communities. In this paper we have introduced a novel feature extraction with dimension reduction technique by using combined signal processing and statistical approach as Discrete Wavelet Transform (DWT) and Multidimensional Scaling (MDS) respectively then Support Vector Machine (SVM) has played a major role for classification of nonlinear, heterogeneous dataset.

Keywords Computer aided diagnosis · Discrete wavelet transform (DWT) · Multidimensional scaling (MDS) · Support vector machines (SVM) · Geographical information system (GIS) · Principal component analysis (PCA) · Factor analysis(FA)

R.C. Barik (✉)
Department of Computer Science and Engineering,
Vikash College of Engineering for Women, Bargarh, Odisha, India
e-mail: ramchbarik@gmail.com

B. Naik
Department of Information Technology, Government (SSD) Intermediate
Science College, Balisankara, Sundergarh, Odisha, India
e-mail: bimalnaik.cs@gmail.com

L.C. Jain et al. (eds.), *Computational Intelligence in Data Mining - Volume 3*,
Smart Innovation, Systems and Technologies 33, DOI 10.1007/978-81-322-2202-6_19

1 Introduction

Nonlinear data creates complex matrix calculation such as covariance, variance, Inverse, high dimensional multiplication etc. for finding a pattern to train a machine in data mining. Extracting proper feature from nonlinear dataset is still a NP hard issue among researchers in machine learning area. So feature extraction or selection is a complex procedure which uses high end mathematics in various domains like Bioinformatics, Image processing, speech processing, Biometric etc. Non-optimized feature extraction from higher dimension of heterogeneous data or nonlinear data has the side effect which may lead to bad interpretation, prediction, classification, optimization, pattern recognition and computer aided Diagnosis etc. Erratic diagnoses depict a fact as correct and timely diagnosis of diseases is an essential matter in medical field. Limited human capability and limitations decrease the rate of correct diagnosis [1]. Therefore Computer Aided diagnosis is a major alternative in medical field. Cancer or tumour diagnosis is not similar with a normal Malaria diagnosis rather a patient can die by taking wrong drugs or pills if the diagnosis of the proper class of disease is not identified. There may be more than one classes of same disease can be possible. So drugs designed for one class of disease cannot be given to other class. Machine learning algorithms such as support vector machine (SVM) can help physicians to diagnose more correctly [1]. Fourier Transform has the limitation of analysing data in Time domain only where as Wavelet Transform analyses the data in both time as well as frequency domain having fixed size window (Mother wavelet). Effective feature extraction and classification methods are crucial to achieve high classification performance in pattern recognition [2]. For pattern recognition, classification in supervised learning the model is to be designed like that there will be a plane or line in between more than one class which is imaginary. For predicting the appropriate class label the machine is trained with the data structure as well as Euclidean distance, Mahalanobis distance of data from the mean data point.

2 Proposed Feed Forward Hybrid Method

2.1 Process of Proposed Method

At first, time domain and frequency domain analysis is done to the original data signal by translating and scaling co-efficient using Discrete Wavelet Transform (DWT) of mother wavelet Daubechies up to level 2 decomposition. Then the resultant transformed data is fed to a statistical method as Multi-Dimensional Scaling (MDS) to find out the similarities and dissimilarities or pattern to classify the correct class or group. Upshot of the hybrid or inter-crossed method is classified by classifier or predictor as Support Vector Machine (SVM). The overall process of proposed method is represented in the following flow diagram in Fig. 1.

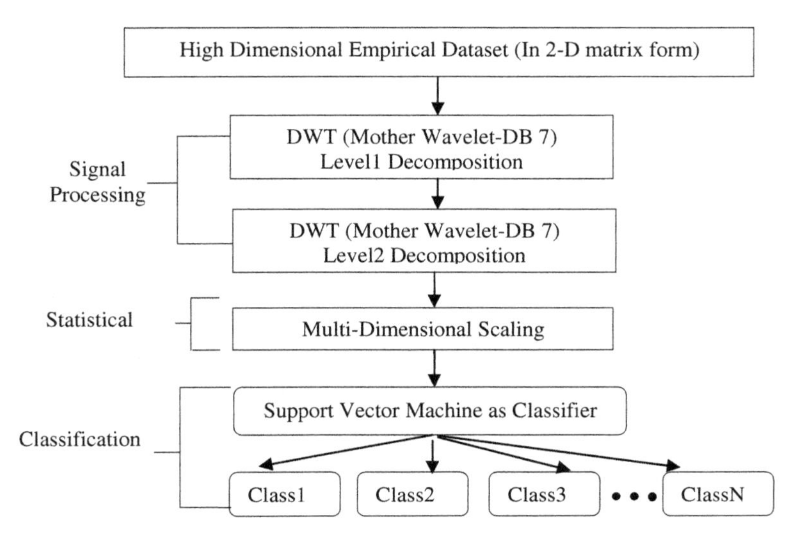

Fig. 1 Overall context flow diagram of proposed method

2.2 *Background of Proposed Method*

2.2.1 Feature Selection

Living being's body is a complex machine in the earth having enormous unexplored areas or issues which inspires various scientists, doctors, biologist to invent new computational algorithm to solve it. Clinical analysis is a traditional approach for any diseases which may have many side effects. Computer aided diagnosis adds a syntactic sugar for treatment of major diseases in medicine or drug design field. For proper diagnosis proper feature selection is required. Discarding or scrutinizing the redundant and irrelevant feature while retaining only the informative feature is the major role of feature selection. In this paper we have analysed wavelet power spectrum in feature selection succeeded by multidimensional scaling analysis.

2.2.2 Wavelet Transform

Wavelets are small waves or windows having fixed length. Wavelet transform (WT) is a Signal processing technique which analyses empirical data signal in both time-frequency domains. It uses a sliding window called as mother wavelet to slice a piece of data signal and analyses it.

The continuous wavelet transform of a signal x (t) is defined as the sum over all time of the signal multiplied by wavelets $\psi_{ab}(t)$ in Eq. (1), the scaled (stretched or compressed) shifted version of mother wavelet $\psi(t)$.

$$\mathrm{CWT}(a,b) = \frac{1}{\sqrt{a}} \int_{-\infty}^{\infty} x(t)\psi^*\left(\frac{t-b}{a}\right)dt \tag{1}$$

$$\psi_{a,b}(t) = \frac{1}{\sqrt{a}}\psi\left(\frac{t-b}{a}\right), a \in R^+, b \in R$$

where, $\psi^*(.)$ represented the conjugated transpose of the wavelet function $\psi(.)$ [3]. The resolution of the signal depends on the scaling parameter 'a' and the translation parameter 'b' determines the localization of the wavelet in time. For smaller 'a', $\psi_{ab}(t)$ have a narrow time-support and therefore a wider frequency support. When parameter 'a' increases, the time support of $\psi_{ab}(t)$ increases as well and the frequency support becomes narrower. The translation parameter 'b' determines the localization of $\psi_{ab}(t)$ in time.

The CWT can be realized in discrete form through the discrete wavelet transform (DWT). To compute Discrete Wavelet Transform (DWT) of each sample signal vector x is passed to low pass filters (scaling functions) as well as high pass filters (Translation function) simultaneously. Decomposition or Down-sampling of original data signal is computed by a factor 2 through filters after individual pass. Two step process for DWT based feature selection from any machine learning dataset One signals (tissue samples) are transformed into time-scale domain by DWT. In the second step features are extracted from the transformed data by correlation coefficients ranking method [4]. The DWT is capable of extracting the local features by separating the components of the signal in both time and scale [4, 24]. The DWT is defined taking discrete values of a and b. The full DWT for signal x (t) can be represented as in Eq. (2).

$$x(t) = \sum_{k \in Z} \mu_{j_0,k}\phi_{j_0k}(t) + \sum_{j=-\infty}^{j_0} \sum_{k \in Z} \omega_{j,k}\psi_{j,k}(t) \tag{2}$$

where ϕ_{j_0k} and $\psi_{j,k}$ are the flexing and parallel shift of the basic scaling function $\phi(t)$ and the mother wavelet function ψ_t and $\mu_{j_0,k}(j<j_0)$ and $\omega_{j,k}$ are the scaling coefficients and the wavelet coefficients respectively.

2.2.3 Multidimensional Scaling

Majoring similar data points and discriminating the dissimilar data points there are a number of statistical solution could be possible like PCA, Factor analysis, LDA, ICA, Isomap, MDS, F-Score, Local Linear Coordination (LLC), Laplacian Eigen maps etc. Among them Multidimensional scaling (MDS) is a unique technique to extract feature from heterogeneous dataset which is used to extract a set of independent variables from a proximity matrix or matrices. Transformation of high dimension data space to low dimension data space is the main aim of Multidimensional Scaling.

Applications of MDS are found in a wide range of areas, including visualization, pattern analysis, data pre-processing, scale development, cybernetics, and localization. The overall rationale behind the paper is to help share innovations across disciplines [5]. It preserves pair wise distances MDS addresses the problem of constructing a configuration of n points in Euclidean space by using information about the distance between the n patterns. MDS provides a visual representation of dissimilarities (or similarities) among objects, cases or, more broadly, observations. In other words, the technique attempts to find structure in data by rescaling a set of dissimilarities measurements into distances assigned to specific locations in a spatial configuration [6].

An $n \times n$ matrix D is called a distance or affinity matrix if it is symmetric, $d_{ii} = 0$ and $d_{ij} > 0$, $i > j$. Given a distance matrix D^X, MDS attempts to find n data points $y_1, y_2 \ldots, y_n$ in D-dimensions such that if $d_{ij}(y)$ denotes the Euclidean distance between yi and y_j. Then D^Y is similar to D^X. Metric MDS minimizes

$$\min_{y} \sum_{i=1}^{n} \sum_{j=1}^{n} \left(d_{ij}^{(X)} - d_{ij}^{(Y)} \right)^2 \quad (3)$$

where

$$d_{ij}^{(X)} = \left\| x_i - x_j \right\|$$

$$d_{ij}^{(Y)} = \left\| y_i - y_j \right\|$$

2.2.4 Classification

Discriminating the distinct classes according to the similarities of data signal is the task of efficient classifier. Pattern recognition and Classification is a major area since last decade and current decade for all Machine Learning Scientist. Neural Network has played a major role in classification or prediction of intelligent dataset but in recent years, SVM has found a wide range of real-world applications, including handwritten digit recognition [7], object recognition [8], speaker identification [9], face detection in images [10], and text categorization [11]. SVM-based approaches are able to significantly outperform competing methods in many applications [12]. SVM achieves this advantage by focusing on the training examples that are most difficult to classify. These “borderline” training examples are called support vectors [13].

We first outline the approach for linear SVMs. Given microarray data with n genes per sample, the SVM outputs the normal to the hyper-plane, w, which is a vector with n components, each corresponding to the expression of a particular gene. Assuming that the expression values of each gene have similar ranges, the absolute magnitude of each element in w determines its importance in classifying a sample, since the following equation holds [14]:

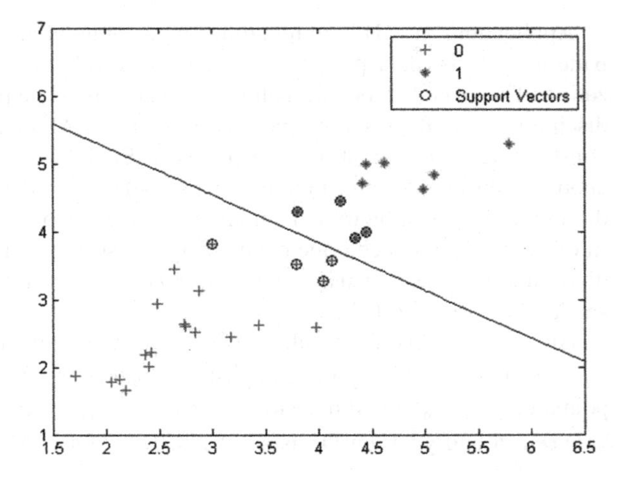

Fig. 2 The rounded circles of + and * are the support vector points which contribute in defining the decision boundary plotted using Matlab (COLON dataset 62 × 2,000)

$$f(x) = w \cdot x + b = \sum_{i=1}^{1} w_i x_i + b \tag{4}$$

Figure 2 explains the support vector points linearly separate binary classes of Colon dataset.

2.3 Dataset Used for Analysis

Cancer gene expression microarray data sets used for analysis in this paper. These datasets are also summarized in Table 1.

3 Result and Discussion

To design an optimized classifier is still a NP hard Problem. In order to compare the efficiency of the proposed method in predicting the class of the cancer microarray data, Six standard datasets such as Leukemia, MLL Leukemia, SRBCT, Prostrate, Lymphoma, Colon have been used for the study which is clearly explained in Table 1. Steps for data decomposition describes below.

First latest signal processing techniques Wavelet transform used for feature extraction methods. Wavelet transform proposed by Grossman and Morlet [15] is an efficient time-frequency representation method which transforms a signal in time domain to a time-frequency domain. Any signal can be decomposed into a series of dilations and compressions of a mother wavelet, resolution of the signal depends on the scaling parameter and the translation parameter determines the localization of

Table 1 Details of the dataset used for analysis

Dataset	Description	Classes			
Leukemia	Consists of two acute leukemia, AML and ALL bone marrow 72 samples. 7,129 probes from 6,817 human genes [18]	AML		ALL	
		Training 38 samples (27 ALL + 11 AML)			
		Testing 34 samples (20 ALL + 14 AML)			
MLL Leukemia	Consists of three types of leukaemia's ALL, MLL and AML with 72 samples from 12,582 genes [19]	ALL	MLL	AML	
		Training 57 samples (20 ALL + 17 MLL + 20 AML)			
		Testing 15 samples (4 ALL + 3 MLL + 8 AML)			
SRBCT	Consists of four types of small round blue cell tumors with 83 samples from 2,308 genes. Burkitt lymphoma (BL), the Ewing family of tumors (EWS), neuroblastoma(NB), rhabdomyosarcoma(RMS) [20]	BL	EWS	NB	RMS
		Training 63 samples (8 BL + 23 EWS + 12 NB + 20 RMS)			
		Testing 20 samples (3 BL + 6 EWS + 6 NB + 5 RMS)			
Prostrate	Consists of prostate tissue samples from 12,600 genes [21]	T		N	
		Training 102 samples 52 prostate tumor tissue and 50 are from normal tissue			
Lymphoma	Consists of three most prevalent adult lymphoid malignancies 62 samples from 4,026 genes	DLBCL	FL	CL	
		Training 62 Samples 42 DLBCL, 9 FL, 11 CL			
Colon	Consists of total 62 samples collected from colon cancer patients 2,000 genes [22]	T		N	
		40 samples are from tumor tissues and 22 are from healthy parts of the colons of the same patients			

the wavelet in time, the wavelets can be realized by iteration of filters with rescaling which was developed by Mallat [16] through wavelet filter banks. Figure 3 shows the time and frequency analysis of Discrete wavelet transform to obtain the SRBCT dataset (83 × 2,308) decomposed signal up to level 2 (83 × 586) using Mother wavelet DB 7 ((1 × 2,308) to (1 × 586)).

Second, after decomposition using wavelet transform still some redundant genes lies with the decomposed data so we have applied it to MDS method (a statistical technique).The details of decomposition is given in Table 2 [5]. Figure 4 describes the decomposition of data signal of SRBCT dataset using Multidimensional Scaling for the first sample among 83 sample (from (1 × 586)) to (1 × 150. Here one axis is SRBCT genes another is MD scale.

After the informative feature selection evaluation of the performance of different classification methods using predictive accuracy, can be defined as:

$$\text{Accuracy} = \frac{\text{TN}_1 + \text{TN}_2 + \cdots \text{TN}_n}{\text{totalnum}} \times 100 \tag{5}$$

Here, TN_1, $\text{TN}_2 \ldots$ TN_n denote the correct classification numbers of the samples belonging to a corresponding class; total num represents total sample numbers. From Tables 3 it is clearly reveals that our proposed method is equivalent to the counterparts with the advantage of reduced computational load [17].

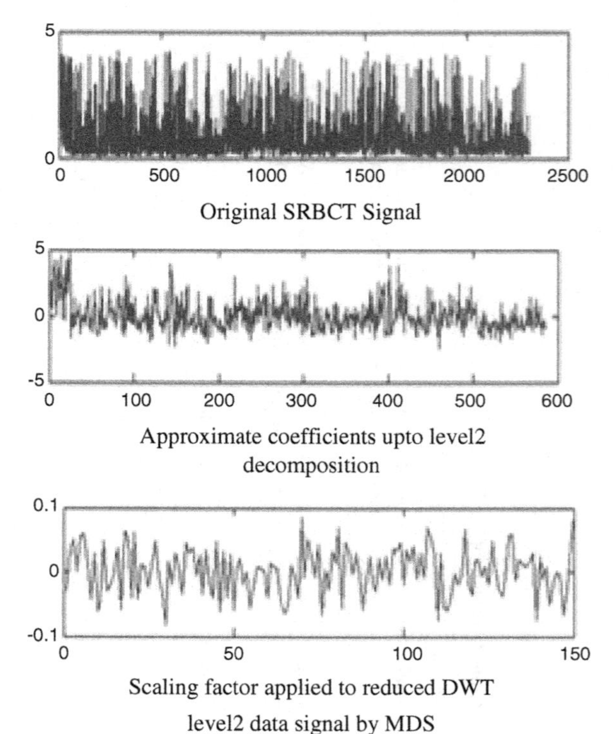

Fig. 3 Shows the time and frequency analysis of DWT and MDS to obtain the SRBCT dataset decomposed signal using mother wavelet Daubechies

Table 2 Describes the feature selection by DWT + MDS of six cancer dataset

Dataset	Original Dimension	DWT(DB 7)		MDS
		Level1	Level2	
Leukemia	72 × 7,129	72 × 3,571	72 × 1,792	72 × 150
MLL Leukemia	72 × 12,582	72 × 6,297	72 × 3,155	72 × 150
SRBCT	83 × 2,308	83 × 1,160	83 × 586	83 × 150
Prostrate	102 × 2,600	102 × 312	102 × 156	102 × 150
Lymphoma	62 × 4,026	62 × 2,019	62 × 1,016	62 × 150
Colon	62 × 2,000	62 × 1,006	62 × 509	62 × 150

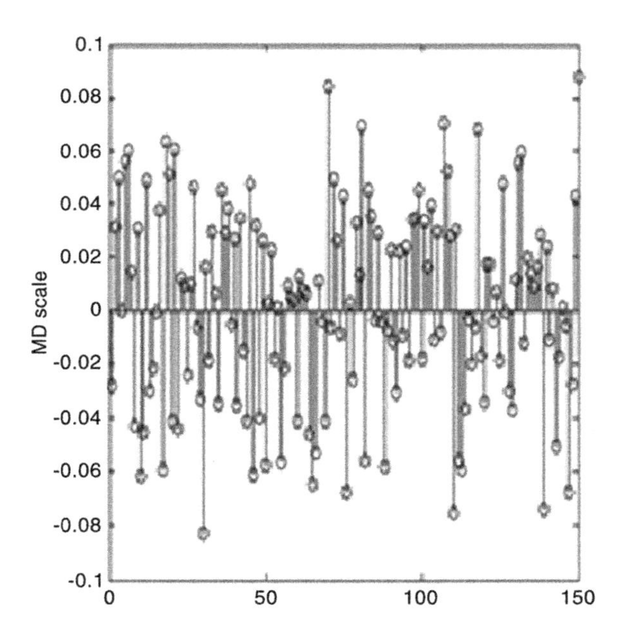

Fig. 4 Describes decomposition of SRBCT data signal using Multidimensional scaling (MDS)

The best classifier based on the average accuracy for the six dataset classification problems used in this experiment is SVM (95.18 %) by analysis of Leave one out cross validation Test (LOOCV) as well in tenfold cross validation test. Table 3 explains all the six benchmark dataset comparison of classification accuracy using SVM (RBF kernel), RBFNN, MLP and LDA methods.

Table 3 Describes the comparative accuracy result by LOOCV test

Dataset	Classes	Method	Cross validation Accuracy %
Leukemia	AML, ALL	SVM (RBF kernel)	98.2
		RBFNN	97.2
		MLP	96.3
		LDA	78
MLL Leukemia	ALL, MLL, AML	SVM (RBF kernel)	95
		RBFNN	93.1
		MLP	89.3
		LDA	71.2
SRBCT	BL, EWS, NB, RMS	SVM (RBF kernel)	96.1
		RBFNN	92.4
		MLP	85.5
		LDA	69
Prostrate	T, N	SVM (RBF kernel)	94.7
		RBFNN	93
		MLP	87.3
		LDA	70.6
Lymphoma	DLBCL, FL, CL	SVM (RBF kernel)	97.7
		RBFNN	96.3
		MLP	93.5
		LDA	81.9
Colon	T, N	SVM (RBF kernel)	95
		RBFNN	93.4
		MLP	89.3
		LDA	66

4 Conclusion

Among more than one alternative if we want to choose only one as best alternative then we must use a hypothesis called as "comparison". For performance comparison, the proposed feature representation method is analyzed with many well-known classifiers like neural network MLP, RBFNN, LDA (statistical classifier) as well as existing feature extractor method like Wavelet, Principal component Analysis (PCA), Independent Component Analysis (ICA), and Factor analysis etc. The success rates of all the classifiers such as PCA + ANN, PCA + LDA, DWT + ANN, DWT + SVM, FA + ANN, DWT + F-Score + ANN are evaluated with all the six benchmark datasets but the proposed method DWT + MDS + SVM is an optimized technique with respect to accuracy as well as complexity. It is not that the proposed method can be applied to bioinformatics domain only it may be applicable to Digital Image Processing, Speech Processing etc. also where heterogeneous data's are used for analysis in Machine Learning.

References

1. Mert, A., Kilic, N., Akan, A.: Breast cancer classification by using support vector machines with reduced dimension. In: IEEE Conference ELMAR. ISSN: 1334-2630
2. Oskoei, M.A., Hu, H.: Support vector machine-based classification scheme for myoelectric control applied to upper limb. IEEE Trans.Biomed. Eng. 55(8) (2008)
3. Sahu, S.S., Barik, R.C., panda, G.: Cancer Classification Using Microarray Gene Expression Data: Approached Using Wavelet Transform and F-Score Method International Conference on Electronic Systems. (ICES-2011). NIT Rourkela, Orissa (2011)
4. Li, T., Zhang, C., Ogihara, M.: A comparative study of feature selection and multiclass classification methods for tissue classification based on gene expression. Bioinformatics **20** (15), 2429–2437 (2004)
5. France, S.L.,Carroll, J.D.: Two-way multidimensional scaling: a review systems, man and cybernetics. part C: applications and reviews. IEEE Trans. **41**(5)
6. Jaworska, N., Chupetlovska-Anastasova, A.: A review of multidimensional scaling (MDS) and its utility in various psychological domains. Tutor. Quant. Methods Psychol. **5**(1), 1–10 (2009)
7. Scholkopf, B., KahKay, S., Burges, C.J., Girosi, F., Niyogi, P., Poggio, T., Vapnik, V.: Comparing support vector machines with Gaussian kernels to radial basis function classifiers. IEEE Trans. Signal Process. **45**, 2758–2765 (1997)
8. Pontil, M., Verri, A.: Support vector machines for 3-D object recognition. IEEE Trans. Pattern Anal. Machine Intell. **20**, 637–646 (1998)
9. Wan, V., Campbell, W.M.: Support vector machines for speaker verification and identification. In: Proceedings IEEE Workshop Neural Networks for Signal Processing, pp. 775–784. Sydney, Australia (2000)
10. Osuna, E., Freund, R., Girosi, F.: Training support vector machines: Application to face detection. In: Proceedings Computer Vision and Pattern Recognition, pp. 130–136. Puerto Rico (1997)
11. Joachims, T.: Transductive inference for text classification using support vector machines. presented at the International Conference Machine Learning, Slovenia (1999)
12. Burges, C.J.: A tutorial on support vector machines for pattern recognition. Knowl. Disc. Data Min. **2**, 121–167 (1998)
13. Issam, E., Yang, Y., Wernick, M.N., Galatsanos, N.P., Nishikawa, R.M.: A support vector machine approach for detection of microcalcifications. IEEE Trans. Med. Imaging **21**(12), 1552–1563 (2002)
14. Mukharjee, S.: Classifying microarray data using support vector machine. Chapter 9.Genome Research and Center for Biological and Computational Learning at MIT
15. Grossmann, A., Morlet, J.: Decomposition of Hardy functions into squareintegrable wavelets of constant shape. SIAM J. Math. Anal. **15**(4), 723–736 (1984)
16. Mallat, S.G.: A theory for multiresolution signal decomposition: the wavelet representation. IEEE Trans. Pattern Anal. Mach. Intell. **11**(7), 674–693 (1989)
17. Sahu, S.S., Panda, G., Barik, R.C.: A hybrid method of feature extraction for tumor classification using microarray gene expression data. Int. J. Comput. Sci. Informatics India **1**(1) (2011)
18. Golub, T.R., Slonim, D.K., Tamayo, P., Huard, C., Gaasenbeek, M., Mesirov, J.P., Coller, H., Loh, M.L., Downing, J.R., Caligiuri, M.A., Bloomfield, C.D., Lander, E.S.: Molecular classification of cancer: Class discovery and class prediction by gene expression monitoring. Science **286**(5439), 531–537 (1999)
19. Armstrong, S.A., Staunton, J.E., Silverman, L.B., Pieters, R., de Boer, M.L., Minden, M.D., Sallan, S.E., Lander, E.S., Golub, T.R., Korsmeyer, S.J.: MLL translocations specify a distinct gene expression profile that distinguishes a unique leukemia. Nat. Genet. **30**(1), 41–47 (2002)
20. Khan, J., Wei, J.S., Ringner, M., Saal, L.H., Ladanyi, M., Westermann, F., Berthold, F., Schwab, M., Antonescu, R., Peterson, C., Meltzer, P.S.: Classification and diagnostic prediction of cancers using gene expression profiling and artificial neural networks. Nat. Med. **7**(6), 673–679 (2001)

21. Singh, D., Febbo1, P.G., Ross, K., Jackson, D.G., Manola, J., Ladd, C., Tamayo, P., Renshaw, A.A., D'Amico, A.V., Richie, J.P., Lander, E.S., Loda, M., Kantoff, P.W., Golub, T.R., Sellers, W.R.: Gene expression correlates of clinical prostate cancer behavior. Cancer Cell **1**, 203–209 (2002)
22. Alon, U., Barkai, N., et al.: Broad patterns of gene expression revealed by clustering analysis of tumor and normal colon tissues probed by oligonucleotide arrays. Proc. Natl. Acad. Sci. **96**, 6745–6750 (1996)
23. Bruce, L.M., Koger, C.H., Li, J.: Dimensionality reduction of hyper spectral data using discrete wavelet transform feature extraction. IEEE Trans. Geosci. Remote Sens. **40**(10), 10 (2002)
24. Suna, G., Ying, X., Dong, Guandong, X.: Tumor tissue identification based on gene expression data using DWT feature extraction and PNN classifier. Neurocomputing **69**, 387–402 (2006) (Elsivier G. Sun et al.)

Implementation of 32-Point FFT Processor for OFDM System

G. Soundarya, V. Jagan Naveen and D. Tirumala Rao

Abstract Due to the advanced VLSI technology, Fast Fourier Transform (FFT) and Inverse Fast Fourier Transform (IFFT) has been applied to wide field in wireless Communication applications. For modern communication systems, the FFT/IFFT is very important in the OFDM. Because of FFT/IFFT is the key computational block to execute the baseband multicarrier demodulation and modulation in OFDM system. An implementation of the 32-point FFT processor with radix-2 algorithm in R2MDC architecture is the processing element. This butterfly- Processing Element (PE) used in the 32-FFT processor reduces the multiplicative complexity by using real constant multiplication in one method and eliminates the multiplicative complexity by using add and shift operation in the proposed method.

Keywords R2MDC · FFT · OFDM · Verilog

1 Introduction

Now a days, FFT processors used in wireless communication system should have faster execution and low power consumption [1]. These are the most important constraints of FFT processor. In FFT/IFFT block the most arithmetic operation is complex multiplication. This is the main issue in processor which consume more time, a large area and power. When large point FFT is to be design it increases the complexity. There are two methods to reduce the multiplication complexity; one method is to perform real and constant multiplication in place of complex

G. Soundarya (✉) · V.J. Naveen · D.T. Rao
Department of ECE, GMR Institute of Technology, Rajam, A.P, India
e-mail: soundarya6020@gmail.com

V.J. Naveen
e-mail: Jagannaveen.n@gmrit.org

D.T. Rao
e-mail: tirumalarao.d@gmrit.org

L.C. Jain et al. (eds.), *Computational Intelligence in Data Mining - Volume 3*,
Smart Innovation, Systems and Technologies 33, DOI 10.1007/978-81-322-2202-6_20

multiplication. The second method is to eliminate the non-trivial multiplication by the twiddle factors and processing with no complex multiplication. Many complicated design of communication system became feasible. There is a rapidly growing demand in communications for high quality video and voice etc. Orthogonal Frequency Division Multiplexing technology is an effective modulation scheme to meet the demand.

According to the literature survey various designs and implementation of FFT/IFFT have been done for OFDM systems. A novel 8-point FFT processor based on pipeline architecture is discussed [1]. High speed data transmission in ultra wide band spectrum by dividing the spectrum band into multiple bands and provides high efficiency is described [2]. A 16-point FFT butterfly PE reduces the multiplication complexity by using real and constant multiplication [3]. The circuit complexity is reduced by means of pipeline FFT/IFFT architecture and provides better performance in terms of area and speed at low frequency [4]. The present works focus on implementation of 32-point FFT, which is used as processing element in R2MDC architecture and non-trivial multiplications are eliminated by using add and shift method for high throughput.

2 OFDM

The Orthogonal Frequency Division Multiplexing is a wideband wireless digital communication technique that is based on block modulation. OFDM is a subset of frequency division multiplexing in which a single channel utilizes multiple sub-carriers on adjacent frequencies. In addition the sub-carriers in an OFDM system are overlapping to maximize the spectral efficiency. Sub-carriers in OFDM system are orthogonal to one another, thus they are able to overlap without interfering. It can support high-speed video communication along with audio with elimination of ISI and ICI.

3 FFT Algorithm

The FFT algorithms are based on fundamental principle of decomposing the computation of the discrete Fourier transform of a sequence of length N into successively smaller DFT's.

$$X[k] = \sum_{n=0}^{N-1} x[n] \cdot W_N^{nk} \quad \text{where } W_N^{nk} = e^{-j\frac{2\pi nk}{N}} \qquad 0 \le \mathrm{k} \le \mathrm{N} - 1 \tag{1}$$

Direct DFT calculations requires a computational complexity of $O(N^2)$.

3.1 Cooley-Tukey Algorithm

By using most popular Cooley-Tukey FFT algorithm, the complexity can be reduced to $O(N.\log_r N)$ [4, 5]. It is most universal of all FFT algorithms. Cooley-Tukey FFTs are those were the transform length is power of a basis r, i.e., $N = r^S$ S—stages. These algorithms are referred to as radix-r algorithms. The most commonly used are base r = 2 and r = 4. Decomposition is important role in FFT. There are two decomposed types of FFT. One is decimation-in-time and other is decimation-in-frequency. In addition there is no difference in computational complexity between these two types of FFT. Since the low computational complexity of FFT algorithms is desired for high speed consideration in VLSI implementation, here we discuss the computational complexity of different algorithms (Fig. 1).

Basic Radix-2 butterfly processor shown in Fig. 2 consists of adder and complex subtraction. Besides that, an additional complex multiplier for twiddle factor W_N is implemented. The complex multiplication with the twiddle factor requires four real multiplications and two add/subtract operations [1, 3].

3.2 Complex Multiplication

Since complex multiplication is an expensive operation, we tend to reduce the multiplicative complexity of the twiddle factor inside the butterfly processor by calculating only three real multiplications and three add/subtract operations as in (3) and (4).

The twiddle factor multiplication:

$$R + jI = (X \ + jY) \cdot (c + jS) \tag{2}$$

However complex multiplication can be simplified:

$$R = (C - S) \cdot Y + Z \tag{3}$$

$$I = (C + S) \cdot X - Z \tag{4}$$

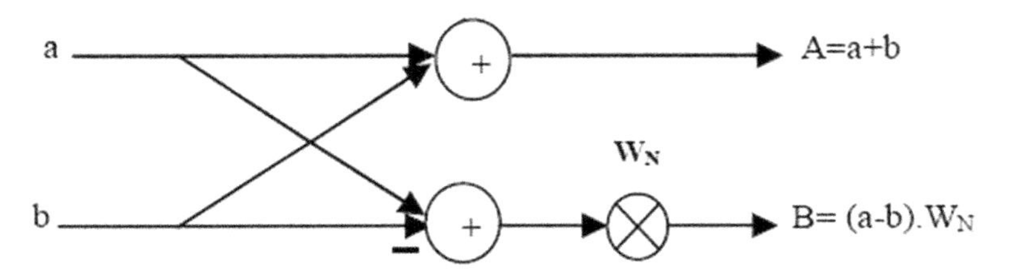

Fig. 1 Signal flow graph of DIF of FFT

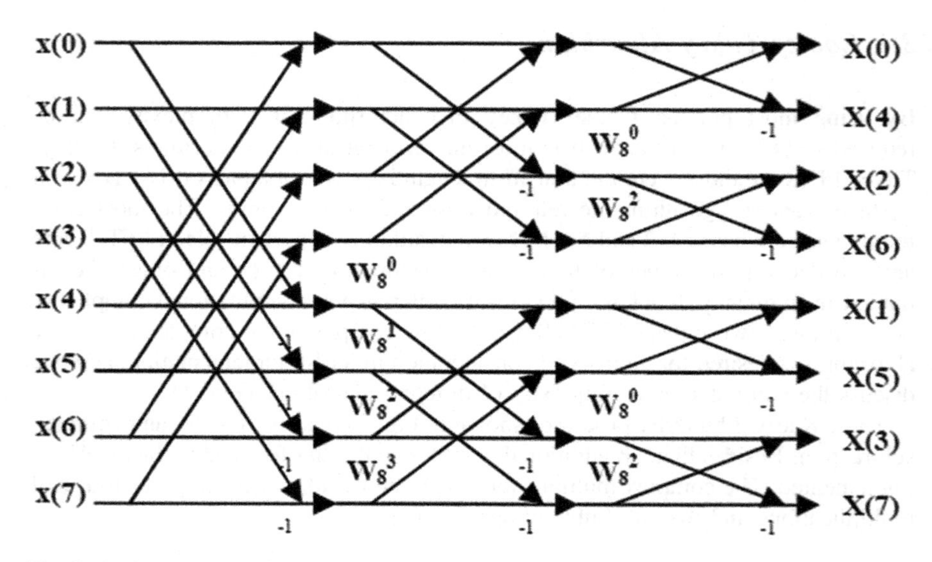

Fig. 2 Basic Butterfly computation

$$\text{with } Z = C \cdot (X - Y) \tag{5}$$

C and S are pre-computed and stored in a memory table. Therefore it is necessary to store the following three coefficients C, C + S and C − S. The implemented algorithm of complex multiplication used in this is three multiplications, one addition and two subtractions as show in Fig. 3.

In the 8-point FFT with radix 2 algorithm, the multiplication with $W_8^2 = -j$ and W_8^0 factors is trivial, the multiplication with W_8^2 simply can be done by swapping from real to imaginary part and vice versa, followed by changing the sign [6, 7]. The number of complex multiplication in this scheme of FFT is two: W_8^1, W_8^3, these are non-trivial complex multiplication was implemented with two multiplications for W_8^1 and three multiplications for W_8^3. Therefore, the number of real multiplications is 5. However this solution was not suitable in practice, because the first

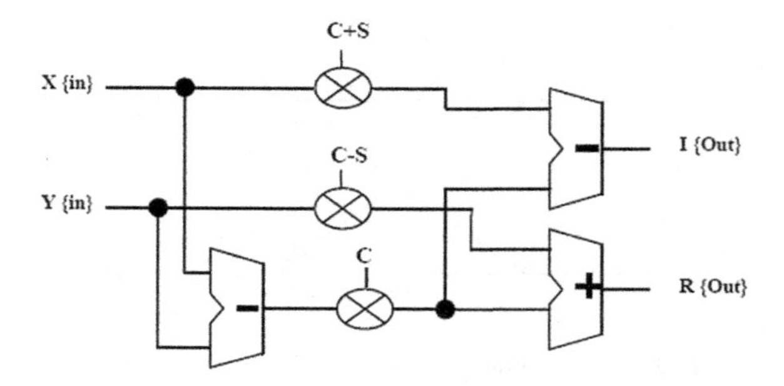

Fig. 3 Implementation of complex multiplication

stage to process four different twiddle factors (trivial and Non-trivial multiplications) in pipeline architecture with one complex multiplier. Therefore it will require more elements in the structure to implement the two complex multiplications in addition to the two trivial multiplications in one block. This can be done by first method i.e., R2MDC. Another architecture solution was proposed in order to eliminate the complex multiplication inside the butterfly processor completely.

4 Pipeline FFT/IFFT Processor Architecture

4.1 R2MDC Architecture

Simplicity, modularity and high throughput are required for FFT/IFFT processors in communication systems. The pipeline architecture is suitable for those ends [5]. The sequential input stream in pipeline architecture unfortunately doesn't match the FFT/IFFT algorithm since the bloc FFT/IFFT requires temporal separation of data. In this case data memory is required in the pipeline processor to be rearranged according to FFT/IFFT algorithm as shown in Fig. 1. One of the straightforward approaches for pipeline implementation of radix-2 FFT algorithm is Radix-2 Multi-path Delay Commutator architecture which is shown in Fig. 4. It is a simplest way to rearrange data for the FFT/IFFT algorithm. The input data sequence are broken into two parallel data elements flowing forward, with correct distance between data elements entering the butterfly scheduled by proper delays. At each stage half of the data flow is delayed via the memory (Registers) and processed with the second half data stream.

4.2 ADD and Shift Method

Another method proposed eliminates the non-trivial complex multiplication with the twiddle factor (W_8^1, W_8^3) and implements the processor without complex multiplication. The proposed butterfly processor performs the multiplication with the

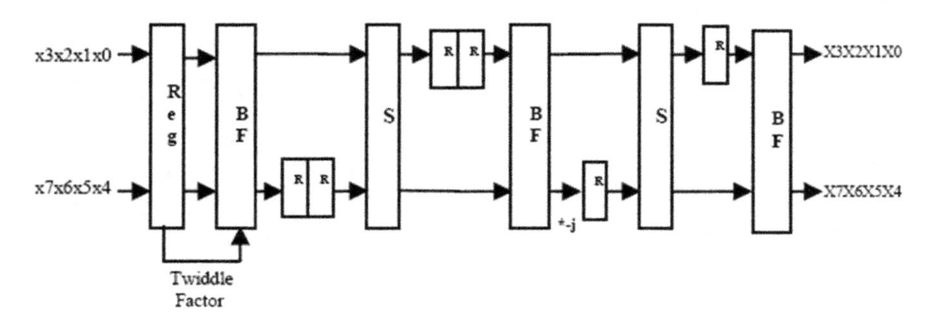

Fig. 4 R2MDC Architecture

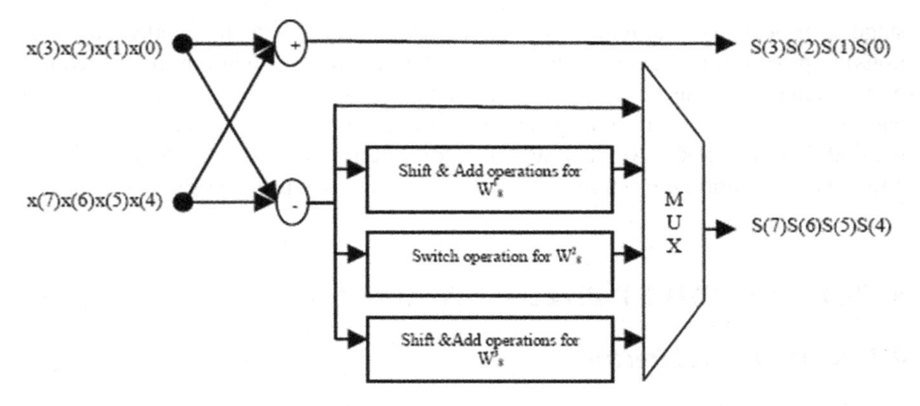

Fig. 5 Butterfly processor with no complex multiplication

trivial factors $W_8^2 = -j$ by switching from real to imaginary and imaginary real, with the factor W_8^0 by a simple cable. With the nom-trivial factors W_8^1, W_8^3 the processor realize the multiplication by factor $1/\sqrt{2}$ using hard wired shift and add operation as shown in Fig. 5.

5 Experimental Results

This simulation shows Figs. 6 and 7 the radix-2 operations. Two inputs are selected with the help of de-multiplexer which is stored inside the buffer block. After addition and subtraction of two inputs with a twiddle factor multiplication final FFT will come.

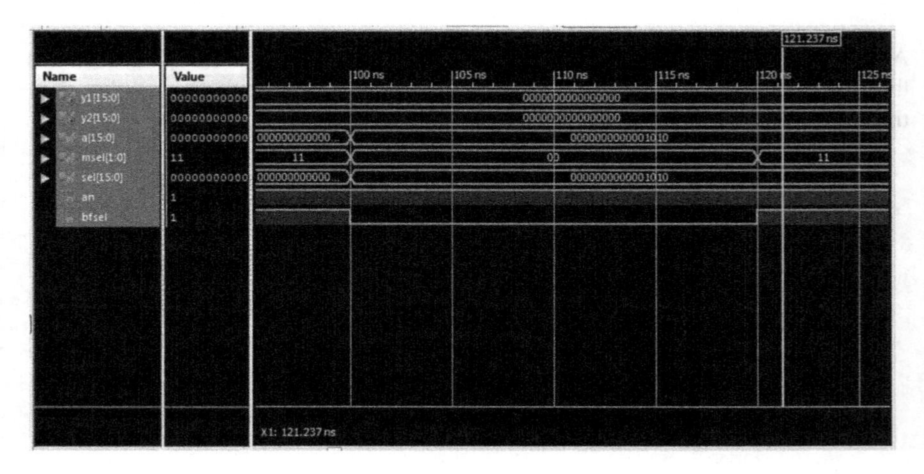

Fig. 6 Simulation result of R2MDC 32-point FFT

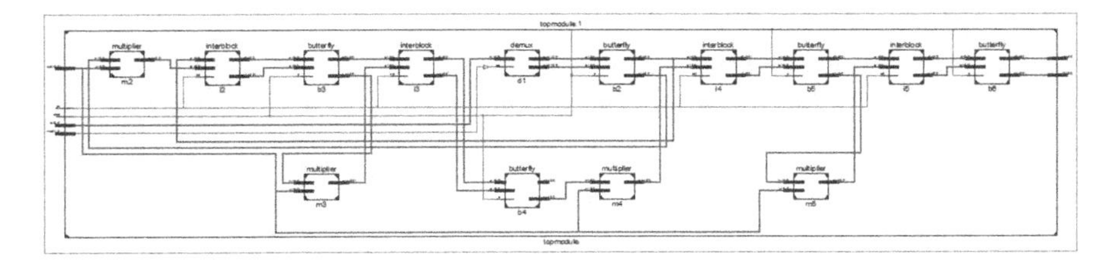

Fig. 7 RTL of R2MDC 32-point FFT

The verilog program is compiled in Xilinx 13.4 to generate a butterfly processor operation with the help of R2MDC architecture, Add and shift method.

The add and shift method simulation shows, that the input after addition and subtraction of radix-2 process multiplied a real number of complex variable, and produce a final FFT.

5.1 Synthesis Report of R2MDC and ADD/Shift Method

The area analysis can be done by using Xilinx power analyzer and synthesis tool in Xilinx 13.4. The 32-point FFT computation with radix-2 in R2MDC was coded in verilog using Xilinx tool simulated and synthesized on vertex4. These results show the better performance when compared to 8-point FFT and speed will be more.

Tables 1 and 2: Shows the device utilization summary of the 32-point FFT for vertex-4 family and number of slice registers, number of slice LUTs, number of and bonded IOBs used. The RTL of R2MDC and Add and shift method are shown in Figs. 8 and 9.

Table 1 R2MDC 32-point FFT

Device Utilization Summary			
Logic Utilization	**Used**	**Available**	**Utilization**
Number of 4 input LUTs	120	10,944	1%
Number of occupied Slices	60	5,472	1%
Number of Slices containing only related logic	60	60	100%
Number of Slices containing unrelated logic	0	60	0%
Total Number of 4 input LUTs	120	10,944	1%
Number of bonded IOBs	67	240	27%
Number of DSP48s	4	32	12%
Average Fanout of Non-Clock Nets	2.47		

Table 2 Add and shift 32-point

Device Utilization Summary			
Logic Utilization	**Used**	**Available**	**Utilization**
Number of 4 input LUTs	932	10,944	8%
Number of occupied Slices	513	5,472	9%
Number of Slices containing only related logic	513	513	100%
Number of Slices containing unrelated logic	0	513	0%
Total Number of 4 input LUTs	983	10,944	8%
Number used as logic	932		
Number used as a route-thru	51		
Number of bonded IOBs	83	240	34%
Average Fanout of Non-Clock Nets	2.54		

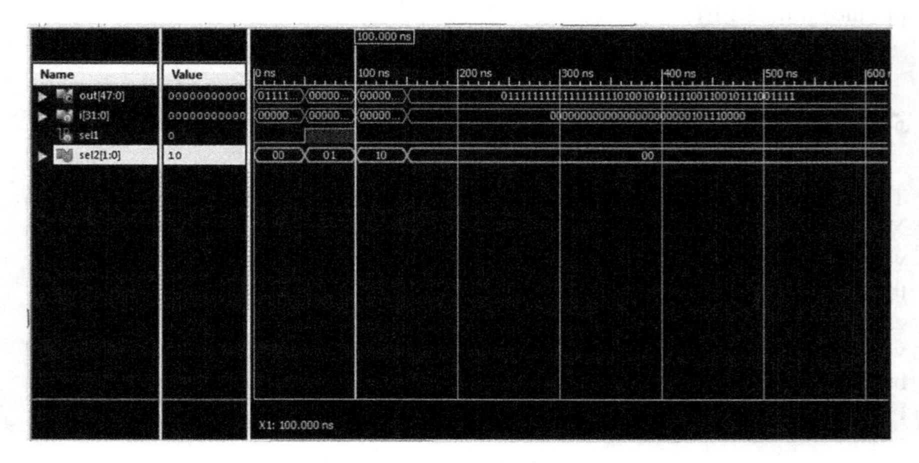

Fig. 8 Simulation result of add and shift method 32-point FFT

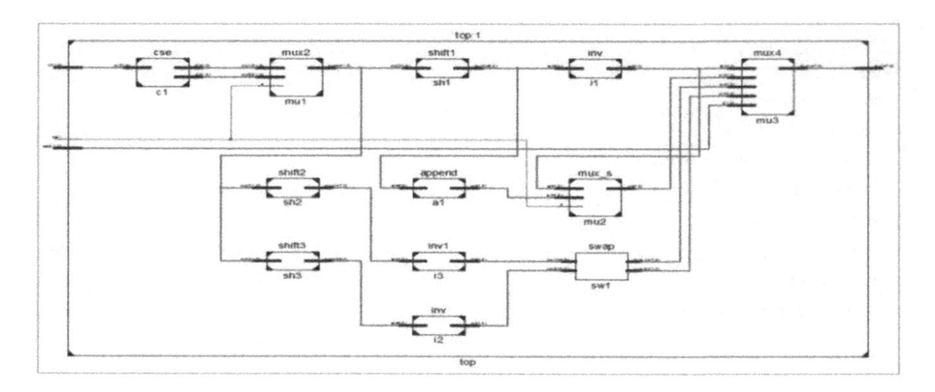

Fig. 9 RTL of ADD and shift method 32-point FFT

6 Conclusion

In this paper, two pipeline-based FFT Architectures are proposed. Both methods are applied on 8-point FFT and 32-point FFT. The 8-point FFT can perform limited operations where as 32-point FFT can perform large complex operations. To reduce computational complexity and increase hardware utility, we adopt different radix FFT algorithms and multiple-path delay commutates FFT architecture in our processors. The multi-path delay commutator FFT architecture requires fewer delay elements and different radix FFT algorithms require fewer complex multiplication. The proposed FFT processor architectures are suitable for various MIMO OFDM-based communication systems, such as IEEE802.11n and IEEE802.16 WiMAX, etc.

References

1. Verma, P., Kaur, H., Singh, M.: VHDL implementation of FFT/IFFT blocks for OFDM. In: Proceedings of International, Conference on Advances in Recent in Communication and Computing, PI 978-1-2244-51-4-3, Kerala, pp. 186–188 (2009)
2. Jinsing, X., Xiaochun, L., Haitao, W., Yujing, B., Decai, Z., Xiaolong, Z., Chaogang, W.: Implementation of MB-OFDM transmitter baseband based on FPGA. In: 4th IEEE (ICCSC 2008), 26–28 May 2008
3. Li, W., Wanhammar, L.: Complex multiplication reduction in FFT processor. In: SSoCC'02 Falkenberg, Sweden, Mar 2002
4. Preeti, G.B., Uma Reddy N.V.: Implementation of area efficient OFDM transceiver on FPGA. Int. J. Soft Comput. Eng. **3**(3), 144–147 (2013). ISSN 2231-2307
5. Prajapati, K., Sharma, A., Thakor, H.: Implementation of an optimized 8 point FFT in pipeline architecture for FPGA's system, vol. 1, Issue 4, (2014)
6. U.M. Baese: Digital Signal Processing with FPGA, 3rd edn. Springer, Berlin (2007)
7. Petrov, M., Glenser, M.: Optimal FFT architecture selection for OFDM receiver on FPGA. In: Proceedings of 2005 IEEE International Conference on Field Programmable technology, PI. 0-7803-9407-0, Singapore, pp. 313–314 (2005)
8. Cooley, J.W., Tukey, J.W.: An algorithm for the machine calculation of complex fourier series. Math. Comput. **19**, 297–301 (1965)
9. Li, W., Wanhammar, L.: An FFT processor based on 16 point module. In: Proceedings of Norchip Conference, Stckholm, Sweden, pp. 125–130 (2001)
10. Wang, B., Zang, Q., Ao, T., Huang, M.: Design of pipelined FFT processor based on FPGA. In: Proceedings of 2nd International Conference on (ICCMS'10), pp. 432–435 (2010)

Performance Analysis of Microstrip Antenna Using GNU Radio with USRP N210

S. Santhosh, S.J. Arunselvan, S.H. Aazam, R. Gandhiraj and K.P. Soman

Abstract Radiation pattern and power spectrum measurements become inevitable while analysing performance of an antenna. Current test bench for these measurements are considerably costly. In this paper, the performance of microstrip antenna is evaluated using GNU (GNU is Not Unix) Radio companion (GRC) and Universal Software Radio Peripheral (USRP). The radiation pattern obtained is used in the measurement of Half Power Beam Width and Front to Back Ratio of the antenna. In addition to these measurements, received signal power is also calculated using the Frobenius norm method. Finally, the economic advantage of the proposed method is compared with conventional measurements.

Keywords Radiation pattern · Power spectrum · GNU radio companion · Universal software radio peripheral · Half power beam width · Front to back ratio · Received signal power · Frobenius norm

1 Introduction

Impressive improvement in the field of communication increases due to its reduced size and compactness. The microstrip antenna is one among those advancements. This advancement makes it compatible with many small devices [1, 2]. In this paper, the micro strip antenna for radio telescope is analysed with the software defined platform and its performance is compared with standard measurements.

S. Santhosh (✉) · S.J. Arunselvan · S.H. Aazam · K.P. Soman
Center for Excellence in Computational Engineering and Networking,
Amrita School of Engineering, Coimbatore, India
e-mail: Santhoshsharma27@yahoo.com

R. Gandhiraj
Department of Electronics and Communication Engineering,
Amrita Vishwa Vidyapeetham University, Coimbatore, India
e-mail: r_gandhiraj@cb.amrita.edu

L.C. Jain et al. (eds.), *Computational Intelligence in Data Mining - Volume 3*,
Smart Innovation, Systems and Technologies 33, DOI 10.1007/978-81-322-2202-6_21

Software Defined Radio (SDR) is a new radical concept in the physical layer of communication where the operation mainly rests on software. In conventional techniques many communication operations involves expensive and specifically designed hardware devices. Whereas SDR allows better reliability of same underlying hardware for many applications [3–5].

USRP is a software controlled communication hardware device which helps in realising many communication problems. The GNU Radio Companion is an open source software platform provides control over USRP by means of software. For this reason, this paper uses the above available platform for analysing antenna's performance [6–8]. This USRP and GNU Radio Companion setup is used for the measurement of radiation pattern and the power spectrum of the test antenna. In order to measure received signal strength a new block is created using the Frobenius norm method [9]. Moreover front and back end ratio of the designed antenna are also calculated [6].

The rest of the paper is organized as follows: Design of Microstrip antenna, GNU Radio and Universal Software Radio Peripheral and other two sections explain experimental procedure with results analysis.

2 GNU Radio Companion and USRP

USRP is a communication hardware which can be tuned to various frequencies through an open source software package called GRC. The following will discuss about GNU Radio and USRP.

2.1 Universal Software Radio Peripheral

Universal Software Radio Peripheral (USRP) is a software hosted radio device. USRP product is comparatively low cost when compared to other communication devices. Among these USRP's, USRPN210 provides dynamic connectivity with high bandwidth. It also provides connectivity to a network through an inbuilt MIMO and a NIC (Network Interface Controller) port. The device can be controlled by open source UHD (USRP Hardware Driver) driver software. USRP is commonly used with the open source GNU Radio software. USRP allows both transmission and reception through SMA (Sub Miniature version A) connector [10]. For these reasons USRP and GNU Radio is used in analysing performance of the antenna. This paper uses USRP incorporated with WBX (Wide Bandwidth) daughter board (operating frequency range 50 MHz–2.2 GHz) for analysing antenna's performance.

2.2 GNU Radio Companion

An open source, software defined tool kit helps in the visualization of many communication problems. In general GNU Radio holds Python and C++ coded GUI (Graphical User Interface) interface for the users. SWIG (Simplified Wrapper and Interface Generator) provides an interface among these two modules. This software package is used to control USRP for the realization of many communication problems [3–9].

3 Microstrip Patch Antenna Design

A Microstrip transmission line consists of a conductive strip etched on a FR-4 (Dielectric material with $\varepsilon_r = 4.4$) substrate of 60 mil thickness with a ground plate at the bottom. The ground plane thickness of the antenna is 0.01 mm. The basic microstrip path consists of a conducting patch printed on the top of a grounded substrate. The design of the antenna has a single patch that can be made to radiate circular polarization (if two orthogonal patch modes are simultaneously excited with equal amplitude and out of phase with determining the sense of rotation) [11]. The patch is selected to be very thin such that $t \ll \lambda_\circ$ where $\lambda_\circ$ is the free space wavelength which is 0.3 m. The resonant frequency of the patch is chosen as 1 GHz. The antenna is designed with 1.42 GHz as the operating frequency.

3.1 Design Equations

The antenna is designed for wavelength 100.7177 mm. Following equations are employed in the calculations of length, width of an antenna.

$$W = \frac{C}{2f_0\sqrt{\frac{\varepsilon_r+1}{2}}} \tag{1}$$

where c is the velocity of light, f_0 is equal to 1.42 GHz, $\varepsilon_r = 4.4$.

The length of the antenna can be calculated by using following equations.

$$L = L_{eff} - \Delta \tag{2}$$

$$L_{eff} = \frac{C}{2f_0}\left(\sqrt{\frac{1}{\varepsilon_{eff}}}\right) \tag{3}$$

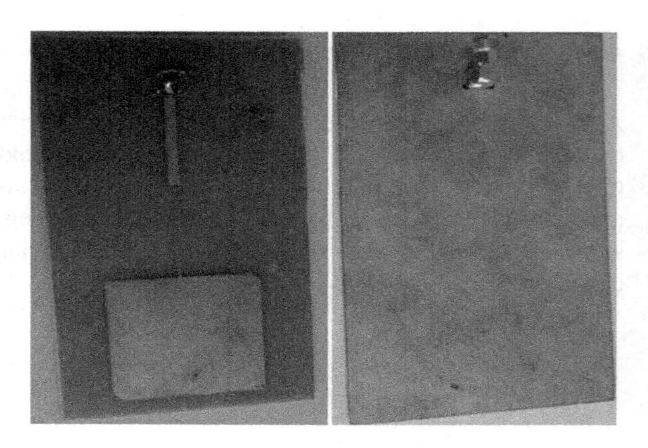

Fig. 1 Depicts fabricated micro strip patch antenna both front and back side [11]

$$\varepsilon_{eff} = \left(\frac{\varepsilon_r + 1}{2} + \frac{\varepsilon_r - 1}{2}\right)\left(1 + \frac{10h}{W}\right)^{-ab} \quad (4)$$

$$a = 1 + \frac{1}{49}\ln\left(\frac{u^4 + \left(\frac{u}{52}\right)^2}{u^4 + 0.432}\right) \quad (5)$$

$$b = 0.564\left(\frac{\varepsilon_r - 0.9}{\varepsilon_r + 3}\right)^{0.093} \quad (6)$$

$$u = \frac{W}{h} \quad (7)$$

The antenna is designed with 1.42 GHz as the operating frequency. Antenna's length and width are determined using (1), (2) by above parameters as the reference and is found to be the length (L) = 48.4667 mm and the width (W) = 64.288 mm with input impedance of 247.5 Ω [11, 12]. To produce circular polarization the design is assumed to have equal length and width (i.e. L, W = 48.4667). The designed antenna is shown in Fig. 1.

4 Proposed Method for Analysing Antenna's Performance

In this paper performance of the antenna is analysed using GNU Radio and USRP. Transmitter part consists of a computer with a USRP from which, the signal is transmitted using an isotropic antenna as shown in Fig. 2. Test antenna is mounted on a stand which is under the control of the Arduino board (Arduino is a programmable open source microcontroller board, intended to make application of interactive objects). This microcontroller device is used to control the rotation of servo motor by 1° is depicted in Fig. 2. This automated rotation helps in accurate

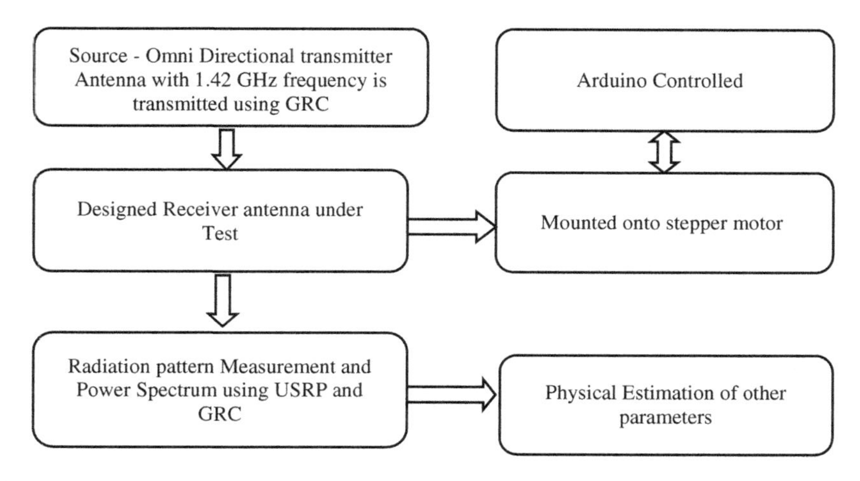

Fig. 2 Illustration of designed methodology

measurement of the radiation pattern over 360°. Then the receiver antenna is connected with a computer which is followed by the analysis using GNU Radio companion.

4.1 Explanation of Proposed Methodology

The transmitter block is designed in GNU Radio and then the input signal of 1.42 GHz is pumped out via the transmitter omni directional antenna using USRP and is shown in Fig. 2. The receiver block is designed for the test antenna for receiving the signal from the transmitter. The rotation of the receiver antenna (Test microstrip antenna) is automated using Arduino controlled stepper motor as stated earlier is shown in Fig. 3a. By the combined action of the transmitter and receiver

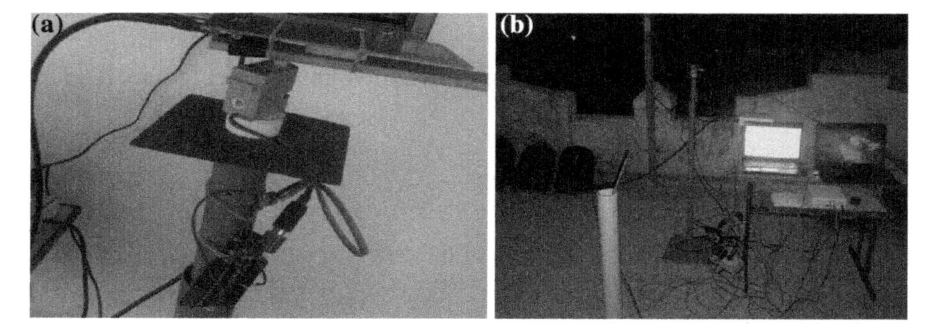

Fig. 3 **a** Test antenna mounted on to the test bed with Arduino controlled servo motor. **b** Experimental setup that depicts receiving and transmitting antenna with the control unit

the radiation pattern [6–8] and Power Spectrum of the antenna is calculated. The Power spectrum of the received signal is measured by taking the magnitude square of the received signal and is shown in Fig. 5. Received signal power is calculated using newly designed Frobenius norm block. The FFT (Fast Fourier Transform) of the received signal is taken in GRC. Observed FFT's of the received signal over 360° is used in plotting the radiation pattern of the test antenna by using of Radiation Pattern Plotter [13]. For better accuracy experiment should perform in open space at proper far field distance.

5 Antenna's Parameter Estimation Using Universal Software Radio Peripheral and GNU Radio Companion

5.1 Radiation Pattern Measurement

The radiation pattern of an antenna determines the direction along which strength of a radio wave is the maximum (or minimum). This also defines the strength of the wave directed by the antenna. The radiation pattern is measured in the far field region with constant radial distance and frequency. Typical radiation pattern leaves main lobe at 3 dB beam width and rest lies in minor lobes [6–8].

Experimental setup as shown in Fig. 3 is used in measuring the radiation pattern of the test antenna in both E and H planes. This is accomplished be properly choosing the mode of excitation beneath the patch. The TM_{11} mode radiates strongest in the broadside direction while the TM_{21} has a null in the broadside direction. The radiation patterns of the higher order modes TM_{02} and TM_{30} also exhibit null in the broadside direction. In this paper, the antenna is analysed in TM_{11} mode. The observed results are compared with the standard anechoic chamber measurement.

5.2 Half Power Beam Width and Front to Back Ratio

From the radiation pattern plot, Half Power Beam Width and front to back ratio of the received signal is calculated.

$$HPBW = \frac{k\lambda}{d} \tag{8}$$

where k is the factor depends on shape of the reflector.

5.3 Power Spectrum Measurement

Power spectrum analysis is necessary while processing the signals at a higher level. In this paper a new way of measuring the power spectrum is proposed. The process involves in acquiring received signal in .DAT format followed by the conversion of the signal from vector to stream. After such conversion Fast Fourier Transform (FFT) of the stream is taken. Then the magnitude of the signal is squared. This square of the magnitude will give the power spectrum of the received signal.

5.4 Received Signal Power

A new method to find the received signal power is proposed in this paper. In this paper Frobenius norm method is used for the calculation of receiving signal power. This way of calculating received signal power is computationally efficient and fast. In this method each value of the received signal is squared and added, by this way the power of the signal can be calculated.

$$\|A\|_F = \sum_{IJ} |A|^2 \tag{9}$$

where A is the recorded data (.dat format), I and J corresponds to rows and columns of the recorded data.

6 Results and Discussion

Microstrip antenna's performance is analysed with the above framework efficiently. Radiation pattern, Power spectrum and Received signal power are calculated and is compared with standard anechoic chamber measurements. The main purpose of this proposed system is to replace the expensive and hard handle devices for analysing antenna's performance.

6.1 Economic Advantage of the Proposed Method Over Standard Anechoic Chamber Measurement

This paper discusses the standard measurement with the proposed method of antenna's performance analysis. From the radiation pattern plot it is obvious that USRP and GNU Radio based measurement is exactly coinciding with the standard anechoic chamber measurement is shown in Fig. 4. Moreover this USRP and GNU Radio platform is not restricted with antenna's performance analysis with this

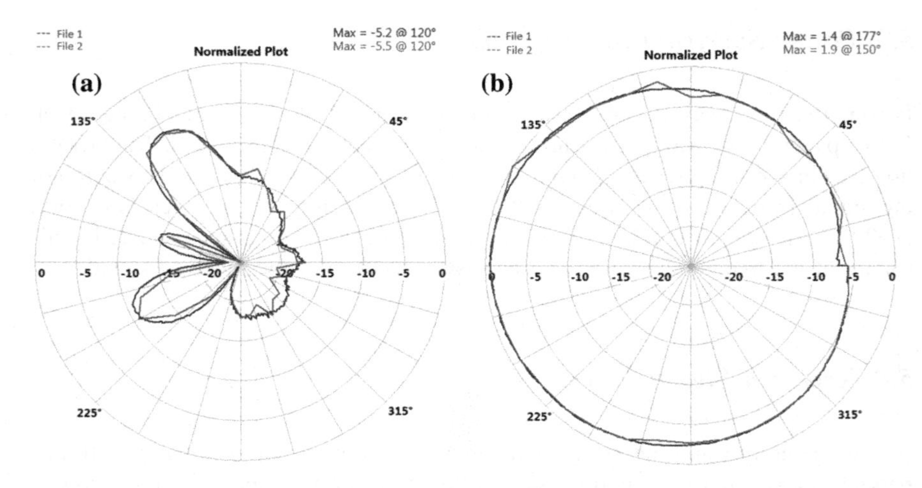

Fig. 4 **a** Radiation pattern along horizontal plane. **b** Radiation pattern along vertical plane, File 1 corresponds to anechoic chamber's reading, File 2 corresponds to GNU Radio and USRP based measurement's reading

Table 1 Performance comparison of GNU Radio based measurement with anechoic chamber measurement

Sl.No	GNU Radio and USRP based measurement possible experiments and research	Anechoic chamber measurement
1.	Antenna's performance analysis	Radiation pattern measurement of an antenna
2.	Personal area network establishment	
3.	Real time communication like AM, FM etc.	
4.	Linear algebra related problems	
5.	Image processing tasks	
6.	Own antenna network design	
7.	Open BTS establishment	
8.	Online video streaming	

platform a researcher can perform an N number of computational and communication experiments apart from the text is shown in the Table 1.

Conventional methods for measuring the radiation pattern and power spectrum of the antenna are expensive when compared with the proposed method.

6.2 Parameter Measurement and Discussion

The radiation pattern along E and H plane of the microstrip antenna is measured with USRP and GNU Radio. The results obtained are then compared with anechoic chamber measurement. These results show that the radiation pattern measurement

Table 2 Half power beam width and received signal power measurement using GNU radio

Sl.No	Half power beam width in degree	Half power beam width in degree	Received signal power in watts
	Horizontal plane using anechoic chamber based setup	Horizontal plane using USRP and GNU radio based setup	Horizontal plane
1.	29.04	28.32	35.65

Table 3 Front to back ratio in both horizontal and vertical plane

Sl.No	Front to back ratio of anechoic in dB		Front to back ratio of GNU radio and USRP setup measurement in dB	
	Horizontal plane	Vertical plane	Horizontal plane	Vertical plane
1.	0.9987	0.704730	0.70366	0.9815

in both the ways is almost same and is depicted in Fig. 4. The plot also shows that the antenna can radiate maximum along a direction with minimum side lobes. Front to Back end ratio, Half Power Beam Width is also calculated from the graph using (8).

Calculation of received signal power is done with the help of Frobenius norm block.

Table 2 shows the measured HPBW value of both anechoic chamber and USRP and GNU Radio setup measurement. Front to Back end ratio measurement is shown in Table 3.

6.3 Radiation Pattern Measurement and Power Spectrum Curve

Figure 4 shows the radiation pattern of the test microstrip antenna. From the radiation pattern it is observed that in vertical plane antenna can able to radiate in all the directions equally. But in horizontal plane antenna can produce maximum radiation a 120°. Moreover from the Fig. 4 it is visible that anechoic chamber measurement is exactly coincides with the proposed system of measurement. In Fig. 4 blue lines corresponds to anechoic chamber measurement whereas the red line corresponds to the proposed system's measurement. The power spectrum of the antenna is also calculated is shown in Fig. 5. From these observations, this paper suggests an alternative way of analysing antenna's performance apart from conventional methods at low cost and better accuracy. From Fig. 5 it is shown that the antenna's operating frequency is set at 1.42 GHz and that is operated at −46.2 dB power. On the either side of the operating frequency point the power is decreasing and it is clear that the antenna can perform better at −46.2 dB operating power.

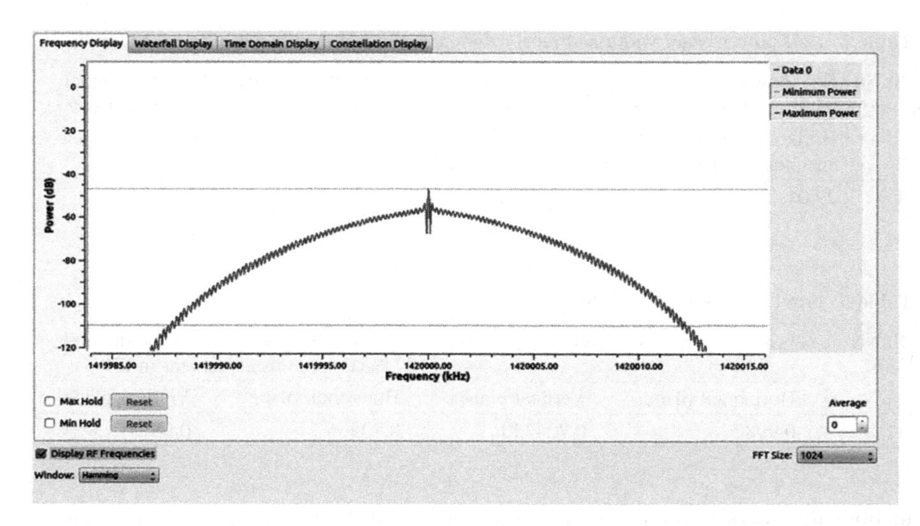

Fig. 5 Power Spectrum curve shows the maximum power at the operating frequency 1.42 GHz

7 Conclusion

The performance of the micro strip antenna is analysed using USRP and GNU Radio and parameters are calculated. The performance of the antenna is mainly restricted itself to the radiation pattern measurement. This paper provides a better and less expensive way of measuring antenna's radiation pattern using USRP and GNU Radio platform. From the radiation pattern measured, Half Power Beam Width and Front to Back Ratio of the antenna is calculated. The Received signal power of the signal is calculated using the Frobenius norm method. The further improvement of the paper involves in automating all the manual operations involved during the course of measurement. Moreover future work aims at developing a virtual lab setup for antenna's performance analysis.

References

1. Hugine, A.L.: Antenna selection for public safety cognitive radio (2006)
2. Chougale, N.A., Magdum, S.K., Patel, S.D., Dongale, T.D., Vasmbekar, P.N.: Design and development of UHF Moxon antenna. Nat. J. Sci. Inf. 93–95 (2012). ISSN: 2229-5836
3. Abirami, M., Hariharan, V., Sruthi, M.B., Gandhiraj, R., Soman, K.P.: Exploiting GNU radio and USRP: an economical test bed for real time communication systems. In: Fourth IEEE International Conference on Computing Communication and Networking Technologies, 4–6 (2013)
4. Fähnle, M.: Software—defined radio with GNU radio and USRP2 hardware front end: setup and FM/GSM applications (2010)

5. Sruthi, M.B, Gandhiraj, R., Soman, K.P.: Realization of wireless communication system in SDR using GNU radio. Int. J. Adv. Res. Comput. Sci. Softw. Eng. **3**(6) (2013)
6. Ajitha, T., Joy, E., Joyce, A., Gandhiraj, R.: Radiation pattern measurement of log- periodic antenna on GNU raio platform. In: IEEE International Conference on Green Computing. Communication and Electrical Engineering (ICGCCEE'14) (2014)
7. Sreelakshmi, S., Ashitha, P.S., Pradeep, A., Gandhiraj, R.: Measurement of radiation pattern of quadrifilar helix antenna (QHA) using GNU radio. In: IEEE International Conference on Green Computing. Communication and Electrical Engineering (ICGCCEE'14) (2014)
8. Sreethivya, M., Dhanya, M.G., Nimisha, C., Gandhiraj, R., Soman, K.P.: Radiation pattern of Yagi-Uda antenna using GNU radio platform. In: International Conference on Trends in Technology for Convergence (TITCON '14) (2014)
9. Kokila, M., Nandhini, K., Vishnu: Minor project thesis—frobenius norm block creation (2014)
10. http://code.ettus.com/redmine/ettus/. Accessed on May 2014
11. Rajendran, J., Peter, R., Soman, K.P.: Design and optimization of band pass filter design for software defined radio telescope. Int J Inf. Electron. Eng. **2**(4), 649–651 (2012)
12. Park, J., Na, H., Baik, S.H.: Design of a modified L—probe fed microstrip patch antenna. IEEE Antennas Wirel. Propag. Lett. **3** (2004)
13. http://www.softpedia.com/get/Science-CAD/Antenna-Radiation-Diagram-Plotter.shtml

Selection of Control Parameters of Differential Evolution Algorithm for Economic Load Dispatch Problem

Narendra Kumar Yegireddy, Sidhartha Panda, Umesh kumar Rout and Rama Kishore Bonthu

Abstract Differential evolution (DE) is a population based heuristic search algorithm for global optimization capable of handling non differentiable, nonlinear and multi-modal objective functions. The performance of this type of heuristic algorithms is heavily dependent on the setting of control parameters as proper selection of the control parameters is very important for the success of the algorithm. In this paper, a study of control parameters on the performance of DE algorithm for economic load dispatch problem has been addressed. The effectiveness of the proposed method is illustrated on a standard *IEEE* 30 bus test system. The results of the effect of the variation of different strategy, control parameters are presented. It is observed that the DE algorithm may fail in finding the optimal value if the strategy and control parameters are not chosen carefully.

Keywords Differential evolution · Optimal power flow · Control parameters · Economic dispatch · IEEE 30 bus test system

1 Introduction

With the development of modern power systems, economic load dispatch (ELD) problem has received an increasing attention. The main objective of ELD is to schedule the committed generation unit outputs so as to meet the load demand at

N.K. Yegireddy (✉) · S. Panda · U.k. Rout
Department of Electrical and Electronics Engineering,
Veer Surendra Sai University of Technology (VSSUT),
Burla 768018, Odisha, India
e-mail: Narenyegireddy@gmail.com

S. Panda
e-mail: panda_sidhartha@rediffmail.com

R.K. Bonthu
Department of Electrical and Electronics Engineering,
Lendi Institute of Engineering and Technology,
Vizianagaram, Andhra Pradesh, India
e-mail: ramkishore.bonthum@gmail.com

L.C. Jain et al. (eds.), *Computational Intelligence in Data Mining - Volume 3*,
Smart Innovation, Systems and Technologies 33, DOI 10.1007/978-81-322-2202-6_22

minimum operating cost, while satisfying all unit and system equality and inequality constraints [1]. Over the years, many efforts have been made to solve the problem, incorporating different kinds of constraints or multiple objectives, through various mathematical programming and optimization techniques. The input-output characteristics of modern generators are nonlinear and highly constrained. Hence, conventional methods may fail to provide satisfactory results. Evolutionary computational approaches, on the other hand, gained remarkable importance in this area since they can provide solutions leading to a considerable reduction of generator fuel consumption and also provide these solutions at low computational costs [2].

Differential evolution (DE) is a branch of evolutionary algorithms developed by Stron and Price in 1995 [3] is an improved version of genetic algorithm for faster optimization. DE is a population based direct search algorithm for global optimization capable of handling non differentiable, nonlinear and multi-modal objective functions, with few, easily chosen, control parameters. DE differs from other evolutionary algorithms in the mutation and recombination phases. DE uses weighted differences between solution vectors to change the population whereas in other stochastic techniques such as perturbation occurs in accordance with a random quantity. In view of the above, many researchers have applied DE to various engineering problems [4–7]. The general convention used in mutation process, is DE/x/y, where DE stands for differential evolution, x represents a string denoting the type of the vector to be perturbed (whether it is randomly selected or it is the best vector in the population with respect to fitness value) and y is the number of difference vectors considered for perturbation of x. It has been reported in literature that there are ten possible working strategies of DE and proper selection of a particular strategy is very much necessary [8]. The success of DE is also heavily dependent on setting of control parameters namely; mutation strategy, step size F and crossover probability of CR. One of the main problems in evolution strategies of DE is to choose the control parameters and mutation strategy such that it exhibits good behavior.

In this paper, tests for various mutation strategy and parameter settings of DE are conducted for the economic load dispatch with generator constraints. The influence of parameter settings of DE is tested on an IEEE 30 bus test system and results are presented. Based on the results obtained, recommendations are suggested for the mutation scheme and suitable range of control parameters of DE for the present problem.

2 Problem Formulation

The primary objective of ELD is to allocate the most optimum real power generation level for all the available generating units in the power station that satisfies the load demand at the same time meeting all the operating constraints. The fuel cost characteristics of each generator unit, is represented by a quadratic equation and the valve point effect is modeled. Mathematically, the problem is represented as:

$$\text{Minimize} \quad \sum_{i=1}^{n} F_i(P_i) \tag{1}$$

where $F_i(P_i)$ is the fuel cost equation of the i-th plant and is expressed as continuous quadratic equation

$$F_i(P_i) = a_i P_i^2 + b_i P_i + c_i \tag{2}$$

Here $a_{i,}$ b_i and c_i are the fuel consumption cost coefficients of the i-th unit. P_i is the real power output (MW) of i-th generator, n is the number of online thermal generating units to be dispatched. The constants $a_{i,}$ b_i and c_i are generally obtained by curve fitting.

The cost is minimized with the following generator capacities and active power balance constraints.

$$P_{i,\min} \leq Pi \leq P_{i,\max} \quad \text{for} \quad i = 1, 2, \ldots n. \tag{3}$$

$$\sum_{i=1}^{n} P_i = P_D + P_L \tag{4}$$

where, $P_{i,\min}$, $P_{i,\max}$ are the minimum and maximum power generation by ith unit respectively, P_D is the total power demand and P_L is the total transmission loss. To calculate the transmission loss, the load flow solutions are obtained by Newton Raphson method and losses are calculated as:

$$P_L = real(\sum_{j}^{n} V_i Y_{ij}^{*} V_j) \quad \text{for} \quad i = 1, 2, \ldots n. \tag{5}$$

The constrained optimization problem is converted to an unconstrained problem by penalty function method as:

$$\text{Minimize} \sum_{i=1}^{n} F_i(P_i) + 1000 * abs\left(\sum_{i=1}^{n} P_i - P_D - P_L\right) \tag{6}$$

With all these functions, the main objective of this study is to find the optimal values of strategy and control parameters for DE based combined economic emission dispatch with above mentioned constraints.

3 Differential Evolution

Differential Evolution (DE) algorithm is a stochastic, population-based optimization algorithm introduced by Storn and Price in 1996 [3]. DE works with two populations; old generation and new generation of the same population. The size of the population is adjusted by the parameter N_P. The population consists of real valued vectors with dimension D that equals the number of design parameters/control variables. The population is randomly initialized within the initial parameter bounds. The optimization process is conducted by means of three main operations: mutation, crossover and selection. In each generation, individuals of the current population become target vectors. For each target vector, the mutation operation produces a mutant vector, by adding the weighted difference between two randomly chosen vectors to a third vector. The crossover operation generates a new vector, called trial vector, by mixing the parameters of the mutant vector with those of the target vector. If the trial vector obtains a better fitness value than the target vector, then the trial vector replaces the target vector in the next generation. The evolutionary operators are described below [4–7].

3.1 Initialization

Chromosome For each parameter j with lower bound X_j^L and upper bound X_j^U, initial parameter values are usually randomly selected uniformly in the interval $[X_j^L, X_j^U]$.

3.2 Mutation

For a given parameter vector $X_{i,G}$, three vectors $(X_{r1,G}\, X_{r2,G}\, X_{r3,G})$ are randomly selected such that the *indices i,* $r1$, $r2$ and $r3$ are distinct. A donor vector $V_{i,G+1}$ is created by adding the weighted difference between the two vectors to the third vector as:

$$V_{i,G+1} = X_{r1,G} + F.(X_{r2,G} - X_{r3,G}) \tag{7}$$

where F is a constant from (0, 2).

3.3 Crossover

The Three parents are selected for crossover and the child is a perturbation of one of them. The trial vector $U_{i,G+1}$ is developed from the elements of the target vector $(X_{i,G})$ and the elements of the donor vector $(X_{i,G})$.Elements of the donor vector enters the trial vector with probability CR as:

$$U_{j,i,G+1} = \begin{cases} V_{j,i,G+1} & if \quad rand_{j,i} \leq CR \quad or \quad j = I_{rand} \\ X_{j,i,G+1} & if \quad rand_{j,i} > CR \quad or \quad j \neq I_{rand} \end{cases} \tag{8}$$

With $rand_{j,\,i} \sim U$ (0,1), I_{rand} is a random integer from (1, 2, ….D) where D is the solution's dimension i.e. number of control variables. I_{rand} ensures that $V_{i\,,\,G+1} \neq X_{i,G}$.

3.4 Selection

An The target vector $X_{i,G}$ is compared with the trial vector $V_{i\,,\,G+1}$ and the one with the better fitness value is admitted to the next generation. The selection operation in DE can be represented by the following equation:

$$X_{i,G+1} = \begin{cases} U_{i,G+1}\, f(U_{i,G+1}) < f(X_{i,G}) \\ X_{i,G} \quad otherwise. \end{cases} \tag{9}$$

where $i \in [1, N_P]$.

DE offers several strategies for optimization. They are classified according to the following notation such as *DE/x/y/z*, where *x* refers to the method used for generating parent vector that will form the base for mutated vector, *y* indicates the number of difference vector used in mutation process and *z* is the crossover scheme used in the cross over operation to create the offspring population [8]. The symbol *x* can be 'rand' (randomly chosen vector) or 'best' (the best vector found so far). The symbol *y*, number of difference vector, is normally set to be 1 or 2. Each mutation strategy was combined with either the exponential (notation: 'exp') type crossover or the binomial (notation: 'bin') type crossover. This yielded 5 × 2 = 10 DE strategies. The mutation process given by Eq. (7) has to be changed for each strategy.

4 Results and Discussions

The proposed method has been tested on IEEE 30 bus system. The system consists of 6 generators, 4 transformers, 41 lines and 2 shunt reactors. The study on the effect of different parameter settings of differential evolution (DE) is performed for a demand of 283.4 MW.

Table 1 Effect of variation of strategy with $N_P = 20$, $F = 0.8$, $CR = 0.8$, Iterations = 200

Sl. no.	Strategy	Minimum cost	Maximum cost	Mean cost
1	DE/best/1/exp	801.84359	801.84359	**801.84359**
2	DE/rand/1/exp	801.84359	801.84371	801.84365
3	DE/rand-to-best/1/exp	801.84359	801.84359	**801.84359**
4	DE/best/2/exp	801.84359	802.0537	801.94865
5	DE/rand/2/exp	801.8518	802.59232	802.22206
6	DE/best/1/bin	801.84359	801.84359	**801.84359**
7	DE/rand/1/bin	801.84359	801.84374	801.84366
8	DE/rand-to-best/1/bin	801.84359	801.84359	**801.84359**
9	DE/best/2/bin	801.84359	801.84977	801.84668
10	DE/rand/2/bin	801.84652	802.18891	802.01771

4.1 Effect of Variation of Strategy

To check the effect of variation strategy, all ten possible strategies are applied. The population size, step size and crossover probability are set to 20, 0.8 and 0.8 respectively following the recommendation from the literature [4–7]. Ten (10) independent runs are performed for every strategy and the maximum number of iteration is set at 200. Table 1 shows the minimum cost, worst cost and mean cost obtained in the 10 runs for all strategy. It is clear from Table 1 that strategies 1, 3, 6 and 8 give the minimum cost of 801.84359 $/hr. Hence these strategies are chosen for further investigation.

4.2 Effect of Variation of Crossover Probability and Step Size

In order to obtain the effect of the variation of crossover probability (CR) and step size, the CR value is change 0.2–0.8 in steps of 0.2 and for each value of CR, the step size is changed from 0.4 to 1.6 in steps of 0.4. The number of iterations and population size is fixed to 200 and 20 respectively for all cases. The results of above investigations are presented in Tables 2, 3, 4 and 5 for strategies 1, 3, 6 and 8 respectively obtained in the 10 runs. It can be seen from the Tables 2, 3, 4 and 5 that the minimum cost is obtained 7 times for strategy 1, 6 times for strategy 3, 8 times for strategy 6 and 5 times for strategy 8. It can also be seen from the Tables 2, 3, 4 and 5 that with a step size of 0.4–0.8, minimum cost is obtained for almost all strategies. Further, it can be observed that a cross over probability of 0.2–0.4 gives the optimum result in maximum number of times.

Table 2 Effect of variation of CR and F with N_P = 20, Iterations = 200 for **Strategy 1**

Sl. no.	Value of CR and F		Minimum cost	Maximum cost	Mean cost
1	CR = 0.2	F = 0.4	801.84359	801.84359	**801.84359**
		F = 0.8	801.84359	801.84371	**801.84359**
		F = 1.2	801.84359	801.84376	801.84368
		F = 1.6	801.84367	801.84536	801.84452
2	CR = 0.4	F = 0.4	801.84359	801.84359	**801.84359**
		F = 0.8	801.84359	801.84359	**801.84359**
		F = 1.2	801.84359	801.8436	801.843595
		F = 1.6	801.84379	801.84652	801.84515
3	CR = 0.6	F = 0.4	801.84359	801.84359	**801.84359**
		F = 0.8	801.84359	801.84359	**801.84359**
		F = 1.2	801.84359	801.86234	801.85297
		F = 1.6	801.84383	801.85833	801.85108
4	CR = 0.8	F = 0.4	801.8436	801.8436	801.8436
		F = 0.8	801.84359	801.84359	**801.84359**
		F = 1.2	801.84359	801.84811	801.84585
		F = 1.6	801.84522	802.6156	802.21041

Table 3 Effect of variation of CR and F with N_P = 20, Iterations = 200 for **Strategy 3**

Sl. no.	Value of CR and F		Minimum cost	Maximum cost	Mean cost
1	CR = 0.2	F = 0.4	801.84359	801.84359	**801.84359**
		F = 0.8	801.84359	801.84371	**801.84359**
		F = 1.2	801.8436	801.84782	801.84571
		F = 1.6	801.84802	801.93006	801.88904
2	CR = 0.4	F = 0.4	801.8436	801.8436	801.8436
		F = 0.8	801.84359	801.84359	**801.84359**
		F = 1.2	801.84359	801.84625	801.84492
		F = 1.6	801.84508	802.12369	801.98438
3	CR = 0.6	F = 0.4	801.84359	801.84359	**801.84359**
		F = 0.8	801.84359	801.84359	**801.84359**
		F = 1.2	801.8436	801.85103	801.84731
		F = 1.6	801.88534	802.40428	802.14468
4	CR = 0.8	F = 0.4	801.8436	802.6946	802.2691
		F = 0.8	801.84359	801.84359	**801.84359**
		F = 1.2	801.84375	801.98065	801.9122
		F = 1.6	802.07429	804.11937	803.09688

Table 4 Effect of variation of CR and F with N_P = 20, Iterations = 200 for **Strategy 6**

Sl. no.	Value of CR and F		Minimum cost	Maximum cost	Mean cost
1	CR = 0.2	F = 0.4	801.84359	801.84359	**801.84359**
		F = 0.8	801.84359	801.84371	**801.84359**
		F = 1.2	801.84359	801.84363	801.84361
		F = 1.6	801.8437	801.84646	801.84508
2	CR = 0.4	F = 0.4	801.84359	801.84359	**801.84359**
		F = 0.8	801.84359	801.84359	**801.84359**
		F = 1.2	801.84359	801.84359	**801.84359**
		F = 1.6	801.84362	801.84683	801.845225
3	CR = 0.6	F = 0.4	801.84359	801.96631	801.90495
		F = 0.8	801.84359	801.84359	**801.84359**
		F = 1.2	801.84359	801.84359	**801.84359**
		F = 1.6	801.84367	801.85022	801.84695
4	CR = 0.8	F = 0.4	801.8436	801.9663	801.905
		F = 0.8	801.84359	801.84359	**801.84359**
		F = 1.2	801.84359	801.84361	801.8436
		F = 1.6	801.84377	802.00616	801.942965

Table 5 Effect of variation of CR and F with N_P = 20, Iterations = 200 for **Strategy 8**

Sl. no.	Value of CR and F		Minimum cost	Maximum cost	Mean cost
1	CR = 0.2	F = 0.4	801.84359	801.84359	**801.84359**
		F = 0.8	801.84359	801.84371	**801.84359**
		F = 1.2	801.84362	801.84476	801.84419
		F = 1.6	801.8447	801.91231	801.87851
2	CR = 0.4	F = 0.4	801.8436	801.8436	801.8436
		F = 0.8	801.84359	801.84371	**801.84359**
		F = 1.2	801.84363	801.84514	801.84439
		F = 1.6	801.84742	802.00788	801.92765
3	CR = 0.6	F = 0.4	801.84359	801.94509	801.84934
		F = 0.8	801.84359	801.84359	**801.84359**
		F = 1.2	801.84361	801.85242	801.84802
		F = 1.6	801.84954	802.18703	802.01829
4	CR = 0.8	F = 0.4	801.8436	801.9797	801.9116
		F = 0.8	801.84359	801.84359	**801.84359**
		F = 1.2	801.84387	801.91292	801.8784
		F = 1.6	802.0519	804.0572	803.036355

Table 6 Effect of variation of iterations and N_p with CR = 0.4, F = 0.8 for **Strategy 1**

Sl. no.	Value of iteration and NP		Minimum cost	Maximum cost	Mean cost
1	Iter = 50	NP = 20	801.84380	801.84940	801.84660
		NP = 40	801.84368	801.84627	801.84497
		NP = 60	801.84374	801.84412	801.84393
		NP = 80	801.84368	801.84448	801.84408
2	Iter = 100	NP = 20	801.84359	801.84359	**801.84359**
		NP = 40	801.84359	801.84359	**801.84359**
		NP = 60	801.84359	801.84359	**801.84359**
		NP = 80	801.84359	801.84359	**801.84359**
3	Iter = 150	NP = 20	801.84359	801.84359	**801.84359**
		NP = 40	801.84359	801.84359	**801.84359**
		NP = 60	801.84359	801.84359	**801.84359**
		NP = 80	801.84359	801.84359	**801.84359**

4.3 Effect of Variation of Iterations and Populations

In the above analysis we found strategy-1 and strategy-6 are giving better results for optimal operation. Further the analysis has been done for the strategies 1 and 6 by keeping the crossover probability (CR) 0.4 and step size (F) 0.8 constant and changing the population size from 20 to 80 in steps of 20, iterations from 50 to 150 in steps of 50. The results of above investigations are presented in Tables 6 and 7 for strategies 1 and 6 respectively obtained in the 10 run. It also observed from Tables 6 and 7 that the minimum cost is obtained 8 times for both strategies 1 and 6.

Table 7 Effect of variation of iterations and N_p with CR = 0.4, F = 0.8 for **Strategy**

Sl. no.	Value of iteration and NP		Minimum cost	Maximum cost	Mean cost
1	Iter = 50	NP = 20	801.84368	801.84616	801.84492
		NP = 40	801.84372	801.84440	801.84406
		NP = 60	801.84367	801.84418	801.84392
		NP = 80	801.84370	801.84434	801.84402
2	Iter = 100	NP = 20	801.84359	801.84359	**801.84359**
		NP = 40	801.84359	801.84359	**801.84359**
		NP = 60	801.84359	801.84359	**801.84359**
		NP = 80	801.84359	801.84359	**801.84359**
3	Iter = 150	NP = 20	801.84359	801.84359	**801.84359**
		NP = 40	801.84359	801.84359	**801.84359**
		NP = 60	801.84359	801.84359	**801.84359**
		NP = 80	801.84359	801.84359	**801.84359**

5 Conclusion

The success of DE is heavily dependent on setting of mutation strategy and control parameters. In this paper, selection of mutation strategy and the parameter study of control parameters DE are conducted for economic load dispatch problem applied to an *IEEE* 30- bus power system with generator constraints and transmission loss. The effects of variation of strategy and control parameters (crossover probability and step size) on the performance of DE are investigated and the results are presented. It is observed that premature convergence and even stagnation can occur due to wrong selection of strategy and control parameters. Based on the results obtained, recommendations are suggested for the mutation strategy and range of control parameters of DE.

References

1. Wood, J., Wollenberg, B.F.: Power Generation, Operation and Control, 2nd edn. Wiley, New York (1996)
2. Padhy, N.P.: Artificial Intelligence and Intelligent Systems. Oxford University Press, Oxford (2005)
3. Storn, R., Price, K.: Differential evolution—a simple and efficient adaptive scheme for global optimization over continuous spaces. J. Global Optim. **11**, 341–359 (1997)
4. Panda, S.: Differential evolutionary algorithm for TCSC-based controller design. Simul. Model. Pract. Theory **17**, 1618–1634 (2009)
5. Panda, S.: Robust coordinated design of multiple and multi-type damping controller using differential evolution algorithm. Int. J. Electr. Power Energy Syst. **33**(4), 1018–1030 (2011)
6. Panda, S., Swain, S.C., Baliarsingh, A.K.: Differential evolution algorithm for simultaneous tuning of excitation and FACTS-based controller. Int. J. Bio-Inspired Comput. **2**, 404–412 (2010)
7. Panda, S.: Differential evolution algorithm for SSSC-based damping controller design considering time delay. (Article in press). J. Franklin Inst. doi:10.1016/j.jfranklin.2011.05.011
8. Das, S., Abraham, A., Konar, A.: Particle swarm optimization and differential evolution algorithms: technical analysis. Applications and hybridization perspectives. Stud. Comput. Intell. **116**, 1–38 (2008)

Wireless QoS Routing Protocol (WQRP): Ensuring Quality of Service in Mobile Adhoc Networks

Nishtha Kesswani

Abstract Quality of Service ensures certain level of parameters that need to be fulfilled while transmitting data through a network. In this paper, implementation of Quality of Service constraints such as bandwidth, delay, jitter and propagation loss in wireless adhoc networks has been described. Taking into consideration these constraints, a unique QoS routing protocol, Wireless QoS Routing Protocol (WQRP) has been designed.

Keywords Wireless QoS routing algorithm (WQRA) · Propagation loss · Delay · Jitter · Bandwidth · QoS guarantee

1 Introduction

Providing reliable network performance has been a challenge not only in wired but in wireless networks as well. Also it is essential that the network resources are utilized in an optimum manner. Several challenges associated with providing quality of service include delivering packets in minimum delay, reducing packet loss and increasing the throughput. Though challenges exist for wired networks as well but providing quality of service in wireless networks incorporates additional challenges. These additional challenges as well as several techniques for overcoming these are described in this paper.

N. Kesswani (✉)
Central University of Rajasthan, Ajmer, Rajasthan, India
e-mail: nishtha@curaj.ac.in

L.C. Jain et al. (eds.), *Computational Intelligence in Data Mining - Volume 3*,
Smart Innovation, Systems and Technologies 33, DOI 10.1007/978-81-322-2202-6_23

2 Quality of Service (QoS)

There are several mechanisms that are used for providing the desired quality of service. Amongst these, resource reservation and traffic prioritization are mostly used. In resource reservation, the resources are allocated as per the request while in traffic prioritization; preferential treatment is given to certain traffic as compared to others [1]. Different applications or data flows are given different priorities.

3 QoS Guarantees

3.1 QOS Service Guarantees

Three basic levels of service that are provided include Best effort, differentiated and guaranteed QoS. Differentiated QoS provides Soft Quality of Service while guaranteed QoS provides hard quality of service [2]. A certain level of service is guaranteed to a particular data flow or application. For instance certain level of packet loss or delay or bit error rate may be guaranteed.

Service guarantees are typically made for one or more of the following four characteristics [1]. A guarantee of delay assures the sender and receiver that it will take no more than a specified amount of time for a packet of data to travel from sender to receiver. A guarantee of loss assures the sender and receiver that no more than a specified fraction of packets will be lost during transmission. A guarantee of jitter assures the sender and receiver that the delay will not vary by more than a specified amount. Finally, a guarantee of throughput assures the sender and receiver that in some specified unit of time no less than some specified amount of data can be sent by the sender to receiver.

3.2 QoS Schemes

QOS schemes can work on various parts of the network architecture [1]. The focus has been placed on the data link layer of the protocol stack and the layers below. Schemes at data link layer include 802.11e, Adaptive inter-frame spacing in 802.11, 802.16, INSIGNIA [1]. Other schemes at Network layer include SWAN and Integrated Adaptive QoS [1].

4 Wireless QoS Routing Algorithm (WQRA)

Due to the characteristics of wireless networks, using on-demand routing protocols is a better choice. Most on-demand QoS routing protocols in the literature can be roughly divided into three phases: route discovery, route selection, and route maintenance.

- **Route Discovery**

When a source node has data to transmit, it first checks whether the QoS requirements (i.e., bandwidth or delay or jitter) can be satisfied in its neighbor or not. If yes, the source node broadcasts a QoS Route Request (QRREQ) packet to its neighbors. Whenever a node other than the destination receives a QRREQ packet, it will do the same operations as the source node.

After the QRREQ packet arrives at the destination node, a QoS Route Reply (QRREP) packet will be sent back to the source node then the route discovery completes.

- **Route Selection**

Determine which route will be used for data transmission by some metrics, such as end-to-end delay, or bandwidth etc.

- **Route Maintenance**

Upon detecting a link breakage or violation of the QoS requirements possibly due to mobility or congestion, the upstream node of this link sends a Route Error (RERR) packet to the source node for rediscovering a new QoS route, and the downstream node sends a Release packet to the destination to release the resources reserved before.

In this paper, a QoS algorithm that takes into consideration the degree of nodes, transmission power, mobility, bandwidth and battery power of the nodes has been proposed. Each node in the MANET needs to keep track of its two-hop neighbors' information. This can be done by periodical exchange of Hello messages.

Hello packet is also needed to acquire available bandwidth for admission control. Connecting the nodes is an important issue since the nodes need to communicate with each other and connectivity is the prime concern of the Internet users. Finally, detailed simulation experiments have been conducted that demonstrate our algorithm yields better results as compared to the existing Best effort protocols in terms of Quality of Service.

4.1 Design Philosophy

4.1.1 QoS Route Discovery

All nodes in the network broadcast a Hello packet periodically. In a Hello packet, the estimated available bandwidth is included. Once the source node receives a QoS call request from its application layer, it first checks whether it has enough link resources available to anyone of its neighbors. If the available bandwidth cannot meet the QoS requirement of this application, the source node denies this request. If there is sufficient bandwidth, the source node broadcasts a QRREQ packet to search for a QoS path.

When an intermediate node receives a QRREQ packet, it also checks its available bandwidth as the source node does. If the available bandwidth is enough, it continues broadcasting a QRREQ packet.

After receiving the QRREQ packet at the destination node, it sends a QRREP packet back to the source node along the reverse path established. Whenever a node receives the QRREP packet, it sets a forward path to the node that it receives the QRREP packet from. Note that the forward path is the same as the reverse path with the direction reversed.

The procedure that is invoked after initialization is QRoute_discovery as shown in Fig. 1. This procedure is also invoked whenever a Route error message is received by the source. In this procedure the nodes in the network that fall on the selected path send a QOS route request QRREQ to their neighbors. As soon as the

Fig. 1 Procedure QoS route discovery

```
Procedure QRoute_discovery
Begin
  1.  If node = src then
  2.  For each node in network do
  3.   For each neighbor do
  4.      If avail_bandwidth >= min_required then
            1.  Send RREQ
            2.  QRoute_select
  5.      End If
  6.   End For
  7.  End For
  8.  End If
  9.  If node = dest then
  10.       Send QRREP on reverse_path
  11. End If
End
```

destination node receives the QRREQ, it sends the QoS route reply QRREP along the reverse_path. The reverse_path is the same as the forward_path with the directions reversed.

Compute the available bandwidth Bv of the node. During transmission a node (denoted A) continues monitoring the neighbor information. If it finds that there exists a neighboring node (denoted C) satisfying the following two conditions, it will change its original next hop (denoted B) to this neighboring node C. The bandwidth requirement can be satisfied from A to C and from C to B.

$$\frac{1}{\text{Data\_Rate}(A,C)} + \frac{1}{\text{Data\_Rate}(C,\ B)} < \frac{1}{\text{Data\_Rate}(A,B)}$$

4.1.2 QoS Route Selection

Here we simply select the shortest path from the source to the destination.

Procedure QRoute_select
Begin

1. Select forward_path between src and dest
2. QRoute_maintain

End
Procedure QoS Route Selection

4.1.3 QoS Route Maintenance

If a node fails to receive a Hello packet from one of its neighbors for a period of time, the link to this neighbor is considered broken, and any route passing this broken link is considered disconnected. When this happens, a Route Error (RERR) packet and a Release packet will be sent to the source node and the destination node respectively. The source node will rediscover a new QoS path after it receives the QRERR packet, and any node within the partial path from the broken link to the destination node will free the resources reserved earlier.

As the search for better heuristics for this problem continues, which takes into account several system parameters like the ideal node-degree, transmission power, mobility and the battery power of the nodes along with QoS parameters such as bandwidth, delay, jitter and packet loss as shown in Fig. 2.

Fig. 2 Procedure QoS route maintenance

```
Procedure QRoute_maintain
Begin
 1.   For each node in network do
 2.     Send Hello to neighbor
 3.     If neighbor->Hello = NULL then
 4.       Send QRERR to src and dest
 5.     End If
 6.     If node = dest then
              If delay > threshold_delay or
                 jitter >threshold_jitter or
                 packet_loss > threshold_loss then
 7.           discard packet
 8.       End If
 9.     End If
 10.  End For
End
```

4.2 System Activation and Update Policy

When a system is initially brought up, every node v broadcasts Hello packet that is registered by all other nodes lying within v's transmission range. It is assumed that a node receiving a broadcast from another node can estimate their mutual distance from the strength of the signal received. GPS (Global Positioning System) can be another solution since it is mainly used to obtain the geographical location of nodes. GPS might make the problem relatively simple, but there is a cost associated with the deployment of GPS since every mobile node must be a GPS receiver. Based on the received signal strength, every node is made aware of its neighboring nodes and their corresponding distances. Once the neighbors' list for each node is ready, our algorithm chooses the initial route. It can be noted that the mobility factor and available bandwidth is the same for all the nodes when the system is initialized.

Due to the dynamic nature of the system considered, the nodes tend to move in various directions, thus impacting the stability of the configured system. The system must be updated from time to time. The update may result in formation of new routes and possibly change the point of attachment for a node from one node to another. This is called rearrangement. The frequency of update and hence rearrangement is an important issue. If the system is updated periodically at a high frequency, then the latest topology of the system can be used to find new routes. However, this will lead to high computational cost resulting in the loss of battery power or energy. If the frequency of updates is low, there are chances that the current topological information will be outdated resulting in sessions terminated midway.

All the nodes continuously monitor their signal strength as received from the neighbors. When the mutual separation between the nodes increases, the signal strength decreases. In that case, the mobile node has to notify its current source that it is no longer able to attach itself to that source. The source tries to hand-over the data to a neighboring node.

If the destination discovers that the data has been received but QoS parameters such as delay and jitter are not within the threshold limits then the packets are discarded. Also the packet delivery ratio, i.e. the number of packets received divided by the number of packets sent is calculated. It should be ensured that these QoS parameters are within their threshold values otherwise the packets are discarded.

The objective of our QoS algorithm is to minimize the number of changes in the route. If the destination discovers that the data has been received but QoS parameters such as delay and jitter are not within the threshold limits then the packets are discarded. Also the packet delivery ratio, i.e. the number of packets received divided by the number of packets sent is calculated. It should be ensured that these QoS parameters are within their threshold values otherwise the packets are discarded.

5 Experiments and Results

The simulations have been performed on open source simulator ns2 [3] and ns3. In addition to building QoS routes, the protocol also builds best effort routes. A mobile adhoc network of 50 nodes has been setup in an area of 1,000 m by 1,000 m. Random movement of nodes has been modeled using Random2DMobility model. Each node chooses a random point and starts moving towards it. The speed of movement follows a uniform distribution. The simulation has been carried out for 200 s. Constant bit rate is used to generate traffic so that the source and destination are chosen randomly. The proposed Wireless QoS routing protocol has been compared to Best Effort Qos routing protocol AODV. The protocol has been compared to AODV due to the on-demand nature of the proposed protocol. Other reactive protocols such as DSR can also be used for comparison. The two protocols have been compared on different QoS metrics such as packets delivered. Figure 3

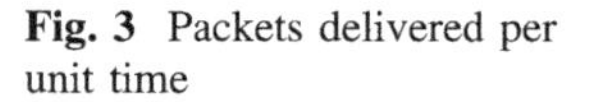
Fig. 3 Packets delivered per unit time

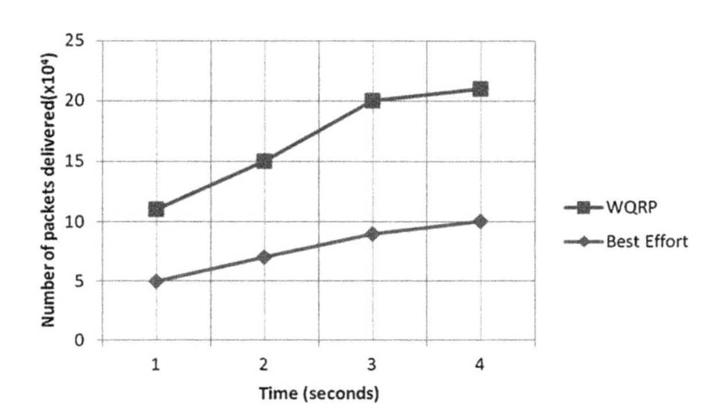

shows how significant improvement has been achieved in WQRP with respect to the number of packets delivered. Also, flooding of packets is considerably reduced as the QoS routes have been established on top of the best effort routes.

Due to the bandwidth reservation requirement, it is difficult to maintain routes especially for longer routes and is thus more time consuming as compared to the Best effort counterpart. Though average packet delay initially may be high in WQRP due to the overhead in maintenance of QoS routes on top of Best effort routes, but later on this delay becomes constant as the routes once detected can be used until the nodes change the routes again. The average packet delay has been shown in Fig. 4.

Simulations were performed for different values of threshold delays. The results indicated that higher value of threshold delay(td) ensured that more packets are delivered to the destination as for low threshold values, the packets are dropped if not delivered in time. The results have been shown in Fig. 5.

The proposed algorithm tries to ensure the QoS constraints but has the limitation in case multiple QoS routes are established. In that case several overlapping QoS routes may reserve the bandwidth and no policy for contention resolution has been used.

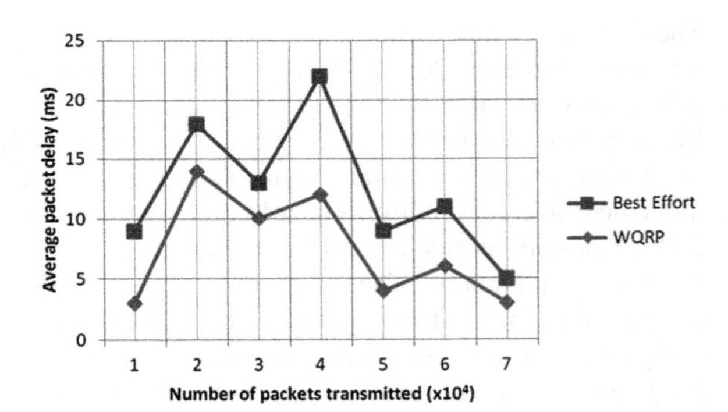

Fig. 4 Average packet delay

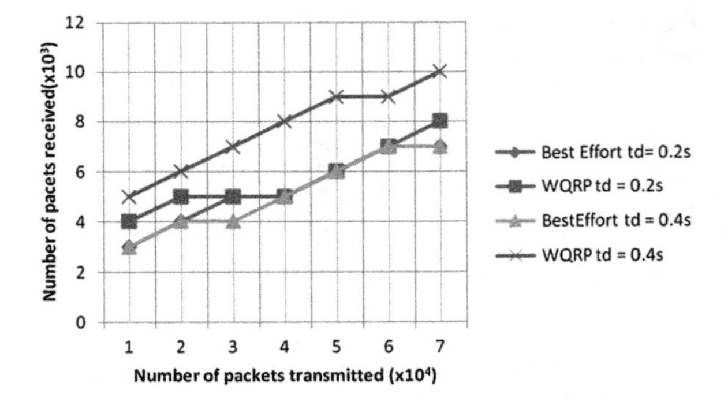

Fig. 5 Packet delivery ratio for td = 0.2 and 0.4 s

6 Conclusions

In this paper, A Qos routing algorithm Wireless QoS Routing Algorithm (WQRA) that can dynamically adapt itself with the ever-changing topology of ad hoc networks. It takes into account the combined effect of the degree of nodes, transmission power and mobility of the nodes. The algorithm is executed only when there is a QoS demand. The algorithm tries to provide Quality of Service by fulfilling constraints such as providing proper bandwidth and minimizing delay, jitter and propagation loss.

References

1. Petcher, A.: QoS in wireless networks. http://www-docs/cse574-06/ftp/wireless_qos/index.html
2. PrincetonUniversity. http://www.princeton.edu/~achaney/tmve/wiki100k/docs/Quality_of_service.html
3. The official site of NS2. www.nsnam.org
4. The 6net consortium Ipv6 deployment guide September. http://www.6net.org
5. Chung, J., Claypool, M.: NS by Example (2006)
6. Fall, K., Floyd, S.: Improving simulation for network research. USC Computer Science Department Technical Report (2007)
7. Baker, D.J., Ephremides, A.: The architectural organization of a mobile radio network via a distributed algorithm. IEEE Trans. Commun. COM-**29**, 1694–1701 (2001)
8. Basagni, S., Turgut, D., Das, S.K.: Mobility-adaptive protocols for managing large ad hoc networks. In: Proceedings of IEEE International Conference on Communications (ICC), Helsinki, Finland, pp. 1539–1543 (2011)
9. Farrel, A.: Network Quality of Service Know it All. Morgan Kaufmann, Amsterdam (2008)
10. Elliott, C., Heile, B.: Self-organizing, self-healing wireless networks. In: IEEE International Conference on Personal Wireless Communications, pp. 355–362 (2000)
11. Amokrane, A., Langar, R.: Energy efficient management framework for multi-hop TDMA based wireless networks. Comput. Netw. **62**, 29–42 (2014)
12. Frodigh, M., et al.: Wireless ad hoc networking: the art of networking without a network. Ericsson Rev. **77**(4), 248–263 (2000)
13. Haas, Z.J., et al.: Special issue on wireless adhoc networks. IEEE J. Sel. Areas Commun. **17** (8), 1454–1465 (1999)

CS-ATMA: A Hybrid Single Channel MAC Layer Protocol for Wireless Sensor Networks

Sonali Mishra, Rakesh Ranjan Swain, Tushar Kanta Samal and Manas Ranjan Kabat

Abstract In wireless sensor network (WSN), the main problem in medium access control are sleep-awake scheduling and its overhead, idle listening, and the energy consumed in transmitting the collided packets. In this paper, we propose a new energy efficient MAC protocol which combines the mechanism of CSMA/CA along with the recently developed advertisement based time division MAC protocol. In the proposed protocol, the time frame is divided into very small slots before sending the data to its receiver. The node that wants to transmit occupies a time slot at the beginning of the time frame. The node wakes up at its own slot and sense the medium in order to find whether the channel is busy or idle. The simulation results show that the proposed protocol outperforms the existing protocols in terms of packet delivery ratio, latency and energy consumption.

Keywords Wireless sensor network · MAC protocol · Advertisement based MAC · CSMA

1 Introduction

The Wireless Sensor Networks (WSNs) have a wide range of potential applications, including surveillance applications [1], habitat monitoring [2], health care [3], elder fall detection, multi-media sensor applications, and military applications. A number

S. Mishra (✉) · R.R. Swain · T.K. Samal · M.R. Kabat
Department of Computer Science and Engineering,
Veer Surendra Sai University of Technology, Burla 768018 Odisha, India
e-mail: sonaliabc.mishra@gmail.com

R.R. Swain
e-mail: rakeshswain89@gmail.com

T.K. Samal
e-mail: samaltushar1@gmail.com

M.R. Kabat
e-mail: kabatmanas@gmail.com

L.C. Jain et al. (eds.), *Computational Intelligence in Data Mining - Volume 3*,
Smart Innovation, Systems and Technologies 33, DOI 10.1007/978-81-322-2202-6_24

of MAC Protocols [4] designed for these applications perform well in single channel scenario whose primary goal is to improve energy efficiency, scalability and throughput. However, many of these WSN applications generate bursty traffic that requires high throughput and high delivery rate. During bursty traffic the large no. of packets generated within a short period leads to a high degree of channel contention and thus a high probability of packet collision.

The major cause of energy waste in conventional MAC protocols is idle listening. In MAC protocols such as IEEE 802.11 or CDMA, the nodes have to keep listening to the channel because they do not know when they might receive a message. The next cause of energy waste is collisions. If a node receives two different packets that coincide partially, then the packets are corrupted. As a result, these packets are discarded and energy is wasted in the process. The third cause is overhearing. This happens when a node receives a packet that is destined for another node. The last cause of energy waste is protocol overhead. Several MAC protocols have been proposed for WSNs that have been designed with the objective of minimizing energy consumption, often at the cost of other performance metrics such as latency and packet delivery ratio.

The Low-Energy Adaptive Clustering Hierarchy (LEACH) [5] is a cluster based hierarchical MAC layer protocol. All the sensor nodes arrange themselves to form a cluster and there is a cluster head (CH) within a cluster. All the nodes within the cluster send their data to the CH. The CH creates slotted TDMA to transmit their data. This leads to increase in energy consumption as the data aggregation and communication becomes an overhead to the CH. The increase in energy consumption leads to lowering network lifetime. The Power Aware Clustered TDMA (PACT) [6] uses a passive clustering like LEACH to create a network with CHs and gateway nodes for communication. PACT uses TDMA super frame that is composed of mini slots followed by longer data slots. In the control slots, nodes broadcast their destination addresses for the data slots. Nodes only awake in their respective data slots and turn off their transceivers in other slots, saving energy. The small control mini-slots require precise synchronization, which may pose problems for large and dynamic networks.

The Lightweight Medium Access Protocol (L-MAC) [7] uses a TDMA mechanism to provide a collision-free communication environment for the nodes. The main drawback of L-MAC is that idle-listening overhead is substantial. The Traffic-Adaptive Medium Access (TRAMA) [8] is a distributed TDMA-based protocol. In TRAMA, the owner of a slot is determined by a hash function that uses node IDs and the slot number as its parameter. When a node transmits its schedule, all nodes in its neighborhood must be awake. Energy consumption is thus heavily dependent on the number of one-hop neighbors. The S-MAC [9] is a contention based protocol which is used to save the energy wasted in idle listening. The nodes can transmit their data in the active period and nodes those do not have data to send can turn off their radio in the sleep period to reduce energy. However, in case of S-MAC, only one node can transmit data in each frame and the solution within a fixed duty cycle is also not optimal. In T-MAC [10], the active period ends when there are no activation events. When none of these events occur for a specific time-out period, the nodes undergo sleep mode. Thus the use of timeouts makes the active period in T-MAC robust to various traffic loads. However, whenever an activation event occurs, all the nodes hear the event and renew their timer which leads to unnecessary energy waste.

The advertisement based MAC (ADV-MAC) [11] and its variant advertisement-based Time-division Multiple Access (ATMA) [12] use an advertisement (ADV) period. In ADV-MAC, the SYNC period is used for synchronization followed by a short fixed period known as ADV period. In the ADV period, the nodes transmit the ADV packet which contains the id of the intended receiver. After ADV period there is a data period and a sleep period present in this protocol. The nodes having data contend for the medium in the ADV period and reserve a slot. Then in the data period they again contend the medium to send their data. In ADV-MAC protocol, the contention happen twice i.e. both in ADV period and data period which leads to more energy consumption. In Z-MAC [13], the nodes are assigned time slots and the nodes those don't own the time slot can also utilize the slots through CSMA with prioritized back-off times. The Z-MAC behaves as TDMA in high contention level (HCL) state and as CSMA in low contention level (LCL) state. The HCL can be detected by sending explicit congestion notification (ECN) message to its two-hop neighborhood. However, the ECN message becomes a burden in a busy network which further results in packet losses.

In this paper, we propose a carrier sense advertisement based time division MAC (CS-ATMA) protocol which combines the features ATMA with CSMA. The proposed protocol divides the WSN into *k* groups and the time frame is also divided into *k* sub-frames. Each sub-frame is dedicated for each group and then the sub-frame is further divided into minor slots called advertisement (ADV) slots. The sensors in each group try to occupy minor slots at the beginning of the major slot. Then, a node wakes up at its allotted time slot to sense the channel. If the channel is busy then the node goes to sleep mode and takes another time slot. Otherwise it transmits the data to the receiver.

The rest of the paper is organized as follows. In Sect. 2, the working of the proposed CS-ATMA protocol is presented. In Sect. 3, we present the performance of our proposed protocol with the existing protocols. The summary of conclusions is presented in Sect. 4.

2 Our Proposed CS-ATMA Protocol

In this section, we present the design and working of our proposed CS-ATMA protocol.

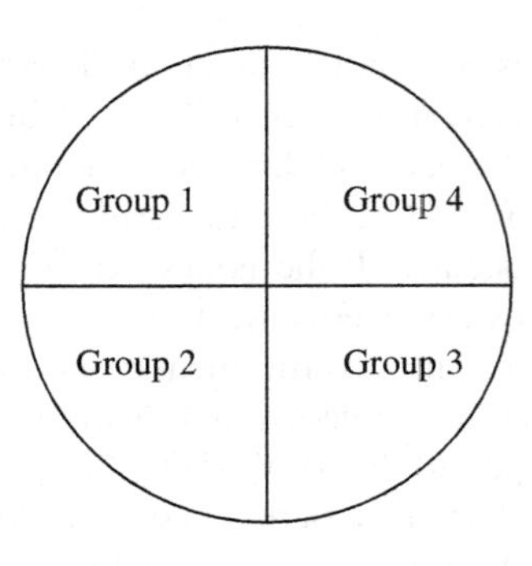

Fig. 1 Arrangement of sensors nodes in a network

2.1 System Model

We consider the sensor network consisting of N sensor nodes (v_1, $v_{2,}$ v_3, …, v_N) and a sink node v_0 in a circular area of radius *r*. The nodes (v_1, $v_{2,}$ v_3, …, v_N) are distributed along the network in a particular fashion and the sink node is at the center of the circle. We divide the total number of nodes into *k* groups and there are N/*k* numbers of nodes in each group. The value of *k* depends on the sensing range of sensor nodes. In our protocol, we have assumed the value of *k* as four. We assume that the data packets sent by all the nodes have the same data size. The traffic is always assumed to be distributed among all the nodes. At MAC layer the sensor nodes use CSMA/CA for channel access. The protocol makes use of a back up timer. When the timer of a particular node gets zero, the node wakes up at that time to sense the medium. The provision of making the group is due to the fact that in CS-ATMA all the nodes present in the network don't remain in wake up state all the time. Due to this, when nodes of one group will sense the medium for reserving the time slot at that time all the other nodes of other groups remain in sleep state. Due to which they consume less energy than other single channel MAC layer protocols. The arrangements of sensor nodes in a network are shown in Fig. 1.

2.2 Protocol Structure

In CS-ATMA, a node can transmit in different states like initialization in the network, synchronization with different nodes, discovery of time slot and last but not the least contention of the medium.

When a network is being deployed or a new node joins the network, it is said to be in initialization state. In this state the nodes only sample the medium in order to synchronize themselves in the network. In the synchronization state, the node checks for any empty slot in which it can transfer its data if any. In this state, the node also checks about its intended receiver to which it will send data. Then in the discovery of time slot state, the node tries to find a free slot in order to transmit the data by contending the medium. If the node gets success in finding a slot

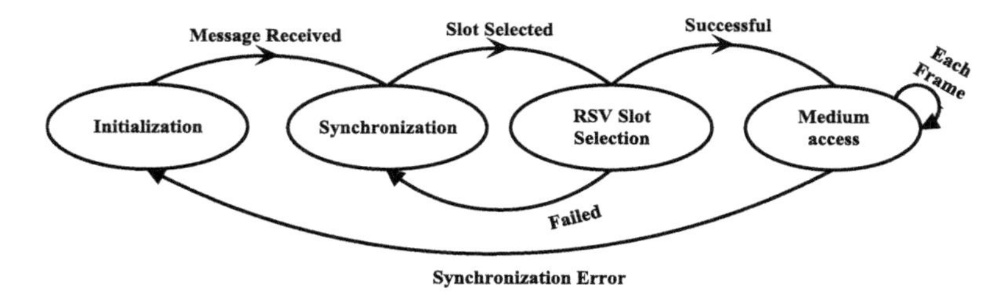

Fig. 2 States of a sensor node while transmitting data

number, it jumps into medium access state in order to send data. In the medium access state, all the nodes that have selected a time slot wait for their timer to get zero. When the timer becomes zero that particular node wakes up and sense the channel for sending the data. If it finds the medium idle then transmits the data or else it chooses a time slot and resets its timer to sense the channel. If a node failed in getting the access to channel then it waits tries to access the channel in the next frame. The states of a node are shown in Fig. 2.

2.3 Design Overview of CS-ATMA

In CS-ATMA, the time is divided into frames. Each frame begins with a fixed SYNC period and a DATA period. The method of synchronization in CS-ATMA is same as in S-MAC [9].

The DATA period is divided into k equal sub-frames and each sub-frame is dedicated for each group. Then each sub-frame is further divided into a number of small time slots known as reservation slot. The Fig. 3 shows the DATA period of our proposed protocol. The number of sub-frames is equal to the number of groups present in the network. Each sub frame is assigned to a single group so that all the nodes that have data to send can contend for the medium and can get a fair chance to send their data in that particular frame.

At the beginning of the frame, the nodes of a particular group contend for the medium to occupy for a slot. The size of the small slots depends upon the clock resolution. When the node occupies a slot in a particular sub frame, it sets a timer

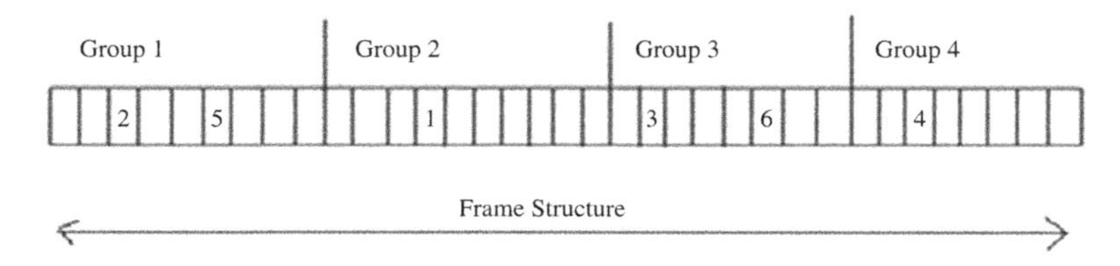

Fig. 3 DATA period of CS-ATMA

whose value is equal to the *slot number* × *slot duration*. At the beginning of the frame all nodes are in sleep states and they transmit their data in CSMA/CA fashion. When the timer of a node becomes zero, it wakes up and senses the channel to find out whether it is busy or idle. If it found that the channel is idle, it immediately sends its data.

When one node is transmitting its data, the timer of other nodes of that group doesn't get freeze. Therefore, if the timer of the any other node reaches to zero then that node wakes up to sense the channel and doesn't find the channel to be idle. Thus, it again chooses another time slot, resets its timer and goes to the sleep state till its timer expires. If a node fails to get the medium after a fixed number of trials then it tries in the next frame.

Figure 3 shows an example of the operation of our proposed CS-ATMA protocol. In this example nodes numbered 1–6 select random slots and initialize their timers. Let us assume that node 2 and 5 of group 1, node 1 of group 2, node 3 and 6 of group 3 and node 4 of group 4 have the data to transmit in the current frame. The frame is divided into four sub-frames and the nodes of those groups select the time slots from their respective groups. Suppose the duration of each reservation slot is set to 0.1 ms and a node selects slot number 6 then the timer is set at 0.6 ms. When the timer for that node becomes zero, it wakes up and senses the medium for a small period and finds it to be idle then it sends its data to the sink. In sub-frame1, if node 2 is transmitting its data and the timer of node 5 becomes zero then node 5 wakes up and found the channel to be busy. Therefore, it goes to the sleep mode without transmitting its data.

3 Simulation Results and Analysis

We implement our proposed CS-ATMA protocol in Network Simulator NS 2.34 [14] along with ADV-MAC [11], ATMA [12] and Z-MAC to study and compare the performance of our proposed protocol along with these existing protocols. In this simulation, We consider 150 nodes in an area of 700*700 m^2. All the nodes are deployed uniformly in the area. We vary the average number of sources in the carrier sense range from 1 to 10. The simulation parameters considered during simulation is shown in Table 1. We investigate the effect of number of sources on the performance of single channel protocols in multi hop scenario.

The Fig. 4a shows the average energy per packet versus the number of sources. It is observed that as the number of sources increases the number of packets transmitted and the energy consumption also increases. However, the energy consumption per packet decreases with the increase in the number of sources for ADV-MAC, ATMA and CS-ATMA. The energy consumption per packet of our protocol is less than that of ADV-MAC and ATMA. This is because, in case of ADV-MAC and ATMA, all the sensors wake up during the ADV period to know the nodes allowed transmitting in the DATA period. However, our CS-ATMA only wakes up at the time of transmission. The Fig. 4b shows that CS-ATMA takes less

Table 1 Simulation parameters

Parameter	**Value**
Duration of ADV period in ATMA	5 ms
Duration of frames	236.4 ms
Duration of one ADV slot	0.1 ms
Duration of data slots of ATMA	12 ms
Node number	150
Duration of SYNC period	8.4 ms
Duration of control packet	0.9 ms
Duration of data packet	8.5 ms
Rx/Idle listening power	59.1 mW
Tx Power	52.2 mW
Total simulation time	307 ms
Transmission rate	250 kbps
Transmission range	100 m
Carrier sense range	200 m

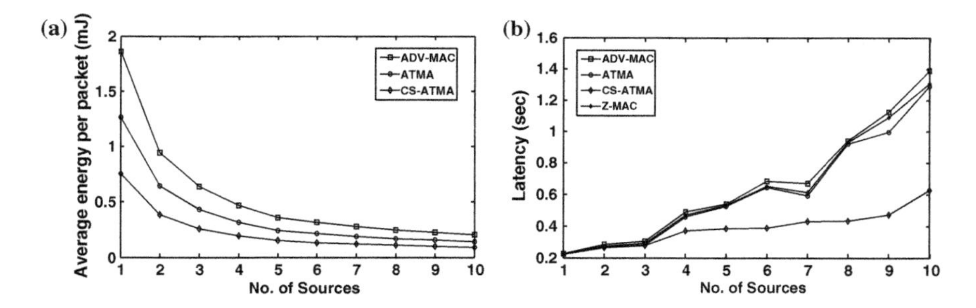

Fig. 4 **a** Average energy per packet. **b** Latency versus number of sources

time to transmit the data from source to destination. The latency of ADV-MAC and Z-MAC increases due to the hidden terminal problem which can be avoided in CS-ATMA. In case of Z-MAC, a non optimal slot selection and allocation will result in collisions and would turn in increase in delay. In ATMA, a fixed time slot is assigned to a source and as the number of sources increase it may not be possible to transmit the whole data in that period. Therefore, the data is transmitted in the next frame and the delay increases. The packet delivery ratio (PDR) of ATMA and ADV-MAC also decreases with the increase in number of sources. The PDR of Z-MAC decreases because of the preamble sampling of the nodes to send their data.

The PDR of our CS-ATMA with the other algorithms is shown in Fig. 5a. It is clear from the figure that the PDR of CS-ATMA is better that others. Furthermore, we investigate the effect of energy consumption, latency, PDR by varying the burst length in multi hop scenario. The interval between the bursts is exponentially

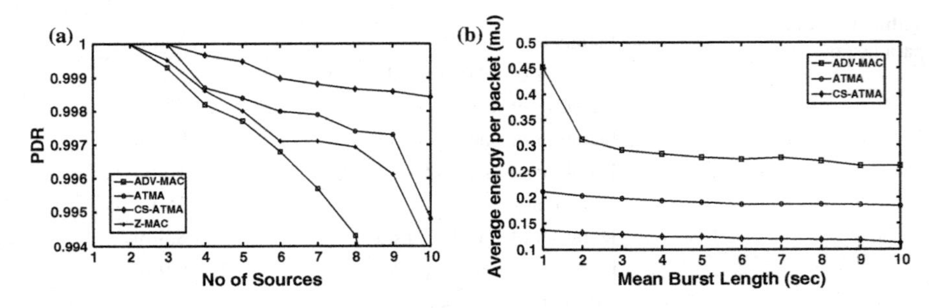

Fig. 5 **a** Packet delivery ratio (PDR) versus no of sources. **b** Average energy per packet versus mean burst length

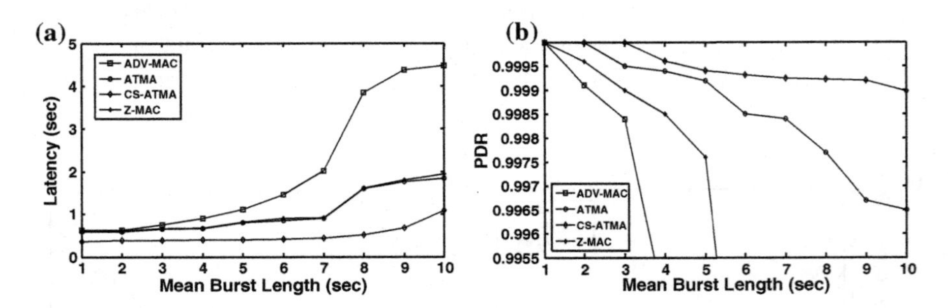

Fig. 6 **a** Latency. **b** Packet delivery ratio (PDR) versus mean burst length

distributed with a mean of 15 s. The average energy per packet versus the burst length is shown in Fig. 5b.

The latency and PDR versus the burst length is shown in Fig. 6a, b respectively. It is observed from the figures that our protocol outperform ATMA and ADV-MAC and Z-MAC in terms of energy consumption, throughput and latency. Since the contention occurs during ADV period, ADV-MAC and ATMA consume more energy than CS-ATMA. Due to the longer ADV period the latency increases and the PDR decreases in case of ADV-MAC and ATMA.

4 Conclusion

This paper presents a hybrid protocol named CS-ATMA which uses both CSMA and ATMA approach for wireless sensor network. In this protocol, nodes randomly select a time slot to set its timer. When the timer of a node becomes zero, the node senses the medium in order to send its data. Therefore it provides less idle listening time and collision. The performance of our proposed protocol is compared with other existing algorithms through simulation. The simulation results reveal that our

CS-ATMA outperforms the existing algorithms in terms of energy efficiency, latency and PDR with respect to the number of sources. The performance of the proposed protocol can further be studied with varying the network density and data rate.

References

1. He, T., Luo, L., Yan, T., Gu, L., Cao, Q., Zhou, G., Stoleru, R., Vicaire, P., Cao, Q., Stankovic, J.A., Son, S.H., Abdelzaher, T.F.: An overview of the VigilNet architecture. In: Proceedings on IEEE International Conference on Embedded and Real-Time Computing Systems and Applications, pp. 109–114 (2005)
2. Mainwaring, A., Culler, D., Polastre, J., Szewczyk, R., Anderson, J.: Wireless sensor networks for habitat monitoring. In: Proceedings of the 1st ACM International Workshop on Wireless Sensor Networks and Applications, pp. 88–97 (2002)
3. Hanjagi, A., Srihari, P., Rayamane A.S.: A public health care information system using GIS and GPS: a case study of Shiggaon. In: Lai, P.C., Mak, S.H. (eds.) GIS for Health and the Environment. Springer-Verlag, New York (2007)
4. Demirkol, Ersoy, C., Alagoz, F.: MAC protocols for wireless sensor networks. In: a survey. IEEE Communications Magazine. vol. 44. no. 4, pp. 115–121 (2006)
5. Heinzelman, W.R., Chandrakasan, A., Balakrishnan, H.: Energy-Efficient communication protocol for wireless microsensor networks. In: Proceedings of the HICSS '00 (2000)
6. Pei, G., Chien, C.: Low power TDMA in large wireless sensor networks. In: Communications for Network-Centric Operations: Creating the Information Force. MILCOM 200, pp. 347–351 (2001)
7. Hoesel, L.V., Havinga, P.: A lightweight medium access protocol (LMAC) for wireless sensor networks. In: International Conference on Networked Sensing System. INSS (2004)
8. Mao, J., Wu, Z., Wu, X.: A TDMA scheduling scheme for many-to-one communications in wireless sensor networks. Comput. Commun. **30**(4), 863–872 (2007)
9. Ye, W., Heidemann, J., Estrin.: An energy-efficient MAC protocol for wireless sensor networks. In: Proceedings of the IEEE INFOCOM, pp. 1567–1576 (2002)
10. Dam, T.V., Langendoen, K.: An adaptive energy-efficient MAC protocol for wireless sensor networks. In: ACM SenSys, '03, pp. 171–180 (2003)
11. Ray, S.S., Demirkol, I., Heinzelman, W.: ADV-MAC: analysis and optimization of energy efficiency through data advertisements for wireless sensor networks. Ad Hoc Netw. **9**(5), 876–892 (2011)
12. Ray, S.S., Demirkol, I., Heinzelman, I.: Supporting bursty traffic in wireless sensor networks through a distributed advertisement-based TDMA protocol. Ad Hoc Netw. **11**(3), 959–974 (2013)
13. Rhee, I., Warrier, A., Aia, M., Min, J., Sichitiu, M.L.: Z-MAC: a hybrid MAC for wireless sensor network. IEEE/ACM Trans Netw **16**, 511–524 (2008)
14. The Network Simulator NS-2. http://www.isi.edu/nsnam/ns/

Energy Efficient Reliable Data Delivery in Wireless Sensor Networks for Real Time Applications

Prabhudutta Mohanty, Manas Ranjan Kabat and Manoj Kumar Patel

Abstract Wireless sensor networks (WSNs) are inherently unreliable and experience excessive delay due to congestion or in-network data aggregation. Furthermore, this problem is more aggravated due to the redundant data in WSNs. Therefore, an energy efficient and reliable data reporting scheme is essential for the delay sensitive applications. In this paper, we propose an Energy efficient Delay Sensitive Reliable Transport (EDSRT) protocol for intelligent data aggregation and forwarding. The EDSRT protocol computes the admissible waiting time of data in each intermediate node for convergence of similar data packet at the same node. Our protocol also uses the explicit and implicit acknowledgement scheme. Moreover, the delay sensitive packets are prioritized at near-sink nodes during congestion to increase the on-time delivery ratio. The proposed protocol is evaluated trough extensive simulations. The simulation results reveal that it outperforms the existing delay sensitive transport protocols in terms of energy efficiency, reliability and on-time delivery ratio.

Keywords EDSRT · Eenergy efficient · Delay-sensitive applications · Reliability · WSN

P. Mohanty (✉) · M.R. Kabat · M.K. Patel
Department of Computer Science and Engineering, VSS University of Technology, Burla, Sambalpur, Odisha, India
e-mail: prabhudutta.mohanty@gmail.com

M.R. Kabat
e-mail: manas_kabat@yahoo.com

M.K. Patel
e-mail: patel.mkp@gmail.com

L.C. Jain et al. (eds.), *Computational Intelligence in Data Mining - Volume 3*,
Smart Innovation, Systems and Technologies 33, DOI 10.1007/978-81-322-2202-6_25

1 Introduction

A wireless sensor network (WSN) [1] consists of a large number of ad hoc networked, low-power, short lived and unreliable multifunctional micro-sensors, which are capable of sensing a phenomenon and transmitting them to one or more base stations (BS). Sensors are limited in computation power, memory capacity and radio range [2]. The WSNs are widely applied to applications [3–6] such as military operation, factory automation, process control, controlling the vehicle traffic in highways require the reliable event transport to be achieved within a certain application-specific delay bound. The late detection of such an event at the BS leads to the failure of the ultimate objectives of the deployed WSN. The main objective of real-time sensor network is to provide feature for the critical applications such as timeliness, design for peak load, fault tolerance, predictability and maintainability. The factors that mostly influence the delay-sensitive applications in WSN are low power radio transmission, limited power supply and processing capacity, node failure and data redundancy. The redundant data drains the energy of the node, increases congestion, communication and computational overhead. The sensor network mostly senses redundant data due to multiple sensors sensing same phenomena and slow change in phenomena. The performance of the real-time WSN can be improved by reducing data redundancy. The data aggregation plays a vital role in reducing data redundancy. In structured WSN, such as cluster-based [7] or tree based [8] the data aggregation is done at the cluster head of each cluster or at the intermediate nodes in a tree topology. In structured data aggregation [9] multiple sources send their data to the aggregation point which eliminates redundant data using various methods [10] such as statistical approaches [11], probabilistic approaches [12], artificial intelligence [13] etc. In structure free approaches [14] the redundant data sent to the same upstream node so that the redundant data can be aggregated at that node and the energy spent to build a structure can be saved. However, the energy spent due to the data transmission by the sensor nodes and data aggregation at the aggregation point cannot be avoided.

Furthermore, many researchers have worked significantly for supporting real-time communication in WSNs. The SPEED [15] is the most well-known real-time routing protocol that implements the end-to-end transmission delay control. It finds out the neighbors' information using a beaconing mechanism and chooses the next hop based on transmission velocity and local geographical information.

The Delay Sensitive Transport (DST) protocol [16] was proposed to ensure reliable on-time data delivery. The DST protocol introduces Time Critical Event First (TCEF) scheduling policy for reducing the buffering delay at intermediate sensor nodes. The TECF scheduling algorithm serves the real-time packets on the principle of Earliest Deadline First (EDF) service discipline at each sensor node. The DST protocol implements a combined congestion control mechanism in which if the buffer overflows at any sensor nodes or the average node delay is above a certain delay threshold value then the congestion is notified to the sink node by the Congestion Notification (CN) bit in the header of the event packet transmitted from the sensor to the sink. The sink can detect the congestion if the CN bit is marked and adjusts the reporting frequency of the sensors. It achieves high performance in terms of reliable event detection, communication latency and energy consumption. In DST, the whole functionality is totally dependent on the sink to regulate the flow of data, which results in increase in traffic and more energy consumption. This does not address multiple events happening in parallel which is usually case in the real world. The Real-Time and Reliable Transport (RT^2) [17] protocol is a variation of the DST protocol that provide event-based reliability by exploiting spatial-temporal correlation of generated packets by source nodes in wireless sensor actuator network (WSAN). It transports the event from the sensor nodes to the actor nodes with minimum energy dissipation and timely reacts to sensor information with a right action.

In this paper, we propose an energy efficient delay sensitive reliable transport (EDSRT) protocol to assure not only reliable but also timely detection of event in WSN. The EDSRT protocol addresses spatial redundancy by using a timeout mechanism so that the base station can assume the redundant data sensed by the sender without requiring any transmission from sender. The temporal redundancy is handled by transmitting the redundant packets to the same upstream nodes for aggregation. The waiting time of packets at each intermediate node is calculated very sensibly so that data can be aggregated efficiently without affecting the on-time delivery of packets. Furthermore, the packets are promoted to the upper classes service level priority during congestion detection. The explicit and implicit ACK mechanisms are used for achieving reliability.

The paper is organized as follows. The proposed Energy Efficient Delay Sensitive Reliable Transport (EDSRT) Protocol is presented in Sect. 2. In Sect. 3, we analyze the performances of the proposed protocol through simulation and finally Sect. 4 concludes the paper.

2 Energy Efficient Delay Sensitive Reliable Transport (EDSRT) Protocol

In this section, we present the operation of our proposed energy efficient delay sensitive reliable transport protocol (EDSRT) protocol. The EDSRT is a distributed protocol that provides structure free data transmission over WSN. The proposed protocol builds a logical hierarchy of nodes to control the topology. A sensor node comes to know about its own position, position of the neighborhood nodes and base station during topology control phase. In the topology construction phase, each node selects its neighbor with minimum hop-count value that received during neighbor discovery phase and sets it own hop-count as received minimum hop-count +1. In our approach, the BS initiates topology set by broadcasting a HELLO message. The HELLO message broadcast by the BS contains hop-count = 0 (Zero). The sensor nodes those hear the HELLO message transmitted by the BS sets its hop-count as one to the BS. These nodes are leveled as first level. The nodes of first level broadcasts HELLO messages with hop-count one and the nodes that hear these set their level and hop-count as two. This process is repeated till the topology is constructed. After logical topology setup, the sensor nodes sense the phenomena and forward the data by intelligently selecting the next-hop node. The next hop data forwarding node is selected either with the help of a neighbor information table maintain at the nodes in the previous level or with probability. When a node senses data or receives a data packet from the nodes of lower level for forwarding it to BS, the node first searches its neighbor information table (Tables 1 and 2) for the presence of same type of packet information. If same type of packet information is present, it finds the next node ID of that having same data packet and transmit the data to that node. If that type of packet is not present or the next node is not a neighbor of the forwarding node then it chooses the next node with a probability computed using Eq. (1).

$$p(n) = \frac{e^{\alpha} \times AR_{TP}^{\gamma}}{\sum_{i=1}^{k} e_i \times AR_{Tp}} \times min\left[\sqrt{\sum_{l=1}^{k}\left(x_i^{(w)} - x_j^{(w)}\right)^2}\right] \tag{1}$$

Where *p(n)* is probability of the next node to be chosen. e is the residual energy of the node *n*, AR_{TP} is the required transmission power of the node *n*, *k* is all neighbor node of *n*, *α*, *γ* two constant to give weightage to energy and data transmission power respectively. x_i and x_j are two distance vector with w dimension that finds the neighbor with shortest minimum distance of node *n*.

The node having largest probability has to be chosen as the next-hop node. The EDSRT minimizes both spatial and temporal data redundancy during data

Table 1 Neighbour information table of a node during topology

NID	S	CP	HC	RTP (nj/bits)	LD	TF	TTE
N42	A	50,80	3	60	0	0	0
N45	A	75,94	3	74	0	0	0
N48	A	30,43	3	82	0	0	0

(In the above two tables *NID* neighbor ID, *S* status, *CP* co-ordinate position of the node, *HC* hop-count, *RTP* required transmission power, *LD* last data, *TF* time of forward and *TE* time to elapse)

Table 2 Neighbour information table of a sensor node during data communication

NID	S	CP	HC	RTP (nj/bits)	LD	TF	TTE
N42	A	50,80	3	60	40	10:03	0:20
N45	A	75,94	3	74	42	10:10	0:18
N48	A	30,43	3	82	39	9:58	0:22

transmission. The proposed protocol minimizes temporal data redundancy with the co-ordination of resource wealthy base-station. The base station maintains a data table containing node id, previous data received, time at which previous data received and a time counter to check the timeout for receiving data from a node. The base station updates the data received value for the redundant data when timeout occurs for a node without being transmitted by node. Thus, it reduces number of redundant data transmission and conserves significant amount energy. The proposed protocol minimizes spatial data redundancy by aggregating redundant data packets received from multiple nodes in the intermediate nodes. The data aggregation is maximized by adopting dynamic waiting policy for a packet at each node along the way to the base station. The EDSRT protocol implements the waiting policy that computes the waiting time of packets at the intermediate nodes so that the data aggregation and on time end-to-end delivery ratio can be maximized.

The total time remain for deadline (*TTRD*) is computed by (*TTD-CT*), where *TTD* is the time remain for delivery which is assigned by the last transmitted node after processing and *CT* is the time at which the current node receive the data.

The maximum approximate waiting time (*MWT*) that can be allowed for a node is computed by calculating $(TTRD-(T_{pd} + T_{td})/hc)-\alpha$, Where *hc* is the hop count of the current received node and α is a constant factor used to leave some remaining time as a safety margin to ensure the deadline meet, T_{pd} is the processing delay which is calculated as (*time to process a data packet* × *No of incoming packet at that instance*), T_{td} is transmission delay which is the time spent by a sensor node to transmit the data packet over the wireless channel and calculated using formula (*channel access delay* + *propagation delay*)

The end-to-end delivery ratio of time constrained packets is maximized by prioritized packet scheduling. The packets are served at each node with respect to their priorities. There are three different priorities assigned to the packets i.e. min-priority, normal-priority and max-priority. A packet sets a min-priority when it is sensed and promoted to next higher level priority in intermediate nodes during congestion to meet their deadline. The congestion is detected at each node by comparing buffer availability field with predefined minimum threshold (th_{min}). If buffer availability < th_{min}, then notify the source node about the congestion. The congestion is notified by assigning a value one (1) in congestion notification bit (CN). The proposed protocol minimizes congestion both by increasing or decreasing reporting frequency during congestion as well as increasing priority of the data packet. The network congestion increases packet loss and delay. The data packet forwarding is ensured with the implicit acknowledgement (IACK) at intermediate node. The base station forwards an explicit acknowledgement (EACK) to the node from which it receives the data packet. After data aggregation when data is forwarded to the next hop by the aggregated intermediate node by overhearing this transmission source node ensures that their data is forwarded this is treated as implicit acknowledgement to the source node. This process continues till the aggregated data packet reaches as Base station. The base station sends an explicit acknowledgement to the last for wording node to ensure that packet is received properly as there is no further data forwarding by the base station. Therefore, the proposed protocol increases the energy efficiency in addition to reliability and on time delivery ratio in WSN, in comparison to the existing protocols.

In the `next_hop_node_selection` algorithm, a node first checks that it has sensed data or received data from downstream nodes to forward through `sense_data()` and `send_data()` function. The node set a flag `get_next_node` as false then search its neighbor information table for similar data packet (`NI_table_packet_type`). The node selects the next hop data forwarding node form neighbor information table if data packet type matches and sets `get_next_node` as true . If next hop data forwarding node not found through neighbor information table then a probability (`p(n)`) is computed for each neighbor node and a node with maximum `p(n)` is selected as next hop data forwarding node. The algorithm computes `max_AWT` (maximum average waiting time) of a data packet in the next hop node, that plays a vital role in data aggregation.

Algorithm: Next Hop node selection algorithm

```
next_hop_node_selection( ){
if (sense_data( )= =TRUE|| send_data( )= = TRUE)
// Nodei has data to send to base station.
get_next_node = FALSE
for ( row=1 to n)
// search neighbor information table of nodei
if( NI_table_packet_type[row]= = sdata_packet_type)
then
            set
next_hop_node_ID=next_hop_node_ID[row]
  endif;
  if ( node_i_rechable (next_hop_node_ID) = = TRUE)
then
            calculate max_AWT for next
next_hop_node_ID
            set get_next_node=TRUE
  endif;
endfor;
if (get_next_node == FALSE) then
   for ( k=1 to n) //for all neighbor node of node i
      calculate p(n) for each neighbor of node i
      select next_hop_node_ID=max(p(n)_node_ID
      calculate max_AWT for next next_hop_node_ID
   endfor;
endif;
send_datapacket(next_hop_node_ID)
update (NI_table) of node i
```

3 Performance Evaluation

In this section, we evaluate the performance of our proposed protocol through simulation in NS-2.30 [18]. The results are compared with SPEED [15], the most well-known real-time routing protocol RAG [14] and DST [16] a novel transport solution for reliable and timely event detection with minimum possible energy consumption. The simulation parameters for our model are mentioned in Table 3. We run the simulation with parameters like different data rate and an average data redundancy of 40 % generated randomly. The same type of data packets could be aggregated in the intermediate nodes along the way to the sink. To identify the aggregated packets an extra field containing information about the number of data aggregated is added to the data packets. We considered up to 65 numbers of data packets can be aggregated during data transmission. The packet generation rate

Table 3 Simulation parameters

Area of sensor field	200×200 m2	Radio range	40 m
Number of sensor nodes	100	Propagation model	Two ray
Packet length	30 bytes	Eelec	50 nJ/bit
Buffer length	65 packets	Esense	0.083 J/s
Initial node energy	0.6 J	Eagg	5 nJ/bit/signal
Bandwidth	200 Kb/s	Eamp	10 pJ/bit/m2

increased step by step from 1 to 100 pps (packets/sec). We analyze our proposed protocol in terms of average energy consumption, miss ratio, end-to-end delay and on-time data delivery ratio through the simulation. The deadline miss ratio is computed as the percentage of packets that do not meet the dead line and discarded in delay sensitive applications. End-to-end delay is the time from the packet generated by the source till it delivered to the destination. The percentage of packets delivered to the sink within its time is on-time data delivery ratio.

The Fig. 1 shows that EDSRT conserves more energy than SPEED, DST and RAG with the increase of data rate. The SPEED doesn't implement any data aggregation during data transmission. Thus energy consumption increases when more packets are injected into the network. On the other hand DST adjusts it reporting frequency according to data rate and adapts congestion control mechanism to avoid energy wastage due to congestion but like SPEED, DST does not adapt any data aggregation policy. RAG adapts a Judiciously Waiting policy for efficient aggregation of data packets to conserves more energy and tries to eliminating the inherent redundancy of raw data. EDSRT out performs all these different proposed protocols due to the use of neighbor information table that helps to decide which node can perform better aggregation in time and saves the burden of broadcasting the data packet to all its neighbor. It minimizes the transmission of temporal data with the co-ordination of resource rich base station and spatial redundancy by data aggregation. With the increase of data rate both temporal and spatial redundancy also increases due to this our protocol saves more energy as compare to others.

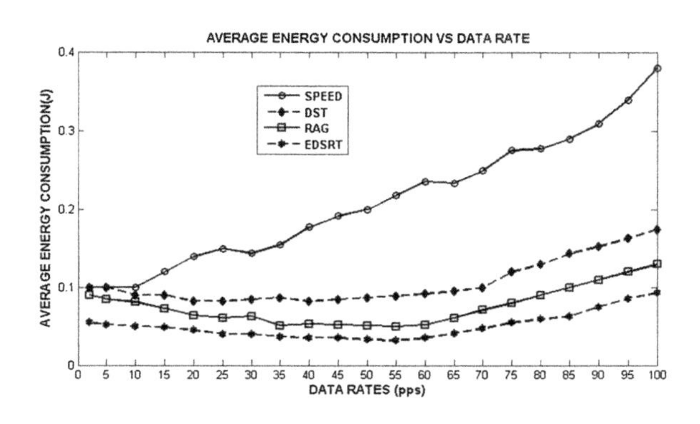

Fig. 1 Average energy consumption versus data rate

The Figs. 2, 3 and 4 evaluates the performance in terms of miss ratio, end-to-end delay and on time delivery ratio. The miss ratio and end-to-end delay increases with the increase of traffic load. This happens due to more contention, longer queuing delays and congestion in intermediate nodes during increase in data rate. Due to the lost packets the retransmissions increases that leads to higher energy consumption

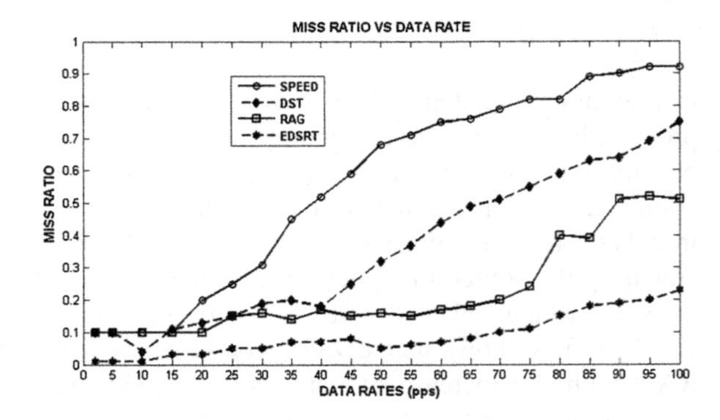

Fig. 2 Miss ratio versus data rate

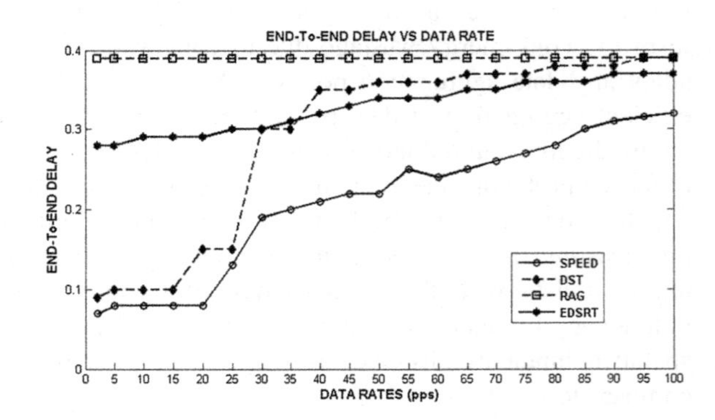

Fig. 3 End-to-end delay versus date rate

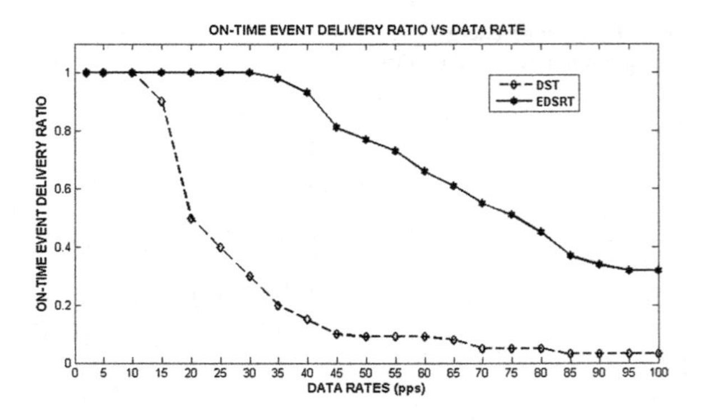

Fig. 4 On-time event delivery ratio versus data rate

in heavy load. Our proposed protocol performs aggregation as well as minimizes the redundant data transmission with the co-ordination with base station so there is reduction of traffic load in case of data redundancy. The prioritized scheduling of data transport adds a step to minimize the miss ratio in our proposed protocol. Our waiting policy dynamically adjusts the waiting time limit in each intermediate and increases/decreases the priority to achieve better on-time data delivery with minimum end-to-end delay where as equal waiting time is assigned for each intermediate nodes for data aggregation in the RAG. DST and SPEED do not follow waiting policy for data aggregation and suffers from congestion with increase in traffic load. We also consider different delay strategies near sink node as compare to near source node as the traffic load is unequal rough out the network.

4 Conclusions

In this paper, we considered the problem of real-time data aggregation in WSNs and proposed a probabilistic next node selection method for non redundant data forwarding and a neighbor information table is used to select data during redundant data forwarding. We address both spatial and temporal data redundancy during communication and minimized the same by the co-ordination of the base station and data aggregation. Priority based data transmission adds a step to achieve timely data delivery ratio in delay sensitive applications. Results evaluated in simulation demonstrated a significant performance improvement in terms of energy consumption, miss ratio, end-to-end delay and on-timely data delivery.

References

1. Mohanty, P., Kabat, M.R.: Transport protocols in wireless sensor networks, Chapter 10 In: Ibrahiem, E.I., Emay, M.M., Ramkrishan, S., (eds.) Wireless Sensor Networks: From Theory to Applications, pp. 265–305 CRC Press, New York (2013)
2. Hadim, S., Mohamed, N.: Middleware: Middleware challenges and approaches for wireless sensor networks. IEEE Distrib. Syst. **7**(3) (2006)
3. Hadjidj, A., Souil, M., Bouabdallah, A., Challal, Y., Owen, H.: Wireless sensor networks for rehabilitation applications: challenges and opportunities. J. Netw. Comput. Appl. **36**, 1–15 (2013)
4. Jiang, S.F., Zhang, C.M., Zhang, S.: Two-stage structural damage detection using fuzzy neural networks and data fusion techniques. Expert Syst. Appl. **38**(1), 511–519 (2011)
5. Roantree, M., Shi, J., Cappellari, P., O'Connor, M.F., Whelan, M., Moyna, N.: Data transformation and query management in personal health sensor networks. J. Netw. Comput. Appl. **35**(4), 1191–1202 (2012)
6. Soro, S., Heinzelman, W.: A survey of visual sensor networks. Adv. Multimedia. 1–22 (2009)
7. Younis, O., Fahmy, S.: HEED: A hybrid energy–efficient distributed clustering approach for ad hoc sensor networks. IEEE Trans. Mob. Comput. **3**, 366-379 (2004)
8. Ding, M., Cheng, X., Xue, G.: Aggregation tree construction in sensor networks in. In: Proceedings of the 58th IEEE Vehicular Technology Conference, pp. 2168-2172 (2003)

9. Du, H., Hu X., Jia, X.: Energy efficient routing and scheduling for real-time data aggregation in WSNS. Compu. Commun. **29**, 3527-3535 (2006)
10. Faouzi, N.E.E.I., Leung, H., Kurian, A.: Data fusion in intelligent transportation systems progress and challenges survey. Inf. Fusion. 4-10 (2010)
11. Faouzi, N.E.E.I.: Heterogeneous data source fusion for impedance indicators. In: IFAC Symposium on Transportation Systems. vol. **3**, pp. 1375–1380, Greece, Chania (1997)
12. Okutani, I.: The kalman filtering approaches in some transportation and road traffic problems. In: Wilson, G, (eds.) Proceedings of the 10th ISTTT, pp. 397–416 (1987)
13. Han, J., Kember, J.: Data Mining Concepts and Techniques. Morgan Kaufmann Publishers, New York (2000)
14. Yousefi, H., Yeganeh, M.H., Naser, A., Ali M.: Structure-free real-time data aggregation in wireless sensor networks. Comput. Commun. **35**, 1132-1140 (2012)
15. He, T., Stankovic, J.A., Lu, C., Abdelzaher, T.: SPEED: A stateless protocol for realtime communication in sensor networks. In: Proceedings of the 23rd International Conference on Distributed Computing Systems, pp. 46–55 (2003)
16. Gungor, V.C., Akan, O.B.: DST: delay sensitive transport in wireless sensor networks. In: Proceedings of the 7th International Symposium on Computer Networks (ISCN), pp. 116–122, Istanbul, Turkey (2006)
17. Gungor, V.C., Akan, O.B., Akyildiz, I.F.: A real-time and reliable transport protocol for wireless sensor and actor networks. IEEE/ ACM Trans. Netw. **16**(2), 359–370 (2008)
18. The network simulator, http://www.isi.edu/nsnam/ns

A Rule Induction Model Empowered by Fuzzy-Rough Particle Swarm Optimization Algorithm for Classification of Microarray Dataset

Sujata Dash

Abstract Selection of small number of genes from thousands of genes which may be responsible for causing cancer is still a challenging problem. Various computational intelligence methods have been used to deal with this issue. This study introduces a novel hybrid technique based on Fuzzy-Rough Particle Swarm Optimization (FRPSO) to identify a minimal subset of genes from thousands of candidate genes. Efficiency of the proposed method is tested with a rule based classifier MODLEM using three benchmark gene expression cancer datasets. This study reveals that the hybrid evolutionary Fuzzy-Rough induction rule model can identify the hidden relationship between the genes responsible for causing the disease. It also provides a rule set for diagnosis and prognosis of cancer datasets which helps to design drugs for the disease. Finally the function of identified genes are analyzed and validated from gene ontology website, DAVID, which shows the relationship of genes with the disease.

Keywords Particle swarm optimization · Population based algorithm · Induction rule · Rough-set · Fuzzy-rough · Minimal reduction

1 Introduction

Feature selection (FS) is to determine a minimal feature subset from a problem domain while retaining a suitably high accuracy in representing the original features [1]. In real world problems FS is essential due to the abundance of noisy, irrelevant or misleading features. Only a small fraction of training samples from the large pool of genes involved in the experiment are effective in performing a task. Also a small subset of genes is desirable in developing gene-expression-based diagnostic tools for delivering precise, reliable, and interpretable results. With the gene selection

S. Dash (✉)
North Orissa University, Baripada, Odisha, India
e-mail: Sujata_dash@yahoo.com

L.C. Jain et al. (eds.), *Computational Intelligence in Data Mining - Volume 3*,
Smart Innovation, Systems and Technologies 33, DOI 10.1007/978-81-322-2202-6_26

results, the cost of biological experiment and decision can be greatly reduced by analyzing only the marker genes. Hence, identifying a reduced set of the most relevant genes is the goal of gene selection [2].

Gene selection method can be divided into three sub-sections: Filter, Wrapper and Embedded approaches. Wrappers utilize learning machine and search for the best features in the space of all feature subsets. Despite being optimal in simplicity and performance, wrappers rely heavily on the inductive principle of the learning model. Due to this they suffer from excessive computational complexity as the learning machine has to be retrained for each feature subset considered [3]. The filter method is generally inferior to the wrapper method as the filter approach usually employs statistical methods to collect the intrinsic characteristics of genes in discriminating the targeted phenotype class, such as statistical tests, Wilcoxon's rank test and mutual information, to directly select feature genes [4] whereas, the wrapper method involves inter-correlation of individual genes in a multivariate manner, and can automatically determine the optimal number of feature genes for a particular classifier.

In the gene selection process, an optimal gene subset is always relative to a certain criterion. Several information measures such as entropy, mutual information [5] and f-information [6] have successfully been used in selecting a set of relevant and non redundant genes from a microarray data set. An efficient and effective reduction method is necessary to cope with large amount of data by most techniques. Growing interest in developing methodologies that are capable of dealing with imprecision and uncertainty is apparent from the large scale research that are currently being done in the areas related to fuzzy [7] and rough sets [8]. The success of rough set theory is due to three aspects of the theory. First, only hidden facts in the data are analyzed. Second, additional information about the data is not required for data analysis. Third, it finds a minimal knowledge representation for data. Due to this Rough set theory is used in complement with other concepts such as, fuzzy set theory. The two fields may be considered similar in the sense that both can tolerate inconsistency and uncertainty. The only difference among these two fields is the type of uncertainty and their approach to it. While fuzzy sets are concerned with vagueness, rough sets are concerned with indiscernibility. The fuzzy-rough set-based approach considers the extent to which fuzzified values are similar.

This paper presents a Particle Swarm Optimization (PSO)-based Fuzzy-Rough algorithm to address the problem of gene selection. Particle Swarm Optimization (PSO), developed by Kennedy and Eberhart [10], is a population-based meta-heuristic on the basis of stochastic optimization, inspired by the social behavior of flocks of birds or schools of fish [11]. A swarm of particles move toward the optimal position along the search path that is iteratively updated on the basis of the best particle position and velocity in PSO [12]. Here we consider a rule induction algorithm called MODLEM, introduced by Stefanowskiin [13]. MODLEM is a rule based specific search induction algorithm, whose rules are one of the most popular type of knowledge used in practice. This is mainly because they are both very expressive and human-readable at the same time. This algorithm is most suitable for analyzing data containing a mixture of numerical and qualitative attributes [14].

The proposed approach in this paper is an integration of PSO searching algorithm with Fuzzy-Rough subset evaluator called FRPSO. Combining PSO with Fuzzy-Rough as an evaluator has rarely been investigated by previous researchers. The performance of our proposed method will be evaluated by 4 microarray datasets collected from Bioinformatics Laboratory, University of Ljubljana, http://www.biolab.si/supp/biancer/projections/info/lungGSE1987.htm, [15]. In Sect. 2 of this paper, the population based Fuzzy-Rough feature selection algorithm has been investigated to identify the minimal set of responsible genes. In Sect. 3 of this paper, the application of MODLEM on the microarray datasets is discussed along with experimental results and validation of the rule sets and Genes. In the concluding section of this paper, discussions have been done on this novel approach followed by list of references.

2 Selection of Attributes

DNA microarray (commonly known as DNA chip or biochip) is a collection of microscopic DNA spots attached to a solid surface. It allows the researchers to measure the expression levels of thousands of genes consecutively in a single experiment. This is operated by classifier approaches to compare the gene expression levels in tissues under different conditions [16]; for example, the study of Jiang et al. [17] devised an RF-based method to classify real pre-miRNAs using a hybrid feature set for the wild type versus mutant, or healthy versus diseased classes. Wang et al. [18] presented a hybrid method combining GA and SVM to identify the optimal subset of microarray datasets. This method was claimed to be superior to those obtained by microPred and miPred. Furthermore, Nanni et al. [19] recently devised a support vector machine (SVM) as classifier for microarray gene classification. This method combines different feature reduction approaches to improve classification performance of the accuracy and area under the receiver operating characteristic (ROC). Appropriate gene selection procedures are required to improve classification performance to reduce high computational cost and memory usage.

For some learning problems, some attributes may be irrelevant or even redundant. In case there are too many irrelevant attributes in the initial description of examples, the complexity of a learning process might increase and a reverse trend is visible for the classification performance of the induced classifier [20]. Thus, the aim should be to find the smallest subset of attributes leading to higher classification accuracy than the set of all attributes.

In this paper we have discussed how the feature selection can be reduced to a search problem, where one selects a good feature subset based on a selected evaluation measure [20]. Each state in the search space represents a subset of possible features. As per Kohavi [20], to design a feature selection algorithm one has to define three following components: search algorithm (technique that looks through the space of feature subsets), evaluation function (used to evaluate examined subsets of features), and classifier (a learning algorithm that uses the final subset of features). These elements can be integrated in two ways: (a) filter approach and (b) wrapper approach [20]. In the filter approach, features are selected based on properties of the data independently of the learning algorithm used as a classifier.

To provide instances of studies which have proposed to use PSO algorithm for gene selection problems, the following stand out. Alba [9] stated that a modified PSO (geometric PSO) for high-dimensional microarray data. Both augmented SVM and GA were proposed for comparison on six public cancer datasets. Mohamad et al. [21] presented an improved binary PSO combined with an SVM classifier to select a near-optimal subset of informative genes relevant to cancer classification. Zhao et al. [22] lately presented a novel hybrid framework (NHF) for gene selection and cancer classification of high dimensional microarray data by combining the information gain (IG), F-score, GA, PSO, and SVM. Their method was compared to PSO-based, GA-based, ant colony optimization-based, and simulated annealing (SA)-based methods on five benchmark data sets: leukemia, lung carcinoma, colon, breast, and brain cancers. Chen et al. [24] used PSO+ 1NN for feature selection and tested their algorithm against 8 benchmark datasets from UC Irvine Machine Learning Repository as well as to a real case of obstructive sleep apnea.

3 Fuzzy-Rough Data Reduction

Fuzzy-Rough feature selection (FRFS) method is devised on the basis of the fuzzy lower approximation to enable reduction of datasets containing real-valued features. This method becomes similar to the crisp approach when dealing with nominal well-defined features. The crisp positive region in Rough Set Theory (RST) is defined as the union of the lower approximations. By the extension principle, the membership of an object $x \in U$, belonging to the fuzzy positive region can be defined by

$$\mu_{POS_{F(Q)}(x)} = \sup_{X \in \frac{U}{Q}} \mu_{P_X}(x) \tag{1}$$

Object x will belong to the positive region only if the equivalence class it belongs to is a constituent of the positive region. This is similar to the crisp version where objects belong to the positive region only if their corresponding equivalence classes do so. The new dependency function can be defined using the definition of the fuzzy positive region as,

$$\gamma_P(Q) = \frac{\left|\mu_{POS_P(Q)}(x)\right|}{|U|} = \frac{\sum_{x \in U} \mu_{POS_P(Q)}(x)}{|U|} \tag{2}$$

Here the dependency of Q on P is the proportion of objects that are discernible out of the entire dataset, similar to the crisp rough sets. This corresponds to determining the fuzzy cardinality of $\mu_{POSP(Q)}$ (x) divided by the total number of objects in the universe. This may give rise to a problem if compared to the crisp approach. In conventional rough set-based feature selection, a reduct is a subset R of the features which have the same information as the full feature set A. In terms of the dependency function this means that the values γ (R) and γ (A) are identical and equal to 1, provided the dataset does not contain any contradictory information. However, in the fuzzy-rough approach this is not necessarily true just because the uncertainty encountered when objects belong to many fuzzy equivalence classes results in a reduced total dependency. This can be overcome by determining the degree of dependency of a set of decision features D upon the full feature set and use this as the denominator rather than |U|, allowing γ' to reach 1.

4 MODLEM Algorithm

It is a sequential covering algorithm, which treats nominal and numeric data in a similar way. It generates heuristically a minimal set of rules for every decision concept by finding best rule condition during rule induction from the attribute search space. It handles directly numerical attributes during rule induction when elementary conditions of rules are created, without using any preliminary discretization process. The rules induced from objects are represented as logical expressions of the following form:

IF (conditions) THEN (decision class);

where conditions are specified as conjunctions of elementary conditions on values of attributes, and decision part of the rule assigns a given class to an object which satisfies the condition. Due to the restriction in size of the paper, formal presentation of the algorithm is skipped and only the main objective of the algorithm is discussed. For more information [21, 23, 25] can be referred. The set of rules tries to cover all positive examples of the given concept and not to cover any negative examples. The procedure of rule induction starts by creating a first rule, i.e., choosing sequentially the best elementary conditions according to the criteria chosen (we have used class entropy in our experiments). Then the rule is matched with all positive learning examples. Moreover, when a rule is formed, all positive learning examples are required to be matched with the rule and the matching examples are needed to be removed from further consideration. The process is repeated iteratively as long as some positive examples of the decision concept remain still uncovered.

Then, the procedure is repeated sequentially for each decision class. The elementary conditions on numerical attributes are represented as either $(a < v_a)$ or $(a \leq v_a)$, where a denotes an attribute and v_a its value. Selection of the threshold v_a is made from checking several possible candidates and then all temporary conditions are evaluated over uncovered examples by means of the chosen evaluation measure. But, if the same attribute is selected twice while framing a single rule, one can obtain the condition ($a = [v_1; v_2]$) which results from an intersection of two conditions ($a < v_2$) and ($a \geq v_1$) such that $v_1 < v_2$. In case of nominal attributes, the equivalent conditions are represented as ($a = v_a$). For more details about the function finding best elementary conditions see, e.g., [21, 25].

When the input data contains inconsistent examples, the rough sets theory can be used to handle them. In this theory inconsistencies are neither removed from consideration nor aggregated by probabilistic operations. Instead, lower and upper approximations of each class are computed. Basing on that two sets of rules corresponding to lower and upper approximations are induced. When the set of induced rules are applied to classify new examples, a classification strategy introduced by Grzymala in LERS system is used. This strategy considers completely matched rules but in some cases allows partially matching rules when it does not find the rules matching the description of tested examples [26].

5 Experimental Methods

We integrated PSO algorithm with Fuzzy-Rough algorithm to address the gene selection problem. The important genes were identified by PSO algorithm employing Fuzzy-Rough as fitness function to verify the efficiency of the selected genes. We used a random function to initialize the particle population of PSO. Good initial seed value leads to a better result.

5.1 Representation and Initialization of Particle in PSO

A particle represents a potential candidate solution (i.e., gene subset) in an n-dimensional space. Each particle is coded as a binary alphabetical string. We updated the dimension d of particle i by

$$x_{id}^{new} = \begin{cases} 1, & if\ sigmoid\left(v_{id}^{new}\right) > u(0,1) \\ 0, & \end{cases} \tag{3}$$

where, the sigmoid function

$$\left(v_{id}^{new}\right) \quad is \quad {}^{1}\!/\!\left(1 + e^{-v_{id}^{new}}\right) \tag{4}$$

The fitness function of PSO is measured by Fuzzy-Rough evaluation algorithm. Figure 1 shows the process of FRPSO on gene selection process.

```
input
   c1, c2, r1, r2, w, v min, v max
output
   GB
   The fitness value of GB
begin
   Initialize Population
   while (max iteration or convergence criteria is not met)
   dc
      for i=1 to numbers of particles
       Evaluates fitness value of the particle by PSO
      if the fitness value of Xi is greater than that of PBi
         then PBi = Xi
          end if
              if the fitness value of Xi is greater than that
of GB
                 then GB = Xi
              end if
                 for d=1 to no of genes
                 v^new_id = w * v^old_id + c1r1 (pb^old_id- x^old_id) +(c2r2
(pb^old_id- x^old_id)
                 if v^old_id> vmax then v^new_id = vmax
       if v^old_id < vmax then v^new_id = vmin
       if sigmoid (v^new_id) > U (0,1)
          then
             x^new_id =1
              else
              x^new_id=0
       end if
            next d
          next i
    end while
end
```

Fig. 1 Fuzzy-rough-PSO (FRPSO) algorithm

5.2 Fuzzy-Rough-PSO (FRPSO) Algorithm

See Fig. 1.

6 Results and Discussion

6.1 Experimental Studies

In this study, 4 microarray cancer datasets with different sizes, features, and classes have taken for experimental analysis to evaluate the effectiveness of our proposed method. The datasets were collected from "Bioinformatics Laboratory, University

Table 1 Microarray data sets used in this experiment

Data set	No. of features	No. of samples	No. of classes	Particle size PSO	No. of selected genes/attributes
MLL	12,534	72	03	130	14
Childhood tumours	9,946	23	03	99	6
Lung tumour	12,601	203	05	135	20
DLBCL tumour	7,130	77	02	73	8

of Ljubljana" [15], which include MLL, Lung Tumor, Childhood Tumor and DLBCL Tumor. Table 1 summarizes the characteristics of those microarray datasets. The PSO parameters were selected referring many research articles and optimized by using the literature [10]. Moreover, many trials were conducted to test the parameter setting to obtain the best objective value. The parameter values selected for FRPSO are as follows: number of particles in the population was set to the one-tenth number of genes (field of "particle size" in Table 1). The parameter, c1 and c2, were both set at 2, whereas the parameter, lower (vmin) and upper bounds (vmax), were set at −4 and 4, respectively. The algorithm was repeated until either the fitness was 1.0 or the number of the repetitions was reached by the default value of T = 100. Table 1 summarizes the characteristics of microarray dataset.

To avoid generating random results, this study has adopted a tenfold cross-validation statistical strategy by dividing dataset into two segments for evaluating and comparing learning algorithms. One segment of data is used to learn or train a model and the other is used to validate the model. Cross-validation can be used as a means for choosing proper parameters of the model rather than using purely for estimating model performance.

Figure 2 shows, 10 certain rules which are generated from lower approximation involving 14 genes in which some are identified as strongly associated with 3 cancers i.e., ALL, AML and MLL. Rules 3, 5 and 9 strongly identify all the 3 types of cancer than the other 7 induced rules. Gene 33589_at is strongly associated with all three cancers i.e., rule one: (38989_at >= 582.5)&(33589_at < 13)&(32189_g_at < 1046.5) => (class = ALL) (17/17, 70.83 %), classify more than 70 %

Rule 1. (38989_at >= 582.5)&(33589_at < 13)&(32189_g_at < 1046.5) => (class = ALL) (17/17, 70.83%)
Rule 2. (41651_at >= 5066.5)&(32850_at < 1516) => (class = ALL) (7/7, 29.17%)
Rule 3. (32138_at < 199)&(34449_at < 135.5) => (class = ALL) (4/4, 16.67%)
Rule 4. (38627_at < -58) => (class = ALL) (3/3, 12.5%)
Rule 5. (41664_at >= 586)&(33589_at in [13, 543])&(32581_at < 31) => (class = MLL) (15/15, 75%)
Rule 6. (32799_at >= 2713)&(34449_at >= 888) => (class = MLL) (3/3, 15%)
Rule 7. (32799_at >= 2713)&(33589_at < -1134) => (class = MLL) (2/2, 10%)
Rule 8. (38866_at < 13)&(33589_at >= -286) => (class = MLL) (3/3, 15%)
Rule 9. (38989_at < 133)&(36045_at < 3.5) => (class = AML) (25/25, 89.29%)
Rule 10. (41651_at < 1672)&(33589_at < -286) => (class = AML) (15/15, 53.57%)

Fig. 2 Certain rules generated from lower approximation for MLL by MODLEM

ALL cancer data, rule 5: (41664_at >= 586)&(33589_at in [13, 543])&(32581_at < 31) => (class = MLL) (15/15, 75 %) classify 75 % MLL cancer and rule 9: (38989_at < 133)&(36045_at < 3.5) => (class = AML) (25/25, 89.29 %) classify more than 89 % AML cancer dataset. Gene 38989_at is also strongly associated with ALL and MLL. Conditional entropy is used as a condition measure to find best elementary conditions to form rules. Class approximation is natural type of rules while certain rules are generated from lower approximation and possible rules are generated from upper approximation in terms of Rough-Set Theory.

Similarly, 6 important genes out of 9,946 are identified by 5 rules in Childhood Tumor which is shown in Fig. 3. Rule 1 classifies more than 81 % EWS, more than 66 % eRMS by rule 3 and 100 % aRMS by rule 5.

Figure 4, demonstrates 12 induced rules generated from lower approximation for Lung Tumor which classify all 5 cancers from the dataset identifying 20 genes out of 12,601. Gene 38138_at is strongly associated with all five cancers.

Figure 5 shows the classifier model of DLBCL which contains 7 rules which classify DLBCL and FL by identifying 8 significant genes from the dataset. Gene L38810_at is strongly associated with DLBCL and FL.

Rule 1. (2017_s_at >= 0.09) => (class = EWS) (9/9, 81.82%)
Rule 2. (32334_f_at >= 0.66) => (class = EWS) (6/6, 54.55%)
Rule 3. (38522_s_at >= 0.67) => (class = eRMS) (2/2, 66.67%)
Rule 4. (33500_i_at >= 0.58) => (class = eRMS) (1/1, 33.33%)
Rule 5. (32334_f_at < 0.54)&(37287_at < 0) => (class = aRMS) (9/9, 100%)

Fig. 3 Certain rules generated from lower approximation for childhood tumor by MODLEM

Rule 1. (38138_at >= 377.86)&(37141_at >= 7.64)&(41549_s_at < 92.88)&(33693_at < 30.04) => (class = AD) (100/100, 71.94%)
Rule 2. (41151_at < 258.56)&(38138_at < 1788.99)&(37797_at >= -16.47)&(37187_at >= 23.28)&(37148_at < 177.38) => (class = AD) (87/87, 62.59%)
Rule 3. (35090_g_at >= 124.49)&(38442_at >= 298.22)&(38138_at >= 328.33) => (class = AD) (14/14, 10.07%)
Rule 4. (39805_at >= 62.22) => (class = AD) (8/8, 5.76%)
Rule 5. (749_at < 54.89)&(37148_at >= 102.74)&(33667_at >= 1792.99) => (class = NL) (16/16, 94.12%)
Rule 6. (32052_at >= 4253.91) => (class = NL) (5/5, 29.41%)
Rule 7. (36572_r_at >= 592.41)&(39023_at < 136.77) => (class = SMCL) (6/6, 100%)
Rule 8. (33693_at >= 106.75)&(38138_at >= 512.27) => (class = SQ) (14/14, 66.67%)
Rule 9. (38968_at < 44.24)&(36702_at < 76.58) => (class = SQ) (6/6, 28.57%)
Rule 10. (33667_at < 1870.71)&(547_s_at < 4.33)&(33843_g_at >= -58.08) => (class = SQ) (4/4, 19.05%)
Rule 11. (38138_at < 119.98) => (class = COID) (19/19, 95%)
Rule 12. (34265_at >= 390.66) => (class = COID) (18/18, 90%)

Fig. 4 Certain rules generated from lower approximation for lung tumor by MODLEM

Rule 1. (X61123_at < 1444)&(S82024_at < 265.5) => (class = DLBCL) (20/20, 34.48%)
Rule 2. (L38810_at >= 512)&(U47101_at < 2662.5)&(D42073_at >= -93.5) => (class = DLBCL) (39/39, 67.24%)
Rule 3. (L38810_at >= 1164.5) => (class = DLBCL) (24/24, 41.38%)
Rule 4. (D28423_at < 444) => (class = DLBCL) (12/12, 20.69%)
Rule 5. (L38810_at < 585)&(X61123_at >= 1444)&(U57721_at < 479.5) => (class = FL) (10/10, 52.63%)
Rule 6. (U47101_at >= 2662.5)&(L38810_at < 1164.5) => (class = FL) (6/6, 31.58%)
Rule 7. (U65402_at >= 2030) => (class = FL) (3/3, 15.79%)

Fig. 5 Certain rules generated from lower approximation for DLBCL by MODLEM

Table 2 Classification accuracy (%) validated with tenfold CV

Data set	Percentage of accuracy	Mean absolute error	No. of rules	Model building time (in s)
MLL	68.0556	0.213	10	27.54
Childhood tumours	60.8696	0.2609	5	8.24
Lung tumour	86.2069	0.0552	12	120.95
DLBCL tumour	74.026	0.2597	7	4.85

The results of these experiments are given in Table 2. For each data set, the first column shows the average classification accuracy obtained by the classifier over ten-fold-cross-validations. Mean absolute error and model building time in seconds is also given. The next column represents the number of rules generated by the classifier for each datasets.

6.2 Validations of the Obtained Results

The validation of the result is shown in two ways. One is mathematical validation and another one real life functional classification of genes. The mathematical validation is based on the induced rules generated from the reducts obtained from Fuzzy-Rough-PSO and classification of whole dataset on the basis of that generated rules which have only few responsible genes. The accuracy of prediction of the diseases is verified by applying rule sets generated by MODLEM classifier on the datasets. The cross validated result, as shown in Table 2 indicates that these rules can accurately predict the data with total coverage 1. In all the datasets only few marker genes classify the entire datasets which validate our results.

To find biological relevance of the method, next part of validation need to be applied to find actual functional classification of those genes in human body. This is obtained from a Gene Ontology website called DAVID (http://david.abcc.ncifcrf.gov/) where it is available. If the lists of marker genes are provided as input with

Table 3 Functional classification of marker genes for leukemia

Gene identifiers	Gene name	Gene symbol	Gene functions obtained from DAVID
34449_at	Caspase 2, apoptosis-related cysteine protease (neural precursor cell expressed, developmentally down-regulated 2)	CASP2	This gene encodes a protein which is a member of the cysteine-aspartic acid protease (caspase) family. Sequential activation of caspases plays a central role in the execution-phase of cell apoptosis
38866_at	GRB2-related adaptor protein 2	GRAP2	This gene encodes a member of the GRB2/Sem5/Drk family. This member is an adaptor-like protein involved in leukocyte-specific protein-tyrosine kinase signaling. polyA sites exist
32189_g_at	Myelin transcription factor 1	MYT1	The protein encoded by this gene is a member of a family of neural specific, zinc finger-containing DNA-binding proteins. The protein binds to the promoter regions of proteolipid proteins of the central nervous system and plays a role in the developing nervous system
32850_at	Nucleoporin 153 kDa	NUP153	Nuclear pore complexes are extremely elaborate structures that mediate the regulated movement of macromolecules between the nucleus and cytoplasm. These complexes are composed of at least 100 different polypeptide subunits, many of which belong to the nucleoporin family

appropriate gene identifier, the website gives the function of these genes or proteins in human body. In addition to this, it is also possible to find genetic disease which happens due to variation in gene expressions. Table 3 shows a sample of functional classification of marker genes for Leukemia dataset.

7 Conclusion

In this study, the genetic information available in 4 microarray datasets such as MLL, Childhood Tumor, Lung Tumor and DLBCL Tumor have been used to diagnose human disease. It is also considered that the set of genes will behave in a general way in this investigation. With this constraint, a model has been proposed to provide optimize solution for two different purposes. First one is predicting disease using genetic information from microarray data and second one, identification of marker genes responsible for causing the disease. The algorithm of the model is developed using an intelligent Fuzzy-Rough population based method, Fuzzy-Rough-PSO (FRPSO) as a gene selection tool. The results obtained here is quite dependable. The validation of the predicted results has been carried out in two ways. The result shows more than 60 % accuracy in prediction in the form of diagnosis of disease. The result also predicted marker genes for causing cancers. Then using these responsible genes, induction rules can be generated to classify the datasets. This may help in drug design for cancer with regulating the function of the marker genes or proteins by injecting drugs which will stop proliferation of cancer in human body. The real life validation can be carried out in the DAVID website.

References

1. Golub, T.R., Slonim, D.K., Tamayo, P., Huard, C., Gaasenbeek, M., Mesirov, J.P., Coller, H., Loh, M.L., Downing, J.R., Caligiuri, M.A., Bloomfield, C.D., Lander, E.S.: Molecular classification of cancer. Class discovery and class prediction by gene expression monitoring. Science **286**(5439), 531–537 (1999)
2. Dash, M., Liu, H.: Feature selection for classification. Intell. Data Anal. **1**(3), 131–156 (1997)
3. Ahmad, A., Dey L.: A feature selection technique for classificatory analysis. Pattern Recognit. Lett. **26**, 43–56 (2005)
4. Su, Y., Murali, T.M.: RankGene: identification of diagnostic genes based on expression data. Bioinformatics **19**, 1578–1579 (2003)
5. Ding, C., Peng, H.: Minimum redundancy feature selection from microarray gene expression data. J. Bioinf. Comput. Biol. **3**(2), 185–205 (2005)
6. Maji, P.: f-information measures for efficient selection of discriminative genes from microarray data. IEEE Trans. Biomed. Eng. **56**(4) 1063–1069 (2009)
7. Zadeh, L.A.: Fuzzy sets. Inf. Control **8**, 338–353 (1965)
8. Pawlak, Z.: Rough Sets. Theoretical Aspects of Reasoning About Data. Kluwer, Norwell (1991)
9. Alba, E.: Gene selection in cancer classification using PSO/SVM and GA/SVM hybrid algorithms. IEEE C Evol. Comput. **9**, 284–290 (2007)
10. Kennedy, J., Eberhart, R.: Particle swarm optimization. In: IEEE International Conference on Neural Networks—Conference Proceedings, vol. 4, pp. 1942–1948 (1995)
11. Chen, LF.: Particle swarm optimization for feature selection with application in obstructive sleep apnea diagnosis. Neural Comput. Appl. **21**(8), 2087–2096 (2011)
12. Mohamad, M.S., et al.: Particle swarm optimization for gene selection in classifying cancer classes. In: Proceedings of the 14th International Symposium on Artificial Life and Robotics, pp. 762–765 (2009)
13. Stefanowski, J.: Changing representation of learning examples while inducing classifiers based on decision rules. In: Symposium on Artificial Intelligence Methods, 5–7 Nov 2003
14. Stefanowski, J.: Algorithms of rule induction for knowledge discovery. (In Polish), Habilitation Thesis published as Series Rozprawy no. 361. Poznan University of Technology Press, Poznan (2001)
15. Bioinformatics Laboratory, University of Ljubljana. http://www.biolab.si/supp/bi-ancer/projections/info/lungGSE1987.htm (1987)
16. Dudoit, S., Fridlyand, J.: Speed TP. Comparison of discrimination methods for the classification of tumors using gene expression data. J. Am. Stat. Assoc. **97**, 77–86 (2002)
17. Jiang, P., et al.: MiPred. Classification of real and pseudo microRNA precursors using random forest prediction model with combined features. Nucleic Acids Res. **35**, 339–344 (2007)

18. Wang, Y., et al.: Predicting human microRNA precursors based on an optimized feature subset generated by GA-SVM. Genomics **98**, 73–78 (2011)
19. Nanni, L., Brahnam, S., Lumini, A.: Combining multiple approaches for gene [8] R. Kohavi, D. Sommerfield, feature subset selection using the wrapper method: overfitting and dynamic search space topology. In: Proceedings of the First International Conference on Knowledge Discovery and Data Mining, pp. 192–197. AAAI Press, Montreal (1995)
20. Kohavi, R., George, H., John.: Wrappers for feature subset selection. AIJ special issue on relevance. http://robotics.stanford.edu/~fronnyk,gjohng (1996)
21. Mohamad, M.S., et al.: Particle swarm optimization for gene selection in classifying cancer classes. In: Proceedings of the 14th International Symposium on Artificial Life and Robotics, pp. 762–765 (2009)
22. Zhao, W., et al.: A novel framework for gene selection. Int. J. Adv. Comput. Technol. **3**, 184–191 (2011)
23. Chen, L.F., et al.: Particle swarm optimization for feature selection with application in obstructive sleep apnea diagnosis. Neural Comput. Appl. **21**, 2087–2096 (2011)
24. Chen, et al.: Gene selection for cancer identification: a decision tree model empowered by particle swarm optimization algorithm BMC Bioinf. **15**(49) (2014). doi:10.1186/1471-2105-15-49
25. Stefanowski, J.: The rough set based rule induction technique for classification problems. In: Proceedings of 6th European Conference on Intelligent Techniques and Soft Computing EUFIT'98, pp. 109–113. Aaachen (1998)
26. Grzymala-Busse, J.W.: Managing uncertainty in machine learning from examples. In: Proceedings of 3rd International Symposium in Intelligent Systems, pp. 70–84. IPI PAN Press, Wigry, Poland (1994)

Bird Mating Optimization Based Multilayer Perceptron for Diseases Classification

N.K.S. Behera, A.R. Routray, Janmenjoy Nayak and H.S. Behera

Abstract Classification of data is a significant method of data analysis that can be used for intelligent decision making and neural networks are vital contrivances of such classification. Several meta-heuristic algorithms based neural network models such as genetic algorithm (GA) and differential evolution (DE) based MLP are efficiently implemented for this task. However these methods are trapped at local optima. To overcome such limitations, firefly algorithm (FA) and bird mating optimization (BMO) based MLP techniques have been proposed in the paper and tested over several bio-medical datasets like thyroid, hepatitis and heart diseases for classification. The result shows efficient classification of different patients into their diseases categories according to the data obtained from different pathological test.

Keywords Bird mating optimization · Firefly algorithm · Multilayer perceptron · Liver disease · Indian Pima diabetes disease

1 Introduction

Data classification is a two-step predictive data mining technique used for discovering classes of unknown data, by building a classifier. Before applying classification techniques, data pre-processing is carried out, which eliminates the

---

N.K.S. Behera (✉) · J. Nayak · H.S. Behera
Department of Computer Science and Engineering and Information Technology, Veer Surendra Sai University of Technology, Burla 768018, Odisha, India
e-mail: nksbehera@gmail.com

J. Nayak
e-mail: mailforjnayak@gmail.com

H.S. Behera
e-mail: mailtohsbehera@yahoo.co.in

A.R. Routray
Department of Computer Science and Applications, Fakir Mohan University, Balasore 756019, Odisha, India
e-mail: ashanr@gmail.com

L.C. Jain et al. (eds.), *Computational Intelligence in Data Mining - Volume 3*, Smart Innovation, Systems and Technologies 33, DOI 10.1007/978-81-322-2202-6_27

noisy data, missing data, and irrelevant data from the raw dataset? which takes place by four steps (1) data integration (2) data cleaning (3) discretization (4) attribute selection. After that two-step classification techniques are applied to the training set. First step is the learning step that builds the classifier by analyzing a training set made up of attribute vector and class label attribute. There are three different learning approaches: supervised, unsupervised and semi-supervised. In the second step test data are used to evaluate the accuracy of the classification rule. Some basic data classification techniques are Bayesian classifier, rule based classifier, decision tree classifier, Bayesian belief network, support vector machine, k-nearest-neighbour classifier, case based reasoning, fuzzy logic techniques, rough sets, artificial neural network and genetic algorithms.

A multilayer perceptron is a feed-forward neural network which consists of multiple layers of nonlinear-activating neurons in a directed graph. Each neuron has a linear activating function and utilizes different supervised learning techniques. Each activation function maps the weighted input to the output of the neuron. Connection weights are updated according to amount of error in the output compared to the expected result after each piece of data is processed and this process is known as learning of neural network. In the recent years several nature inspired meta-heuristic algorithms are used as leaning techniques to train neural network. Firefly algorithm is incorporated into multilayer perceptron is used to train the neural network and to optimize the performance index [1]. This nature inspired meta-heuristic optimization techniques, which are inspired by the flashing light of fireflies that occurs due to the bioluminescence process, and it is used to update weight, bias and other parameters of the multilayer perceptron.

Bird mating optimization based multilayer perceptron is an optimum searching technique, which implements different mating strategies to train each neurons of the neural network [2]. Mating between different birds follows different strategies to breed a brood having high quality genes which similar to the optimization problems [3]. This technique is tested over three different biomedical datasets such as thyroid, hepatitis and heart diseases for early and correct classification of patients suffering in these diseases. Thyroid is one of the largest glands of endocrine system which secrete some hormones such as triodothyronine and thyroxine which metabolize food, use energy, temperature preferences and gain or lose weight. Thyroid diseases occurs due to imbalance in secretion of these hormones and according to a survey report of Indian Journal of Endocrinology and Metabolism, 10.95 % of the 5,376 people surveyed were diagnosed with hypothyroidism. This is an alarming number and needed more awareness about the thyroid gland and thyroid diseases. Proposed methods are used for correct and early detection of diseases from its symptoms to avoid harmful situations such thyroid related cancer and this technique is also tested over hepatitis and heart diseases dataset to classify its patients suffering from these diseases.

2 Related Work

Feature selection is one of the important task in classification problems. The main purpose feature selection is to select a number of features used in the classification. Several variables are stored in a multi dimensional data sets are sometimes are called features. Several efficient classifiers such as k-nearest neighbor, C4.5 (Quinlin 1993) and back-propagation classifier. The decision tree method shows that feature selection can improve case based learning [4]. A well known tree-growing algorithm for generating decision trees based on univariate splits in Quinlan's ID3 with an extended versions called C4.5 search methods like Greedy search is employed in these algorithms to explore the exponential space of possible modes, which involve growing and pruning decision tree structure. KNN-GA feature selection is one of the data reduction technique introduced by Siedlecki and Sklansky (1989). The feature selection for classification by using a GA-based neural network approach proposed by Li [5], the results are compared with the k-nearest neighbor (KNN) classification rule. GA-based classifiers are robust and effective in finding optimal subsets of features from large data sets [5]. Evolutionary algorithm (EAs) inspired by biological evolution and swarm intelligence algorithms proposed by Askarzadeh [6].

Bird mating optimization (BMO) is a population based optimization algorithm which employs mating process of birds as a framework [3]. And it is proved that search capability of BMO is better than other evolutionary algorithms like FEP, FES, CES and CEP [4, 7]. This algorithm is also tested over several other algorithms like PSO and GA using a number of multimodal functions. Firefly algorithm (FA) is one of most powerful nature inspired algorithms used for multimodal optimization proposed by Yang [2]. It is observed that FA optimization is more promising than particle swarm optimization, when tested using multimodal functions [1, 2]. This paper proposed a new meta-heuristic approach named MLP-BMO for feature selection, which uses bird mating optimization algorithm based multilayer perceptron for classification of several real world biomedical datasets such as thyroid, Indian Pima diabetes, heart and liver disease.

2.1 Firefly Based Multilayer Perceptron

Firefly algorithm is incorporated with multilayer perceptron to update the weight, bias and other parameters of the neural network. In each step the calculated output is compared with the expected output to find the error rate such as mean square error and sum of squared error and upgrade the parameters in subsequent steps.

2.1.1 Behavior of Fireflies

There are about two thousand fireflies species exists in the world. The pattern of rhythmic flashing light of fireflies is a unique characteristic of a particular fireflies species. A process of bioluminescence produces the flashing light. Fireflies use this flashing light [8] as the signaling system to attract their prey. This rhythmic flash, the rate of flashing and the duration of flashing light may also serve as a protective warning mechanism. The flashing light can be formulated in such a way that it is associated with the objective function to be optimized. This makes it possible to formulate a new optimization algorithm.

2.1.2 Attractiveness and Movement of Fireflies

Flashing characteristics of fireflies can be used to develop fireflies-inspired algorithm. In fireflies algorithm there are 3 basic assumptions.

(1) All fireflies are unisex, so that one firefly will be attracted to other regardless of their sex.
(2) Attractiveness is proportional to their brightness, and they both increases as the distance between two fireflies' decreases.
(3) The brightness of firefly is determined by the value of the objective function.

2.1.3 Pseudocode of MLP-FFA

Attractiveness of a firefly is proportional to the distance from adjacent fireflies [1].

$$\beta = \beta 0 e^{-\gamma r^2} \tag{1}$$

where $\beta 0$ is the attractiveness at $r = 0$, and γ is the light absorption constant. Attractiveness function $\beta(r)$ can be any monotonically decreasing function. The distance between any two fireflies i and j at x_i and x_j, respectively is the Cartesian distance. Where $x_{i,k}$ is the spatial coordinate x_i of ith firefly.

Begin :
Define objective function f
Initialize all parameter like light absorption coefficient , beta, number of fireflies(n).
Generate initial firefly population ' ff_i ' and assign weight bias of MLP.
Calculate Mean squared Error (MSE) for each weight
Calculate light intensity I_i at x_i is determined by f.
do
Calculate MSE and find brighter firefly f_i .
do
Choose any less brighter firefly f_j than the brightest firefly.
If $(f_i<f_j)$
Calculate the distance between two firefly .
Move less brighter towards the more.
Update weight and bias value.
end
Recalculate MSE of the updated weight and bias.
Correctly classify the features.
Repeat until last firefly
If (classification rate > Expected value)
Stopping condition =true
End
Repeat until last fireflies
Repeat Until Stopping Condition=true.

---

3 Proposed Techniques

3.1 Bird Mating Optimization Based Multilayer Perceptron

Bird Mating Optimization is one of the most recent and efficient meta-heuristic global optimization approach proposed by Askarzadeh [3]. This approach is inspired by the mating strategies followed by different bird species to produce efficient broods. During mating season birds employ a variety of intelligent behaviors such as singing, tail drumming or dancing to attract potential mates [9]. Birds mating strategies are classified into five different categories like monogamy, polygamy, promiscuity, parthenogenesis and polyandry. Birds society is classified in male and female. The main responsibility of a male bird is to provide security and search food sources, while the main task of a female bird of the society is to produce egg and supply them. Again male birds are divided into monogamous, polygynous, and promiscuous. Female birds are classified into polyandrous and partheogenetic [5]. Before breeding each bird search for a qualitative partner of opposite sex according to produce a healthy brood. The quality of each is determined by the ability to fly long distance, to get more food and memorize paths. Proposed approach MLP-BMO uses this BMO algorithm to train a multilayer layer perceptron used to select features for classification. This approach is tested with three different datasets like thyroid data, Indian pima diabetes dataset, liver and heart diseases datasets and then its performance is compared with MLP-FFA.

3.2 Pseudocode of BMO Based MLP

Begin:
 Initialize MLP parameters, input and percentage of bird's category.
 Specify bird society size, Maximum generation.
 Generate initial population of birds and assign weight and bias of the MLP.
 Calculate the Mean Squared Error (MSE) of the weights.
 do
  Sort the birds according to their performance.
  Partition birds society into polyandrous, parthenogenesis, monogamous Polyandrous, and promiscuous respectively according to their percentage.
  Regenerate promiscuous birds randomly.
  for i=1 to ***society size***
   Compute fitness value of each bird
   S_1 = Calculate MSE value.
  end
  Produce brood of each species by performing corresponding mating.
  for i=1 to ***society size***
   Compute fitness value of each brood.
   S_2 = Calculate MSE value
  end
  for i=1 to ***society size***
   If $S_2 < S_1$
    Replace old birds by their brood.
   end
  end
 Repeat Until termination criterion is met.

---

4 Experimental Set Up and Result Analysis

All computations of FFA based MLP and BMO based MLP classification were performed on a personal computer with 2.00 GHz Intel Pentium-Dual core processor and 1 GB memory, running under Windows-8.1 operating system. Above classification techniques are implemented and trained in MATLAB-8.1 (MATLAB-2014a) using a combination of Neural Network Toolbox and Statistical Toolbox.

4.1 Case-1: Liver Diseases Dataset

This dataset is archived from UCI respiratory which contains 416 liver patients and 167 non-liver patients. This dataset was collected from north east of Andhra Pradesh, India. This dataset contains 10 variables that are age of patient, gender of patient, total biluribin, direct biluribin, alkalin phosphate, alamine aminotransferace. Figure 1 shows the comparison of MSE between MLP-FFA and MLP-BMO for liver diseases dataset.

4.2 Case-2: Thyroid Dataset

Generally diseases diagnosis is considered as a classification task. Proposed approaches were employed in this field to classify whether an individual has a thyroid diseases. Figure 2 shows the comparison of MSE between MLP-FFA and MLP-BMO for thyroid diseases dataset.

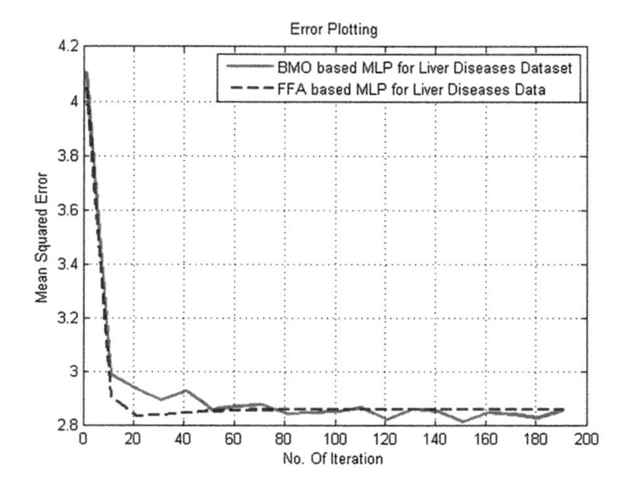

Fig. 1 Comparison of MSE between MLP-FFA and MLP-BMO (Liver diseases dataset)

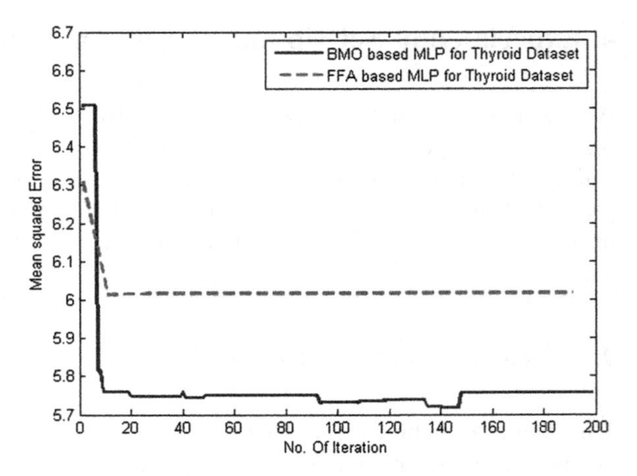

Fig. 2 Comparison of MSE between MLP-FFA and MLP-BMO (Thyroid disease data)

Thyroid diseases arise due to imbalance in production of thyroid hormones from dysfunction of thyroid gland and pituitary gland. Thyroid hormones like stimulating hormone or hypothalamus regulates the pituitary gland via thyroprotein releasing hormones. Several classical diagnosis methods are exists like blood tests, ultrasound, radioiodine scanning, biopsy. But early and correct diagnosis depends upon the correct classification of diseases into their classes. Thyroid diseases dataset contains several attributes like age, sex, thyroxin, anti-thyroid medication, sick, pregnant, thyroid surgery, I131 treatment, TT4 measure, Lithium, goiter, tumor, hypo pituitary, psych, TSH measured, T3 measured, TT4, T4U, FTI measured, TBG measured etc. with these attributes individuals are classified in three different classes like positive, discordant and negative.

4.3 Case-3: Indian Pima Diabetes Dataset

Hepatitis disease is a medical condition defined by the inflammation of liver and characterized by the presence of inflammatory cells in the tissue of organ. The main causes of hepatitis are both infectious and non-infectious. Hepatitis dataset is archived from UCI respiratory, donated by G. Gong (Carnegie-Mellon University) via Bojan Cestnik, Jozef Stefan Institute. This dataset contains several attributes like age, sex, steroid, antivirals, fatigue, malaise, anorexia, liver size, spleen palpable, spiders, ascites, varices, sgot, albumin and protime. Early detection of hepatitis diseases depends upon correct classification of symptoms into different classes. This can be archived by using multilayer neural networks based on metaheuristics algorithm. Figure 3 shows the comparison of MSE between MLP-FFA and MLP-BMO for Indian Pima diabetes diseases dataset.

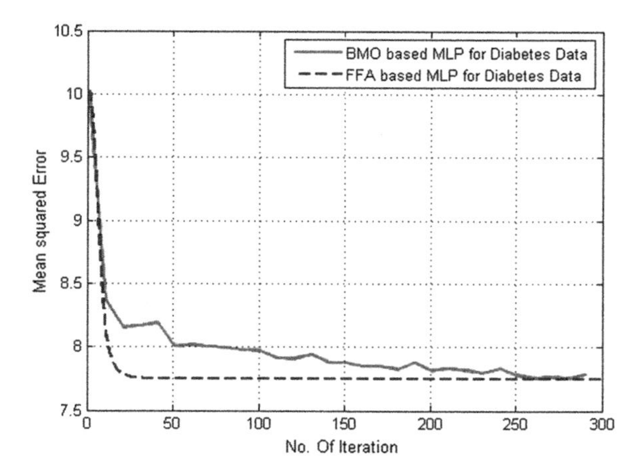

Fig. 3 Comparison of MSE between MLP-FFA and MLP-BMO (Diabetes disease data)

4.4 Case-4: Heart Diseases Dataset

Heart diseases are also known as cardiovascular diseases, which involves heart and blood vessels. This diseases basically affects older adults, there are several causes like age, tobacco consuming, excess alcohol consumption, sugar consumption, diabetes milieus, streptococcal bacterial infections, air pollution, lack of physical exercise and other psychological factors. Several preventions methods can be followed to decrease the possibility of this diseases, this include avoidance of alcohol consumption, tobacco cessation, and decrease body fat. Early and correct detection of heart diseases depends upon correct classification of disease data. The proposed work MLP-BMO and MLP-FFA is applied for heart diseases classification. Figure 4 shows the comparison of MSE between MLP-FFA and MLP-BMO for heart diseases dataset.

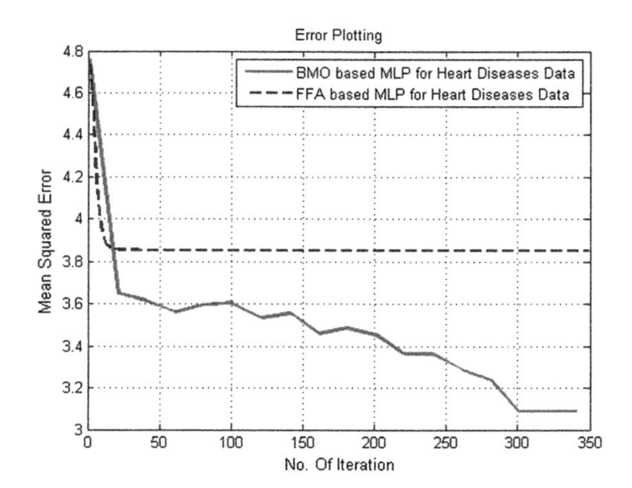

Fig. 4 Comparison of MSE between MLP-FFA and MLP-BMO (Heart diseases dataset)

Table 1 Comparison of results between MLP-FFA and MLP-BMO

Dataset	Methods	MSE				Exec. time(s)
		Min	Max	Avg.	St. D	
Liver diseases	MLP-FFA	2.8892	4.1277	2.9456	0.4425	51.1094
	MLP-BMO	2.8002	4.0714	2.9802	0.3534	24.6754
Thyroid diseases	MLP-FFA	6.0128	6.3012	6.1567	0.5432	48.1689
	MLP-BMO	5.7321	6.5217	6.3632	0.3201	38.2745
Diabetes diseases	MLP-FFA	8.8031	10.0132	8.5634	0.2334	22.0528
	MLP-BMO	7.7732	10.1145	8.4563	0.3453	15.5684
Heart diseases	MLP-FFA	3.8743	4.7832	3.9456	0.2234	45.4556
	MLP-BMO	3.1132	4.8123	3.1534	0.1256	32.6545

4.5 Result Analysis

Table 1 shows the comparison of performance between MLP-FFA and MLP-BMO in-terms of minimum, maximum and average MSE, Standard deviation is also considered as the performance parameter for four bio-medical dataset such as liver, heart, thyroid and Indian Pima diabetes diseases dataset.

5 Conclusion

Performance index of the proposed technique BMO and FFA based multilayer perceptron is measured in terms of mean squared error MSE) and sum of squared error (SSE). It is observed that, MLP-BMO produces early convergence then MLP-FFA. However, BMO based MLP can continue up to a large number of iteration as compared to other techniques and provide a global optimum result. These techniques can classify patients suffering from diseases such as thyroid, liver and heart diseases into three different classes positive, negative and discordant. Classification result shows that MLP-BMO outperforms then the MLP-FFA in the case of heart and liver diseases datasets.

References

1. Yang, X.S.: Firefly algorithm, stochastic test functions and design optimization. Int. J. Bio-inspired Comput. **2**(2), 78–84 (2010)
2. Yang, X.S.: Firefly algorithm for multimodal optimization. In: Stochastic Algorithm: Foundation and Applications, SAGA 2009, Lecture Notes in Computer Sciences, vol. 5792, pp. 169–178 (2009)
3. Askarzadeh, A.: Bird mating optimizer: an optimization algorithm inspired by bird mating strategies. Commun. Nonlinear Sci. Numer. Simul. **19**(4), 1213–1228 (2014)

4. Price, K.V., Glover, F., Corne, D., Dorigo, M.: An introduction to differential evolution new ideas in optimization, pp. 79–108. McGraw-Hill, London (2009)
5. Li, T.-S.: Feature selection for classification by using a GA-based neural network approach. J. Chin. Inst. Ind. Eng. **23**(1), 55–64 (2006)
6. Askarzadeh, A., Rezazadeh, A.: Artificial neural network training using a new efficient optimization algorithm. Appl. Soft Comput. **13**, 1206–1213 (2013)
7. Storn, R., Price, K.: Differential evolution—a simple and efficient heuristic for global optimization over continuous spaces. J. Global Optim. **11**, 341–359 (2004)
8. Nayak J., Nanda, M., Nayak, K., Naik, B., Behera, H.S.: An Improved Firefly Fuzzy C-Means (FAFCM) Algorithm for Clustering Real World Data Sets, Smart Innovation, Systems and Technologies. Springer, Switzerland, vol. 27, pp. 339–348 (2014)
9. Charbonneau, P.: An Introduction to Genetic Algorithms for Numerical Optimization. CAR Technical Note, NCAR
10. Schwefel, H.P.: Evolution and Optimum Seeking: The Sixth Generation. Wiley, New York (1999)
11. Eiben, A.E., Smith, J.E.: Introduction to Evolutionary Computing. Springer, Berlin (2003)
12. Rezazadeh, A., Askarzadeh, A., Sedighizadeh, M.: ANN-based PEMFC modeling by a new learning algorithm. Int. Rev. Modell. Simul. **3**(2), 187–193 (2010)
13. Askarzadeh, A., Rezazadeh, A.: Artificial immune system-based parameter extraction of proton exchange membrane fuel cell. Int. J. Electr. Power Energy Syst. **33**, 933–938 (2011)
14. Blum, C., Socha, K.: Training feed-forward neural networks with ant colony optimization: An application to pattern classification. In: Proceedings of the Fifth International Conference on Hybrid Intelligent Systems, pp. 233–238 (2005)
15. Shi, Y.-J., Teng, H.-F., Li Z.-Q.: Cooperative co-evolutionary differential evolution for function optimization. Adv. Nat. Comput. **3611**, 1080–1088 (2005)
16. Yang, X.S., Deb, S.: Cuckoo search via L'evy flights. In: Proceedings of World Congress on Nature and Biologically Inspired Computing. IEEE Publications, USA pp. 210–214 (2009)

Comparison and Prediction of Monthly Average Solar Radiation Data Using Soft Computing Approach for Eastern India

Sthitapragyan Mohanty, Prashanta Kumar Patra and Sudhansu S. Sahoo

Abstract The paper presents a comparative study of monthly average solar radiation data using different soft computing approaches for Eastern regions of India. Soft computing tools like Multi layer Perceptron (MLP), Artificial Neuro Fuzzy Inference System (ANFIS), and Radial Basis Function (RBF) are used for the said analysis for three cities, i.e. Kolkata, Bhubaneswar and Vishakhapatnam. The input parameters used in these tools are ratio of sunshine hours (S/S_0), average temperature ratio (T/T_0), relative humidity ratio (R/R_0). The monthly average data recorded during the period 1984–1999 for the region mentioned has been considered as training data. The purpose of the study is to compare the accuracy of measured radiation value with the output parameter of solar radiation through different models and using different techniques. The performance of the model is evaluated by comparing the predicted and measured parameters in terms of absolute relative error among ANFIS, MLP and RBF.

Keywords Solar radiation · MLP · ANFIS · RBF

1 Introduction

Solar radiation is the most important parameter in the design and evaluation of solar energy devices. Many developing nations' solar radiation measurements data are not easily available. Therefore it is necessary to elaborate some methods to estimate

S. Mohanty (✉) · P.K. Patra
Department of Computer Science & Engineering, C.E.T, Bhubaneswar, India
e-mail: msthitapragyan@gmail.com

P.K. Patra
e-mail: pkpatra@cet.edu.in

S.S. Sahoo
Department of Mechanical Engineering, C.E.T, Bhubaneswar, India
e-mail: sudhansu@cet.edu.in

L.C. Jain et al. (eds.), *Computational Intelligence in Data Mining - Volume 3*,
Smart Innovation, Systems and Technologies 33, DOI 10.1007/978-81-322-2202-6_28

the solar radiation on the basis of meteorological data. Global solar radiation (H_G) is the most important component of solar radiation as it gives the total solar availability at a particular given place. Solar radiation data provide information on how much of the sun's energy strikes a surface at a location on earth during particular time period. The accurate information of the solar radiation intensity at a given location is essential for solar energy system design, cost analysis, and efficiency calculations of a project in sizing a photovoltaic cell. Mostly, monthly mean daily data are needed for the estimation of long-term performance of solar systems.

Solar radiation data is available for most parts of the world, but is not available for many countries which cannot afford the measurement equipment and techniques involved. Several empirical and regression models have been developed by researchers [1, 2] to calculate solar radiation based on sunshine hours. The applications of MLP network in optimization, prediction, forecasting, estimation of solar radiation performed through different regression method is found in literature review [3–6]. As the weather condition is uncertain some researchers proposed fuzzy based models [7–9] for predicting solar radiation based on different meteorological and climatological parameters. Traditional regression approaches also have been employed in literature [10] for predicting solar radiation with several time scales i.e. hourly, daily and monthly.

Everybody is aware that solar energy is considered to be the most effective and economic alternative resource. India is not far behind as the interest in solar energy application has been growing for providing electricity especially to rural areas and solar heating, cooling processes for domestic and industrial process applications. However, it has been observed in last few years that four States (West Bengal, Andhra Pradesh, Odisha and Tamil Nadu) on the East Coast and One State (Gujarat) on the West Coast are more vulnerable to cyclone disasters. Hence solar radiation insolation in these areas is haphazard in nature as the solar radiation is directly or indirectly generates high wind flow. Again for setting up any renewable energy systems such as solar PV or solar thermal, wind machines, the solar radiation data analysis in these areas becomes crucial. Although traditional approaches have been used to measure solar radiation in almost all Indian cities, it has been observed that most of time no measuring equipments are used in these coastal regions due to high wind effects which cause damage of the solar radiation measuring equipments. Hence, the model based solar radiation measuring techniques can predict the solar radiation data in these coastal areas.

In this paper, a feasibility of various soft computing tools like Neural Network (MLP), ANFIS, RBF have been employed to compare and predict the solar irradiance for three cities on the east coast region of India i.e. Kolkata, Bhubaneswar and Vishakhapatnam.

2 Methodology

Here we have considered three important stations such as Bhubaneswar, Kolkata and Vishakhapatnam for predicting average monthly global solar radiation from 1984–1999. The input parameters used in this model are sunshine per hour (S/S_0) where S is the sunshine duration and S_0 is the maximum sunshine duration, average temperature ratio (T/T_0) where T is air temperature and T_0 is maximum temperature, and relative humidity (R/R_0) where R stands for relative humidity and R_0 is the maximum relative humidity. The output parameter is the clearness index (H/H_0). The estimated output (H, solar radiation) is obtained by multiplying the estimated clearness index with H_0, where H_0 is the extraterrestrial solar radiance measured on the 15th of every month. H is the monthly average solar radiance falls on horizontal surface. Following are the four models have been devised to get the required output.

Model I: Estimates the global solar radiation from average Temperature, Relative Humidity and Sunshine duration. $\frac{H}{H_0} = f\left(\frac{S}{S_0}, \frac{R}{R_0}, \frac{T}{T_0}\right)$

Model II: Estimates the global solar radiation from Relative Humidity and Sunshine duration. $\frac{H}{H_0} = f\left(\frac{S}{S_0}, \frac{R}{R_0}\right)$

Model III: This model estimates the global solar radiation from average Temperature, Relative Humidity. $\frac{H}{H_0} = f\left(\frac{R}{R_0}, \frac{T}{T_0}\right)$

Model IV: This last model estimates the global solar radiation from sunshine duration and temperature. $\frac{H}{H_0} = f\left(\frac{S}{S_0}, \frac{T}{T_0}\right)$

Following sub sections gives the brief insight into the different tools used in this paper for prediction of solar radiation data.

2.1 Neural Networks Model (MLP)

The MLP network in Fig. 1 consist of an input layer having three inputs, output layer having one output and one or more hidden layer. MLP have the ability to mode both linear and non-linear systems without making any assumptions as in most traditional statistical approaches.

NN can be classified into two categories: feed-forward and feedback (recurrent) networks. In the feed-forward network doesn't form any loops by the network connections, But one or more loops may exist in the feedback networks.

To obtain the optimal structure of the MLP neural network, we conducted several training by varying the network parameters such as the activation function, number of hidden layers, number of neurons in each layer, the learning function, the number of iteration.

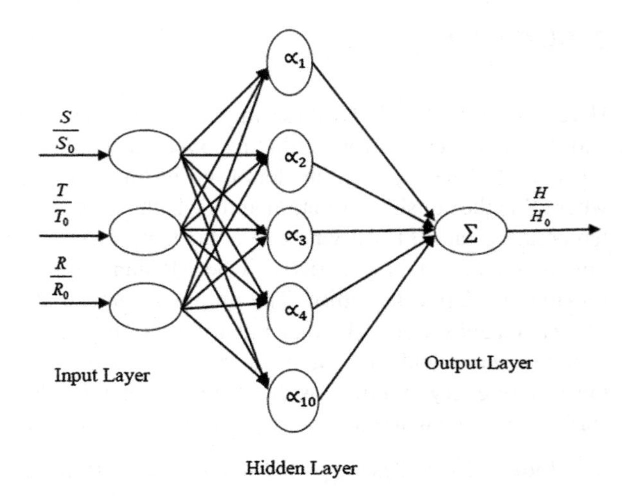

Fig. 1 Neural network (MLP) model

2.2 ANFIS

ANFIS is the implementation of fuzzy Inference system (FIS) to adaptive networks for developing fuzzy rules having some membership functions. The basic structure of a FIS consists of three components: a rule base, which contains a selection of fuzzy rules; a database, which defines the membership functions (MF) used in the fuzzy rules and a reasoning mechanism, ANFIS is a multilayer feed-forward network which uses neural network learning algorithms and fuzzy reasoning to map inputs into output. A general adaptive neuro-fuzzy network shown in Fig. 2. This network consists a number of nodes connected through directional links. Hybrid learning algorithm is used and model is run till the error is minimized.

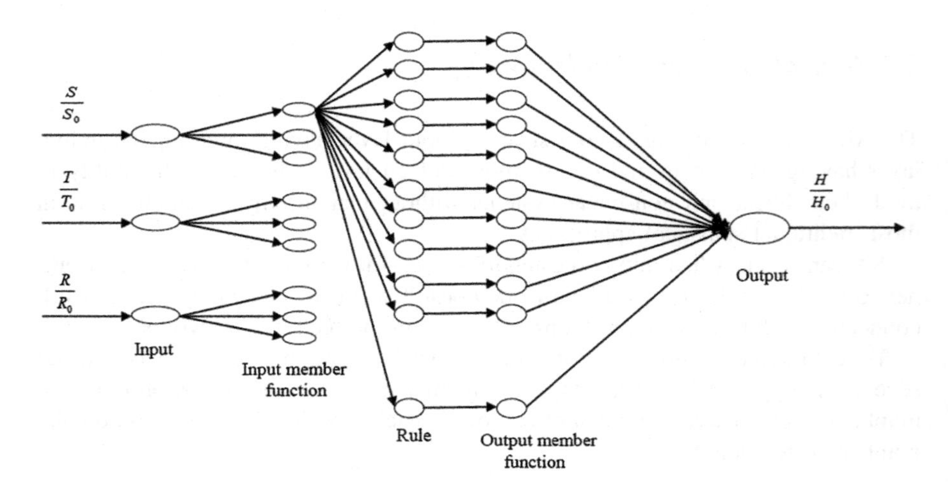

Fig. 2 ANFIS

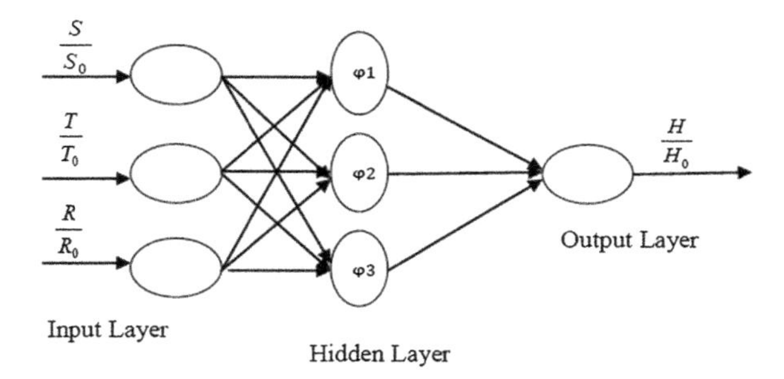

Fig. 3 RBF model

2.3 Radial Basis Function Networks (RBF)

A radial basis function network is an artificial network whose activation function is radial basis function. The output of the network is a combination of radial basis functions of the inputs and neuron parameters. A radial basis function (RBF) network contains three layers: an input layer, a hidden layer with a non-linear RBF activation function and a linear output layer. The second layer/hidden layer performs non-linear mapping from the input space into a higher dimensional space by using a Gaussian kernel function. Output layer/the final layer performs a weighted sum with a linear output. The model is shown in Fig. 3.

3 Results and Discussion

Here in this paper measured monthly average data for 16 years at Kolkata, Bhubaneswar and Vishakhapatnam from (1984–1999) have been taken from solar radiation handbook and National Renewable Energy Laboratory [11, 12]. Computer codes for different models were developed in the MATLAB software (Version 10).

The measured monthly averaged data such as sunshine per hour, temperature ratio, relative humidity and clearness index for the mentioned three cities are mentioned in Table 1. The model is trained and tested by using different soft computing technique such as MLP, ANFIS and RBF. Using MLP, number of epoch, number of hidden layer nodes, learning rate and momentum were varied for back propagation algorithm. We found the best performance; maximum regression (0.96886) with Mean Square Error (MSE) (0.000650) is obtained at epoch 2000, iteration 6 using 10 hidden layer neurons. For Bhubaneswar the maximum regression coefficient (0.82582) and MSE (0.00868) is found at epoch 2000 and iteration 6 and for Vishakhapatnam the regression coefficient (0.97908) and MSE (0.00113) is found at epoch 2000 and iteration 6. Similarly the RBF, after several simulation it was found that the best performance is obtained at epoch 2, and spread = 2.0. The MSE for Kolkata is 0.00414128. Similarly for Bhubaneswar and Vishakhapatnam MSE is 0.0061278 and 0.0075529. In case of ANFIS, after several

Table 1 Training file of monthly average data of S/S_0, T/T_0, R/R_0, H/H_0 for mentioned cities

Month	Kolkata				Bhubaneswar				Vishakhapatnam			
	S/S_0	T/T_0	R/R_0	H/H_0	S/S_0	T/T_0	R/R_0	H/H_0	S/S_0	T/T_0	R/R_0	H/H_0
Jan	0.78	0.751	0.641	0.58	0.747	0.728	0.508	0.617	0.877	0.884	0.621	0.622
Feb	0.8	0.791	0.843	0.533	0.741	0.825	0.517	0.616	0.962	0.894	0.681	0.634
Mar	0.682	0.825	0.536	0.557	0.647	0.935	0.496	0.588	0.79	0.916	0.683	0.617
Apr	0.557	0.845	0.843	0.556	0.637	0.999	0.553	0.599	0.821	0.938	0.706	0.603
May	0.679	0.879	0.711	0.521	0.601	0.996	0.67	0.559	0.77	0.951	0.749	0.568
Jun	0.635	0.907	0.731	0.428	0.355	0.974	0.779	0.575	0.628	0.959	0.766	0.448
Jul	0.551	0.922	0.838	0.38	0.263	0.942	0.819	0.67	0.648	0.959	0.784	0.412
Aug	0.556	0.929	0.667	0.406	0.328	0.937	0.825	0.36	0.723	0.949	0.779	0.428
Sep	0.677	0.918	0.869	0.424	0.416	0.921	0.803	0.415	0.654	0.941	0.795	0.472
Oct	0.734	0.886	0.626	0.498	0.623	0.874	0.734	0.516	0.829	0.926	0.744	0.542
Nov	0.808	0.827	0.767	0.528	0.66	0.802	0.581	0.574	0.87	0.908	0.643	0.571
Dec	0.743	0.768	0.737	0.525	0.762	0.72	0.492	0.486	0.877	0.881	0.583	0.606

Table 2 Absolute relative error for given three cities using different soft computing techniques

Month	Kolkata			Bhubaneswar			Vishakhapatnam		
	MLP	ANFIS	RBF	MLP	ANFIS	RBF	MLP	ANFIS	RBF
Jan	3.0165	0.0016	0.6961	5.9449	0.00502	3.8024	0.1117	0.0135	1.709
Feb	4.2324	0.0381	3.5534	5.3647	0.0033	6.4643	2.5307	0.0044	0.006
Mar	3.476	0.0053	5.6596	5.6702	0.0077	4.0527	0.86201	0.0049	8.4653
Apr	2.7738	0.0426	7.5107	0.9199	0.0012	3.4472	0.3502	0.0355	4.6453
May	2.8363	0.0158	6.1261	3.3133	0.002	6.5792	1.1997	0.0064	5.4589
June	1.1139	0.0059	7.4124	0.8544	0.0047	3.7042	6.1806	0.0066	4.4648
July	1.5638	0.0036	6.193	8.7102	0.004	6.8342	6.0207	0.0089	6.7951
Aug	4.2362	0.0091	3.6189	3.4637	0.0253	9.0297	6.7013	0.0119	8.1804
Sep	4.3099	0.0044	6.9943	8.9104	0.0063	8.3468	1.112	0.0232	6.036
Oct	4.5103	0.0082	0.8281	6.4035	0.0089	0.3072	3.7027	0.0004	5.8702
Nov	0.3375	0.0066	2.3547	0.9799	0.0062	3.7557	9.0403	0.0067	6.293
Dec	1.8914	0.0066	2.7157	6.7407	0.00053	5.1809	2.3607	0.0065	0.5159

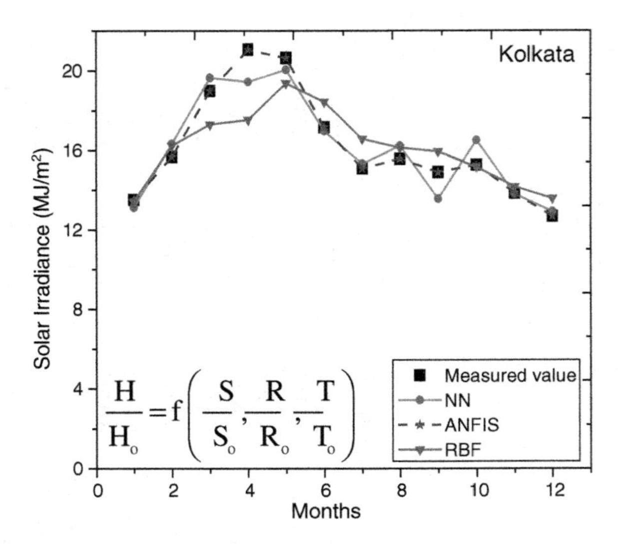

Fig. 4 Measured data and predicted data using different soft computing approaches for Kolkata

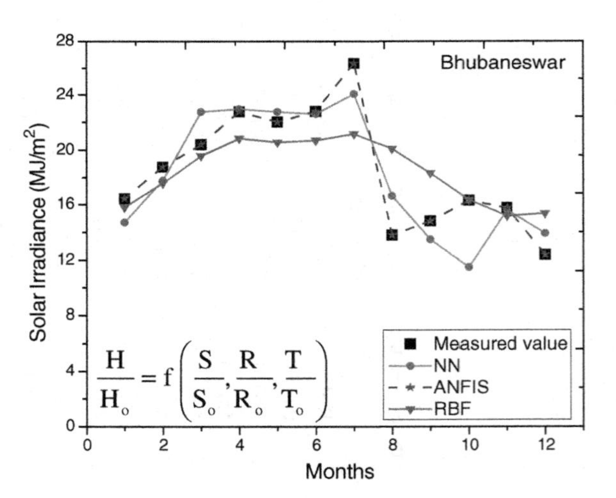

Fig. 5 Measured data and predicted data using different soft computing approaches for Bhubaneswar

simulations the model gave better accuracy at epoch 3. The model uses trimf (triangular membership function) for input and linear type function for output. The overall performances of the models are calculated on the basis of absolute relative error in percentage which is given in Table 2. By calculating absolute relative percentage error, ANFIS is more effective compared to other soft computing technique (MLP and RBF)

Figures 4, 5 and 6 depict the monthly averaged solar data comparison of different soft computing tools for Kolkata, Bhubaneswar and Vishakhapatnam respectively. It has been observed that results from ANFIS matches quite well with the measured

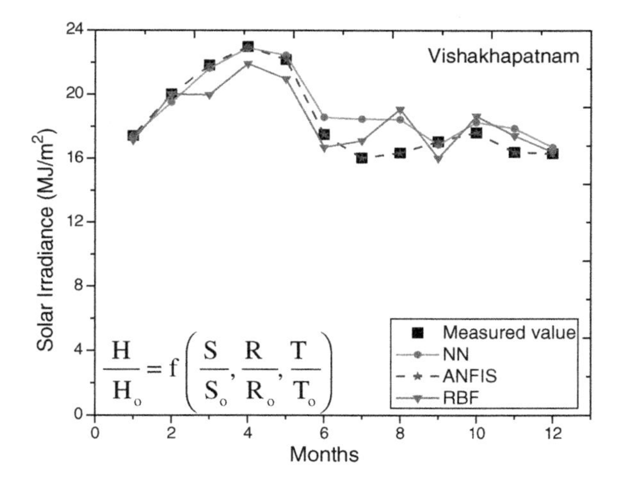

Fig. 6 Measured data and predicted data using different soft computing approaches for Vishakhapatnam

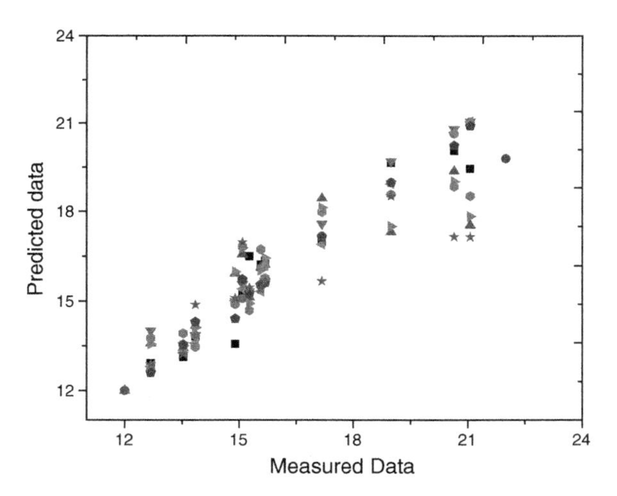

Fig. 7 Parity plot showing measured and predicted solar irradiance

values. However, the data from NN and RBF are within 10 % error band with that of measured value. The Parity plot showing measured data and predicted data using different soft computing techniques is shown in Fig. 7.

4 Conclusion

This paper presented a comparison of mean monthly global solar radiation data of Eastern India using different soft computing techniques like NN, ANFIS and RBF. The measured average monthly global solar radiation was compared with measured values of Kolkata, Bhubaneswar and Vishakhapatnam. Due to frequent cyclonic effects in these places are considered much useful to study the feasibility of use of soft computing approaches for measurement of solar data as most of time no equipments are used for measuring solar data. It was seen that the mentioned soft computing tools predicts much well with measured values. It can be concluded that with sunshine duration data, temperature data, and/or humidity data, solar radiation can be estimated with soft computing techniques. Hence this tool seems to be promising to estimate solar data for the places where costly equipments for measuring solar data cannot be installed. Furthermore this tool can be used for offline designing renewable energy systems such as solar PV, solar thermal, wind machines where solar irradiation data are required.

References

1. Medugu, D.W., Yakubu, D.: Estimation of mean monthly global solar radiation in Yola. Nigeria using angstrom model. Adv. Appl. Sci. Res. **2**, 414–421 (2011)
2. Kumar, R., Aggrawal, R.K., Sharma, J.D.: New regression model to estimate global solar radiation using artificial neural network. Adv. Energy Eng. **1**, 66–73 (2013)
3. Ahmed, A., El-Nouby, M.: Estimate of global solar radiation by using artificial neural network in Qena, Upper Egypt. J. Clean Energy Technol. **1**, 148–150 (2013)
4. Benghanem, M., Mellit, A.: Radial basis function network-based prediction of global solar radiation data: application for sizing of a stand-alone photovoltaic system at AI-Madinah, Saudi Arabia. Energy **35**, 3751–3762 (2010)
5. Rehman, S., Mohandes, M.: Artificial neural network estimation of global solar radiation using air temperature and relative humidity. Energy Policy **36**, 571–577 (2008)
6. Mohandes, M., Balghonaim, A., Rehman, S.: Use of radial basis functions for estimating monthly mean daily solar radiation. Sol. Energy **68**, 161–169 (2000)
7. Mellit, A., Hadj Arab, A., Shaari, S.: An ANFIS based prediction for monthly clearness index and daily solar radiation. Application for sizing of a stand-alone photovoltaic system. J. Phys. Sci. **18**, 15–35 (2007)
8. Sumithira, T.R.: An adaptive neuro-fuzzy inference system (ANFIS) based prediction of solar radiation. J. Appl. Sci. Res. **8**, 346–351 (2012)
9. Rahoma, W.A., Hassan, A.H.: Application of neuro-fuzzy techniques for solar radiation. J. Comput. Sci. **7**, 1605–1611 (2011)
10. Falayi, E.O., Adepitan, J.O., Rabiu, A.B.: Empirical models for the correlation of global solar radiation with meteorological data for Iselin. Nigeria. Int. J. Phys. Sci. **3**, 210–216 (2008)
11. Solar Radiation Handbook 2008
12. http://rsd.gsfc.nasa.gov/rsd/remotessensing.html

Effect of Thermal Radiation on Unsteady Magnetohydrodynamic Free Convective Flow in Vertical Channel in Porous Medium

J.P. Panda

Abstract The aim of this paper is to study the Magnetohydrodynamic free-convective flow in a vertical channel with thermal radiation in porous medium. The Rosseland and approximation is used to describe radiative heat flux in energy equation. The channel flow is described by momentum and energy equations. The expression for temperature and velocity are obtained analytically using perturbation method. There is also remarkable influence of magnetic parameter, radiation parameter, temperature difference parameter and Prandtl number in the flow.

Keywords Thermal radiation · Magnetohydrodynamic · Free-convective flow · Vertical channel · Porous medium

1 Introduction

The interests of many researchers have been acknowledged in the study of unsteady laminar free convection flow. This area continues to drag its recognition due to its vast scientific and technological applications, such as in the early stage of melting and in transient heating of insulating air gap by heat input at the start-up of furnace, solar heating, ventilating passive systems and electric equipment in construction of vertical circuit boards which can be modeled by parallel heated plates with upward flow in the intervening space as reported by Jha et al. [1]. In deed more interest have been in the study of radiation on MHD free convection flows. Sahoo et al. [2] investigated convective boundary layer flow and heat transfer past a stretching porous wall embedded in a porous medium. Ghaly [3] proposed an investigation on the radiation effect on steady MHD free convection flow near isothermal stretching sheet with mass transfer. Makinde [4] used numerical superposition methods to

J.P. Panda (✉)
Department of Mathematics, Veer Surendra Sai University of Technology, Burla 768018, India
e-mail: jppanda123@yahoo.co.in

L.C. Jain et al. (eds.), *Computational Intelligence in Data Mining - Volume 3*, Smart Innovation, Systems and Technologies 33, DOI 10.1007/978-81-322-2202-6_29

solved boundary layer flow with thermal radiation and mass transfer past a permeable moving plate. Ogulu [5] obtained an analytical solution on radiation heat absorption and mass transfer on the flow of polar fluid in the presence of a uniform magnetic field. Das et al. [6] showed the influence of mass transfer on MHD flow through a porous medium. Abdul Hakeem and Sathiyathan [7] presented an analytical solution for nonlinear oscillatory flow past a porous medium with radiation effect. Panda et al. [8] studied heat and mass transfer on MFD flow through porous media over an accelerating surface in presence of suction and blowing. Nandkeolyar et al. [9] discussed unsteady hydromagnetic radiative flow of a dusty fluid past a porous plate with ramped wall temperature. In this paper the role of radiation on MHD free-convection flow in a vertical channel in porous medium using analytical method is investigated.

2 Governing Equations

We consider coupled time-dependent heat transfer by unsteady MHD free-convective flow of a viscous incompressible and electrically conducting fluid under the influence of transverse magnetic field in a vertical channel formed by two infinite parallel plates. The schematic representation of the flow is shown in Fig. 1. The two vertical infinite parallel plates are assumed to be separated by a distance H. Initially, the temperature of the both vertical plates and the fluid is assumed to be T_0. At $t' \geq 0$, the imposed temperature on plate $y' = 0$ is raised to uniform temperature T_w, causing flow of free convection current. The (x', y') are taken such that the flow is assume to be in the x'-direction, vertically upward along the channel walls and y'-axis normal to the plate. The fluid is assumed to obey the Boussinesq approximation. Under this assumption the momentum and energy equations in dimensional form are

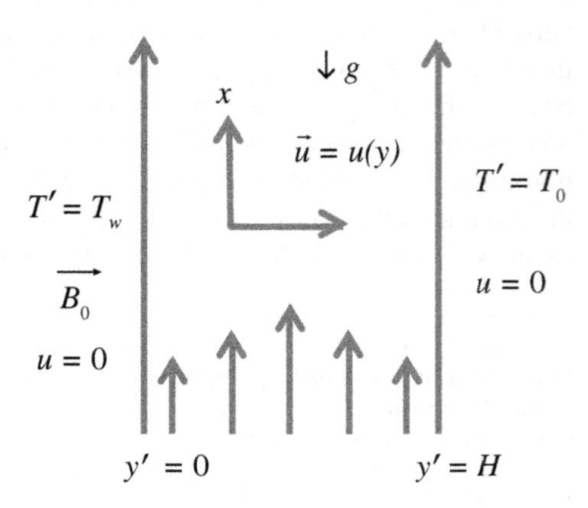

Fig. 1 Schematic diagram of the problem

$$\frac{\partial u'}{\partial t'} = v\frac{\partial^2 u'}{\partial y'^2} + g\beta(T' - T_0) - \frac{\sigma_1 B_0^2 u'}{\sigma} - \frac{vu'}{k} \tag{1}$$

$$\frac{\partial T'}{\partial t'} = \alpha\left[v\frac{\partial^2 T'}{\partial y'^2} - \frac{\partial q_r}{K\partial y'}\right] \tag{2}$$

where T_0 is the initial temperature of the fluid and plates, T', is the dimensional temperature of the fluid, α is the thermal diffusivity, K is the thermal conductivity, ρ is the density of the fluid, β is the coefficient of the thermal expansion, σ_1 is the fluid electrical conductivity, k is permeability and B_0 is the strength of applied magnetic field. The flow is assumed laminar.

The radiation heat flux term in the problem is simplified by using the Rosseland approximation.

$$q_r = \frac{4\sigma\partial T'^4}{3k^*\partial y'} \tag{3}$$

where σ is Stefan-Boltzmann constant and k^* the mean absorption coefficient.

The physical quantities used in the above equations are defined in the nomenclature, hence the initial and boundary conditions to be satisfied are

$$t' \leq 0 : u' = 0 \quad T' = T_0 \quad 0 \leq y' \leq H$$

$$t' > 0 : \begin{cases} u' = 0 \quad T' = T_w \; at \quad y' = 0 \\ u' = 0 \quad T' = T_0 \; at \quad y' = H \end{cases} \tag{4}$$

The solutions of Eqs. (1), (2) and (3) subject to the condition (4) in dimensionless form can be obtain by introducing the following dimensionless quantities

$$t = \frac{t'v}{H^2}; \quad t = \frac{y'}{H}; \quad Pr = \frac{v}{\alpha}; \quad R = \frac{4\sigma_1(T_w - T_0)^3}{k^*K}; \quad u = u'v\left[g\beta H^2(T_w - T_0)\right]^{-1};$$
$$M^2 = \frac{\sigma_1 B_0^2 H^2}{\rho v}; \quad C_T = \frac{T_0}{T_w - T_0}; \quad \theta = \frac{T' - T_0}{T_w - T_0} \tag{5}$$

Using expression (5) in (1), (2) and (3), we have dimensionless momentum and energy equations as

$$\frac{\partial u}{\partial t} = \frac{\partial^2 u}{\partial y^2} + \theta - \left(M^2 + \frac{1}{K_p}\right)u \tag{6}$$

$$\Pr\left[\frac{\partial \theta}{\partial t}\right] = \left[1 + \frac{4R}{3}(C_T + \theta)^3\right]\frac{\partial^2 \theta}{\partial y^2} + 4R[C_T + \theta]^2\left(\frac{\partial \theta}{\partial y}\right)^2 \tag{7}$$

with the dimensionless initial and boundary conditions as

$$t \leq 0: \quad u = 0, \quad \theta = 0, \quad for \; 0 \leq y \leq 1$$

$$t > 0: \begin{cases} u = 0, & \theta = 1 \quad at \;\; y = 0 \\ u = 0, & \theta = 0 \quad at \;\; y = 1 \end{cases} \tag{8}$$

3 Solutions

Equations (6) and (7) are coupled and nonlinear, hence provide no analytical solution. But to be realistic to the numerical result such a solution is useful. Since analytical solution is often an opportunity to validate computer routines of complicated problems and comparisons with data as well as inspecting the internal consistency of mathematical models and the approximation adopted and developed new theoretical result [19]. The mathematical model representing the steady-state MHD free-convection can be obtained by letting $\partial u/\partial t = 0$ and $\partial \theta/\partial t = 0$ in (6) and (7) to have

$$\frac{d^2u}{dy^2} - \left(M^2 + \frac{1}{K_p}\right)^2 u + \theta = 0 \tag{9}$$

$$\left[1 + \frac{4R}{3}(C_T + \theta)^3\right]\frac{d^2\theta}{dy^2} + 4R[C_T + \theta]^2\left(\frac{d\theta}{dy}\right)^2 = 0 \tag{10}$$

with the following boundary conditions

$$\begin{cases} u = 0, & \theta = 1 \quad at \;\; y = 0 \\ u = 0, & \theta = 0 \quad at \;\; y = 1 \end{cases} \tag{11}$$

To construct an approximated solution to Eqs. (9) and (10) subject to (11), It is assumed that the radiation parameter is small and taking a power series expansion in the radiation parameter R employs a regular perturbation method.

$$\begin{aligned} \theta(y) &= \theta_0 + R\theta_1 + \mathrm{O}(R^2) \\ u(y) &= u_0 + Ru_1 + \mathrm{O}(R^2) \end{aligned} \tag{12}$$

Substituting Eq. (12) into Eqs. (9) and (10) and equating the coefficient of like powers of R, the required solution of the velocity and energy equations are

$$
\begin{aligned}
u(y) = {} & \frac{\sinh(M_1 y - M_1)}{\sinh(M_1)} + \frac{4\sinh(M_1 y)}{M_1^4 \sinh(M_1)} \\
& + R\left\{\frac{A\sinh(M_1 - M_1 y)}{\sinh(M_1)} + \frac{4\sinh(M_1 y)}{M_1^4\sinh(M_1)}\left[B^2 - 2B + 1 + 2M_1^{-2}\right]\right. \\
& \quad - 4M_1^{-4}\lfloor B^2 + 2M_1^{-2}\rfloor + yM_1^{-2}\lfloor 2B^2 - 4B/3 + 8BM_1^{-2} + 1/3\rfloor \\
& \quad - 2y^2 M_1^{-2}\lfloor B^2 + 2M_1^{-2}\rfloor \\
& \quad \left. + 4y^3 BM_1 - y^4 M_1^{-2}/3\right\}
\end{aligned} \tag{13}
$$

$$
\theta(y) = 1 - y + \frac{R}{3}\left[6B^2\left(y - y^2\right) - 4B\left(y - y^3\right) + \left(y - y^4\right)\right] \tag{14}
$$

From (13) the steady-state skin frictions on the boundary are:

$$
\begin{aligned}
\tau_0 = \left.\frac{du}{dy}\right|_{y=0} = {} & \frac{M_1\cosh(M_1)}{\sinh(M_1)} + \frac{4M_1^{-3}}{\sinh(M_1)} \\
& + R\left\{\frac{-AM_1}{\sinh(M_1)}\frac{4M_1^{-3}\cosh(M_1)}{\sinh(M_1)}\left[B^2 - 2B + 1 + 2M_1^{-2}\right]\right. \\
& \quad \left. M_1^{-2}\lfloor 2B^2 - 4B/3 + 8BM_1^{-2} + 1/13\rfloor\right\}
\end{aligned} \tag{15}
$$

$$
\begin{aligned}
\tau_1 = \left.\frac{du}{dy}\right|_{y=1} = {} & \frac{M_1}{\sinh(M_1)} + \frac{4M_1^{-3}\cosh(M_1)}{\sinh(M_1)} \\
& + R\left\{\frac{-AM_1}{\sinh(M_1)} + \frac{4M_1^{-3}\cosh(M_1)}{\sinh(M_1)}\left[B^2 - 2B + 1 + 2M_1^{-2}\right]M_1^{-2}\right. \\
& \quad \left[2B^2 - 4B/3 + 8BM_1^{-2} + 1/3\right] \\
& \quad \left. - 4M_1^{-2}\lfloor B^2 + 2M_1^{-2}\rfloor + 12BM_1^{-2} - 4M_1^{-2}/3\right\}
\end{aligned} \tag{16}
$$

where $A = 4B^2M_1^{-4} + 8M_1^{-6}$, $B = C_T + 1$ and $M_1 = \sqrt{M^2 + \frac{1}{K_p}}$

4 Results and Discussion

The dimensionless system of the unsteady flow problems are solved analytically. The essential dimensionless parameters that govern the flow are: the magnetic parameter (M), the radiation parameter®, the temperature difference (C_T), and the Prandtl number Pr, which is inversely proportional to the thermal diffusivity of the working fluid. The numerical computation of the solutions are performed using different values of magnetic parameter (M = 1, 2, 3), the radiation parameter

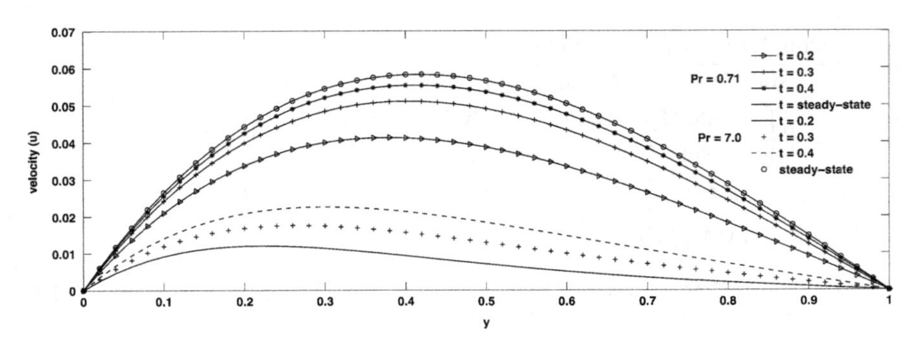

Fig. 2 Velocity profile (C_T = 0.01, R = 0.001, M = 1, Kp = 1)

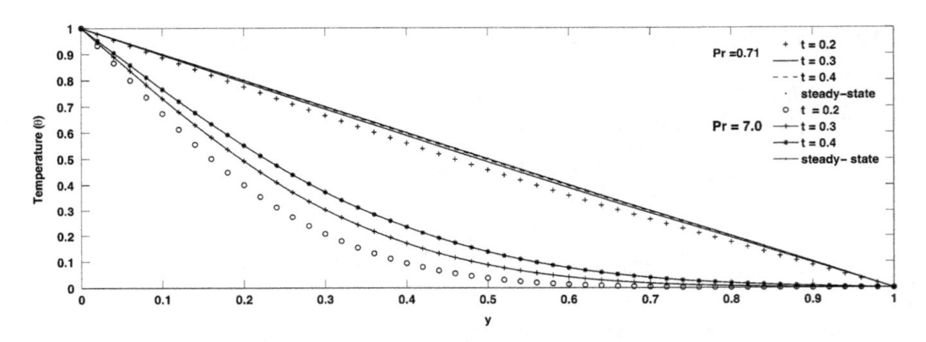

Fig. 3 Temperature profile (C_T = 0.01, R = 0.001, Kp = 1)

$(R = 0.001,\ 0.2,\ 0.4)$, temperature different parameter $(C_T = 0.01,\ 0.2,\ 0.4)$ and Prandtl number for air (Pr = 0.71) and water (Pr = 7) on velocity field, temperature distributions, skin friction and Nusselt number. The unsteady and steady state profiles behaviours as a function of time with default values of magnetic parameter $(M = 1)$, radiation parameter $(R = 0.001)$ and temperature difference parameter $(C_T = 0.01)$ with Prandtl number $(Pr = 0.71,\ 7.0)$ are shown on Figs. 2 and 3.

It is observed from these two figures that velocity and temperature increases with increase in time and finally reaches their steady state. These two figures revealed that time required to reach steady-state in case of water (Pr = 7) is high than that of air (Pr = 0.71), even though both later reached equal steady state with fixed default values. This is true since thermal diffusivity reduces at high value of Prandtl number.

Figures 4, 5 and 6 depict comparison between unsteady and steady state velocity profile for different dimensionless values of time (t) and varying magnetic parameter (M), radiation parameter (R) and temperature difference parameter C_T under two working fluid numbers for air (Pr = 0.71) and water (Pr = 7) respectively. In Fig. 4, the behaviour of varying magnetic parameter (M) with adjusted radiation parameter (R) and temperature difference parameter are studied. The figures show increase in the velocity as time increases. Also, as M increases the velocity

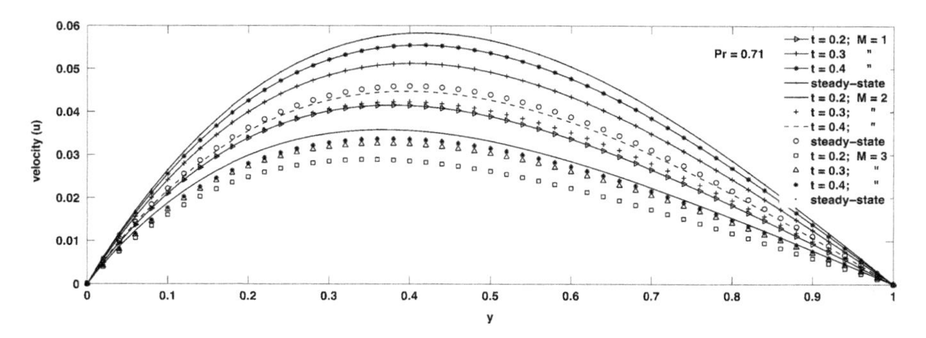

Fig. 4 Velocity profile ($C_T = 0.01$, $R = 0.001$, $Kp = 1$)

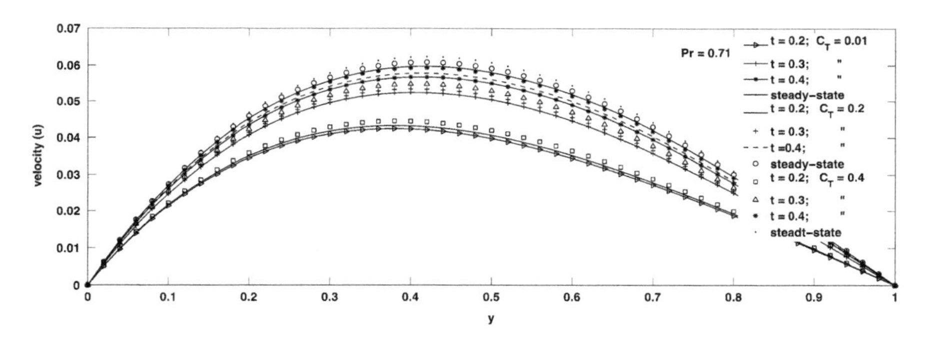

Fig. 5 Temperature profile ($C_T = 0.01$)

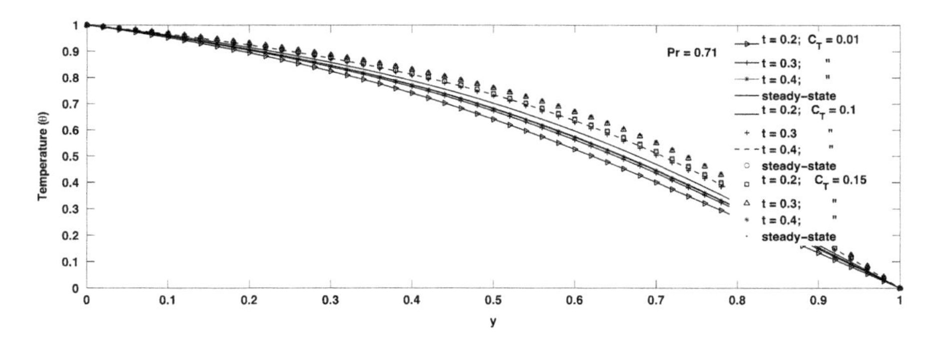

Fig. 6 Temperature profile ($R = 1$)

decreases for both unsteady and steady state. It equally observed that time required to reach steady state is decreases as M increases. In Figs. 5 and 6 the comparative behaviour between unsteady and steady state temperature as a function of time (t) and radiation parameter (R) for fixed temperature difference parameter (C_T) and temperature difference parameter (C_T) for fixed radiation parameter (R) are reported. It is noted that the temperature increases with time and gradually attain its steady state.

5 Conclusion

The role of radiation on time dependent Magnetohydrodynamic free-convection flow in a vertical channel formed by two infinite vertical plates has been studied. The velocity field, temperature field and skin-friction for the nonlinear partial differential equation are obtained analytically by the well-known perturbation series method. The effects of the pertinent parameter on velocity and temperature are presented on figures and discussed. From the result obtained, it is seen that velocity and temperature increases with increase in time.

References

1. Jha, B.K., Samaila, A.K., Ajibade, A.O.: Transient free-convection flow of reactive viscous fluid in a vertical channel. Int. Commun. Heat Mass Transf. **38**, 633–637 (2011)
2. Sahoo, S.N., Panda, J.P., Dash, G.C.: MHD convective boundary layer flow and heat transfer past a stretching porous wall embedded in a porous medium. J. Energy Heat Mass Transf. **33**, 131–142 (2011)
3. Ghaly, A.Y.: Radiation on a certain MHD free-convection flow. Chaos Soliton. Fract. **13**, 1843–1850 (2002)
4. Makinde, O.D.: Free convection flow with thermal radiation and mass transfer past a moving vertical porous plate. Int. Commun. Heat Mass Transf. **32**, 1411–1419 (2005)
5. Ogulu, A.: The influence of radiation absorption on unsteady free convection and mass transfer flow of a polar fluid in the presence of a uniform magnetic field. Int. J. Heat Mass Transf. **48**, 5078–5080 (2005)
6. Das, S.S., Satapathy, J.K., Panda, J.P.: Mass transfer effect on MHD and heat transfer past a vertical porous plate through a porous medium under oscillatory suction and heat sourcent. Int. J. Heat Mass Transf. **52**, 5962–5969 (2009)
7. Abdul Hakeem, A.K., Sathiyanathan, K.: An analytical solution of an oscillatory flow through a porous media with radiation effect. Nonlinear Anal. Hybrid Syst **3**, 288–295 (2009)
8. Panda, J.P., Dash, N., Dash, G.C.: Heat and mass transfer on MFD flow through porous media over an accelerating surface in presence of suction and blowing. J. Eng. Thermophys. Russia **21** (2), 119–130 (2012)
9. Nandkeolyar, R., Sibanda, P., Ansari, S.: Unsteady hydromagnetic radiative flow of a dusty fluid past a porous Plate with ramped wall temperature. ASME **7A**, 70 (2013)

A Fuzzy Based Approach for the Selection of Software Testing Automation Framework

Mohd Sadiq and Fahamida Firoze

Abstract In literature, we have identified different software test automation frameworks (STAF) like modular framework, data driven framework, keyword driven framework, and hybrid framework. Selection of these frameworks on the basis of different criteria creates a multi-criteria decision making (MCDM) problems. Therefore, this paper presents a fuzzy based approach for the selection of STAF to test the software. Finally, the utilization of the proposed method is demonstrated with the help of an example.

Keyword Fuzzy approach and their implementation in the selection of software testing automation framework

1 Introduction

Software testing is the process of evaluating the software behavior to check whether it operates as expected in order to improve its quality and reliability. Software testing process is highly time and resource consuming [1]. Software testing is divided into different phases, i.e., Modeling the software environment, selecting test scenarios, running and evaluation test scenarios, and measuring testing process. Test automation is a kind of development activity. It refers to anything that streamlines the testing process and facilitates things to move along more quickly and with less delay. Test automation involves automating a manual process already

M. Sadiq (✉)
Faculty of Engineering and Technology, Computer Engineering Section, University Polytechnic, Jamia Millia Islamia (A Central University), New Delhi 110025, India
e-mail: sadiq.jmi@gmail.com

F. Firoze
Department of Computer Science and Engineering, Al-Falah School of Engineering and Technology, Dhauj, Faridabad, Haryana, India
e-mail: fahamida_firoze@yahoo.com

L.C. Jain et al. (eds.), *Computational Intelligence in Data Mining - Volume 3*,
Smart Innovation, Systems and Technologies 33, DOI 10.1007/978-81-322-2202-6_30

in place that uses formalize testing process. Test automation involves three phases, i.e., Test planning, test preparation, and test execution [2–5].

In test planning phase following are documented: test strategy, scope of testing, concepts and the practices to be followed. The outcome of the phase is the master test plan. Test preparation phase includes activities such as gathering the functional requirements and translating them into test requirement, writing the test cases, studying the application under test (AUT). Performing feasibility analysis with various automation tools and selecting the appropriate tools. The third and last phase of test automation is Test execution. In this phase, the automated test scripts and library files are implemented and executed against the AUT to generate test report and problems or defects reports.

In literature we have identified the following frameworks which are independent of automation tool [3, 6–9]:

1.1 Modular Framework

1.1.1 Test script modular framework
1.1.2 Test library modular framework

1.2 Data-Driven Framework
1.3 Keyword Driven Framework
1.4 Hybrid Framework

Selection of Software Testing Automation Framework (STAF) on the basis of several criteria creates a multi-criteria decision making (MCDM) problem. Different stakeholders participate in requirements elicitation process and they may have different preferences for the same requirements because stakeholders may have different levels of experience, knowledge, and understanding [10]. In general most of the people use linguistic terminology instead of exact numbers in describing their preferences. Theirs is vagueness in human judgments and often difficult to represent in exact numbers. The weights assigned to requirements by stakeholders are often described qualitatively [11]. Fuzzy logic handles imprecise and linguistic variables mathematically; the term as low cost, high performance, good quality etc. can be expressed in well defined manner using fuzzy logic [12, 13].

In 2012, Jain and Sharma proposed an efficient keyword driven test automation framework for web application. Keyword driven framework reduces the complex script programming knowledge problem for the automation testing. In [3], Patwa provides an overview of how the hybrid test automation framework can be implemented using Quick Test Professional (QTP). In a technical report of IBM global business services, Rashmi Mascarenhas [9] developed and implemented an automation framework. In 2011, Shahamiri et al. proposed an automated framework for software test oracle. Test Oracles are used as a complete and reliable source of expected outputs and a tool to verify the correctness of actual outputs [14]. The objective of this paper is to propose a fuzzy based approach for the selection of STAF. Rest of the paper is organized as follows: In Sect. 2, we present the proposed method for the selection of STAF. Implementation part is given in Sect. 3; and finally, we conclude the paper in Sect. 4.

2 Proposed Method

In this section we present a method for the selection of STAF using fuzzy based approach. In this method we have used different notation of fuzzy set theory. Detailed descriptions about these notations is given in [11, 15] The proposed method is presented as below: (i) Identify the existing framework; (ii) Identify the criteria; (iii) Collect decision making (DM) fuzzy assessment to establish the relationship between framework and criteria; (iv) Construct comparison performance and weight matrix and then apply the following steps: (a) aggregate fuzzy performance ratings with fuzzy weights (b) Define each sub-goal/STAF as a fuzzy number (c) Define extended average (EA) and (d) define the extended difference; (v) calculate the ranking values (rv) for each STAF.

Step 1: **Framework identification** is an important activity of software testing process. Therefore first step is to identify the existing framework. We have identified the following framework, i.e., (i) Modular Framework: (a) Test script modular framework and (b) Test library modular framework; (ii) Data-Driven Framework; (iii) Keyword Driven Framework; (iv) Hybrid Framework.

Step 2: **Identify the criteria**. Before the selection of any automation framework, software tester should identify the criteria's for the selection of an STAF. On the basis of our literature review, we have identified the following criteria's which influence the decision of choosing an automation framework methodology:

(a) Clarity of the requirements; (b) Cost and development time; (c) Incorporation of requirements change; (d) System complexity; (e) Communication among different stakeholders; (f) Size of development team etc.

Step 3: **Collect decision maker's fuzzy assessment**. In this step, expert's opinions regarding the importance of each STAF are obtained in the form of linguistic variable such as, very good, good, medium etc. In this step, we collect the expert's fuzzy assessments and express their opinions on the importance of each STAF.

Step 4: **Compute fuzzy group preference from the fuzzy individual preferences**. For the selection of STAF on the basis of various criteria's we aggregate fuzzy performance rating through all decision maker by means of extended addition and scalar multiplication to form a comprehensive performance matrix P, in which performance rating:

$$P_{ij} = \frac{1}{n} \odot \left(P_{ij}^{1} \oplus P_{ij}^{2} \oplus P_{ij}^{3} \oplus, \ldots, \oplus P_{ij}^{n} \right)$$

Is a triangular fuzzy number of the form:

$$\left(P1_{ij}, P2_{ij}, P3_{ij}\right) = \frac{1}{n} \sum_{k=1}^{n} P1_{ij}^{k}, \frac{1}{n} \sum_{k=1}^{n} P2_{ij}^{k}, \frac{1}{n} \sum_{k=1}^{n} P3_{ij}^{k}$$

Now calculate the fuzzy weight through all decision makers (DM) by means of extended addition and scalar multiplication to form a comprehensive weight vector (WV). Once we have obtained the comprehensive performance and weight matrix then apply the following steps [4]:

Step 4.1: Aggregate fuzzy ratings with fuzzy weights by means of extended multiplication to form a weighted, comprehensive decision matrix D, in which

$$d_{ij} = P_{ij} \odot W_j$$

Is a fuzzy number with parabolic membership functions in the form of:

$$\delta_1 ij, \delta_2 ij, \delta_3 ij|\overline{d_{ij}}|\Delta_1 ij, \Delta_2 ij, \Delta_3 ij$$

where

$$\delta_1 ij = (W_2 j - W_1 j)(P_2 ij - P_1 ij)$$

$$\delta_2 ij = W_1 ij(P_2 ij - P_1 ij) + P_1 ij(W_2 ij - W_1 ij)$$

$$\delta_3 ij = W_1 ijP_1 ij$$

$$\Delta_1 ij = (W_3 j - W_2 j)(P_3 ij - P_2 ij)$$

$$\Delta_2 ij = W_3 j(P_3 ij - P_2 ij) + P_3 ij(W_3 j - W_2 j)$$

$$\Delta_3 ij = W_3 ijP_3 ij$$

$$\overline{d_{ij}} = W_2 ijP_2 ij$$

Step 4.2: Define each STAF as a fuzzy number $A_i; i = 1, 2 \ldots m$ by means of extended addition and scalar multiplication through the following criteria:

$$A_i = \frac{1}{m} \odot (d_{i1} \oplus d_{i2} \oplus d_{i3} \oplus \cdots \oplus d_{iC})$$

With parabolic membership function in the form of

$$\left(\delta_1 i, \delta_2 i, \delta_3 i|\overline{EA_i}|\Delta_1 i, \Delta_2 i, \Delta_3 i\right)$$

where

$$\delta_{Ii} = \frac{1}{m} \sum_{j=1}^{m} \delta_I ij; \quad I = 1, 2, 3 \ldots$$

$$\Delta_{Ii} = \frac{1}{m}\sum_{j=1}^{m} \Delta_I ij; \quad I = 1, 2, 3\ldots$$

$$EA_i = \frac{1}{m}\sum_{j=1}^{m} d_{ij}$$

Step 4.3: Define EA by means of extended addition and scalar multiplication through all alternatives (STAF).

$$EA = \frac{1}{n} \odot (g_1 \oplus g_2 \oplus g_3 \oplus \cdots \oplus g_h)$$

With parabolic membership function in the form of

$$(\delta_1, \delta_2, \delta_3 | sum\_EA | \Delta_1, \Delta_2, \Delta_3)$$

where

$$\delta_I = \frac{1}{m}\sum_{ij=1}^{m} \delta_I ij; \quad I = 1, 2, 3\ldots$$

$$\Delta_I = \frac{1}{m}\sum_{i=1}^{m} \Delta_I i; \quad I = 1, 2, 3\ldots$$

$$Sum\_EA = \frac{1}{m}\sum_{i=1}^{m} EA_i; \quad I = 1, 2, 3..$$

Step 4.4: Define the extended difference, $EA_i \ominus Sum - EA$, for each $A_i \varepsilon R$, with parabolic membership function in the form of:

$$((\delta_{1i-}\Delta_1), (\delta_{2i} + \Delta_2), (\delta_{3i} - \Delta_3) | EA_i - Sum\_EA | (\Delta_{1i-}\delta_1), (\Delta_{2i} - \delta_2), (\Delta_{3i} - \delta_3))$$

Step 4.5: Calculate rv of each STAFF

In this step, we calculate the ranking values (rv_i) for each STAF A_i by means of F-preference relation R:

if $((\delta_{3i} - \Delta_3) < 0, (\Delta_{3i} - \delta_3) \geq 0, EA_i \geq Sum\_EA)$ then
$rv_i = \mu R(A_i \ominus EA, 0) = \prod^+ \ / \ (\prod^+ + \prod^-);$
Else *if* $((\delta_{3i} - \Delta_3) \leq 0, (\Delta_{3i} - \delta_3) > 0, EA_i \leq Sum\_EA)$ then
$rv_i = \mu R(A_i \ominus EA, 0) = \Psi^+ \ / \ (\Psi + \Psi^-);$
Else *if* $((\delta_{3i} - \Delta_3) = 0, (\Delta_{3i} - \delta_3) = 0, EA_i = Sum\_EA)$ then
$rv_i = \mu R(A_i \ominus EA, 0) = 0.5);$
Else *if* $((\delta_{3i} - \Delta_3) \geq 0, (\Delta_{3i} - \delta_3) > 0, EA_i \geq Sum\_EA)$ then

$rv_i = \mu R(A_i \ominus EA, 0) = 1)$;

Else *if* $((\delta_{3i} - \Delta_3) < 0, (\Delta_{3i} - \delta_3) \leq 0, EA_i \leq Sum\_EA)$ then $rv_i = \mu R(A_i \ominus EA, 0) = 0)$;

where

$$\Pi^+ = \begin{bmatrix} \frac{1}{4}(\Delta_1 i - \delta_1) - \frac{1}{3}(-\Delta_2 i - \delta_2) + \frac{1}{2}(\Delta_3 i - \delta_3) \\ +\frac{1}{4}(\delta_1 i - \Delta_1)(1 - \mu_1^4) + \frac{1}{3}(\delta_2 i + \Delta_2)(1 - \mu_1^3) \\ +\frac{1}{2}(\delta_3 i - \Delta_3)(1 - \mu_1^2) \end{bmatrix}$$

$$\Pi^- = -\left[\frac{1}{4}(\delta_1 i - \Delta_1)\mu_1^4 + \frac{1}{3}(\delta_2 i + \Delta_2)\mu_1^3 + \frac{1}{2}(\delta_3 i - \Delta_3)\mu_1^2\right]$$

$$\mu_1 = \frac{-(\delta_2 i + \Delta_2) + \sqrt{(\delta_2 i + \Delta_2)^2 - 4(\delta_1 i - \Delta_1)(\delta_3 i - \Delta_3)}}{2(\delta_1 i - \Delta_1)}$$

$$\Psi^+ = \left[\frac{1}{4}(\Delta_1 i - \delta_1)\mu_2^4 + \frac{1}{3}(-\Delta_2 i - \delta_2)\mu_2^3 + \frac{1}{2}(\Delta_3 i - \delta_3)\mu_2^2\right]$$

$$\Psi^- = -\begin{bmatrix} \frac{1}{4}(\delta_1 i - \Delta_1) + \frac{1}{3}(\delta_2 i + \Delta_2) + \frac{1}{2}(\delta_3 i - \Delta_3) \\ -\frac{1}{4}(\Delta_1 i - \delta_1)(1 - \mu_2^4) - \frac{1}{3}(\Delta_2 i + \delta_2)(1 - \mu_2^3) \\ +\frac{1}{2}(\Delta_3 i - \delta_3)(1 - \mu_2^2) \end{bmatrix}$$

$$\mu_2 = \frac{(\Delta_2 i + \delta_2) - \sqrt{(-\Delta_2 i - \delta_2)^2 - 4(\Delta_1 i - \delta_1)(\Delta_3 i - \delta_3)}}{2(\Delta_1 i - \delta_1)}$$

3 Implementation

In this section we present the implementation of proposed method to test the software, i.e., Institute Management System. In this paper, we use the following linguistic variable, i.e., Very Low (0, 0, 0.25); Low (0, 0.25, 0.5); Middle (0.25, 0.5, 0.75); High (0.5, 0.75, 1); Very High (0.75, 1, 1). For the relationship between STAF and the criteria we use the following linguistic variables: Very Weak (2, 2, 4); Weak (2, 4, 6); Medium (4, 6, 8); Strong (6, 8, 10); Very Strong (8, 10, 10). In Table 1, we present the comprehensive performance matrix; and the values of the quadratic membership functions are given in Table 2.

Calculations for membership function are given below:

$$1.95_{QF} = 0.5, 1.05, 0.4|1.95|0.5, 3.05, 4.5$$

$$3.3_{QF} = 0.5, 1.6, 1.2|3.3|0.25, 2.55, 5.6$$

Table 1 Comprehensive performance

Model	C_1	C_2	C_3
HD	(3.2, 5.2, 7.2)	(2.4, 4.4, 6.4)	(2, 4, 6)
MD	(4, 6, 8)	(4.4, 6.4, 8.4)	(4.8, 6.8, 8.8)
KD	(5.2, 7.2, 9.2)	(5.6, 7.6, 9.6)	(6, 8, 10)
DD	(7.6, 9.6, 10)	(7.2, 9.2, 10)	(8, 10, 100

Table 2 Quadratic membership functions

STAF	Criteria		
	C_1	C_2	C_3
HD	1.95_{QF}	3.3_{QF}	3.5_{QF}
MD	2.25_{QF}	4.8_{QF}	5.95_{QF}
KD	2.7_{QF}	5.7_{QF}	7_{QF}
DD	3.6_{QF}	6.9_{QF}	8.75_{QF}

$$3.5_{QF} = 0.5, 1.75, 1.25|3.5|0.125, 2.25, 5.625$$

$$2.25_{QF} = 0.5, 1.25, 0.5|2.25|0.5, 3.25, 5$$

$$4.8_{QF} = 0.5, 2.1, 2.2|4.8|0.25, 2.8, 7.35$$

$$5.95_{QF} = 0.5, 2.45, 3|5.95|0.125, 2.425, 8.25$$

$$2.7_{QF} = 0.5, 1.55, 0.65|2.7|0.5, 3.55, 5.75$$

$$5.7_{QF} = 0.5, 2.4, 2.8|5.7|0.25, 2.95, 8.4$$

$$7_{QF} = 0.5, 2.75, 3.75|7|0.125, 2.5, 9.375$$

$$3.6_{QF} = 0.5, 2.15, 0.95|3.6|0.1, 2.75, 6.25$$

$$6.9_{QF} = 0.5, 2.8, 3.6|6.9|0.1, 1.95, 8.75$$

$$8.75_{QF} = 0.5, 3.25, 5|8.75|0.625, 9.375$$

We have got the following values after applying steps 4.2 and 4.3

(HD) $2.9166_{QF} = 0.5, 1.4666, 0.95\ |2.9166|\ 0.29166, 2.7166, 5.29166$

(MD) $4.3333_{QF} = 0.5, 1.933, 1.9\ |4.3333|\ 0.2916, 2.825, 6.866$

(KD) $5.133_{QF} = 0.5, 2.233, 2.4\ |5.133|\ 0.2916, 3, 7.8416$

(DD) $6.416_{QF} = 5, 2.733, 3.183\ |6.416|\ 0.0666, 1.775, 8.125$

After executing 4.4 and 4.5, we get the following ranking values of STAFs:

HD = 1.93077; MD = 0.4386; KD = 0.87315; and DD = 0.50033

In our case study, HD has the highest priority and MD has the lowest priority. Among all the models, Hybrid Framework model has the highest priority, so this framework would be used to test the software.

4 Conclusion

This paper presents a method for the selection of STAF using fuzzy approach for the testing of Institute Management System. Proposed method is a 4 steps process, i.e., (i) Identify the existing framework, (ii) identify the criteria, (iii) Collect decision making (DM) fuzzy assessment to establish the relationship between framework and criteria, (iv) Construct comparison performance and weight matrix and then apply the following steps: (a) aggregate fuzzy performance ratings with fuzzy weights. (b) Define each sub-goal/STAF as a fuzzy number. (c) Define extended average (EA) (d) define the extended difference (v) calculate the ranking values (rv) for each STAF. On the basis of our analysis, we identify that Hybrid framework (HD) has highest priority, i.e., 1.93077. Therefore, this framework would be used to test the Institute Management System. Future research agenda includes the following:

1. To generate the test cases of Institute Management System using hybrid framework
2. To proposed a fuzzy based approach for the selection and prioritization of STAF.

References

1. Khan, M.A., Sadiq, M.: Analysis of black box software testing techniques: a case study. IEEE International Conference and Workshop on Current Trends in Information Technology, December 2011. Dubai, UAE, pp. 1–5
2. Khan, M.A., Bhatia, P., Sadiq, M.: BBTool: a tool to generate the test cases. Int. J. Recent Technol. Eng. **1**(2), 192–197 (2012)
3. Patwa, P.: Hybrid test automation frameworks implementation using QTP. Technical Article
4. Li, R.J.: Fuzzy method in group decision making. Comput. Math. Appl. **38**, 91–101 (1999)
5. Trivedi, S.H.: Software testing techniques. Int. J. Adv. Res. Comput. Sci. Softw. Eng. **2**(10) (2012)
6. Divya, A., Mahalakshmi, S.D.: An efficient framework for unified automation testing: a case study on software industry. Int. J. Adv. Res. Comput. Sci. Technol. **2**(1) (2014)
7. Jain, A., Sharma, S.: An efficient keyword driven test automation framework for web applications. Int. J. Eng. Sci. Adv. Technol. **2**(3) (2012)
8. Laukkanen, P.: Data-driven and keyword-driven test automation frameworks. M. Tech thesis, Department of Computer Science and Engineering Software Business and Engineering Institute, Helsinki University of Technology (2006)
9. Rashmi, M.: Developing and Implementing an Automation Framework. IBM Global Business Services (2008)

10. Wang, X.-T., Xiong, W.: An integrated linguistic-based group decision making approach for quality function deployment. Experts Syst. Appl. **38**, 14428–14438 (2011)
11. Sadiq, M., Jain, S.K.: Applying fuzzy preference relation for requirements prioritization in goal oriented requirements elicitation process. Int. J. Syst. Assur. Eng. Manage. (2014) (Springer)
12. Zadeh, L.A.: Fuzzy sets. Inf. Control **8**, 338–353 (1965)
13. Zadeh, L.A.: The concept of a linguistics variable and its application to approximate reasoning. Inf. Sci. **8**, 199–249 (1975)
14. Shahamiri, S.R., et al.: An automated framework for software test oracle. Inf. Softw. Technol. **53**, 774–788 (2011)
15. Sadiq, M., Jain, M.: A fuzzy based approach for requirements prioritization in goal oriented requirements elicitation process. In: 25th International Conference on Software Engineering and Knowledge Engineering, Boston, USA, 27–29 June 2013

Time Frequency Analysis and Classification of Power Quality Events Using Bacteria Foraging Algorithm

S. Jagadeesh and B. Biswal

Abstract This paper proposes a novel method Modified Hilbert Huang Transform which is the combination of empirical-mode decomposition (EMD) and Hilbert transform with an equivalent window for Time frequency analysis. Initially the Non-stationary power signal is decomposed using EMD to get Intrinsic Mode functions (IMFs) and then Hilbert transform with an equivalent window is applied to all the IMFs to obtain instantaneous amplitude and frequency for Modified Hilbert Energy Spectrum. Different features are extracted from the Modified Hilbert Energy Spectrum and these features are applied to the Bacteria Foraging algorithm for automatic classification.

Keywords Non-stationary power signals · EMD (Empirical mode decomposition) · Modified Hilbert Huang transform (MHHT) · Bacteria foraging algorithm

1 Introduction

For the Electrical Industry precise analysis of power-quality disturbances is required to overcome the problem of equipment malfunction due to voltage and current fluctuations in electrical power systems. Such an analysis is derived from understanding power quality and its effects on the industry. Consequently, many power quality upgrading efforts has either fallen short or have been in excess of the need. This paper delivers data that relates power quality and time-frequency analysis to detection, localization and classification of non-stationary power signal disturbances. The disturbances presented in non-stationary power signals are classified

S. Jagadeesh (✉) · B. Biswal
Department of ECE, GMR Institute of Technology, GMR Nagar, Rajam,
Srikakulam District 532127, Andhra Pradesh, India
e-mail: Jagadeeshsambangi5890@gmail.com

B. Biswal
e-mail: birendra.biswal@gmrit.org

L.C. Jain et al. (eds.), *Computational Intelligence in Data Mining - Volume 3*,
Smart Innovation, Systems and Technologies 33, DOI 10.1007/978-81-322-2202-6_31

as transients, Momentary interruption, Voltage sag, oscillatory transient, Swell, harmonics etc. [1]. These power quality disturbances unveil distinct features that can be used for feature extraction and classification.

It is significant to understand that there are other sources of disturbances which are not associated with the external power supply which include electrostatic ejections, emitted electromagnetic interference and operator errors and also mechanical, ecological issues can play a part in system disturbances. These can include excessive heat, pollution, extreme vibration and loose connections [2].

The time–frequency analysis gives visual detection and localization of the frequency components which are present at particular time sample [3]. FT states the frequency components present in the signal, however do not express when it occurs and for how long. FT is inadequate for non-stationary signals. Power quality events are non-stationary in nature, therefore require not only frequency information and also the timing of the existence of the disturbance. STFT provides frequency as well as time information, however the fixed window length is the disadvantage, because different type of disturbances requires windows of different length. Wavelet Transform [4] and Modified Frequency Slice Wavelet Transform [5] also provides time and frequency information. The main drawback of wavelet transform is its corrupted performance in the presence of noise.

Present developments in signal processing leads to the improvement of a new method for non-stationary power signal analysis called Hilbert Huang Transform (HHT) [6, 7]. In the Hilbert energy spectrum unexpected noise is still present as reported in [8]. To reduce the scattering of Hilbert Energy Spectrum a novel method called 'Modified Hilbert Huang Transform' is proposed with an equivalent window. Different features are extracted from Modified Hilbert energy spectrum and given as input to Bacteria Foraging algorithm [9, 10] for automatic classification.

2 Empirical Mode Decomposition

In this paper, a new method MHHT is proposed, which is a combination of Empirical Mode Decomposition (EMD) method, in which various Intrinsic Mode functions are extracted from the Non stationary power signal by repeatedly applying the shifting process to satisfy the conditions of IMFs and then applied to HT with an equivalent window to get the instantaneous amplitude and frequency.

An IMF is a function that satisfies these conditions:

1. For a dataset, the no. of extrema (local maxima or local minima) and the no. of zero crossings must be either equal or differ at most by one.
2. At any given time sample, the mean value of the envelope defined by the local maxima and the envelope defined by the local minima is zero.

The steps followed in sifting process are given as:

(1) In the decomposition method first envelopes are identified using local maxima and minima distinctly.

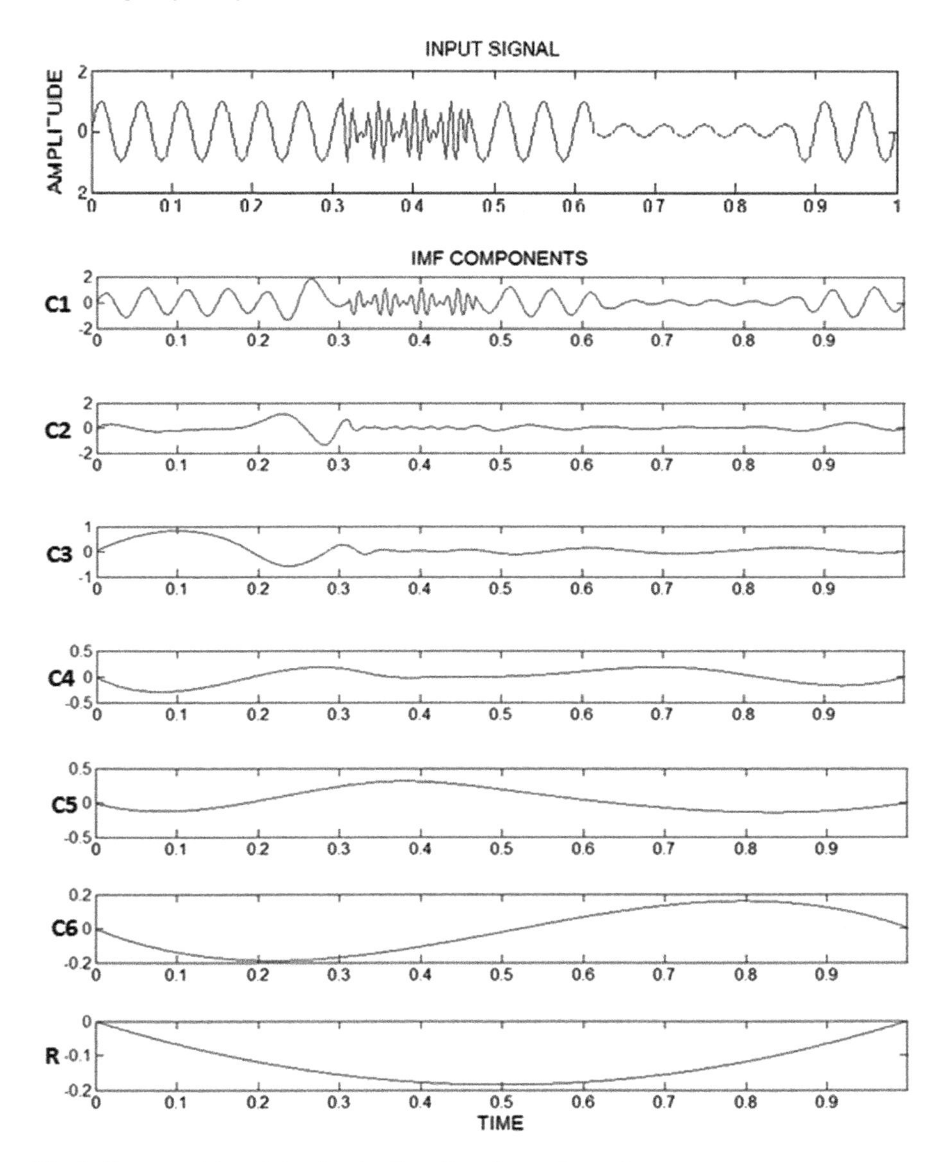

Fig. 1 Decomposition of the signal with voltage harmonics and sag

(2) The mean value of envelopes is defined as m_1. The difference between the original data and mean m_1 is the first component Q_1.

$$x(t) - m_1 = Q_1 \tag{1}$$

(3) If Q_1 satisfies the two IMF conditions, then Q_1 is the first IMF, else if Q_1 is not an IMF then it is treated as original signal and Steps (1) and (2) are repeated to get component Q_{11}.

$$Q_{11} = Q_1 - m_{11} \tag{2}$$

(4) After repeated sifting i.e. up to n times, Q_{1n} becomes an IMF

$$Q_{1(n-1)} - m_{1n} = Q_{1n} \tag{3}$$

$$\text{Then it is designated as} \quad P_1 = Q_{1n} \tag{4}$$

(5) P_1 is the first IMF component from the original data. Separate P_1 from $x(t)$

$$Re_1 = x(t) - P_1 \tag{5}$$

(6) Now treating R_{e1} as the original data and repeating the above processes second IMF can be obtained.
(7) The above procedure is repeated 'r' times and 'r' IMFs of the signal $x(t)$ are obtained.

The Fig. 1 output shows the IMF Components which are obtained by the decomposition of a given non stationary signal input.

3 Modified Hilbert Huang Transform

Hilbert energy spectrum gives visual detection of frequencies which are obtained during the signal, and also shows where most of the signal energy is concentrated in time and frequency axis. After performing the Hilbert transform on each IMF component, provides Hilbert energy spectrum. However the Hilbert energy spectrum often suffered from the ripple phenomenon effect and scattering of the energy spectrum. To conquer this, a new method Modified Hilbert Huang Transform is proposed, which consists of an equivalent window in between EMD and Hilbert Transform. This equivalent window reduces the scattering of the energy spectrum and removes the short time disturbances.

$$MHHT(x(t)) = \check{x}(t) = \frac{1}{\pi} \int_{-\infty}^{\infty} \frac{W_{\Delta}(\tau - t)x(\tau)}{t - \tau} d\tau. \tag{6}$$

Then, calculate the MHHT of $W_K[n-k]x[n]$ which is represented as $\check{x}[k]$, i.e.

$$\check{x}[k] = \hat{x}_{K,k}\left[\frac{K-1}{2}\right] \tag{7}$$

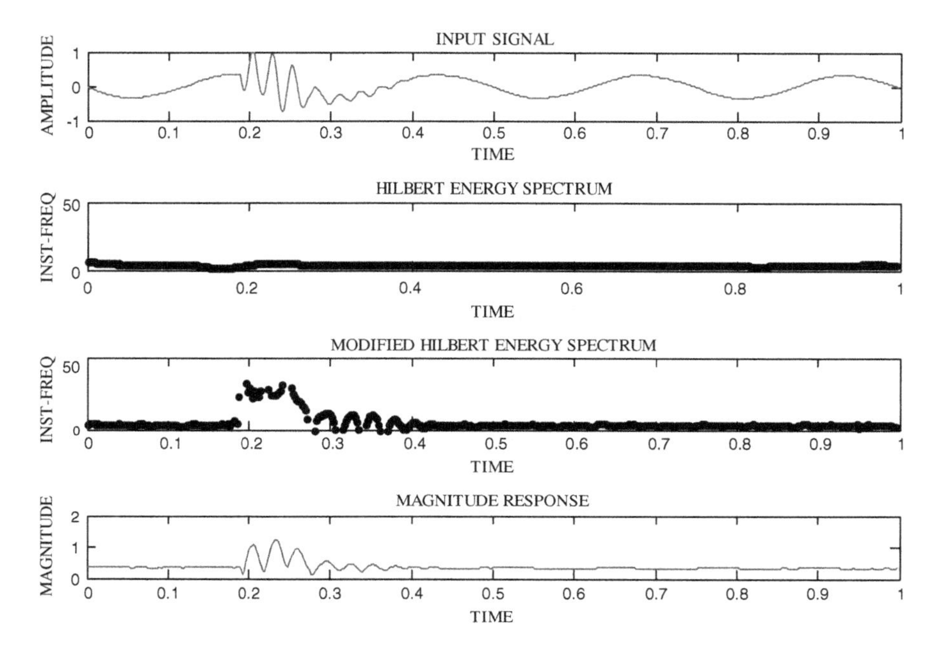

Fig. 2 Detection and visual localization of the signal with transient

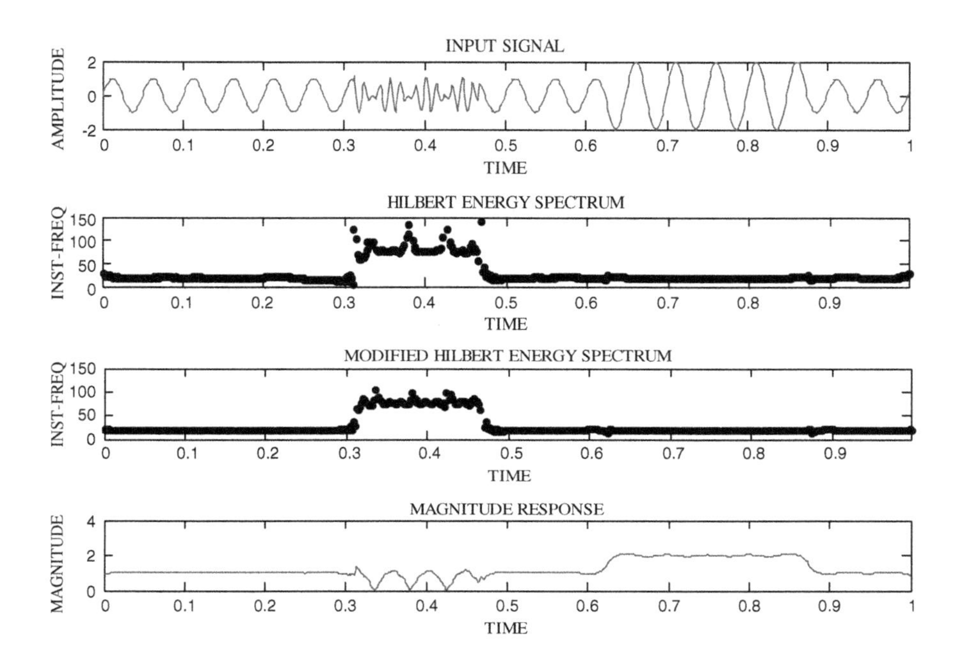

Fig. 3 Detection and visual localization of the signal with swell and harmonics

$$\text{where} \quad \hat{x}_{K,k} = HT[W_K(n-k)x(n)] \tag{8}$$

This section presents the MHHT which reduces the scattering presented in conventional Hilbert energy spectrum and it is shown in the plot of the Modified Hilbert Energy spectrum plotted in a time-frequency plane where the energy concentration is represented in terms of intensity shown in Figs. 2 and 3.

4 Bacteria Foraging Algorithm

The features that are derived from the Modified Hilbert Energy spectrum, such as entropy, energy, normalized value, etc. are given as input to the Bacteria Foraging algorithm. This technique determines the minimum of a function $J(i,j)$ where the variables under consideration constitute the high-dimensional vector $\theta \in R$. Here θ determines the location of a bacterium in high dimensional space. In chemotaxis steps each bacterium either experience a tumble followed by another tumble or a tumble followed by a run. Let $\theta(i,j,k,l) \in R$ represent the ith bacterium $(i = 1, 2, 3\ldots s)$ in the jth chemotactic, kth reproduction and lth elimination dispersal step. Let $J(i,j,k,l)$ represent the cost associated with this position of the bacterium, although.

$$\theta(i,j+1,k,l) = \theta(i,j,k,l) + c(i)\varphi(j) \tag{9}$$

where $\varphi(j)$ denotes a unit length random direction to represent a tumble and determine a future direction of movement and $C(i)$ denotes the run length. This combined cell to cell attraction and repelling effect is given by

$$\sum Jcc(\theta, \theta(i,j,k,l)) = \sum_{i=1}^{s}\left[-d_{attract}\exp\left(-w_{attract}\sum_{m=1}^{p}\left(\theta m - \theta m^{i}\right)^{2}\right)\right] + \sum_{i=1}^{s}\left[h_{repellant}\exp\left(-w_{repellant}\sum_{m=1}^{p}\left(\theta m - \theta m^{i}\right)^{2}\right)\right] \tag{10}$$

Which is added to the actual cost function J to be minimized and this presents a time varying nature for J. The $d_{attract}$, $w_{attract}$, $h_{repellant}$ and $w_{repellant}$ parameters will determine the shape of these attraction and repulsion signals. After every N_c chemotactic steps are completed, one reproduction step is undertaken and after N_{re} reproduction steps are completed, one elimination—dispersal step is undertaken. Hence a bacterium will undergo several chemotactic steps before it is allowed to

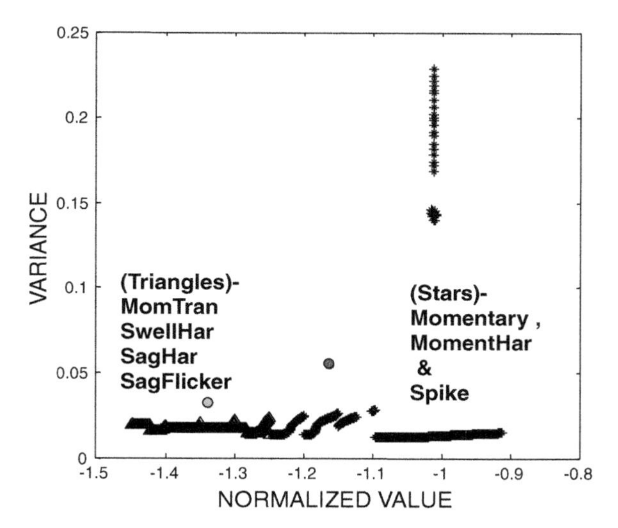

Fig. 4 Classification of total dataset

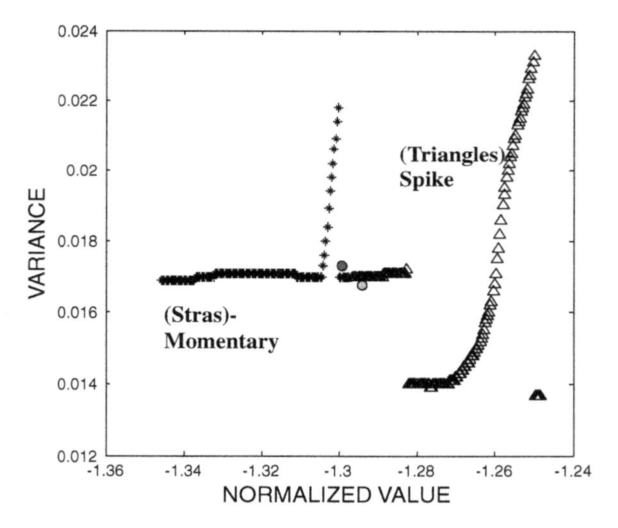

Fig. 5 Classification of dataset into momentary interruption and spike

reproduce and the system will undergo several generations before an elimination-dispersal step is executed. Figure 4 describes the classification of the entire power quality events. Figure 5 shows the classification of events into Momentary Interruption and Spike.

5 Conclusion

The Modified Hilbert Huang transform is found to be a powerful exploration tool for recognition, localization and classification of non-stationary power signals. Comparing with former Hilbert transform, the MHHT gives better visualization of Hilbert energy spectrum in which the scattering of the energy spectrum is reduced. The relevant features that take out from the modified Hilbert energy spectrum apply to Bacterial Foraging Optimization Algorithm (BFOA). Hence, EMD with Modified Hilbert Huang transform provides a practically improved time-frequency representation for the non-stationary power signal.

References

1. Sejdic, E., Djurovic, E., Jiang, J.: Time–frequency feature representation using energy concentration: an overview of recent advances. Digit. Signal Proc. **19**, 153–183 (2009)
2. Edward Reid, W.: Power quality issues-standards and guidelines. IEEE Trans. Ind. Appl. **32** (3) (1996)
3. Biswal, B., Dash, P.K., Panigrahi, B.K. : Power quality disturbance classification using fuzzy C-means algorithm and adaptive particle swarm optimization. IEEE Trans. Ind. Electron. **56**(1) (2009)
4. Santoso, S., Grady, W.M., Powers, E.: Characterization of distribution power quality events with fourier and wavelet transforms. IEEE Trans. Power Deliv. **15**(1), 247–254 (2000)
5. Biswal, B., Mishra, S.: Detection and classification of disturbances in non-stationary signals using modified frequency slice wavelet transform. Gen. Trans. Distrib. (IET) **7**(9) (2013)
6. Jayasree, T., Devaraj, D., Sukanesh, R.: Power quality disturbance classification using hilbert transform and RBF networks. Neurocomputing **73**, 1451–1456(2010)
7. Biswal, B., Biswal, M., Mishra, S., Jalaja, R.: Automatic classification of power quality disturbances with balanced neural tree. IEEE Trans. Ind. Electron. **61**(1) (2014)
8. Sun, S., Jiang, Z., Wang, H., Yu, F.: Automatic moment segmentation and peak detection analysis of heart sound pattern via short-time modified Hilbert transform. Comput. Methods Programs Biomed. **114**(3), 219–230 (2014)
9. Mishra, S.: A hybrid least square-fuzzy bacteria foraging strategy for harmonic estimation. IEEE Trans. Evol. Comput. **9**(1), 61–73 (2005)
10. Passino, K.M.: Biomimicry of bacterial foraging for distributed optimization and control. IEEE Control Syst. Mag. **22**(3), 52–67 (2002)

Non-Stationary Signal Analysis Using Time Frequency Transform

M. Kasi Subrahmanyam and Birendra Biswal

Abstract In this paper, visual detection and classification of the non-stationary power signals are demonstrated by well-known transform called generalized synchrosqueezing transform.The wavelet based time frequency representation gives poor quality and understandability, hence the proposed synchrosqueezing transform is an effective method to get better quality and readability of the wavelet-based TFR by summarizing along the frequency axis. Different feature vectors have been extracted from the frequency contour of the generalized synchrosqueezing transform and these feature vectors applied as input to the Reformulated Fuzzy C-Means algorithm for automatic classification.

Keywords Continuous wavelet transform (CWT) · Generalized synchrosqueezing transform (GST) · Reformulated fuzzy C-Means algorithm

1 Introduction

The time or frequency domain description only may not give sufficient information for a non-stationary power signal [1]. Time-frequency representation is an effective technique which shows frequency information as well as time information, to study the non-stationary signals [1]. In general, for time-frequency representation different techniques are used such as wigner–ville distribution, Short Time Fourier Transform (STFT), and wavelet transform [2–4]. In STFT, the constant window length leads to degraded time–frequency (TF) resolution and Wigner-Ville distribution leads cross-terms in between various signal components [5]. Recently power quality disturbance classification using Hilbert Transform [6] and modified

M. Kasi Subrahmanyam (✉) · B. Biswal
Department of ECE, GMR Institute of Technology, Razam, A.P, India
e-mail: kasisubrahmanyammaturi@gmail.com

B. Biswal
e-mail: birendra_biswal1@yahoo.co.in

L.C. Jain et al. (eds.), *Computational Intelligence in Data Mining - Volume 3*,
Smart Innovation, Systems and Technologies 33, DOI 10.1007/978-81-322-2202-6_32

frequency slice wavelet transform [7] has been addressed. Wavelet transform (WT) also useful for extracting the instantaneous attributes by detecting the edge of the WT. But, the wavelet transform gives blurred TFR in the presence of noise and therefore reduces the readability. However a novel method based on wavelet transform for TFR known as "Synchrosqueezing" is proposed by Daubechies et al. [8, 9]. The synchrosqueezing technique concentrates on the TFR, $T(t, f)$ of the signal $x(t)$ by squeezing its value along the frequency f (scale p) axis. However, this transform is restricted to scale the variables of the continuous wavelet transform (CWT).

In this paper, a new method known as Generalized Synchrosqueezing Transform (GST) is proposed to determine the two-dimensional (p and q) diffusion problem. This GST method provides much more reduced TFR by an arrangement of several TF plane calculations (such as translation, revolving, and distortion) in unique operation. In this work the TFR thus obtained by the CWT is sharpened to get visual detection, localization of power quality disturbances presented in the non-stationary power signals. Distinct features are derived from the Time frequency representation of GST and are given as input to the Reformulated Fuzzy C- Means algorithm for automatic classification [10].

2 Generalized Synchrosqueezing Transform (GST)

To provide the synchrosqueezing of a non-stationary power signal, the IF trajectory mapping is first introduced in this section. Figure 1 represents there is no blur at the time-dimension. Hence it is necessary to squeeze the horizontal edge only in the frequency dimension. The GFT of the signal $x(t)$ is given by

$$x_G(t) = F_G(x(t)) = \int_{-\infty}^{\infty} x(t)e^{-i2\pi(f_t+s_0(t))}dt \tag{1}$$

Here $s_0(t)$ is actual transform function specifying the phase transform behavior of the signal. The signal might be also calculated from the inverse GFT, i.e.

$$x(t) = F_G^{-1}(x_G(f)) = e^{i2\pi s_0(t)} \int_{-\infty}^{\infty} x_G(f) \cdot e^{i2\pi f_0} df \tag{2}$$

If $x_G(f) = \delta(f - f_0)$, then $x(t) = e^{i2\pi(f_0 t+s_0(t))}$ where f_0 is the required horizontal GFT frequency trajectory represented as

$$f_0 = f(t) - \frac{ds_0(t)}{dt} = f(t) - s_0^{-1}(t) \tag{3}$$

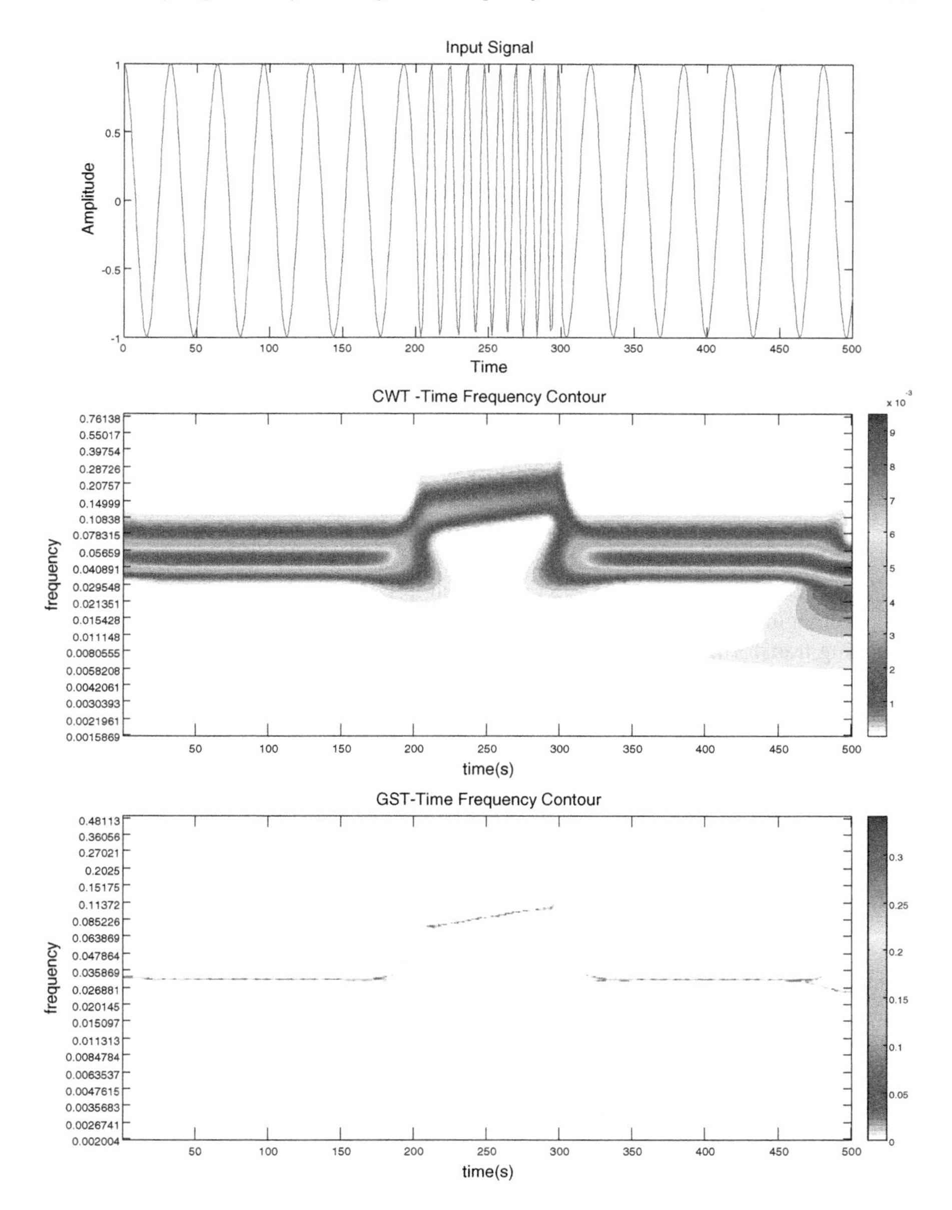

Fig. 1 Comparision of TFR for chirp (increasing frequency) signal

Since both the GFT and the CWT gives the equal IF trajectory, the horizontal GFT frequency ridge is related to a horizontal IF for the same signal. To apply the IF trajectory plotting in the complex domain, $x(t)$ is first transformed to an analog signal.

$$u(t) = x(t) + iH(x(t)) \tag{4}$$

where $H(x(t))$ gives the Hilbert transform of $x(t)$, i.e.

$$H(x(t)) = x(t) * \left({}^{1}\!/_{\pi t}\right) \tag{5}$$

From Eq. (4), the analog signal $u(t)$ is then transformed with the plotting function $e^{-i2\pi s_0(t)}$ as

$$v(t) = u(t)e^{-i2\pi s_0(t)} \tag{6}$$

$$w(t) = v(t) + iH(v(t)) \tag{7}$$

This time-scale result is then transformed as

$$w_y(p, q) = C_w(p, q)e^{-i2\pi s_0(t)} \tag{8}$$

The main reason behind this GST approach is $|w_y| = w_w$ with respect to plotted real & imaginary part for enhance TFR of the signal. Therefore the synchro-squeezing transform computed gives the TF convergence quality of analog signal $w(t)$ that has a frequency f_0 can be preserved by the IF trajectory plotting, i.e.

$$x_x(f_x, q) \Leftrightarrow x_y(f_y, q) = x_w\left(f_w - \frac{f(t) - s_0^{-1}(t)}{f(t)}, q\right) \tag{9}$$

The above equation indicates that the IF of $x(t)$ has been restored and the TFR has been further condensed by the proposed GST. Hence the TFR obtained by the GST is given by the following equation

$$x(q) = R(C_\varphi^{-1} \sum x_y(f_y, q)(\Delta f)) \tag{10}$$

Figure 1 shows squeezing of the frequency contour using GST versus CWT.

3 Reformulated Fuzzy C-Means Algorithm

In this paper for clustering of feature vectors, earlier Fuzzy C-means algorithm is replaced with Reformulated Fuzzy C-Means algorithm. Different features extracted from Generalized Synchrosqueezing Transform are given as input to the Fuzzy C—Means algorithm for automatic classification. It's essentially based on minimization of the following objective function.

$$J_q = \sum_{m=1}^{P} \sum_{n=1}^{C} u_{mn}^{q} \|y_m - c_n\|^2 \tag{11}$$

where q is the number of clusters, u_{mn} is the degree of membership of y_m in cluster n, y_m is the mth of n-dimensional measured data, and C_n is the n-dimensional center of the cluster.

$$C_n = \frac{\sum_{m=1}^{P} u_{mn}^{q} m}{\sum_{m=1}^{P} u_{mn}^{q}}, \; u_{mn} = \sum_{p=1}^{c} \left[\frac{\|y_m - C_n\|}{\|y_m - C_p\|} \right]^{\frac{-2}{q-1}} \tag{12}$$

Due to the parameter 'q', it clusters the total feature vectors into a definite amount of clusters. The "fuzziness" of the grouping is measured by the parameter 'q'. As this value of factor 'q' moves to 1, the grouping of the feature vectors is nearly a hard decision-making process. As the value of 'q' increases resulting in lower membership to extremely fuzzy classification.

To overcome the problem due to 'q' Reformulated Fuzzy C-Means Algorithm is reformulated and is based on the minimization of the objective function in Eq. (11) given as,

$$Re_q = \sum_{m=1}^{Q} \left[\sum_{n=1}^{c} (\| y_m - C_n \|^2)^{\frac{1}{(1-q)}} \right]^{1-q} \tag{13}$$

$$Re_q = \sum_{m=1}^{Q} C^{\frac{1}{k}} B_k(y_m, C) \tag{14}$$

where $k = \frac{1}{1-q}$ and $B_k(y_m, C)$ is the generalized mean (or weighted p-norm) of the 'C' positive real numbers.

$$B_k = \left[\frac{1}{C} \sum_{n=1}^{c} (\| y_m - C_n \|^2)^k \right]^{\frac{1}{k}} \tag{15}$$

As 'k' reaches to '0', the generalized mean $B_k(y_m, C)$ approaches to the geometric mean

$$B_G(y_m) = \lim_{k \to 0} B_k(y_m, C) = \left(\prod_{n=1}^{c} \|y_m - C_n\|^2 \right)^{\frac{1}{c}} \tag{16}$$

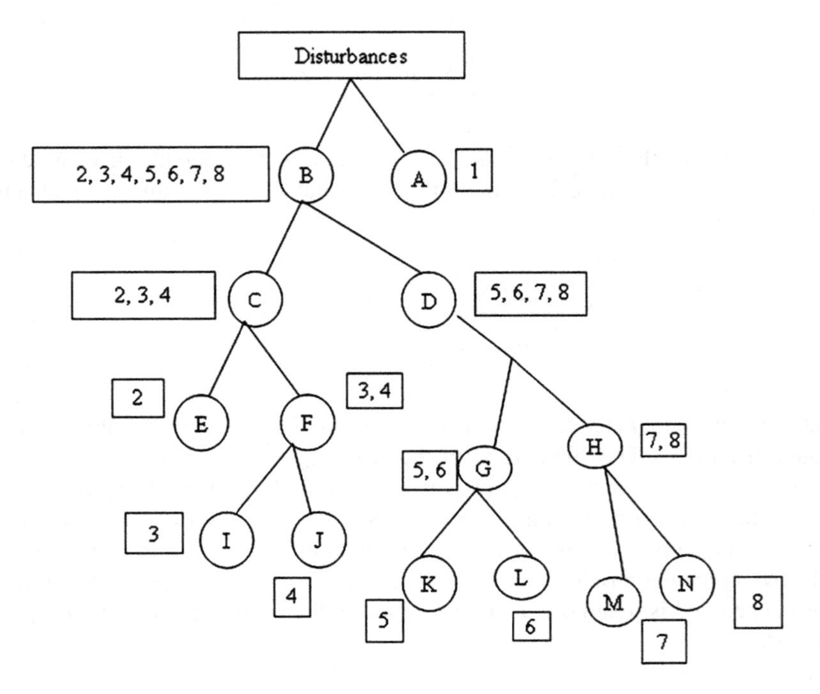

Fig. 2 Fuzzy C-Means tree

As 'q' reaches towards ∞ 'k' moves to '0' from the left. Thus

$$\lim_{q \to \infty} Re_q = \lim_{k \to 0_-} \sum_{m=1}^{Q} C^{\frac{1}{k}} B_G(y_m) = 0 \tag{17}$$

The generalized mean $B_k(y_m, C)$ moves towards the minimum of $\{\| y_m - C_n \|^2\}$ as 'k' tends to $-\infty$. As 'k' tends to $-\infty$, 'q' moves to 1from the right. Thus

$$\lim_{q \to 1} Re_q = \lim_{k \to -\infty} \sum_{m=1}^{Q} C^{\frac{1}{k}} B_k(y_m, C) = \sum_{m=1}^{Q} \sum_{n=1}^{c} U_{mn} \|y_m - C_n\|^2 \tag{18}$$

For classification of Non-stationary power signal disturbances Reformulated Fuzzy C-Means tree is used. The disturbances such as 1. Transient, 2. Momentary with Harmonics, 3. Momentary interruption, 4. Spikes, 5. Sag with Harmonic, 6. Momentary with Transient, 7. Swell with Harmonic, 8. Sag with Flicker are classified in the same order as shown in the Reformulated Fuzzy C-means tree is shown in Fig. 2.

The feature vectors that are extracted from generalized synchrosqueezing transform are clustered using Reformulated Fuzzy C-Means algorithm. Figures 3 and 4 shows the simulated results of non-stationary power signal disturbances classification.

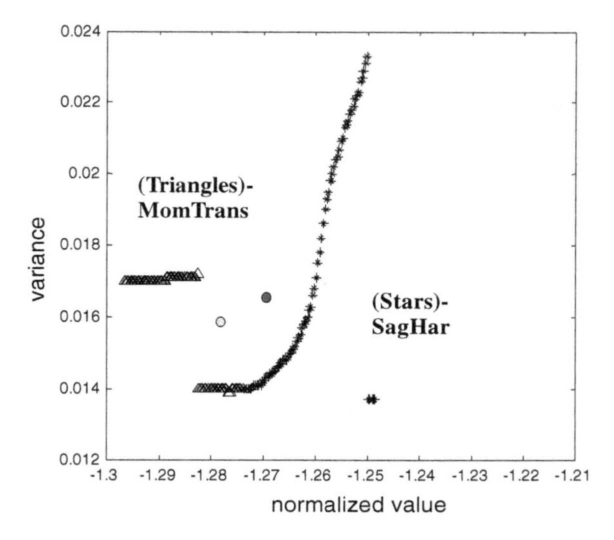

Fig. 3 Classification of dataset into Momentary interruption with transient and Voltage sag with harmonics

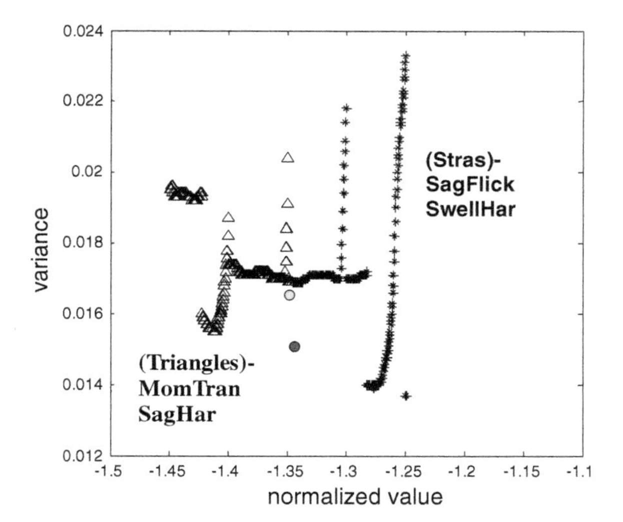

Fig. 4 Classificication of dataset into Momentary with Transient, Sag with Harmonics and Sag with flicker, Swell with harmonics

4 Conclusion

A Generalized Synchrosqueezing Transform is proposed to get better time-frequency representation for non-stationary power signals. This technique is very efficient to perform the broad range of TF plane transforms such as rendition, revolve and distortion within unique operation. It is observed that diffusions occurred in the frequency dimension of wavelet transform for the non-stationary signal. The GST approach is used to remove the diffusions of the frequency, then IF trajectory mapping function is used to suppress the extension of the time dimension with the help of IF trajectory mapping operation. Therefore GST is capable to give sharpened TFR in both time and frequency axes with the combination of signal and time frequency improvement.

References

1. Li, C., Liang, M.: A generalized synchrosqueezing transform for enhancing signal time frequency representation. Signal Process. **92**(9), 2264−2274 (2012)
2. Sejdic, E., Djurovic, E., Jiang, J.: Time-frequency feature representation using energy concentration: an overview of recent advances. Digit. Signal Proc. **19**, 153–183 (2009)
3. Tantibundhit, C., Pernkopf, F., Kubin, G.: Joint time–frequency segmentation algorithm for transient speech decomposition and speech enhancement. IEEE Trans. Audio. Speech. Lang. Process. **18**, 1417–1428 (2010)
4. Padovese, L.R.: Hybrid time–frequency methods for non-stationary mechanical signal analysis. Mech. Syst. Signal Process. **18**, 1047–1064 (2004)
5. Li, C., Liang, M.: Separation of the vibration-induced signal of oildebris for vibration monitoring. Smart. Struct. Struct. **20**, 045016 (2011)
6. Biswal, B., Biswal, M., Mishra, S., jalaja, R.: Automatic classification of power quality disturbances with balanced neural tree" IEEE. Trans. Ind. Electron. **61**(1), 521–530 (2014)
7. Biswal, B., Mishra, S.: Detection and classification of disturbances in non-stationary signals using modified frequency slice wavelet transform". IET Gener. Transm. Distrib. **7**(9), (2013)
8. Daubechies, I., Lu, J.F., Wu, H.T.: Synchrosqueezed wavelet transforms: an empirical mode decomposition-like tool. Appl. Comput. Harmonic. Anal. **30**, 243–261 (2011)
9. Daubechies, I., Maes, S.: A nonlinear squeezing of the continuous wavelet transform based on auditory nerve models. In: Aldroubi, A., Unser, M. (eds.) Wavelets in Medicine and Biology, pp. 527−546. CRC Press, Boca Raton (1996)
10. Biswal, B., Dash, P.K., Panigrahi, B.K.: Power quality disturbance classification using Fuzzy C-means algorithm and adaptive particle swarm optimization. IEEE Trans. Ind. Electron. **56** (1), 212–220 (2009)

Multivariate Linear Regression Model for Host Based Intrusion Detection

Sunil Kumar Gautam and Hari Om

Abstract Computer security is an important issue for an organization due to increasing cyber-attacks. There exist some intelligent techniques for designing intrusion detection systems which can protect the computer and network systems. In this paper, we discuss multivariate linear regression model (MLRM) to develop an anomaly detection system for outlier detection in hardware profiles. We perform experiments on performance logfiles taken from a personal computer. Simulation results show that our model discovers intrusion effectively and efficiently.

Keywords Intrusion detection system · Multivariate linear regression model · Mean square error · Host based intrusion detection system · Network based intrusion detection system

1 Introduction

Data security has become a challenging task for researchers since the suspicious attacks are increasing day by day. Companies are worried about the security of their valuable information, which is not safe due to malicious activities. Researchers have discussed some statistical techniques for data security, which are helpful for entire systems from a global perspective [1, 2]. The namely, signature based and anomaly based intrusion detection systems. The signature based systems can recognize the pattern by matching with the pre-stored patterns in the database. The anomaly based systems recognize both the known and unknown patterns of intrusion. These both types of intrusion detection systems follow models, which are created by using several statistical approaches such as data mining algorithms, genetic algorithms, artificial neural networks, and fuzzy logic, principal component analysis (PCA), to

S.K. Gautam (✉) · H. Om
Department of Computer Science and Engineering, Indian School of Mines, Dhanbad, India
e-mail: gautamsunil.ism@gmail.com

H. Om
e-mail: hariom4india@gmail.com

L.C. Jain et al. (eds.), *Computational Intelligence in Data Mining - Volume 3*,
Smart Innovation, Systems and Technologies 33, DOI 10.1007/978-81-322-2202-6_33

name some [3]. In this paper, we discuss multivariate linear regression model in host based intrusion detection system to find the residual error for an intrusion attack and the efficacy of the model is evaluated by using experimental results.

The rest of the paper is organized as follows. Section 2 discusses the related works and in Sect. 3 we present our multivariate linear regression model. Section 4 discusses performance logfile and related attributes and the experimental methodology is discussed in Sect. 5. Finally, results and discussions are presented in Sect. 6.

2 Related Works

The Intrusion Detection Systems (IDS) have been categorized in two types: (1) Host based Intrusion Detection System (HIDS) monitors only indiviadaul existing sytem attacks and gives an alert to the user system upon the detection of these attacks. In other words, HIDS can only provide protection to inividual user system (i.e. only local data). (2) Network based Intrusion Detection System (NIDS) monitors the suspicious activity on entire network. This type of system analyzes the data packets of entire network (i.e. Wide Area Network). Furthermore, NIDS detects Denial of Services (DoS), User to Root, Remote to Local and Probing attacks. However, HIDS only detects DoS [4–6]. The IDS are recognized as an efficient tool for security to a system administrator due to several vulnerabilities in a system. There exist several works related to IDS; but none is ideal IDS for network.

Many researchers have applied several statistical methods for intrusion detection such as Principal Component Analysis (PCA), Chi–square statistic, T2 test statistical technique and X2 test, Gaussian mixture distribution, etc. The PCA, also known as Karhunen-Loeve transform, is used for dimensionality reduction and compression of the data. The dimensionality reduction is generally done by computing eigenvalues and corresponding eigenvectors of the sample covariance matrix by using singular value decomposition (SVD) [7, 8]. The PCA mainly extracts the useful information from the large dataset and transforms it into lower dimensional space. It detects two types of outliers (abnormal behavior) from the data. One type of outliers affects the variance and they are detected by major principle components (PCs) and the other type of outliers that violate the structure are identified by minor PCs. The simplicity and effectiveness of PCA make it more popular than the other existing state of art statistical techniques [9]. The papers [10, 11] have used Chi-square distribution for outlier detection. It is applied on the sample data to know that the sampled data is in asymmetric distribution and the minimum value of this test is numeric zero. It is a multivariate statistical technique that deals with multiple activities in an information system. The X2 test gives better performance if the computer audit dataset is small. It can finds out 75 percent of intrusive events and never gives false alarms in normal sessions. It however does not provide correlated structure of the variables. The T2 test is suitable for large audit dataset and is reliable for individual variables. It uses statistical distance that incorporates the multivariate variance-covariance matrix. It determines the shift and counter relationship.

3 Proposed Model

The multivariate linear regression model is a statistical technique that provides a way to understand the structure of relationship between the dependent and independent variables. In this model, we analyse those attributes of data that influence some other attributes of data. The influencing attributes are categorized into several independent and dependent variables. The multivariate linear regression falls into a class of linear models that is based on the statistical model known as the general linear model and it is given as.

$$Y = X\beta + \varepsilon. \tag{1}$$

$$Y = \beta_0 + \beta_1 x_1 i + \beta_2 x_2 i + \cdots + \beta_q x_{qi} + \varepsilon_i \tag{2}$$

where,

Y	is dependent variable
$X_1, X_2, \ldots, X_q$	are independent variables
β_0	is intercept
$\beta_1, \beta_2, \ldots, \beta_q$	are coefficients of independent variables
ε_i	is regression residual

We now calculate the different statistical parameters to get the best statistical relationship between the independent and dependent variables. For this purpose, the various performance measures such as coefficient of determination (R^2), coefficient of correlation (CC), standard error of estimate (SSE), and mean square error (MSE) are calculated [14]. In a regression model, we express each y in a sample of n observations as a linear function of x as follows:

$$\left.\begin{aligned} y_1 &= \beta_0 + \beta_1 x_{11} + \beta_2 x_{12} + \cdots + \beta_q x_{1q} + \varepsilon_1 \\ y_2 &= \beta_0 + \beta_1 x_{21} + \beta_2 x_{22} + \cdots + \beta_q x_{2q} + \varepsilon_2 \\ &\cdots \\ y_n &= \beta_0 + \beta_1 x_{n1} + \beta_2 x_{n2} + \cdots + \beta_q x_{nq} + \varepsilon_n \end{aligned}\right\} \tag{3}$$

where y_i is ith observation, $1 \leq i \leq n$. In order to make computation simpler we impose the following constraints:

$$\left.\begin{aligned} &E(\varepsilon_i) = 0, \forall i = 1, 2, \ldots, n \\ &Var(\varepsilon_i) = \sigma^2, \forall i = 1, 2, \ldots, n \\ &Cov(\varepsilon_i, \varepsilon_j) = 0, \forall i \neq j. \end{aligned}\right\} \tag{4}$$

The first constraint in (4) indicates that if the average error is zero, the model does not need constraints (4) and extra terms are needed to predict y because the variation in y is purely random and unpredictable.

The second constraint in (4) says that the variance of each ε_i is same which means that $var(y_i) = \sigma^2$ as x_i's are fixed.

The third constraint in (4) says that the error terms are uncorrelated which means that y_i's are also uncorrelated, i.e., $Cov(y_i, y_j) = 0$

$$\left.\begin{array}{l} E(y_i) = \beta_0 + \beta_1 x_{i1} + \beta_2 x_{i2} + \cdots + \beta_q x_{iq}, \\ i = 1, 2, \ldots, n. \\ Var(y_i) = \sigma^2, \ i = 1, 2, \ldots, n. \\ Cov(y_i, y_j) = 0, \text{ for all } i \neq j \end{array}\right\} \quad (5)$$

In matrix notation, the model for n observations may be written in the following form;

$$\begin{pmatrix} y_1 \\ y_2 \\ \cdots \\ y_n \end{pmatrix} = \begin{pmatrix} 1 & x_{11} & x_{12} & x_{1q} \\ 1 & x_{21} & x_{22} & x_{2q} \\ \cdots & \cdots & \cdots & \cdots \\ 1 & x_{n1} & x_{n2} & x_{nq} \end{pmatrix} \begin{pmatrix} \beta_1 \\ \beta_1 \\ \cdots \\ \beta_q \end{pmatrix} + \begin{pmatrix} \varepsilon_1 \\ \varepsilon_2 \\ \cdots \\ \varepsilon_n \end{pmatrix}$$

or

$$Y = X\beta + \varepsilon$$

The least squares estimates of $\beta_0, \beta_1, \ldots, \beta_q$ minimize the sum of squares of deviations of the n observed y_i's from their "modeled" values, that is, from the values predicted by the model.

$$\left.\begin{array}{l} SSE = \sum_{i=1}^{n} \hat{\varepsilon}(y_t - \hat{y}_t)^2 \\ \quad = \sum_{i=1}^{n} (y_t - \hat{\beta}_0 - \hat{\beta}_1 x_{i1} - \hat{\beta}_2 x_{i2} - \cdots - \hat{\beta}_q x_{iq})^2 \end{array}\right\} \quad (6)$$

The value of $\hat{\beta} = (\hat{\beta}_0, \hat{\beta}_1, \ldots, \hat{\beta}_q)'$ that minimizes SSE above is given by

$$(X'X)^{-1} X^{-1} y$$

The regression residual ε_i is calculated by $\varepsilon_i = (Y - \widehat{Y})$

We assume that X'X is a non-singular, which generally for $n > q + 1$ and no x_j is a linear combination of other x's. Assigning ith row of X as $x_i' = (1, x_{i1}, x_{i2}, x_{i3}, \ldots, x_{iq})$, we may write SSE as

$$SSE = \sum_{i=1}^{n} (y_i - x_i')\,\hat{\beta} \text{ or}$$

$$SSE = (y - X\hat{\beta})' - (y - X\hat{\beta})$$

It is not difficult to show that

$$\begin{aligned} \mathrm{SSE} &= \sigma^2[\mathrm{n} - (\mathrm{q} + 1)] \\ &= \sigma^2(\mathrm{n} - \mathrm{q} - 1) \end{aligned}$$

Thus, we obtain an unbiased estimator of σ^2 as follows:

$$\begin{aligned} \mathrm{s}^2 &= \mathrm{SSE} \div (\mathrm{n} - \mathrm{q} - 1) \\ &= 1 \div (\mathrm{n} - \mathrm{q} - 1)(\mathrm{y} - \mathrm{X}\hat{\beta})' - (\mathrm{y} - \mathrm{X}\hat{\beta}) \end{aligned}$$

The SSE can be expressed in the following form:

$$\mathrm{SSE} = \mathrm{y}'\mathrm{y} - \mathrm{X}'\mathrm{y}$$

4 Performance Logfile

We discuss the accuracy of our regression model on offline performance log files data that we have created with intrusion and without intrusion from our PC. The hardware details of the PC are given below.

- Intel(R) Core (TM) i3
- 4.00 GB RAM
- Microsoft Windows 7
- 64-bit Operating System

In this paper, we use host based anomaly detection technique to show abnormal system behavior. Normal behavior of the system is created based on the processing running in the system. When any intrusion occurs in the system, it means that an error is generated in the system and the system behaves abnormally. The performance of a personal computer can be measured by using the performance log. We consider the following attributes in our experiments on the performance log analysis

(a) **Demand Zero Faults/sec**: The template is designed so that author affiliations are not repeated each time for multiple authors of the same affiliation. Please keep your affiliations as succinct as possible. This template was designed for two affiliations.
(b) **Cache Bytes**: It defines the sum of the values of system cache resident bytes, system driver resident bytes, system code resident bytes, and pool paged resident bytes.
(c) **Page Faults**: It refers to the average number of pages faulted per second. It is measured in numbers of pages faulted because only one page is faulted in each fault operation. It is also equal to the number of page fault operations.

(d) **Transition Faults/sec**: It refers, in incidents per second, at which the page faults are resolved by recovering pages without additional disk activity, including pages that are being used by another process sharing the page.
(e) **Pool Paged Allocations**: It refers to the number of calls to allocate space in the paged pool. It is measured in numbers of calls to allocate space regardless of the amount of space allocated in each call.
(f) **Committed byte in use**: Committed bytes in use are the ratio of memory or committed byte to the memory or commit limit. The committed memory is the physical memory in use for which the space has been reserved in the paging file
(g) **Available Mbytes**: It is the amount of physical memory in megabytes available to processes running in the computer. It is calculated by summing up the space of the zeroed, free, and standby memory lists.
(h) **Cache Faults/sec**: It is the rate, in incidents per second, at which faults occur when a page that is sought in the file system cache is not found and it is to be retrieved either from elsewhere in memory (a soft fault) or from disk (a hard fault).
(i) **Write Copies/sec**: It defines the rate, in incidents per second, at which page faults are caused by attempts to write that are satisfied by copying the page from elsewhere in physical memory.
(j) **Pages/sec**: It is the rate, in incidents per second, at which pages are read from or written to disk to resolve hard page faults. It is a primary indicator for the kinds of faults that cause system-wide delays.

5 Experiment Methodology

In this section, we briefly describe the process of generating data from system performance log file of the personal computer. The performance logs are resources of data and which are saved in MS-Excel format. The steps involved in data generation from the system logfiles are explained while referring to the Fig. 1.

Step 1: Go to Start Menu → Settings → Control Panel → Administrative Tools and then double click on Computer Management (See Fig. 1)
Step 2: Explore the performance and then select Data Collector Sets → right click on User Defined and finally choose new → Data Collector Set (See Fig. 1)
Step 3: User must provide a name for New data Collector Set and then click on OK (See Fig. 1)
Step 4: Finally user can add or remove the counters by going through properties of the created Data Collector Set (See Fig. 1)
Step 5: The generated logfile can be found by going through the following action: Performance → Reports → User Defined and then select create logfile (See Fig. 1)

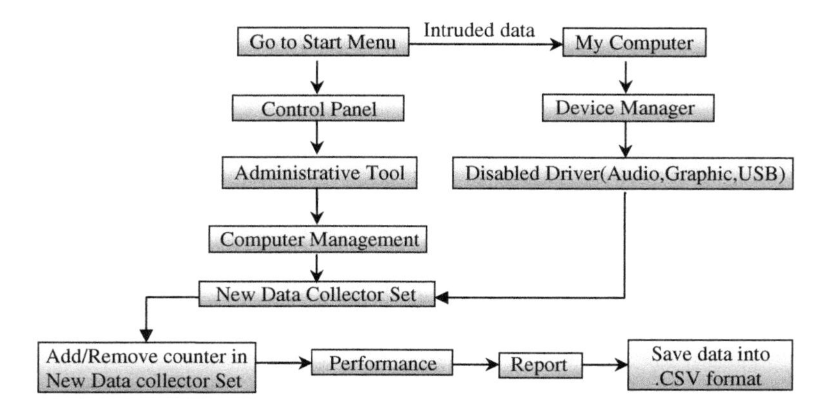

Fig. 1 Flow chart for data generation

Step 6: Now, we go to My Computer → select Propeties → select Device Manager and disable the following drivers: Audio driver, Graphics driver, and USB driver (See Fig. 1)
Step 7: Now, we go to My Computer → select Propeties → select Device Manager and disable the following drivers: Audio driver, Graphics driver, and USB driver (See Fig. 1)
Step 8: Repeat the steps from step 2 through step 8 for intruded data (See Fig. 1).

We have collected performance log for every day and then used these values as the normal data set because these data are generated without any intrusion (disabled driver). After, this we have disabled driver i.e. data are considered as intruded data. For our experiments, we have taken the normal dataset and the mixed dataset (normal and intrusion), which are given in Tables 1 and 2, respectively.

6 Results and Discussions

In our experiments, we have evaluated the residual error between the tested (abnormal) and normal datasets as shown in Fig. 2. In this experiment, we have taken generated performance log files for 30 s. for normal data which is saved into excel format. Then after, we have taken generated performance log files for 30 s. after disabling the aforesaid drivers of the system. Here, disabling of these drivers is considered as intrusion has occurred in this system. In the Fig. 2, Series 1 represents the normal behavior of the system with respect to time (in sec) using normal data while Series 2 indicates the abnormal behavior of the system. When the number of records in tested (abnormal) dataset increases, the residual error decreases (see Fig. 3 and Table 2). However, when the percentage of abnormal dataset instances increases the residual error decreases.

Table 1 Normal dataset with 10 selective attributes

Demand zero faults/s	Cache bytes	Page faults/s	Transition faults/s	Pool paged allocs	Committed bytes in use	Available MByte	Cache faults/s	Write copies/s	Pages/s
5864.13	73928704.00	10505.40	4562.89	236531.00	35.41	1698.00	67.32	24.12	6.03
7219.62	73949184.00	7502.15	277.48	236102.00	35.40	1692.00	5.05	20.18	1.01
2988.74	73793536.00	3134.29	135.58	236137.00	35.40	1693.00	5.98	19.94	39.88
496.95	73777152.00	695.12	184.09	236129.00	35.39	1693.00	14.08	20.12	1.01
544.57	73760768.00	1246.57	649.52	236159.00	35.39	1693.00	52.48	19.80	25.74

Table 2 Tested dataset with 10 selective attributes

Demand zero faults/s	Cache bytes	Page faults/s	Transition faults/s	Pool paged allocs	Committed bytes in use	Available MByte	Cache faults/s	Write copies/s	Pages/s
5864.13	73928704.00	10505.40	4562.89	236531.00	35.41	1698.00	67.32	24.12	6.03
7219.62	73949184.00	7502.15	277.48	236102.00	35.40	1692.00	5.05	20.18	1.01
2988.74	73793536.00	3134.29	135.58	236137.00	35.40	1693.00	5.98	19.94	39.88
496.95	73777152.00	695.12	184.09	236129.00	35.39	1693.00	14.08	20.12	1.01
544.57	73760768.00	1246.57	649.52	236159.00	35.39	1693.00	52.48	19.80	25.74

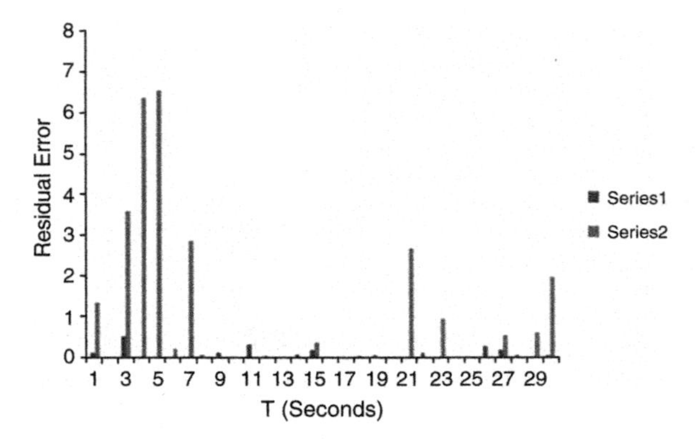

Fig. 2 Residual error between normal and mixed data with respect to time

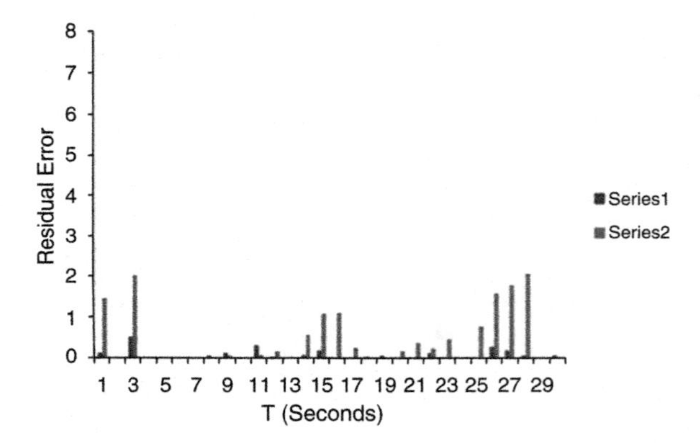

Fig. 3 Residual error between normal and mixed data with respect to time

7 Conclusion

In this paper, we have studied the performance of intrusion detection by using multivariate linear regression model that can detect intrusion with high accuracy for system performance. Furthermore, it reduces the residual error on the performance logfiles. We conclude that this model is suitable for high speed of intrusion detection.

References

1. Tomášek, M., Čajkovský, M., Madoš, B.: Intrusion detection system based on system behaviour. In: IEEE Jubilee International Symposium Applied Machine Intelligence and Informatics, pp. 271–275 (2012)
2. Ali, F., Ali, B.H., Len, Y.Y.: Development of host based intrusion detection system for log files. In: IEEE Symposium on Business, Engineering and Industrial Applications-ISBEIA, pp. 281–285 (2011)

3. Om, H., Hazra, T.: Design of anomaly detection system for outlier detection in hardware profile using PCA. Int. J. Comput. Sci. Eng. **4**(9), 1623–1632 (2012)
4. Yeung, D.Y., Ding, Y.: Host-based intrusion detection using dynamic and static behavioral models. Pattern Recogn. **36**(1), 229–243 (2003)
5. Ou, C.M.: Host-based intrusion detection systems adapted from agent-based artificial immune systems. Neurocomputing **88**, 78–86 (2012)
6. Laureano, M., Maziero, C., Jamhour, E.: Protecting host-based intrusion detectors through virtual machines. Comput. Netw. **51**(5), 1275–1283 (2007)
7. Mechtri, L., Djemili, F., Ghoualmi, N.: Intrusion detection using principal component analysis. Eng. Syst. Manage. Appl.vol no 1−6 (2010)
8. Shyu, M., Chen, S., Kanoksri, S., Chang, L.: A novel anomaly detection scheme based on principal component classifier. In: IEEE Foundations and New Directions Data Mining Workshop. , pp. 172−179 (2003)
9. Badr, N., Noureldien, N.A.: Examining outlier detection performance for principal components analysis method and its robustification method. int. J. Adv. Eng. Technol. **6**(2), 573–582 (2013)
10. Ye, N., Emran, S.M., Chen, Q., Vilbert, S.: Multivariate statistical analysis of audit trails for host based intrusion detection. IEEE Trans. Comput. **51**(7), 810–820 (2002)
11. Om, H., Hazra, T.: Statistical techniques in anomaly intrusion detection system. int. J. Adv. Eng. Technol. **5**(1), 378–398 (2012)
12. Filzmoser, P., Garrett Robert, G., Reimann, C.: Multivariate outlier detection in exploration geochemistry. Comput. Geosci. **31**(5), 579–587 (2005)
13. Sayed, A., Aziz, A., Dayem, A., Darwish, G.: Network intrusion detection system applying multivariate control charts: In INFOS. (2008)
14. Rencher, A.C.: Methods of Multivariate Analysis. A Wiley Inc. Publication. (2002)
15. Gupta, B.B., et al.: Estimating Strength of a DDoS Attack Using Multiple Regression Analysis: Advanced Computing. Springer, Heidelberg (2011)
16. Jobson, J.D.: Applied Multivariate Data Analysis Categorical and Multivariate Methods. Springer, New York (1992)

CBDF Based Cooperative Multi Robot Target Searching and Tracking Using BA

Sanjeev Sharma, Chiranjib Sur, Anupam Shukla and Ritu Tiwari

Abstract Path planning is a problem where the objective to reach up to target from source without collide with obstacle This problem would be complex when it is considered with multi robot and unknown environment. Reaching up to the target is considered as an optimization problem where the objective to minimize the distance, time and energy. This paper use the Bat algorithm (BA) for the movement of robot form one location to next location with optimizes the time, distance and energy. Here the direction of the movement is given by clustering based distribution factors (CBDF) that guide the robot to move in different direction. Different parameters are calculated during the moving of robots that help to analyze the process of target searching and tracking. Simulation is done with both simple and complex environment and results shows that the method works in both cases in searching and tracking.

Keywords Multi robot path planning · Target tracking and searching · Bat algorithm

1 Introduction and Literature Review

Robotics is the field that are now closely related to the human life, because it can help in many way like cleaning [1], fire source detection [2], reaching up to target in hazardous area [3] where human is not able to work safely. All of these work are

S. Sharma (✉) · C. Sur · A. Shukla · R. Tiwari
ABV- Indian Institute of Information Technolgy and Management, Gwalior, India
e-mail: sanjeev.sharma1868@gmail.com

C. Sur
e-mail: chiranjibsur@gmail.com

A. Shukla
e-mail: dranupamshukla@gmail.com

R. Tiwari
e-mail: tiwariritu2@gmail.com

L.C. Jain et al. (eds.), *Computational Intelligence in Data Mining - Volume 3*,
Smart Innovation, Systems and Technologies 33, DOI 10.1007/978-81-322-2202-6_34

related to the movement of robot in environment reach up to the target without collide with obstacle. Multi Robot path planning is a very important and challenging problem in multi robot system. It guides the robots to work together in cooperative manner without colliding with each other and obstacle. Multi robot path planning can be defined as follows.

Suppose there is n robot of k dimension, and each robot has initial configuration that will define the starting and ending point for robot. The objective is to determine the path for each robot that will avoid the obstacle as well as the other robot. So here multi robot is enabling to navigate collaboratively to achieve special position goal. In multi robot collaboratively path planning the objective are to minimized the total path, minimize the total path length and to minimize the energy to reach goal [4]. Multi robot cooperative path planning can be divided into two categories. It may be either coupled approach or decoupled approach. In the coupled centralized approach, all robots consider as a single team, coordination and path planning can be done in one phase. In the decoupled approach the path planning and coordination will be done in two different phase. In the first phase path to be generated and in the second phase the same coordination happens to avoid the potential conflicts. Different robot architecture like centralized, distributed and master slave can be chosen for the different type of problem [5].

Previously many works has been done that focus on the study of multi robot cooperative path planning.

In [6], probabilistic multi robot path planning is described. They used A* algorithm for path planning in dynamic environment. The process is divided into two phases. In the first phase the workspace converted into configuration space. Then create the probabilistic road map for the graph and find the optimal path. Second phase is called moving phase where the move in prioritized manner without collide with obstacle. They also show the comparison with depth first search method with A*.

In [7], the authors given the new method for plan the multi robot path planning efficiently. They use the concept of abstraction, which partition the map into sub graphs of known structure with entry and exit points. After dividing the planning become easy in small parts. Once author find the abstract plan, it is easy to resolved concrete plan.

In [8], multi robot path planning is solving by co-evolutionary genetic programming. Authors consider the different source and goal for each robot. Here each robot used grammar based genetic programming to find out the optimal path in environment, while master evolutionary caters to the need of overall path optimally. Co operation among the robots leads to the optimally path. Here authors used the local optimization using memory based look up.

In [9], this paper present new artificial potential field method to support the multi robot path planning. They name this new potential field method as swarm. Here APF uses the priority scheme between multi robots and calculate new potential function. This is able to solve the local minimum problem. This approach is base on the distributed multi robot path planning.

In [10], an author gives the new method for multi robot path planning using differential evolution. Here both centralized and decentralized approach is realized and results are compared with previous methods. The distributed approach to this problem out-per forms its centralized version for multi-robot planning. Results are tested for varying number of robots and obstacles.

In [11], to solve the multi robot path planning authors use the well known RRT method. During planning they considered both constraints conflictions among the robots. RRT* algorithm by considering the motion constraints and the sensor ranges of the robots and calculate the path of the robots in real time. Simulation studies are performed to show the effectiveness of our algorithm.

The rest of the paper is organized as follows. Section 2 discusses the Bat algorithm and it modified version for the robotics path planning problem. Section 3 discusses the directional movement and flow graph. Section 4 is for experimental parameters setup and results. At last we are concluding the things in Sect. 5.

2 Bat Algorithm for Robotics

2.1 Classical Bat Algorithm

Bat Algorithm (BA) [12–16] is the new meta heuristic search algorithm, which is developed by Xin-She Yang in 2010. Here BA is used for finding the location for movement of the robot from one location to another location. Bat algorithm is inspired by the echolocation of the microbats. Bats are the fascinating animals. These bats use the type of sonar, called echolocation that can be use to avoids obstacle, detect pray, and locate their roosting crevies in darkness. These bats emit a sound and listen for echo, when it will be return after collide with the objects. Bats echolocation property can be idealized in following way. All bats use the echolocation to find out the distance and they also know the difference between food/prey and background barrier in some way.

Bats can be randomly fly with some velocity V_i at position X_i. Here frequency f_{min} is fixed and wavelength $\lambda()$, and loudness A_0 can be varied to find out the prey. Depend on the proximity of target they can be varied.

Loudness can be very in particular range from A_0 to A_{min}, where A_o is the large positive and A_{min} is the minimum cost value. Movements of the bats— following rule are use to update the position x_i and velocity v_i in d-dimensional search space. Here equation can be defined by like this

$$f_i = f_{min} + (f_{max} - f_{min})\beta \tag{1}$$

$$V_i^t = V_i^{t-1}(X_i^t - X_*)f_i \tag{2}$$

$$X_i^t = X_i^{t-1} + V_i^t \quad (3)$$

where β is the random vector and X_* is current global best.

Based on the current based solution, a new solution for bat is generated locally using random walk.

$$X_{new} = X_{old} + \varepsilon\, A_t \quad (4)$$

where ε is a random number between 0 and 1. And A_t is the average loudness of all the bats at this step.

2.1.1 Loudness and Pulse Emission

As the iteration proceeds, the loudness A_i and the rate of emission r_i is updated by the following rule.

When the bats are close to the prey, loudness decrease and emission increase. Equation can be defined as:

$$A_i^{t+1} = \alpha A_i^t \quad (5)$$

$$r_i^{t+1} = r_i^0\, [1 - \exp(-\gamma t)] \quad (6)$$

where α and γ are constant weightage.

2.2 Modified Bat Algorithm for Robotics

In robot agent path generation the role of Bat Algorithm is to introduce randomness in local exploration through better frontier selection. Frontier selection through a combination of deterministic and stochastic process yield better result than the individuals alone. Hence we have introduced the following bat algorithm for better optimization of robot movement. The basic equations that are used in Bat algorithm for robot movement and their implications with respect to robotics application are discussed below.

Where we have

$$x_f = x_i + dx \quad (7)$$

where we have

$$dx = V_1 + (p_i - X_i) * f \quad (8)$$

and

$$f = f_{min} + (f_{max} - f_{min}) * \beta \tag{9}$$

Here β is the weightage parameter that need to be considered for contribution of the difference for the next iteration, f_{min} and f_{max} depends on range of environment. It must be kept in mind that as the movement or rather the step of movement depends on the global solution and as any of the solution is not complete, hence the p_i (global best) cannot be determined, but can be replaced by an equivalent. Here we have used the remaining distance as the criteria for determination of the best solution, and the more near the agent is, the better is its fitness. The rest of the variables have the same implications as the classical Bat Algorithm, but the range depends on the search space, and partly determined and converged through experimentation.

3 Methodology and Flow Graph

3.1 Directional Movement

Here the path planning is considered based on the local search. Robot has no previous information of environment. It senses the environment and move according to that. In the case of multi-robot the process with be want when the robot move in different direction to sense the different portion. For the scattering of robot a new concept clustering based distribution factor is introduced. In this concept the environment is divided into x region, so each robot has x direction to move. So when the process of target searching and tracking is start robot, the process will be like that so each robot will cover different direction. For this purpose we have two methods.

Scheme 1— Directional scattering effect
Directional Scattering Effect (DSE) is provided to guide the robot to go pass a certain explored area and when there is requirement of changing a region, come out of a trap or an enclosure etc. The direction factor is derived out of the cluster head and normally selected randomly to explore new direction and outings.

Scheme 2— Zig zag effect
Zig-Zag Search Effect (ZSE) is like the Directional Scattering Effect in all respect but mostly used in local search. Here the direction changes more frequently and thus helping more hovering over the workspace. Here also the cluster heads based direction is chosen dynamically and randomly.

Both of these is represent in Figs. 1 and 2 respectively.

Fig. 1 Directional movement

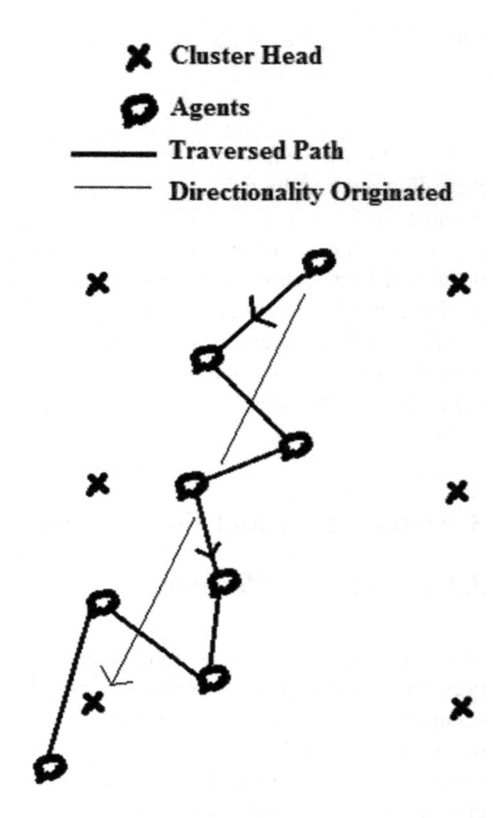

3.2 Processing Diagram

Figure 3 helps to understand the path planning process. Here a map or environment is given to input. After applying the methodology we get explored map. Here in output map the colored section showing the explored area that the robot traverses to reach up to target.

3.3 Flow Diagram

Figure 4 showing the flow graph for over all process. Here first the parameters are to be initialized. Then we check if the robot is reach up to target if we not find the target we use either directional movement or BA according to the condition. If next selection target is free move onto other otherwise find the suitable target by using

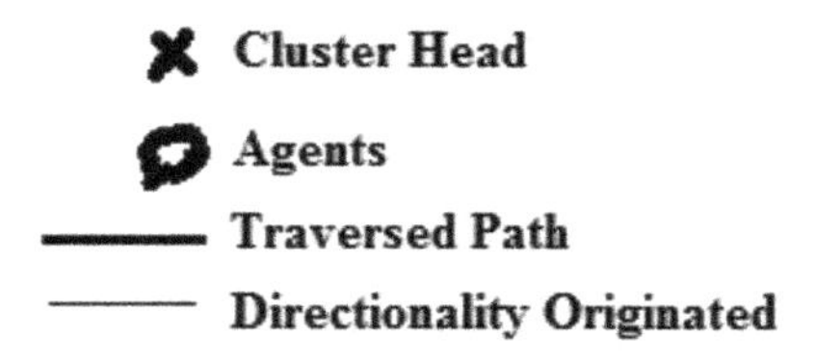

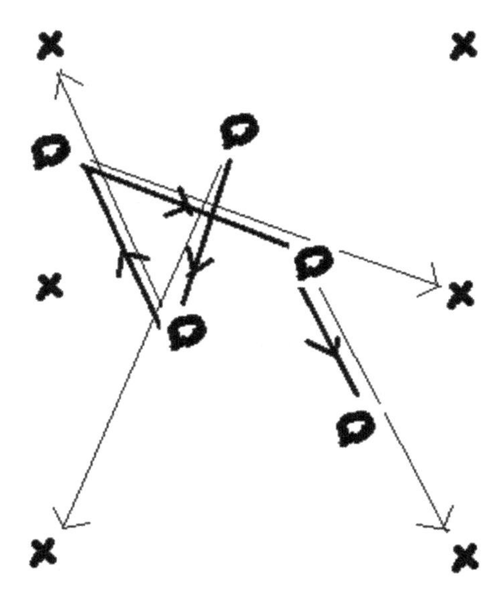

Fig. 2 Zig zag effect

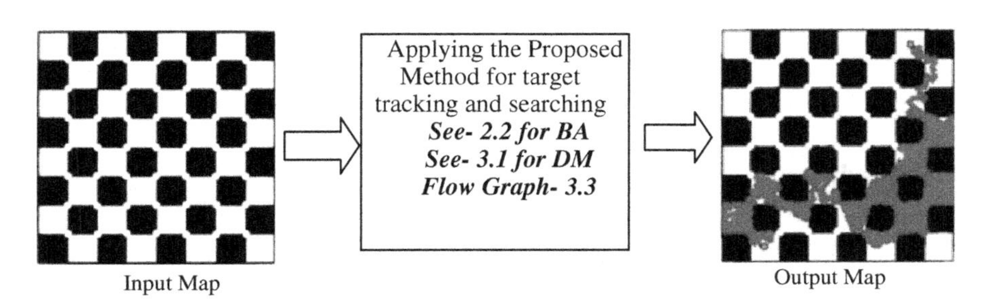

Fig. 3 Path planning process

BA or directional movement. Once we are able to choose the next target we update the correction position of robot. We repeat this process until we reach up to target or termination condition meets.

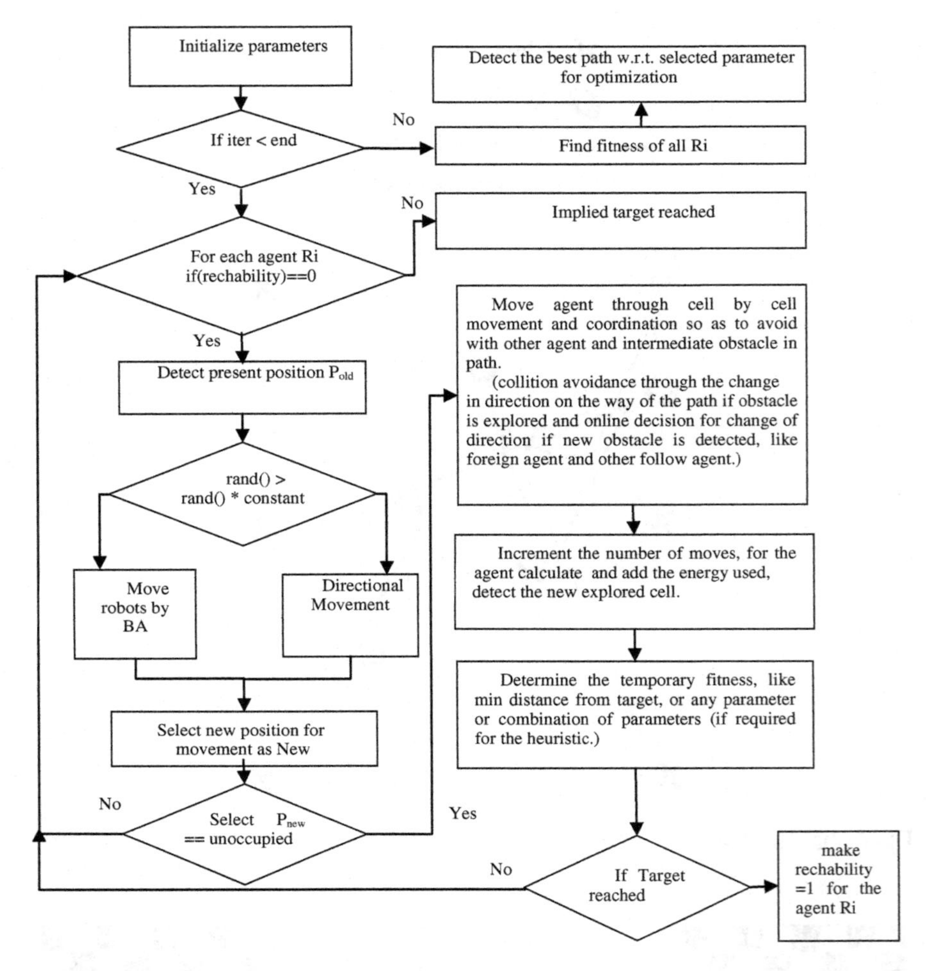

Fig. 4 Flow graph

4 Experimental Setup and Results

The search based movement is tested on two kinds of scenario that is both where direction of the destination is known to the robot and also where the direction is not known and the robot has to move to discover the destination until and unless it goes near the target with its detection range and can detect the target.

Known target is one kind of target tracking problem which involves reaching the target in efficient way and with optimized expenditure of the resources.

But in the case the no idea of direction for the target so robots have to be explore area to find the target.

4.1 Environment Description

The scenario or environment, we have considered, have static obstacles and tend to move robots through NIA based position selection probabilistically so that all agent moves differently and one of them comes up with the same path.

Here two environments are considered that has different structure and complexity. These environments are represents in Fig. 5.

4.2 Parameters Set up

Matlab 2011a is using for simulation purpose. The workspace is taken as an input in the form of an image. All simulation experiments were performed on a Duo Core 2.00 GHz machine running Windows 7 with 2 GB of RAM. Parameters setup for experiments are define in Table 1.

4.3 Explored Map to Reach up to Target

4.3.1 Unknown Target

Here we are showing the traverse path to reach up to target from source to target for unknown target. For movement from one point to another we are using BA. Figure 6 shows the results.

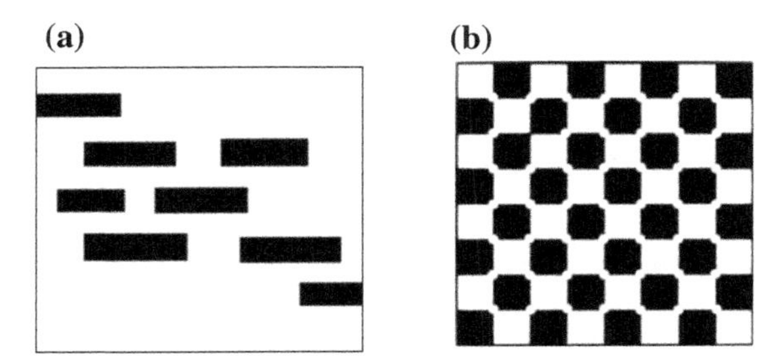

Fig. 5 Environment map. **a** Multi obstacle environment. **b** Chess like environment

Table 1 Experimental parameters

Parameters	Value
Map size	600 × 600
Sensor range	4 cell
Target	Known/unknown
Number of robots	5
Start point	Any point set as start

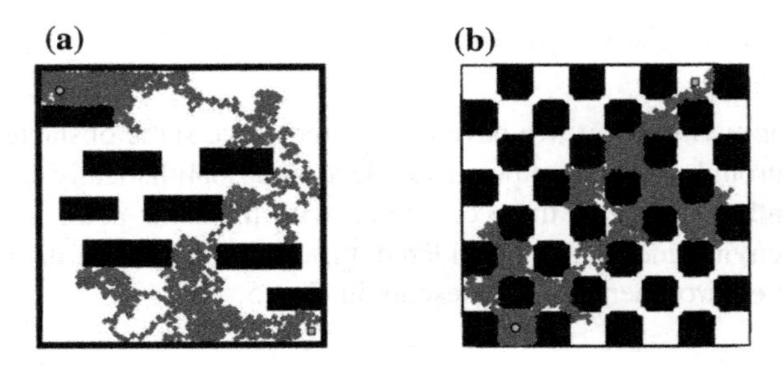

Fig. 6 Explored area to reach unknown target using BAT. **a** Multi obstacle environment. **b** Chess like environment

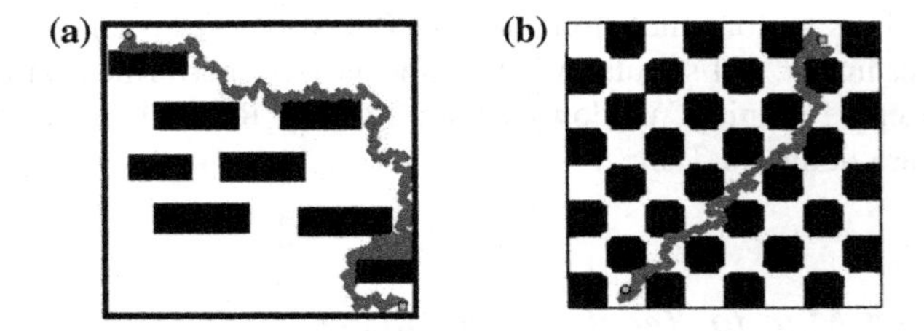

Fig. 7 Explored area to reach known target using BAT. **a** Multi obstacle environment. **b** Chess like environment

4.3.2 Known Target

Here we are showing the path planning simulated result for known target. In Fig. 7 we can see that robot reach the target smoothly, because it known the direction.

4.4 Comparison of Parameters

Here four parameters move, time coverage and energy are calculated for comparison purpose.

Move. Number of intermediate selected target form source to goal. For the given example Fig. 8 the number of move is 5.

Energy. Similarly for energy we consider the number of unit shifts in any direction (only horizontal and vertical). In example Fig. 8 this value is 27.

Time. Here time is simulation time.

Coverage. It can visualize from the nodes it has covered. The coverage area depends on the visibility or sensing range of the agents. Here for the given Fig. 8 this value is 46.

Figure 8 demonstrate the calculation of these parameters.

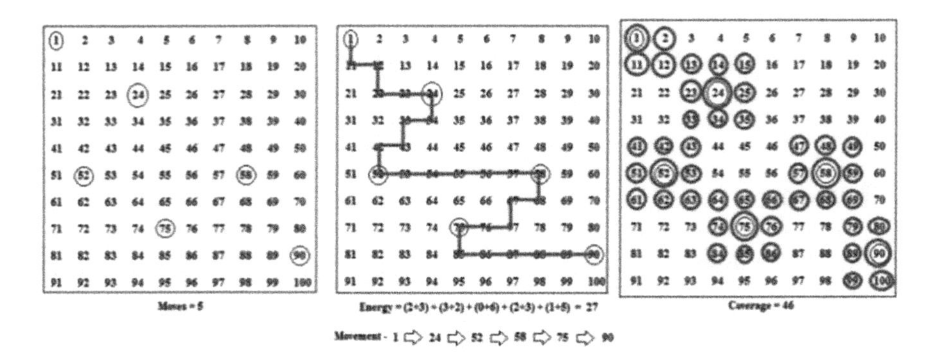

Fig. 8 Parameters calculations

Table 2 Calculated values for parameters

	Map1		Map2	
	Known target	Unknown target	Known target	Unknown target
Move	5,290	8,236	2,949	4,064
Time	0.43165	2.5865	0.2566	2.6616
Coverage	22,240	29,916	12,265	12,933
Energy	25,305	36,903	12,997	17,748

4.5 Calculated Values

Table 2 shows the results for both known and unknown environment. Here form the results we can conclude that the unknown target takes more time, move, energy and coverage, because robots need to search randomly in all directions.

5 Conclusion

Multi robot path planning is the problem where objective is to reach from source to target without collide the obstacle and other robots. This problem can be considered as optimization problem. Here we are considering the two cases of path planning one is path planning for known target, this can be considered as target searching problem, another is for known target: it can be considered as target searching problem. So here multi robot system is used for target searching and target tracking. A new concept clustering based distribution factor (CBDF) is used here for dividing the area. So the robot can move in all area to search the target in the case of target searching. So in this CBDF based division of environment if the direction of the target is known it is known as target tracking if the direction is not known it is known as target searching.

For the movement of robots we are using Nature inspired algorithm BAT algorithm. Results show that this method is able to both target tracking and searching.

References

1. Prassler, E., Ritter, A., Schaeffer, C., Fiorini, P.: A short history of cleaning robots. Auton. Robots **9**(3), 211–226 (2000)
2. Marjovi, A., Nunes, J.G., Marques, L., de Almeida, A.: Multi-robot exploration and fire searching. In: International Conference on Intelligent Robots and Systems, IEEE/RSJ, pp. 1929–1934 (2009)
3. Lee, C.H., Kim, S.H., Kang, S.C., Kim, M.S., Kwak, Y.K.: Double-track mobile robot for hazardous environment applications. Adv. Robot. **17**(5), 447–459 (2003)
4. Kumar Das, P.: D* lite algorithm based path planning of mobile robot in static Environment. Int. J. Comput. Commun. Technol. (IJCCT) **2**, 32–36 (2011)
5. Marzouqi, M.A.: Efficient path planning for searching a 2-D grid-based environment map. In: IEEE Conference and Exhibition (GCC) Dubai, United Arab Emirates, pp. 19–22 (2011)
6. Shwail, S.H., Karim, A., Turner, S.: Article: probabilistic multi robot path planning in dynamic environments: a comparison between A* and DFS. Int. J. Comput. Appl. **82**(7), 29–34 (2013)
7. Ryan, M.R.K.: Exploiting subgraph structure in multi-robot path planning. J. Artif. Intell. Res. (JAIR) **31**, 497–542 (2008)
8. Kala, R.: Multi-robot path planning using co-evolutionary genetic programming. Expert Syst. Appl. **39**(3), 3817–3831 (2012)
9. Kim, U.-H., Lee, G., Hong, I., Kim, Y.-J., Kim, D.: New potential functions for multi robot path planning: SWARM or SPREAD. In: The 2nd International Conference on Computer and Automation Engineering (ICCAE), vol. 2, pp. 557–561 (2010)
10. Chakraborty, J., Konar, A., Jain, L.C., Chakraborty, U.K.: Cooperative multi-robot path planning using differential evolution. J. Intell. Fuzzy Syst. **20**(1), 13–27 (2009)
11. Li, Rongxin, C., Chenguang, Y., Demin, X.: Multi-robot path planning based on the developed RRT* algorithm. In: 32nd Chinese Control Conference (CCC), pp. 7049–7053 (2013)
12. Yang, X.S.: A new metaheuristic bat-inspired algorithm. In: Nature Inspired Cooperative Strategies for Optimization (NICSO 2010), Springer, pp. 65–74 (2010)
13. Gandomi, A.H., Yang, X.S., Alavi, A.H., Talatahari, S.: Bat algorithm for constrained optimization tasks. Neural Comput. Appl. **22**(6), 1239–1255 (2013)
14. Tsai, P.W., Pan, J.S., Liao, B.Y., Tsai, M.J., Istanda, V.: Bat algorithm inspired algorithm for solving numerical optimization problems. Appl. Mech. Mater. **148**, 134–137 (2012)
15. Wang, G., Guo, L., Duan, H., Liu, L., Wang, H.: A bat algorithm with mutation for UCAV path planning. Sci. World J. doi:10.1100/2012/418946.2012
16. Khan, K., Sahai, A.: A comparison of BA, GA, PSO, BP and LM for training feed forward neural networks in e-learning context. Int. J. Intell. Syst. Appl. (IJISA) **4**(7), 23–29 (2012)

A Highly Robust Proxy Enabled Overload Monitoring System (P-OMS) for E-Business Web Servers

S.V. Shetty, H. Sarojadevi and B. Sriram

Abstract E-business applications play significant role in present day's business operations. Due to the surge in browsing requests, the Web-servers associated with E-business web systems suffer from unpredictable overloads that adversely affects revenue generating requests. In this research, a Proxy Enabled Overload Monitoring System (P-OMS) is developed that provides a flexible and robust overload control for E-business Web-servers. The P-OMS paradigm incorporates the combination of LIFO and FIFO scheduling for overload control. The implemented P-OMS is tested on a testbed with TPC-W benchmark setup and found to be very efficient in enhancing the throughput as well as revenue generating requests in E-business systems.

Keywords Overload control · E-business · Web-Server · TPC-W · LIFO scheduling

1 Introduction

Web-servers play vital role in E-business applications. The efficiency of Web-servers is estimated in terms of its rate of request/response generation per second. In case of higher request rates, the servers suffer overloads that results into extremely increased response time and request timing out. Such abandonments cause retries that ultimately becomes cause for more overloads. In such circumstances, the Web-servers

S.V. Shetty (✉) · B. Sriram
NMIT, Govindapura, Yelahanka Bangalore 560064, India
e-mail: vijus_shetty@yahoo.com

B. Sriram
e-mail: sriramshridhar@gmail.com

H. Sarojadevi
NMAMIT, Nitte Karkala Taluk, Udupi District 574110, Karnataka, India
e-mail: hsarojadevi@gmail.com

L.C. Jain et al. (eds.), *Computational Intelligence in Data Mining - Volume 3*,
Smart Innovation, Systems and Technologies 33, DOI 10.1007/978-81-322-2202-6_35

exhibit poor performance and the overall throughput and productivity of the system decreases. In general, Web-servers suffer from overload due to higher bursty behavior of web traffic. An illustration of bursty behavior could be the behavioral dynamicity of Web-server during closing time of sale. In highly competitive operational scenario, as web servers are expected to give good performance even during overload conditions, there is the necessity to enhance Web-servers with a goal to make them functional even with overload conditions. Therefore, there is a requirement of a scheme that can avoid servers from being forced to any infertile state during overload. In this research, the proposed P-OMS for E-business Web-servers is implemented and tested using TPC-W benchmark. The P-OMS takes into consideration of multiple optimization aspects like enhancing overall throughput and distinguishing the direct transaction request for revenue generation and browsing requests which cause revenue generation indirectly. In shopping or online retail shopping sites, the load caused because of browsing request gets exceedingly higher as compared to revenue-generating requests. In such scenario it becomes imperative that the requests for browsing do not put off requests of revenue-generation from getting finished.

Presently, the overload control strategies have gained immense interest of scientific society. The related work studied and a survey prepared advocates that even though numerous schemes exist for overload monitoring none of the existing works emphasize on the indispensable dissimilarity between revenue-generating and browsing requests, which is the significant characteristic of an E-business web system. In this work, we have developed a system paradigm that distinguishes revenue-generating (transaction) and browsing requests, monitors the load on the server and provides a dynamic overload control for E-business Web-servers. The objective of the research is to accomplish maximum revenue-generating requests. We have used a combination of LIFO and FIFO scheduling schemes for overload monitoring and control.

2 Related Work

The overload control paradigms are basically admission control [1] or certain sophisticated scheduling paradigm [2] or even a hybrid of these two. Some other overload control schemes are based on session defined policies. In research [1], an approach that does not admit any new session during overload was proposed. In [2], a dynamic weighted sharing scheme was developed for processing requests from those kinds of sessions which are likely to finish and it was accomplished by adjusting the weights of queues in dynamic manner as estimated by optimizing productivity function. Elnikey et al.[3] developed a robust admission control and request scheduling paradigm where the requirements of resources were calculated with certain external entity and on this basis the author performed admission control. A shortest job primary scheduling for enhancing the response times was presented in [4]. In their work, a feedback control loop based scheme was employed for avoiding overload and it was accomplished by means of supervising resource utilization on server. Cherkasova and Phaal [5] advocated a paradigm based on session workload to estimate the performance of web-server and defined session as the queue of revenue generation requests. To optimize the QoS of the E-business web servers they developed a session-based admission control (SBAC) mechanism along with an adaptive admission control approach. In research [6], a scheduling scheme called self-adaptive scheduling system was developed that exhibited better results for faster system recovery during transient overload. Wang et al. [7] developed a QoS-enabled work-manager scheme (WMQ) that reduces complexity of QoS schemes. Authors developed a hybrid QoS mechanism on the basis of self-optimization and fine-grained QoS control facility with a TPC-W workload generator allied with an E-commerce web application. Menasce and Akula [8] proposed a business-oriented framework to dispatch requests to a number of web-servers and incorporated a controller that could exhibit shifting of loads across varied types of servers as the workload changes. Yue and Wang [9] developed a profit-aware admission control scheme for overload avoidance for E-commerce Web sites and developed two hash tables possessing full IP address and network ID prefix that maintains the records of consumers in fine-grain and coarse-grain manners, respectively. In their work they classified clients on the basis of purchase category and estimate effectiveness of the profit-aware mechanism using the TCP-W workloads. Chen et al. [10] worked for request scheduling and focused on performance enhancement for web systems. They developed the multi-objective dynamic priority optimization scheduling scheme for E-commerce web systems. Iwasa et al. [11] presented a dynamic scaling process for distributed session control servers and developed a cluster based model to achieve both scalability and dynamic scaling by session control server. Pateriya [12] enhanced overload control using an adaptive SSL policy and session handling scheme. Kuzminykh [13] developed a model for message service system in SIP server in the standard conditions and estimated server's congestion status using a local overload control mechanism. The author employed FIFO model in its normal condition and LIFO with precedence service scheme in overload situation. To the best of our knowledge none of the research used proxy based Web-server overload control mechanism. Proxy based approach isolates overhead of overload monitoring and control from the Web-server.

3 Queuing Based P-OMS Scheme for Overload Control

The objective of the P-OMS overload control scheme is to optimize the overall throughput of revenue generating requests on E-business web sites along with preserving an enhanced performance during overload situation on web servers. In order to accomplish this goal, we have tried to minimize dropping probability of transaction requests to zero. In P-OMS there are separate dynamic queues for browsing requests and transactional requests. The browsing request joins the browsing queue while transaction request joins transaction queue. Both types of queues are set with predefined priorities. It is ensured that the transaction queue has the higher priority than the browsing queue in case of pending transaction requests. The priorities change dynamically depending on the type of request. Transaction requests are always scheduled in FIFO manner because transaction queue has higher priority with a dropping probability of zero. Browsing requests are processed when there are no pending transaction requests. If the browsing requests are served in FIFO queuing manner, during server overload requests which are neither served nor dropped get queued at the last of queue and thus wait for a long time which ultimately results into dropping of requests. On the contrary, we have used LIFO queuing model during server overload that facilitates requests to get the server access before it is dropped. Employing developed LIFO queuing during server overload, the response time as well as throughput can be improved. The algorithm for LIFO queuing and scheduling paradigm is given in Table 1.

In the proposed LIFO scheduling, the browsing request processing relies on the estimation of request volume per second; which is the browsing rate control (B_{RC}). If the B_{RC} increases beyond certain upper browsing rate control threshold (B_{RCUTh}), then it initiates serving the browsing requests in LIFO queuing order and the serving continues till the B_{RC} comes down the lower browsing rate control threshold (B_{RCLTh}). B_{RCUTh} and B_{RCLTh} values are estimated from maximum and minimum CPU utilization time.

The P-OMS approach decides which queue is to be processed on the basis of queue priority. Queue priority represents a dynamic quantity that depends on the queue length and varies as per the request being made. In case of pending request, the developed LIFO priority scheduling estimates respective priority of the queues and the queue with maximum priority value is served next. In general priority of the queue tells which queue would be served next and the B_{RC} factor decides LIFO or FIFO schedule of browsing requests. On the other hand, FIFO order is followed for transaction request serving.

In the overloaded state, we change the scheduling policy of only browsing requests from FIFO to LIFO. The decision of when to switch the queuing policy depends upon B_{RC}. Whenever B_{RC} exceeds B_{RCUTh}, we treat this as server in an unsafe state and change the policy of browsing queues from FIFO to LIFO.

Table 1 Algorithm for LIFO Queuing and Scheduling

Algorithm: LIFO Queuing and scheduling paradigm

Define Control:
Initialize system
  while alive **do**
    $B_{RC} \leftarrow$ Estimation of transactions volume per second
    If $B_{RC} > B_{RCUTh}$ **AND** with FIFO as browsing scheme B_r **then**
        $B_r \leftarrow$ LIFO
    end if
    if $B_{RC} < B_{RCLTh}$ **AND** with LIFO as browsing scheme B_r **then**
        $B_r \leftarrow$ FIFO
    end if
  end while

Similarly, whenever B_{RC} drops below B_{RCLTh}, we consider it as server is safe from the overload. Hence, we switch back to FIFO policy. This overload detection mechanism is based on assumption that B_{RC} is the true indicator of the server load and there is no other bottleneck resource in the system.

4 Experimental Setup

A TPC-W benchmark is used for emulating proposed overload model with E-business web server. A fundamental model of the overall system on test with P-OMS is illustrated in Fig. 1. The implemented P-OMS model is integrated with the TPC-W benchmark for the test purpose. In TPC-W benchmark certain clients possess privileges for accessing servers from web console and then apply their requests in terms of workload to the server.

The Remote Browser Emulators (RBEs) in the benchmark construct the client side and server is formed by the unit known as system under test (SUT). SUT encompasses varied components such as web servers, servers for databases and image servers. RBE transmits a request to Web-server and accordingly Web-server executes those requests and as per the needs of details to be served, it also contacts database server and image server.

The workload of the benchmark is submitted to SUT with its associated load parameters. The benchmark then evaluates system metrics which is to be estimated by performance monitoring. In P-OMS system, we have implemented a proxy server in request flow path lying between client and server, where it exhibits the role of an admission controller which additionally avoids upstream server from being overloaded.

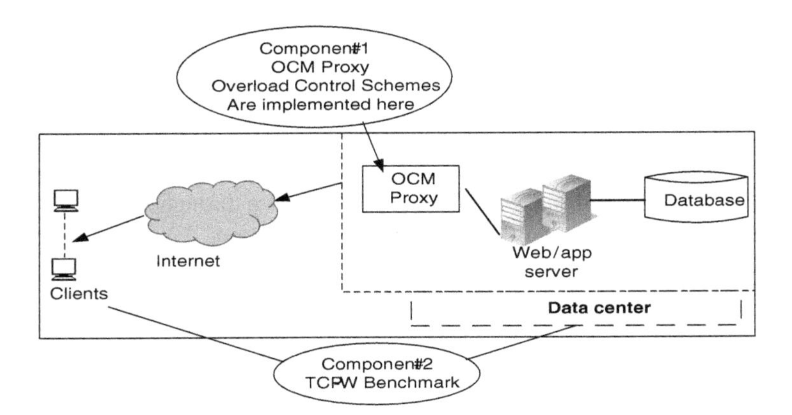

Fig. 1 P-OMS implementation model

4.1 TPC-W Benchmark

In order to evaluate the efficiency and performance of transaction processing systems, a standard association named Transaction Processing Council (TPC) has specified the benchmark TPC-W. In this research we have employed TPC-W benchmark for overload control realization and performance evaluation of E-business web systems. TPC-W benchmark encompasses a combination of functions for exhibiting robust web server system for E-business web application. The considered environment has been characterized in terms of its various synchronized sessions of on-line browser, service delivery for static as well as dynamically generated pages and numerous patterns for access and update for certain database comprising multiple tables with varied sizes, attributes and allied relationships.

The constituents of TPC-W are categorized into a combination of emulated web browsers, a web server and certain unrelenting storage. The first category performs simulation for the activities of various synchronized web browsing users, individually forming self-directed requests to the web server for various requirements like data, detail or images. On the basis of requested web pages it might be important to communicate with the unrelenting storage mechanism for web servers and to generate response pages dynamically. The relentless storage paradigm exhibits the recording of required data needed for an online E-business web servers such as inventory, customer records, order records, etc.

TPC-W benchmark defines 14 web oriented interactions where those are different from each other in terms of needed amount of server-side processing. Few of them are light weight, seeking solitary web serving for HTML pages, details or images and additional intercommunication for databases. Others may require the construction of dynamic pages.

A general TPC-W benchmark starts functioning by initializing certain count of emulated browsers where each initiates a browsing session at the E-business web site. These defined emulated browsers keep traversing pages, ensuing to varied interaction links and inflowing towards user data with dynamic probabilities. The browser counts are employed for estimating maximum throughput. There exists a random session spent being idle amid succeeding entity browser requests, for simulating the think time for certain user. Thus the browser continues accessing system for long time and as soon as it approaches its steady state the system performance analysis starts. In our work, the factors for performance with TPC-W are system throughout in terms of web interactions counts per second (WIPS) and CPU Utilization.

5 Results and Analysis

The developed P-OMS is employed with TPC-W benchmark. To validate our mechanism, we set up a TPC-W Web site that emulates the distinctiveness of a typical E-business Web site. In our experiment TPC-W was employed with the browsing mix of 95 % of browsing request and 5 % of revenue generating request. The varied sizes of browsers have been populated and the respective CPU utilization as well as throughput has been obtained. The throughput is estimated for varied emulated browser (EBs) counts. Unlike traditional approaches, in the developed testbed we have not incorporated any modification in server algorithms and even the basic infrastructure of TPC-W has not been altered. In spite of any modification we have used a proxy between server and clients; classified overall requests in two kinds, browsing requests and revenue generating (transaction) requests. The revenue generating requests have been scheduled with FIFO priority scheme and LIFO has been employed for browsing requests only in case of overload situations. The simulation results are obtained in terms of throughput, CPU utilization and emulated browsing (EB) counting. Table 2 shows the statistics of Throughput and CPU utilization obtained for different EB counts. EB count represents number of clients sending requests and throughput is an indication of web interactions per second processed by the server.

As depicted in Fig. 2, the CPU utilization for the developed system is found to be 0.45 ms when the EB count is 10. In the very next increase in EB count CPU utilization reduces/increases drastically. This happens due to generation of varying workloads by emulated browsers which results in reduction/increase of loads on the server. Similar behavior is observed with respect to throughput parameter, as illustrated in the Fig. 3.

Table 3 shows simulation results of the benchmark with implemented P-OMS and without P-OMS. Figures 4 and 5 depict the comparison and analysis of the system with P-OMS and without P-OMS. Being in uncertain user behavior scenario, simulation results fluctuate a little in the original system and P-OMS. The P-OMS scheme proved to be optimal solution for efficient web-servers both in terms of throughput and CPU Utilization. The implemented priority scheduling scheme causes the reduction of overloads for effective transaction oriented browsing functions and thus it makes the system effective to achieve more transactions/sec. Taking into consideration of the implemented proxy enabled overload

Table 2 Statistics of throughput and CPU utilization

Number of EB's	Throughput (WIPS)	CPU utilization (ms)
10	0.92	0.45
20	3.13	0.60
30	4.34	0.31
40	2.98	1.65
50	6.45	2.00
60	4.98	1.96
70	6.65	2.15
80	7.78	0.38
90	5.98	0.59
100	6.30	1.13

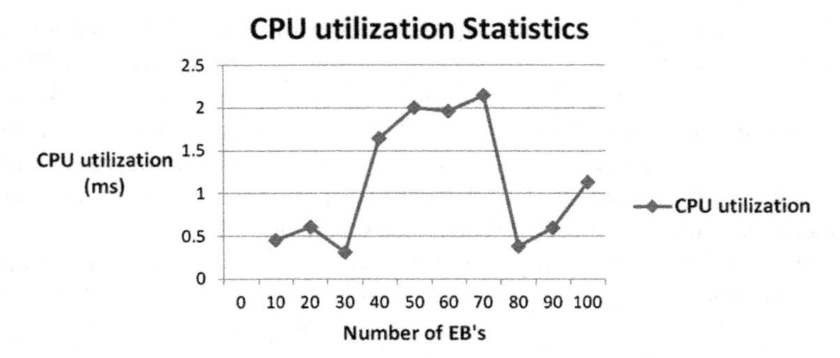

Fig. 2 CPU utilization statistics

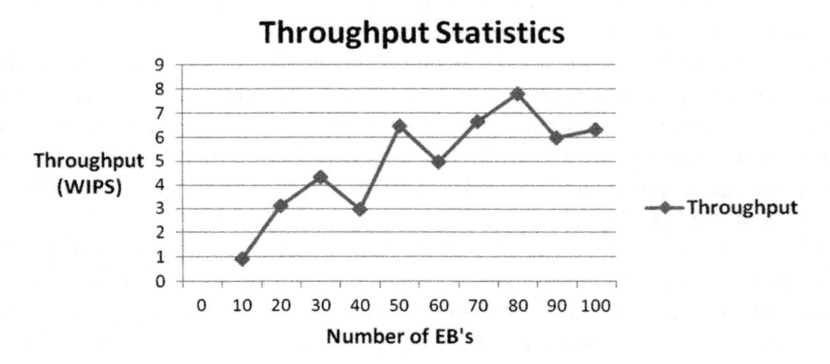

Fig. 3 Throughput statistics

Table 3 Throughput and CPU utilization comparison

Number of EB's	Throughput (P-OMS)	Throughput (without P-OMS)	CPU utilization (P-OMS)	CPU utilization (without P-OMS)
10	0.92	0.54	0.45	0.57
20	3.13	2.36	0.60	0.75
30	4.34	4.00	0.31	2.80
40	2.98	2.18	1.65	1.73
50	6.45	5.18	2.00	3.32
60	4.98	4.54	1.96	2.33
70	6.65	5.90	2.15	6.80
80	7.78	7.36	0.38	7.70
90	5.98	5.60	0.59	3.20
100	6.30	5.20	1.13	2.83

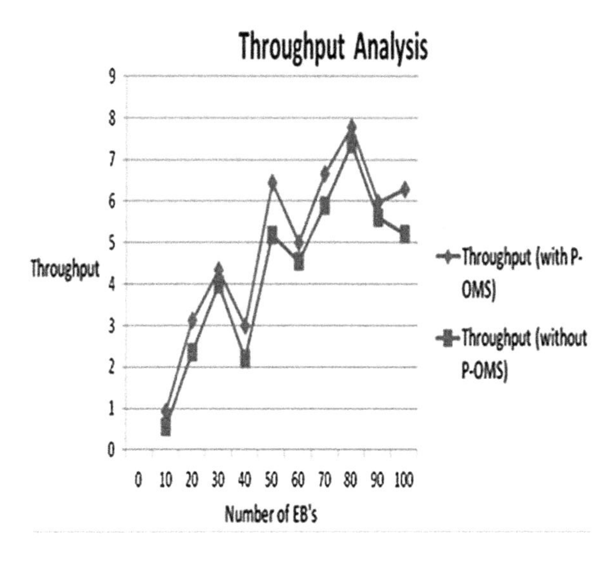

Fig. 4 Throughput analysis

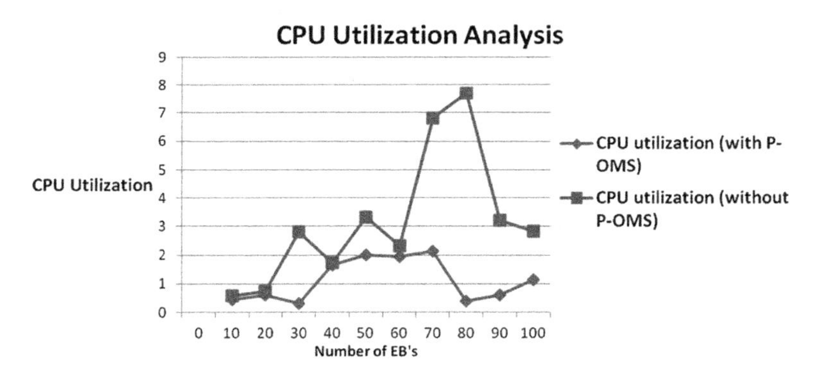

Fig. 5 CPU utilization analysis

monitoring system and the respective results obtained in terms of efficient CPU utilization and throughout achieved, the overall efficiency of web servers can be enhanced optimally.

6 Conclusions

Considering the requirement of an efficient system for overload avoidance in E-business systems, a robust overload control scheme called Proxy Enabled Overload Monitoring System (P-OMS) has been implemented and tested for

E-business Web-servers. The developed system uses LIFO paradigm for browsing queues during server overload and it employs a dynamic priority scheme to serve the browsing and transaction requests. In the developed system, the transaction queue is given higher priority than the browsing queue. The developed system has exhibited significant enhancement in throughput of the system and CPU utilization. The proposed system also enhances the processing of revenue-generating requests. The combination of LIFO and FIFO scheduling for overload control proved to be an efficient technique to improve overall performance of E-business systems. The developed P-OMS system can further be extended to optimize other bottleneck resources in real-time E-business systems.

Acknowledgments Our special thanks to our Management, Principal, Dean R&D and Head of the department for their continuous research encouragement and motivating guidelines.

References

1. Cherkasova, L., Phaal, P.: Session-based admission control-A mechanism for peak load management of commercial web sites. IEEE Trans. Comput. **51**(6), 669–685 (2002)
2. Chen, H., Mohapatra, P.: Session-based overload control in qos-aware web servers. IEEE INFOCOM (2002)
3. Elnikety, S., Nahum, E., Tracey, J., Zwaenepoel, W.: A method for transparent admission control and request scheduling in e-commerce web sites. In: Proceedings of the 13th international conference on World Wide Web. ACM Press pp. 276–286 (2004)
4. Abdelzaher,T.F., Shin, K.G. Bhatti N.: Performance guarantees for web server end systems-a control-theoretical approach. IEEE. Trans. Parallel. Distrib. Sys. **13**, 80–96 (2002)
5. Cherkasova, L., Phaal, P.: Session-based admission control: a mechanism for peak load management of commercial Web sites. Comput. IEEE. Trans. **51–6**, 669–685 (2002)
6. Zhang, Q., Riska, A., Riedel, E.: Workload propagation—overload in bursty servers. Quantitative Evaluation of Systems. Second International Conference 179–188 (2005)
7. Wang, W., Zhang, W., Wei, J., Huang, T.: A QoS-enabled work manager model for web application servers. Qual. Softw. Seventh International Conference 40–49 (2007)
8. Menasce, D., Akula V.: A business-oriented load dispatching framework for online auction sites. Quantitative Evaluation of Systems. QEST. Fourth International Conference 249–258 (2007)
9. Yue, C., Wang, H.: Profit-aware admission control for overload protection in E-commerce web sites. Qual. Serv. Fifteenth IEEE Int. Workshop 188–193 (2007)
10. Chen, M., Gao, Z., Wang, W.: Research on multi-objective dynamic priority request scheduling of business web system. Wireless Communications. Networking and Mobile Computing, WiCom .International Conference 1954–1957 (2007)
11. Iwasa, E., Irie, M, Kaneko, M, Fukumoto, T., Iio, M., Ueda, K.: Methods for reducing load of dynamic scaling for distributed session control servers. Information and Telecommunication Technologies (APSITT). 9th Asia-Pacific Symposium 1–5 (2012)
12. Pateriya, R.K.:Web server load management with adaptive SSL and admission control mechanism. Computer Science and Education (ICCSE). 7th International Conference 1178 –1183 (2012)
13. Kuzminykh, I.: A combined LIFO-Priority algorithm for overload control of SIP Server. Modern Problems of Radio Engineering Telecommunications and Computer Science (TCSET). International Conference 330–330 (2012)

Short Term Hydro Thermal Scheduling Using Invasive Weed Optimization Technique

A.K. Barisal, R.C. Prusty, S.S. Dash and S.K. Kisan

Abstract This paper presents an efficient and reliable new invasive weed optimization algorithm to solve scheduling of hydro-thermal systems with cascaded reservoirs. A multi chain cascaded hydro-thermal systems with non-linear relationship between water discharge rate, power generation and net head is considered. The water transport delay between cascaded reservoirs, prohibited operating zones, valve point loading effects and transmission losses are also considered. The feasibility of the proposed method is demonstrated in one standard test system consisting of four cascaded hydro units and an equivalent thermal unit. The findings of the proposed method are better than the results of other established methods reported in literature in terms of quality of solution and convergence characteristics.

Keywords Hydrothermal scheduling · Cascade reservoir · Invasive weed optimization · Valve point loading effect

1 Introduction

Over the last few decades, we experience energy crisis and ill effects of pollution to save nation, so it is very important to utilize energy in an efficient manner. Besides, cost must be as less as possible and this possesses a requirement to develop

A.K. Barisal · R.C. Prusty · S.S. Dash (✉) · S.K. Kisan
Department of Electrical Engineering, Veer Surendra Sai University of Technology, Burla, Odisha, India
e-mail: subhenduwithu@gmail.com

A.K. Barisal
e-mail: a_barisal@rediffmail.com

R.C. Prusty
e-mail: ramesh.prusty82@gmail.com

S.K. Kisan
e-mail: kisansamit@gmail.com

L.C. Jain et al. (eds.), *Computational Intelligence in Data Mining - Volume 3*,
Smart Innovation, Systems and Technologies 33, DOI 10.1007/978-81-322-2202-6_36

scheduling methods that accommodate generation diversity and line flow limitations to produce accurate scheduling results. Again the power generation is much lesser as compared to power demand in our country, so the main aim of power system operation is to generate and transmit power to meet the system load demand and losses at minimum fuel cost and minimum environmental pollution. Hence, a mixed of hydrothermal scheduling is necessary. The short range hydrothermal problem has usually as an optimization interval of one day or week. This period is normally subdivided into subintervals for scheduling purposes. Here the load demand, water inflow and discharge of water are known. A set of starting condition, that means the reservoirs storage levels are being given. The optimal power schedule can be prepared that can minimize the desired objective while meeting the system constraints successfully. The head of water level is assumed to be constant. The objective of short term hydrothermal scheduling of power system is to determine the optimal hydro and thermal generations in order to meet the load demands over a scheduling horizon of time while satisfying the various operational constraints imposed on the hydraulic and thermal power system network. The optimal scheduling of hydrothermal power system is usually more complex than that for all thermal system. It is basically a non-linear problem involving nonlinear objective function and a mixture of linear and non-linear constraints on thermal plants as well as hydroelectric plants. The test system consists of a multi-chain cascade of four hydro units and one thermal unit.

Dynamic programming (DP) [1, 2], employed in solving most economical hydro thermal generation schedule under the practical constraints is suffering from the curse of dimensionality and local optimality. Many stochastic search algorithms like artificial neural network [3], simulated annealing technique [4], a coordinated approach [5], genetic algorithm (GA) [6], evolutionary programming (EP) [7], two phase neural network approach [8], fast evolutionary programming [9], fuzzy interactive EP [10], simulated annealing based goal attainment method [11], an efficient real coded GA [12], modified DE [13], improved PSO [14, 15], biogeography based optimization [16], clonal selection algorithm [17], an adaptive artificial bee colony optimization [18], a differential evolution based optimizer [19], teaching learning based optimization [20] and improved DE [21] have applied successfully to solve hydrothermal problems. The neural network [3, 8] based approaches may suffer from excessive numerical iterations. Simulated annealing (SA) [4, 10] requires appropriate annealing schedule otherwise achieved solution will be of locally optimal. Genetic algorithm [6, 12] is commonly used evolutionary technique and is based on selection, crossover and mutation operation. However, it suffered from premature convergence which may lead to a local optimal solution. In differential evolution [13, 19, 21] algorithm, it is difficult to properly choose the control parameter. The literature survey of improved particle swarm optimization (IPSO) [14, 15] reveals that this technique is able to generate high quality solutions with less computational time, but sometimes trapped by local optima. This paper proposes invasive weed optimization technique for short term optimal scheduling of generation in a hydro thermal system which involves the allocation of generations among the multi reservoir cascaded hydro plants having prohibited operating zones and thermal units with valve point loading so as to minimize the total fuel cost of thermal plants while satisfying the various constraints imposed on hydraulic and thermal network. To verify the superiority of the proposed approach, one hydro-thermal system (HTS) has been considered in this study and results are compared with other algorithms.

2 Objective Function

As the fuel cost of hydropower plants are insignificant in comparison with that of thermal power plants, so the main objective of the HTS is to minimize the fuel cost of thermal units while making use of availability of hydro resource as much as possible. The objective function of HTS problem for thermal units is given by

$$\mathrm{FC(PT)} = \sum_{j=1}^{z}\sum_{i=1}^{N_s}\left(a_i PT_{j,i}^2 + b_i PT_{j,i} + c_i\right) \tag{1}$$

where N_s is the number of thermal generating units and z is the number of time intervals.

The fuel cost function of each thermal generating unit considering the valve point effects is expressed as the sum of quadratic and sinusoidal function. Thus the fuel cost function with valve point effects can be expressed as

$$\begin{aligned}\mathrm{FC(PT)} = {} & \sum_{j=1}^{m}\sum_{i=1}^{N_s}\left(a_i PT_{j,i}^2 + b_i PT_{j,i} + c_i\right) \\ & + \left|d_i \times \sin(e_i \times (PT_{\min i} - PT_{j,i})\right|\end{aligned} \tag{2}$$

The total fuel cost for running the thermal system to meet the load demand in scheduling horizon m is given by f_{total}. The objective function is expressed mathematically, as

$$Min\ F = \sum_{j=1}^{m}\sum_{i=1}^{N_s} FC\left(PT_{j,i}\right) \tag{3}$$

where, N_s is the number of thermal plants and $PT_{i,j}$ represents the ith thermal plant generation in MW at jth interval.

The various constraints are discussed as below.

(a) *Demand constraints*: This constraint is based on the principle of equilibrium between the total generation from hydro and thermal plants and the total system demand plus the system losses in each hour of scheduling j.

$$\sum_{i=1}^{N_s} PT_{i,j} + \sum_{i=1}^{N_h} PH_{i,j} = PD_j + PL_j, \text{ for } j = 1, 2, \ldots, \text{m} \tag{4}$$

(b) *Thermal generator constraints*: The operating limit of *i*th thermal generator has a lower and upper bound so that it lies in between these bounds.

$$PT_{i\,\min} \le PT_{ij} \le PT_{i\,\max} \tag{5}$$

(c) *Hydro generator constraint*: The operating limit of *i*th hydro plant must lie in between its upper and lower bounds.

$$PH_{i\,\min} \le PH_{i,j} \le PH_{i\,\max} \tag{6}$$

(d) *Reservoir capacity constraint*: The operating volume of *i*th reservoir storage must lie in between the minimum and maximum capacity limits.

$$V_{i\,\min} \le V_{i,j} \le V_{i\,\max} \tag{7}$$

(e) *The water discharge constraint*: The variable net head operation is considered and the physical limitation of water discharge of turbines in m^3, q_{ij}, must lie in between its maximum and minimum operating limits, as given by

$$q_{i\,\min} \le q_{ij} \le q_{i\,\max} \tag{8}$$

(f) *Hydraulic continuity constraint*: The storage reservoir volume limits are expressed with given initial and final volumes as:

$$\mathrm{V}_{\mathrm{i(j+1)}} = \mathrm{V}_{\mathrm{ij}} + \sum_{\mathrm{u}=1}^{\mathrm{R_u}} \left[\mathrm{q}_{\mathrm{u(j-\tau)}} + \mathrm{s}_{\mathrm{u(j-\tau)}}\right] - \mathrm{q}_{\mathrm{i(j+1)}} - \mathrm{s}_{\mathrm{i(j+1)}} + \mathrm{r}_{\mathrm{i(j+1)}} \tag{9}$$

where, τ is the water delay time between reservoir i and its up-stream u at interval j and R_u is the set of upstream units directly above hydro-plant i.

(g) The *hydro power generation*, PH_{ij}, is assumed to be a function of discharge rate and storage volume.

$$PH_{ij} = c_{1i}V_{ij}^2 + c_{2i}q_{ij}^2 + c_{3i}\left(V_{ij}q_{ij}\right) + c_{4i}V_{ij} + c_{5i}q_{ij} + c_{6i} \tag{10}$$

where, c_{1i}, c_{2i}, c_{3i}, c_{4i}, c_{5i} and c_{6i} are the coefficients.

3 Invasive Weed Optimization

Invasive weed optimization is a novel population based numerical stochastic, derivative free optimization algorithm inspired from the biological growth of weed plants. It was first developed and designed by Mehrabian and Lucas [22]. This technique is based on the colonizing behavior of weed plants [23]. Some of the interesting characteristics of weed plants that are invasive, fast reproduction and distribution, robustness and self adaptation to the changes in climate conditions.

Some important properties of IWO algorithm are described below for which it is considered as a sophisticated tool to solve complex optimization problems. One important property of the IWO algorithm is that it allows all of the plants to participate in the reproduction process. Fitter plants produce more seeds than less fit plants, which tends to improve the convergence of the algorithm. Furthermore, it is possible that some of the plants with the lower fitness value carry more useful information in iteration process as compared to the fitter plants. This IWO algorithm gives a chance to the less fit plants to reproduce and if the seeds produced by them have good finesses in the colony, they can survive [22]. The other important feature of IWO algorithm is that weeds reproduce without mating. Each weed can produce new seeds, independently. This property adds a new attribute to the algorithm that each agent may have different number of variables during the optimization process. Thus, the number of variables can be chosen as one of the optimization parameters in this algorithm. This IWO algorithm has the more chance to avoid local minima points compared to GA and PSO due to its continuous and normally distributed dispersal structure over search space which has a decreasing variance parameter centered on each parent plant [23].

4 Invasive Weed Optimization Algorithm

In this section, the computational procedure of the proposed algorithm to solve hydrothermal scheduling problem are described below.

4.1 Implementation of IWO Algorithm in Hydrothermal Scheduling Problems

The flow chart of the IWO for solving HTS problem is shown in Fig. 1. The computational procedure for proposed IWO technique to solve hydrothermal problems can be described in the following steps.

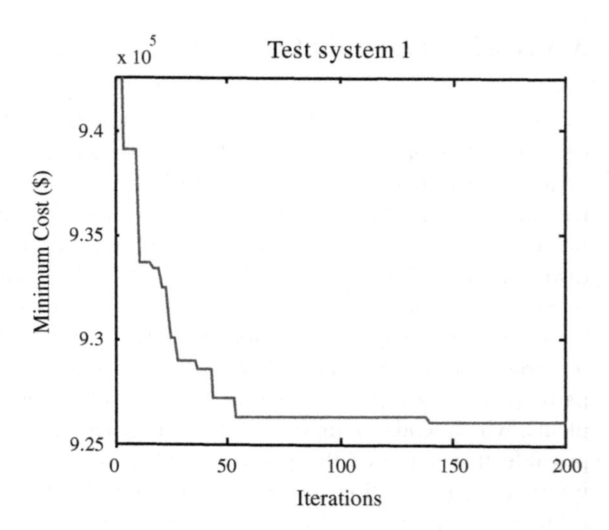

Fig. 1 Convergence characteristics for system

Step-1: Input parameters of the system and specify the upper and lower boundaries of each variable. Read the value of $Pop_{\max}$, d, z $q_{\min}$, $q_{\max}$, cost coefficients, B-coefficient matrix, $V_{\min}, V_{\max}, PT_{\min}, PT_{\max}, PH_{\min}, PH_{\max}, P_D$, $N_{\min}, N_h, N_s, N_{\max}$, σ_0, σ_f, n and $iter_{\max}$.

Step-2: Initialize randomly the seeds of the population. These seed set must be feasible candidate solutions that satisfy the given constraints.

Step-3: Let, $Q_j = [q_1, q_2 \ldots q_d]$ be the trial vector denoting the current seed set of population $(Pop_{\max} \times d)$ to be evolved. The trial random current seed matrix (S) is as follows:

$$Q = \begin{bmatrix} q_{11} & q_{12} & \cdots & \cdots & q_{1d} \\ q_{21} & q_{22} & \cdots & \cdots & q_{2d} \\ \cdots & \cdots & \cdots & \cdots & . \\ \cdots & \cdots & q_{ij} & \cdots & q_{id} \\ q_{pop_{\max},1} & \cdots & q_{pop_{\max},j} & \cdots & q_{pop_{\max},d} \end{bmatrix} \tag{11}$$

The elements of Q are the hydro discharges of the considered $d = N_h$ hydro generating units subjected to their respective capacity constraints in [4].

The discharge of ith hydro plant at jth interval is randomly generated as $q_{ij} \sim U(q_i^{\min}, q_i^{\max})$. Let, q_{id} be a dependent hydro discharge rate randomly selected to satisfy the constraints of initial and final reservoir storage limits. Let, $Q_p = [q_{11}, q_{12}, \ldots q_{1(d-1)}, q_{1(d+1)}, \ldots q_{1m}; q_{21}, q_{22}, \ldots,,$ $q_{2(d-1)} q_{2(d+1)}, \ldots, q_{2m}, \ldots; q_{i1}, q_{i2}, \ldots, q_{i(d-1)}, q_{i(d+1)}, \ldots, q_{im}]$ be the trial vector designating the pth individual of the population. The hydro discharge at the dependent interval q_{id} is calculated from Eq. 9 with zero spillage, as

$$q_{id} = V_{i1} - V_{i25} - \sum_{\substack{j=1 \\ j \neq d}}^{m} q_{ij} + \sum_{j=1}^{m} r_{ij} + \sum_{u=1}^{R_u} \sum_{j=1}^{m} q_{u(j-\tau)} \tag{12}$$

After knowing the water discharges, the reservoir volumes at different intervals are determined. Then, the hydro generations are calculated from Eq. 10. Knowing the calculated hydro generations and the given load demand P_{Dj}, for j = 1, 2…., m, thermal generations PT_j can be calculated as

$$\sum_{j=1}^{Ns} PT_j = P_{Dj} + P_{Loss} - \sum_{j=1}^{Nh} PH_j \tag{13}$$

Step-4: Go to step-3 for few iterations and then, go to step-5.

Step-5: Each seed set (each row) must be a feasible candidate solution that satisfies the given constraints. So, the seed set of population (S) is created. The fitness value (fuel cost) of all individuals (each weed) of the current seed set matrix (S) is calculated according to the objective function Eq. 3. The cost function, which is the sum of the thermal cost and violation of final storage of reservoirs with a penalty multiplier of 10,000, is taken in this problem.
The volume of water in the reservoirs at the end of each interval is then calculated using Eq. 9. All the generation levels, discharges, reservoir water volumes must be checked against their limiting values as per Eqs. 5–10. Each individual (plant) is ranked based on its fitness with respect to other weeds. Subsequently, each weed produces new seeds depending on its rank in the whole population including the quasi opposite seed set. Starting with the maximum number of seeds ($N_{\max}$) produced by the best fit plant.

Step-6: Update the iteration as $iter = iter + 1$.

Step-7: All of the plants (candidate solutions) are participating in the reproduction process. The numbers of new seeds are generated using the Eqs. 8 and 9. Each individual (plant) should satisfy both equality and inequality constraint. The infeasible solutions are replaced by randomly generated new seed set. The current seed set matrix (S) is updated with increasing individuals which are feasible candidate solutions.

Step-8: The fitness value of each individual is calculated in the entire population as per Eq. 3. They are ranked in descending order according their fitness. The best plant is identified with highest fitness value in the colony (the best fuel cost) and corresponding seed set is the optimal solution (hydro discharges) stored in a separate memory location for comparison and updated in every iteration.

Step-9: Plants with lower ranking are eliminated in order to limit the maximum number of plants in a colony to $Pop_{\max}$.

Step-10: If the stopping criterion as maximum iterations $iter_{\max}$ is satisfied then go to step-11, Otherwise, go to step-6.

Step-11: The best plant is identified with highest fitness value in the colony (the best fuel cost) and the corresponding seed set is the optimal solution (hydro discharges).

5 Simulation Results and Analysis

The proposed IWO algorithm has been applied to solve hydrothermal problems of one benchmark test system to demonstrate its performance in comparison to several established optimization techniques reported in literature. The data of the test system considered here are the same as in [6] and the additional data with valve point loading effect and with prohibited discharge zones of turbines are also same as in [9].

The various control parameters like Number of seeds, population size and maximum iteration number have been tested and finally the optimal setting of parameters is provided in Table 1.

In this case, the valve point loading of thermal units and prohibited operating zones of hydro units are taken into consideration to verify the robustness of proposed approach. The optimal hourly water discharges of reservoirs obtained by the proposed IWO algorithm are furnished in Table 2. The obtained fuel cost in this case is compared to that of IFEP [9], DE [13], IPSO [14], TLBO [20] and IDE [21] methods and reported in Table 3. From this Table 3, one can see the superiority of the proposed IWO method over other reported techniques. The cost convergence characteristic and the trajectories of hourly reservoirs volumes of hydro plants by IWO algorithm are shown in Figs. 1 and 2, respectively.

Table 1 Optimal parameter setting of different test cases

Test systems	Parameters					
	$S_{\max}$	$S_{\min}$	n	σ_0	σ_f	$Pop_{\max}$
1	5	0	3	0.7	0.001	20

Table 2 Hourly plants discharge ($\times 10^4$ m^3) of test system

Time	Q1	Q2	Q3	Q4
1	10.6418	8.3268	30.0000	13.0000
2	12.0406	6.9225	30.0000	13.0000
3	9.0634	6.1595	30.0000	13.0001
4	6.1508	6.3110	29.9991	13.0000
5	7.9918	6.2457	15.7921	13.0088
6	7.9014	6.0189	17.5326	13.0032
7	7.2692	6.3724	17.2258	13.0063
8	6.4386	6.0000	14.4533	13.0361
9	7.9181	6.9488	15.2817	13.0021
10	9.0446	6.0009	14.8901	13.0000
11	7.5095	9.2512	11.5517	13.0000
12	7.8830	10.5182	15.4248	13.0014
13	9.8317	6.0000	11.0317	13.0061
14	5.0611	6.0042	15.1822	13.0677
15	9.7554	6.6860	15.3604	13.0174
16	11.1972	8.5657	19.1742	13.8676
17	7.9412	9.2072	16.7676	13.2697
18	7.9712	10.5095	14.7497	13.0415
19	5.0089	9.7622	17.0283	15.4577
20	9.0288	10.9988	16.0086	21.6471
21	9.5177	12.5509	11.3445	19.5648
22	7.1683	11.9402	10.3054	15.7349
23	5.0415	13.0447	10.0076	24.9827
24	7.6243	11.6549	10.8829	21.5386

Table 3 Comparison of optimal costs for test system (VPL and PDZ)

Method	Best fuel cost ($)	Calculated best fuel cost ($) from discharges	Average fuel cost ($/day)	Worst fuel cost ($/day)	CPU time (s)
IFEP [9]	933949.25	933949.25	938508.87	942593.02	1450.90
DE [13]	935617.76	935617.76	–	–	
IPSO [14]	925978.84	925978.84	–	–	–
TLBO [20]	924550.78 without VPL	930761.58	924702.43	925149.06	–
IDE [21]	923016.29	927814.96	923036.28	923152.06	547.07
IWO	925390.461	925390.461	925399.427	925414.287	398.99

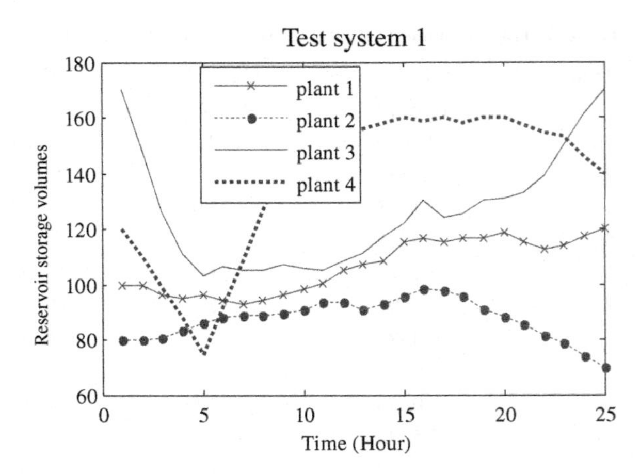

Fig. 2 Hydro reservoir storage volumes of IWO technique with VPL and PDZ by IWO method

6 Conclusion

This paper proposes an efficient invasive weed optimization algorithm (IWO) method successfully implemented to solve one hydrothermal system. The results have been compared with those obtained by other heuristic optimization techniques reported in literature. It is found that the execution time and total production costs produced by the proposed IWO method over the scheduling horizon is less than other existing optimization techniques in almost all cases considered in this study.

References

1. Yang, J.S., Chen, N.: Short term hydro thermal coordination using multi-pass dynamic programming. IEEE Trans. Power Syst. **4**(3), 1050–1056 (1989)
2. Chang, S., Chen, C., Fung, I., Luh, P.B.: Hydroelectric generation scheduling with an effective differential dynamic programming. IEEE Trans. Power Syst. **5**(3), 737–743 (1990)
3. Liang, R., Hsu, Y.: Scheduling of hydroelectric generation units using artificial neural networks. IEE Proc. Gener. Transm. Distrib. **141**(5), 452–458 (1994)
4. Wong, K.P., Wong, Y.W.: Short term hydrothermal scheduling: part 1. Simulated annealing approach. IEE Proc. Gener. Transm. Distrib. **141**(5), 497–501 (1994)
5. Nenad, T.: A coordinated approach for real-time short term hydro scheduling. IEEE Trans. Power Apparatus Syst. **11**(4), 1698–1704 (1996)
6. Orero, S.O., Irving, M.R.: A genetic algorithm modeling framework and solution technique for short term optimal hydro thermal scheduling. IEEE Trans. Power Syst. **13**(2), 501–518 (1998)
7. Hota, P.K., Chakrabarti, R., Chattopadhyay, P.K.: Short term hydro thermal scheduling through evolutionary technique. EPSR **52**, 189–196 (1999)
8. Naresh, R., Sharma. J.: Short term hydro scheduling using two phase neural network. Int. J. Electr. Power Energy Syst. **24**(7), 583–590 (2002)
9. Sinha, N., Chakrabarti, R., Chattopadhyay, P.K.: Fast evolutionary technique for short term hydrothermal scheduling. IEEE Trans. Power Syst. **18**(1), 214–219 (2003)

10. Basu, M.: An interactive fuzzy satisfying method based on evolutionary programming technique for multi-objective short term hydro thermal scheduling. Electr. Power Syst. Res. **69** (2–3), 277–285 (2004)
11. Basu, M.: A simulated annealing–based goal attainment method for economic emission load dispatch of fixed head hydro thermal power system. Int. J. Electr. Power Energy Syst. **27**(2), 147–153 (2005)
12. Kumar, S., Naresh. R.: Efficient real coded genetic algorithm to solve the non-convex hydro thermal scheduling problem. Int. J. Electr. Power Energy Syst. **29**(10), 738–747 (2007)
13. Lakshminarasimman, L., Subramanian, S.: Short term scheduling of hydro thermal power system with cascaded reservoirs by using modified differential evolution. IEE Proc. Gener. Transm. Distrib. **153**(6), 693–700 (2006)
14. Hota, P.K., Barisal, A.K., Chakrabarti, R.: An improved PSO technique for short term optimal hydrothermal scheduling. Electr. Power Syst. Res. **79**(7), 1047–1053 (2009)
15. Mandal, K.K., Chakraborty, N.: Short term combined economic emission scheduling of hydro thermal system with cascaded reservoir using particle swarm optimization technique. Appl. Soft Comput. **11**(1), 1295–1302 (2011)
16. Roy, P.K., Mandal, D.: Quasi-oppositional biogeography-based optimization for multi-objective optimal power flow. Electr. Power. Compont. Syst. **43**(1), 1443–1452 (2011)
17. Swain, R.K., Barisal, A.K., Hota, P.K., Chakrabarti, R.: Short term hydro thermal scheduling using clonal selection algorithm. Int. J. Electr. Power Syst. **33**(3), 647–656 (2011)
18. Liao, X., Zhou, J., Zhang, R., Zhang, Y.: An adaptive artificial bee colony algorithm for long term economic dispatch in cascaded hydropower systems. Int. J. Electr. Power Syst. **43**(1), 1340–1345 (2012)
19. Mandal, K.K., Chakraborty, N.: Differential evolution technique-based short term economic generation scheduling of hydrothermal systems. Electr. Power Syst. Res. **78**(11), 1972–1979 (2008)
20. Roy, P.K.: Teaching learning based optimization for short term hydro thermal scheduling problem considering valve point effect and prohibited discharge constraint. Int. J. Electr. Power Energy Syst. **53**, 10–19 (2013)
21. Basu, M.: Improved differential evolution for short term hydro thermal scheduling. Int. J. Electr. Power Energy Syst. **58**, 91–100 (2014)
22. Mehrabian, A.R., Lucas, C.: A novel numerical optimization algorithm inspired from weed colonization. Elsevier Ecol. Inf **1**(4), 355–366 (2006)
23. Sepehri, R.H., Lucas, C.: A recommender system based on invasive weed optimization algorithm. In: IEEE congr. evol. comput. CEC (2007)
24. Wood, A.J., Wollenberg, B.F.: Power Generation Operation and Control. Wiley, New York (1984)
25. Carvalloh, M.F., Soares, S.: An efficient hydrothermal scheduling algorithm. IEEE Trans. PWRS **4**, 537–542 (1987)
26. Rao, R.V., Savsani, V.J., Vakharia, D.P.: Teaching-learning based optimization: an optimization method for continuous non-linear large scale problems. Inf. Sci. **183**(1), 1–15 (2012)

11. Basu, [illegible] technique for [illegible] (2004) 277–285 [illegible]
12. [illegible] dispatch [illegible]
13. [illegible] 147(6) [illegible]
12. Kumar, S., Naresh, R.: [illegible] hydro-thermal [illegible] Energy Syst. 29(10) [illegible]
15. Lakshminarasimman, L., [illegible] system with cascaded reservoirs [illegible] Transm. Distrib. [illegible]
16. [illegible] Electr. Power Syst. Res. [illegible]
18. [illegible] thermal system with cascaded reservoirs [illegible] Sci. Comput. [illegible]
16. Roy, P.K., [illegible] Electr. Power Energy Syst. [illegible]
17. Swain, R.K., [illegible] using clonal selection algorithm. [illegible] Electr. Power Energy Syst. [illegible]
18. Liao, X., Zhou, J., Zhang, R., Zhang, Y.: [illegible] hydrothermal [illegible] Electr. Power Energy Syst. [illegible]
19. [illegible] scheduling of hydrothermal [illegible] (2008)
20. Roy, P.K.: [illegible] scheduling problem considering [illegible] Electr. Power Energy Syst. [illegible]
21. [illegible] Electr. Power Energy Syst. [illegible] (2013)
22. [illegible]
23. [illegible] algorithm. [illegible]
24. Wood, A.J., Wollenberg, B.F.: Power generation operation and control. Wiley, New York (1996)
25. [illegible] Trans. [illegible]
26. [illegible] (2012)

Contourlet Transform Based Feature Extraction Method for Finger Knuckle Recognition System

K. Usha and M. Ezhilarasan

Abstract Hand based Biometric systems are considered to be more advantageous due to its high accuracy rate and rapidity in recognition. Finger knuckle Print (FKP) is defined as a set of inherent dermal patterns present in the outer surface of the Proximal Inter Phalangeal joint (PIP) of a person's finger back region which serves as a distinctive biometric identifier. This paper contributes a Contourlet Transform based Feature Extraction Method (CTFEM) which initially decomposes the captured finger knuckle print image that results in low and high frequency contourlet coefficients with different scales and various angles are obtained. Secondly, the Principle Component Analysis (PCA) is further used to reduce the dimensionality of the obtained coefficients and finally matching is performed using Euclidean distance. Extensive experiments are carried out using PolyU FKP database and the obtained experimental results confirm that, the proposed CTFEM approach shows an high genuine acceptance rate of 98.72 %.

Keywords Finger knuckle print · Contourlet transform · Knuckle contours · Principal component analysis · Euclidean distance based classifier · Matching score level fusion

1 Introduction

Personal recognition based on hand biometric traits has been widely used for most of the modern security applications due to its low cost in acquiring the data, its reliability in identifying the individuals and its degree of acceptance by the user [1].

K. Usha (✉)
Department of Computer Science and Engineering, Pondicherry Engineering College, Pillachavady, Puducherry 605014, India
e-mail: ushavaratharajan@gmail.com

M. Ezhilarasan
Department of Information Technology, Pondicherry Engineering College, Pillachavady, Puducherry 605014, India
e-mail: mrezhil@pec.edu

L.C. Jain et al. (eds.), *Computational Intelligence in Data Mining - Volume 3*,
Smart Innovation, Systems and Technologies 33, DOI 10.1007/978-81-322-2202-6_37

In hand based biometric authentication, most of the research works proposed in the past decade used different modalities viz., fingerprint, palm print, hand geometry, hand vein patterns, finger knuckle print and palm side finger knuckle print [2]. Among these biometric traits, finger print is considered to be a very old trait and known as the first modality used for personal identification. Apart from the various beneficial aspects, finger print also possesses some limitations such as its vulnerability towards intrusion of acquired image and its feature like minutia, singular points, delta points etc., are highly distracted by means of wounds and injuries created on the finger surfaces [3].

On the other hand, palm print recognition system captures large area for identification, while contains limited number of features like principal lines, wrinkles etc., [4]. In case of finger geometry and hand geometry, the features extracted were not highly distinctive to identify the individuals, when the number of users grows exponentially [5]. In hand vein system, the vein structures present in the dorsum of the hand and in the palm area of the hand are captured by means of high resolution devices, which are found to be more costly [6].

In this paper, we recommend a new approach for personal recognition using Finger Knuckle Print (FKP) based on texture analysis using Contourlet Transform. The proposed Contourlet Transform based Feature Extraction Method (CTFEM) initially decomposes the captured finger knuckle print image by which low and high frequency contourlet coefficients with different scales and various angles are obtained. Further, to reduce the dimensionality of the obtained coefficients, Principle Component Analysis is used. Finally, the Eigen values obtained through the above process are represented in the form of Eigen Vectors, which represents the feature information of the captured FKP image. The matching between the test image and the registered images is done by finding the Euclidean distance between their corresponding feature information vectors. The fusion process is carried out to fuse the matching scores obtained from various samples of finger knuckle print images of a person in order to improve the performance of the recognition system.

The rest of the paper is organized as follows. Section 2 briefs out some of the existing works in hand based biometric recognition with their limitations. Section 3 presents the proposed Contourlet Transform based Feature Extraction Method with its matching and fusion process. Section 4 illustrates the experimental analysis and results discussions in a detailed manner. Section 5 concludes the paper with future research directions.

2 Related Work

Woodard and Flynn [7] were first to propose the finger knuckle print (FKP) as a biometric trait in 2005. In their work, the FKP image is acquired by means of a 3D sensor and the feature extraction process is done by means of geometrical analysis by exploiting the curvature shape features of FKP. The limitation of this work is that its complexity towards 3D data processing which is found to be computationally expensive. Later in 2009, Kumar et al., have proposed number of techniques for personal authentication using hand biometric traits. In the first work Kumar et al. [8] proposed a new personal authentication system using finger knuckle surface. The feature extraction from the finger knuckle surface is done by means of both texture and geometrical feature analysis methods. Kumar et al., in the second work [9], introduced a new modality known as hand vein structure for personal authentication. In this dorsum surface of the hand is captured using infra-red imaging system. Feature extraction using texture based methods were first explored by Kumar and Ravikanth [8] work. In their work, the finger knuckle surface is captured by means of 2D sensor device and line orientation coding algorithm is used to extract feature information. In their further work [10] using finger knuckle print based on texture analysis, the matching is done by means of band limited phase only correlation.

Further, Zhang et al. [11] in their work, they hierarchically coded the finger knuckle print features using competitive coding scheme based on Riesz Transform. Meraoumia et al. [12] incorporated Fourier transform functions for deriving the feature vectors from the finger knuckle and palm print images. Matching process is based on phase correlation algorithm which identifies linear phase shift in frequency. Hedge et al. [13] incorporate Random transform in order to derive feature vector information. When a preprocessed FKP image is subjected to Gabor Wavelet transform, the FKP image was divided into various sub regions and from each regions feature vector information is derived which are termed as Eigen vector. Authors in their further work implemented a real time personal authentication using finger knuckle print in which features of the finger knuckle surface were extracted using Radon transform [14]. Meroumia et al. [15] incorporated 2D block based Discrete Cosine Transform for extracting finger knuckle features.

The detailed investigation of existing methods list show some limitations such as, The existing textural analysis based feature extraction methods like Gabor filters, Fourier transform, Wavelet transform etc., fails to handle higher dimensional singularities and not effective enough in representing curves and smooth contours of an image. Since, the image patterns of FKP images mainly consist of curved lines and contours, the implementation of Multi-scale and directional transform such as Contourlet Transform is required to extract reliable feature information from the finger knuckle print.

3 Contourlet Transform Based Feature Extraction Method (CTBFEM)

3.1 Problem Statement

The proposed CTBFEM approach identifies the feature in the captured FKP image using Discrete Contourlet Transform. The Coefficient obtained through Discrete Contourlet Transform is considered as the feature information. The Eigen values from feature information are obtained through dimensionality reduction technique known as Principle Component Analysis. Further, matching process is done by means of Euclidean Distance Classifier.

3.2 The Contourlet Transform

Wavelet offers time frequency representation of an image and its multi-scale representation [16]. But, the finger knuckle image contains curved lines and smooth contours in different directions. Wavelet transform fail to represent such image structures in an effective manner. The contourlet transform addresses this problem by representing the curved lines and contours using the properties of directionality and anisotropy [17]. Contourlet transform is a multi-scale directional transform to represent the image which initially incorporates a wavelet like structure for edge detection and then a local directional transform for contour segment detection. When a FKP image is subjected to contourlet transform, the implementation of contourlet transform happens in two folds viz., (1) Laplacian pyramid Decomposing (LPD) and Directional Filter Banks (DFB).

The captured FKP images is subjected to Contourlet transform, which divides the FKP image into low pass FKP image and band pass FKP image. Then, each obtained band pass image is further decomposed by DFB. The same steps are processed with low band pass FKP image which results in multi-scale and multi-directional decomposition

The following Fig. 1 illustrates the double filter bank structure of the Contourlet in order to obtain the sparse expansions of FKP images which possess curved line structures or contours. In the double filter bank structure, Laplacian pyramid Decomposing is used to capture the point discontinuities and then the obtained result is further subjected to Directional Filter Bank (DFB), which when implemented links these discontinuities into linear structures [18].

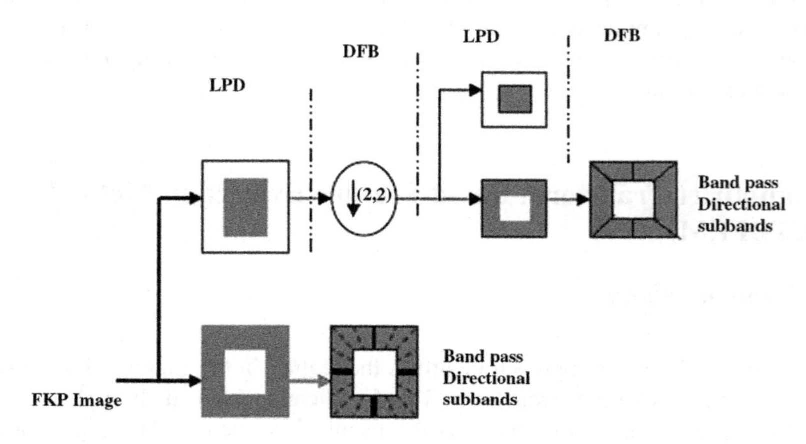

Fig. 1 Illustration of directional filter banks of contourlet transform

The Contourlet transform can also be enhanced to support different scales, directions, and aspect ratio. This further allows Contourlet to efficiently approximate a smooth contour at multiple resolutions.

3.3 Principle Component Analysis

Principal component analysis is the linear dimensionality reduction technique based on the mean-square error. It is also a second-order method that computes the covariance matrix of the variables.

The obtained contourlet coefficients are reduced to a lower dimension by identifying the orthogonal linear combinations with greater variance exists among the coefficients. The first principle component is the linear combination with largest variance. The second principle component is the linear combination with second largest variance and orthogonal to first principle component.

Then, the computation of covariance matrix for the obtained contourlet coefficients is obtained through the following steps

(1) Calculate the algorithmic means of all the feature information vectors viz., $F_1, F_2, F_3, \ldots, F_n$ containing contourlet coefficients through (1)

$$\chi = \frac{1}{n}\sum_{j=1}^{n} F_j \tag{1}$$

(2) The difference between each feature information vector and the calculated mean is computed through (2)

$$\sigma_j = F_j - \chi \tag{2}$$

(3) The covariance matrix is computed through (3)

$$\varphi = \frac{1}{n}\sum_{j=1}^{n} \sigma_j \sigma_j^T \tag{3}$$

(4) The Eigen value viz., $\zeta_1, \zeta_{2,}\zeta_3 \ldots, \zeta_n$ from the obtained covariance matrix is derived through (4)

$$\zeta_j = \frac{1}{n}\sum_{j=1}^{n} \left(v_j^T \boldsymbol{\sigma}_j^T\right)^2 \tag{4}$$

In this context, Principal Component analysis are used for the following two main reasons. Firstly, it is used to reduce the dimensions of the obtained contourlet coefficients to the lower dimension which enables computationally easier for further processing. Secondly, even the reduced contourlet coefficients could able to represents most reliable feature information for identification.

3.4 Feature Extraction and Matching Process

From the captured finger knuckle print image, the feature extraction using the proposed CTBFEM approach is achieved by the following steps:

1. If the obtained image is an RGB image, it is converted to gray scale image and the original size of the image, 120 × 270 is retained
2. The obtained FKP image is subjected to Multidimensional filtering and sharpening techniques for preprocessing.
3. The preprocessed FKP image is subjected to Discrete Contourlet Transform which results with the coefficients of lower and higher frequencies with different scales and multiple directions.
4. Decompose the obtained coefficients with the same size as $C_1, C_{2-2}..., C_{n-d}$, in which d represents the number of directions
5. The contourlet coefficients are incorporated to construct image vector F_i, by column values are reordered with the coefficient values.
6. The obtained feature vector is transformed to lower dimensional sub-band using Principle Component Analysis.
7. Matching between test image and registered image is performed by calculating Euclidean distance between test image vector information and registered image vector information.

3.5 Fusion Process

Matching Score level fusion scheme is adopted to consolidate the matching scores produced by the knuckle surfaces of the four different fingers. In the matching score level, different rules can be used to combine scores obtained by from each of the finger knuckle. All these approaches provide significant performance improvement. In this paper, weighted rule has been used.

In the weighted rule, say that S_1, S_2, S_3 and S_4 represents the normalized score obtained from finger back knuckle surface of index finger, middle finger, ring finger and little finger respectively. The final score S_F is computed using (5).

$$S_F = S_1 w_1 + S_2 w_2 + S_3 w_3 + S_4 w_4 \tag{5}$$

where w_1, w_2, w_3 and w_4 are the weights associated with each unit given by (6).

$$w_i = \frac{EER_i}{\sum_{k=1}^{i} EER_k} \tag{6}$$

4 Experimental Analysis and Results Discussion

The performance of the proposed finger knuckle print recognition system was examined by means of the PolyU Finger Knuckle Print Database [19]. The PolyU FKP database consists of finger knuckle images collected from 165 individual in a peg environment. This database consists of knuckle images collected from four different knuckle surface of a person viz, left index, right index, left middle and right middle finger knuckle surfaces. Totally, this database consists of 7,920 images with the resolution of 140 × 200. In this experimental analysis, images collected from 100 different subjects are used to train the proposed finger knuckle print recognition and images collected from 65 different subjects are used to test the system.

The extensive experiments were conducted to assess the performance of the proposed finger knuckle print recognition system. The genuine acceptance rate is determined by calculation the ratio of the number of genuine matches and imposter matches obtained to the total number of matches made with the system. False acceptance rate and False Rejection rates are computed by tracking the number of invalid matches and invalid rejections made by the system. The equal error rate is

Table 1 Performance of the proposed personal authentication system in terms of GAR values obtained from the corresponding FAR values

FPK features	Genuine acceptance rate		
	FAR = 0.5 %	FAR = 1 %	FAR = 2 %
Left index finger knuckle (LIFK)	74	76	79
Left middle finger knuckle (LMFK)	75	73	74
Right index finger knuckle (RIFK)	76	73	78
Right middle finger knuckle (RMFK)	72	74	75
LIFK + LMFK	82.5	84.5	86.3
LIFK + RIFK	83.4	84.7	86.5
LIFK + RMFK	84.8	85.9	86.3
LMFK + RIFK	85.2	87.3	89.7
LMFK + RMFK	86.1	88.2	89.3
RIFK + RMFK	87.1	88.5	89.6
LIFK + LMFK + RIFK	91.5	94.7	96.5
LIFK + LMFK + RMFK	92.6	94.8	96.7
LIFK + RIFK + RMFK	91.8	93.7	95.9
LMFK + RIFK + RMFK	93.5	95.6	96.9
All four finger knuckles	97.4	98.2	98.7

computed based on the point at which the false acceptance rate and false rejection rate become equal and the decidability threshold is computed by finding the distributions of genuine and imposters matching scores. The following Table 1 illustrates the values of the genuine acceptance rate obtained for the different values of false acceptance rate from the individual finger knuckle print feature and also from their various combinations using PolyU dataset.

The tabulated results prove that the combined performance of all the four finger knuckle print features using sum-weighted rule of matching score level fusion shows higher GAR of 98.72 % for the lower FAR values.

The ROCs illustrating the individual performance of FKP features, combined performance of two, three and all four FKP features are shown in Fig. 2a–d respectively. The ROC shown in Fig. 2a, illustrates that the performance of the system is considerably good even when features of single FKP image is used. Further, the performance gets increases as when fusion of two, three finger knuckle print image features are used as shown in Fig. 2b, c respectively. The best performance 98.7 % of GAR is achieved by fusing the entire four fingers knuckle print features as shown in Fig. 2d.

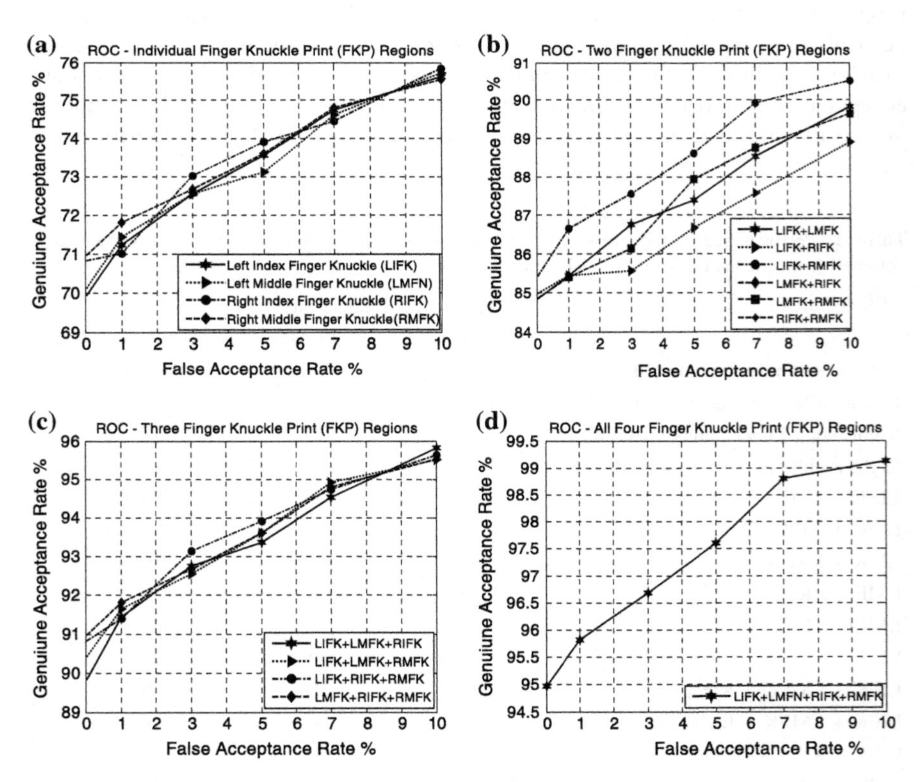

Fig. 2 Performance of the proposed system when experimented with **a** individual FKP features, **b** fusion of two FKP features, **c** fusion of three FKP features, **d** fusion of all four FKP features

Table 2 Comparative analysis of proposed method with some of the existing feature extraction methods in hand based biometrics

Reference	Data set	Feature extraction methodology	Results obtained (EER %)
Meraoumia et al. [12]	PolyU database for palm print. PolyU database for knuckle print	Generation of phase correlation function based on discrete fourier transform	1.35
Saigaa et al. [15]	PolyU FKP Database with 165 subjects and 7,440 images	Measurement of average gray scale pixel values and frequency values of the block subjected to 2D DCT and matching using correlation method	1.35
Hedge et al. [14]	PolyU FKP Database with 165 subjects and 7,440 images	Computation of Eigen values by subjecting the FKP image to random transform and matching by calculating the minimal distance value	1.28
This paper	PolyU FKP Database with 165 subjects and 7,440 images	Contourlet transform based feature extraction method	0.82

The proposed finger knuckle recognition system is compared with the existing personal recognition systems based on texture analysis methods which has been implemented on various hand based biometric traits such as finger knuckle print, palm print, hand vein structure and finger prints. The following Table 2 illustrates the summary of reported results of existing systems and comparative analysis of those results with the performance of proposed system.

In the comparative analysis as shown in the Table 2, it has been found that, the existing finger knuckle print authentication system based on coding method and appearance based method produces accuracy which is purely dependent upon the correctness of the segmentation techniques and quality of the image captured respectively. But in the case of the proposed Contourlet Transform based Feature Extraction Method which is based on texture analysis produces the lower equal error rate (EER) of 0.82 % with less computational complexity.

5 Conclusion

This paper has presented a robust approach for feature extraction from finger knuckle print using Contourlet transform. The proposed CTFEM approach extracts reliable feature information from finger knuckle print images is very effective in achieving high accuracy rate of 99.12 %. From the results analysis presented in the paper, it is obvious that the finger back knuckle print offers more features for personal authentication. In addition, it requires less processing steps as compared to the other hand traits used for personal authentication and hence it is suitable for all types of access control applications. As a future work, first we plan to incorporate shape oriented features of FKP along with this CTFEM texture feature extraction method. Secondly, analyze the performance of the recognition methods using two different finger knuckle datasets. Finally, derive the computational complexity of the recognition methods in order to analyze its space and time requirements.

References

1. Rao, R.M., Bopardikar, A.S.: Hand-based biometrics. Biometric Technol. Today **11**(7), 9–11 (2003)
2. Ribaric, S., Fratric, I.: A biometric identification system based on Eigen palm and Eigen finger features. IEEE Trans. Pattern Anal. Mach. Intell. **27**(11), 1698–1709 (2005)
3. Sun, Z., Tan, T., Wang, Y., Li, S.Z.: Ordinal palm print representation for personal identification. In: Proceedings of CVPR 2005, vol. 1, no. 1, pp. 279–284 (2005)
4. Jain, A.K., Ross, A., Pankanti,S.: A prototype hand geometry based verification system. In: Proceedings of AVBPA. Washington, DC, vol. 1, no. 1, pp. 166–171 (1999)
5. Kumar, A., Zhang, D.: Improving biometric authentication performance from the user quality. IEEE Trans. Instrum. Measur. **59**(3), 730–735 (2010)
6. Zhang, L., Zhang, L., Zhang, D.: Finger-knuckle-print: a new biometric identifier. In: Proceedings of IEEE International Conference on Image Processing, Cairo, Egypt, vol. 1, no. 1, pp. 76–82 (2009)
7. Woodard, D.L., Flynn, P.J.: Finger surface as a biometric identifier. Comput. Vis. Image Underst. **100**(1), 357–384 (2005)
8. Kumar, A., Ravikanth, Ch.: Personal authentication using finger knuckle surface. IEEE Trans. Inf. Secur. **4**(1), 98–110 (2009)
9. Kumar, A., Venkataprathyusha, K.: Personal authentication using hand vein triangulation and knuckle shape. IEEE Trans. Image Process. **18**(9), 640–645 (2009)
10. Zhang, L., Zhang, L., Zhang, D.: Finger-knuckle-print verification based on band-limited phase-only correlation. LNCS 5702, vol. 1. No. 1, pp. 141–148. Springer, Berlin (2009)
11. Zhang, L., Zhang, L., Zhang, D.: MonogenicCode: a novel fast feature coding algorithm with applications to finger-knuckle-print recognition. In: IEEE International Workshop on Emerging Techniques and Challenges (ETCHB), vol. 1. No. 1, pp. 222–231 (2010)
12. Meraoumia, A., Chitroub, S., Bouridane, A.: Fusion of finger-knuckle-print and palm print for an efficient multi-biometric system of person recognition. In: Proceedings of IEEE International Conferences on communications (ICC), vol. 1, No. 1, pp. 1–5 (2011)
13. Hegde, C., Phanindra, J., Deepa Shenoy, P., Patnaik, L.M.: Human Authentication using finger knuckle print. In: Proceedings of COMPUTE'11 ACM, Bangalore, Karnataka, India, vol. 1. No. 1, pp. 124–131 (2011)
14. Hegde, C., Deepa Shenoy, P., Venugopal, K.R., Patnaik, L.M.: Authentication using finger knuckle prints, signal, image and video processing, vol. 7. No. 4, pp. 633–645. Springer, Berlin (2013)
15. Saigaa, M., Meraoumia, A., Chitroub, S.B.: A efficient person recognition by finger-knuckle-print based on 2D discrete cosine transform. In: Proceedings of ICITeS, vol. 2. No. 1, pp. 1–6 (2012)
16. Yang, L., Guo, B.L., Ni, W.: Multimodality medical image fusion based on multiscale geometric analysis of contourlet transform. Neurocomputing **72**(1–3), 203–211 (2008)
17. Lu, Y., Do, M.N.: A new contourlet transform with shape frequency localization. IEEE Int. Conf. Image Process. **1**(1), 1629–1632 (2009)
18. Hu, H., Yu, S.: An image compression scheme based on modified contourlet transform. Comput. Eng. Appl. **41**(1), 40–43 (2005)
19. PolyU Finger Knuckle Print Database. http://www.comp.polyu.edu.hk/biometrics/FKP.htm

CDM Controller Design for Non-minimum Unstable Higher Order System

T.V. Dixit, Nivedita Rajak, Surekha Bhusnur and Shashwati Ray

Abstract The major problem in the field of control system design is to develop a control design procedure which is simple to implement and reliable in performance for complex control problems. Yet, the design of an effective controller for highly complicated, Non-minimum phase, higher order and unstable system is a challenging problem in the control community. The classical control procedure using PID controller is only effective for ordinary control problems and fails, when it is applied to some complex. In the present work, an algorithm is proposed to calculate the PID parameters. Coefficient Diagram Method (CDM) of controller design has been proposed that keeps a good balance of stability, speed of response and robustness. In this work, a CDM controller is designed and implemented to a fourth order non-minimum unstable system. The results of proposed CDM method depict a better disturbance rejection property, stability and speed of response as compared to the PID controller.

Keywords Cart-inverted pendulum (CIP) · PID controller · Coefficient diagram method (CDM) · Stability indices

T.V. Dixit (✉)
Sarguja University, Ambikapur (C.G.), India
e-mail: tvdixit@gmail.com

N. Rajak · S. Bhusnur · S. Ray
Bhilai Institute of Technology, Durg (C.G.), India
e-mail: nivedita.5rajak@gmail.com

S. Bhusnur
e-mail: sbhusnur@yahoo.co.in

S. Ray
e-mail: shashwatiray@yahoo.com

L.C. Jain et al. (eds.), *Computational Intelligence in Data Mining - Volume 3*,
Smart Innovation, Systems and Technologies 33, DOI 10.1007/978-81-322-2202-6_38

1 Introduction

In this paper, a classical problem of Cart Inverted Pendulum (CIP) is considered. CIP is a fourth order unstable non-minimum system. Since, the CIP is inherently unstable [1, 2] the pendulum will not remain upright without the external forces. The problem involves a cart, which moves back and forth, and a pendulum, hinged to the cart at the bottom of its length such that the pendulum can move in the same plane as the cart [3]. Thus, the pendulum mounted on the cart is free to fall along the cart's axis of motion. The system is to be controlled so that the pendulum remains balanced and upright. Like any other physical system, the CIP also exhibits non-linear behaviour as the system parameters may change during operation [4, 5]. Therefore, designing a sufficiently robust controller to the aforesaid CIP model is a challenging task [6, 7].

1.1 CIP: A Fourth Order System and Its Transfer Function

In the state model of CIP, the position, velocity of the cart, the angular position and change of angular position of the pendulum are taken as a set of state variables.

This system state-space equation can be written as follows

$$\dot{X} = Ax + Bu, Y = CX + Du$$

$$\begin{bmatrix} \dot{x} \\ \ddot{x} \\ \dot{\varphi} \\ \ddot{\varphi} \end{bmatrix} = \begin{bmatrix} 0 & 1 & 0 & 0 \\ 0 & -(I + ml^2 b)/I(M+m) + Mml^2 & m^2 gl^2/I(M+m) + Mml^2 & 0 \\ 0 & 0 & 0 & 1 \\ 0 & -mlb/I(M+m) + Mml^2 & mgl(M+m)/I(M+m) + Mml^2 & 0 \end{bmatrix} \begin{bmatrix} x \\ \dot{x} \\ \varphi \\ \dot{\varphi} \end{bmatrix} + \begin{bmatrix} 0 \\ I + ml^2/I(M+m) + Mml^2 \\ 0 \\ I + ml^2/I(M+m) + Mml^2 \end{bmatrix} \quad (1)$$

$$Y = \begin{bmatrix} x \\ \varphi \end{bmatrix} = \begin{bmatrix} 1\,0\,0\,0 \\ 0\,0\,1\,0 \end{bmatrix} \begin{bmatrix} x \\ \dot{x} \\ \varphi \\ \dot{\varphi} \end{bmatrix} + \begin{bmatrix} 0 \\ 0 \end{bmatrix} u \quad (2)$$

The state equations that describe the behavior of the inverted pendulum are given in (2).

1.2 Parameters of the Cart-Inverted Pendulum

In this paper the parameters of the Cart-inverted pendulum for simulation are listed in Table 1.

Table 1 Parameters of the cart-inverted pendulum

M	Mass of the cart	0.5 kg
b	Friction of the cart	0.1 N/m/s
l	Length of the pendulum	0.3 m
m	Mass of the pendulum	0.2 kg
I	Inertia of the pendulum	6E-4 kg m^2
g	Gravity	9.8 m^2

By considering the parameters (from Table 1) the transfer functions for pendulum angle and cart position control are obtained as given in (3) and (4).The transfer function for pendulum angle control is given by

$$\frac{\varphi(s)}{U(s)} = \frac{4.5455s^2}{s^4 + 0.18181s^3 - 31.1818s^2 - 4.4545s} \tag{3}$$

$$\frac{X(s)}{U(s)} = \frac{1.8182s^2 - 44.5455}{s^4 + 0.18181s^3 - 31.1818s^2 - 4.4545s} \tag{4}$$

2 PID Controller Design

The PID controller TF is given by

$$\frac{U(s)}{E(s)} = K_p + \frac{K_I}{s} + K_D s = \frac{K_D s^2 + K_p s + K_I}{s} \tag{5}$$

To ensure better performance of PID controller, poles other than the dominant poles of the desired Characteristic Equation (CE) are selected 2, 3 and 5 times farther than the dominant poles (Pendulum Angle and cart position). The dominant poles are decided based on desired settling time t_s and peak overshoot M_P. Where, the unknown PID parameters of controller are obtained by comparing it with desired CE.

3 A Brief Review on CDM

The CDM is fairly new and its basic philosophy has been known in control community since four decades [8]. The power of CDM lies in that it generates not only non-minimum phase controllers but also unstable controllers when required. Unstable controllers are shown to be very effective in controlling such unstable plants as CIP [9, 10]. The detailed description of CDM design method can be found in [11–13].

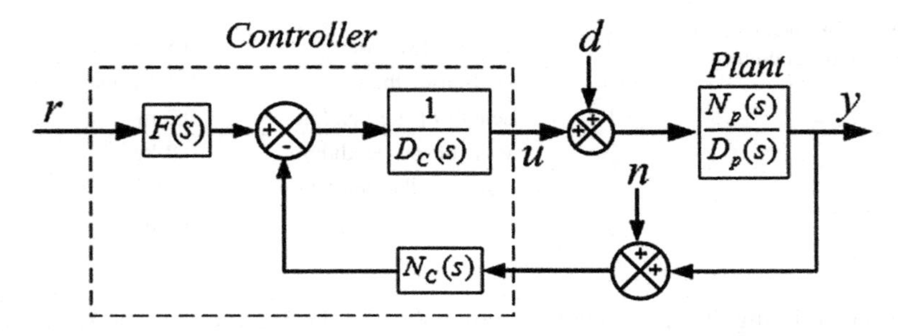

Fig. 1 Mathematical model of CDM control system

Block diagram of CDM is shown in Fig. 1. Where, *r* is the reference input, *y* is the output, *u* is the control and *d* is the external disturbance signal. Further, $N_c(s)$ and $D_c(s)$ are controller polynomials designed to meet the desired transient response and the pre-filter $F(s)$ is used to provide the steady-state gain. The output of this closed loop system is

$$y = \frac{N_c(s)F(s)r}{P(s)} + \frac{D_c(s)N_c(s)d}{P(s)} \tag{6}$$

where P(s) is the characteristic polynomial and is given by.

$$P(s) = D_p(s)D_c(s) + N_p(s)N_c(s) \tag{7}$$

$$= a_n s^n + \cdots + a_1 s^1 + a_0 = \sum_{i=0}^{n} a_i s^i \tag{8}$$

The mathematical relations involving coefficients of characteristic equation and stability indices, limits and time constant are detailed in [8, 14].

The characteristic polynomial is expressed using a_0, τ and γ_i as follows:

$$a_0 \left[\left\{ \sum_{i=2}^{n} \left(\prod_{j=1}^{i-1} \frac{1}{\gamma_{i-j}^{j}} \right) (\tau s)^i \right\} + \tau s + 1 \right] \tag{9}$$

The details of basic stability conditions involved are discussed in detail in the earlier papers on CDM [11, 12, 15].

3.1 Proposed Algorithm for CDM Controller

Let the degree of the numerator and denominator polynomial of the plant TF be assumed as m_p and n_p respectively. So, the TF of an LTI system can be expressed in the form [4]:

$$\frac{N_p(s)}{D_p(s)} = \frac{N_{m_p}s^{m_p} + N_{m_{p-1}}s^{m_{p-1}} + \cdots N_1 s + N_0}{D_{n_p}s^{n_p} + D_{n_{p-1}}s^{n_{p-1}} + \cdots D_1 s + D_0} \tag{10}$$

Let the degree of the denominator polynomial $D_c(s)$ of the controller TF be n_c and that of the numerator polynomial of the controller transfer function $N_c(s)$ be m_c. The controller polynomials are suitably chosen so that the effect of disturbance and noise is decreased to the minimum. Let the polynomials be given by

$$D_c(s) = \sum_{i=0}^{n_c} l_i s^i, N_c(s) = \sum_{i=0}^{m_c} k_i s^i \tag{11}$$

Considering only input r, the pre-filter $F(s)$ is chosen as

$$F(s) = \frac{[P(s)_{s=0}]}{N_p(s)} \tag{12}$$

In CDM, τ is chosen as

$$\tau = \frac{t_s}{2.5 \sim 3} \; or \; t_s = (2.5 \sim 3)\tau \tag{13}$$

where t_s is user-specified settling time [5].

By using the standard values of stability indices and the value of equivalent time constant τ calculated from (13), the target characteristic polynomial and coefficients are calculated. The closed loop characteristic equation is obtained with the help of (10)–(11) in terms of unknown controller parameters and plant parameters. From (7),

$$D_p(s)D_c(s) + N_p(s)N_c(s) = P_{target}(s) \tag{14}$$

The controller parameters are obtained by solving the Diophantine Eq. (14).

To control the selected design problem following steps are adopted in the proposed design algorithm:

1. Define $N_p(s)$, $D_p(s)$ as in (10) and select. $N_c(s)$, $D_c(s)$ of the controller as in (11).
2. Derive the Closed Loop Characteristic Polynomial (CLCP) in terms of unknown controller parameter and defined polynomials of plant.
3. Derive the target Characteristic Polynomial (CP) from (9) (order of target CP is selected on the basis of order of CLCP from (14).
4. To derive the target CP, select a_0, τ, γ_i. The value of τ is obtained on the basis of desired settling time from (13).
5. Draw the Coefficient Diagram and compare the target characteristic polynomial with closed loop characteristic polynomial by using Diophantine Eq. (14).
6. Derive the controller parameters and calculate Pre-filter from (12).
7. Simulate the plant with proposed controller.

Table 2 PID controller results of pendulum angle of CIP

t_s (s)	$\%M_p$	ζ (rad/s)	Dominant poles	Other poles	Parameters K_p, K_I, K_D
2	10	0.591	-2 ± 2.73	–4, –6	23.456, 63.889, 3.039
2	10	0.591	-2 ± 2.73	–4	12.896, 13.562, 1.719

(1st *row*) WOPZC-without, (2nd *row*) WPZC-with pole-zero cancelation

4 Results and Discussion

The simulation results of CIP control system are obtained using MATLAB/SIMULINK 7.1 (R2010a) environment.

4.1 Calculation of K_p, K_I, K_D for Pendulum Angle of CIP

The location of dominant poles, other assumed poles as well as parameters of PID Controller of pendulum angle for desired settling time and percentage overshoot are listed in Table 2.

4.2 Calculation of K_p, K_I, K_D for Cart Position of CIP

Substituting the plant and proposed controller transfer functions from (4) and (5), the characteristic equation of the overall system is obtained as:

$$b_5 s^5 + b_4 s^4 + b_3 s^3 + b_2 s^2 - b_1 s - b_0 = 0 \quad (15)$$

where,
$b_5 = 1, b_4 = 0.1818 + 1.8182K_D, b_3 = 1.8182K_P - 31.1818,$
$b_2 = 1.8182K_I - 44.5455K_D - 4.4545, b_1 = 44.5455K_P, b_0 = 44.5455K_I.$

The location of dominant poles, other assumed poles as well as parameters of PID Controller of cart-position for desired settling time and percentage overshoot are listed in Table 3.

Table 3 PID controller results of cart-position of CIP

t_s (s)	$\%M_p$	ζ (rad/s)	Dominant poles	Other poles	Parameters K_p, K_I, K_D
12.5	10	0.591	-0.32 ± 0.44	–0.64, –0.96, –1.59	19.927, 52.088, 1.957
15	10	0.591	-0.27 ± 0.36	–0.54, –0.81, –1.35	19.239, 44.349, 1.659

Table 4 CDM controller results of pendulum angle of CIP

Sys/$F(s)$		TOC	t_s (s)	τ	P_{target} ref	Controller parameters $k_2, k_1, k_0, l_2, l_1, l_0$
WOPZC	$\frac{1}{4.544s^2}$	2/2	5	2	(16)	0.032, 0.23, 0.36, 6E-4, 0.013, 0
WPZC	$\frac{1}{4.544s}$	2/2	5	2	(17)	0.223, 0.1226, 0.560, 0.013, 0.126, 0
WPZC	$\frac{0.4}{4.544s}$	2/2	2	0.8	(18)	0.004, 0.031, 0.072, 1E-4, 12E-4, 0

4.3 Calculation of Controller Parameters for Proposed CDM Controller

The procedure involved in the CDM controller design is described as follows: The target polynomial (pendulum angle) WOPZC with $t_s = 2\,\text{s}$ is given in (16) and the corresponding values of controller parameters are listed in Table 4. Similarly, the target polynomials WPZC with $t_s = 2\,\text{s}$ as well as $t_s = 5\,\text{s}$ are presented in (17) and (18) respectively.

$$P_{trg}(s) = 0.0006s^6 + 0.0128s^5 + 0.128s^4 + 0.64s^3 + 1.6s^2 + 2s + 1 \quad (16)$$

$$P_{trg}(s) = 0.0128s^5 + 0.128s^4 + 0.64s^3 + 1.6s^2 + 2s + 1 \quad (17)$$

$$P_{trg}(s) = 0.0001s^5 + 0.0013s^4 + 0.0164s^3 + 0.1024s^2 + 0.32s + 0.4 \quad (18)$$

4.4 For Cart-Position of Cart-Inverted Pendulum

The seventh order target polynomial (cart-position) with $t_s = 12.5$ s and $t_s = 15$ s are given in (19) and (20) respectively and their controller parameters are listed in Table 5.

Table 5 CDM controller results of cart position of CIP

Sys/$F(s)$	TOC	t_s (s)	τ	P_{trg} ref	Controller parameters $k_3, k_2, k_1, k_0, l_3, l_2, l_1, l_0$
$(a_0 = 3)$ $\frac{3}{1.82s^2-44.55}$	3/3	12.5	5	(19)	14.263,80.541, 10.961, −0.067
					0.029, 0.464, −21.357, −112.94
$(a_0 = 2.5)$ $\frac{2.5}{1.82\,s^2-44.55}$	3/3	15	6	(20)	30.452, 167.068, 23, −0.056
					0.088, 1.150, −45.073, −233.38

$$P_{trg}(s) = 0.0293s^7 + 3.75s^6 + 0.0128s^5 + 15s^4 + 30s^3 + 30s^2 + 15s + 3 \quad (19)$$

$$P_{trg}(s) = 0.0875s^7 + 1.1664s^6 + 7.7760s^5 + 25.92s^4 + 43.2s^3 + 36s^2 + 15s + 2.5 \quad (20)$$

The Coefficient diagrams for (16)–(20) are shown in Fig. 2.

4.5 Simulation Results of Proposed PID Controller

The calculated values of K_p, K_I, K_D for the proposed PID controller (pendulum angle), from Table 2 are simulated with the plant and the result is shown in Fig. 3a. Both achieve the reference condition of pendulum angle i.e. $\theta = 0°$ (vertically upright position). But, the system WOPZC depicts better performance with respect to both, disturbance rejection time and effect of step disturbance as compared to the system WPZC.

The simulation results of cart position with PID controller parameters of Table 2 are shown in Fig. 3b. It is observed that for both the conditions viz., WOPZC and WPZC the system shows unstable behavior i.e., the cart is unable to achieve the rest position and it keeps on moving forward.

4.6 Simulation Results of Proposed CDM Controller

The calculated values of controller parameters of proposed CDM controller WOPZC and WPZC for pendulum angle are simulated with the plant and the results are shown in Fig. 4a. The system WOPZC shows better performance in disturbance rejection as compared to the system WPZC.

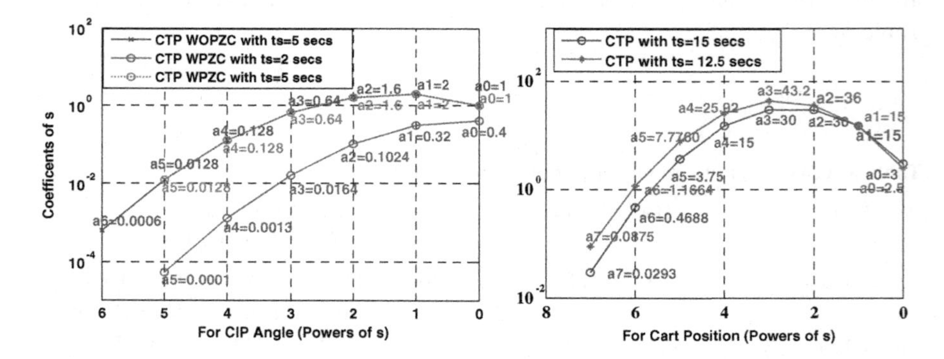

Fig. 2 Coefficient diagram for pendulum angle WOPZC, WPZC and for cart-position

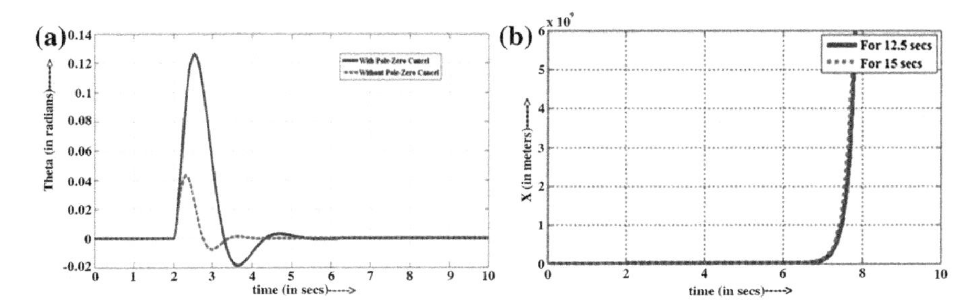

Fig. 3 Response due to PID controller with **a** for pendulum angle **b** for cart position

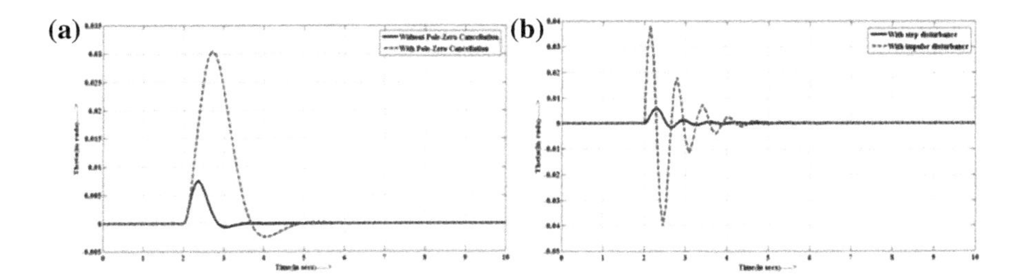

Fig. 4 Response due to CDM controller for **a** $\theta = 0^0$, with step disturbance for pendulum angle **b** pendulum angle WPZC

The values of controller parameters of proposed controller obtained from Table 5 for pendulum angle of CIP WPZC are simulated with plant and the simulation result is shown in Fig. 4b. The system shows more transient behavior when it is subjected to impulse disturbance as compared to step disturbance.

In Fig. 5a, the cart is initially at rest position x = 0, the system is subjected to a step disturbance and an impulse disturbance. The cart moves 1.6 m and takes 11.94 s to reject the step disturbance which is less than its desired disturbance rejection time of 12.5 s and settles into rest position x = 0. The cart moves 0.65 m and takes 11.93 s to reject the impulse disturbance which is less than its desired disturbance rejection time of 12.5 s and gets into rest position x = 0. The cart moves 3.4 m and takes 17.5 s to reject the step disturbance which is greater than its desired disturbance rejection time of 15 s and settles into rest position x = 0. The cart moves 1.35 m and takes 17 s to reject the impulse disturbance which is greater than its desired disturbance rejection time of 15 s and gets into rest position x = 0. The controller designed for $t_s = 12.5$ s gives better performance when the system is given an impulse disturbance as compared to the case of a step disturbance. The controller designed for $t_s = 12.5$ s also rejects the disturbances (step and impulse) but shows more transient effect on the system as compared to the controller designed for $t_s = 12.5$ s.

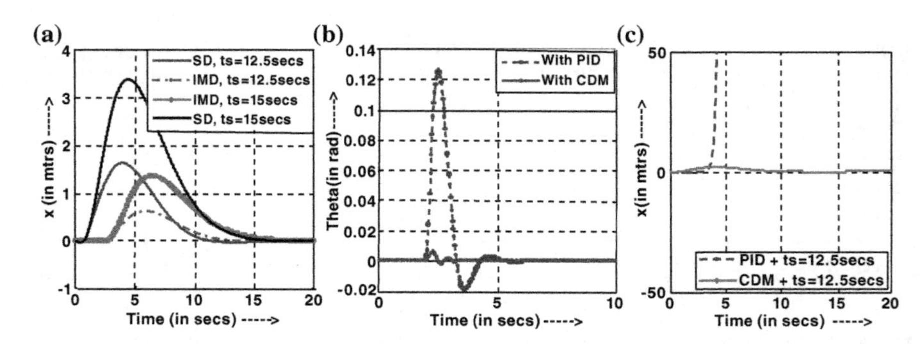

Fig. 5 Response of **a** CDM controlled system for cart position (x = 0), **b** due to proposed PID controller and CDM controller for (**b**) pendulum angle, and **c** for cart-position (SD)

The proposed PID and CDM controller WPZC (pendulum angle) are simulated with the plant and the simulation results are shown in Fig. 5b. It is observed that the PID controller does not acquire $\theta = 0°$ within the desired rejection time and transient effect of disturbance is more in both the cases. CDM gives better performance in terms of both disturbance rejection time and its effect. The simulation results of PID and CDM controller for cart position are shown in Fig. 5c.

4.7 *Simulation Results of Proposed PID and CDM Controller for Pendulum Angle and Cart Position*

The performance of three controllers, viz., PID controller, CDM controller to control pendulum angle and cart-position of CIP are summarized in Tables 6 and 7.

Table 6 The disturbance rejection time (DRT) for pendulum angle of CIP

Proposed controller	PID		CDM controller			
	WOPZC	WPZC	WOPZC	WPZC	SD	IMD
DDRT (in s)	2		5		2	
ODRT (in s)	2.06	2.35	1.5	3	0.809	0.648

where, *DDRT* desired DRT, *ODRT* obtained DRT, *SD* step disturbance, *IMD* impulse disturbance

Table 7 The disturbance rejection time (DRT) for cart-position of CIP

Proposed controller	PID		CDM controller with			
			SD	IMD	SD	IMD
DDRT (in s)	12.5	15	12.5		15	
ODRT (in s)	∞	∞	11.94	11.93	17.5	17

5 Conclusion

In this paper, CIP model was considered and an algorithm was proposed in place of trial-error method to calculate PID controller parameters to achieve desired performance. Nevertheless, the proposed PID controller only was able to achieve stability of pendulum angle but cart-position remained unstable. An alternative controller design approach known as Coefficient Method Diagram (CDM) was proposed in this paper. A simple design procedure of CDM was developed and implemented on CIP; the CDM controller gave better results in terms of stability, robustness, speed of response and easy design procedure.

The proposed PID controller and CDM controller were implemented under two conditions one WOPZC at origin and the other, WPZC. It can be concluded from the results that a better performance is obtained WOPZC. The objective of keeping pendulum in vertical upright and cart at reference position x = 0 are the problems of disturbance rejection. Generally, CDM controller has 2 DOF structure but in this case reference input was considered as zero. Hence, pre-filter has no effect on system performances.

References

1. Udhayakumar, K., Lakshmi, P.: Design of robust energy control for cart-inverted pendulum. Int. J. Eng. Technol. **4**(1), 66–76 (2007)
2. Chanchareon, R., Kananai, J.,Chantranuwathana, S.: Stabilizing of an inverted pendulum using computed feedback linearization technique. In: 19th Conference of Mechanical Engineering Network of Thailand, Phuket, Thailand, pp. 19–21 (2005)
3. Mahadi, H., Chanchal, S., Mostafizur, R., Islam, M.R.S., Subrata, K.A.: Balancing of an inverted pendulum using PD controller. Dhaka Univ. J. Sci. **60**(1), 115–120 (2012)
4. Hamamci, S.E., Koksal, M.: Robust control of A DC motor by coefficient diagram method. In: MED'01 The 9th Mediterranean Conference on Control and Automation, Dubrovnik, Crotia. (2001)
5. Hamamci, S.E., Kaya, I., Koksal, M.: Improving performance fora class of processes using coefficient diagram method. In: 9th IEEE Mediterranean Conference on Control and Automation (2001)
6. Razzaghi,K., Jalali, A.A.: A new approach on stabilization control of an inverted pendulum, using PID controller. In: International Conference on Control Robotics and Cybernetics (2011)
7. Marholt, J., Gazdos F., Dostal, P.: Control of the unstable system of the inverted pendulum using the polynomial approach. Cybern. Lett. (2011)
8. Manabe, S.: Importance of coefficient diagram in polynomial method. In: 42nd IEEE Conference on Decision and Control, Maui, Hawaii, USA (2003)
9. Manabe, S.: A solution of the ACC benchmark problem by coefficient diagram method. In: ISAS 6th Workshop on Astrodynamics and Flight Mechanics, ISAS Sagamihara, Japan (1996)
10. Erkam IMAL: CDM based controller design for nonlinear heat exchanger process. Turk. J. Electr. Comp. Sci. **17**(2), TUBITAK (2009)
11. Manabe,S.: Brief tutorial and survey of coefficient diagram method. In: 4th Asian Control Conference. TA1-2, Singapore, pp. 1161–1166 (2002)
12. Manabe, S.: Coefficient diagram method. In: 14th IFAC Symposium on Automatic Control in Aerospace, Seoul (1998)

13. Bhusnur, S., Ray, S.: Robust control of integrating systems using CDM-based two-loop control structure. Int. J. Reliab. Saf. **5**(3/4), 250–269 (2011)
14. Bhusnur, S.: Robust control using two loop structure and coefficient diagram method. Ph.D. Thesis, CSVTU, Bhilai (C.G.) (2013)
15. Manabe, S., Kim, Y.C.: Recent development of coefficient diagram method. In: 3rd Asian Control Conference, Shanghai, pp.2055–2060 (2000)

Model Based Test Case Prioritization Using Association Rule Mining

Arup Abhinna Acharya, Prateeva Mahali and Durga Prasad Mohapatra

Abstract Regression testing has gained importance due to increase in frequency of change requests made for software during maintenance phase. The retesting criteria of regression testing leads to increasing cost and time. Prioritization is an important procedure during regression testing which makes the debugging easier. This paper discusses a novel approach for test case prioritization using Association Rule Mining (ARM). In this paper, the system under test is modelled using UML Activity Diagram (AD) which is further converted into an Activity Graph (AG). A historical data store is maintained to keep details about the system which revealing more number of faults. Whenever a change is made in the system, the frequent patterns of highly affected nodes are found out. These frequent patterns reveal the probable affected nodes i.e. used to prioritize the test cases. This approach effectively prioritizes the test cases with a higher Average Percentage of Fault Detection (APFD) value.

Keywords Regression testing · Association rule mining · Test case prioritization and test case

A.A. Acharya (✉) · P. Mahali
School of Computer Engineering, KIIT University, Bhubaneswar 751024, India
e-mail: aacharyafcs@kiit.ac.in

P. Mahali
e-mail: prateevamahali@gmail.com

D.P. Mohapatra
Department of Computer Science Engineering, National Institute of Technology, Rourkela 769008, India
e-mail: durga@nitrkl.ac.in

L.C. Jain et al. (eds.), *Computational Intelligence in Data Mining - Volume 3*,
Smart Innovation, Systems and Technologies 33, DOI 10.1007/978-81-322-2202-6_39

1 Introduction

Software Testing is a time, effort and cost consuming phase in the software development life cycle. Retesting of new versions of the product during maintenance phase is carried out using previously developed test suite to identify the issue related to the modifications [1]. The biggest challenge for software testers is to complete this retesting process of regression testing within time and budget. Researchers have proposed various techniques for regression testing such as test case selection, test suite reduction and test case prioritization [2]. Out of these techniques, test case prioritization is the most effective one to have higher percentage of fault detection [2].

Test Case Prioritization is a methodology used for scheduling of test cases with the intention of detecting maximum faults with minimum time and cost [3]. Prioritization also focuses on increase in rate of high risk faults detection and ensures reliability of software products at a faster rate. Researchers had proposed several criteria for scheduling of test cases during prioritization such as, coverage-based, requirement-based, risk factor based etc. [4, 5]. But, test cases with maximum coverage do not confirm maximum fault detection. Hence we have to propose a technique which should give maximum coverage with high fault detection capability.

Average Percentage of Fault Detection (**APFD**) metric is available to check the efficiency of test suites [6].

In this paper we have tried to overcome this problem by generating frequent patterns of nodes revealing more faults which help in identifying the test cases which detects maximum faults, collected from last few releases of the software. If maximum priority is given to retest those test cases first in the next release of software, then we can ensure that the probability of getting defects or faults early will be enhanced and the quality of software will be also improved by the time it is delivered. Hence the software analyst can be able to reduce the estimated budget for the new release of software. The frequent pattern of nodes is discovered by using data mining mechanism.

Data Mining is a process of discovering or extracting data patterns which will help in identifying the frequently occurring faults [7]. A frequent pattern can be generated by using association rule, apriori algorithm or vertical data format [7]. In this paper, association rule is used to generate frequent patterns for mostly affected nodes.

Association rule is a popular method for discovering interesting relations between variables in large database. The main goal of this rule is to discover regularities between products in large scale database [7]. In an association rule mining, a rule is defined as an implication form $X \Longrightarrow Y$ where $X, Y \subseteq I$ and $X \cap Y = \phi$, where I defines the itemset. Here, it requires to satisfy a user-specified minimum support and user-specified minimum confidence simultaneously.

The rest of the paper is organized as follows: Sect. 2 discusses the related work and its analysis and Sect. 3 describes the proposed approach. Section 4 presents the implementation of proposed approach using a case study of Shopping Mall Management System (SMMS) and Sect. 5 compares the current approach with the related work. Section 6 discusses the conclusion and future work.

2 Related Work

Many researchers [8–11] have proposed different mechanisms for test case prioritization using model based approach. Khandai et al. [3] proposed an approach for prioritizing the test cases according to Business Criticality Test Value (BCTV) of the test cases. In this case functionalities which incorporate more risks to the business are given more priority. Miller et al. [8] proposes a set of new techniques for functional test case prioritization based on the inherent structure of dependencies between tests i.e. Dependency Structure Prioritization (DSP). The objective of this technique is to verify the fault detection capability of DSP relative to existing coarse grained and random fine grained techniques and efficiency of this technique is compared with existing techniques.

Mahali et al. [12] proposed another method for test case prioritization using model based approach. In that paper, they have carried out the prioritization on an optimized test suite. The authors have optimized the generated test suite using genetic algorithm and that optimized test suite is used for prioritization by comparing the current version of software with the previous version of corresponding software. Acharya et al. [13] proposed an approach for test case prioritization by assigning priority to user requirement concerning parameter like database parameter, networking parameter and Graphical User Interface (GUI) parameter.

Muthusamy et al. [14] proposed an approach for test case prioritization process by identifying the severe faults. Here test case prioritization is performed based on four practical weight factor such as customer allotted priority, developer observed code execution complexity, change in requirements, fault impact, completeness and traceability.

3 Proposed Approach

In this paper a new heuristic approach for test case prioritization is presented. In the proposed approach, UML activity diagram is used as system model and test case prioritization is performed by analysing the modification history of different version of that SUT. First the system model is converted to a model dependency graph called Activity Graph (AG) and test cases are generated from AG for the new version of SUT. Then, the proposed forward slicing algorithm is used on the AG to track the modified node and trace out the affected nodes of all modified nodes.

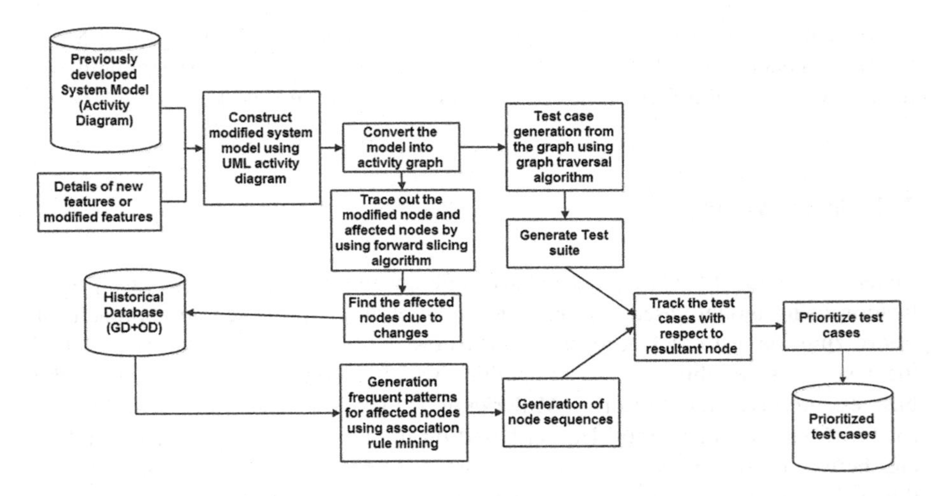

Fig. 1 Proposed framework for test case prioritization

Simultaneously historical data of affected nodes of the previous version of SUT are analysed. The historical data contains Graph Data (GD) and Observation Data (OD). Graph data defines information or data collected from AG without knowing the behaviour of system. The tester only knows the objective and proposed technique of testing process. Observation data defines information or data collected from the low level design of the system. At that time, software tester knows behaviour and required technical knowledge to implement the proposed technique by using an automation tool to show the actual result. GD and OD for each modified node are used to find out a common pattern of affected nodes revealing more number of faults. Then, test case prioritization is done by using the patterns of affected nodes generated using ARM. The proposed framework for test case prioritization is shown in Fig. 1 and elaborated in Sect. 4 with a case study of Shopping Mall Management System (SMMS).

4 Implementation of Proposed Approach

4.1 Test Case Generation Using Activity Diagram

In this paper the UML Activity Diagram (AD) is used for system modelling. The AD of Shopping Mall Management System (SMMS) is shown in Fig. 2. Then AD is converted into a model dependency graph called as Activity Graph (AG) (shown in Fig. 3), where each activity is represented as a node and activity flow between two activities is represented as an edge between the two nodes. The graph contain

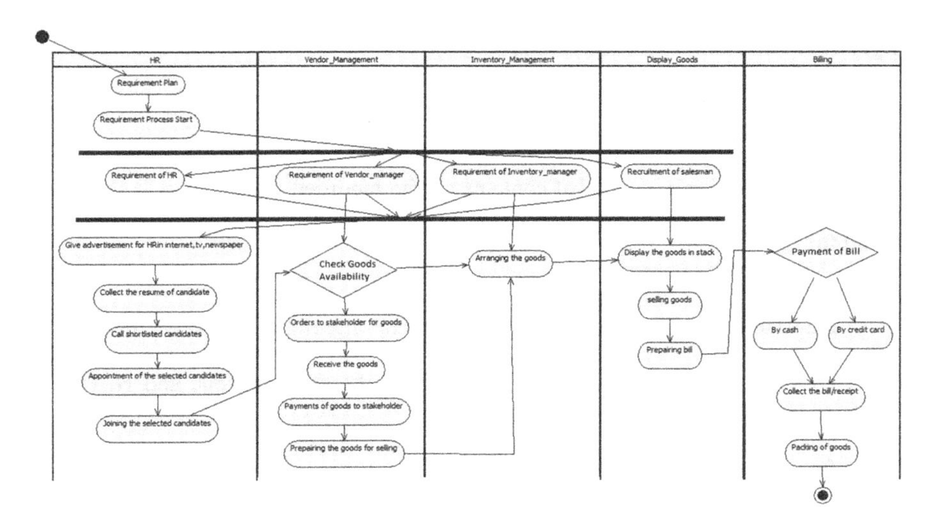

Fig. 2 Activity diagram of shopping mall management system (SMMS)

two decision nodes, one fork and join node, five control nodes and eighteen normal nodes. Test cases are generated from the activity graph by using Breadth-First-Search (BFS) algorithm [15]. Each individual path in the graph is considered as a test scenario. There are 38 paths in the activity graph and hence 38 test scenarios are possible for this SMMS, which are shown in Table 1. Due to space constrained seven number test cases are represented.

4.2 Creation of Historical Data Table Using Forward Slicing

Historical database schema representation is one of the goal of the proposed approach. The database schema contains all information about the developed system like Project ID, Project Name, Project Type, Total No. of Nodes, Total No. of Modified Nodes, and Total No. of Affected Nodes for each modified node. Database containing information about oldest version of system is called as historical database and it is updated for each latest version of that system. It contains two types of data such as Graph Data (GD) and Observation Data (OD).

As per the proposed framework, modified nodes are collected from the AG and affected nodes are obtained from proposed forward slicing algorithm (given in Algorithm 1). The affected nodes with respect to the modified nodes are shown in Table 2.

Algorithm 1 Forward slicing algorithm

Input: Activity Graph (AG)
Output: Affected nodes for each modified node
1: **for** p = 0 **to** n **do**
2: scanf("%d",A[p]); // A[p] contains all the modified nodes and n is the total number of modified nodes.
3: **end for**
4: **for** i = 0 **to** n **do**
5: tnode = A[p]; // tnode = target node
6: Traverse the graph from tnode using Kruskal algorithm to find the depended nodes;
7: AN ← {N1,N2,N3,− − − − −Nn}; // AN is the affected node and Ni is the depended node.
8: printf("%d",AN);
9: **end for**

4.3 *Finding Common or Frequent Patterns Using Association Rule Mining*

After collecting the required data for the system, ARM [7] is used to generate a common pattern or frequent pattern. To generate association rule, first the items are collected and stored in database called as *itemset*. Here items are the affected nodes for a particular modification done in different versions of the system. Several types of modifications can be possible in one system like requirement modification, design modification, functionality modification etc. In this paper, we have considered requirement modification for each version of the system and affected nodes due to the modification are collected. These affected nodes are stored in the historical database. Historical data for last four versions (C1, C2, C3, C4) due to requirement modification are given below.

C1:	GD - {2, 3, 8, 9, 13, 14, 15}
	OD - {3, 4, 5, 8, 9, 15}
C2:	GD - {4, 5, 6, 8, 10, 11, 17, 18, 19}
	OD - {4, 6, 7, 11, 12, 15, 18, 19}
C3:	GD - {6, 7, 8, 10, 12, 14, 15, 16, 19, 20, 21}
	OD - {4, 5, 8, 11, 14, 15, 16, 17, 18, 19, 20, 21, 22}
C4:	GD - {3, 4, 5, 6, 7, 8, 10, 11, 12, 13, 14, 15, 16, 17, 18, 19, 20, 21, 22, 23, 24, 25, 26, 27}
	OD - {4, 5, 6, 7, 10, 14, 16, 20, 21, 22, 26}

These affected nodes are called as itemsets. Then first step of association rule is applied i.e. generation of frequent itemsets from the whole itemset. To generate a frequent itemset, support of individual items and support of combination of items

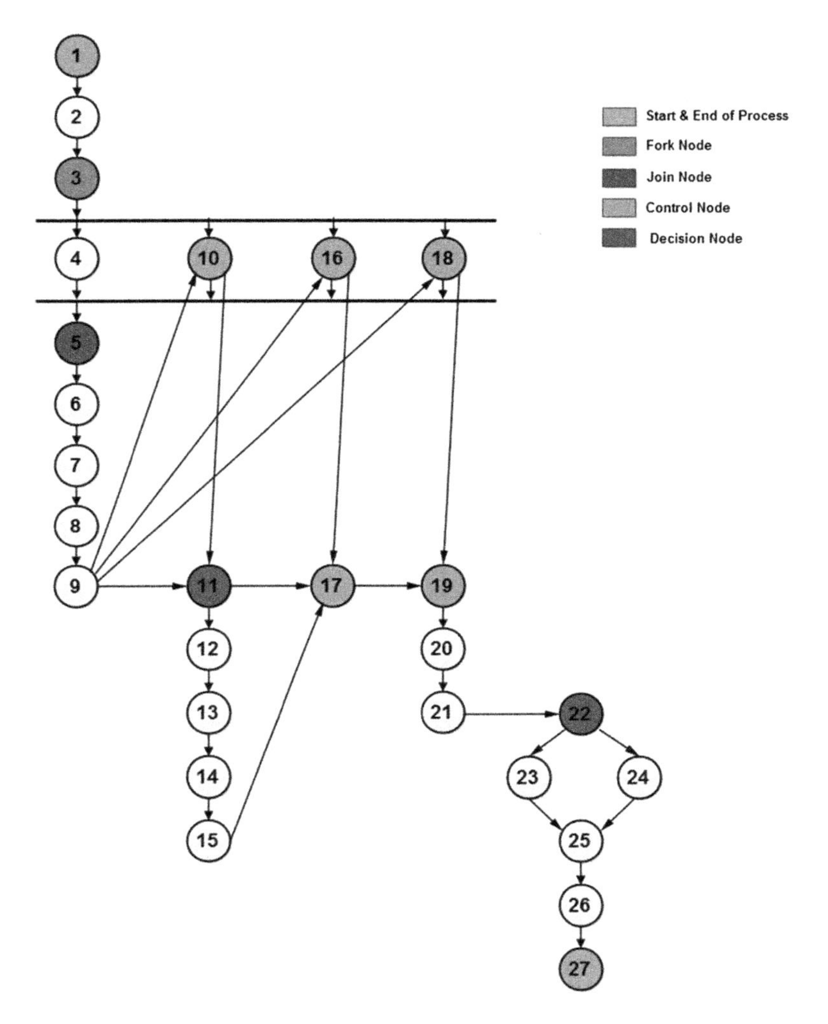

Fig. 3 Activity graph of shopping mall management system

are calculated and compared with user-specified minimum support value. Then the generated frequent pattern is used for calculation of confidence value. After calculating the confidence value for each item, the frequent pattern is compared with user-specified minimum confidence value. The resultant pattern is called as common pattern or frequent pattern for frequently occurring itemset. Here, we have used MATLAB R2012a(7.14.0.739) to implement the itemsets by taking minimum support value as 40 % and minimum confidence value as 85 % and the resultant frequent pattern graph found from the implementation is shown in Fig. 4. The nodes which affect the other nodes maximum time are collected from the frequent pattern graph and the node sequence (let N) is given by

Table 1 Test scenarios of SMMS

Sl no (path)	Test case id	Independent path
1	T1	1→2→3→4→5→6→7→8→9→11→12→13→14→15→17→19→20→21→22 →23→25→26→27
2	T2	1→2→3→4→5→6→7→8→9→11→12→13→14→15→17→19→20→21→22 →24→25→26→27
3	T3	1→2→3→4→5→6→7→8→9→11→17→19→20→21→22→23→25→ 26→27
4	T4	1→2→3→4→5→6→7→8→9→11→17→19→20→21→22→24→25→26→27
5	T5	1→2→3→4→5→6→7→8→9→10→11→12→13→14→15→17→19→20 →21→22→23→25→26→27
⋮	⋮	⋮
37	T37	1→2→3→18→19→20→21→22→23→25→26→27
38	T38	1→2→3→18→19→20→21→22→24→25→26→27

Table 2 Modified nodes and affected nodes

Modified nodes	Affected nodes
2	3,4,5,6,7,8,9,10,11,12,13,14,15,16,17,18,19,20,21,22,23,24,25,26,27
3	4,5,6,7,8,9,10,11,12,13,14,15,16,17,18,19,20,21,22,23,24,25,26,27
4	5,6,7,8,9,11,12,13,14,15,17,19,20,21,22,23,24, 25,26,27
5	6,7,8,9,10,11,12,13,14,15,16,17,18,19,20,21,22,23,24,25,26,27
6	5,7,8,9,10,11,12,13,14,15,16,17,18,19,20,21,22,23,24,25,26,27
7	5,6,8,9,10,11,12,13,14,15,16,17,18,19,20,21,22,23,24,25,26,27
8	5,6,7,9,10,11,12,13,14,15,16,17,18,19,20,21,22,23,24,25,26,27
9	6,7,8,10,11,12,13,14,15,16,17,18,19,20,21,22,23,24,25,26,27
10	6,7,8,9,11,12,13,14,15,16,17,18,19,20,21,22,23,24,25,26,27
15	17,19,20,21,22,23,24,25,26,27
16	5,6,7,9,10,11,12,13,14,15,17,19,20,21,22,23,24,25,26,27
18	5,6,7,9,10,11,12,13,14,15,16,17,19,20,21,22,23,24,25,26,27
19	20,21,22,23,24,25,26,27
26	27

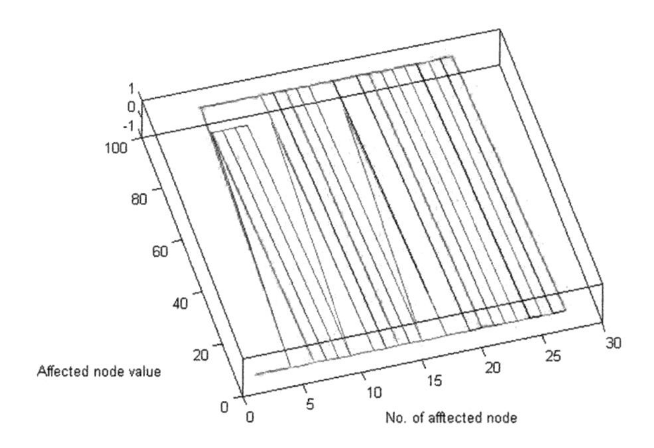

Fig. 4 Frequent pattern graph found for SMMS with mini-support = 40 %, min-confidence = 85 %

N = {6, 7, 8, 9, 11, 12, 13, 14, 15, 16, 17, 18, 19, 20, 21, 22, 23, 24, 25, 26, 27}

The prioritized node sequence (Np) is

Np = {27, 26, 25, 24, 23, 22, 21, 20, 19, 17, 15, 14, 13, 12, 11, 9, 8, 7, 6, 16, 18}

This prioritized node sequence is used for test case prioritization. Test case prioritization is done as per early detection of affected nodes due to a particular modification in the system. The modified information was collected from last few configuration of the system. Hence maximum priority is given to that node which was affected maximum time with maximum confidence value. Here, we have assigned a priority level from 1 to 21 to each node present in Np by giving highest priority to first node of Np and subsequently assigning the priority level to other nodes.

Test case prioritization is done as per the total priority value of test cases. Total priority value of test scenarios is the summation of priority levels of all nodes present in the test scenario (test scenarios are shown in Table 1), given in Eq. 1.

Mathematically, this can be expressed as

$$\text{Total priority value} = \sum_{i=1}^{n} \text{PL}_i \quad (1)$$

where PLi is the total priority level of node present in the test scenario. If node is present in Np then priority level value is consider, otherwise it is zero.

For example, Total Priority value for T1 = 0 + 0 + 0 + 0 + 0 + 3 + 4 + 5 + 6 + 7 + 8 + 9 + 10 + 11 + 12 + 13 + 14 + 15 + 16 + 17 + 19 + 20 + 21 = 210

Similarly the Total Priority value of all test cases are found put. Here the author has implemented this proposed metric/formula to find out the prioritized test suite. Hence the prioritized test suite (T_P) is T24, T23, T32, T2, T6, T12, T16, T31, T1, T5, T11, T15, T20, T19, T26, T25, T34, T4, T14, T18, T33, T3, T13, T17, T28, T27, T8, T7, T22, T36, T10, T21, T35, T9, T30, T29, T38 and T37. The prioritized test suite found using the proposed approach has a higher APFD value as compared to the non-prioritized test cases. APFD value of prioritized and non-prioritized test cases is shown in Fig. 5.

5 Comparison with Related Work

The approaches proposed by Khandai et al. [3], Mahali et al. [10] and Acharya et al. [13] are not efficient enough as far as high risk functions, scalability of the project and frequency of change are concerned. In the real industrial practice, the volume of

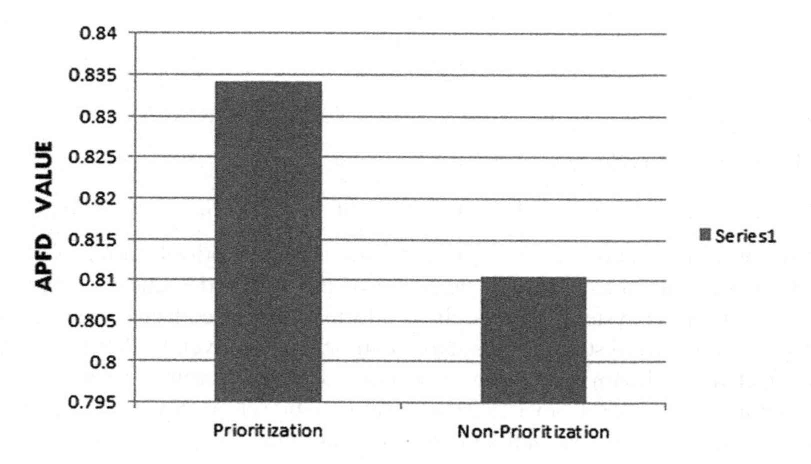

Fig. 5 APFD Value of prioritized and non-prioritized test cases

historical database increases due to increase in complexity of projects and subsequently the number of faults and affected nodes are also increasing. There is no such technique is available which will search the number of modification, faults and affected nodes from the large database and will be able to give highly affected test cases. In our approach, we have taken one attempt to overcome this problem by mining the historical data of projects using ARM and identified a frequent pattern containing highly affected test scenarios.

6 Conclusion and Future Work

In this paper the author has used historical data on the occurrence of past faults to prioritize the test cases. Frequent patterns generated using association rule mining helped effectively in prioritization. This approach gives a better result w.r.t. the average percentage of fault detection. Here the author has only considered the functional requirements of the system. A change during the maintenance phase also affects the non-functional behaviour of the system. Future research can be carried out on prioritizing test cases for testing the non-functional aspects of the system.

References

1. Mall, R.: Fundamental of Software Engineering, 3rd edn. PHI Learning Private Limited, New Delhi (2009)
2. Chauhan, N.: Software Testing Principles: Practices, 3rd edn. Oxford University Press, New Delhi (2010)
3. Khandai, S., Acharya, A.A., Mohapatra, D.P.: Prioritizing test cases using business test criticality value. Int. J. Adv. Comput. Sci. Appl. **3**(5), 103–110 (2011)
4. Askarunisa, A., Shanmugariya, L., Ramaraj, N.: Cost and coverage metrics for measuring the effectiveness of test case prioritization techniques. INFOCOMP J. Comput. Sci. **9**(1), 43–52 (2010)
5. Aggrawal, K.K., Singh, Y., Kaur, A.: Code coverage based technique for prioritizing test cases for regression testing. ACM SIGSOFT Softw. Eng. Notes **29**(5), 1–4 (2004)
6. Srivastava, P.R.: Test case prioritization. J. Theor. Appl. Inf. Technol. 178–181 (2008)
7. Han, J., Kamber, M.: Data Mining: Concepts and Techniques. 2nd edn. Morgan Kaufmann Publishers, San Francisco (2010)
8. Haidry, S.Z., Miller, T.: Using dependency structures for prioritization of fundamental test suites. IEEE Trans. Softw. Eng. **39**(2), 258–275 (2013)
9. Kaushik, N., Salehie, M., Tahvildari, L., Li, S., Moore, M.: Dynamic prioritization in regression testing. In: Fourth International Conference on Software Testing. Verification and Validation Workshops, pp. 135–138 (2011)
10. Malhotra, R., Tiwari, D.: Development of a framework for test case prioritization using genetic algorithm. ACM SIGSOFT Softw. Eng. Notes **38**(3), 1–6 (2013)
11. Yoo, S., Harman, M.: Regression testing, minimisation, selection and prioritisation: a survey. Softw. Test. Verif. Reliab. 1–60 (2007)
12. Mahali, P., Acharya, A.A.: Model based test case prioritization using UML activity diagram and evolutionary algorithm. Int. J. Comput. Sci. Inf. **3**, 42–47 (2013)

13. Acharya, A.A., Budha, G., Panda, N.: A novel approach for test case prioritization using priority level technique. Int. J. Comput. Sci. Inf. Technol. **2**, 1054–1060 (2011)
14. Muthusamy, T., Seetharaman, K.: A new effective test case prioritization for regression testing based on prioritization algorithm. Int. J. Appl. Inf. Syst. **6**(7), 21–26 (2014)
15. Cormen, T.H., Leiserson, C.E., Rivest, R.L., Stein, C.: Introduction to algorithms, 3rd edn. PHI Learning Private Limited, New Delhi (2010)

Rank Based Ant Algorithm for 2D-HP Protein Folding

N. Thilagavathi and T. Amudha

Abstract Bio-inspired computing is the field of studying emergence and social behaviors of social insects. Bio-inspired algorithms are a branch of nature inspired algorithms which are termed as swarm intelligence, focused on insect behavior in order to develop some meta-heuristics which can mimic insect's problem solving abilities. In this Paper Rank Based Ant Colony Optimization (RBACO) was applied to solve 2D-HP protein folding problem. The objective is minimizing the energy to obtain the best 2D structure of protein. The RBACO is one of the bio-inspired algorithms, which could obtain better solutions in 2D-HP protein folding process when compared with other algorithms. The problem instances were taken from the HP benchmarks. The 2D-HP lattice model was used for implementation with HP benchmark instances for 2D-HP protein folding.

Keywords Bio-inspired algorithms · 2D-HP protein folding · Rank based ACO · Misfolding of proteins

1 Introduction

Biologically Inspired Computing also known as Bio-inspired computing (BIC) is an exciting and most popular field in research. Bio-inspired computing is a major subset of Nature Inspired Computing (NIC). BIC is a field, devoted to tackling complex problems using computational methods modeled after design principles encountered in nature [1, 4]. The functioning of the social insects such as Ants, Bees, Termites, Wasps, Birds and school of fishes are inspired the scientist's to design advanced problem solving techniques [20]. 2D-HP protein folding model

N. Thilagavathi (✉) · T. Amudha
Department of Computer Applications, Bharathiar University,
641 046 Coimbatore, Tamil Nadu, India
e-mail: thilagamca86@gmail.com

T. Amudha
e-mail: amudhaswamynathan@buc.edu.in

L.C. Jain et al. (eds.), *Computational Intelligence in Data Mining - Volume 3*,
Smart Innovation, Systems and Technologies 33, DOI 10.1007/978-81-322-2202-6_40

was introduced by Lau and Dill at 1989. The HP model starts with classifying the twenty amino acids as H (hydrophobic or non-polar) or P (hydrophilic or polar). For the HP model, an approximation corresponds to obtaining a folding in which the number of H–H contacts is at least a fraction of that in the optimal, since the energy is inversely proportional to the number of H–H contacts [6].

2 Protein Folding Problem

Lattice Protein Folding has two types, on lattice and off lattice. Proteins are sequenced as H (Hydrophobic) and P (Polar) monomers configured in the 3D simple cubic lattice. In a lattice, each sequence pattern is self-avoiding walk and each monomer occupies one lattice site connected to its chain neighbors. Links between H–H monomers are constructive. The native condition of a HP sequence is defined as the set of conformations with the biggest possible number of H–H contacts [6]. The 2D-HP model consists of 2D square and 3D-HP model consists of 3D cube lattice. The energy E of a sequence pattern is determined by the number of H–H contacts $E(c) = n(-1)$. In the equation, c is conformation and n is number of H–H contacts [2, 6]. Most lattice models apply to the folding process up to the final stage (which involves side-chain packing), whereas the molecular dynamics simulations yield a detailed picture of the initial stages of unfolding up to the molten globule state [7, 8].

2.1 2D-HP Model

2D-HP protein folding model was introduced by Lau and Dill at 1989.The HP model starts with classifying the twenty amino acids as H (hydrophobic or non-polar) or P (hydrophilic or polar). For the HP model, an approximation corresponds to obtaining a folding in which the number of H–H contacts is at least a fraction of that in the optimal, since the energy is inversely proportional to the number of H–H contacts (Fig. 1) [7]. In Hydrophobic-polar model all amino acids are classified as either hydrophobic or polar and the folding of a protein chain is defined as a *self avoiding walk* in a 2 to 3-dimensional lattice. This model imitates the hydrophobic effect by conveying a negative (or favorable) weight to interactions between adjacent, non-covalently bound hydrophobic amino acids. And proteins that have minimum free energy are assumed to be in their native state.

2.2 Self Avoiding Walks

Self avoiding walks were first introduced by the chemist Paul Flory. Self avoiding walk is a series of moves on a lattice or a grid that does not visit the same point more than once. This is because the components of such things cannot live in the

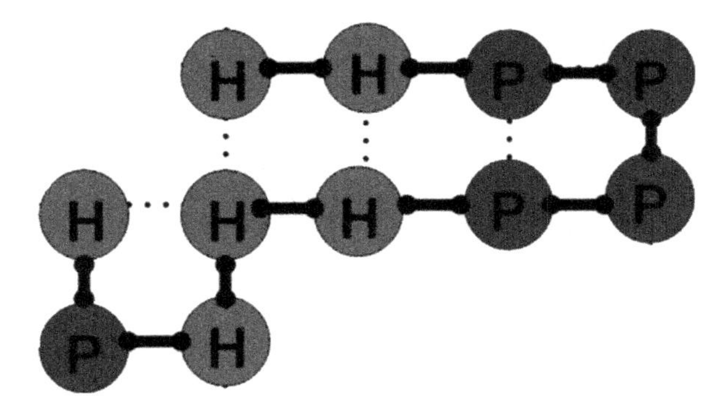

Fig. 1 Protein 2D-HP model [23]

same spatial point. In computing a self-avoiding walk is a sequence like structure in 2D or 3D with a certain number of nodes that typically have a fixed step length and never crosses itself or another walk. They act a lot like random walks.

2.3 3D-HP Model

3D-HP model (Fig. 2) consists of 3D cube lattice. 3D-HP protein model is designed by a common back bone or side chain of a Hydrophobic or Hydrophilic. 3D-HP model is decoded with different letters or symbols. The final conformation is called 3D model, which has minimum energy. Many bio-inspired algorithms are used in the 3D-HP model [19].

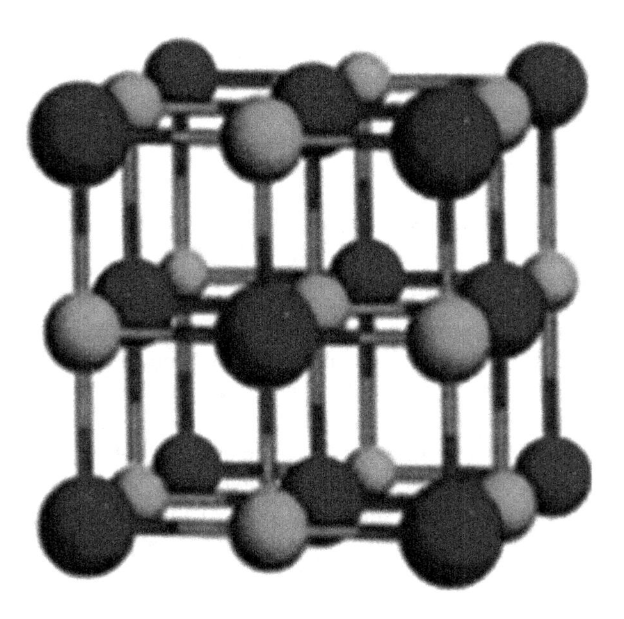

Fig. 2 Protein 3D HP model [23]

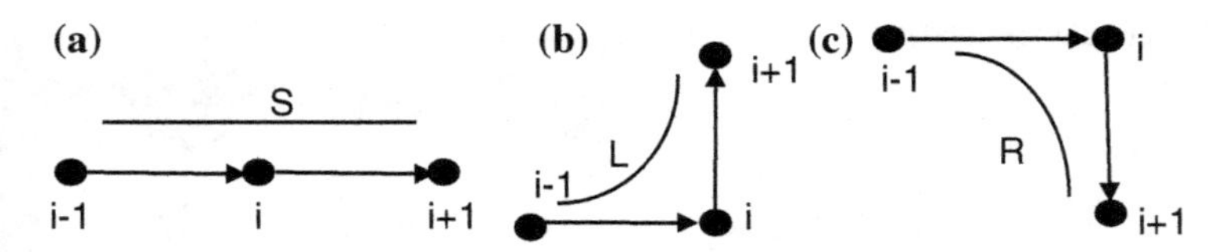

Fig. 3 **a** Straight [18]. **b** Left [18]. **c** Right [18]

2.4 Protein Encoding

Two other isomorphic encoding strategies for HP Lattice models are available, one is relative encoding and another one is absolute encoding. In relative encoding the move direction is relative to the direction of the previous move. In absolute encoding the sequence is replaced by numeric or letters to represent the direction [6]. In 2D square lattice moves the alpha representations are straight denoted by S, Left denoted by L, and Right denoted by R as shown in Fig. 3a–c.

2.5 Energy Calculation

Energy is calculated from the native state. The native state is final Conformation state which has minimal energy. In the HP model, the energy of a conformation is defined as a number of topological contacts between hydrophobic amino acid. Specifically a conformation c with exactly n such H–H contacts has free energy. In this equation n is number of H–H Contacts.

$$E(c) = n \cdot (-1) \tag{1}$$

E Energy, c Conformation state, n Number of H–H contacts

3 Bio-inspired Algorithm for Protein Folding

Ant Colony Optimization (ACO) is a Bio-inspired technique for solving combinatorial optimization problems. ACO takes inspiration from the foraging behavior of ant colony. In the real world, ants (initially) walk randomly, and upon finding food return to their colony while laying down pheromone trails. The ant solution was applied for protein folding problem. The Rank Based Ant Colony Optimization (RBACO) was used to solve the Protein Folding Problem. RBACO is one of the variant of ACO. The RBACO gave better solution than other algorithms in protein folding problem. Ants construct feasible solutions and the pheromone trails are updated. At each step ants compute a set of feasible moves and select out the rest of the tour. The transition probability is based on the heuristic information and pheromone trail level of the move. In the beginning stage the pheromone level is set to a small positive constant value and then the ants update pheromone values after completing the construction stage. The pheromone updations have two stages

i. Local Update
ii. Global Update.

3.1 Local Update

While ants construct their solutions, at the same time they locally update the pheromone level of the visited paths by applying the local update rule. The aim of the local update rule is to make better use of the pheromone information by dynamically changing the desirability of edges.

$$\tau_{ij} \leftarrow (1-\rho)\tau_{ij} + \rho\tau_0 \tag{2}$$

where 'τ_{ij}' is an amount of the pheromone on the arc (i, j) of the 2D square lattice model, 'ρ' is the persistence of the trail and the term $(1-\rho)$ can be interpreted as trail evaporation. 'τ_0' is the pheromone initialization value, set as a small positive constant value.

3.2 Global Update

Global updating is performed after all ants have completed their tours. The pheromone level will be reduced and this will reduce the chance that the other ants will select the same solution and consequently the search will be more diversified.

$$\tau_{ij} \leftarrow (1-\rho)\tau_{ij} + \Delta\tau_{ij} \tag{3}$$

$$where \quad \Delta\tau_{ij} = \begin{cases} -E_{gb} & if(i,j)\in bestsolution \\ 0 & Otherwise \end{cases}$$

'E_{gb}' is the global energy of the best folding. The global update rule is used to provide a greater amount of pheromone on the paths of the best solution.

3.3 Probability to Select Next Move

There are four possible moves in 2D-HP Lattice model. During the construction phase, ants fold a protein from the left end of the sequence adding one amino acid at a time based on the two sources of information; pheromone matrix value and heuristic information. The heuristic values are calculated for each residue before it is placed on the lattice. The Maximum heuristic value for each position is 4. In 2D-HP model the positions are represented as L (Left), R (Right), U (Up) and D (Down).

$$p_{ij} \leftarrow \frac{\tau_{ij}^{\alpha} \eta_{ij}^{\beta}}{\sum_{k \in Unused} \tau_{ik}^{\alpha} \eta_{ik}^{\beta}} \tag{4}$$

The 'τ_{ij}' is the intensity of the pheromone deposited by each ant on the path (i, j). The 'i' is the start position and 'j' is the next movable position, 'α' is the intensity control parameter, 'η_{ij}' is the heuristic information equal to the number of new H–H contacts if the position 'j' is chosen, 'β' is the heuristic parameter. When the next amino acid is polar, the Probability $P_{ij} = 0$.

The Ant Colony Optimization Pseudo code starts with initialization. The initialization has size of population, pheromone value, 'α', 'β' and 'ρ' values. The 'α' is intensity control parameter, 'β' is the heuristic parameter and 'ρ' is the persistence of the trail. The stopping criterion for this folding process is to occupy the unique lattice point by each residue in the given sequence. The starting position for the sequence is selected randomly. Each ant gives unique structure with different energy. At the end of each iteration, the pheromone values are updated locally. The iterations will vary based upon the optimal solution. The best optimal solution is identified after completion of all iterations. The best energy path is updated by the global pheromone updating rule. The energy is calculated, after all iterations are completed.

Pseudocode for RBACO

```
[1]   foreach colony do
[2]   foreach ant do
[3]          generate route
[4]          evaluate route
[5]end for
[6]verify for global or local optimum
[7]   evaporate pheromone in all trails
[8]   generate an elite from current colony
[9]   deposit a Y amount of pheromone on elite trails
[10]  deposit an X amount of pheromone on best   global route
[11]  end for
```

AS_{rank} is a variant of Ant Colony Optimization (ACO) that was proposed by Bullnheimer in 1999. The general procedures of AS_{rank} are similar to Ant System (AS). AS_{rank} draws on the concept of ranking and adopts it into the pheromone update procedure. Rank is defined by the quality of the solution compared to solutions found by other ants in the same iteration [20]. In the Rank Based Ant System algorithm each ant deposits a quantity of pheromone that decreases with rank. The 'n' ants are ranked according to the quality of their solutions in decreasing order. The amount of pheromone deposited is then weighted for each solution, such that solutions with shorter paths deposit more pheromone than the solutions with longer paths.

$$\tau_{ij} \leftarrow \tau_{ij} + \sum_{r=1}^{w-1} (w - \mathbf{r}) \Delta\tau_{ij}^{r} + w\Delta\tau_{ij}^{bs} \tag{5}$$

4 Implementation Results and Discussion

Benchmark Problems are gathered to test the robustness of the algorithm. In the domain of 2D-HP protein folding, there are several famous 2D-HP benchmarks, which are often utilized to test the algorithms. In this research work, the 2D-HP problem instances are taken from the [21, 22], which are used to test the efficiency of proposed Rank Based Ant Colony Optimization (RBACO).

In 2D-HP protein folding problem, the energy calculation is very important to finalize the optimal structure. The energy is calculated after finding the confirmation state or final state. The energy function is defined by number of Hydrophobic (H) nodes, which are adjacent in the 2D-HP lattice and not consecutive in the chain. The path designation implies that the conformation is both connected and self-avoiding, i.e., no amino acids can collide in the same cell of the lattice. The objective is to find a confirmation state that minimizes the total energy of the given amino acid sequence, which therefore corresponds to maximizing the number of H–H contacts. The given two diagrams are represented for a single sequence with different structure and energy. The energy got minimized when the number of H–H contacts is maximized.

The movements are encoded in several formats, like symbols, numbers or alphabets. Sequences have a number of possibility structures,

$$Possibility\ structure \leftarrow n^4 \tag{6}$$

n—Sequence Length, 4—Possible movements in square lattice

In Table 1 the values of parameters used for implementing this research work are listed. Parameters are listed with symbol, range from minimum to maximum and description for the parameter. The parameter values vary for each sequence to get

Table 1 Parameter settings for proposed RBACO

Parameter	Value	Description
α	7–8	Intensity control parameter
β	0.3–0.4	Heuristic parameter
ρ	0.3–0.5	Persistence of the trail
m	100–500	Number of ants in RBACO
τ_0	0.5	Pheromone initialization

Table 2 Comparison of RBACO results with benchmark optimum

S. no	Sequence length	Sequence	Optimal energy	RBACO energy
1	20	HHHPPHPHPHPPHPHPHPPH	**−10**	**−10**
2	50	HHPHPHPHPHHHHPHPPPHPPPHPPPPHPPPHPPPHPHHHHPHPHPHPHH	**−21**	**−21**
3	60	PPHHHPHHHHHHHHPPPHHHHHHHHHHPHPPPHHHHHHHHHHHHPPPPHHHHHHPHHPHP	**−36**	**−36**
4	64	HHHHHHHHHHHHPHPH PPHHPPHHPPHPPHHPPHHPPHPPHHPPHHPPHPHPHHHHHHHHHHHH	**−42**	**−42**
5	85	HHHHPPPPHHHHHHHHHHHHPPPPPPHHHHHHHHHHHHPPPHHHHHHHHHHHHPPPHHH HHHHHHHHHPPPHPPHHPPHHPPHPH	**−53**	**−50**
6	100	PPPPPPHPHHPPPPPHHHPHHHHHPHHPPPPHHPPHHPHHHHHPHHHHHHHHHHPHHPH HHHHHHPPPPPPPPPPPHHHHHHHPPHPHHHPPPPPPHPHH	**−48**	**−46**
7	100	PPPHHPPHHHHPPHHHPHHPHHPHHHHPPPPPPPPHHHHHHPPHHHHHHPPPPPPPPPHPHHP HHHHHHHHHHHPPHHHPHHPHPPHPHHHPPPPPPHHH	**−50**	**−48**

Table 3 Performance comparison of RBACO with other algorithms

S. no.	Sequence length	Optimal energy	Proposed RBACO	ACO* [13]	ABC* [18]	TS* [10]	CI* [9]	IA* [17]	GA* [12]	MC* [12]	EMC* [3]	BB* [11]
1	18	**-4**	**-4**	-	-	-	-	-	-	-	-	-
2	18	**-8**	**-8**	-	-	-	-	-	-	-	-	-
3	18	**-9**	**-9**	-	-	-	-	-	-	-	-	-
4	20	**-9**	**-9**	**-9**	**-9**	**-9**	**-9**	**-9**	**-9**	**-9**	**-9**	**-9**
5	20	**-10**	**-10**	**-10**	-	-	-	**-10**	-	-	-	-
6	24	**-9**	**-9**	**-9**	-	**-9**	**-9**	**-9**	**-9**	**-9**	**-9**	**-9**
7	25	**-8**	**-8**	**-8**	-	**-8**	**-8**	**-8**	**-8**	-7	**-8**	**-8**
8	36	**-14**	**-14**	**-14**	**-14**	**-14**	**-14**	**-14**	**-14**	-12	**-14**	**-14**
9	48	**-23**	**-23**	**-23**	**-23**	**-23**	**-23**	**-23**	**-23**	-18	**-23**	-22
10	50	**-21**	**-21**	**-21**	-	**-21**	**-21**	**-21**	**-21**	-19	**-21**	-21
11	60	**-36**	**-36**	**-36**	-	-	-35	-35	-34	-31	-35	-34
12	64	**-42**	**-42**	**-42**	-40	**-42**	-40	-39	-37	-	-39	-38
13	85	**-53**	-50	-51	-	-	-	-	-	-	-52	-52
14	100	**-48**	-46	-47	-	-	-	-	-	-	-	-
15	100	**-50**	-48	-47	-	-	-	-	-	-	-	-48

Note cells marked with '–' denotes that no results could be obtained from literature

* *ACO* ant colony optimization, *ABC* artificial bee colony, *TS* tabu search, *CI* contact interaction, *IA* immune algorithm, *GA* genetic algorithm, *MC* Monte Carlo, *EMC* evolutionary Monte Carlo, *BB* branch and bound

optimal structure. As a general rule, the best results were found using high 'α' values, low '$\boldsymbol{\beta}$' values, '$\boldsymbol{\rho}$' values between 0.3 and 0.5 and 'm' between 100 and 500, depending on sequence length. '$\boldsymbol{\tau_0}$' is used for pheromone initialization.

From the Table 2, it was clearly understandable that proposed RBACO algorithm gave best optimal solutions for all sequences except sequences 5–7. The longer sequences have taken more space to execute the implementation.

Table 3 has shown the performance comparison of the proposed algorithm RBACO with other algorithms used for protein folding. The proposed RBACO was found to be competitive and produced optimal protein conformations in most of the benchmark problems taken for this research work, except few sequences, compared to other algorithms.

5 Conclusion and Future Work

In this research work, the Rank Based Ant Colony Optimization (RBACO) algorithm was applied to 2D-HP protein folding to solve the problem in an efficient way. The problem instances for testing the algorithm were taken from the standard HP Benchmarks in the literature. It was proven that the RBACO algorithm could obtain the best-so-far optimal solutions (minimal energy) for most of the problem cases and the resultant energy was also compared with the energy obtained by other algorithms. In future work, Bio-inspired algorithms can be applied to perform 3D protein folding. In this research work the RBACO was applied for 2D-HP benchmarks; in future, RBACO can be applied for real proteins to predict the structure and to analyze the protein misfolding diseases. The longer protein sequences used in this research have occupied more space during execution, and hence space complexity was identified as a major limitation of this research which is a critical issue and should be addressed in the future. This research has obtained promising results with protein folding whereas structure visualization is to be carried out as future work.

References

1. Alazzam, A., Lewis, H.W. III: A new optimization algorithm for combinatorial problems. (IJARAI) Int. J. Adv. Res. Artif. Intell. **2**(5) (2013)
2. Soto, C.: Protein misfolding and disease; protein refolding and therapy. Elsevier Sci. **498**, 204–207 (2001)
3. Liang, F., Wong, W.H.: Evolutionary Monte Carlo for protein folding simulations. J. Chem. Phy. **115**(7), 3374–3380 (2001)
4. Forbes, N.: Biologically inspired computing. Comput. Sci. Eng. **2**(6), 83–87 (2000)
5. Chakravarthy, H., Proch, P.B., Rajan, R., Chandrasekharan, K.: Bio inspired approach as a problem solving technique. Netw. Complex Syst. **2**(2) (2012)

6. Yue, K., Fiebig, K.M., Thomast, P.D., Chan, H.S., Shakhnovicht, E.I., Dill, K.A.: A test of lattice protein folding algorithms. Pharm. Chem. **92**, 325–329 (1995)
7. Lau, K., Dill, K.A.: A lattice statistical mechanics model of the conformation and sequence spaces of proteins. Macromolecules **22**, 3986–3997 (1989)
8. Bianchi, L., Dorigo, M., Gambardella, L.M., Gutjahr, W.J.: A survey on metaheuristics for stochastic combinatorial optimization. Nat. Comput. **8**, 239–287 (2009)
9. Toma, L., Toma, S.: Contact interactions method: a new algorithm for protein folding simulations. Protein Sci. **5**, 147–153 (1996)
10. Milostan, M., Lukasiak, P., Dill, K.A., Blazewicz J.: A tabu search strategy for finding low energy structures of proteins in HP-model. In: Proceedings of the Annual Internatinal Conference on Computational Molecular Biology. Berlin. Poster No. 5-108. (2003)
11. Chen, M., Huang, W.Q.: A branch and bound algorithm for the protein folding problem in the HP lattice model. Geno. Prot. Bioinfo. **3**(4) (2005)
12. Unger, R., Moult, J.: Genetic algorithms for protein folding simulations. J. Mol. Biol. **231**, 75–81 (1993)
13. Shmygelska, A., Hoos, H.H.: An improved ant colony optimization algorithm for the 2D HP protein folding problem. In: Proceedings of the 16th Canadian Conference on Artificial Intelligence. LNCS 2671. Springer (2003) 400–417
14. Fidanova, S., Lirkov, I.: Ant colony system approach for protein folding. In: Proceedings of the International Multiconference on Computer Science and Information Technology. (2008) pp. 887–891, ISSN:1896-7094
15. Thalheim, T., Merkle, D., Middendorf, M.: Protein folding in the HP-model solved with a hybrid population based ACO algorithm. IJCS. (2008)
16. Mohan, U.: Bio Inspired Computing. Division of computer science SOE. CUSAT. (2008)
17. Cutello, V., Nicosia, G., Pavone, M., Timmis, J.: An immune algorithm for protein structure prediction on lattice models. IEEE Trans. Evol. Comput. **11**(1) (2007)
18. Zhang, Y., Wu, L.: Artificial bee colony for two dimensional protein folding. Adv. Electr. Eng. Syst. **1**(1) (2012)
19. Zhang, Y., Skolnick, J.: Tertiary structure predictions on a comprehensive benchmark of medium to large size proteins. Biophys. J. **87**, 2647–2655 (2004)
20. http://www.wisegeek.com/what-is-bio-inspired-computing.htm
21. http://www.cs.sandia.gov/tech_reports/compbio/tortilla-hp-benchmarks.html
22. http://www.tamps.cinvestav.mx/~mgarza/HPmodel/HPBenchmarks.pdf
23. http://cpsp.informatik.uni-freiburg.de:8080/faq.jsp

Boolean Operations on Free Form Shapes in a Level Set Framework

V.R. Bindu and K.N. Ramachandran Nair

Abstract Shape modeling is an important constituent of both computer graphics and computer vision research. This paper proposes an efficient shape modeling scheme using the active geometric deformable models, for implementing solid modeling techniques on free form shapes. A fast and computationally efficient narrow band level set algorithm is proposed for reducing the overall computational cost. When compared to traditional geometric modeling, the proposed level set model can easily handle topology changes, is free of edge connectivity and mesh quality problems and provides the advantages of implicit models, supporting straightforward solid modeling operations on complex structures of unknown topology. Using this model boolean operations have been implemented on arbitrary shapes extracted from both synthetic and real image data including some low contrast medical images, with promising results.

Keywords Shape modeling · Level sets · Narrow band · Boolean operations · Geometric deformable models

1 Introduction

Solid modeling technology has significantly outgrown its original scope of computer aided mechanical design and manufacturing automation. Nowadays, it plays an important role in many domains including medical imaging and therapy planning, architecture and construction, animation and digital video production for

V.R. Bindu (✉)
School of Computer Sciences, Mahatma Gandhi University, Kottayam, Kerala, India
e-mail: binduvr@mgu.ac.in

K.N. Ramachandran Nair
Department of Computer Science and Engineering, Viswajyothi College of Engineering, Muvattupuzha, Kerala, India
e-mail: knrn@hotmail.com

L.C. Jain et al. (eds.), *Computational Intelligence in Data Mining - Volume 3*,
Smart Innovation, Systems and Technologies 33, DOI 10.1007/978-81-322-2202-6_41

entertainment and advertising. With the advancement of scanning and imaging technology, an increased amount of sampled biological data is being generated. The data come from sources like Magnetic Resonance Imaging (MRI), Computed Tomography (CT) etc. In this context, it is appropriate to have some mechanism for accurately extracting and manipulating desired shapes from medical imagery. We have developed a level set framework for implementing solid modeling operations on scanned volume data sets. The proposed model can be effectively used for the analysis and extraction of information from biomedical images and further computations of geometric measurements such as area, volume, intensity of a detected contour, surface or region respectively. These information can be later used to study the evolution in time of a growing tumour or to compare two different subjects, usually a normal one and an abnormal one.

The application of geometric deformable models has brought tremendous impact to medical imagery due to its capability of topology preservation and fast shape recovery. Geometric active contours implemented as level set methods have been addressing a wide range of problems in image segmentation and computer vision. Level set methods were first introduced by Osher and Sethian [20] for tracking interfaces moving under complex motions. They offer a highly robust and accurate method for capturing moving fronts, work in any number of space dimensions and handle topological merging and breaking naturally.

Active contour models or snakes were introduced by Kass, Witkins and Terzopoulos [11] for addressing the domain of image segmentation problems using dynamic curves. Based on the representation and implementation, the active contour models can be either parametric active contour models or geometric active contour models. Parametric active contours [8, 11] are represented explicitly as parameterized curves in a Lagrangian formulation. Geometric active contours were introduced more recently and has many advantages over the former, such as computational simplicity (the level set function always remains a function on a fixed grid that allows efficient numerical schemes) and the ability to accommodate changes in the topology during curve evolution (the contours represented by the level set function may break or merge naturally during the evolution, thus automatically handling the topological changes). They are represented implicitly as level sets and were introduced by Caselles et al. [3, 4] and Malladi et al. [14]. These early models, based on curve evolution theory [12] implemented via level set techniques [18, 19] are typically derived using a Lagrangian formulation that yields a certain evolution partial differential equation (PDE) of a parameterized curve. This PDE is then converted to an evolution PDE for a level set function using the related Eulerian formulation from level set methods. At any time during evolution, contours are represented by the zero level set of the level set function, which is an implicit function defined in a higher dimension. An alternative is to define a certain energy functional on the level set function and then directly derive the evolution PDE of the level set function from the problem of minimizing this energy functional. These variational level set methods [5, 13, 25, 28] are more convenient and natural for incorporating additional information, such as region-based information [5, 13] and shape-prior information [25], into energy functionals that are directly formulated in the level set domain. These methods produce more robust results and have larger convergence range. They allow flexible initialization of the level set function and can also eliminate the costly re-initialization procedure [13].

Level set methods as computational techniques have been widely used in solving many scientific and engineering problems. Level set methods have been extensively used in medical image segmentation. However, in the conventional straight forward approach, for evolving the level set function Ø, the initial value partial differential equation is solved in the entire computational domain. This together with the frequent need for reinitialization of Ø make the overall computation complex and expensive. One alternative is the Narrow Band approach, where the value of Ø is updated only within a specified narrow band close to the zero level set.

The Narrow Band Method was introduced by Chopp [6] and used in recovering shapes from images by Malladi, Sethian and Vemuri [14]. These methods were extensively analysed by many [1] and several variants were implemented. A fairly wide narrow band of width twelve grid points was suggested as a reasonable balance between reinitialization costs and update costs, by Sethian [24] for standard narrow band implementation. The narrow band algorithms suggested by [14] and [26] use narrower bands but still require signed distance reinitialization and upwind finite difference scheme. Many other approaches exist [10, 21, 22] that target a variety of level set segmentation problems on unstructured point sets. Some of the recent works [15, 17] combine the idea of narrow band with different models of computational grids for handling high resolution level sets.

Level set methods have been successfully applied in computer graphics, computer vision and visualization [18, 24] specifically in areas like medical image segmentation [14, 27]. There exists little work utilizing level set methods for geometric modeling. The most efficient and prominent among these is the set of algorithms developed by Museth et al. [16] for interactive editing of level set models. This work is claimed to be the first to utilize level set methods to perform a variety of interactive editing operations on complex geometric models. In this system every object is represented as a signed distance volume, which is a volume data set where each voxel stores the shortest distance to the surface of the object being represented by the volume.

Our work differs from these earlier works in many ways. In order to reduce the overall computational overhead, our proposed level set framework uses the variational narrow band formulation presented in Sect. 3. Since our approach does not require the level set function to be initialized to a signed distance function, the complicated calculations involved in finding the signed distance function is completely avoided. A significantly larger time step can be used for numerically solving the evolving partial differential equation which speeds up the curve evolution. Moreover, the computational efficiency is further enhanced by avoiding the re-initialization step and complex upwind finite difference schemes.

In this paper, we present an efficient shape modeling scheme for implementing solid modeling techniques [23] on free form shapes set in a level set frame work implemented using a fast narrow band level set algorithm. Since level set models are volumetric, constructive solid geometry operations of union, difference and intersection may be applied to them [2, 16, 23, 24]. For extracting shapes from various types of image data, the proposed narrow band algorithm is applied on a variational level set approach without re-initialization. The following sections deal with the level set formulations, the proposed model and its implementation, and the experimental results showing the performance of this model.

2 Standard Level Set Formulation

The basis of level set methods is to view a moving interface as the zero level set of a time-dependent level set function Ø. The evolution of this function Ø is linked to the propagation of the front itself through a time-dependent initial value problem. An equation of the motion for this level set function Ø is derived as follows [20].

In order to match the zero level set of ϕ with the evolving front, the level set value of a particle on the front with trajectory $x(t)$ must be zero. Hence

$$\phi(x(t), t) = 0$$

The speed of the particle

$$\frac{\partial x}{\partial t}$$

in the direction n normal to the level set is given by the speed function F.

So

$$\mathrm{F} = \frac{\partial x}{\partial t} \cdot n$$

where the normal vector $n = \frac{\nabla \phi}{|\nabla \phi|}$.

By the chain rule,

$$\phi t + \frac{\partial \boldsymbol{x}}{\partial \boldsymbol{t}} \cdot \nabla \phi = 0$$

This gives an evolution equation for ϕ

$$\phi t + F|\nabla \phi| = 0 \tag{1}$$

given $\phi(x, t = 0)$.

This is the level set equation given by Osher and Sethian [20].

There are several advantages to this level set perspective. The evolving function Ø remains a function as long as the speed function F is smooth. The surface corresponding to the propagating hypersurface, may change topology, merge, break or form sharp corners. Finite differences lead to a numerical scheme to approximate the solution, and intrinsic geometric properties are easily determined from the level set function. The key in level set methods is to approximate the gradient in the level set equation in a way that satisfies the correct entropy condition. The formulation has no significant differences for propagating interfaces in three dimensions.

The initial value level set formulation [24] allows for both positive and negative speed functions F. The front can move forward and backward as it evolves. The speed function F may be defined according to the area of application. For example, while applying the level set methods to problems in the geometric evolution of curves and surfaces, the motion depends on geometric properties such as normal direction and curvature.

Since their introduction, level set techniques have been used in a wide collection of problems [24] involving moving interfaces, the generation of minimal surfaces, singularities and geodesics in moving curves and surfaces, flame propagation, etching, deposition and lithography calculations, crystal growth, and grid generation. In recent years, level set methods have been used in a variety of settings for problems in computer vision, graphics and image processing [8, 15, 16, 19].

2.1 Reinitializing the Level Set Function

Reinitialization, used with the traditional level set methods [4, 14, 24] is a technique for periodically reinitializing the level set function to a signed distance function during the evolution. In many situations, the level set function will develop steep or flat gradients leading to problems in numerical approximations. It is then needed to reshape the level set function to a more useful form, while keeping the zero location unchanged. One way to do this is to perform distance reinitialization.

But from a practical viewpoint, reinitialization of the level set function is obviously a disagreement between the theory of the level set method and its implementation [7]. Many proposed reinitialization schemes have an undesirable side effect of moving the zero level set away from its original location.

2.2 Variational Level Set Formulation Without Reinitialization

While implementing the traditional level set methods, it is important to keep the evolving level set function as an approximate signed distance function during the evolution and a signed distance function [19] must satisfy a desirable property of $|\nabla \phi| = 1$. As a numerical remedy to ensure stable curve evolution, reinitialization, a technique for periodically reinitializing the level set function to a signed distance function is commonly used. But as mentioned earlier, reinitialization poses many problems [7]. More than being computationally expensive and complicated, the procedure has many side effects on evolution of the curve.

In view of the above facts, we have adopted a variational level set formulation [13], as our basic formulation for level set evolution that can be easily implemented by simple finite difference scheme, without the need for reinitialization. This variational formulation of the level set function consists of an internal energy term and an external energy term, respectively. The internal energy term penalizes the deviation of the level set function from a signed distance function, whereas the external energy term drives the motion of the zero level set to the desired image features such as object boundaries. The formulation is given by

$$\varepsilon(\phi) = \mu\rho(\phi) + \varepsilon_m(\phi) \tag{2}$$

where $\mu > 0$ is a parameter that controls the effect of penalizing the deviation of Ø from a signed distance function and $\varepsilon_m(\phi)$ is the energy that would drive the motion of the zero level curve of Ø. The integral

$$\rho(\phi) = \int_{\Omega} \frac{1}{2}(|\nabla\phi| - 1)^2 dxdy \tag{3}$$

is a metric that characterizes how close a function Ø is to a signed distance function in the image domain. The evolution equation

$$\frac{\partial\phi}{\partial t} = -\frac{\partial\varepsilon}{\partial\phi} \tag{4}$$

is the gradient flow that minimizes the functional ε. During the evolution of Ø (4) that minimizes the functional (2), the zero level curve will be moved by the external energy ε_m towards the object boundaries. For achieving this, the external energy is defined as

$$\varepsilon_{g,\lambda,\alpha}(\phi) = \lambda L(\phi) + \alpha A(\phi) \tag{5}$$

where g is the edge indicator function defined by

$$g = \frac{1}{1 + |\nabla(G_\sigma * I)|}$$

G_σ is the Gaussian kernel with standard deviation σ and I, the image. In (5), $\lambda > 0$ and α are constants and the terms $L(\phi)$ and $A(\phi)$ are defined by

$$L(\phi) = \int_{\Omega} g\delta(\phi)|\nabla\phi|dxdy \tag{6}$$

and

$$A(\phi) = \int_{\Omega} gH(-\phi)dxdy \tag{7}$$

respectively, where δ is the univariate Dirac function, and H is the Heaviside function. The total energy functional is defined as

$$\varepsilon(\phi) = \mu\rho(\phi) + \varepsilon_{g,\lambda,\alpha}(\phi) \tag{8}$$

The external energy $\varepsilon_{g,\lambda,\alpha}$ drives the zero level set towards the object boundaries, while the internal energy $\mu\rho(\phi)$ penalizes the deviation of ϕ from a signed distance function during its evolution.

The Gateaux derivative [9] of the functional ε in (8) can be written as

$$\frac{\partial \varepsilon}{\partial \phi} = -\left[\Delta\phi - div\left(\frac{\nabla\phi}{|\nabla\phi|}\right)\right] - \lambda\delta(\phi)div\left(g\frac{\nabla\phi}{|\nabla\phi|}\right) - \alpha g\delta(\phi) \tag{9}$$

where Δ is the Laplacian operator. Therefore, Ø satisfies the Euler-Lagrange equation $\frac{\partial\varepsilon}{\partial\phi} = 0$. Hence the gradient flow

$$\frac{\partial \phi}{\partial t} = \mu\left[\Delta\phi - div\left(\frac{\nabla\phi}{|\nabla\phi|}\right)\right] + \lambda\delta(\phi)div\left(g\frac{\nabla\phi}{|\nabla\phi|}\right) + \alpha g\delta(\phi) \tag{10}$$

gives the evolution equation of the level set function. All the spatial partial derivatives $\frac{\partial\phi}{\partial x}$ and $\frac{\partial\phi}{\partial y}$ are approximated by the central difference and the temporal partial derivative $\frac{\partial\phi}{\partial t}$ is approximated by the forward difference. The difference equation is expressed as the iteration

$$\phi_{i,j}^{k+1} = \phi_{i,j}^{k} + \tau G(\phi_{i,j}^{k}) \tag{11}$$

where $G(\phi_{i,j})$ is the approximation of the right hand side in (10).

3 Narrow Band Approach

While dealing with volumetric representations, the amount of computation time and memory requirements needed to process level set models may become considerably high. Techniques have been developed such as the narrow band approach to accelerate the level set computations by limiting computations to only a narrow band around the level set of interest [24]. The central idea of the narrow band method is to solve the PDE only in the neighbourhood of the tracked level set instead of the whole volume.

In this scheme, the evolving front is moved by updating the level set function at a small set of points in the neighbourhood of the zero level set instead of updating it at all the points on the grid. During a given time step, the value of ϕ is not updated at points outside the narrow band.

In standard level set formulations, if a reinitialization procedure is not applied, the level set function can degrade gradually during evolution making the evolution unstable. And in standard narrow band approach, this degradation is even faster, since the zero level set approaches the narrow band boundaries periodically. This will demand extra reinitialization of ϕ. As mentioned before, many proposed reinitialization schemes have an undesirable side effect of moving the zero level set away from its original location resulting in further error in the location of the zero level set [7, 18]. Moreover, this frequent degradation and re-initialization result in causing significant error in the location of the zero level set. When considering these facts, an evolution model that can avoid the reinitialization procedure is more suitable for an efficient narrow band implementation.

Hence for an efficient narrow band implementation of level set evolution, the variational level set formulation discussed in the previous section, that can be easily implemented by simple finite difference scheme and which avoids the need for re-initialization, has been adopted in this work. In this formulation of the level set evolution model, every update of the level set function not only drives the zero level set towards the target boundaries but also maintains the level set function as an approximate signed distance function.

3.1 Building the Narrow Band

During a given time step, the value of ϕ is not updated at points lying outside the narrow band. One issue in the implementation of the narrow band approach is the choice of the appropriate width for the narrow band. The width of the narrow band determines the number of grid points that should fall within the narrow band. During evolution, it has to be ensured that the evolving curve always remains inside the band. To achieve this, the narrow band needs to be updated.

If the narrow band is too wide, it is never rebuilt and the procedure becomes equivalent to the full matrix approach. No computational speedup is achieved. It is desirable to have the band as narrow as possible to reduce the computational cost of updating the level set function on the band at each iteration. But if the band is too narrow, the zero level set will evolve outside the band faster. This results in more frequent rebuilding of the narrow band. Hence in order to reduce the overall computational cost, our proposed algorithm uses a five grid point wide narrow band.

3.2 Faster Narrow Band Level Set Evolution Algorithm

Step 1 Initialize the iteration count n to zero. Initialize level set function ϕ to ϕ_0. Construct the initial narrow band

$$NB_0 = \bigcup_{x_i \in L_0} B(x_i) \tag{12}$$

L_0 is the zero level set of ϕ_0 $B(x_i)$ is the 5X5 neighbourhood of all $x_i \in L_0$.

Step 2 Update the level set on the narrow band.
Update $\phi^{n+1} = \phi^n + \tau G(\phi^n)$ for all pixels on NB_n according to (11).

Step 3 Update the narrow band.
Identify the zero level set of ϕ^{n+1} on NB_n denoted by L_{n+1}. Then update the narrow band

$$NB_{n+1} = \bigcup_{x_i \in L_{n+1}} B(x_i) \tag{13}$$

Step 4 For all new pixels $p_i \in NB_{n+1}$ which were not in NB_n, determine the sign and assign the ϕ value. Set n = n + 1.

Step 5 If either the difference between ϕ^n and ϕ^{n-1} is less than a prescribed threshold or n exceeds a predefined maximum number of iterations, stop the iteration, otherwise go to *step 2*.

In this formulation, the level set function ϕ need not be initialised as a signed distance function. Instead ϕ is initialized as

$$\phi_0(\mathrm{x}, \mathrm{y}) = \begin{cases} -c(x, y) \in R \\ cotherwise \end{cases} \tag{14}$$

where R is an arbitrary region in the image domain and c > 0 is a constant. The constant c should be chosen as greater than the width in the definition of the regularized Dirac function

$$\delta\varepsilon(x) = \begin{cases} 0, & |x| > \varepsilon \\ \frac{1}{2\varepsilon}\left[1 + \cos\left(\frac{\pi x}{\varepsilon}\right)\right] & |x| \le \varepsilon \end{cases} \tag{15}$$

The arbitrary region R in the image domain can be specified by the user or can be generated automatically. This region based initialization of the level set function is computationally very efficient and allows for flexible applications in many situations.

For updating the level set in each iteration, performing the calculations over the entire computational domain requires O(N^2) operations per time step in two dimensions and order of O(N^3) operations in three dimensions, where N is the number of grid points along a side. In the narrow band approach, if it is assumed that the front has roughly Mpoints along the boundary, the number of operations comes down to O(mM) where m is the number of cells in the narrow band.

4 Boolean Operations on Solids in a Level Set Framework

Level set models can be very effectively used as a geometric model for creating and manipulating complex and closed objects. The efficiency of these models lie in the low level volumetric representation combined with the mathematics of deformable implicit surfaces. They offer very powerful numerical techniques for developing a novel and efficient shape modeling approach. In comparison to other types of geometric models, level set models offer many advantages. They can generate simple closed surfaces without self intersection. It may appear to be inefficient to define a surface with a volume data set, but level set models are guaranteed to generate simple closed surfaces without self intersection.

Since level set models are topologically flexible, they are ideal for representing complex structures of unknown topology. That is, they can easily represent complicated shapes that can form holes, split to form multiple objects or merge with other objects to form a single structure. Moreover, they are free of the edge connectivity and mesh quality problems associated with surface mesh models.

They also provide the advantages of implicit surface models, a powerful surface modeling paradigm supporting solid modeling operations in a straightforward manner. Implicit representations offer many advantages over the explicit surface representations. One such advantage is the ease with which constructive solid geometry (CSG) operations can be performed on implicit surfaces. Performing a CSG operation on polygon meshes, subdivision surfaces or NURBS is a complex and error prone task. On the other hand performing these operations on a pair of implicit surfaces is straight forward.

Constructive Solid Geometry (CSG) has been a useful tool in computer graphics for many years [23]. Usually CSG is applied to primitive shapes like spheres, cylinders and cubes to produce more complex geometric shapes. However, the set theoretic operations of union, intersection and difference are efficient for editing arbitrary solid shapes also. Boolean operations are a natural way of constructing complex objects out of simpler primitives. Adding, subtracting and intersecting shapes enable us to create more complex models. Since level set models are volumetric, CSG operations of union, difference and intersection may be applied to them [2, 24]. In the level set framework, with a positive-inside/negative-outside sign convention for volumes, these operations may be efficiently implemented as min/max operations as shown in Table 1.

Table 1 Implementation of boolean operations on two level set volume models V1 and V2

Boolean operation	Implementation
Intersection $V1 \cap V2$	Min (V1, V2)
Union $V1 \cup V2$	Max (V1, V2)
Difference $V1 - V2$	Min (V1, −V2)

5 Experimental Results

Implementing the proposed narrow band approach, we have extracted shapes from a variety of synthetic and real image data with promising results. In all the experiments, the level set function ϕ has been initialized as in (12) with c = 2.

Table 2 shows the CPU time taken by the narrow band algorithm and the full domain algorithm for different images in Figs. 1a, 2a and 3a respectively. Values given in the table are the averages of ten runs for each of the images. It is evident that the proposed narrow band algorithm runs faster than the full domain algorithm. It is also observed that as the image size increases the computational efficiency achieved by using the narrow band algorithm also increases. It is clear that while dealing with volumetric representations, where the computational time and memory requirements are high, the proposed narrow band algorithm is highly efficient.

Figure 1 shows the result on a particular slice of a 128X128X27 MRI data of human brain. For this image we used the parameters $\lambda = 5$, $\mu = 0.04$, $\alpha = -1.5$

Table 2 Comparison of computation time

Image and size	Iterations	Full domain (s)	Narrow band (s)
Brain (128X128)	200	1.5132	0.9672
Fruits (225X300)	340	15.4909	6.3942
Coins (221X425)	340	22.6201	8.7673

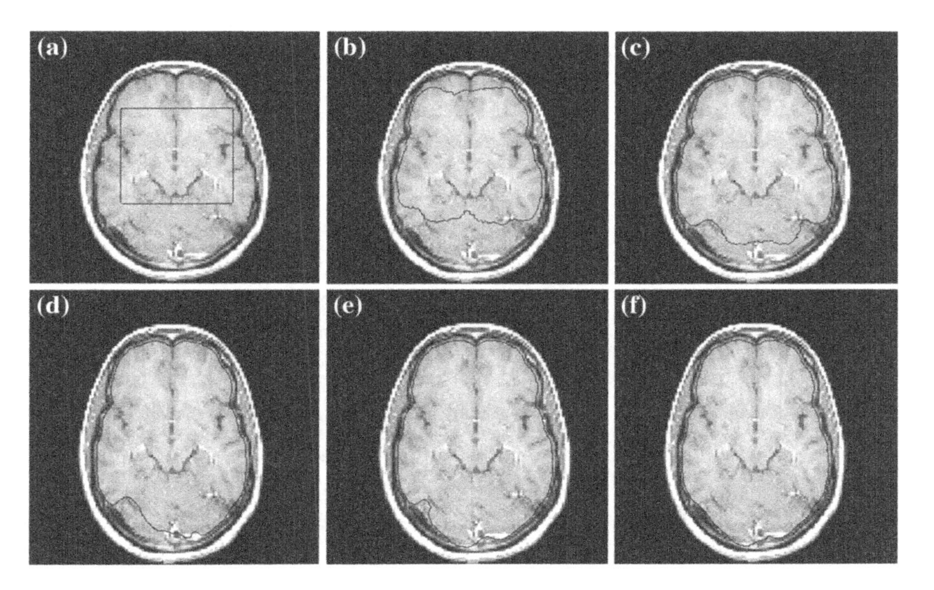

Fig. 1 Result for a 128X128 MRI slice of human brain with parameters $\lambda = 5$, $\mu = 0.04$, $\alpha = -1.5$ and time step $\tau = 5.0$. **a** Initial contour, **b** 40 iterations, **c** 80 iterations, **d** 120 iterations, **e** 160 iterations, **f** 200 iterations

and time step $\tau = 5.0$. The boundary of the white matter is successfully extracted with 200 iterations, even though a portion of the lower part of the boundary is blurred. This result shows the performance of the algorithm even when the object boundaries are weak. Figures 2 and 3 show the result of the proposed algorithm on two real images. For these examples we used the parameters $\lambda = 5$, $\mu = 0.04$, $\alpha = 1.5$ and time step $\tau = 5.0$. Here, α, the co-efficient of the weighted area term in (5) takes a positive value since the initial contour is placed outside the object and during evolution the contour is made to shrink faster. These results show how the

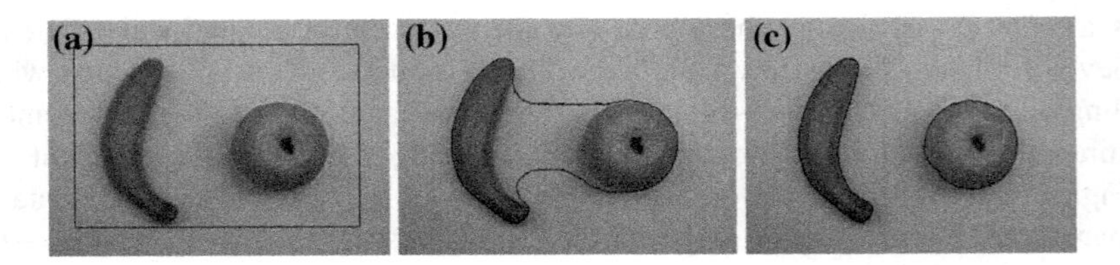

Fig. 2 Result for a real image of fruits with parameters $\lambda = 5$, $\mu = 0.04$, $\alpha = 1.5$ and time step $\tau = 5.0$. **a** Initial contour, **b** 160 iterations, **c** 340 iterations

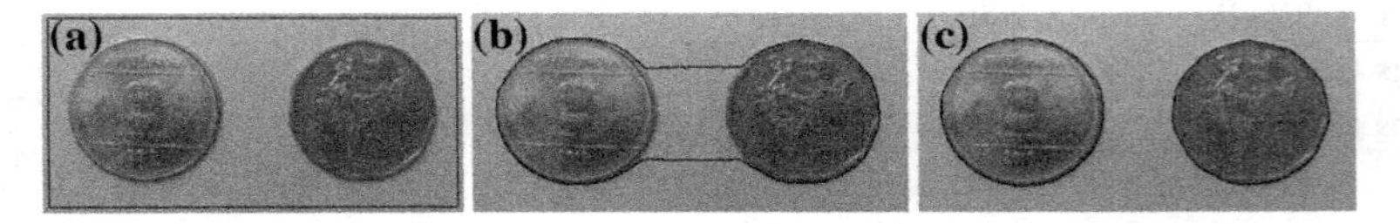

Fig. 3 Result for a real image of coins with parameters $\lambda = 5$, $\mu = 0.04$,$\alpha = 1.5$ and time step $\tau = 5.0$. **a** Initial contour, **b** 160 iterations, **c** 340 iterations

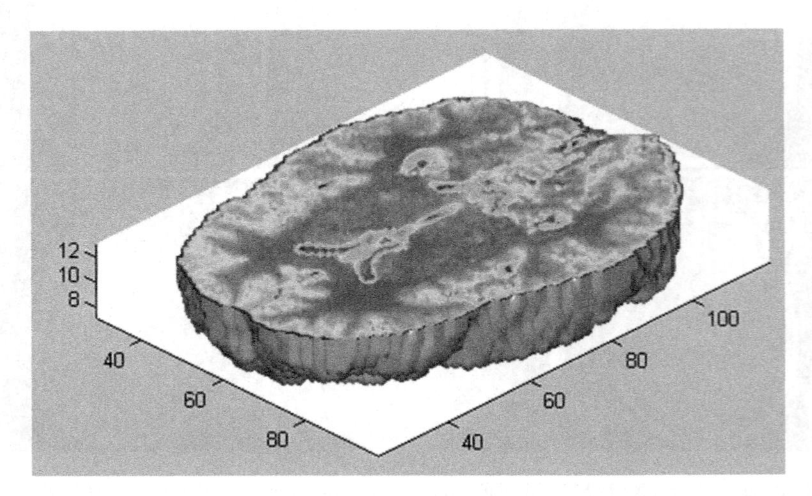

Fig. 4 Result of a portion of the shape extracted from 128X128X27 MRI data of human brain showing the internal volume data underlying the bounding surface

initial curve, during evolution, breaks and finally locks around the boundaries of the two objects. In all the examples, the initial value of the level set function is constructed from a rectangular shaped arbitrary region in the image domain according to (12).

Implementing the proposed shape modeling scheme, we have applied the CSG operations of union, intersection and difference on shapes extracted from a variety of 3D image data with good results. The result of a particular portion of the shape extracted from a 128X128X27 MRI data of human brain is shown in Fig. 4. Using the proposed narrowband approach, the entire shape of the white matter has been successfully extracted. Figure 4 illustrates a portion of the extracted shape showing the underlying volumetric representation. Figure 5 shows the result of the boolean operations on two sections extracted from the same 3D image data.

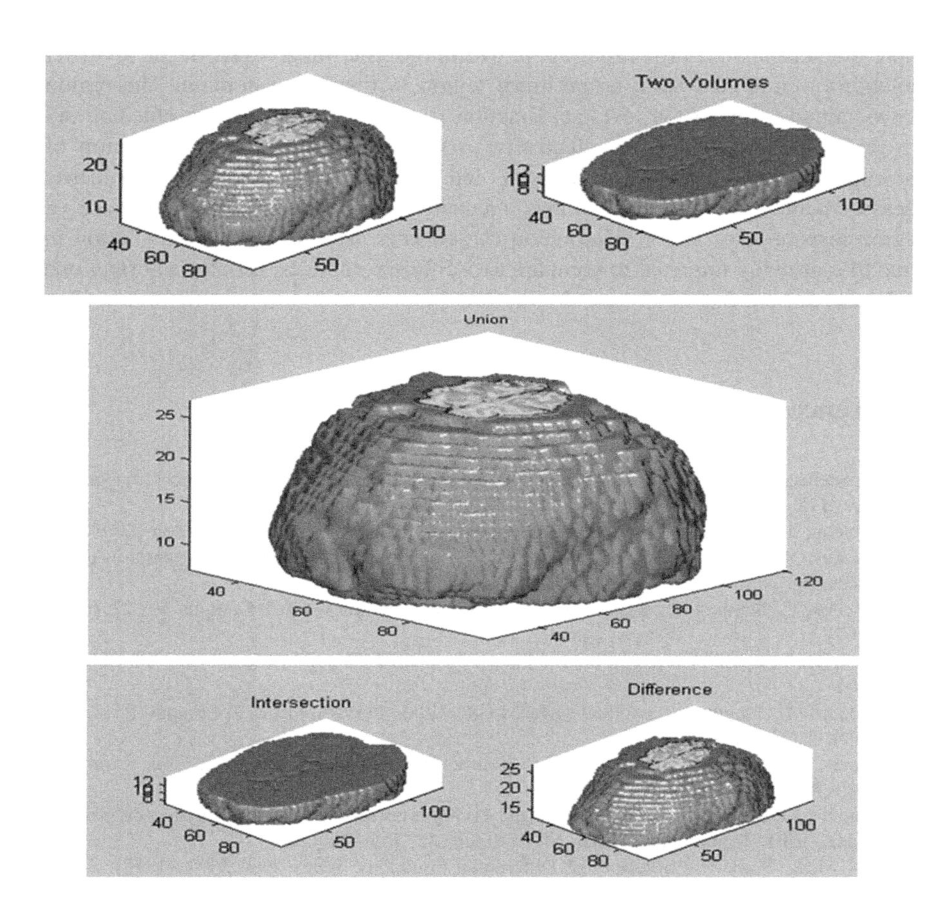

Fig. 5 Result of the boolean operations on two sections extracted from the 3D brain data

6 Conclusion and Future Work

Volume modeling plays an important role in the visualization of scientific data, a major field in computer graphics. Medical imaging was one of the first applications of volume modeling. In this paper, we have presented an efficient shape modeling scheme for manipulating arbitrary shapes retrieved from different types of image data applying a fast narrow band level set evolution algorithm. The proposed narrow band level set algorithm forms the basis of the shape modeling scheme for implementing solid modeling techniques on free form shapes set in a level set frame work. The overall computational cost is reduced by using a five grid point wide narrow band based on a variational level set formulation that can be easily implemented by simple finite difference scheme and avoids the need for re-initialization. Implementation by simple finite difference scheme makes computations more efficient and ensures faster curve evolution. The initial value of the level set function is computed from an arbitrary region in the image domain. The region based initialization of the level set function is computationally more efficient and flexible. This model can be effectively used for the analysis and extraction of information from bio-medical images and further computations of geometric measurements such as area, volume, intensity of a detected contour, surface or region respectively. These information can be later used to study the evolution in time of a growing tumor or to compare two different subjects, usually a normal one and an abnormal one.

References

1. Adalsteinsson, D., Sethian, J.A.: A fast level set method for propagating interfaces. J. Comput. Phys. **118**, 269–277 (1995)
2. Agoston, M.K.: Computer Graphics and Geometric Modeling. Springer, London (2005)
3. Caselles, V., Catte, F., Coll, T., Dibos, F.: A geometric model for active contours in image processing. Numer. Math. **66**, 1–31 (1993)
4. Caselles, V., Kimmel, R., Sapiro, G.: Geodesic active contours. Int. J. Comput. Vis **22**, 61–79 (1997)
5. Chan, T., Vese, L.: Active contours without edges. IEEE Trans. Image Process. **10**, 266–277 (2001)
6. Chopp, D.L.: Computing minimal surfaces via level set curvature flow. J. Comput. Phys. **106**, 77–91 (1993)
7. Gomes, J., Faugeras, O.: Reconciling distance functions and level sets. J. Vis. Commun. Image Represent. **11**, 209–223 (2000)
8. Han, X., Xu, C., Prince, J.: A topology preserving level set method for geometric deformable models. IEEE Trans. Pattern Anal. Mach. Intell. **25**, 755–768 (2003)
9. Jain, M.K.: Numerical Solutions of Differential Equations. Wiley, New York (1979)
10. Kachouie, N.N., Fieguth, P.: A narrow band level set method with dynamic velocity for neural stem cell cluster segmentation, image analysis and recognition. Lect. Notes. Comput. Sci. **3656**, 1006–1013 (2005)
11. Kass, M., Witkin, A.: Terzopoulos, D.: Snakes: active contour models. Int. J. Comput. Vis. **1**, 321–331 (1987)

12. Kimia, B.B., Tannenbaum, A.R., Zucker, S.W.: Shapes, shocks, and deformations I: the components of two dimensional shape and the reaction-diffusion space. Int. J. Comput. Vis. **15**, 189–224 (1995)
13. Li, C., Chenyang, X., Gui, C., Fox, M.D.: Level set evolution without re-initialization: A New Variational Formulation. IEEE Conf. Proc. Comput. Vis. Pattern Recognit.1063-6919/05 (2005)
14. Malladi, R., Sethian, J.A., Vemuri, B.C.: Shape modeling with front propagation: A level set approach. IEEE Trans. Pattern Anal. Mach. Intell. **17**, 158–175 (1995)
15. Museth, K.: An efficient level set toolkit for visual effects. In: SIGGRAPH'09: SIGGRAPH Talks, pp. 1–1. ACM, New York (2009)
16. Museth, K., Breen, D.E., Whitaker, R.T., Mauch, S., Johnson, D.: Algorithms for interactive editing of level set models. Int. J. Eurographics Assoc. Comput. Graph. Forum. **24**(4), 821–841 (2005)
17. Nielsen, M.B., Museth, K.: Dynamic tubular grid: an efficient data structure and algorithms for high resolution level sets. J. Sci. Comput. **26**(3), 261–299 (2006)
18. Osher, S., Fedkiw, R.: Level Set Methods and Dynamic Implicit Surfaces. Springer, New York (2002)
19. Osher, S., Paragios, N.: Geometric Level Set Methods in Imaging. Vision and Graphics. Springer, New York (2003)
20. Osher, S., Sethian, J.A.: Fronts propagating with curvature-dependent speed: Algorithms based on Hamilton-Jacobi formulations. J. Comput. Phys. **79**, 12–49 (1988)
21. Rosenthal, P., Molchanov, V., Linsen, L.: A narrow band level set method for surface extraction from unstructured point-based volume data. In: Proceedings of WSCG, The 18th international conference on computer graphics, Visualization and Computer Vision. pp. 73–80. UNION Agency—Science Press, Plzen, Czech Republic (2010)
22. Rosenthal, P., Linsen, L.: Smooth surface extraction from unstructured point-based volume data using PDEs. IEEE Trans. Vis. Comput. Graph. **14**(6), 1531–1546 (2008)
23. Rossignac, T.R., Requicha, A.A.G.: Solid modeling. In: Webster, J. (ed.) Encyclopedia of Electrical and Electronics Engineering. Wiley, New York (1999)
24. Sethian, J.A.: Level Set Methods and Fast Marching Methods. Cambridge University Press, Cambridge (1999)
25. Vemuri B., Chen Y.: Joint image registration and segmentation. Geometric Level Set Methods in Imaging, Vision, and Graphics pp. 251–269. Springer, New York (2003)
26. Whitaker, R.T.: A level-set approach to 3D reconstruction from range data. Int. J. Comput. Vis. **3**, 203–231 (1998)
27. Whitaker, R.T., Breen, D., Museth, K., Soni, N.: Segmentation of biological volume datasets using a level-set framework. In: Chen, M., Kaufman, A. (eds.) Volume Graphics. pp. 249–263. Springer, Vienna (2001)
28. Zhao, H., Chan, T., Merriman, B., Osher, S.: A variational level set approach to multiphase motion. J. Comput. Phys. **127**, 179–195 (1996)

Feature Extraction and Recognition of Ancient Kannada Epigraphs

A. Soumya and G. Hemantha Kumar

Abstract Optical Character Recognition finds numerous applications and one among them is in the field of Epigraphy, which is the study of inscriptions. Expert epigraphers who read ancient inscriptions are nowadays less in number. Also it is found that preserving these ancient records is important, hence lot of scope for digitization of these historical records and automatic decipherment of the same is important to the mankind. This paper addresses mainly on Segmentation, Feature Extraction and Character Recognition of Ancient Kannada script of Ashoka and Hoysala periods. Initially, input epigraph image is segmented to obtain sampled characters using Nearest Neighbor clustering Algorithm. Statistical Features such as Mean, Variance, Standard Deviation, Kurtosis, Skewness, Homogeneity, Contrast, Correlation, Energy, and Coarseness are extracted to store as training set and for comparison at the later stage of testing. Finally Mamdani Fuzzy Classifier is used in recognition; as a result, output is displayed in modern Kannada form.

Keywords Ancient Kannada script · Nearest neighbor clustering · Statistical features · Optical character recognition · Fuzzy classifier

1 Introduction

An inscription is carved information on stones, coins, pillars and walls in the ancient language used. The use of inscription in the ancient world has importance in helping to preserve information on those ancient cultures for which artifact can be

---

A. Soumya (✉)
Department of Computer Science and Engineering, R V College of Engineering, Bangalore, India
e-mail: soumyaa@rvce.edu.in

G. Hemantha Kumar
Department of Studies in Computer Science, University of Mysore, Mysore, India
e-mail: ghk2007@yahoo.com

L.C. Jain et al. (eds.), *Computational Intelligence in Data Mining - Volume 3*,
Smart Innovation, Systems and Technologies 33, DOI 10.1007/978-81-322-2202-6_42

found. These ancient inscriptions provide insights into the history of the cultures that used them and into the development of doctrinal thought and they also act as a means of transmission of poetic and literary writings from the ancient to the modern world. Optical Character Recognition (OCR) can be applied in the field of epigraphy and paleography for deciphering ancient scripts [1, 2].

High accurate OCR's are available for foreign languages which have very minimal character set and also are less complex. Recognition of Indian scripts is relatively complex because of the existence of compound characters and huge character set. There are very few works reported in literature on ancient script recognition (specifically on Kannada scripts) which is much more complex as the characters have transformed to present form over years. In the proposed system, the input ancient Kannada epigraphs of Ashoka and Hoysala periods are recognized and transformed to text of present form.

Creation of precise alphabet fonts of early Brahmi script from photographic data of ancient Srilankan inscription [1]. The paper scans and separates sample letters from image using adobe Photoshop CS3. Then addresses the creation of font image using majority algorithm which isolates noise.

An offline ancient Tamil script recognition from temple wall inscription using Fourier and wavelet features [2]. Fourier–Wavelet transform is implemented for feature extraction. In final stage, classification method is implemented and tested on a database of cursive ancient Tamil script and is proven as an efficient classifier.

A Random Forest (RF) classifier is designed [3] to date ancient Kannada epigraphs of the renowned dynasties: Ashoka, Satavahana, Kadamba, Chalukya, Rastrakuta and Hoysala. A prediction rate of 85 % is obtained for classifying the Kannada epigraphs into respective periods using RF.

The paper [4] demonstrates the recognition of base characters of ancient Kannada Script pertaining to three periods or dynasties—Ashoka, Badami Chalukya and Mysore Wodeyars using Zernike moment features. SVM classifies the input ancient characters and then the ancient character is mapped to the present form.

The Statistical Analysis of the Indus Script using n-grams, discuss the advantage of using Statistical Feature Extraction methodologies in feature extraction process [5]. As per the analysis statistical features provides 75 % accuracy in the results.

This paper is organized as follows: Sect. 2 provides a mathematical background of feature extraction technique used in the current work. The proposed system and the Methodology are elaborated in Sect. 3. Experimental results are covered in Sects. 4, and 5 provides concluding remarks.

2 Mathematical Background

Feature extraction is an important step in Pattern Recognition and Classification. Texture feature play an important role in feature analysis. Texture patterns are analyzed using statistical approach and the spatial relationship between the pixel values. Texture feature extraction has two approaches: First order statistics and second order statistics [6].

2.1 First Order Statistical Features

Considers only individual pixel and not a neighbor pixel and the methods are:

- **Statistical grey level features**: The statistical features are used to characterize distribution of gray level in the given image. Some of the features are Min, Max, Entropy, Skewness, Kurtosis, Variance, Standard deviation and are given by the following equations:

$$\text{Min } I_{min} = \text{Min}\{I(X, Y)\} \tag{1}$$

$$\text{Max } I_{max} = \text{Max}\{I(X, Y)\} \tag{2}$$

$$\begin{aligned} \text{Variance}\,(\sigma &= \frac{1}{mn - 1}\sum_{x=0}^{m-1}\sum_{y=0}^{n-1}(I(x, y) - \text{Mean})^2 \\ \text{where, Mean} &= \frac{1}{mn - 1}\sum_{x=0}^{m-1}\sum_{y=0}^{n-1}(I(x, y) \end{aligned} \tag{3}$$

$$\text{Standard deviation} = \sqrt{\text{Variance}} \tag{4}$$

$$\text{Skewness} = \frac{1}{mn(\sigma^3)}\sum_{x=0}^{m-1}\sum_{y=0}^{n-1}(I(x, y) - \text{Mean})^3 \tag{5}$$

$$\text{Kurtosis} = \frac{1}{mn - 1(\sigma^4)}\sum_{x=0}^{m-1}\sum_{y=0}^{n-1}(I(x, y) - \text{Mean})^4 \tag{6}$$

  where, m and n are the number of rows and columns respectively in the given input image.
- **Histogram Features**: it describes grey level histogram moments of an image, histogram informs about the content of image, and the features are average intensity, average contrast, smoothness, third moment, uniformity and entropy and are given by the following mathematical expressions:

$$\text{Smoothness} = 1 - \frac{1}{1 + \sigma^2} \tag{7}$$

$$\text{Third moment} = \sum_{i=0}^{L-1} \mathrm{p}(\mathrm{z_i} - \mathrm{m})^3 \mathrm{p}(\mathrm{z_i}) \quad (8)$$

$$\text{Entropy} = \sum_{i=0}^{L-1} \mathrm{p}(\mathrm{z_i}) \log_2(z_i) \quad (9)$$

$$\text{Uniformity} = \sum_{i=0}^{L-1} \mathrm{p}^2(\mathrm{z_i}) \quad (10)$$

$$\text{Average histogram} = \frac{1}{L} \sum_{i=0}^{L-1} N(i) \quad (11)$$

2.2 Second Order Statistics

Finds features based on the pixels relationship between neighboring pixels.

- **GLCM Features**: Provides information regarding relative position of two pixels with respect to each other [6]. It is formed by counting the number of occurrences of pixel pairs at a given distance. GLCM matrix is computed using distance vector d = (x, y), and computes the probability of occurrence of gray levels in the given distance d and angle Θ. Some of the GLCM features are:

$$\text{Contrast} = \sum_{i=0}^{G-1} \sum_{j=0}^{G-1} p(i, j)(i - j)^2 \quad (12)$$

$$\text{Energy} = \sum_{i=0}^{G-1} \sum_{j=0}^{G-1} p(i, j)^2 \quad (13)$$

$$\text{Homogeneity} = \sum_{i=0}^{G-1} \sum_{j=0}^{G-1} \frac{p(i, j)}{1 + |i - j|} \quad (14)$$

$$\text{Dissimilarity} = \sum_{i=0}^{G-1} \sum_{j=0}^{G-1} |i - j| p(i, j) \quad (15)$$

The methods of both first order and second order statistics forms hybrid features and extracts unique features of an image, hence used in pattern extraction and classification.

3 Proposed System and Methodology

Figure 1 shows Structure Chart for Feature Extraction and Character Recognition System for Ancient Kannada Script. The main module is divided into three major modules Segmentation, Feature Extraction and Character Recognition and it is supported by user interfaces to handle input and output images. Input to the main module is an ancient Kannada script image browsed from the host machine and also the era of the text in the script image.

Segmentation of the input image: The segmentation module responsibility is to divide the script image into small recognizable characters.

Statistical Feature Extraction: The feature extraction module responsibility is to compute the statistical features of the given character image.

Classification and Recognition: In recognizer module considering ten features from the previous module, model value is computed using Mamdani fuzzy

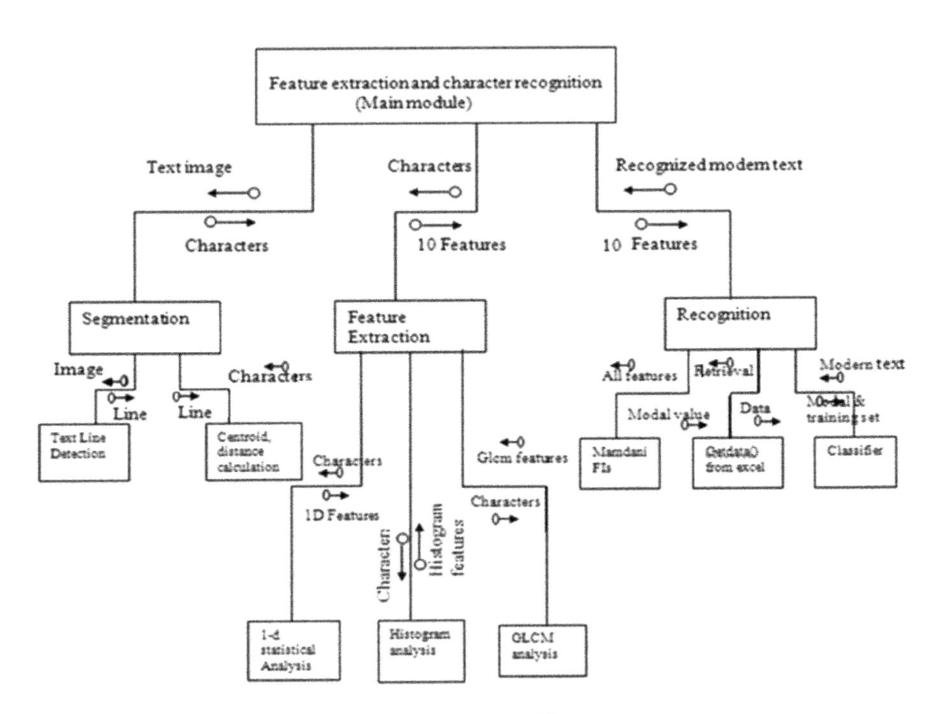

Fig. 1 Structure chart of epigraphical character recognition system

function. Each of the input is set with certain rules and Gaussian function is used to get model value. The data stored as training sample for all the characters of Brahmi and Hoysala are retrieved. Model value is compared with training sample and relevant result is fetched. The result of the classifier is displayed in modern form as stored in training set.

The algorithmic steps towards the stages of character recognition of epigraphs are discussed below.

3.1 Segmentation of the Input Image

The nearest neighbor clustering based method [7] is a novel approach that segments the line and character and works even on the skewed document. The proposed method scans the given input image from the left corner. When it encounters the first black pixel, it identifies the complete character through connected component. This character is segmented and Centroid is computed. Similarly the second character and its centroid are computed. The Euclidian distance between the Centroid is computed to know whether the character belongs to the same line or next line and hence segments into characters.

Input

Noiseless Kannada epigraph image of either Brahmi or Hoysala dynasty.

Output

Segmented characters of the given input image.

Method

Step 1: while (It is not the end of the image) do

Step 1.1: find the first black pixel
Step 1.2: Find the complete character using connected component
Step 1.3: Compute the Centroid of the character and store
Step 1.4: Find the next character and compute the Centroid
Step 1.5: Compute the threshold
Step 1.6: compare the distance between current and previous character using the distance formula

$$\text{Distance} = \sqrt{(x2 - x1)^2 + (y2 - y1)^2}$$

where (x1, y1) and (x2, y2) are the coordinates of centroids of characters
Step 1.7: Segment the character based on the Distance Method ends.

3.2 Extracting Statistical Features of the Character

The feature extraction module responsibility is to compute the statistical features of the given character image. The statistical features are extracted by three methods, each of these methods provide few more extraction methods. The one dimensional statistical analyzer computes Mean, Variance, Standard deviation of an image. Histogram analyzer computes Skewness, kurtosis, Entropy values of an image and GLCM analyzer computes coarseness, smoothness, of an image. Overall ten features per characters are extracted.

Input
Segmented character of Brahmi or Hoysala script will be considered as input.
Output
The values of One Dimensional Statistical Features like Mean, Variance, Standard deviation, the values of Histogram Features like Entropy, Skewness, Kurtosis and the values of GLCM features like Energy, Contrast, Correlation, Homogeneity.
Method

Step 1: segmented character of Brahmi or Hoysala as input

Step 1.1: Compute Mean
Step 1.2: Compute variance
Step 1.3: Compute Standard deviation
Step 1.4: Compute Skewness
Step 1.5: Compute Kurtosis
Step 1.6: Compute Entropy
Step 1.7: Compute GLCM features Energy, Homogeneity, Correlation, Contrast
Step 1.8: Return all 10 features of the character Method ends.

3.3 Classification and Recognition of the Segmented Character

The classifier used in the project is Fuzzy based classifier which employs if-then rules for analysis. The quantitative analysis of features obtained by feature extraction module, qualitative reasoning and human knowledge will convert the ten features into one single model value. Fuzzy models are of two types, Mamdani fuzzy model and Sugeno fuzzy model. In this work Mamdani Fuzzy model is adopted which uses Gaussian based computations. Computations are adopted in two stages [8].

First stage computation is to model and classify all the characters of Brahmi and Hoysala script which will be used as training sample to achieve the proper output.

The computed model value is stored as a document in excel worksheet along with relevant modern Kannada language characters.

Second stage computation is to get model value of features extracted after segmentation. The result of the second stage will be further used in classification and comparisons of data with the data set stored in excel worksheet.

Input

Input to this module is extracted ten features from feature extraction module.

Output

Output is recognized Kannada character in modern form.

Method

Step 1: Input (Ten Statistical features)

Step 1.1: Find the model value
Step 1.2: Compare the model value with classes described in dataset
Step 1.3: Compare model value in particular class applying if-then rules
Step 1.4: if (found nearest match)
Retrieve the entire row in the dataset
Display the result of modern Kannada character stored
Step 1.5: else Display "result not found" Method ends.

4 Experimental Results

In this work the reconstructed Brahmi and Hoysala images which are Noise-free are considered. Brahmi training data set includes base as well as compound characters ranging around 280, where Hoysala training data set includes base as well as compound characters ranging 400. A character set of 31 base characters, 112 compound characters of Brahmi script and character bank of 46 base characters, 180 compound characters of Hoysala script are used for recognition.

Figure 2 shows the results of execution when user has selected Brahmi test image. The segmented characters are displayed with the recognized Kannada characters displayed in modern form.

Figure 3 shows the result of execution, considering the Hoysala era script image file and displays the output in the modern Kannada Language. Since there are many cuts and bruises in the input image many tokens are segmented, and it is recognized as unidentified character.

Inferences:

- The segmentation of characters has direct impact on the output. Due to the cuts and broken piece in the epigraphical text image, on an average validates segmentation results up to 90 % and yields 10 % invalid output.
- Writing style practiced in Handwritten Documents are so versatile, which affects the accuracy of the results obtained in Segmentation and Recognition.

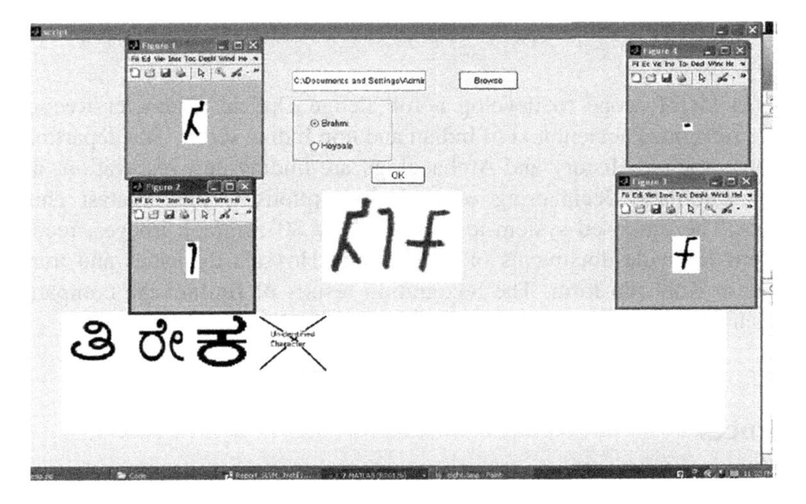

Fig. 2 The result of recognition for Brahmi script

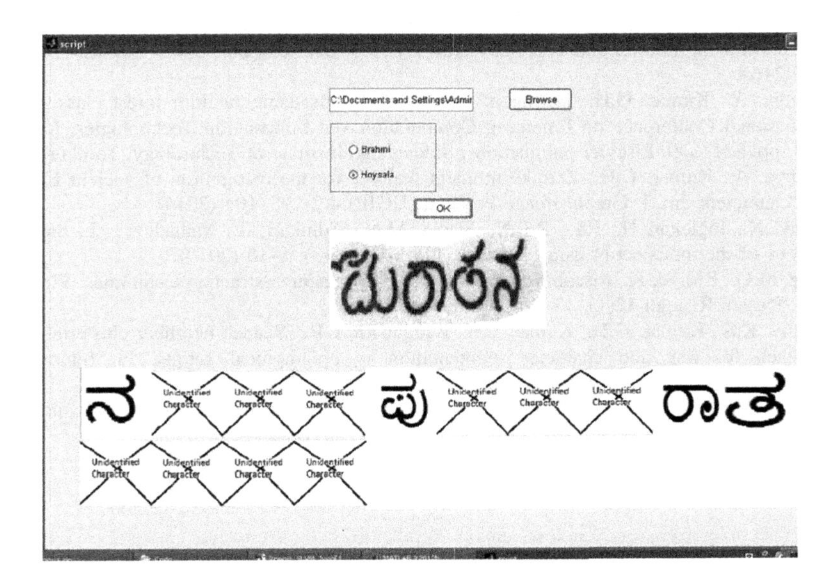

Fig. 3 Result of recognition for Hoysala script

- The results of Segmentation and Recognition are observed to be comparatively better for Brahmi text than Hoysala, because of the disconnectivity of sub-components making compound character in Hoysala text.

5 Conclusion

There is a lot of scope to develop a full fledge Optical Character Recognition system to recognize ancient text of Indian and non Indian script. The departments of Epigraphy, Ancient History and Archaeology are finding new excavations and the need for automatic deciphering of such inscriptions is the greatest challenge nowadays. The proposed system tested on nearly 80 epigraph images, recognizes the ancient Kannada documents of Ashoka and Hoysala dynasties and transform into modern Kannada form. The recognition results of Brahmi are comparatively better than Hoysala script.

References

1. Bandara, D., Warnajith, N., Minato, A., Ozawal, S.: Creation of precise alphabet fonts of early Brahmi script from photographic data of ancient Sri Lankan inscriptions. Can. J. Artif. Intell. Mach. Learn. Pattern Recognit. **3**(3), 33–39 (2012)
2. Kumar, S.R., Bharathi, V.S.: An off line ancient tamil script recognition from temple wall inscription using Fourier and Wavelet features. Eur. J. Sci. Res. **80**(4), 457–464 (2012). ISSN 1450-216X
3. Soumya, A., Kumar, G.H.: Dating of ancient epigraphs using random forest classifier. In: International Conference on Emerging Computation and Information Technologies, ICECIT-2013, pp. 331–339. Elsevier publications, Siddaganga Institute of Technology, Tumkur (2013)
4. Soumya, A., Kumar, G.H.: Zernike moment features for the recognition of ancient Kannada base characters. Int. J. Graph. Image Process. (IJGIP) **4**(2), 99–104 (2014)
5. Yadav, N., Joglekar, H., Rao, R.P.N., Vahia, M.N., Adhikari, R., Mahadevan, I.: Statistical analysis of the indus script using n-grams. PLoS One **5**(3), 1–15 (2010)
6. Yang, M.Q., Kidiyo, K., Joseph, R.: A survey of shape feature extraction techniques, Yin, P.-Y. (ed.). Pattern Recogn **12**(1), 43–90 (2008)
7. Murthy, K.S., Kumar, G.H., Kumar, P.S., Ranganath, P.R.: Nearest neighbor clustering based approach for line and character segmentation in epigraphical scripts. In: International Conference on Cognitive Systems, New Delhi, 14–15 Dec 2004
8. Elamvazuthi, Vasant, P., Webb, J.: The application of Mamdani fuzzy model for auto zoom function of a digital camera. IJCSIS **6**(3), 126–134 (2009)

*e*KMP: A Proposed Enhancement of KMP Algorithm

Nitashi Kalita, Chitra, Radhika Sharma and Samarjeet Borah

Abstract Exact sequence matching is a vital component of many problems, including text editing, information retrieval, signal processing and recently in bioinformatics applications. Various sequence matching or string matching algorithms are found in literature. The reliability of these algorithms constantly depends on the ability to detect the presence of match characters and the ability to discard any mismatch characters. These algorithms are widely used for searching of an unusual sequence in a given DNA sequence. In this paper, a proposed enhancement over KMP algorithm is presented. It is carried out to bring completeness to the parent algorithm.

Keywords Algorithm · Boyer-Moore · DNA sequencing · eKMP · KMP · LCS

1 Introduction

Deoxyribonucleic Acid (DNA) is a nucleic acid that contains genetic instructions. It consists of three components: a five-carbon sugar (Deoxyribose), a series of phosphate groups, and four nitrogenous bases. The four bases in DNA are Adenine (A), Thymine (T), Guanine (G), and Cytosine (C). The four nucleotides are grouped to

N. Kalita (✉) · Chitra · R. Sharma
Department of Computer Science and Engineering,
Sikkim Manipal Institute of Technology, Sikkim, India
e-mail: nitashikalita@gmail.com

Chitra
e-mail: chitra.smit90@gmail.com

R. Sharma
e-mail: sarmah.radhika@gmail.com

S. Borah
Department of Computer Applications,
Sikkim Manipal Institute of Technology, Sikkim, India
e-mail: samarjeetborah@gmail.com

L.C. Jain et al. (eds.), *Computational Intelligence in Data Mining - Volume 3*,
Smart Innovation, Systems and Technologies 33, DOI 10.1007/978-81-322-2202-6_43

form words and these words make sentences. These sentences are called as DNA sequences. The occurrence of unwanted or mutated words causes the disease. Every disease will have its own words or sequences and the intensity of the disease depends on the occurrence of the mutated sequence in the DNA. Genes are made up of DNA; therefore the DNA sequence of each human is unique.

In bioinformatics, matching of a DNA sub-sequence to a larger set of DNA string plays a very important role. It is used for the purpose of disease detection. When a particular sequence is known as the cause for a disease, the trace of the sequence in the DNA and the number of occurrences of the sequence defines the intensity of the disease. There are a handful number of works available in literature on the techniques of DNA sequence analysis [1–3].

2 DNA Sequence Matching

The complex task of DNA sequencing and annotation is computationally intensive. Exponential growth of gene databases enforces the need for efficient information extraction methods and sequencing algorithms exploiting existing large amount of gene information available in genomic data-banks.

There is a wide range of algorithms for DNA sequencing. Some of the popular and widely used algorithms are Knuth-Morris-Pratt (KMP) algorithm [4], Boyer-Moore algorithm [5] and Longest Common Subsequence (LCS) algorithm [6]. These algorithms are briefly explained in the following section. Since, the prime objective of this paper is to discuss a proposed enhancement of the popular KMP algorithm, detailed discussions over rest of the algorithms are not considered here.

3 Previous Work

KMP [4] is a popular algorithm used for string matching. This algorithm was first proposed by Donald Knuth and Vaughan Morris and independently by Morris in 1977, but the three published it jointly. Detail discussion on this algorithm is carried out in Sect. 3.1.

The objective of Boyer-Moore string-matching algorithm [5] is to figure out if or not a match of a specific string exists inside an alternate string. A specific DNA sequence may be a cause for a specific problem. The hint of this problematic DNA arrangement characterizes the particular illness. It is particularly suitable if the sickness DNA grouping is extensive. The matching of the two groupings is carried out in a right to left manner. This computation is a quick grouping matching calculation. Its productivity infers from the way that with every unsuccessful attempt to discover a match between the pursuit groupings, it utilizes the data picked up from the modules utilized as a part of this calculation to skip whatever number positions of the succession as could reasonably be expected where the target grouping can't match.

The simplest form of a DNA sequence similarity analysis is the Longest Common Subsequence (LCS) problem [6], where substitution operations are eliminated and only insertions and deletions are allowed. The longest common subsequence (LCS) problem is used to find the longest subsequence of DNA common to all sequences in a set of DNA sequences. This technique has useful applications in DNA sequencing.

LCS is not necessarily unique; for example the LCS of "AGC" and "ACG" is both "AG" and "AC". The LCS problem is often defined to be finding all common subsequences of DNA of a maximum length.

3.1 KMP Algorithm

KMP algorithm is linear sequence matching algorithm and plays an important role in bioinformatics, in DNA sequence matching, disease detection, finding intensity of disease etc. It is a forward linear sequence matching algorithm. It scans the strings from left to right. This algorithm computes the match in two parts: computes KMP table and a linear sequential search.

In case of making comparisons, if the current character doesn't matches with that of the sequence, it doesn't throw away all the information gained so far. Instead the algorithm calculates the new position from where to begin the search. The KMP algorithm uses a partial match table, often termed as failure function. It preprocesses the diseased DNA sequence and helps in computation of new position from where to begin the search, avoiding unnecessary search.

3.2 Analysis of KMP Algorithm

KMP algorithm is having comparatively lesser time complexity, because it is not required to start the search from the scratch due to the use of partial match table maintains by KMP. The preprocessing phase has time complexity O(m) and searching phase has complexity O(m + n) where n is DNA sequence and m is DNA sub sequence. Complexity in case of KMP algorithm does not depend upon the reparative pattern. It remains the same. Details can be found in [7, 8].

KMP is a simple algorithm. It is fast and easy to implement. The algorithm is very effective in multiple detections of DNA sub sequences. Use of KMP table is a great beneficial feature of this algorithm. The table decreases the unnecessary comparisons and backtracking.

While experimenting, it was observed that the KMP algorithm is unable to indicate to what extend a match was found. Even then algorithm has nothing to do if a match is not found.

The use of partial match table may increase the space complexity of the KMP algorithm for relatively longer patterns. The time complexity in worst case is O(m + n). To hold the partial match table the algorithm requires O(m) more space.

4 *e*KMP: The Proposed Algorithm

In order to detect a disease in a human DNA sequence, the disease DNA sequence is to be searched in the human DNA sequence. Again, the number of occurrences of the disease DNA sequence determines the intensity of the disease. But sometimes it may happen that a trace of the disease DNA might be present in the Human DNA sequence i.e. a mutated disease DNA sequence might be present. In such a case it may happen that in a particular window (selected for sequence matching with disease DNA sequence) of the human DNA sequence there can be mismatches of 50 %, <50 % or >50 % with the disease DNA sequence. This information can be of great importance to medical sciences, so it cannot be overlooked. The KMP algorithm searches for the exact sequence (short sequence) that can be present in the longer sequence and overlooks the pattern that can have few mismatches. To overcome this problem KMP algorithm is enhanced to make it more efficient and complete.

In the proposed enhancement, initially, the window size is set to the length of the disease DNA sequence. Comparison of the disease DNA sequence is performed with the characters of the selected window in the longer sequence (human DNA sequence). If a match occurs, index of the window is incremented and of the of disease DNA sequence. If a mismatch occurs, a value "m" is computed which gives the starting of the new window to be searched (similar to KMP algorithm). In the previous window percentage of mismatch and penalty is calculated and comparison in a new window is started. This process will be repeated until the end of the longer DNA sequence.

4.1 Basic Working Methodology

Let, take an example to understand the working methodology of the proposed enhancement. Array S holds the DNA Sequence and W holds the disease DNA sequence, to be searched; i is the index for disease DNA sequence and m denotes the starting index of DNA sequence from where to begin our comparison search.

Search begins at $m = 0$ and $i = 0$.

When comparison matches we increment the indexes, and move forward, but at S [3] is a space and W [3] is 't'. So new value for m through KMP table. $m = m + i - T[i]$, initially $m = 0 + 3 - 0$ and later $T[0] = -1$, and set $m = 4$. In the first window we compute the number of mismatches, penalty = −2, match = 71.42 % (Fig. 1a–e).

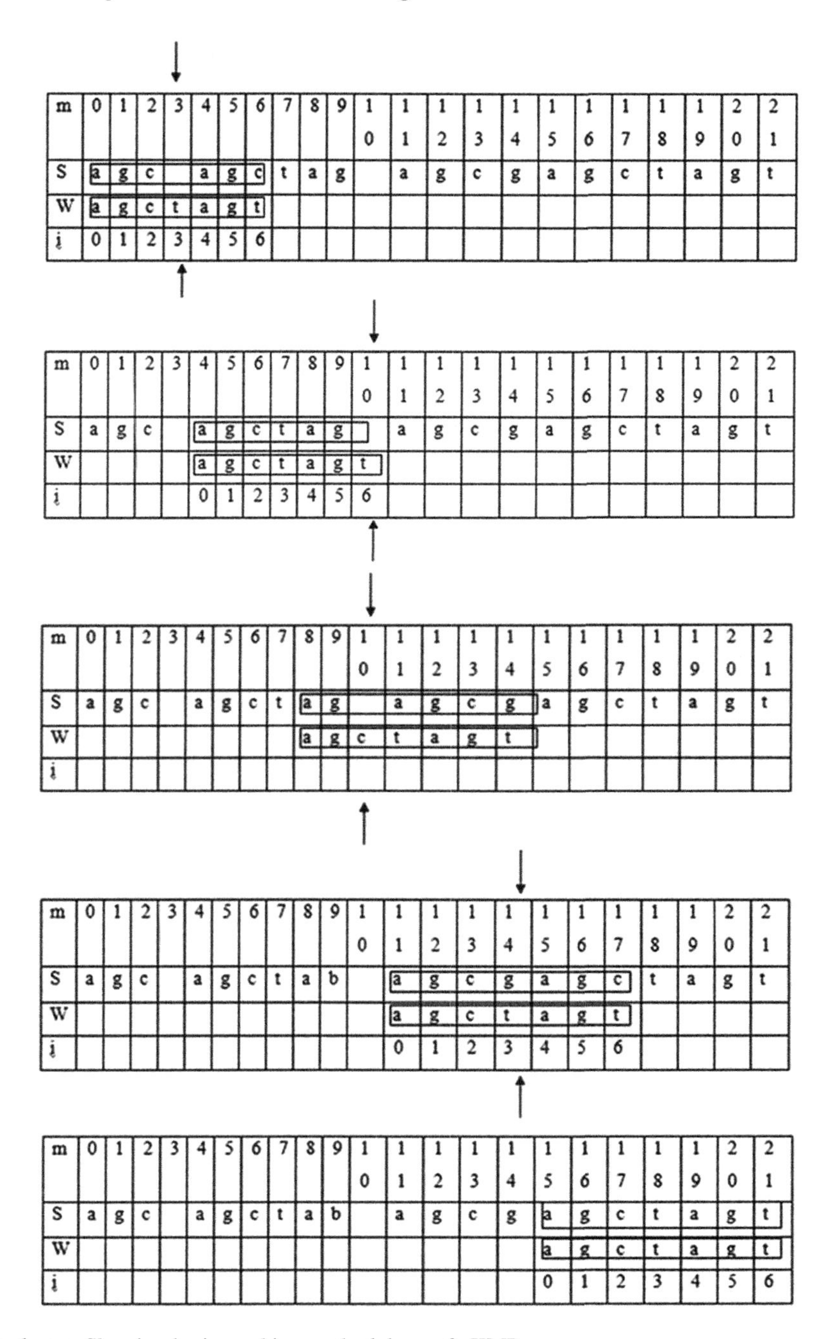

Fig. 1 a–e Showing basic working methodology of eKMP

A nearly complete match is found, but there is a mismatch again at W [6] and S[10], then m = 8 is set. In the second window we compute the number of mismatches, penalty = −1, match = 85.71 %.

A mismatch is found, again m = 11 is set. In the third window we compute the number of mismatches, penalty = −5, match = 28.57 %.

A mismatch and reset value of m = 15, and begin our search. In the fourth window we compute the number of mismatches, penalty = −2, match = 71.42 %.

Finally, a complete match is found at m = 15. In the fifth window the number of mismatches is computed, penalty = 0, match = 100 %.

4.2 Algorithm for eKMP

```
Begin
m←0
i←0
While  m+i <(length of S)-1 do
 if  W[i]=S[m+i]
  if i=(length of w)-1
    Return pos=m+1
    y=m
  End if
    i=i+1
 Else
   m←m+i-T[i]
   if T[i]>-1
      i←T[i]
   Else
      i←0
       for l←y to 1<w_length do
       if word[l] = sentence[k+1]
          Increment match counter
       Else
          Increment mismatch counter
           g=mismatchcounter/w_length*100
       if g==0
          Write "Match 100%"
       else if g<50
          Write "Match less than 50%"
        else
          Write "Match greater than 50%"
       End if
   End while
End
```

5 Implementation and Results

The algorithms were implemented using Dev C++ in Windows environment. Both synthetic and real DNA sequences were used for testing purpose.

5.1 Datasets Used

Different DNA sequence datasets were used to perform this experiment. Some of these are given below:

DNA sequence (500 *characters*):
atcgtcccgcatatccgcagaacacggaatacttactaaagcaacctagtgacctatgggactggaatcagatgttacg
gggccccaagatgcttacgaaagaaatccagaggcccggtcgggttatgaaccgtagtcccatccctcaccctattct
ggagctctggaggcgaggtgcacatgccgcttggcaatcgtcccgcatatccgcagaacacggaatacttactaaagc
aacctagtgacctatgggactggaatcagatgttacggggccccaagatgc atcgtcccgcatatccgcagaacacgga
atacttactaaagcaacctagtgacctatgggactggaatcagatgttacggggccccaagatgcttacgaaagaaatcca
gaggcccggtcgggttatgaaccgtagtcccatccctcaccctattctggagctctggaggcgaggtgcacatgccgcttg
gcaatcgtcccgcatatccgcagaacacggaa

Disease DNA sequence (15 *characters*):
atcgtcccgcatatc

DNA sequence (700 *characters*):
Atcgtcccgcatatcaagagattgatggaagaaggaacggcgtcatacaacgagtaaacggctcacgcgggtgg
ggtgcaaatcgcggtggagattccgatcgtcccgcatatcagtggcttaatgctcatatcgcaaatcttttgaactcaacaaa
aactgccccagcggggactcaccgtttcgccaatctAtcgtcccgcatatcaagagattgatggaagaaggaacggcgtc
atacaacgagtaaacggctcacgc cacccacgtcagtaaaccgatgggaaagtacaagtaattagttctggtcctacgtag
cccatagtagtccgactctacgggtcgaagagacgtaccctgatctgtcgccgtgcatttcaccgataactttacagggagt
cacataccaacctaaagagcgtgaaacttgcgaggcgtaaggcggttgcgagga□cacccacgtcagtaaaccgatggg
aaagtacaagtaattagttctggtcctacgtagcccatagtagtccgactctacgggtcgaagagacgt cacccacgtcag
taaaccgatgggaaagtacaagtaattagttctggtcctacgtagcccatagtagtccgactctacgggtcgaagagacgta
ccctgatctgtcgccgtgcatttcaccgataactttacagggagtcacatacc

Disease DNA sequence (15 *characters*):
atcgtcccgcatatc

5.2 Analysis

A comparative analysis of the various algorithms studied as part of this work is summarized in Table 1. The table also contains the proposed enhancement of the KMP algorithm. Since, the enhancement is introduced to bring completeness to KMP algorithm, computational overhead increased in this case. Therefore, KMP is superior to eKMP in terms of time complexity. Scalability issues are not considered in this work.

Table 1 Comparative analysis of the algorithms

Algorithms				
Parameters	KMP	LCS	Boyer-Moore	*e*KMP
Input parameters	A human DNA sequence and a diseased DNA sequence	Two human DNA sequence	A human DNA sequence and a diseased DNA sequence	A human DNA sequence and a diseased DNA sequence
Data structures used	KMP table	LCS table	Skip table	KMP table
Direction of scan	Left to right	Both left to right and right to left	Right to left	Left to right
Backtracking used	No	Yes	Yes	No
Result	Shows the occurrence of the diseased DNA sequence along with its position	Shows the longest common sub sequences of the input human DNA sequences	Shows the occurrence of the diseased DNA sequence	Shows the multiple occurrence of the diseased DNA sequence along with its position. Also gives the penalty incurred and the match percentage

6 Conclusion

DNA Sequence analysis needs great understanding, matching and comparison concepts of DNA sequence cannot be deployed directly with one algorithm. The program codes developed are simple to understand, easily compatible and executable. A comparative study of different algorithms is provided. The studied standard algorithms were having less time complexity but lacked completeness in extracting information from DNA Sequence from different point of views. It is now possible to find genetic similarities and dissimilarities with better completeness.

It has been found from the experiments that, the time complexity of the proposed enhancement is comparatively higher than the other algorithms under study. This is due to consideration of all the characters for search even when a mismatch occurs in order to calculate the mismatch percentage. It may be improved by introducing heuristic. The proposed enhancement over KMP algorithm can be further improved by introducing gaps.

References

1. Li, H., Homer, N.: A survey of sequence alignment algorithms for next-generation sequencing. Brief. Bioinform **11**, 473–483 (2010)
2. Johnstone, J.: A survey of sequence matching and alignment algorithms. url:https://people.ok.ubc.ca/rlawrenc/teaching/404/Project/Samples/Report/SequenceMatching_Report.pdf
3. Das, M.K., Dai, H.K.: A survey of DNA motif finding algorithms. Proceedings of the fourth annual MCBIOS conference. Computational frontiers in biomedicine. BMC Bioinform. **8**(Suppl 7), S21 (2007). doi:10.1186/1471-2105-8-S7-S21
4. Knuth, D.E., Morris, J.H., Pratt, V.R.: Fast pattern matching in strings. SIAM J. Comput. **6**(2), 323–350 (1977)
5. Boyer, R.S., Smoore, J.: A fast string-searching algorithm. Commun. Assoc. Comput. Mach. **20**, 762–772 (1977)
6. Hirschberg, D.S.: A linear space algorithm for computing maximal common subsequences. Commun. ACM **18**(6), 341–343 (1975)
7. Knuth–Morris–Pratt Algorithm. url:http://en.wikipedia.org/wiki/Knuth–Morris-Pratt_algorithm. Accessed 12 Sept 2014
8. Cao, M.: Time complexity of Knuth-Morris-Pratt string matching algorithm. Course Project Report for COMP-160. Tufts University, Medford, USA. url:http://www.eecs.tufts.edu/~mcao01/2010f/COMP-160.pdf (2010)

Acquisition and Analysis of Robotic Data Using Machine Learning Techniques

Shivendra Mishra, G. Radhakrishnan, Deepa Gupta and T.S.B. Sudarshan

Abstract A robotic system has to understand its environment in order to perform the tasks assigned to it successfully. In such a case, a system capable of learning and decision making is necessary. In order to achieve this capability, a system must be able to observe its environment with the help of real time data received from its sensors. This paper discusses certain experiments to highlight methods and attributes that can be used for such a learning. These experiments consider attributes recorded in different virtual environments with the help of different sensors. Data recording is performed from multiple directions with multiple trials for obtaining different views of the environment. Clustering performed on these multimodal data results in clustering accuracies in the range of 84.3–100 %.

Keywords Clustering · Multimodal · Robotic data · Time series

1 Introduction

Intelligence in robotic systems has been a key requirement when these systems are deployed in a dynamic or uncertain environment. It is important for the system to learn the parameters of the environment, or more precisely, it should have knowledge of the objects in order to achieve successful navigation through the

S. Mishra (✉) · G. Radhakrishnan · D. Gupta · T.S.B. Sudarshan
School of Engineering, Amrita Vishwa Vidyapeetham, Bangalore Campus,
Bangalore 560035, India
e-mail: shivendra_mishra@comsoc.org

G. Radhakrishnan
e-mail: g_radhakrishnan@blr.amrita.edu

D. Gupta
e-mail: g_deepa@blr.amrita.edu

T.S.B. Sudarshan
e-mail: tsb_sudarshan@blr.amrita.edu

L.C. Jain et al. (eds.), *Computational Intelligence in Data Mining - Volume 3*,
Smart Innovation, Systems and Technologies 33, DOI 10.1007/978-81-322-2202-6_44

environment. It should also keep track and recall the environment in which it has been navigating.

Generally, exploration of environment using robotic system has wide range of applications including search and rescue, monitoring, defense and general servicing. There have been attempts of urban search and rescue operations using robotic systems in situations such as natural disasters, earth quakes and nuclear disasters. These robotic systems generate time series data collected from environment which can be analyzed to get insight of environment. There has been a few attempts of using time series data analysis for health assessment of industrial machines [1]. These methods can also be used for trajectory modeling in any environment, since such analysis can extract patterns from real time data streams. In all these experiments, multiple sensors are interfaced to robot for observing the environment. These sensors return huge amount of multimodal time series data which have to be preprocessed and analyzed for desired patterns. It is also feasible for these type of analysis to support decision making process in order to achieve intelligent actuation in the environment.

The rest of this paper is organized as follows. Section 2 discusses related past work. Section 3 discusses the experimental setup and simulated robotic environment. Section 4 discusses robotic system and sensors used for recording. Section 5 discusses algorithms and techniques followed for pre-processing of raw data. Section 6 discusses clustering and analysis of the result. Finally, Sect. 7 presents conclusions and proposes future work.

2 Related Past Work

Time series data recorded from multiple sources have been considered for analysis in multiple works, which are discussed further in this section. In [1], the authors use K-means clustering algorithm for detecting unhealthy gearbox. Analysis starts with setting up classes of the given time series data which are used for searching of a particular pattern in data. Experimental setup of demonstration considers recording the current consumed by a motor (Time Series Data) attached to gearbox. This data is further given to Fast Fourier Transform (FFT) to perform frequency analysis. Finally, results are shown based on the different symptoms detected in different gearboxes. The method used in [2] uses experimental setup to record time series data from Pioneer-I robot sensors. They use dynamic time warping (DTW) to measure the dissimilarity between time series and to cluster them. This experiment shows accuracy in the range of 82–100 %. The work described in [3] presents neural data mining technique to analyses robot's sensor data for imitation learning. This technique demonstrates a hybrid approach of differential ratio data mining to perform analysis on spatio-temporal robot behavior data. It shows three layered time series analysis, which consists of normalization, differential ratio data mining, and classification. Normalized times series data are given for covariance calculation which is further used for calculating pair-wise ratios of elements and differential ratio (dFr) with respect to time is obtained. This dFr is classified using multi-layer perceptron (MLP) neural network. In [4], the authors propose an extension to the method proposed by Oates et al. [2]. Experimental method considers sensor weightage, which defines the sensors or pair of sensors to be recorded at a time. These sensor data go through a clustering algorithm chosen based on the nature of data recorded for analysis. Inter-cluster distance measure has been considered as stopping criteria for further clustering and sensor weightage has been used for improving inter and intra cluster distance. The evaluation of the method is conducted using typical trajectory in an office environment by recording sensor readings of Pioneer-2 Robot. Finally, the results have been shown for different values of K. In [5], the authors introduce a predictor-corrector filter model for time series prediction. In this method, environment has a temporal process and an observable vector which is used to reduce the relative noise present in data. This data is further given to self-organized map (SOM) and simple recurrent network (SRN) classifiers. This framework is further evaluated on robotic system where music mixed with speech is given as input to microphone and a visual camera is rotated if certain phrases are classified. In [6], thermal hydraulic computer code output curves are clustered using a k-means scheme, with a projection onto a lower dimensional space. It also has the properties of the empirically optimal cluster centers found by the clustering method based on projections compared to the true ones. The choice of the projection basis is discussed, and an algorithm is implemented to select the best projection basis among a library of orthonormal bases. The approach is illustrated on a simulated example and then applied to an industrial problem. In [7], visualization method of heat transfer distributions over the human body in the indoor environment under cross ventilation conditions has been considered for experiment. The time series images are recorded at 30 samples/s sampling rate. The thermal distributions are analyzed taking cross ventilation and air condition into account. The distributions of sensible heat on thermal mannequin's surfaces were also visualized by classifying the pixels of power spectrum images by the cluster analysis and calculating the correction coefficient that was defined on each cluster for multiplying to roughly estimated value of sensible heat using the mean heat transfer coefficient. A survey of techniques applied for time series data mining is presented in [8].

The experiments described in [9, 10] have robotic environment recorded with the help of different sensors. Experiment shown in [9] has a robotic system which records different views of object with the help of ultrasonic and light sensors. These data are recorded at different speeds. Further, data have been clustered using agglomerative clustering techniques after applying DTW to some of the trials. This experiment yields accuracy in the range of 73–98 %. In [10], authors perform similar experiments with the help of an array of sonar sensors interfaced to a robotic system. These sensors are installed at different angles, and record the environment surrounding its path. Data recorded in different paths go through different phases; such as pre-processing, distance matrix calculation using DTW and clustering. Results obtained through this method fall in the range of 88–97 %. The work discussed in [11] explains an experiment conducted on the time series data recorded from sonar sensors. Experiment uses back propagation neural network (BPNN) for classification of robotic environments. This experiment shows result accuracy of about 100 %.

The work explained in [9–11] have very limited simulation of robotic environments. It considers only a small number of objects in the environment. Work shown in [2–4] are close to our interest. It is also observed that work done in [1, 3, 4] have a real time data recording setup which is close to our work. However, it is concluded that none of these experimental setups considers data collected from multiple time series source and correlating them. The primary aim of our work is to achieve higher learning outcome in addition to taking more parameters of environment into account. In order to achieve that, our work conducts experiments in a virtual robotic environment. The environment has about 32 objects of different shapes in a square, even surface with an area of 210 × 210 cm. Some of these objects have temperatures higher than air temperature as well as lower than air temperature. Some of these objects may also emit different sound tones. A firebird V platform based mobile robot interfaced with various sensors is used for recording the environment. The next section discusses the same in detail.

3 Experimental Setup

The experimental setup used for simulating the robotic environment consists of a robotic system which records data while moving in a straight path. This robot is interfaced with different sensors. The environment has different sets of objects to be sensed. The following sub-section discusses the robotic environment and the robotic system configuration in detail.

3.1 Specification of Objects

Objects used in the design of environment mainly comprised of three different sets, and about 32 objects. The first set of objects are normal objects that have rectangular, cylindrical, or a mix of different shapes. These objects are made of materials which are generally found around us, such as wood, brick, stone, card board etc. These objects may be in any color, as color is not taken into account for experiment. All objects are wide enough, so that robot does not miss any objects for chosen sampling rate. Second set of objects are thermal objects. These objects are in different shapes, and maintain temperatures over or below the air temperature. These objects have a constant temperature anywhere between 15 and 70 °C. The air temperature throughout experiment is considered close to 30 °C. Third set of objects consist of sound sources. This set of objects are mostly intersection of normal or thermal objects and sound sources. These objects keep emitting tone of 1 or 2 kHz.

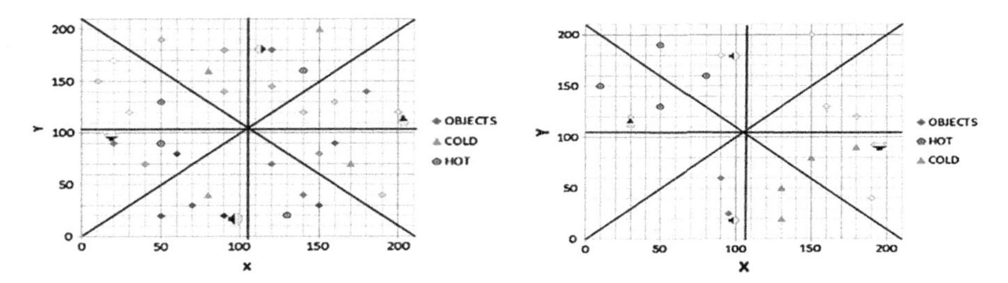

Fig. 1 The scenario on the *left* shows placement of objects for scenarios of set 1. Whereas the one on the *right* is a diversely dense scenario with thermal objects

3.2 Robotic Environment

Seven different scenarios have been created to simulate robotic environment for conducting experiments. Each scenario has a unique set of 32 spots for placement of objects. These seven scenarios are categorized into three types. First three scenarios are uniformly dense scenarios; i.e. all objects are uniformly placed throughout the environment. However, these three scenarios do not intersect each other in terms of objects placed at particular spot. The next three scenarios are diversely dense scenarios i.e., one quadrant of environment may be highly dense, whereas another may have very few objects. These scenarios also do not share any spot and object with each other. A sample of two scenarios of set 1 and set 2 are shown in Fig. 1. The seventh scenario is also a diversely dense scenario. This scenario does not have normal objects in high density; it has thermal objects in higher density instead.

As shown in Fig. 1, there are eight paths for the robot to move inside the environment to record observations. These paths are named as path1, path2 ... path8. Path2 is the returning direction of path1. Similarly, path4 is the returning direction of path3 and so on. The following section discusses the robotic system used for data recording.

4 Robotic System

Robotic system used for this experiment is based on firebird V robotic platform, which is an open source hardware-software robotic platform. This system is based on AVR architecture microcontrollers and uses ATmega 2560 as master and ATmega 8 as slave microcontroller. The block diagram of overall robotic system with sensors is shown in Fig. 2. The sensors used in this experiment are IR-proximity sensors, thermal imaging sensors and microphones. The following subsections discuss about interfacing of sensors and data recording in detail.

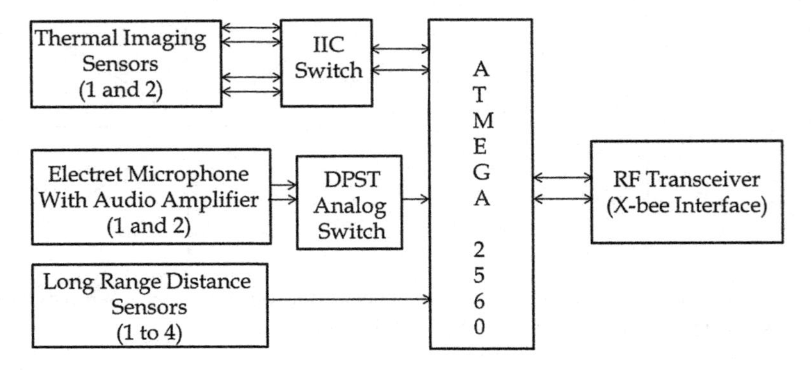

Fig. 2 Block diagram of robotic system

4.1 Sensor Interfacing

The robotic system has a pair of thermal imaging sensors, two microphones and four Infrared (IR) proximity sensors. Thermal imaging sensor used in this experiment returns a thermal image of 4 × 4 resolution every 250 ms on an inter-integrated circuit (I2C) interface. In addition to image, it also returns the value of air temperature. It can sense temperatures in the range of 0–70 °C with a field view of about 45° on horizontal and vertical axes. Similarly, two electret microphones have been interfaced for recording audio signals from the environment. Signal received from these microphones are amplified and positively clamped to make suitable for processing. Third type of sensors used is IR proximity sensor. It gives an analog output, which is further given to different channels of analog to digital convertor (ADC) of microcontroller to get equivalent digital value. Data received from thermal imaging sensor, microphone and IR proximity sensor are transmitted through a transceiver to a computer. This transceiver is also responsible for controlling the robot by issuing commands for movement and for controlling the sensor recordings.

4.2 Configuration of Data Recording

The apparatus and environment explained in earlier sub-section is used for data recording. Ten trials are recorded for each path in each scenario with different sampling rates for different types of sensors. The robot runs on straight even surface at a speed of about 19 cm/s. The data received on computer terminal is 8 bit raw hexadecimal data. The 8 bit data received from IR proximity sensors can be directly used for calculating distance of object from sensor. Each sensor is recorded at different sampling rate. IR proximity sensors are recorded at 10 samples/s, thermal imaging sensor at 5 samples/s, and electret microphone at 4,000 samples/s. These sampling frequencies are calculated by taking environment and apparatus limitation into consideration. Ten trials are recorded for each path for travel period of about 9–10 s. It makes the total number of trials for all paths in all scenarios to 560. These files are further taken for clustering.

5 Clustering of Robotic Environments

As discussed in previous section, the robot returns a stream of hexadecimal data while recording the environment. These streams have to be preprocessed to get the set of actual parameters of environment. The next phase calculates distance between different trials and performs clustering based on distance between trials. Following sub-sections discuss the same in details.

5.1 Pre-processing

The data recorded by the robot with the help of thermal imaging sensors, microphones and IR proximity sensors are received as a continuous stream of hexadecimal data by the computer. The pre-processing of this stream yields a time series data of sensor readings. The sensor readings are recorded with a sampling frequency of 200 ms. Hence, a trial run of length 210 cm will typically have 50 snapshots of sensor readings. The steps followed for pre-processing of data are:

Step 1: Scan file for calculating number of iterations recorded.
Step 2: Perform parsing and remove erroneous new lines/spaces.
Step 3: Compute thermal image and distance from IR proximity sensor data.
Step 4: Compute amplitude of audio signal from Microphone data.
Step 5: Repeat steps 1 to 4 for all trials.

The data after pre-processing will have 36 attributes, which include 4 distance measurements from 4 IR-proximity sensors, and 16 pixels values of 4×4 images obtained from each of the two thermal imaging sensors. Hence, the time series data for a typical trial will have 50 samples of 36 attributes each. A sample of 16 attributes of a thermal sensor is plotted in Fig. 3. The amplitude of audio signal is given to FFT algorithm to get the frequency spectrum of audio signal. A sample of processed audio data is plotted in Fig. 4. Finally, data considered for further processing consist of data from thermal imaging sensor and IR proximity sensor along with frequency spectrum of audio data.

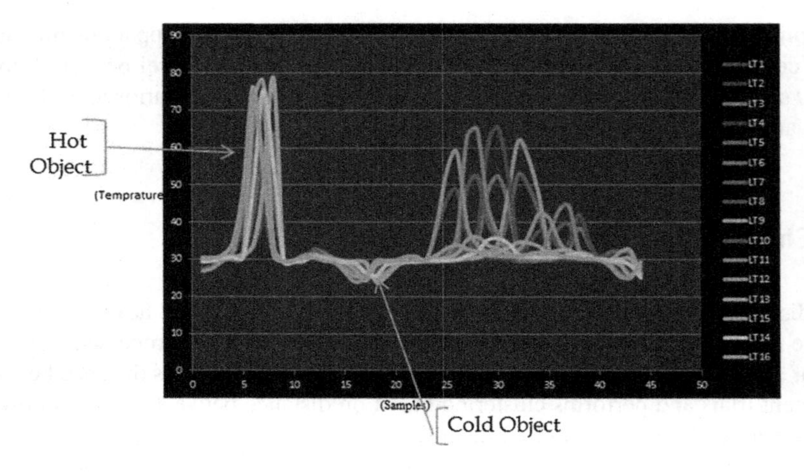

Fig. 3 Plot of thermal data. The x axis is the number of samples recorded and y axis is the temperature observed

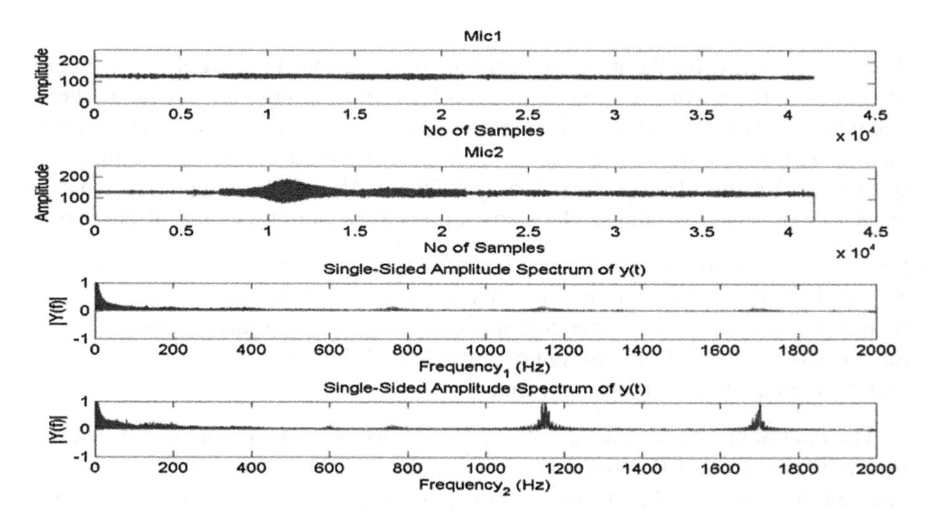

Fig. 4 Frequency spectrum of microphone data. First two graphs are the plot of data in time domain. Second two graphs are the plot for same data in frequency domain

5.2 Distance Matrix Calculation and Clustering

Distance matrix is calculated using 36 attributes, i.e., 32 from thermal imaging sensors, and 4 from IR proximity sensors. Manhattan distance is calculated between the trials and stored in a distance matrix. The following expression shows how the distance is calculated between two trials T1 and T2:

$$d(T1, T2) = \sum_{i=1}^{50} \sum_{j=1}^{36} (T1_{ij} - T2_{ij}) \tag{1}$$

where, $T1_{ij}$ and $T2_{ij}$ represent the value for jth attribute of ith sample of trials T1 and T2 respectively. As there are 10 trials per scenario, the distance matrix is a 70 × 70 matrix consisting of distances between trials taken from all 7 scenarios using a particular path. This square distance matrix is given as input to the clustering algorithm. Distance matrices for path1 to path8 are created separately.

K-medoid clustering algorithm is used for clustering. The input given to the algorithm is the distance matrix for a single path. It is desired that algorithm should be able to put all trials from a particular scenario into the same cluster. This procedure is repeated for all eight paths of robotic environment.

6 Experimental Result

Table 1 shows the result of clustering for each path with attributes selected varying from 4 to 40. The second column of the table shows the results obtained using four IR-Proximity sensor readings. The third column shows the results when all 40 attributes, i.e., 32 thermal Image attributes, 4 IR proximity sensor attributes and four attributes from audio frequency spectrum are used. It can be seen that the clustering result has improved when thermal and audio attributes are added.

The overall time complexity of the algorithm is $O(k.l.n^2)$, where, k is the number of samples per trial, l is the number of attributes and n is the number of trials. In order to reduce the execution time of the clustering algorithm, the number of thermal attributes has been reduced from 32 to 2 by using the average thermal values of each of the two thermal images. The results from this experiment are shown in the last column of the table. It has been observed that the results with 10 attributes are slightly better than the results with 40 attributes. This could be due to the fact that random noise gets eliminated when averaging of thermal values is done.

Table 1 Results of clustering

Path	4 attributes (%)	40 attributes (%)	10 attributes (%)
1	97.14	98.57	98.57
2	88.57	98.57	98.57
3	100	98.57	100
4	97.14	98.57	98.57
5	97.14	97.14	97.14
6	98.57	97.14	98.57
7	75.71	84.29	84.29
8	85.17	87.14	87.14

7 Conclusions and Future Work

This work summarizes the experimental setup created for recording the virtual robotic environment. The data recorded is clustered using k-medoid clustering algorithm and using Manhattan distance. It is observed that clustering accuracy in the range of 75–100 % is obtained with IR proximity sensor values. The results showed an improvement when thermal images and audio data are added to IR sensor data. The running time and clustering accuracy have been improved by pruning the thermal attributes.

In this work, data sets obtained from different path are clustered separately. Further efforts are required to perform clustering analysis by combining the data collected from different paths.

References

1. Olsson, E.: Identifying discriminating features in time series data for diagnosis of industrial machines. In: The 24th Annual Workshop of the Swedish Artificial Intelligence Society (2007)
2. Oates, T., Schmill, M.D., Cohen, P.R.: A method for clustering the experiences of a mobile robot that accords with human judgments. In: Proceedings of the Seventeenth National Conference on Artificial Intelligence, pp. 846–851 (2000)
3. Malone, J., Elshaw, M., McGarry, K., Bowerman, C., Wermter, S.: Spatio-temporal neural data mining architecture in learning robots. In: Proceedings of 2005 IEEE International Joint Conference on Neural Networks, IJCNN '05, vol. 5, pp. 2802–2807, 31 July–4 Aug 2005
4. Großmann, A., Wendt, M., Wyatt, J.: A semi-supervised method for learning the structure of robot environment interactions. Adv. Intell. Data Anal. Lect. Notes Comput. Sci. **2810**, 36–47 (2003)
5. Swarup, S., Lakkaraju, K., Smith, N.A., Ray, S.R.: A recurrent anticipatory self-organized map for robust time series prediction. In: The Workshop on Self-Organizing Maps (WSOM), Paris, France (2005)
6. Auder, B., Fischer, A.: Projection-based curve clustering. J. Stat. Comput. Simul. **82**(8), 1145–1168 (2012)
7. Iino, A., Annaka, T., Iino, Y., Ohba, M.: Image Analysis of Characteristics of Fluctuations of Thermal Mannequin's Surface Temperature Distributions Using Time Series Infrared Images, vol 13, pp. 71–79 (2008–2012). ISSN:1342-792X
8. Esling, P., Agon, C.: Time series data mining. ACM Comput. Surv. **45**(1), Article 12, 1–34 (2012)
9. Radhakrishnan, G., Gupta, D., Sudarshan, T.S.B.: Experimentation and analysis of time series data for rescue robotics. In: Thampi, S.M., Abraham, A., Pal, S.K., Manuel, J., Rodriguez, C. (eds.) Recent Advances in Intelligent Informatics. Advances in Intelligent Systems and Computing, vol. 235, pp. 443–453. Springer, Zofingen (2014)
10. Radhakrishnan, G., Gupta, D., Abhishek, R., Ajith, A., Sudarshan, T.S.B.: Analysis of multimodal time series data of robotic environment. In: 2012 12th International Conference on Intelligent Systems Design and Applications (ISDA), pp. 734–739, 27–29 Nov 2012
11. Gopalapillai, R., Vidhya, J., Gupta, D., Sudarshan, T.S.B.: Classification of robotic data using artificial neural network. In: 2013 IEEE Recent Advances in Intelligent Computational Systems (RAICS), pp. 333–337, 19–21 Dec 2013

Maximum Likelihood DOA Estimation in Wireless Sensor Networks Using Comprehensive Learning Particle Swarm Optimization Algorithm

Srinivash Roula, Harikrishna Gantayat, T. Panigrahi and G. Panda

Abstract Direction of arrival (DOA) estimation is one of the challenging problem in wireless sensor networks. Several methods based on maximum likelihood (ML) criteria have been established in literature. Generally, to obtain the ML solutions, the DOAs must be estimated by optimizing a complicated nonlinear multimodal function over a high-dimensional problem space. Comprehensive learning particle swarm optimization (CLPSO) based solution is proposed here to compute the ML functions and explore the potential of superior performances over traditional PSO algorithm. Simulation results confirms that the CLPSO-ML estimator is significantly giving better performance compared to conventional method like MUSIC in various scenarios at less computational costs.

Keywords Wireless sensor networks · Maximum likelihood DOA estimation · Comprehensive learning particle swarm optimization

---

S. Roula (✉) · H. Gantayat · T. Panigrahi
Department of ECE, National Institute of Science and Technology,
Berhampur 761008, India
e-mail: srinivashroula@gmail.com

H. Gantayat
e-mail: harifranky2006@gmail.com

T. Panigrahi
e-mail: tpanigrahi80@gmail.com

G. Panda
School of Electrical Sciences, Indian Institute of Technology Bhubaneswar,
Odisha, India
e-mail: ganapati.panda@gmail.com

L.C. Jain et al. (eds.), *Computational Intelligence in Data Mining - Volume 3*,
Smart Innovation, Systems and Technologies 33, DOI 10.1007/978-81-322-2202-6_45

1 Introduction

Direction of arrival (DOA) estimation is an important problem in wireless sensor networks (WSNs) to estimate the source localization and tracking. This is an well-known problem in the fields of radar, sonar, radio-astronomy, underwater surveillance and seismology etc. One of the simplest versions of this problem is the estimation of the DOA of narrow-band sources where the sources are located in the far field of the sensor array [1].

Many high resolution suboptimal techniques have been proposed and analyzed, such as multiple signal classification (MUSIC) [2], the minimum variance distortionless response (MVDR) method of Capon [3], estimation of signal parameters via rotational invariance technique (ESPRIT), and more [4]. The ML technique is used here because of its superior statistical performance compared to spectral based methods [5, 6]. The ML method is a standard technique in statistical estimation theory. A likelihood function can be formulated easily if we know the observed parametric data. The ML estimate is computed by maximizing the log-likelihood function or minimizing the negative log-likelihood function with respect to all unknown parameters. Since the ML function is multimodal, so direct optimization is seems to be unrealistic due to large computational burden.

There are different optimization techniques available in literature for optimization of ML function like AP-AML [1], fast EM and SAGE algorithms [7] and a local search technique e.g. Quasi-Newton methods. All these techniques have several limitations because of multidimensional cost function which need extensive computation, good initialization is also crucial for global optimization and we can not guarantee that these local search techniques always have global converge.

The evolutionary algorithms like genetic algorithm [8], particle swarm optimization and simulated annealing can be designed to optimize the ML function. Genetic algorithm [9], particle swarm optimization [6, 10, 11] and also PSO algorithm has been applied in distributed way in [12, 13] had already used as a global optimization technique to estimate the DOA. Genetic Algorithm (GA) is one of the most powerful and popular global search tools; however, its implementation is somewhat cumbersome due to slow convergence.

As an emerging evolutionary algorithm, PSO has attracted a lot of attention in recent years, and has been successfully applied in many fields. Most of the applications demonstrated that PSO could give competitive or even better results in a faster and cheaper way, compared with other heuristic methods such as GA. In this paper a variant of PSO known as comprehensive learning PSO (CLPSO) algorithm [14] is used to ML criterion functions for accurate DOA estimation.

Due to the multimodal, nonlinear, and high-dimensional nature of the parameter space, the problem seems to be a good application area for CLPSO, by which the excellent performance of ML criteria can be fully explored. Via extensive simulation studies, we demonstrate that with properly chosen parameters, CLPSO achieves fast and robust global convergence over PSO and MUSIC.

2 Data Model and ML Estimation Problem

Let us consider a planner array of M sensors are distributed in an arbitrary geometry and received signals form N narrow band far-field signal sources at unknown locations. We have considered this arbitrary geometry for the application of DOA finding in sensor network. The output of sensors modeled by standard equation as [15]

$$\mathbf{x}(i) = \mathbf{A}(\boldsymbol{\theta})\mathbf{s}(i) + \mathbf{n}(i), \qquad i = 1, 2, \ldots, L \tag{1}$$

where $\mathbf{s}(i)$ is the unknown vector of signal waveforms of size $N \times 1$, $\mathbf{n}(i)$ is unpredicted noise process of size $M \times 1$, L denotes the number of data samples (snapshots) taken. The matrix $\mathbf{A}(\boldsymbol{\theta})$ has the following special structure defined as

$$\mathbf{A}(\boldsymbol{\theta}) = [\mathbf{a}(\theta_1), \ldots \mathbf{a}(\theta_N)] \tag{2}$$

where $\mathbf{a}(\theta)$ is called steering vector of the array and it depends on the geometry of the array, and $\boldsymbol{\theta} = [\theta_1, \theta_2, \ldots, \theta_N]^T$ are the parameters of interest or true DOA's. The exact form of $\mathbf{a}(\theta)$ depends on the position of the sensors.

Further, the vectors of signals and noise are assumed to be stationary, temporarily white, zero-mean complex Gaussian random variables with the following second-order moments given by Panigrahi et al. [10, 11]

$$\begin{aligned} E[\mathbf{s}(i)\mathbf{s}(j)^{\mathrm{H}}] &= \mathbf{S}\delta_{ij} \qquad E[\mathbf{s}(i)\mathbf{s}(j)^{\mathrm{T}}] = 0 \\ E[\mathbf{n}(i)\mathbf{n}(j)^{\mathrm{H}}] &= \sigma^2\mathbf{I}\delta_{ij} \qquad E[\mathbf{n}(i)\mathbf{n}(j)^{\mathrm{T}}] = 0 \end{aligned} \tag{3}$$

where δ_{ij} is the Kronecker delta, $(\cdot)^{\mathrm{H}}$ denotes complex conjugate transpose, $(\cdot)^{\mathrm{T}}$ denotes transpose, $E(\cdot)$ stands for expectation.

The observation process, $\mathbf{x}(i)$, constitutes a stationary, zero-mean Gaussian random process having second-order moments

$$E[\mathbf{x}(i)\mathbf{x}(i)^{\mathrm{H}}] = \mathbf{R} = \mathbf{A}(\boldsymbol{\theta})\mathbf{S}\mathbf{A}^{\mathrm{H}}(\boldsymbol{\theta}) + \sigma^2\mathbf{I} \tag{4}$$

The problem addressed herein is the estimation of $\boldsymbol{\theta}$ along with the parameter in $\mathbf{S}$ and σ^2 (noise power) from a batch of L measured data $\mathbf{x}(1), \ldots, \mathbf{x}(L)$. Further the dimension of the problem is reduced by solving the noise power and signal covariance matrix in terms of source DOA. Under the assumption of additive Gaussian noise and complex Gaussian distributed signals we can have negative log-likelihood function [6] is given as

$$\ell(\boldsymbol{\theta}, \mathbf{S}, \sigma^2) = \log|\mathbf{R}| + \mathrm{tr}\{\mathbf{R}^{-1}\hat{\mathbf{R}}\} \tag{5}$$

where $\hat{\mathbf{R}}$ is the sample covariance matrix of size $M \times M$ and it defined as

$$\hat{\mathbf{R}} = \frac{1}{N}\sum_{i=1}^{N}\mathbf{x}(i)\mathbf{x}(i)^{\mathrm{H}} \quad (6)$$

The ML criterion function can be concentrated with respect to $\mathbf{S}$ and σ^2 by following [16]. The stochastic maximum likelihood (SML) estimates of the signal covariance matrix and the noise power are obtained by inserting the SML estimates of $\boldsymbol{\theta}$ in the following expressions

$$\hat{\mathbf{S}}(\boldsymbol{\theta}) = \mathbf{A}^{\dagger}(\boldsymbol{\theta})\big(\hat{\mathbf{R}} - \hat{\sigma}^2\mathbf{I}\big)\mathbf{A}^{\dagger\mathrm{H}}(\boldsymbol{\theta}) \quad (7)$$

$$\hat{\sigma}^2(\boldsymbol{\theta}) = \frac{1}{M-N}\mathrm{Tr}\{\mathbf{P}_{\mathbf{A}}^{\perp}(\boldsymbol{\theta})\hat{\mathbf{R}}\} \quad (8)$$

where $\mathbf{A}^{\dagger}$ is the pseudo-inverse of $\mathbf{A}$ and $\mathbf{P}_{\mathbf{A}}^{\perp}$ is the orthogonal projection onto the null space of $\mathbf{A}^{\mathrm{H}}$ and are defined as

$$\mathbf{A}^{\dagger} = \left(\mathbf{A}^{\mathrm{H}}\mathbf{A}\right)^{-1}\mathbf{A}^{\mathrm{H}} \quad (9)$$

$$\mathbf{P}_{\mathbf{A}} = \mathbf{A}\mathbf{A}^{\dagger} \quad (10)$$

$$\mathbf{P}_{\mathbf{A}}^{\perp} = \mathbf{I} - \mathbf{P}_{\mathbf{A}} \quad (11)$$

Therefore the concentrated form of the EML function now can be obtained by using (2) in (5) as

$$f_{ML}(\boldsymbol{\theta}) = \log\left|\mathbf{A}(\boldsymbol{\theta})\hat{\mathbf{S}}(\boldsymbol{\theta})\mathbf{A}^{\mathrm{H}}(\boldsymbol{\theta}) + \hat{\sigma}^2(\boldsymbol{\theta})\mathbf{I}\right| \quad (12)$$

The statistical performances of CML and EML are similar, but for highly correlated or coherent sources, the performance of EML is significantly superior.

3 Comprehensive Learning Particle Swarm Optimizer

The swarm is typically modeled by particles in multidimensional space that have a position and a velocity. Consider a D dimensional problem in which a swarm of J particles are trying to find the optimum solution. The position vector of the particles are represented by the vector $X = [x^1, x^2, \cdots, x^D]^T$. The individual position of each particle j is denoted by the D dimensional vector $X_j = \left[x_j^1, x_j^2, \cdots, x_j^D\right]^T$.

Similarly, the velocity vector of each particle j is given by the vector $V_j = \left[v_j^1, v_j^2, \cdots, v_j^D\right]^T$.

Each particle updates its position based on its own best recorded position (pbest) as well as the best recorded position (gbest) of the whole swarm. Let the best individual position vector of each particle and the best global position vector be denoted by the vectors $P_j = \left[p_j^1, p_j^2, \cdots, p_j^D\right]^T$ and $G_j = \left[g^1, g^2, \cdots, g^D\right]^T$, respectively. Members of a swarm communicate good positions to each other and adjust their own position and velocity based on these good positions. So a particle has the following information to make a suitable change in its position and velocity.

The velocity update equation for the PSO algorithm is given by Kennedy and Eberhart [17]:

$$v_j^d(i+1) = v_j^d(i) + c_1 r_1 \left(p_j^d - x_j^d(i)\right) + c_2 r_2 \left(g^d - x_j^d(i)\right), \tag{13}$$

where, $j = 1, \cdots, J$, $d = 1, \cdots, D$ and $i = 1, 2, \ldots$ denotes is the particle index, the dimension index and the time index respectively, c_1 and c_2 are positive constants known as the acceleration coefficients, and r_1 and r_2 are uniformly distributed random numbers within the range $[0, 1]$. The global best is generated from the neighborhood of each particle. In general, the velocity and the position for each particle are bounded within a predefined limit, $[-V_{max}, V_{max}]$ and $[X_{min}, X_{max}]$, respectively. These bounds ensure that the particles do not diverge from the solution space.

In CLPSO algorithm the unknown estimate divided into sub-vectors [14]. For each sub-vector, a particle chooses two random neighbor particles. The neighbor particle which gives the best fitness value for that particular sub-vector is chosen as an exemplar. The combined result of all sub-vectors gives the overall best vector, which is then used to perform the update of the velocity vector. If the particle is not improving for a certain number of iterations then the neighbor particles for the sub-vectors are changed. The velocity update equation for the CLPSO algorithm is given by Liang et al. [14]:

$$v_j^d(i+1) = w_j(i).v_j^d(i) + c_j r_j \left(\hat{p}_j^d - x_k^d(i)\right), \tag{14}$$

where $w_j(i)$ is a time-varying weighting coefficient, $\hat{p}_j^d$ gives the overall best position value for dimension d of particle j, c_j is the acceleration coefficient and r_j is a uniformly distributed random number. The overall CLPSO algorithm is thus defined by the following set of equations [14]:

$$V_j(i+1) = w_j(i).V_j(i) + c_j r_j \left(\hat{P}_j - X_j(i)\right) \tag{15}$$

$$X_j(i+1) = X_j(i) + V_k(i+1), \tag{16}$$

4 Maximum Likelihood DOA Estimation Using CLPSO Algorithm

Here we describe the formulation of the CLPSO algorithm for ML optimization to estimate the source DOA. At first initialize a population of particles in the search space with random positions and random velocities constrained between 0 and π in each dimension [6]. The N-dimensional position vector of the jth particle takes the form $x_j = [\theta_1, \ldots, \theta_N]$, where θ represents the DOAs. A particle position vector is converted to a candidate solution vector in the problem space through a suitable mapping. The score of the mapped vector evaluated by a likelihood function f_{ML} which is given in (12), is regarded as the fitness of the corresponding particle.

To evaluate the likelihood function f_{ML} required the data from all the nodes for K number of snapshots. During the evolution of algorithm, in every iteration each particle update their velocity and position, then evaluate the global best. The manipulation of a particles velocity according to (14) is regarded as the central element of the entire optimization. Since there was no actual mechanism for controlling the velocity of a particle, it is necessary to define a maximum velocity to avoid the danger of swarm explosion and divergence. The velocity limit can be applied along each dimension at every node as

$$v_j^n = \begin{cases} V_{\mathrm{MAX}}, & if\ v_j^n > V_{\mathrm{MAX}} \\ V_{\mathrm{MIN}}, & if\ v_j^n < V_{\mathrm{MIN}} \end{cases} \tag{17}$$

In this work, we keep the limitation of V_{MAX} is set to the half value of the dynamic range. The new particle position is calculated using (14).

The optimization iteration will be terminated if the specified maximum iteration number is reached. The final global best position p_g is taken as the ML estimates of source DOA.

5 Simulation Results and Discussions

In this section a numerical example is presented to demonstrate the performance of CLPSO based DOA estimation using (against adaptive PSO and MUSIC algorithm. MUSIC is the best known and well investigated algorithm for DOA estimation. The performances of those methods are compared in two ways i.e. DOA estimation root mean square error and probability of resolution. These two parameters are defined as follows.

(a) the root-mean-squared error (RMSE):

$$\mathrm{RMSE} = \sqrt{\frac{1}{NN_{\mathrm{run}}} \sum_{l=1}^{N_{\mathrm{run}}} \sum_{n=1}^{N} \left(\hat{\theta}_n(l) - \theta_n\right)^2} \tag{18}$$

where N is the number of sources, $\hat{\theta}_n(l)$ is the estimate of the nth DOA achieved in the lth run, θ_n is the true DOA of the nth source; and

(b) Probability of resolution: The ability to resolve closely spaced sources known as probability of resolution (PR). By definition, two sources are said to be resolved in a given run if both $|\hat{\theta}_1 - \theta_1|$ and $|\hat{\theta}_2 - \theta_2|$ are smaller than $|\theta_1 - \theta_2|/2$.

Let us assume that two equal-power, uncorrelated signals impinge on distributed wireless sensor network with sixteen sensors from 130° to 138°. The signal-to-noise ratio (SNR) varies from −20 to 30 dB with the step size of 2 dB taken for simulation. The function f_{ML} is optimized using PSO and CLPSO algorithms. The number of snapshots are 20, in case of PSO and CLPSO, but 1,000 snapshots are taken for MUSIC to get better result.

The Fig. 1 gives the the DOA estimation RMSE values obtained using PSO-EML, APSO-EML, and MUSIC as a function of SNR. Figure 2 shows the resolution probabilities (RP) for the same methods. Two sources are considered to be resolved in an experiment if both DOA estimation errors are less than the half of their angular separation.

As can be seen from Figs. 1 and 2, APSO-EML yields significantly superior performance over PSO-EML as a whole, by demonstrating lower DOA estimation RMSE and higher resolution probabilities. The accurate DOA estimates are observed because (1) ML criterion functions are statistically optimal although computation-extensive, and (2) the designed CLPSO is a robust and reliable global optimization algorithm. MUSIC, on the other hand, produces less accurate estimates compared to both PSO and CLPSO algorithm.

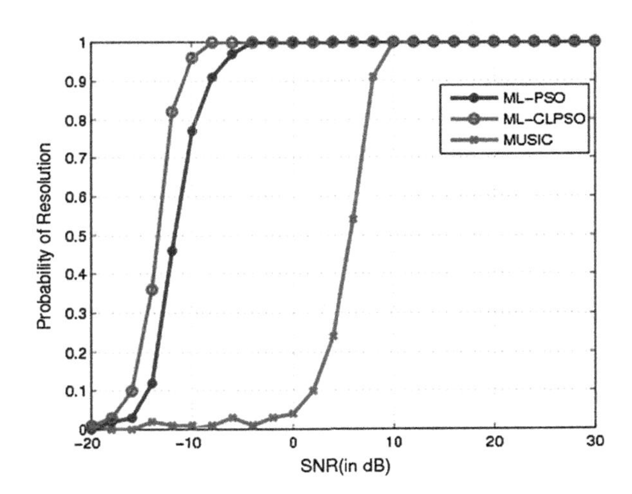

Fig. 1 DOA estimation RMSE values of PSO-EML, APSO-EML, and MUSIC versus SNR. Two uncorrelated sources impinge on WSN with 16 sensors at 130° and 138°

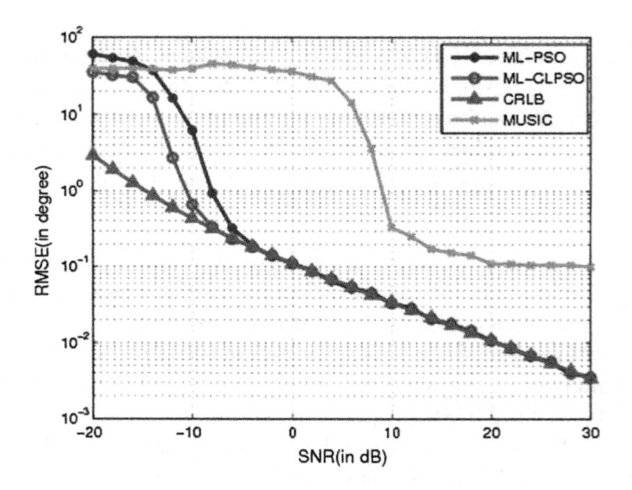

Fig. 2 Resolution probabilities of PSO-EML, APSO-EML and MUSIC versus SNR. Two uncorrelated sources impinge on WSN with 16 sensors at 130° and 138°

6 Conclusion

In this paper, a modified particle swarm optimization algorithm known as comprehensive learning particle swarm optimization (CLPSO) is proposed to estimate the source DOA using maximum likelihood function in distributed wireless sensor networks. With newly introduced features matching scheme and intelligent initialization, carefully selected evolution operators and fine-tuned parameters, the ML-CLPSO estimator achieves best performance. Simulation results demonstrate that the root mean square error (RMSE) performance of the ML-CLPSO algorithm asymptotically attain statistical Cramer-Rao lower bound (CRLB) at lower SNR than the ML-PSO and MUSIC. Similarly, the probability of resolution (PR) performance of the proposed method is much better compared to the existing algorithms.

References

1. Ziskind, I., Wax, M.: Maximum likelihood localization of multiple sources by alternating projection. IEEE Trans. Acoust. Speech Signal Process. **36**, 1553–1560 (1988)
2. Schmidt, R.: Multiple emitter location and signal parameter estimation. IEEE Trans. Antennas Propag. **34**, 276–280 (1986)
3. Capon, J.: High-resolution frequency–wavenumber spectrum analysis. Proc. IEEE **57**, 1408–1418 (1969)
4. Krim, H., Viberg, M.: Two decades of array signal processing research: the parametric approach. IEEE Signal Process. Mag. **13**, 67–94 (1996)
5. Noel, M.M., Joshi, P.P., Jannett, T.C.: Improved maximum likelihood estimation of target position in wireless sensor networks using particle swarm optimization. In: Third International Conference on Information Technology: New Generations vol. 0, pp. 274–279 (2006)

6. Li, M., Lu, Y.: Maximum likelihood DOA estimation in unknown colored noise fields. IEEE Trans. Aerosp. Electron. Syst. **44**, 1079–1090 (2008)
7. Chung, P.J., Böhme, J.F.: Doa estimation using fast EM and SAGE algorithms. Signal Process **82**, 1753–1762 (2002)
8. Rodriguez, J., Ares, F., Moreno, E., Franceschetti, G.: Genetic algorithm procedure for linear array failure correction. Electron. Lett. **36**, 196–198 (2000)
9. Li, M., Lu, Y.: A refined genetic algorithm for accurate and reliable DOA estimation with a sensor array. Wirel. Pers. Commun. **43**, 533–554 (2007)
10. Panigrahi, T., Rao, D.H., Panda, G., Mulgrew, B., Majhi, B.: Maximum likelihood DOA estimation in distributed wireless sensor network using adaptive particle swarm optimization. In: ACM International Conference on Communication, Computing and Security (ICCCS2011) pp. 134–136 (2011)
11. Panigrahi, T., Panda, G., Majhi, B.: Maximum likelihood source localization in wireless sensor network using particle swarm optimization. Int. J. Signal Imaging Syst. Eng. **6**, 83–90 (2013)
12. Panigrahi, T., Panda, G., Mulgrew, B., Majhi, B.: Distributed doa estimation using clustering of sensor nodes and diffusion PSO algorithm. Swarm Evol. Comput. **9**, 47–57 (2013)
13. Panigrahi, T., Panda, G., Mulgrew, B.: Distributed bearing estimation techniques using diffusion particle swarm optimization algorithm. IET Wireless Sens. Syst. **2**, 385–393 (2012)
14. Liang, J., Qin, A., Suganthan, P., Baskar, S.: Comprehensive learning particle swarm optimizer for global optimization of multimodal functions. IEEE Trans. Evol. Comput. **10**, 281–295 (2006)
15. Trees, H.V.: Optimum array processing. 1em plus 0.5em minus 0.4em. Wiley-Interscience Publication, New York (2002)
16. Jaffer, A.G.: Maximum likelihood direction finding of stochastic sources: a separable solution. In: Proceedings of ICASSP, vol. 5, pp. 2893–2896 (1988)
17. Kennedy, J., Eberhart, R.C.: Particle swarm optimization. In: Proceedings of IEEE Conference Neural Network, vol. IV, pp. 1942–1948 (1948)

Performance Evaluation of Some Clustering Indices

Parthajit Roy and J.K. Mandal

Abstract This paper analyzes the performances of four internal and five external cluster validity indices. The internal indices are Banfeld-Raftery index, Davies-Bouldin index, Ray-Turi index and Scott-Symons index. Jaccard index, Folkes-Mallows index, Rand index, Rogers-Tanimoto index and Kulczynski index are the external indices considered. The standard K-Means algorithm and CLARA algorithm has been considered as testing models. Four standard data sets, namely Iris, Seeds, Wine and Flame data sets has been chosen for testing the performance of the indices. The performance of the indices with the increasing number of parameters of the data set is measured. The results are compared and analyzed.

Keywords Data clustering · Internal index · External index · Cluster validity · Jaccard index · Davies-Bouldin index

1 Introduction

Clustering is the grouping of data points by exploiting the underlying structural representation and proximity among the data points [11, 20]. As training data set is not available in the case of clustering, the clustering models learn the underlying dynamics of the data points from the data set itself [7]. To measure the result of clustering, researchers have proposed several indices. These indices are collectively called cluster validity indices. Almost all indices consider either distribution of the

P. Roy (✉)
Department of Computer Science, The University of Burdwan, Burdwan 713104, India
e-mail: roy.parthajit@gmail.com

J.K. Mandal
Department of Computer Science and Engineering, The University of Kalyani, Nadia, Kalyani 741235, India
e-mail: jkm.cse@gmail.com

L.C. Jain et al. (eds.), *Computational Intelligence in Data Mining - Volume 3*,
Smart Innovation, Systems and Technologies 33, DOI 10.1007/978-81-322-2202-6_46

data points or the prior knowledge about the data set. This paper considers some popular indices of both the types.

Based on the knowledge available about the data set, the whole clustering indices have been divided into two sets. Internal and external validity indices. Internal validity indices only exploits distribution of the data set. External indices, on the other hand, assume some external information, like class information, class correspondence of the data points etc. It is obvious that external indices gives better result than the internal indices as the association of the cluster points with the class is assumed to be known in case of external indices.

The present paper considers four internal indices namely Banfeld-Raftery [1], Davies-Bouldin [4], Ray-Turi [16] and Scott-Symons [18] indices. All of these indices are minimum rule indices i.e. the minimum the index value the more accurate the result. Five external indices, namely Jaccard index, Folkes-Mallows index, Rand index, Rogers-Tanimoto index [5] and Kulczynski index [13]. The external indices are all maximum rule based indices i.e. the larger index value is better. In all external cases, the value of the index lies between 0 and 1 where 0 means no match and 1 means perfect match. Some other indices are Xie-Beni index [19], Dunn Index [6], Ratkowsky-Lance index [15] etc. The Present paper assumes four standard data sets for testing purposes. These data sets are Iris data set [8], Seeds data set [3], Wine data set [9] and Flame data set [10]. This type of review is not new. Some early works in this filed has been done by Bezdek et al. [2], Maulik et al. [14] and Saha et al. [17].

The rest of the paper is organized as follows. Section 2 introduces the cluster validity indices with their mathematical backbones. Section 3 describes the material and the methods for the experiments. Section 4 shows the results and analysis. Conclusion is given at Sect. 5 and references are drawn at the end.

2 Clustering Indices and Mathematical Backgrounds

This section discusses the clustering indices and the mathematical backgrounds of them. Let $S = \{e_1, e_2, \ldots, e_N\}$ be the set of points to be clustered where $\forall i \in N, e_i \in R^M$ i.e. each e_i is an M dimensional vector. In clustering, M is called the number of parameters. The elements of S in the present paper will be denoted in a matrix format called $D_{N \times M}$ matrix. Each row of D will be called X_i where $X_i = e_i, \forall i \in 1, 2, \ldots, N$. Each column of D matrix is the value of all data points for a particular parameter. These column vector are of length N and these column vector will be called $P_m \forall m \in 1, 2, \ldots, M$. The present paper also assumes the number of clusters, K, as a priori. Clusters are denoted by $C_k\ \forall k \in 1, 2, \ldots, K$. The clustering in this setup can be defined as a vector V with |V| = N, the number of instances, where each cell of vector V has an integer value between 1 and K indicating the association of points to the cluster. Mathematically $\forall k \in K, C_k = \{X_n : X_n \in D_{N \times M} \,\&\, V_n = k\}$. The instances that belongs to cluster C_k is called $D^{\{k\}}$. This paper assumes the size of

$D^{\{k\}}$ is N_k. Given a cluster C_k, the center of the cluster is defined as, $\mu_k = \frac{1}{N_k}\sum_{i \in N \& V_i = k} X_i$ and the mean of all data instances is defined as $\mu_k = \frac{1}{N}\sum_{i=1}^{N} X_i$.

With these mathematical foundations two important concepts of validity indices will be developed. These two concepts are within cluster scatter and between cluster scatter. A cluster set is a good cluster set if its results is such that the within cluster scatter is low and between cluster scatter is high. Every internal index is based on these two facts on some way or the other. Mathematically these two concepts can be defined as follows.

Within Cluster Scatter (WCS) is a matrix of size $M \times M$ that can be defined as $WCS^{\{k\}} = \left(a_{ij}^{\{k\}}\right)_{M \times M}$ where $a_{ij}^{\{k\}} = N_k \times COV\left(p_i^{\{K\}}, p_j^{\{K\}}\right)$ and $a_{ii}^{\{k\}} = N_k \times \mathrm{var}\left(V_i^{\{K\}}\right)$.

The Between Cluster Scatter (BCS) is the measure scatter among the clusters. Let $X_i^{\{K\}}$ is the set of points that belongs to cluster k. So the between cluster scatter ca be defined as $BCS = B^T B$, a matrix of order K × K (The number of clusters) where each element of the matrix B can be defined as

$$b_{ij} = \sum_{k=1}^{K} N_k \left(\mu_k^{\{i\}} - \mu^{\{i\}}\right) \times \left(\mu_k^{\{j\}} - \mu^{\{j\}}\right) \quad (1)$$

With these definitions the present paper defines the internal indices one after another that it considers.

Banfeld-Raftery Index: Banfeld-Raftery index can be defined as:

$$C = \sum_{k=1}^{K} N_k \log\left(\frac{trace\left(WCS^{\{k\}}\right)}{N_k}\right) \quad (2)$$

where *Trace()* is the Trace of the matrix.

Davies-Bouldin Index: To define Davies-Bouldin index, let us first define the following.

Let us denote $\Delta_{kk'} = \|\mu_{k'} - \mu_k\|$ is the distance between the center of the clusters k and k'. With these, the Davies-Bouldin index can be defined as

$$C = \frac{1}{K}\sum_{k=1}^{K} \max_{k' \neq k}\left(\frac{\delta_k - \delta_{k'}}{\Delta_{kk'}}\right) \quad (3)$$

Ray-Turi Index: Assuming the definition of given above $\Delta_{kk'}$, the Ray-Turi index can be defined as

$$C = \frac{1}{N} \times \frac{\frac{1}{N}\sum_{k=1}^{K}\sum_{i \in N \& V_i = k} \|X_i - \mu\|^2}{\min\limits_{k' < k} \Delta_{kk'}^2} \quad (4)$$

Scott-Symons Index: The Scott-Symons index can be defined as

$$C = \sum_{k=1}^{K} N_k \log\left(Det\left(\frac{WCS^{\{k\}}}{N_k} \right) \right) \tag{5}$$

The external indices consider only the distributions of the points in different clusters which are known, i.e. the sample data for testing. When a labeled data set is given, the clustering algorithm clusters them into number of clusters. The algorithm may say the right thing or may say the wrong thing. Right saying are of two types. A point actually belongs to a cluster and the algorithm may say the same. This is called True Positive or TP. Alternatively the algorithm may say no to a point which is actually not a member of that cluster. This is called True Negative or TN. The Algorithm may say wrong thing also. Like right, wrong are also of two types. False Positive or FP means the point actually doesn't belong to the cluster but the algorithm says that it belongs and the False Negative or FN means that the point actually belongs to the cluster but the algorithm says that it does not. The following part of this subsection illustrates the external indices that this paper considers.

Jaccard Index: Jaccard index is the ratio of TP and sum of TP, FP and FN. i.e. Jaccard index is,

$$C = \frac{TP}{TP + FP + FN} \tag{6}$$

i.e. this is the total number of unique elements common to both the sets by total number unique elements in both the sets.

Folks-Mallows Index: The Folks-Mallows index or FM index is defined as follows

$$C = \sqrt{\frac{TP}{TP + FP}} \times \sqrt{\frac{TP}{TP + FN}} \tag{7}$$

Rand Index: Rand index can be defined as

$$C = \frac{TP + TN}{TP + FP + TN + FN} \tag{8}$$

Rogers-Tanimoto Index: The Rogers-Tanimoto index or RT index can be defined as

$$C = \frac{TP + TN}{TP + TN + 2(FP + FN)} \tag{9}$$

Kulczynski Index: the Kulczynski index can be defined as

$$C = \frac{1}{2} \times \left(\frac{TP}{TP + FP} + \frac{TP}{TP + FN} \right). \tag{10}$$

3 Experimental Setup

For experiment some model needs to be chosen. In the present case, this model is the algorithm. There are numerous algorithms available for data clustering. For all the algorithms, the performance of the indices cannot be checked. Further, the algorithms cannot be checked for all instances of data sets. For these, the present paper has decided to test the performance of the validity indices strategically. The present study is based on K-Means algorithm [21] and CLARA algorithm [12].

This paper considers K-Means and CLARA algorithm and four data with different capabilities. The performance of the indices is measured against these data sets. Table 1 shows the brief description of the data sets.

Flame dataset has two attributes and 240 instances. Iris data set [8] is the most popular data set for data clustering. The number of instances is 150 and the number of attributes is 4. The total number of classes is 3. The Seeds data set has 210 instances of information and 7 attributes. The number of classes is 3. In case of Wine data set, the number of data instances is 178 and the number of attributes is 13. The number of classes is again three. The objective is to observe the performances of the different clustering indices with the increasing number of attributes, against the same algorithm, when the other thing like number of instances, number of classes are almost invariant. The present paper considers the data in its original form. No data refinement or data transformation or data normalization has been done. The data in its raw form has been passed to K-Means algorithm and CLARA algorithm for the clustering and the performance of the results has been measured in different indices. The idea is, as the algorithm is same, the result is same. Then the indexes which will results more closure to the actual is the best index. The Table 1 shows the summary of the data sets.

The external indices are easier to analyze. As the result produced by the algorithm is being compared with the actual results, the performances of the indices can be measured directly. In all the cases of external indices, the 100 % accurate result is indicated by 1 and all other results are between 1 and 0.

Table 1 The details of the standard datasets considered for experiment

Data	Instances	Attributes	Missing values	No. of clusters
Iris	150	4	None	3
Wine	178	13	None	3
Seed	210	7	None	3
Flame	240	2	None	2

4 Result and Discussions

The experimental results are shown in Tables 2, 3, 4, 5 and 6. The Table 2 shows that the performance of Roy-Turi is good as the values are closure to zero but it is not consistent because its values are not at per with the external indices. When an external index says a result is bad as in case of Wine data, the index is saying it is not so good. The Banfeld-Raftery index is not also consistent in that sense. The result of Scott-Symons is not also good because the Scott-Symons index has even failed to produces a valid value in case of Seeds data. This is no wander as the Scott-Symons index finds the determinant of a semi-definite matrix and thereafter the logarithm. So, if the resultant determinant value is zero, the log of zero becomes undefined.

Davies-Bouldin index is a good and consistent. In Table 4, the correlation of the three internal indices with the Jaccard index has been shown. The correlation coefficients in Table 4 show that the Davis-Bouldin index is more close to the external index. As the Jaccard index is computed from the actual class distribution and the K-Means suggested distribution, the Jaccard index, or any other external index can be taken as the benchmark distribution for the analysis of internal indices.

Table 2 The result of the internal indices based on K-Means algorithm

Data	Banfeld Raftery (rule: min)	Davies-Bouldin (rule: min)	Ray-Turi (rule: min)	Scott-Symons (rule: min)
Iris	−104.8013	0.6206659	0.162755	−1604.542
Wine	1653.275	0.5342432	0.1822258	−916.249
Seed	215.4227	0.8025661	0.2094518	NaN
Flame	617.6787	1.121857	0.38538	788.2648

Table 3 The result of the external indices based on K-Means algorithm

Data	Jaccard (rule: max)	Folkes-Mallows (rule: max)	Rand (rule: max)	Rogers-Tanimoto (rule: max)	Kulczynski (rule: max)
Iris	0.6958588	0.8208081	0.8797315	0.8037773	0.8209596
Wine	0.4119676	0.583537	0.7186568	0.5605288	0.5835371
Seed	0.6815294	0.8106152	0.8743677	0.781016	0.8106238
Flame	0.6104777	0.7586289	0.7498257	0.6379602	0.7591257

Table 4 Correlation of internal indices with Jaccard index (produced by K-Means algorithm)

	Banfeld Raftery versus Jaccard	Davies Bouldin versus Jaccard	Ray-Turi versus Jaccard	Scott-Symons versus Jaccard
Correlation value	−0.991943638	0.717191268	0.05075541	−1

Table 5 The result internal indices based on Clara algorithm

Data	Banfeld Raftery (rule: min)	Davies-Bouldin (rule: min)	Ray-Turi (rule: min)	Scott-Symons (rule: min)
Iris	−104.2308	0.6259434	0.1653186	−1613.243
Wine	1648.66	0.4678189	0.2038049	NaN
Seeds	215.8931	0.8054753	0.216418	NaN
Flame	620.3148	1.115713	0.3962113	822.39

Table 6 The result of external indices based on Clara algorithm

Data	Jaccard (rule: max)	Folkes-Mallows (rule: max)	Rand (rule: max)	Rogers-Tanimoto (rule: max)	Kulczynski (rule: max)
Iris	0.7084381	0.8294489	0.885906	0.8109436	0.8295577
Wine	0.4264369	0.597912	0.7295118	0.5712865	0.5979189
Seeds	0.6676616	0.8007185	0.8680793	0.7691997	0.8007211
Flame	0.6113639	0.7592351	0.7498257	0.6347965	0.759655

The external indices, on the other hand are all consistent. Figure 1 shows that all of the external indices produce comparable results. But our conclusion is Rogers Tanimoto index is the best among these four indices. The reason for not choosing the other indices is given below. The Folkes-Mallows and the Jaccard index does not consider the True Negative (TN) in its computations. Rand and Rogers-Tanimoto indices are comparable because they considers all the four cases of data inclusion but present paper prefers Rogers-Tanimoti because there index is more sensitive to False

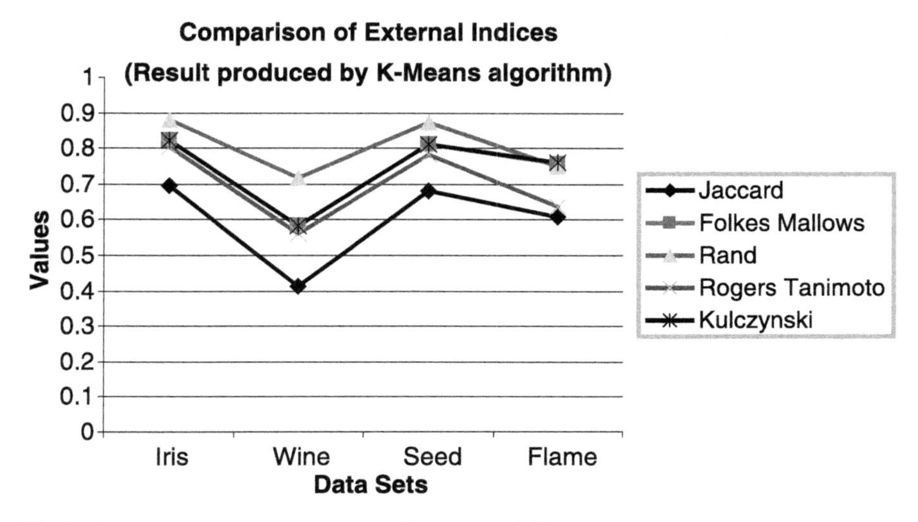

Fig. 1 The comparative performance of the external indices

Positive (FP) and False Negative (FN), by giving them double weight which is not the case of Rand index.

5 Conclusions

This paper endeavors to review the performance of different internal and external cluster validity indices. The careful choice of the standard data sets gives the better idea about the performance of the validity indices. Both external and inter validity indices consideration is the strength of the paper. Nevertheless, there is scope of improvement in these directions. More number of cluster validity indices will definitely strengthen the weight of the research. The variation of the data set is another dimension where there is scope of improvement. The Number of clustering algorithms is another direction where the pros and cons of the cluster validity indices can be understood more clearly.

Acknowledgments The authors express the deep sense of gratitude to the Department of Computer Science, the University of Burdwan, West Bengal, India and the DST, PURSE Program, Government of India running under the University of Kalyani, West Bengal, India, for providing necessary infrastructure and support for the present work.

References

1. Banfeld, J.D., Raftery, A.E.: Model-based gaussian and non-gaussian clustering. Biometrics **49**, 803–821 (1993)
2. Bezdek, J.C., Pal, N.R.: Some new indexes of cluster validity. IEEE Trans. Syst. Man Cybern. Part B Cybern. **28**(3), 301–315 (1998)
3. Charytanowicz, M., Niewczas, J., Kulczycki, P., Kowalski, P., Lukasik, S., Zak, S.: A complete Gradient Clustering Algorithm for Features Analysis of X-Ray Images. Information Technologies in Biomedicine, pp. 15–24 . Springer, Berlin (2010)
4. Davies, D.L., Bouldin, D.W.: A cluster separation measure. IEEE Trans. Pattern Anal. Mach. Intell. PAMI **1**(2), 224–227 (1979)
5. Desgraupes, B.: Clustering indices. Technical Report, University Paris Ouest, Lab Modal'X (2013)
6. Dunn, J.: Well separated clusters and optimal fuzzy partitions. J. Cybern. **4**, 95–104 (1974)
7. Everitt, B.S., Stahl, D., Leese, M., Landau, S.: Cluster Analysis, 5 edn. Wiley, New York (2011)
8. Fisher, R.A.: UCI Machine Learning Repository. University of California, Irvine, School of Information and Computer Sciences. http://archive.ics.uci.edu/ml (1936)
9. Forina, M.: UCI Machine Learning Repository. Institute of Pharmaceutical and Food Analysis and Technologies. http://archive.ics.uci.edu/ml
10. Fu, L., Medico, E.: FLAME, a novel fuzzy clustering method for the analysis of DNA microarray data. BMC Bioinf. **8**(1), 3 (2007). http://cs.joensuu.fi/sipu/datasets
11. Jain, A.K., Murty, M.N., Flynn, P.J.: Data clustering: a review. ACM Comput. Surv. **31**(3), 264–323 (1999)
12. Kaufman, L., Rousseeuw, P.J.: Finding Groups in Data: An Introduction to Cluster Analysis. Wiley, New York (1990)

13. Kulczynski, S., Die Pflanzenassoziationen der Pieninen: Bulletin International de l'Academie Polonaise des Sciences et des Lettres. Classe des Sciences Mathematiques et Naturelles, B, 57–203 (1927)
14. Maulik, U., Bandyopadhyay, S.: Performance evaluation of some clustering algorithms and validity indices. IEEE Trans. Pattern Anal. Mach. Intell. **24**(12), 1650–1654 (2002)
15. Ratkowsky, D.A., Lance, G.N.: A criterion for determining the number of groups in a classification. Aust. Comput. J. **10**, 115–117 (1978)
16. Ray, S., Turi, R.H.: Determination of number of clusters in k-Means clustering and application in colour image segmentation. In: Proceedings of the 4th International Conference on Advances in Pattern Recognition and Digital Techniques, pp. 137–143 (1999)
17. Saha, S., Bandyopadhyay, S.: Performance evaluation of some symmetry-based cluster validity indexes. IEEE Trans. Syst. Man Cybern. Part C Appl. Rev. **39**(4), 420–425 (2009)
18. Scott, A.J., Symons, M.J.: Clustering methods based on likelihood ratio criteria. Biometrics **27**, 387–397 (1971)
19. Xie, X.L., Beni, G.: A validity measure for fuzzy clustering. IEEE Trans. Pattern Anal. Mach. Intell. **13**(8), 841–847 (1991)
20. Xu, R., Wunsch-II, D.: Survey of clustering algorithms. IEEE Trans. Neural Netw. **16**(3), 645–678 (2005)
21. Yu, J.: General c-means clustering model. IEEE Trans. Pattern Anal. Mach. Intell. **27**(8), 1197–1211 (2005)

A New (n, n) Blockcipher Hash Function Using Feistel Network: Apposite for RFID Security

Atsuko Miyaji and Mazumder Rashed

Abstract In this paper, we proposed a new (n, n) double block length hash function using Feistel network which is suitable for providing security to the WSN (wireless sensor network) device or RFID tags. We use three calls of AES-128 (E_1, E_2, E_3) in a single blockcipher E' so that the efficiency rate is 0.33. Surprisingly we found that the security bound of this scheme is better than other famous (n, n) based blockcipher schemes such as MDC-2, MDC-4, MJH. The collision resistance (CR) and preimage resistance (PR) security bound are respectively by $O(2^n)$ and $O(2^{2n})$. We define our new scheme as JAIST according to our institute name.

Keywords AES · Deal cipher model · Collision security · Preimage security

1 Introduction

A cryptographic hash function is a function which maps an input of arbitrary length to an output of fixed length. For any cryptographic hash at least collision resistance and preimage resistance properties should be satisfied.

Cryptographic hash function can be constructed by blockcipher or scratch. Due to adversarial successful attacks on MD4/5 and SHA-family [20, 21] blockcipher based hash functions gained popularity. Another issue is low cost requirement in respect of power and chip size. The AES module requires only a third of the chip area and half of the mean power in compare of SHA. Smaller hash functions like SHA-1, MD5 and MD4 are also less suitable for RFID tags than AES. Inclusively it can be said that the total power consumption of SHA-1 is about 10 % higher than

A. Miyaji (✉) · M. Rashed
Japan Advanced Institute of Science and Technology,
1-1, Asahidai, Nomi-Shi, Ishikawa, Japan
e-mail: miyaji@jaist.ac.jp

M. Rashed
e-mail: rashed@jaist.ac.jp

L.C. Jain et al. (eds.), *Computational Intelligence in Data Mining - Volume 3*,
Smart Innovation, Systems and Technologies 33, DOI 10.1007/978-81-322-2202-6_47

Table 1 Different (n, $2n$) and (n, n) based blockcipher result analysis [6, 10, 17]

Hash type	Comp. function	Eff. rate	No. of E calls	Key sch.	Coll. resistance	Pre. resistance
Weimar	$3n \rightarrow 2n$	1/2	2	2	$O(2^n)$	$O(2^{2n})$
Hirose	$3n \rightarrow 2n$	1/2	2	1	$O(2^n)$	$O(2^{2n})$
ISA-09	$4n \rightarrow 2n$	2/3	3	3	$O(2^n)$	–
MDC-2	$3n \rightarrow 2n$	1/2	2	2	$O(2^{n/2})$	$O(2^n)$
MDC-4	$3n \rightarrow 2n$	1/4	4	1	$O(2^{5n/8})$	$O(2^{5n/4})$
MJH	$3n + c \rightarrow 2n$	1/2	2	1	$O(2^{n/2})$	$O(2^n)$
JAIST (proposed)	$3n \rightarrow 2n$	1/3	3	3	$O(2^n)$	$O(2^{2n})$

the AES [9]. Day by day the uses of RFID tags or WSN's devices has been increased rapidly. So it is a great challenge for the scientists to provide a scheme which can make balance between cost and security issues.

Blockcipher based hash functions are classified into single-block-length (SBL) and double-block-length (DBL). The output length of SBL hash function is equal to the block length and DBL hash function is the twice of block length. Due to birthday attack the collision resistance of hash function can be occurred with the time complexity $O(2^{1/2})$ (l: output length of hash function). So SBL hash function is no longer secure in terms of CR. From the following Table 1 current research status of DBL has been found which categorizes into (n, n) and (n, $2n$) blockcipher. In CT-RSA-09, it is found that (n, n) based blockcipher hash function is 40 % faster than (n, $2n$) blockcipher hash function [13]. For implementation purpose critical issue is to measure the cost of security under RFID tags or WSN's devices. According to [14], AES-128 is more user friendly because of less power consumption, less encryption and decryption time.

2 Related Work

At first we mention some famous (n, n) blockcipher hash function which are based on AES-128. MDC-2 and MDC-4 has been introduced in the late eighties by Bracht et al. but their CR and PR security bound prove have been achieved in Africacrypt-2012 [8]. The CR and PR security bound result of MDC-2 and MDC-4 are respectively by $(O(2^{3n/5}), O(2^n))$ and $(O(2^{5n/8}), O(2^{5n/4}))$. Another famous MJH hash function introduced in CT-RSA (2011). It's CR and PR bound is as $O(2^{n/2})$ and $O(2^n)$. Two things are remarkable here such as no (n, n) based blockcipher hash function's CR and PR security bound are equal to $O(2^n)$ and $O(2^{2n})$. In other hand if we follow the (n, $2n$) based blockcipher then we can find that the security bound of AES-256 based hash function is better than AES-128 based hash function. Weimar-DM [6] double block length hash functions has been proposed in ACISP-2012 where

CR and PR bound is respectively $O(2^n)$ and $O(2^{2n})$. Other famous two schemes of Abreast and Tandem-DM which were proposed Lai and Messy [11]. The CR and PR of Abreast-DM and Tandem-DM was being proved by Lee et al. [15]. The CR and PR bound of these two schemes are respectively $O(2^n)$ and $O(2^{2n})$. In FSE 2006, Hirose [7] proposed another famous construction and shoed that it was bound in $O(2^n)$ for the CR and $O(2^n)$ for the PR. Later this PR security bound has been improved by Armknecht et al. [1].

3 Preliminaries

3.1 Ideal Cipher Model

A blockcipher is a keyed function $E : \{0,1\}^k \times \{0,1\}^n \to \{0,1\}^n$. For each $k \in \{0,1\}^k$ the function $E_k(\cdot) = E(k,\cdot)$ is a permutation on $\{0,1\}^n$. If E is a block cipher the E^{-1} denotes it's inverse, where $E_k(x) = y$ and $E_k(x) = y$ is called forward and backward query respectively. Assume that block (k, n) is the family of all blockciphers $E : \{0,1\}^k \times \{0,1\}^n \to \{0,1\}^n$. A blockcipher based hash function $H : \{0,1\}^* \to \{0,1\}^l$ where $E \in Block(k,n)$ used as round function. An adversary is given access to oracle E/E^{-1} and for the ith query response q_i, adversary keeps the record. In the ICM model the complexity of an attack is measured by the total number of the optimal adversary's queries to the two oracles E and E^{-1}.

3.2 Security Definition

An adversary is a computationally unbounded but always-halting collision-finding algorithm A with resource-bounded access to an oracle $E \in Block(k,n)$ that means in the collision resistance experiment, a computationally unbounded adversary A is given oracle access to a blockcipher E. It is allowed that, A can make query to a both E and E^{-1}.

Definition 1 Collision resistance of a compression function: The adversary A is given oracle access and f be a blockcipher based hash function, then the advantage of A to find collisions in f is:

$$\Pr\left[\begin{array}{l} E \leftarrow B(k,n); (h,g,m),(h',g',m') \leftarrow A^{E,E^{-1}} : (h,g,m) \neq (h',g',m') \\ \wedge f^E(h,g,m) = f^E(h',g',m') \vee f^E(h,g,m) = (h_0,g_0) \end{array}\right]$$

For $q \geq 1$, it can expressed that, $Adv_f^{COMP}(q) = \max_A\left\{Adv_f^{COMP}\right\}$ maximum is taken over all adversaries which can query at best q oracle queries.

Definition 2 Preimage resistance of a compression function: The adversary A is given oracle access to a block cipher $E \in Block(k,x)$ and f be blockcipher based hash function. Adversary A arbitrary selects a value of (h', g') before making any query to oracle either E or E^{-1}. Then the advantage of A is to find preimage in f such as $Adv_f^{pre}(A) = \Pr[H(g,h,m) = (h',g')]$.

For $q \geq 1$, $Adv_f^{pre}(q) = \max_A \left\{ Adv_f^{pre} \right\}$ where the maximum is taken over all adversaries which can query at best q oracle queries.

4 A New (*n, n*) Double Block Length Hash Function

In this section, a new (n, n) double block length hash function has been discussed with diagram which is defined as JAIST scheme according to our institute name. This scheme is created based on Feistel network which is shown in Fig. 1b. In ISA-2009, there was another scheme based on Feistel network which actually inspires us. We should mention here the similarities and dissimilarities of these two schemes. Our scheme is based on (n, n) blockcipher where ISA-09 scheme is based on

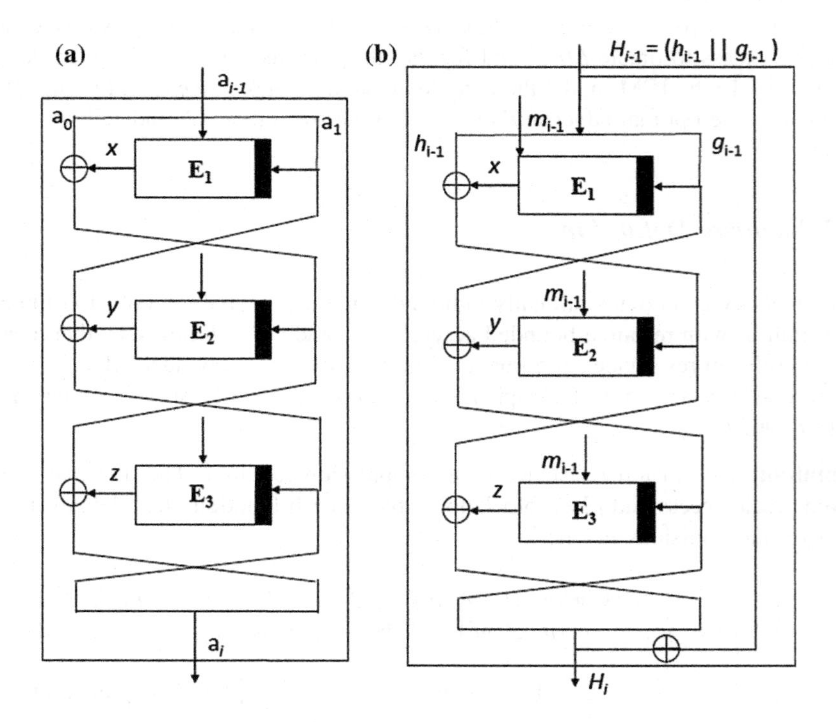

Fig. 1 **a** Compression function (*Source* LNCS 5576, p. 16), 1. **b** JAIST scheme

(n, 2n) blockcipher [8]. In our CR proof technique we used the idea of external and internal collision. Also we provide PR security proof according to Armknecht [1].

Definition 3 Let $E(k, n)$ be a blockcipher taking k: n bit key and n-bit block size such as $E : \{0,1\}^k \times \{0,1\}^n \rightarrow \{0,1\}^n (k = n)$. We define, $E' : \{0,1\}^k \times \{0,1\}^{2n} \rightarrow \{0,1\}^{2n}$ be a DBL cipher using Feistel network created by calling three independent blockcipher of E.

So $E' : \{(a_0, a_1), m_i\} = (y \oplus a_1)||(x \oplus z \oplus a_0)$ where x, y, z is defines as:

$$x : E_{a_1}(m_i),\ y : E_{a_0 \oplus x}(m_i),\ \ z : E_{a_1 \oplus y}(m_i).$$

Definition 4 Let $F : \{0,1\}^n \times \{0,1\}^{2n} \rightarrow \{0,1\}^{2n}$ be a compression function. Then we replace the compression function by our defined blockcipher E' from Definition 3 and Fig. 1a. Such as, $f(h_{i-1}, m_i) = E'_a(m) \oplus c$.

5 Security Analysis of the JAIST Scheme

5.1 Collision Security Analysis

For any collision resistance finding experiment, a computationally unbounded adversary A is given oracle access to a blockcipher E uniformly sampled among all blockciphers of key length n and message length n. A is allowed to query both E and E^{-1}. After q queries to E, the query history of A is the set of triples Q = (Xi, Ki, Yi) such that E(Ki, Xi) = Yi and ith query of adversary A is either E(Ki, Xi) or E^{-1}(Ki, Yi) for $1 \leq i \leq q$. Assume that $Q = \{(X_i, Y_i, K_i)\}_{j=1}^{i}$ be the first i elements of the query history. Then it can be said that A succeeds or finds a collision after its first i queries if there exists distinct $(H_{i-1}, m_i), (H'_{i-1}, m'_i)$ such that,

$$H^{JAIST}(H_{i-1}, m_i) = H'^{JAIST}\left(H'_{i-1}, m'_i\right)$$

Theorem 1 *Let H^{JAIST} be a double-length hash function composed of compression Function F specified in Definition* 1. *Then the advantage of an adversary in finding a collision in H^{JAIST} after q queries can be upper bounded by:*

$$\frac{3q}{2(N-1)^2}$$

Proof It is assumed that the adversary has made any relevant query to E or E^{-1} which can occur collision in the ideal cipher model. Another issue is the adversary never makes a query which is already available at his query database. In formal meaning, one can assume that the adversary never makes a query E(K, X) = Y obtaining an answer Y and then makes the query $E^{-1}(K, Y) = X$ (which will necessarily be answered by X). At first consider, adversary A which is able to make an arbitrary q-query collision. Let A be a collision-finding algorithm of H^{JAIST} with oracles E, E^{-1}. A asks q pairs of queries to E, E^{-1} in total. Since, h' and g' depends both on the plaintext and ciphertext of E/E^{-1}. One of them is fixed by a query and other is determined by randomly from the query-database Q. As a result h', g' selected randomly from the query and query-oracle-database. It can be describe as like tabular form in the Table 2. At first one of the important issue should be raised here, in our scheme we construct E' from three calls of E and then combine the result of $h_i||g_i = H_i$. So at first we find out the security of desire (E_1, E_2, E_3) of three calls of blockcipher under some conditions and assumptions. Under the following three cases we mentioned these.

External Collision

For every j, $[where, j \leq q]$, let C_j be the event that a colliding pair found for F with the jth pair of queries. The event is as like $j' < j$:

$$y \oplus h_{i-1} \oplus g_{i-1} = y' \oplus h'_{i-1} \oplus g'_{i-1}$$
$$x \oplus z \oplus h_{i-1} \oplus g_{i-1} = x' \oplus z' \oplus h'_{i-1} \oplus g'_{i-1}$$

It implies that, $H_{i-1} \neq H'_{i-1} \wedge m_{i-1} = m'_{i-1}$. According to Table 2. described three cases can be occurred for collision resistance. So at first we need to find out this probability of these three cases. This is trivial and also it can be said that from the famous PGV [3] paper for any case the probability is as like $(1/2^n - 1)^2$. So if C_j be the event that A finds a collision pair of the compression function for with the jth pair of queries. Then three queries to the oracle E/E^{-1} are required to compute the output of the compression function for above conditions. It implies that $\Pr[C_j] \leq (j-3)\left(\frac{1}{2^n-1}\right); (j \geq 3)$. Let C be the event that a pair is found for F with q pairs of queries then,

$$\Pr[C] = \Pr[C_3 \vee C_4 \vee C_5 \vee \cdots \vee C_q] = \left(\frac{1}{2^n-1}\right)^2 \cdot \sum_{j=3}^{q} (j-3)$$
$$= \left(\frac{1}{2^n-1}\right)^2 \cdot \frac{(q-2)(q-3)}{2} \tag{1}$$

Table 2 Sub case of $H_{i-1} \neq H'_{i-1} \wedge m_{i-1} = m'_{i-1}$

$h_{i-1} \neq h'_{i-1}, g_{i-1} = g'_{i-1}, m_{i-1} = m'_{i-1}$
$h_{i-1} = h'_{i-1}, g_{i-1} \neq g'_{i-1}, m_{i-1} = m'_{i-1}$
$h_{i-1} \neq h'_{i-1}, g_{i-1} \neq g'_{i-1}, m_{i-1} = m'_{i-1}$

Internal Collision
This is actually for internal collision that means there is a probability to collide such as $(h_i = g_i) \Rightarrow y \oplus h_{i-1} \oplus g_{i-1} = x \oplus z \oplus h_{i-1} \oplus g_{i-1}$. Let C be the event that a pair are found or F with q pairs of queries, then,

$$\Pr[C] = \Pr\left[C_3 \vee C_4 \vee C_5 \vee \cdots \vee C_q\right] = \frac{(q-2)(q-3)}{2(2^n-1)^2} \quad (2)$$

IV Collision
For $3 \le j \le q$, it is the event for collision occur with the IV value such as $(h_i = h_0, g_i = g_0) \Rightarrow (y \oplus h_{i-1} \oplus g_{i-1} = h_0, x \oplus z \oplus h_{i-1} \oplus g_{i-1} = g_0)$. Let C be the event that a pair are found for F with q pairs of queries, then

$$\Pr[C] = \Pr\left[C_3 \vee C_4 \vee C_5 \vee \cdots \vee C_q\right] = \frac{(q-2)(q-3)}{2(2^n-1)^2} \quad (3)$$

Take the result from Eqs. (1), (2) and (3). Then finally it is shown that,

$$\left(\frac{1}{2^n-1}\right)^2 \cdot \frac{(q-2)(q-3)}{2} + \frac{(q-2)(q-3)}{2(2^n-1)^2} + \frac{(q-2)(q-3)}{2(2^n-1)^2} \le \frac{3q^2}{2(N-1)}$$

5.2 Preimage Security Analysis

Let A be an adversary that tries to find a preimage for its input σ which is randomly chosen by A before making any query to oracle. We follow a similar proof strategy of Armknecht and implementation strategy of Armknecht [1], when A selects its queries. The adversary A asks the conjugate queries in pair. Now adversary needs to bind the probability that ith query pair leads to a preimage for σ where σ is defined as $(h'||g') \in \sigma$. So findings is that to calculate the probability that in q queries the adversary finds a point (σ), such that $H^{JAIST}(h, g, m) = \{(h'||g')\}$.

Theorem 2 *Let H^{JAIST} be a double-length compression Function $(E \in block(n, n))$. Then the advantage of an adversary in finding a preimage in H^{JAIST} after q queries can be upper bounded by*

$$= \frac{16q}{2^{2n}}$$

Proof According to definition of adjacent query pair [1], the adversary B maintains an adversary query database Q in the form of $y \oplus h_{i-1} \oplus g_{i-1}, x \oplus z \oplus h_{i-1} \oplus g_{i-1}$ which has been run by adversary A. This is called adjacent query pair. Now need to make and implement super query. It implies that the query contains exactly $N/2$

queries with the same key, all remaining queries under this key are given for free to the adversary. Now an adjacent query pair $y \oplus h_{i-1} \oplus g_{i-1}, x \oplus z \oplus h_{i-1} \oplus g_{i-1}$ can be succeed iff,

$$y \oplus h_{i-1} \oplus g_{i-1} = y' \oplus h'_{i-1} \oplus g'_{i-1}$$
$$x \oplus z \oplus h_{i-1} \oplus g_{i-1} = x' \oplus z' \oplus h'_{i-1} \oplus g'_{i-1}$$

Thus the adversary obtains a preimage of $\left\{(h'||g') \in \{0,1\}^{2n}\right\} \in \sigma$ in particularly if it attains a winning query pair. It can be occurred by any of the following way such as NormalQueryWin(Q) and SuperQueryWin(Q).

Case 1: *Probability* of NormalQueryWin(Q)

Adversary B which has been called by adversary A, can make forward or backward query. Under this section, the goal is to find out the NormalQueryWin (Q). According to super query and adjacent query pair [1] the fresh value of h_i/g_i could be found in the following two ways.

- *Sub-Case* 1.1 The adversary B can make forward or backward query. Assume adversary makes a forward query, where at most $\left(2^n/2 - 1\right)$ queries could be answered previously and earlier for adjacent query it could be answered at most $\left(2^n/2 - 1\right)$ queries. Otherwise super query can be occurred. So the value of h_i and g_i comes uniformly and independently from the set size $2^n/2$. So probability forms as $\left(2/2^n/2\right)$.
- *Sub-Case* 1.2 If $y \oplus h_{i-1} \oplus g_{i-1} = h_i$ then there is a probability for the free query (part of adjacent query pair) to return from the set size $\left(2^n/2 + 1\right)$. So probability could be $(1/2^n/2) = 2/2^n$.

$$\text{So desired probability of NormalQueryWin(Q) is } 8/2^{2n} \tag{4}$$

Case 2: Probability of SuperQueryWin(Q)

In this section the target is to find out the probability of Super query. As for example under the keys, the value of h'/g' already have been known on exactly $2^n/2$ points. So from the definition of super query and adjacent query pair [1] if any pair of the query is the part of super query then the corresponding others query must be the member of the super query domain. From the above discussed points, it can be said that, probability of any query of any blockcipher is either 0 or $2/2^n$. Now the question how it can be found. The probability will be 0 if the h' is not in the range of super query that means it is available in the domain of normal query. Conversely it is assumed that due to super query the result comes from the set size $2^n/2$, so probability is $2/2^n$. For the adjacent query pair following cases can be happened:

$$y \oplus h_{i-1} \oplus g_{i-1} = y' \oplus h'_{i-1} \oplus g'_{i-1}, x \oplus z \oplus h_{i-1} \oplus g_{i-1} = x' \oplus z' \oplus h'_{i-1} \oplus g'_{i-1}$$
$$y \oplus h_{i-1} \oplus g_{i-1} = x' \oplus z' \oplus h'_{i-1} \oplus g'_{i-1}, x \oplus z \oplus h_{i-1} \oplus g_{i-1} = y' \oplus h'_{i-1} \oplus g'_{i-1}$$

- Sub-Case 2.1 For the above first condition, the answer will come from the set size $2^n/2$. So the probability would be $2/2^n$. As well as the probability for the second equation is equal to $2/2^n$. So total probability of sub-Case 2.1 looks like $(2/2^n)^2$.
- Sub-Case 2.2 As like same explanation of sub-Case 2.1, the total probability of sub-Case 2.2 is $(2/2^n)^2$.

Now, analysis the probability of case-1 and case-2 and point that the cost of super query occurs is $2^n/2$. Another important factor is that the probability of super query occurs, which is at most $q/2^n/2$. It implies that, Pr[SuperQueryWin (Q)]:

$$\leq {}^{q}\!/\!_{(2^n/2)} \times (2^n/2) \times 2 \times \left(\frac{2}{2^n}\right)^2 = \frac{8}{2^{2n}} \tag{5}$$

Taking the value of Eqs. (4) and (5), we got the final probability of PR security bound is $16q/2^{2n}$.

6 Conclusion

In this article, a new scheme of DBL hash function has been proposed which is based on (n, n) blockcipher. The CR and PR security bound are respectively $O(2^n)$ and $O(2^{2n})$. The result of this construction is better than existing other (n, n) based blockcipher hash function and also this scheme is suitable for providing security to RFID tags/WSN's because of faster operation of AES-128. In our proof technique we used ICM method which is widely known. But in real life AES does not behave ideally. So there is an open problem to introduce weak cipher model for the security proof. Our scheme's key schedule is more than one which is not cost effective. So there is another challenge to propose a new scheme which obtains single KS and as well as better security bound.

References

1. Armknecht, F., Fleischmann, E., Krause, M., Lee, J., Stam, M., Steinberger, J.: The Preimage Security of Double-Block-Length Compression Functions. LNCS. ASIACRYPT, vol. 7073, pp. 233–251. Springer, Berlin (2011)
2. Black, J.A., Rogaway, P., Shrimpton, T.: Black-Box Analysis of the Block-Cipher-Based Hash-Function Constructions from PGV. LNCS, CRYPTO, vol. 2442, pp. 320–335. Springer, Berlin (2002)
3. Black, J.A., Rogaway, P., Shrimpton, T., Stam, M.: An analysis of the blockcipher-based hash functions from PGV. J. Cryptol. **23**, 519–545 (2010)
4. Bogdanov A., Leander G., Paar C., Poschmann A., Robshaw, M.J.B., Seurin, Y.: Hash Functions and RFID Tags: Mind the Gap. LNCS, CHES, vol. 5154, pp. 283–299. Springer, Berlin (2008)
5. Fleischmann, E., Forler, C., Gorski, M., Lucks, S.: Collision Resistant Double-Length Hashing. LNCS, PROVSEC, vol. 6402, pp. 102–118. Springer, Berlin (2010)
6. Fleischmann, E., Forler, C., Lucks, S., Wenzel, J.: Weimar-DM: A Highly Secure Double Length Compression Function. LNCS, ACISP, vol. 7372, pp. 152–165. Springer, Berlin (2012)
7. Hirose, S.: Some Plausible Constructions of Double-Block-Length Hash Functions. LNCS, FSE, vol. 4047, pp. 210–225. Springer, Berlin (2006)
8. Jesang, L., Seokhie, H., Jaechul, S., Haeryong, P.: A New Double-Block-Length Hash Function Using Feistel Structure. LNCS, ISA, vol. 5576, pp. 11–20. Springer, Berlin (2009)
9. Kaps, J.P., Sunar, B.: Energy Comparison of AES and SHA-1 for Ubiquitous Computing. LNCS, Emerging Directions in Embedded and Ubiquitous Computing, vol. 4097, pp. 372–381. Springer, Berlin (2006)
10. Knudsen, L., Preneel, B.: Fast and Secure Hashing Based on Codes. LNCS, CRYPTO, vol. 1294, pp. 485–498. Springer, Berlin (1997)
11. Lai, X., Massey, X.: Hash Function Based on Block Ciphers. LNCS, EUROCRYPT, vol. 658, pp. 55–70. Springer, Berlin (1993)
12. Lee, J., Kwon, D.: The security of abreast-DM in the ideal cipher model. IEICE Trans. **94-A** (1), 104–109 (2011)
13. Lee, J., Stam, M.: MJH: A Faster Alternative to MDC-2. LNCS, CT-RSA, vol. 6558, pp. 213–236. Springer, Berlin (2011)
14. Lee, J., Kapitanova, K., Son, S.H.: The price of security in wireless sensor networks. Comput. Netw. **54**(17), 2967–2978 (2010)
15. Lee, J., Stam, M., Steinberger, J.: The Collision Security of Tandem-DM in the Ideal Cipher Model. LNCS, CRYPTO, vol. 6841, pp. 561–577. Springer, Berlin (2011)
16. Menezes, A.J., Oorschot, P.C., Vanstone, S.A.: Handbook of Applied Cryptography, 5th edn. CRC Press, Boca Raton (2001)
17. Mennink, B.: Optimal Collision Security in Double Block Length Hashing with Single Length Key. LNCS, ASIACRYPT, vol. 7658, pp. 526–543. Springer, Berlin (2012)
18. Ozen, O., Stam, M.: Another Glance at Double-Length Hashing. LNCS. Cryptography and Coding, vol. 5291, pp. 176–201. Springer, Berlin (2009)
19. Shannon, C.E.: Communication theory of secrecy systems. Bell Syst. Tech. J. **128–134**, 656–715 (1949)
20. Wang, X., Lai, X., Feng, D., Chen, H., Yu, X.: Cryptanalysis of the Hash Functions MD4 and RIPEMD. LNCS, EUROCRYPT, vol. 3494, pp. 1–18. Springer, Berlin (2005)
21. Wang, X., Lai, X., Yu, X.: Finding Collisions in the Full SHA-1. LNCS, CRYPTO, vol. 3621, pp. 17–36. Springer, Berlin (2005)

Efficient Web Log Mining and Navigational Prediction with EHPSO and Scaled Markov Model

Kapil Kundra, Usvir Kaur and Dheerendra Singh

Abstract Web log mining is an important part of web usage mining, which help us to retrieve the important and hidden information from web server logs, for tuning up websites and increase the capabilities of web servers. In this paper we are proposing an enhanced methodology on web log mining process and online navigational prediction to improve the accuracy and stability of all web log mining stages. First, we are introduced some improvements on preprocessing stage. Second, we proposed refined approach for user identification and time based heuristic approach for session identification. Third, we are purposing efficient hierarchical particle swarm optimization clustering algorithm (EHPSO) to find the similarity based user sessions, which reduced the complexity of the Markov Model. Finally, we are suggesting Markov Model for online navigational prediction and we also proposed improved popularity and similarity based page rank algorithm (IPSPR) to solve the Markov Model ambiguous result problems.

Keywords Data clustering · Web log mining · PSO · Markov model · Data preprocessing

K. Kundra (✉) · U. Kaur
Department of Computer Science and Engineering, Sri Guru Granth Sahib World University, Fatehgarh Sahib, India
e-mail: kapil.kundra@hotmail.com

U. Kaur
e-mail: usvirkaur@gmail.com

D. Singh
Department of Computer Science and Engineering, Shaheed Udham Singh College of Engineering and Technology, Tangory, India
e-mail: professordsingh@gmail.com

L.C. Jain et al. (eds.), *Computational Intelligence in Data Mining - Volume 3*,
Smart Innovation, Systems and Technologies 33, DOI 10.1007/978-81-322-2202-6_48

1 Introduction

Today all organizations, business firms, institutions and companies are depending on their websites and online services to deal with clients. The web now becomes very powerful, that provides us many types of online services to interact with each other at any location in the world. Business sectors growing very fast and demands high tech online services that they serve to clients, they are using many attractive ways to make their websites more efficient to attract the users. To make their websites efficient there are some most frequently techniques are using, first one approach is to improve the content and website's structure according to the feedback that returns from the client and second approach is to analyze the clients or users navigation behavior logs that automatically recorded on servers and according to these logs, tune up the websites and online services to improve the efficiency of the client based services. In our proposed work we are working with second option, which also called web log mining, because in second option there is no need to direct interaction with clients, servers automatically records user inputs and we are not depending on the feedbacks that returns from the users. So in this paper we are using web log mining that is the part of web usage mining (WUM) for optimized the performance of websites.

Section 2 of this paper presents the background and related work, which explain the highlights of web mining algorithms. Section 3 describes our proposed web usage mining approach. Sections 4 and 5 describes the refined clustering and online navigation prediction approach. Section 6 provides the discussions and results of our experiments and the last section presents the conclusion of this paper.

2 Background and Related Work

In this section, we present many aspects from literatures that distinguish our work. In general, according to [1] work on web data mining is very challenging task for researchers because the nature of web data is unstructured and heterogeneous. In paper [2] author pointed out, data cleaning of log file is very complex and hard job, it takes 80 % of the total time of web log mining process. In fact it becomes more complicated when researchers dealing with proxy server logs, because most of users share same IP when they browsing websites with the help of proxy servers [3].

According to paper [4] author using DBSCAN algorithm for clustering, that is a density based clustering algorithm. There are two major benefits of this algorithm, first it is detect outliers, second it does not required cluster number at initialization time. But it is very sensitive to input parameters, so in this paper they are using Optics algorithm for appropriate values and give that values to DBSCAN as input data. Approach is good but very time consuming, because they are using two algorithms for a single task called clustering. Currently there are lots of algorithms are available, that takes less time for clustering as comparatively to this approach.

In paper [1] author proposed HPSO clustering algorithm that is based on particle swarm optimization technique and this algorithm works very well with heterogeneous data. The main purpose of this algorithm is to retrieve the sessions on the similarity bases and it is also especially optimized for web log data. So this algorithm solves the problems regarding of clustering that faced by many of researchers, when they are working on web usage mining. But in this algorithm author faced one major problem in many of cases results after clustering still have deficiency of density of similar sessions and algorithm gives cluster results with noisy data without proper distribution and it also decreasing the accuracy of the algorithm.

In paper [2] author suggest Markov model and PSPR algorithm for prediction, in this paper they are directly apply Markov Model on web log data set after preprocessing, that increases the complexity and raised some problems, first problem is low order Markov models that do not use sufficient history for prediction and therefore, they gives lack of accuracy [2, 5]. Second problem is high order Markov Models that increases space complexity due to longer sequences, but gives more accurate results comparative to the low order Markov Models. Third problem is ambiguous results, in the case of ambiguous results we cannot take decisions on the next page prediction. In this paper author also proposed Popularity and Similarity Based Page Rank Algorithm (PSPR) that solve the problem of ambiguous results.

3 Proposed Web Usage Mining Approach

In this section, we demonstrate our view on how we overcome the problems of data preprocessing or we called data cleaning and session identification, what types of data mining techniques we are using to increase the efficiency of clustering, how we can improve and get efficient results of online navigation prediction. All the components of our purposed work are presented in Fig. 1.

3.1 Preprocessing

As shown in Fig. 1, first we received raw log data from server which includes heterogeneous data of the users that frequently visited on the website. But before any formulation on this raw log data, our first task is data cleaning, so that we apply our purposed work on good quality data.

3.1.1 Data Cleaning

In WUM researchers purposed many approaches to clean the data. In paper [4, 6] author suggest a good approach to capture the robot requests, they are using look up list that includes the database of robots and check new request with that list to

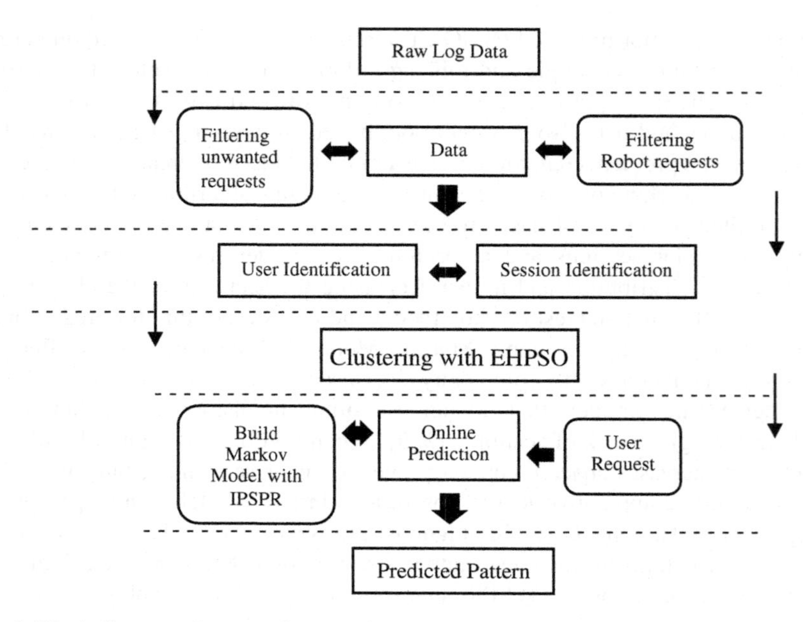

Fig. 1 Block diagram of proposed approach

capture the robot entries. We apply this approach in our purposed work to capture the robot requests and remove the unwanted entries from the raw log data, we are using many filters that are capable to detect and remove the noise, first filter check suffix of URLs for e.g., (*.png, *.gif, *.jpg, *.css) to remove entries that requests graphical contents on the web page, second we filtered entries that has unsuccessful HTTP status codes, here we consider only those entries that has successful HTTP codes, which occurs between the range of 200–299 [7], third we filtered those entries that has not GET and POST request methods. Figure 2 shows some filtered unwanted entries from the web log file.

Fig. 2 Example of some filtered entries

```
anon00000000000000148950|GET
/music/machines/images/title.jpg
HTTP/1.1|image/jpeg|200|1999/01/01-00:09:05|-|16072|-|-|

anon00000000000000148968|GET
/music/machines/images/title.jpg
HTTP/1.1|image/jpeg|200|1999/01/01-01:04:04|-|16072|-|-|

anon00000000000000148969|GET
/music/machines/categories/software/Windows/k4v311.zip
HTTP/1.1|application/zip|200|1999/01/01-01:04:52|-|80798|-|-|

anon00000000000000148970|GET/music/machines/images/bull
et.gif HTTP/1.1|image/gif|200|1999/01/01-01:05:42|-|205|-|

anon00000000000000148978|GET/music/machines/images/bell
.png HTTP/1.1|image/png|200|1999/01/01-01:08:42|-|116|-|
```

3.1.2 User Identification

User identification is a very important task to identify unique users. In paper [8] author purposed a good approach to identify unique users, according to the author we apply two simple assumptions to capture unique users, first if the user IP address is new, then we identify that as a new user, second if the user IP address is available in the data set but operating system and browsing software is different from another entries that has same IP address then we identify that as a new user. We apply this approach in our purposed work to identify the unique users. User identification is very important for session identification, because with the help of user identification we easily capture all possible unique users, which helps the session identification to refine the sessions, which further using for clustering the data. Following given program code shows the pseudo code of user identification technique.

Program code for User Identification process.

```
Input: Cleaned web log data.
Output: List E of identified unique users.
Begin
01 – E ← { }
02 – For each entry do
03 – Check if IP is not available in E then
04 – E ← E U {current entry} /* add as a unique user */
05 – Else if IP is available in E then
06 – Check if operating system and browsing software are not same
07 – E ← E U {current entry} /* add as a unique user */
08 – Else if operating system and browsing software are same then
09 – E ← E U {current entry} /* add user to sub collection, where IP is equal to
current IP */
10 – End For
11 – Return
End
```

3.1.3 Session Identification

According to the literature, session identification is very important task of web usage mining. In session identification we find each frequently accessed path and user access pattern to divide user accessed paths into individual sessions. Most of commercials used 30 min time-out threshold for session identification. We also used 30 min time-out threshold for identify each session, in this time based heuristic approach we check frequently accessed pages by the user and also check how much time user devote on the website for browsing, if the user takes longer time then the threshold interval, than we create a new session for that user. Table 1 shows the top five identified session, which we calculate with the help of session identification process.

Table 1 Identified first five sessions from the purposed model

User ID	Session ID	Transactions
User01	S1	P2, P4, P5, P1
User02	S2	P1, P4, P2
User03	S3	P5, P1, P3, P2
User04	S4	P3, P5, P2, P1, P4
User05	S5	P3, P4, P5, P3, P2

Our contribution for this stage is, first we use 30 min time-out based heuristic approach for the sessions and second we are using trimming approach to trim the sessions, in this process we omit all those pages that coming after 23rd visit of the user. The purpose of trimming is to trim the sessions, so that at clustering time we easily calculate the similarity of the users. As a recommendation, we wanted to have at least 10 pages to be used for efficient sessions, and first 5 pages used to calculate the similarity of the users to formulate the similar sessions based on sessions properties, with the help of this approach we can easily retrieved all possible patterns from the sessions. More detail of trimming process can be found in [1, 9, 10].

4 Refined Clustering Process

We are proposing optimized clustering approach called Efficient Hierarchical Particle Swarm Optimization (EHPSO) algorithm, which we modified from the existing HPSO [1] clustering algorithm. The problem which we tackled in HPSO formulation, in many of cases algorithm not successfully remove the pages, which are not frequently accessed by the users. The results of the clusters still lack the density of similar sessions and it also decreasing the accuracy of the algorithm. Our contribution on this algorithm is, we are purposing the formulation method that modifies the HPSO to overcome the problem of denoising without any duplication of the data. We named this new improved algorithm as EHPSO.

Program code of our purposed formulation for noise reduction method.

```
Input: a session
Output: a denoised session.
Begin
01 – S ← { }
02 – For each URL do
03 – if the URL weight is less than given threshold
04 – S ← S U {current URL} /* add as a valid url */
05 – Else if URL weight is greater than given threshold
06 – Remove URL from the session /* remove as a noise
07 – End For
End
```

We apply previous given formulation method before the calculation of Euclidean distance for each session. In our proposed work we input a session to our formulation method and calculate the weight of each requested page of the session, check if the weight of each requested page is less than given threshold value, add that page as a valid request and if the weight is greater than given threshold value, remove the requested page as a noise. So before measuring the distance of two sessions we apply our formulation on them to remove the noise from each session at each iteration of the algorithm, with the help of this approach, we solved the problem of denoising and we received more accurate clusters with very less error rate of noise. The results and reports are given in Sect. 6.

In our proposed approach we are using two type of similarity measure techniques, which is mentioned in paper [1], first is Euclidean distance and second is Boolean distance, the reason why we are using two techniques for measuring the similarity of sessions, because a session contains heterogeneous type of data and Euclidean distance only works with numeric attributes, so we are using Boolean distance to measure the remaining attributes for e.g. clickstream of the pages (pages that requested by the user), we briefly discussed these two similarity measure techniques as following.

$$d(XS, YS) = \left(\sum_{i}^{n} (XS_{ai} - YS_{ai})^2 \right)^{(1/2)} \tag{1}$$

The Euclidean Distance method is given in Eq. (1), which calculate the similarity of two sessions based on three numeric attribute values, for e.g. the values of session attributes are Data Download = 4,500 Kb, Session Time = 20 min and Session Length = 18 pages. Where *(XS, YS)* are the two sessions and *d (XS, YS)* denotes the dissimilarity of them.

$$b(XS, YS) = \left(\sum_{i}^{n} (XS_i \cap YS_i) \right) \tag{2}$$

The Boolean Distance method is given in Eq. (2), which calculates the similarity between two web pages with the help of AND operator, where XSi and YSi denotes the web pages contains by the two different sessions and to calculate the similarity based Euclidean and Boolean distance measure, we are using distance function which is given as following in Eq. (3).

$$Dist(XS, YS) = d(XS, YS) + b(XS, YS) \tag{3}$$

We are using different type of attributes to calculate the similarity measure and each of attribute has his own impact on the formulation process, some of attributes are numeric and some of attributes has different types like clickstream of the users etc., to balance the impact of values it is must to add some weight to each function result, so with the help of Eq. (4) we assign the value of weight to each given method to balance the results of the functions.

$$Dist(XS, YS) = W_1(d(XS, YS)) + W_2(b(XS, YS)) \quad (4)$$

Previous given equations helps to calculate the similarity and distance between two sessions with using numeric and text based parameters, which are very important because of all sessions contains heterogeneous data. More details about the process of these formulations can be found in paper [1].

5 Efficient Online Navigation Prediction

To predict online navigational patterns, we purpose a scalable Markov model approach, which is working with improved popularity and similarity based page rank algorithm (IPSPR) to predict the next page. Our purposed EHPSO clustering algorithm helps the Markov model to decrease the complexity. Markov model is commonly used for understanding the stochastic processes, and it is well suited for modeling the browsing behavior of the user on a website [2, 5]. Markov property builds the user behavior models that used to predict the web page, which user will most likely access next. Markov model estimate the conditional probabilities by assuming that, the web pages visited by the users follows a Markov process, with the help of Markov model process assumption, the web pages Pi + 1 will be generated next is calculated by the following Eq. (5).

$$\mathrm{P}_{\mathrm{i+2}} = \mathrm{argmax}_{\mathrm{p \in P}}\left\{\mathrm{P}\left(\mathrm{P}_{\mathrm{i}}, \mathrm{P}_{\mathrm{i+1}}, \ldots, \mathrm{p}_{\mathrm{i-(x-1)}}\right)\right\} \quad (5)$$

where, x act as preceding number of pages and indicate the order of Markov model, with the help of this equation we easily create the kth-order Markov model. It is required to know about each sequence of x web pages to use kth-order Markov model.

In our proposed work we are using Markov model to calculate the prediction probability for the web pages and in the condition of ambiguous results, we are using proposed improved similarity based page rank algorithm (IPSPR) to predict the next page. IPSPR algorithm formulates the results based on many important aspects like frequency of page, duration of page etc. In Markov model, the actions consistent to the different pages in the website, and the states consistent to all succeeding pages in the web sessions, which helps to take decisions on the prediction results. So our first work is to identify the states, once states have been identified, then we easily compute the transition probability matrix (TPM), which helps the Markov model to calculate the probabilities. In Table 2, we build the first order transition probability matrix, which is calculated with the help of Table 1.

In Table 2, we evaluate the web session S5(P3, P4, P5, P2, P3) that is given in Table 1. If we build first order Markov model then each state using only one page for formulation, so in Table 2 we can see that first page P3 of session S5 correlates with the state S3(P3) and follows by the page P4, so the entry of TPM is updated with t34. Similarly the next state is S4(P4) which correlates with page P4 and follows by the P5, so the entry of TPM is updated with t45. In the higher order Markov model, we are using second order to build the transition probability matrix, for e.g. in web session S3(P5, P1, P3, P2) of Table 1, contains the first parameter {P5; P1} in which page P3 follows the state {P5; P1}. So the state {P5; P1} and corresponding value of P3 of Table 3 will be updated.

Once we build the transaction probability matrix, it is straight forward for Markov model to making the predictions for web sessions, transition probability matrix is very effective approach to manage the pattern values of web pages, we can see that with the help of TPM we easily predict the next pages for e.g. user accessed the pages first P3 and next P4, now question is what page user will be access next, so with the help of Markov model we first identify the state {P3, P4} and after that just look up the transition probability matrix to find the next page (pi) that has highest probability and predict it to the user. But there are some conditions where we cannot take the decision on prediction, because sometimes in transition probability matrix we have multiple probability values of single state with same impact factor. For e.g. we can see in 2nd order Markov model last and second last entry, where states have multiple probability values with the same impact factor. So we cannot take decisions on these conditions, which is also known as ambiguous problem. So, we are using our proposed algorithm IPSPR to calculate the prediction probability of the next page, when we found ambiguous results. Next given equation shows the formulation of IPSPR algorithm.

Table 2 1st order transition probability matrix

1st order	P1	P2	P3	P4	P5
s1 = P1	0	0	1	2	2
s2 = P2	1	0	2	1	1
s3 = P3	0	1	0	2	1
s4 = P4	0	1	0	0	2
s5 = P5	1	2	1	1	0

Table 3 2nd order transition probability matrix

2nd order	P1	P2	P3	P4	P5
{P1, P3}	0	1	0	0	0
{P2, P4}	0	0	0	0	1
{P2, P1}	0	0	0	1	0
{P3, P5}	0	1	0	0	0
{P3, P4}	0	0	0	0	1
{P4, P5}	1	1	0	0	0
{P5, P2}	1	0	1	0	0

$$IPSPR_i = \varepsilon \times \sum_{xj \in In(x_i)} \left[IPSPR_J \times \frac{w_{j \to i}}{\Sigma_{P_k \epsilon Out(P_j)} w_{j \to k}} \times \frac{(d_{j \to}/s_i)}{\max(d_{m \to n}/s_n} \times tf.idf(t,d) \to \cos(\theta) = \frac{A.B}{||A||\,||B||} \right] + (1 - \varepsilon) \times \frac{W_i}{\Sigma_{P_j \in WS} W_j} \times \frac{d_i/s_i}{\max(d_m/s_m)} \tag{6}$$

Above given Eq. (6) formulate the many aspects to solve the ambiguous problems. Where ϵ denotes the value of damping factor and usually $\epsilon = 0.85$ for balancing the equation. In(xi) denotes the in-links of a particular page and Out(pj) denotes the out-links of a particular page, where *tf.idf* (*t,d*) calculates the term frequency and inverse document frequency of two web pages and $\cos(\theta)$ calculates the cosine similarity between two web pages. The popularity of the page calculates with average duration and page frequency. $\frac{w_{j \to i}}{\Sigma_{P_k \epsilon Out(P_j)} w_{j \to k}} \times \frac{(d_{j \to}/s_i)}{\max(\mathrm{d_{m \to n}}/\mathrm{s_n}}$ denotes the transition popularity based on transition duration and frequency. $\frac{W_i}{\Sigma_{P_j \in WS} W_j}$ calculates the frequency of a particular page and $\frac{d_i/s_i}{\max(d_m/s_m)}$ calculate the average duration for a particular page. This algorithm solves the Markov model ambiguous problem for correct predictions. More details about the formulation can be found in paper [2, 5].

6 Evolution of the Results

This section describes the experimentation and results, which we conducted on our purposed formulations, to verify the efficiency of the algorithms and results of the prediction probabilities and data memberships of the clusters.

6.1 The Dataset

In this research work, we are using 24 h timespan log data file, which we retrieved from the website http://music.hyperreal.org/. In Table 4, we can see the information about the dataset, which we are using for our proposed work. It is must the log file contains real time data, so that we efficiently apply the all formulation methods to retrieve and capture the important and valuable information.

Table 4 Dataset information

Data set	Period	Number of total entries	Number of unique users
Hyper real web log data	1st July of 1995 (24 h)	22,132	1,571

6.2 Results of Clustering Process

In our proposed work, we are using two modified algorithms, first is EHPSO for clustering and second is IPSPR for prediction and both of two algorithms are working very well with our proposed framework. The EHPSO is the improved version of HPSO [1] algorithm, in which we improved the denoising process to increase the efficiency of the algorithm. Further detail of EHPSO process can be found in Sect. 4. In Fig. 3a, b we can see the results of clustering without using our approach with 35 and 30 particles. The figure shows that 90 % data distributed into only two clusters and remaining 10 % is distributed among all the other clusters. This is happening because during the process of clustering HPSO algorithm not accurately detects and removes the unwanted requests. These factors intercept the intra cluster distance from considerably changing in each sequential generation of the swarm.

In our proposed approach, it is not required to apply the denoising method after the clustering process, because we capture and remove the unwanted requests from the sessions within the process of clustering at each iteration level. So now the problem of cluster denoising easily removed from the solution. In Fig. 4a, b we can see that, after using our approach with clustering now data distributed in many clusters and each cluster has data that dissimilar to the another cluster. The boundaries of the clusters now easily identified.

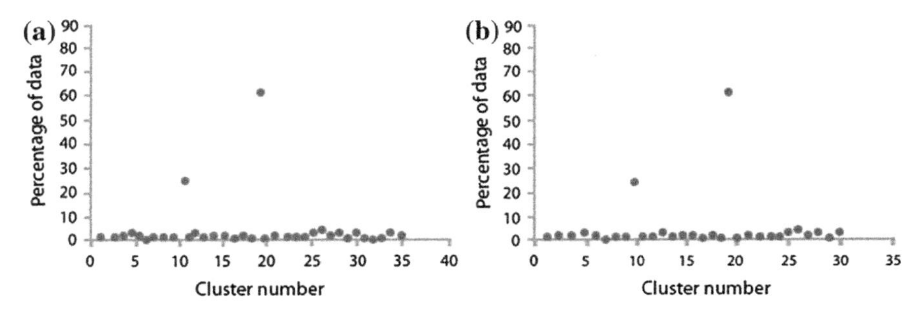

Fig. 3 Percentage of clusters membership without using our approach with **a** 35 and **b** 30 particles

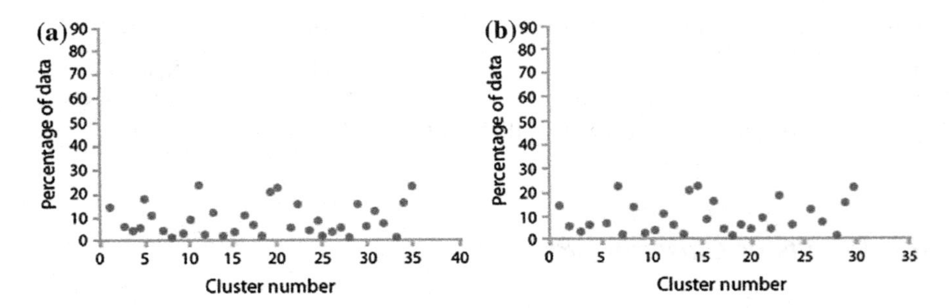

Fig. 4 Percentage of clusters membership with using our approach with **a** 35 and **b** 30 particles

6.3 Results of Navigational Prediction

Markov model is very accurate for prediction, if we solve the problem of complexity of the model, which we solved with the help of EHPSO clustering technique. Table 5 shows the second order Markov model, in which we can see the last three entries, where model easily predict the next page P17 of the user behaviour pattern P5 => P226, because the frequency of the pattern P5 => P226 => P17 is high as comparative to the another same patterns.

In Table 6 we shows the top five pages from our model, which gives the required attribute results. To predict the next page in the condition of Markov model ambiguous problem, we can easily calculates the rank and weight of a particular web page with the help of improved popularity and similarity based page rank algorithm.

Table 5 2nd order Markov model

Pattern	Next page	Frequency
P3 => P167	P849	3
P3 => P4	P5	1
P5 => P226	P312	3
P5 => P226	P320	4
P5 => P226	P17	12

Table 6 Top five attribute results of IPSPR algorithm

Page	Frequency (wi)	Page size in bytes (si)	Duration (di) in milliseconds
P1	3	4,330	21,085
P2	6	15,803	15,209
P3	5	4,042	24,004
P4	6	8,855	18,814
P5	4	6,468	6,373

Table 7 Markov model response time in milliseconds

Number of sessions	Without EHPSO	With EHPSO
200	6.189	1.052
300	7.087	1.287
400	6.291	1.148
500	6.362	1.187
600	8.013	2.079
700	7.472	1.729
800	7.253	1.518

We identify in our results, if we apply Markov model direct on sessions after preprocessing without using EHPSO for predict the next page. It takes lots of extra time due to the complexity of the model and the probability to predict the next page is low as comparative to our proposed approach, where we are using first EHPSO for clustering to find the similarity based sessions and then we give the clustered sessions to Markov model.

Due to the heterogeneous nature of data, preprocessing not detects all noised entries from the data log file, but with the help of EHPSO clustering algorithm we also remove the noised data form the sessions, which decrease the complexity of the Markov model. We can see the difference of response time in Table 7, where Markov model with EHPSO takes very less time as comparative to using without EHPSO.

7 Conclusion

In this paper we proposed a novel approach for web log mining process to increase the efficiency of the model. We are suggesting some aspects for preprocessing to clean the web log file, also we introduce the user and session identification approaches which optimized the sessions and for clustering process we purposed a novel approach called EHPSO and after that we are using Markov model with proposed IPSPR algorithm to predict the next page. Clustering is very important to understand the behavior of the web users but due to the nature of heterogeneous data of web log file, we cannot apply direct clustering algorithms on web log data. So in purposed EHPSO clustering technique, we are using similarity measure to identify the sessions which contains heterogeneous data also we apply our proposed denoising method to increase the density of the clusters and minimized the error rates of noisy clusters. So that Markov model accurately and fast predict the next page. The results of our purposed work verify the efficiency of the similarity measure and prediction response time. We also think that it would be better if we can further improve the efficiency of the overall process and using another hybrid approach in which we apply the cluster and prediction based approaches in a single combined algorithm, so that time and space complexity may be reduced and better results will be considered.

References

1. Alam, S., Dobie, G.: Clustering heterogeneous web usage data using hierarchical particle swarm optimization. In: Proceedings in Symposium on Swarm Intelligence (SIS), IEEE, pp. 147–154 (2013)
2. Thwe, P.: Proposed Approach for web page access prediction using popularity and similarity based page rank algorithm. Proc. Int. J. Sci. Technol. Res. **2**(3), 240–246 (2013)
3. Sha, H.Z., Liu, T.: EPLogCleaner: improving data quality of enterprise proxy logs for efficient web usage mining. Proc. Comput. Sci. **17**, 812–818 (2013)
4. Guerbas, A., Addam, O., Zaarour, O.: Effective web log mining and online navigational pattern prediction. Proc. Knowl.-Based Syst. **49**, 50–62 (2013)
5. Gunel, B.D., Senkul, P.: Investigating the effect of duration paze size and frequency on next page recommendation with page rank algorithm. In: Workshop on Web Search Click Data WSCD (2012)
6. Robot-Pages: http://www.robotstxt.org/db.html. Accessed March 2014
7. HTTP-Status-Codes: http://httperrorcodes.blogspot.in/. Accessed 8 March 2014
8. Suneetha, K.R., Krishnamoorthi, R.: Data preprocessing and easy access retrieval of data through data ware house. In: Proceedings in World Congress on Engineering and Computer Science WCECS, vol. I, San Francisco, USA (2009)
9. Alam, S., Dobbie, G., Riddle, P.: Towards recommender system using particle swarm optimization based web usage clustering. New Frontier in Applied Data Mining, pp. 316–326 (2012)
10. Alam, S.: Intelligent web usage clustering based recommender system. In: Proceedings in Fifth ACM Conference on Recommender Systems, ACM, pp. 367–370 (2011)
11. Velasquez, J.D.: Web mining and privacy concerns: some important legal issues to be consider before applying any data and information extraction technique in web-based environments. Proc Expert Syst. Appl. **40**, 5228–5239 (2013)
12. Eltahir, M.A., Anour, F.A., AllFa, D.: Extracting knowledge from web server logs using web usage mining. In: Proceedings of International Conference on Computing Electrical and Electronic Engineering ICCEEE, IEEE, pp. 413–417 (2013)
13. Kotiyal, B., Kumar, A.: User behavior analysis in web log through comparative study of eclat and apriori. IEEE, pp. 421–426 (2012)
14. Weichbroth, P., Owoc, M.: Web user navigation pattern discovery from WWW server log files. In: Proceedings in Federated Conference on Computer Science and Information Systems, IEEE, pp. 1171–1176 (2012)
15. Tanasa, D.: Web usage mining: contributions to inter sites logs preprocessing and sequential pattern extraction with low support. In: Proceedings in Universite de Nice Sophia Antipolis—UFR Sciences (2005)
16. Nayak, R.: Fast and effective clustering of xml data using structural information. Knowl. Inf. Syst. **14**, 197–215 (2008)
17. Fayyad, U.M., Shapiro, G.P., Smyth, P.: From data mining to knowledge discovery: an overview. In: Proceedings in Advance Knowledge Discover and Data Mining, pp. 1–34 (1996)
18. Pal, N.R., Pal, S.K.: A review on image segmentation techniques. Pattern Recognit. **26**(9), 1277–1294 (1993)
19. BenDor, A., Yakhini, Z.: Clustering gene expression patterns. In: Proceedings in Annual International Conference on Computational Molecular Biology, ser. RECOMB '99. ACM, New York, NY, USA, pp. 33–42 (1999)

20. Fu, Y., Sandhu, K., Shih, M.Y.: A generalization-based approach to clustering of web usage sessions. In: Revised Papers from the International Workshop on Web Usage Analysis and User Profiling ser. WEBKDD '99. Springer, London, UK, pp. 21–38 (2000)
21. Steinbach, M., Larypis, G., Kumar, V.: A comparison of document clustering techniques. In: KDD Workshop on Text Mining (2000)
22. Nassar, O.A., Nedhal, A., Saiyd, A.: The integrating between web mining and data mining techniques. In: Proceedings in 2013 5th International Conference on Computer Science and Information Technology CSITIEEE, pp. 243–247 (2013)

View Invariant Human Action Recognition Using Improved Motion Descriptor

M. Sivarathinabala, S. Abirami and R. Baskaran

Abstract Human Action Recognition plays a major role in building intelligent surveillance systems. To recognize actions, object tracking and feature extraction phases are crucial to detect the person and track its motion correspondingly. A good motion descriptor is to analyze the motion in each and every frame under different views. As a result, this paper addresses the two major challenges such as motion representation and action recognition. Thus a novel motion descriptor based on optical flow has been proposed to describe the activity of the persons though they are in various views. The new motion descriptor has been framed by fusing shape features and motion features together to increase the performance of the system. This has been fed as the input to the SVM classifier to recognize the human actions. Performance of this system has been tested over Weizmann and IXMAX multi-view data sets and results seem to be promising.

Keywords Motion descriptor · Optical flow · Activity recognition

1 Introduction

Human Action Recognition (HAR) is one of the major research areas in the field of video Surveillance. The aim of video surveillance is to collect information from the videos by tracking the people and recognizing their actions involved in the videos.

M. Sivarathinabala (✉) · S. Abirami
Department of Information Science and Technology, Anna University,
Chennai, India
e-mail: sivarathinabala@gmail.com

S. Abirami
e-mail: abirami_mr@yahoo.com

R. Baskaran
Department of Computer Science and Engineering, College of Engineering,
Anna University, Chennai, India
e-mail: baaski@annauniv.edu

L.C. Jain et al. (eds.), *Computational Intelligence in Data Mining - Volume 3*,
Smart Innovation, Systems and Technologies 33, DOI 10.1007/978-81-322-2202-6_49

Video Surveillance has been employed in all public places for Security purpose. Over a long period of time, a human guard would be unable to watch the behavior of the persons from the videos. To avoid the human intervention, automatic video surveillance has been emerged where people/objects could be identified, tracked and their activities have been recognized effectively. Action Recognition of the particular person/object under various views and illumination conditions, occlusion and dynamic environments is still a challenging research. Human Action Recognition includes the following stages: Background modeling, Object detection and tracking, Feature extraction, classification and Action Recognition. Major works in the field of video surveillance have been devoted to the Action Recognition phase.

Numerous tracking approaches have been proposed in the literature to track objects [1–4]. An Object Tracking mechanism starts with the Object detection stage which consists of Background modeling, Object representation and Feature extraction. Background modeling is required to differentiate/segment the foreground pixels from the background. Gaussian Mixture Modeling [5] updates the background frame frequently in order to adapt to the changes in the environment. Particle filters and Kalman filters [6] are widely used for tracking the movement of the interested objects in previous works.

Next to Object detection and tracking, features have been extracted in order to recognize the action of the particular person in every frame over time. From the previous literatures [2–4, 7] Bag Of Words (BOW) and Interest point detectors has been used widely to represent the action. Jain et al. [8] has proposed BOW to represent in a pre-defined model of human actions as a collection of independent codeword in a pre-defined codebook generated from training data. Huang et al. [9] proposed motion context as feature that is sensitive to scale and direction. Motion context has been extracted from motion images. Robertson and Reid [10] proposed the concept of Motion History Volumes (MHV) an extension of Motion History Images (MHI) to capture motion information from a sequence of video frames. Sadek et al. [11] proposed a statistical approach for human activity recognition by fusing local and global features. Hu and Boulgouris [12] described the action by a feature vector comprising both trajectory information and a set of local descriptors. In this, HMM has been used to encode the actions. In comparison with the shape and motion features, Optical flow is a better choice for view and distance invariance. Horn Schunck optical flow algorithm [13] has been used widely because of its smoothness and accurate time derivatives for more than two frames. The major disadvantages of Horn Schunck algorithm are: error due to noise and its discriminative performance. This motivated us to generate new features to improve Horn Schunck algorithm which are view invariant to describe the actions.

A particular action of the person may fail to get identified under various views. While analyzing the various views, the motion based features can depict the approximate moving direction of the body but most of them are not robust in capturing velocity. On the other hand, shape based features can represent the pose information of the body but lacks to detect the movement of the body part from the silhouette region. To solve this problem, a novel motion descriptor has been modeled here with improved motion features fused with the shape information that are extracted from the silhouettes. This new motion descriptor has been provided as input to the SVM classifier to recognize the actions successfully. In addition to motion information, this research also has been motivated to recognize the activity of the person under different views. The main contribution of this work lies in the extension of Horn Schunck optical flow algorithm with enhanced motion features which results in a new motion descriptor to describe the action in a better way.

The Paper has been organized as follows: Sect. 2 describes the proposed approach to recognize the human action Sect. 3 illustrates the results and discussion and Sect. 4 concludes the paper.

2 Proposed Methodology

Activity Analysis application rely heavily on motion detection. Actions are characterized by some specific movement or motion patterns. Figure 1 depicts the View Invariant Human Action Recognition (VIHAR) system. The input video has been considered from the different datasets. Initially, background has been modeled with GMM (Gaussian Mixture Model) to obtain the foreground as blobs. The motion of the person has been estimated using dense optical flow algorithm. Optical flow [13] is a pattern of apparent motion of objects, surfaces and edges in a visual scene caused by the relative motion between an observer and the scene. Understanding and predicting motion is an important part in recognizing human actions in live videos. Motion estimation is difficult when the object is shapeless, in the presence of shadow of the object and motion blur. The parametric motion is limited and cannot describe the motion in arbitrary videos. Motion features has been estimated using dense optical flow algorithm called Horn Schunck algorithm.

Using the contour of the input image, shape features such as blob size, blob width and height have been extracted. Both the shape and motion features have been combined to the single feature vector. Gradient velocity and gradient

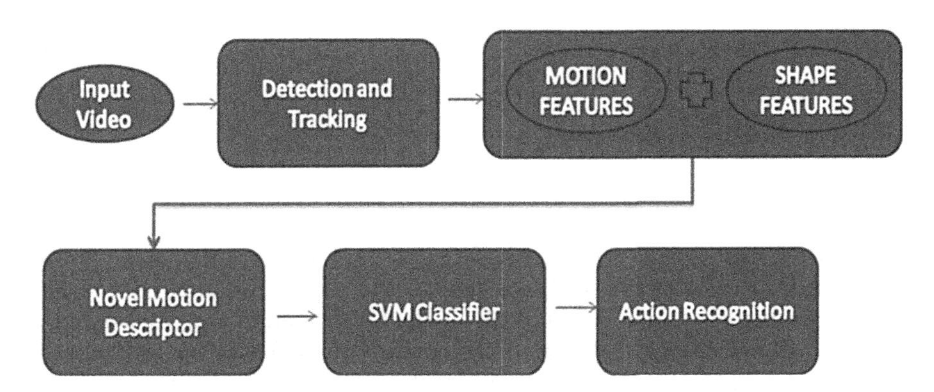

Fig. 1 View invariant human action recognition (VIHAR) system

orientation are the motion features that can be extracted from the ordinary optical flow method. In this work, the new features are extracted in addition to it. Normalized speed of the optical flow for pixel, Euclidean norm of the optical flow vector and Average Angular Error (AAE) are the features that can solve the view invariance problem. Thus extended motion features along with shape information are combined to form a new motion descriptor. The new descriptor describes the actions and provides input to the classifier SVM and recognizes the human actions.

2.1 *Object Detection and Tracking*

To detect/track the objects, frames are extracted from input videos and temporal information are obtained from those sequences. A suitable background model has been selected using GMM mixture of Gaussians and the background information gets updated constantly. To obtain the background model, Gaussian Mixture model (GMM) [6] has been employed here. For outdoor environments, Single Gaussian model is not suitable. Therefore to suit outdoor and indoor environments, mixture of Gaussians has been used here to model the background. Later, Blobs or pixels are segmented as foreground to depict the region of interest (ROI). Further, suitable features are selected from ROI to represent the motion in every frame.

2.2 *Feature Extraction*

Feature Extraction has been considered as crucial step to understand and predict the human actions. In general, features [9] are spatial, motion and spatio-temporal based. Spatial based features also referred as shape based, uses contour or silhouette information. Spatio-temporal based features have been widely used in recent years. Here in this paper, the target person/object has been represented in contour and their motion has been estimated in every frame. The person/object motion has been computed here using Optical Flow.

2.2.1 Optical Flow Estimation

In this paper, our motivation is to recognize the human action even if they are in various camera views. In literatures, several optical flow approaches has been described. In this context, Optical flow can give useful information about spatial arrangement of the object viewed and the rate of change of this arrangement. The approaches that have been used to compute optical flow are: Lucas Kanade algorithm and Horn Schunck algorithm. Lucas Kanade algorithm/Sparse method estimates the displacement for selected number of pixels in the image. The pixels are chosen randomly so there cause error on the boundaries. The major disadvantage of the sparse method is that neighboring surface points of a rigid object have approximately same local displacement vectors. While the Horn Schunck algorithm/Dense optical flow calculates the displacement of each pixel by using global constraints. This Dense method provides accurate time derivatives using more than two frames. This method is smoothened over time- domain. Optical flow estimation assumes that pixel intensities are translated from one frame to the next. Horn Schunck algorithm [13] determines the optical flow as a solution of the following partial differential equation:

$$\frac{\partial E}{\partial x}\frac{\partial x}{dt} + \frac{\partial E}{\partial y}\frac{dy}{dt} + \frac{dE}{dt} = 0 \tag{1}$$

The solution of Eq. 1 is obtained by error function minimization. The error function E is defined in terms of spatial and time gradients of optical flow vector field and consist of two terms shown in Eq. 2.

$$E = E_x u + E_y v + E_t \tag{2}$$

$$E_x = \frac{\partial E}{\partial x}, \quad E_y = \frac{\partial E}{\partial y}, \quad E_t = \frac{dE}{dt} \tag{3}$$

$$u = \frac{dx}{dt} \quad \text{and} \quad v = \frac{dy}{dt} \tag{4}$$

E (x, y, t) denotes luminance of the image in point (x, y) at time instant t. The optical flow algorithm has two main stages. In the first phase, gradient coefficients E_x, E_y, E_t are computed from the given input images. The coefficients represent estimates of the image gradients in space and time, and they are defined by Eq. 3. In the second phase, the optical flow vectors u and v defined by Eq. 4 are computed. Errors and noises in the optical flow computation will provide error when computing the motion of the person in each and every frame. The error will reduce the performance of the algorithm.

2.3 Improved Motion Descriptor

Optical flow algorithm has been improved to overcome the disadvantage of the Horn Schunck algorithm. Motion filtering has been used to filter out the noise in the flow computation. Gradient velocity and gradient orientation are the ordinary features extracted from the optical flow algorithm. In order to improve the performance of the optical flow we have been extracting more features to estimate the motion between the frames.

Optical flow can be normalized in both time and size. In order to improve the discriminative performance of the algorithm, normalized speed has been considered as features invariant to views. In this paper, we have been considered H_{um} and W_{um} as height and width of the union of blobs in (m − 1)th and mth frame. Normalized speed can be represented as in Eq. 5.

$$\rho_{nm}(i,j) = \frac{F \times \rho_m(i,j)}{\sqrt{H_{um^2} + W_{um^2}}} \quad (5)$$

where $\rho_{nm}(i,j)$ indicates the normalized speed of optical flow for pixel $\rho_m(i,j)$ in mth frame. F is the video frame rate 25 frames/s. Threshold [α β] set as 1.5 by choosing walking and running as reference motion for large and small motions respectively.

Euclidean norm of the optical flow vector has been represented by

$$E_{opt} = \sqrt{u^2 + v^2} \quad (6)$$

Average Angular Error (AAE): The Flow vector at certain positions in the image can deviate from the true flow vector at one position in direction and in length. AAE have been estimating the movement from one frame to next frame. So, time component of flow vector has been considered as 1 and yielding 4 dimensional vector v = {u,v,w,1}

$$AAE = \arccos\left(\frac{v_t}{\sqrt{u_{t^2} + v_{t^2} + w_{t^2} + 1}} \cdot \frac{v_e}{\sqrt{u_{e^2} + v_{e^2} + w_{e^2} + 1}}\right) \quad (7)$$

where V_t and v_e represents the true vector and estimated vector with spatial component u_t, v_t, w_t and time component 1 respectively.

2.4 Feature Fusion and Classification

In Optical flow field estimation, the motion part has been considered. The motion features such as gradient velocity, gradient orientations, euclidean norm of optical flow vector, normalized speed, Average Angular Error (AAE) have been extracted from the optical flow algorithm and from the silhouette, shape features such as size of the blob and Aspect ratio has been considered. The motion features and shape features have been fused together to form a new motion descriptor such that this novel descriptor describes the activity involved in the video.

The novel motion descriptor describes the human activity which has been given as input to the SVM classifier. The features extracted have been concatenated to a single feature vector and fed as input to support vector machine. The Radial basis Function has been used here to map the training vectors into the feature space for classification of actions.

3 Results and Discussion

The implementation of this Human Action Recognition system has been done using MATLAB (Version2013a). MATLAB is a high performance language for technical computing. The proposed algorithm have been applied and tested over Weizmann dataset [14] and IXMAX dataset [15].

The Weizmann Dataset consists of 90 videos that are evaluated by leave one out cross validation. Resolution can be given as 180 × 144. The Videos can be recorded using Static camera. Ten data actions can be explored such as walk, run, jump, gallop sideways, bend, one-hand wave, two-hand wave, jump in place, jumping jack, skip. In this work, we have been considered only 7 actions and evaluated. The sample frames from the Weizmann dataset has been shown in Fig. 2

Tracking and computation of optical flow vector is shown in Fig. 3. Thresholded Horn schunck algorithm and the moving human body part alone shown in bounding box. From the flow features size of the blob and aspect ratio has been considered and estimated flow vector has been shown in figure. The classified output has been shown in Fig. 3. Next to the Flow vector calculation, Action classification results

Fig. 2 Shows some of the sample frames from the Weizmann activity dataset for the actions such as side, bend and jump

Fig. 3 Optical flow vectors and action classification

has been carried out in 9 runs using leave-one-out procedure with N = 4. The method correctly classifies 96.33 % of all testing sequences. The above method correctly classifies the action sequences such as bend, side, walk and wave2 action. The confusion matrix of the Weizmann activity dataset has been shown in Table 1.

INRIA Xmas Motion Acquisition Sequences (IXMAS) is a Multi view dataset for view-invariant human action recognition [15]. 13 daily-live motions have been performed 3 times each by 11 actors. The actors choose free position and orientation. Sample frames from INRIA IXMAS dataset has been shown in Fig. 4. Shows the horizontal, diagonal and vertical views of the person the angles varying from 0 to 90°.

Horizontal view can be measured from the angle 0 to 30°. The Actions such as sit down, getup, turn around, walk and punch has been recognized under various views in this work. The overall recognition rate under horizontal view obtained is 98.85 %. Here in the horizontal view, getup and turn around action may get slightly confused. Table 2 shows the confusion matrix for horizontal view. The Diagonal view can be measured from 30 to 45°. In the diagonal view shape features does not

Table 1 Confusion matrix for Weizmann dataset

Bend(B)	1.0	X	X	X	X	X	X
Jump(J)	X	0.89	X	0.11	X	X	X
Pjump(PJ)	X	X	0.90	X	0.10	X	X
Side(S)	X	X	X	1.0	X	X	X
Walk(W)	X	X	X	X	1.0	X	X
Wave1(W1)	X	X	X	X	.X	0.96	0.04
Wave2(W2)	X	X	X	X	X	X	1.0
	B	J	PJ	S	W	W1	W2

Fig. 4 Horizontal view, diagonal view and vertical view

Table 2 Confusion matrix for horizontal view

Horizontal view	Sit down	Get up	Turn around	Walk	Punch
Sit down	1	X	X	X	X
Get up	X	0.98	0.02	X	X
Turn around	X	X	0.97	0.03	X
Walk	X	X	X	1	X
Punch	X	X	X	X	1

Table 3 Confusion matrix for diagonal view

Diagonal view	Sit down	Get up	Turn around	Walk	Punch
Sit down	1	X	X	X	X
Get up	X	0.98	0.02	X	X
Turn around	X	X	0.96	0.04	X
Walk	X	X	X	1	X
Punch	X	X	X	X	1

Table 4 Confusion matrix for vertical view

Vertical View	Sit down	Get up	Turn around	Walk	Punch
Sit down	1.0	X	X	X	X
Get up	X	0.95	0.05	X	X
Turn around	X	0.03	0.93	0.04	X
Walk	X	X	X	1.0	X
Punch	X	X	X	X	1.0

provide more effect. The overall recognition rate in vertical view is 96 %. Getup action and turn around action may get confused in different situations. The Vertical view can be measured from 45 to 90°. Action recognition in the vertical view depends not only on the motion features but also depends on the shape features. The addition of shape features to the motion descriptor, the increases the performance to identify the actions. The overall recognition rate is 94.5 % for the actions under vertical view. Tables 3 and 4 shows the confusion matrix for diagonal and vertical view respectively.

4 Conclusion

In this research, an automated human action recognition system has been developed using novel motion descriptor to describe the human actions in various views. This system has the ability to analyze the motion of the person in each and every frame. The new motion descriptor has been formed by fusing shape features and motion features together to increase the performance of the system. This combined feature vector provides the input to the multi class SVM classifier to recognize the human actions. The proposed algorithm has been tested over Weizmann and IXMAS datasets and the results are promising. The addition of shape features to the motion descriptor increases the performance to identify the actions. The overall recognition rate has been improved to 98.85, 96 and 94.5 % for the actions under horizontal, diagonal and vertical view respectively. In future, this system could be extended along with the detection of occlusion too.

References

1. Yilmaz, A., Javed, O., Shah, M.: Object tracking :a survey. ACM Comput. Surv. **38**(4 Article no. 13), 1–45 (2006)
2. Ryoo, Agarwal, J.K.: Human activity analysis: a survey. ACM Comput. Surv. **43**(3), 16:1–16:43 (2011)
3. Poppe, R.: A survey on vision-based human action recognition. Image Vis. Comput. **28**, 976–990 (2010)
4. Gowshikaa, D., Abirami, S., Baskaran, R.: Automated human behaviour analysis from surveillance videos: a survey. Artif. Intell. Rev. (2012). doi:10.1007/s10462-012-9341-3
5. Gowsikhaa, D., Manjunath, S., Abirami, S.: Suspicious human activity detection from Surveillance videos. Int. J. Internet Distrib. Comput. Syst. **2**(2), 141–149 (2012)
6. Sivarathinabala, M., Abirami, S.: Motion tracking of humans under occlusion using blobs. Adv. Comput. Netw. Inf. (Smart Innov. Syst. Technol. **1**) **27**, 251–258 (2014)
7. Gowsikhaa, D., Abirami, S., Baskaran, R.: Construction of image ontology using low level features for image retrieval. In: Proceedings of the International Conference on Computer Communication and Informatics, pp. 129–134 (2012)
8. Jain, M., J'egou, H., Bouthemy, P.: Better exploiting motion for better action recognition. In: The Proceedings of IEEE International Conference on Computer Vision and Pattern Recognition, pp. 2555–2562 (2013)
9. Huang, K., Wang, S., Tan, T., Maybank, S.: Human behaviour analysis based on new motion descriptor. In: IEEE Transactions on Circuits and Systems for Video Technology (2009)
10. Robertson, N., Reid, I.: A general method for human activity recognition in video. Comput. Vision Image Underst. **104**, 232–248 (2006)
11. Sadek, S., Al-Hamadi, A., Michaelis, B., Sayed, U.: A fast statistical approach for human activity recognition. Int. J. Comput. Inf. Syst. Ind. Manag. Appl. **4**, 334–340 (2012). ISSN 2150-7988
12. Hu, J., Boulgouris, N.V.: Fast human activity recognition based on structure and motion. Pattern Recogn. Lett. **32**, 1814–1821 (2011)
13. Bruhn, A., Weickert, J.: Lucas/Kanade meets Horn/Schunck: combining local and global optic flow methods. Int. J. Comput. Vision **61**(3), 211–231 (2005)
14. http://www.wisdom.weizmann.ac.il/ ~ vision/
15. http://4drepository.inrialpes.fr/public/datasets

Visual Secret Sharing of Color Image Using Extended Asmuth Bloom Technique

L. Jani Anbarasi, G.S. Anadha Mala and D.R.L. Prassana

Abstract Secret image sharing is the technique for securing images that involves the distribution of secret image into several shadow images. Secret sharing for color image is proposed in this article using Asmuth bloom technique which is further extended for accurate reconstruction of given original image. The secret image is reduced in pixel value using a quantization factor resulting in quantized secret image and a difference image. Secret shares based on Asmuth Bloom scheme are generated from the quantized secrets and are distributed among n participants specified at least t of them should gather to reveal the difference secret image and therefore the respective quantized image. The original secret is reconstructed using the inverse quantization process.

Keywords Image processing · Asmuth bloom technique · Secret sharing

1 Introduction

In 1979, Blakley and Shamir introduced the (t, n) threshold secret sharing schemes, where a dealer divides the secret data into n elements which are then distributed to n legitimate participants. At least t participants should be combined to retrieve the secret data. Visual Cryptography, proposed by Naor and Shamir encodes a binary secret into two shares based on white and black pixels and the secret is reconstructed

L. Jani Anbarasi (✉)
Sri Ramakrishna Institute of Technology, Coimbatore, India
e-mail: jani_lj_2000@yahoo.com

G.S. Anadha Mala
Easwari College of Engineering, Chennai, India
e-mail: gs.anandhamala@gmail.com

D.R.L. Prassana
Vasavi College of Engineering, Hyderabad, India
e-mail: prasannadusi@gmail.com

L.C. Jain et al. (eds.), *Computational Intelligence in Data Mining - Volume 3*,
Smart Innovation, Systems and Technologies 33, DOI 10.1007/978-81-322-2202-6_50

by superimposing the two shares. Threshold Secret sharing proposed by Shamir [1–4] uses Lagrange interpolation technique, Asmuth and Bloom [5] and Mignotte's secret sharing scheme are based on the Chinese Remainder Theorem [6–8] is based on hyper plane geometry. In weighted threshold secret sharing scheme, a positive weight is assigned to each participant and the secret can be reconstructed only if the sum of the weights of all the gathered participants is greater than the fixed threshold. This weighted secret sharing is analyzed in [9–13]. In Shamir's, (t, n) threshold approach, all the gray pixels of the secret image that are higher than 250 are truncated to 250, as a result of the prime number chosen is 251. In medical and sensitive images, even tiny distortions don't seem to be tolerable. To overcome these issues two pixels are used to represent the gray values larger than 250. In this paper we have overcome this problem using quantization process. Identification of fake shares and prevention of Cheating has become an important issue. The dishonest participants interpret the secret by pooling a fake share during secret reconstruction. Identification of secret share and detection of cheating participants play vital role in efficient reconstruction of secret.

The rest of the paper is organized as follows: Sect. 2 explains the proposed work. The experimental and simulation results are detailed in Sect. 3; Security analysis is discussed in Sect. 4, and the conclusion is described in Sect. 5.

2 Proposed System

In this paper, we proposed a method by extending the Asmuth bloom technique using a quantization factor which securely shares the secret color images for various participants.

2.1 Secret Sharing Using Extended Asmuth Bloom Technique

Dealer D and a set of n participants, say P_1, $P_2 \ldots P_n$ plays a vital role in this extended Asmuth bloom secret sharing scheme. Dealer D performs the generation and distribution of secret shares to the participants. This phase includes prime number generation, Quantization and Share creation process. Dealer D generates key for each individual participants $\{K = K_1, K_2 \ldots K_n\}$ in the prime generation phase. The pixel value of the secret image is reduced in the quantization phase and secret shares are generated using extended Asmuth bloom technique.

2.1.1 Prime Number Generation

Secret shares are created for each participant P using their secret key K. If the secret keys are chosen or above 255 then it results in partial reconstruction or no image representation respectively. A range of secret keys are chosen by reducing the images SI using a Quantization factor F such that selection of key lies between, $F \leq K_z \leq 255$ where $1 \leq z \leq n$ and n is the number of participants. The Quantization Factor F is inversely proportional to chosen key range, probably between F and 255.

$$F \propto {}^{1}/_{K} \tag{1}$$

The chosen keys should be co primes inorder to results in perfect secret sharing.

2.1.2 Quantization Phase

Quantization [11] is a technique which helps to reduce the pixel value of the secret image $SI = \{SI_{ij} \in [0 - 255]\}$. Quantization process is performed using a quantization factor F $(0 \leq F \leq 25)$ chosen as very small whole integer. To support lossless image secret sharing Quantization factor $F(0 \leq F \leq 25)$ is chosen as very small whole integer. Quantized secret image QI and the respective Difference Image DI are the two different images which results from quantization process. Every pixel in the image is reduced using pixel reduction process. The quantized secret image $QI = \left\{QI_{ij} \in {}^{(0-255)}/_{F}\right\}$ where $1 \leq i,j \leq m$ is calculated using the quantization factor chosen by the dealer D. The *i*th row and the *j*th column pixel value of the Quantized Image QI can be determined as follows

$$QI = \left\lfloor {}^{SI}/_{F} \right\rfloor \tag{2}$$

The quantized difference image $DI = \{DI_{ij} \in [0 - (F-1)]\}$ where $1 \leq i,j \leq m$ is the difference value or the truncation error occurred during processing. The size of Difference Image DI is same as that of SI and DI. The DI is determined as follows

$$DI = SI - F * \lfloor QI \rfloor \tag{3}$$

2.2 *Extended Secret Sharing Using Asmuth Bloom Technique*

The quantized secret image Quantized Image QI is encoded using the Asmuth Bloom scheme which results in n meaningless shares. The difference Image DI replaces the random integer γ in the Asmuth bloom technique which leads to lossless reconstruction of secret Image SI. Unique shares are created for each participants using the participant's key K_z. Meaningless shares are securely distributed among n participants. Extended Asmuth bloom technique generates the secret shares as follows

$$I_z = QI + (DI * F) \bmod K_z \text{ Where } 1 \leq z \leq n \quad (4)$$

2.2.1 Quantized Secret Retrieval Phase

During reconstruction, at least t participants should pool their share to reconstruct the original secret whereas less than t cannot. Chinese remainder theorem is used to reconstruct the unique solution by solving the system of congruence.

$$X_o = I_z * e_z \bmod M \quad (5)$$

where I_z is the pooled share of the chosen participants and $M = \prod_{z=1}^{t} m_z$ The value e_z is found using Bézout's identity followed by Extended Euclidean algorithm.

The Quantized secret QI and Difference secret DI are reconstructed from the above generated results. The reconstruction phase is extended by the dealer, from usual computations through reconstruction the quantized secret QI and difference DI.

Quantized secret QI and Difference Image DI are determined as follows

$$QI = X_o \bmod F \quad (6)$$

The quantized difference secret image DI is determined using the following,

$$DI = {}^{(X_o - QI)}/_{F} \quad (7)$$

Thus the quantized secrets are reconstructed successfully with out loss. Using these values, the original secret is reconstructed through inverse Quantization.

$$SI = QI * F + DI \quad (8)$$

3 Experimental Results and Analysis

The performance and simulation of the proposed system are detailed in this section. The proposed scheme is tested for various color images. The algorithm is tested and coded in Matlab 7.9. The color images are considered as 8-bit depth image of size 256 × 256. A simulation for the (3, 4) scheme is applied for color data set, where the threshold value 't' is chosen as 3 and the number of participants 'n' is 4. The performance evaluation is carried out on various 256 × 256 color images. Secret color image specified in Fig. 1a is partitioned into quantized secret image QI and quantized difference image DI using a quantization factor F is shown in Fig. 1b, c. Generated secret shares using extended Asmuth bloom technique is shown in the Fig. 2.

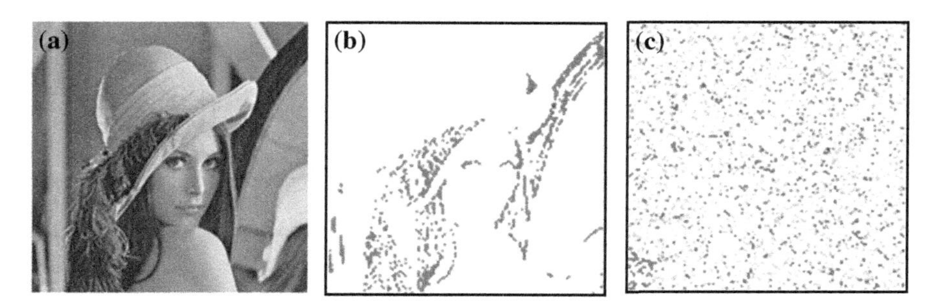

Fig. 1 **a**. Secret image **b**. Quantized secret **c**. Quantized difference secret

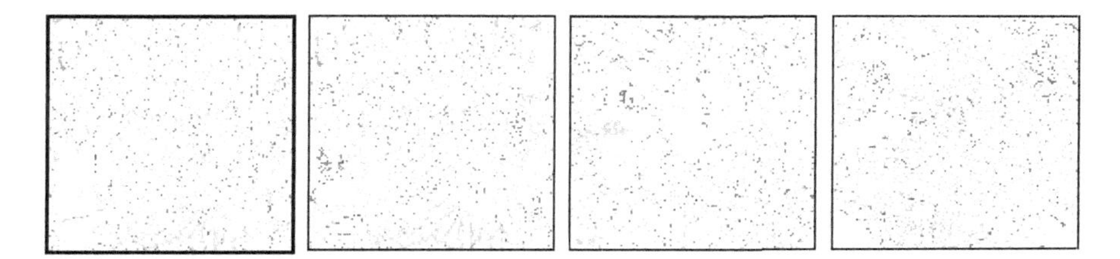

Fig. 2 Share images

4 Security Analysis

4.1 Reconstruction of Secret with Less Than t Shares

Reconstruction of secret with less than t shares leads to meaningless secret. Thus the stated reconstruction is tested for various images and the analysis is given below. Less than t shares generated for secrets are pooled to reconstruct the secret. In (3, 4) secret sharing scheme at least three shares are required to reconstruct the secret, incase two shares are combined irrelevant information is the result. The reconstructed secrets are presented in Fig. 3.

Fig. 3 Reconstructed secret with 2 shares

Fig. 4 Reconstructed secret with fake shares

4.2 Reconstruction of Secret with Fake Shares

During reconstruction a fake share is pooled along with other two shares generated for secrets. Some of the original shares are replaced with fake shares and the system was tested and the results obtained are shown in Fig. 4. The obtained result shows that the reconstructed secret has no meaning.

4.3 Reconstruction of Secret with Tampered Shares

The proposed system was tested with various tampered shares and the results obtained are shown. One of the share is tampered and pooled along with other original shares to reconstruct the secret. The obtained result shows that if any modification is done to the shares, then a loss secret is only reconstructed. Figure 5 shows the tampered shares and the corresponding reconstructed secret is given in Fig. 6.

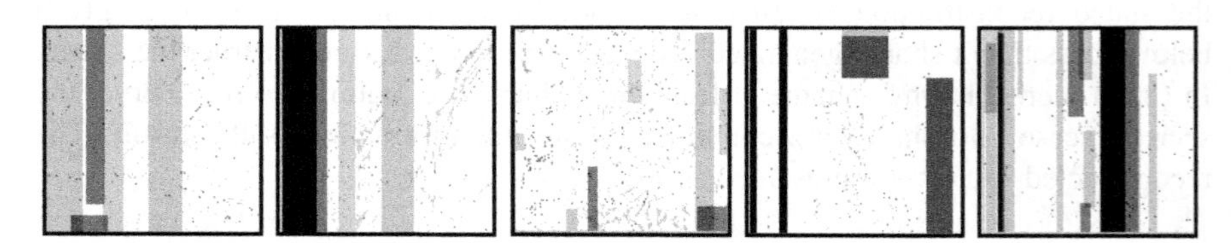

Fig. 5 Tampered shares

Fig. 6 Reconstructed shares

To resist a statistical attack, less correlation among two adjacent pixels is very essential and critical. The correlation coefficient of two adjacent horizontal and diagonal pixels in the original image and encrypted image is examined and given in Tables 1 and 2.

Table 1 Vertical correlation coefficient

Quantization factor	Prime	Secret	R	G	B
			0.9676	0.9595	0.9569
19	31	Shadow1	0.0398	0.0328	0.0362
	37	Shadow2	0.0451	0.0303	0.0318
	41	Shadow3	0.0026	0.0151	0.0136
	43	Shadow4	0.0797	0.0766	0.0758
47	67	Shadow1	0.0727	0.0692	0.0407
	71	Shadow2	0.0113	0.0320	0.0274
	73	Shadow3	0.0108	0.0226	0.0058
	79	Shadow4	0.0249	0.0274	0.0017
53	101	Shadow1	0.0479	0.0467	0.0650
	103	Shadow2	0.1164	0.0651	0.0944
	109	Shadow3	0.0860	0.0837	0.0815
	113	Shadow4	0.0346	0.0192	0.0233

Table 2 Horizontal correlation coefficient

Quantization factor	Prime	Secret	R	G	B
			0.9712	0.9667	0.9587
19	31	Shadow1	0.0232	0.0263	0.0275
	37	Shadow2	0.0229	0.0285	0.0436
	41	Shadow3	0.0183	0.0692	0.0196
	43	Shadow4	0.0145	0.0758	0.0820
47	67	Shadow1	0.0512	0.0396	0.0619
	71	Shadow2	0.0402	0.0380	0.0128
	73	Shadow3	0.0134	0.0223	0.0147
	79	Shadow4	0.0248	0.0088	0.0403
53	101	Shadow1	0.0287	0.0616	0.0682
	103	Shadow2	0.1062	0.0748	0.0745
	109	Shadow3	0.0533	0.0091	0.0023
	113	Shadow4	0.0191	0.0269	0.0171
	131	Shadow4	0.0135	0.0437	0.0145

5 Conclusion

In this paper a secret color image is securely shared using the extended Asmuth Bloom secret sharing. The proposed threshold secret sharing uses Quantization to avoid loss of the secret data. The secret color image is reconstructed without loss. The achieved size of the shadow is the same as that of the secret image. This is a perfect scheme where less than t shadows has no information about the secret.

References

1. Abbas, C., Joan, C., Kevin, C., Paul Mc, Kt.: Digital image steganography. Survey and analysis of current methods. Sig. Process **90**, 727–752 (2010)
2. Thien, C.C., Lin, J.C.: Secret image sharing. Comput. Graph. **26**, 765–770 (2002)
3. Tso, H.K.: Sharing secret images using Blakley's concept. Opt. Eng. **47**, 077001 (2008)
4. Mustafa, U., Güzin, U., Vasif, N.: Medical image security and EPR hiding using Shamir's secret sharing scheme. J. Syst. Softw. **84**, 341–353 (2011)
5. Asmuth, C., Bloom, J.: A modular approach to key safeguarding. IEEE Trans. Inf. Theor. **29**, 208–210 (1983)
6. Blakley, G.: Safeguarding cryptographic keys. In: Proceedings of AFIPS National Computer Conference, pp. 313–317 (1979)
7. Sorin, I., Ioana, B.: Weighted threshold secret sharing based on the Chinese remainder theorem. Sci. Ann. Cuza Univ. **15**, 161–172 (2005)
8. Kamer, K., Ali, A.S.: Secret Sharing Extensions Based on Chinese Remainder Theorem, p. 96. IACR Cryptology, (2010)
9. Beimel, A., Tassa, T., Weinreb, E.: Characterizing Ideal Weighted Threshold Secret Sharing. In: Second Theory of Cryptography Conference, pp. 600–619 (2005)
10. Amos, B., EnavWeinre.: Monotone circuits for weighted threshold functions. In: 20th Annual IEEE Conference on Computational Complexity. pp. 67–75 (2005)
11. Guzin, U., Mustafa, U., Vasif, N.: Distortion free geometry based secret image sharing. Procedia Comput. Sci. **3**, 721–726 (2011)
12. Cheng, G., Chin, C.C., Chuan, Q.: A hierarchical threshold secret image sharing. Pattern Recogn. Lett. **33**, 83–91 (2012)
13. Pei-Yu, L., Jung, S.L., Chin, C.C.: Distortion-free secret image sharing mechanism using modulus operator. Pattern Recogn. **42**, 886–895 (2009)

Enhancing Teaching-Learning Professional Courses via M-Learning

V. Sangeetha Chamundeswari and G.S. Mahalakshmi

Abstract Professional courses impose restrictions on teaching methodologies. Though not very theoretical in nature, following ICT techniques for teaching-learning professional courses will only cater to better presentation to the audience but not with the objectives being understood. Assessment techniques have to be evolved for ensuring the reach of such courses. And teaching-learning-assessing becomes less interesting when the traditional techniques are to be followed during online learning. It might be expected in the near future that learning will take place in versatile environments that are smoothly integrated into everyday life. This paper proposes a course management system for m-learning. This paper also discusses and shares the experiences in using mobile learning and innovative assessment methodologies followed for teaching-learning professional course, say GE9021-Professional Ethics and Engineering, R2008, Anna university, Chennai, India.

Keywords Mobile learning · Course management systems · Flexible learning · Interactive tools · Lifelong learning

1 Introduction

CMS is a course-based system that enables participants to communicate and collaborate, access activities and resources, meanwhile being tracked, assessed and reported on via the web [4]. The Course management systems are used mainly for online learning which includes placing course materials online, course administra-

V.S. Chamundeswari (✉)
Department of Information Technology, Jawahar Engineering College, Chennai 600 093, Tamil Nadu, India
e-mail: sangeetharam26@gmail.com

G.S. Mahalakshmi
Department of Computer Science and Engineering, Anna University, Chennai 600 025, Tamil Nadu, India
e-mail: gsmaha@annuniv.edu

L.C. Jain et al. (eds.), *Computational Intelligence in Data Mining - Volume 3*,
Smart Innovation, Systems and Technologies 33, DOI 10.1007/978-81-322-2202-6_51

tion, associating students with courses, tracking student performances, submission of assignments, assessing student grades, and also providing effective communication between the learner and the instructor [7]. The CMS can provide an instructor with a set of tools and a framework which allows building course content online and to provide various interactions with students taking the course. CMSs provide an integrated set of Web-based tools for learning and course management. Examples of a CMS include Blackboard, Angel, Sakai, Desire2Learn, WebCT, Oncourse and Moodle serves as prominent resources which are nearly ubiquitous in Higher Education and have a positive effect on teaching and learning [2]. This prominence comes from using CMSs to enhance courses, offer distance learning courses, and support hybrid courses, which combine online courses, can further be extended to mobile learning with the tremendous growth of the increased mobile technologies.

This article proposes a model for CMS research that equally considers technical features in these systems and conceptual issues about how people learn. Categories in the model relate to these features and issues. The categories are transmitting information to students, creating class discussions, evaluating students, evaluating courses, and creating computer based instruction [6]. Detailed information about each category will be provided.

2 M-CMS Framework

The CMS framework proposed in this work essentially addresses several important design aspects through a modular design approach and we have mapped the framework with the cognitive domain aspect of the bloom's taxonomy as shown in Fig. 1.

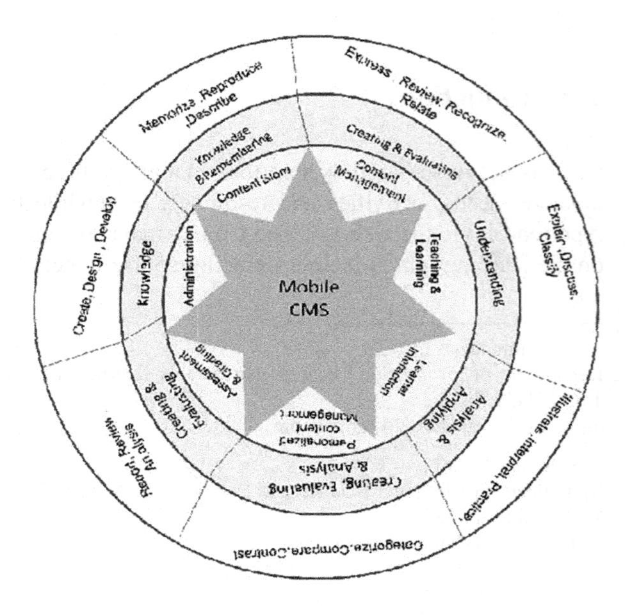

Fig. 1 M-CMS framework

The proposed framework in Fig. 1 will create, store, assemble and deliver personalized e-learning content. We have taken a sample course GE9021—Professional Ethics, which comes under curriculum of VIII Semester Computer Science Engineering students at Anna University, Chennai and provide a case study on how the same can be conducted via mobile devices and social media. The outcome of the course will be the knowledge level of the learner gained through the course and also to fulfill learning objectives of the course.

3 GE9021 Professional Ethics via M-Learning—A Case Study

The aim of Professional Ethics course is to foster in students the competence of applying integrated knowledge with mobile technologies. Therefore, it is important to develop an integrated curriculum for teaching standard operating procedures in physical assessment courses through mobile devices. During the learning activities, each student is equipped with a mobile device; the learning system not only guides individual students to perform each operation of the physical assessment procedure but also provides instant feedback and supplementary materials to them if the operating sequence is incorrect [11]. Students were given a wide array of choice for the type of group activities they should be a part of. Apart from group activities, some marks were set aside in each assessment for evaluating students individually for their creativity in bringing out issues related to ethics. While these activities promised the students a fun learning process, a good effort was taken to ensure that these activities also covered the syllabus as prescribed by the University from an examination point of view. In future, sensing devices can also be used to detect whether the student has conducted the activities correctly for assessing the current status of the student.

3.1 Activities in Professional Ethics Course

Students were given a wide array of choice for the type of activities they should be a part of. The overall findings of the activities indicated that the students generally viewed Mobile Learning as beneficial and useful. This indicated a satisfactory level of perception of the use of mobile devices in learning. In terms of time management, the respondents held the opinion that Mobile Learning will better assist them in managing their time as well as paying much more attention to learning percent [5]. This was because Mobile Learning provides great flexibility for students to manage their learning time [8]. The various activities followed for team assessments were: Ad wars, Surveys, debates, news reporting and corporate ethics, Fig. 2

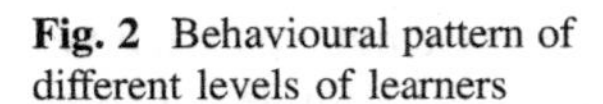

Fig. 2 Behavioural pattern of different levels of learners

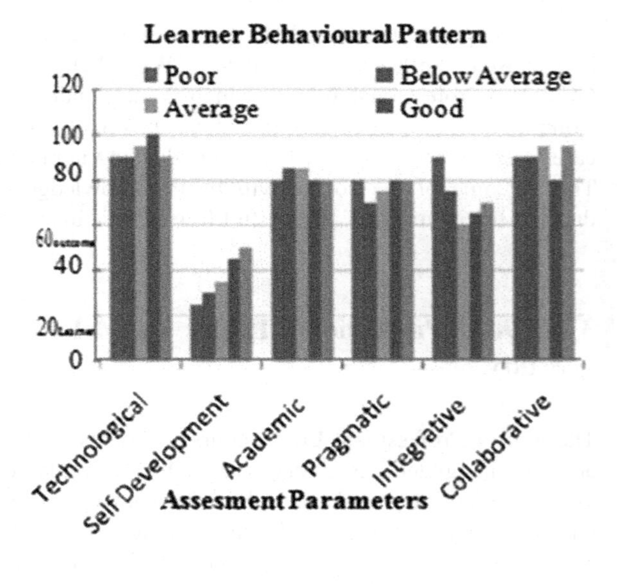

depicts the behavioral pattern of different levels of learners for all the above mentioned activities.

Preparation of news bulleting and movie reviews were performed for individual assessment. As the final leg of the activity-based approach that was adopted for the class “Professional Ethics in Engineering”, students were asked to analyze various ethical issues prevalent in the society using any form of social networking of their choice.

3.1.1 Using Social Media Platforms to Support Group Collaboration

Community opinion is always invaluable. Crowd sourcing enables you to tap into the collective intelligence of others in a community and with social media this can be done at a scale never before possible [10]. The avenues of social networking employed by students were: facebook, twitter and quora.

3.1.2 Using Social Media to Monitor Group Members’ Individual Contributions

Well-defined responsibilities helped in working asynchronously and putting together the combined efforts. The split-up proposed by most teams participating was of the form:

- Deciding the social media platform(s) to be used.
- Deciding the ethical issue to be analyzed.

- Circulating/Promoting the page or post among people via social networking to harvest as many opinions as was possible.

Comparison and contrasting others' reasoning, opinions and conclusions helped endorse higher quality decision making, better problem solving and also increased creativity of each member [9]. There was healthy competition to come up with the best conclusions and thereby contribute to a major portion of the analysis of the ethical issue.

Since all the work was done online (Social Networking sites and Document Sharing platforms) and since the interdependence between the work done by each of the group members required constant data exchange, it was easy and necessary to ensure that there was constant progress in each individual's contribution [12].

3.1.3 Using Social Media to Facilitate the Assessment of Students' Contributions and Learning Outcomes

The activity of each student is effectively logged in social networking websites and teachers can be given access to this information for evaluating each student [13]. The same metrics mentioned for peer evaluation can also be used by teachers to assess the quality of individual posts [1]. Besides this, the number of people following the thread or group or question is an excellent measure to identify the popularity and quality of the discussion as a whole. An example metric that can be devised by teachers for assessing the student can be the ratio of the number of people who followed the question to the number of people who viewed it. The higher the value of this ratio, the better the quality of discussion. The regulation of posts by the students should also be taken into account for their assessment [14].

3.2 Assessment

The various parameters taken into consideration are technological, self development, Academic, pragmatic, integrative and collaborative. Engineers must perform under a standard of professional behaviour that requires adherence to the highest principles of ethical conduct.' In conformance to this, this course is directed at helping students:

- Identify the core values that shape the ethical behaviour of an engineer
- Utilize opportunities to explore one's own values in ethical issues
- Become aware of ethical concerns and conflicts
- Enhance familiarity with codes of conduct
- Increase the ability to recognize and resolve ethical dilemmas.

Irrespective of the different levels of the learner, the activity based approach for teaching and learning professional courses serves to an equal extent which meets the requirements of the learners. As shown in Fig. 2, a below average student could also perform extremely well in an activity compared to a bright student. Thus the technology enhanced learning has got the benefit of retaining any kind of learner for a long time [3].

4 Conclusion

This paper has presented a reflective overview of developments in mobile course management systems from the perspective of pedagogy. The challenge will be to discover how to use mobile technologies to transform learning into a seamless task where it is not recognized as learning at all. Integrated technology-enhanced courses seem to be more demanding for students and also for instructors. In the future, the mobile CMS should cater for easy course administration, tracking facilities, targeted email facilities, reporting functions, assessments and statistics in a mobile friendly manner by hosting on cloud servers.

References

1. Naji Shukri, A., Abdul Razak, Y.: Students awareness and requirements of mobile learning services in the higher education environment. Am. J. Econ. Bus. Adm. **3**(1), 95–100 (2011)
2. Sarrab, M., Elgamel, L., Aldabbas, H.: Mobile learning (M-learning) and educational environments. Int. J. Distrib. Parallel Syst. **3**(4), 78–85 (2012)
3. Norazah, M.N., Embi, M.A., Yunus, M.Md.: Mobile learning framework for lifelong learning. Int. Conf. Learner Divers. 2. Procedia Social and Behavioral Sciences, vol. 7(C), 130–138 (2010)
4. Giemza, A., Bollen, L., Hoppe, H.U.: LEMONADE: A flexible authoring tool with support for integrated mobile learning scenarios. Int. J. Mob. Learn. Organ. **5**(1), 96–114 (2010)
5. Douglas, M., Praul M., Lynch M.J.: Mobile learning in higher education: an empirical assessment of a new educational tool. J. Educ. Technol. **7**(3), Article 2, 15–21 (2008)
6. Goh, T., Kinshuk, D.: Getting ready for mobile learning—adaptation perspective. J. Educ. Multimedia Hypermedia **15**(2), 175–198 (2006)
7. Ha, L., De, J., Holden, H., Rada, R.: Literature trends for mobile learning: word frequencies and concept maps. Int. J. Mob. Learn. Organ. **3**(3), 275–288 (2009)
8. Pachler, N.: Mobile learning in the context of transformation, Pachler, N. (ed). Int. J. Mob. Blended Learn. Special Part1 pp. 129–137 (2010)
9. Jacob, S.M., Issac, B.: The mobile devices and its mobile learning usage analysis. In: Proceedings of the International Multi Conference of Engineers and Computer Scientists (IMECS), Hong Kong, vol. 1, 19–21 (2008)
10. Kim, S.H., Mims, C., Holmes, K.P.: An introduction to current trends and benefits of mobile wireless technology use in higher education. Assoc. Adv. Comput. Educ. J. **4**(1), 77–100 (2006)

11. Singh, D., Zaitun, A.B.: Mobile learning in wireless classrooms. J. Instr. Technol. (MOJIT) **3**(2), 26–42 (2006)
12. Liu, S.P.: A study of mobile instant messaging adoption: within culture variation. Int. J. Mob. Commun. **9**(3), 280–297 (2011)
13. Van der Merwe, H., Brown, T.: Mobile technology: the future of learning in your hands. In: mlearn 4th World Conference on M-learning, pp. 25–28 (2005)
14. Denk, M., Weber, M., Belfin, R.: Mobile learning challenges and potentials. Int. J. Mob. Learn. Organ. **1**(2), 122–139 (2007)

Evaluation of Fitness Functions for Swarm Clustering Applied to Gene Expression Data

P.K. Nizar Banu and S. Andrews

Abstract Clustering problem is being studied by many of the researchers using swarm intelligence. However, the search space is not carried out entirely randomly; a proper fitness function is required to determine the next step in the search space. This paper studies Particle Swarm Optimization (PSO) based clustering with two different fitness functions namely Xie-Beni and Davies-Bouldin indices for brain tumor gene expression dataset. Clustering results are validated using Mean Absolute Error (MAE) and Dunn Index (DI). To analyze function of genes, genes that have similar expression patterns should be grouped and the datasets should be presented to the physicians in a meaningful way. High usability of algorithm and the encouraging results suggests that swarm clustering (PSO based clustering) with Davies-Bouldin index as fitness functions with respect to Dunn index can be a practical tool for analyzing gene expression patterns.

Keywords Swarm clustering · PSO · Xie-Beni · Davies-Bouldin · Gene clustering

1 Introduction

Gene expression datasets comprises thousands of genes and hundreds of conditions for mining functional and class information, is becoming highly significant. Genes that behave similarly may be co-regulated and belong to a common pathway or a cellular structure [1]. Clustering is the process of grouping objects whose characteristics are similar to each other. In the similar way, gene clustering measures gene expression levels for thousand of genes simultaneously and groups related genes

P.K. Nizar Banu (✉)
B. S. Abdur Rahman University, Chennai, India
e-mail: nizarbanu@gmail.com

S. Andrews
Mahendra Engineering College, Namakkal, India
e-mail: andrewsmalacca@gmail.com

L.C. Jain et al. (eds.), *Computational Intelligence in Data Mining - Volume 3*,
Smart Innovation, Systems and Technologies 33, DOI 10.1007/978-81-322-2202-6_52

together. Clustering gene expression patterns can be classified as gene clustering and sample clustering. This paper focuses on gene clustering, in which genes are treated as objects and samples as attributes. Gene clustering reveals the similarity between genes or a set of genes with similar conditions that leads to identify differentially expressed genes.

Hierarchical clustering algorithms [2], Self-Organizing Maps [3] and K-Means clustering algorithms [4] are the most commonly applied full space clustering algorithms on gene expression profiles. Clustering is a most important data mining technique which divides genes of similar expression pattern into a small number of meaningful homogenous groups or clusters. The expression levels of different genes can be viewed as attributes of the samples, or the samples as the attributes of different genes. Clustering can be performed on genes or samples [2].

In literature, swarm optimization algorithms are used for clustering effectively, since it converges to global minima [5]. These algorithms use two basic strategies while searching for the global optimum; exploration and exploitation [6]. While the exploration process succeeds in enabling the algorithm to reach the best local solutions within the search space, the exploitation process tries to reach the global optimum solution that exists around the local solutions obtained so far. Particle Swarm Optimization (PSO), introduced in [7], is a population based stochastic optimization technique inspired by bird flocking and fish schooling that is based on iterations/generations [8]. These methods are efficient, adaptive and robust by producing near optimal solutions. Unlike other optimization problems, gene clustering does not depend on exact optimum solution, but requires robust, fast and near optimal solutions that are produced by swarm clustering efficiently. It is easier to implement and computationally efficient when compared to other evolutionary algorithms. Clustering techniques based on the swarm intelligence tools have reportedly outperformed many classical methods of partitioning a complex real world dataset [9, 10]. Fitness function in swarm clustering is used to evaluate the new coordinates at each time step for finding the compactness and isolation of objects in the cluster [10]. Hybrid rough-PSO technique [11] with Davies-Bouldin clustering validity index is used for grouping the pixels of an image in its intensity space. This paper analyzes swarm clustering for brain tumor gene expression dataset with two different fitness functions and finds the suitable fitness function for swarm clustering and the optimal number of clusters.

The rest of the paper is organized as follows. Section 2 presents a brief review on gene clustering. Section 3 discusses methodology adopted in this paper. Section 4 presents experimental results and its interpretation followed by conclusion in Sect. 5.

2 Gene Clustering Background

This section gives a brief review about gene clustering. Clustering has been shown to be very effective, in associating gene expression patterns with the ligand specificity of neurotransmitter receptors and their functional class. In cancer studies, [12–15] both gene expression, signatures for cell types and signatures for biological processes have been successfully identified by clustering [16]. Gene clustering techniques were also used to analyze temporal gene expression data during rat central nervous system development [15]. Gene expression data are often highly connected and may have interesting and embedded patterns [17].

Gene expression values are presented in a matrix called gene expression matrix; columns represent samples and rows represent genes. K-Means is a well known partitioning method. Genes are classified as belonging to one of K-groups, K chosen a priori. This approach minimizes the overall within cluster dispersion by iterative reallocation of cluster members. In [18], Informative Gene Selection method is applied to identify highly expressed genes and suppressed genes and clustered using K-Means clustering. Self-Organizing Maps partition genes into user-specified K optimal clusters. Full space clustering algorithms like Bayesian network [19] and neural network are also applied on gene expression data. Fuzzy based clustering is applied for clustering gene expression data. Fuzzy modifications of K-Means include Fuzzy C-Means (FCM) [20, 21]. Penalized Fuzzy C-Means approach is applied for brain tumor gene expression dataset in [22]. Fast Genetic K-means Algorithm (FGKA) is applied for clustering genes in [23]. In [24], Kernel density clustering method for gene expression analysis is reported.

In this paper, we examine swarm clustering with different fitness functions for gene expression data for grouping highly expressed and suppressed genes which helps physicians for early diagnosis of disease and drug development.

3 Methodology

Gene expression data is represented by a matrix with rows representing genes and columns corresponding to samples. Methodology adopted in this work is shown in Fig. 1.

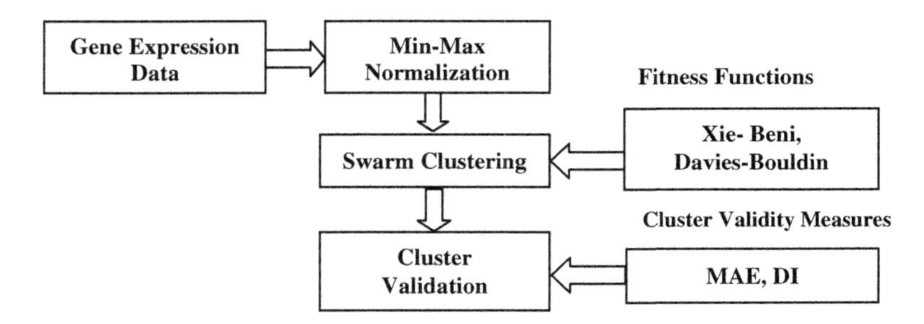

Fig. 1 Framework used in this paper

3.1 Min-Max Normalization

Gene expression dataset comprises huge number of genes with very less number of samples. To make the entire dataset in a specific range, the dataset should be normalized by using any of the normalization techniques available in the literature. We prefer Min-Max normalization technique [25] to normalize the dataset.

$$v' = \frac{v - \min_a}{\max_a - \min_a}(new\_\max_a - new\_\min_a) + new\_\min_a \tag{1}$$

In Eq. 1, v is the original value; $\min_a$ and $\max_a$ are the minimum and maximum values of a column. new_min and new_max are the new range of values to which we want to change. Here, new_min_a is 0 and new_max_a is 1.

3.2 PSO Clustering

Particle Swarm Optimization is a well known swarm intelligence as well as population based method and evolutionary optimization technique used for clustering. As given in [5], first *n* particles are created and distributed randomly in the search space then every particle is made to move towards the optimal solution with updated velocity at every iteration. Fitness of each particle is evaluated using Xie-Beni index and Davies-Bouldin index. All the particles are moved towards the fittest particle and its personal best position with a velocity v_i given by

$$v_i(t+1) = w * v_i(t) + C_1 * r_1 * (p_i - c_i) + C_2 * r_2 * (g - z_i) \tag{2}$$

where p_i is the personal best position of the gene, c_i (centroid) is the current position of the gene, g is the global best of the entire genes, w is the inertial constant, C_1 is the personal best constant and C_2 is the global best constant, $i = 1,2,\ldots, n$. [5]. Each gene moves using

$$z_i(t+1) = z_i(t) + v_i \tag{3}$$

Fitness of each gene is calculated, and then the personal best position and the global best position are found. This process is repeated until maximum number of iteration is met. After finding the global best position, best value of cluster centroid is found [5].

PSO Clustering: Pseudo-code

Input: K – Number of Clusters; D – Dataset containing n objects; tmax – Maximum number of iterations

Output: Set of K clusters

Initialize parameters including population size P, C_1, C_2, r_1, r_2, w and tmax
Create swarms with P particles *(X, P_i, g and V)*
For it = 1: tmax
Calculate cluster centers
Calculate fitness of each particle *f(x)* using XB (Eq. 4) or DB (Eq. 5) indices
Update personal best for each particle and global best for the swarm
Update velocity using Eq. 2 and position for each particle using Eq. 3

End

3.3 Fitness Functions for PSO Clustering

Fitness of a particle to each cluster is evaluated using its compactness and isolation. Compactness measures the uniqueness of genes within the cluster and isolation measures distinctiveness of genes between a cluster and the rest of the clusters [26]. The purpose of the fitness function is to generate partitions that have compact and well-separated clusters. In this paper we have used Xie-Beni index and Davies-Bouldin index as a fitness function for PSO clustering.

3.3.1 Xie-Beni (XB) Index

Xie-Beni index has been proved to detect correct number of clusters in many experiments [27] and hence it is chosen as a fitness function. Xie-Beni index is the combination of two functions. First, it computes the compactness of genes within the cluster and then it finds the separateness of genes in different clusters. Let S be the fitness function, compactness of genes be π and separation of genes be s. Then the Xie-Beni index is expressed as

$$S = \frac{\pi}{s} \tag{4}$$

In Eq. 4, $\pi = \frac{\sum_{i=1}^{K} \sum_{j=1}^{n} \mu_{ij}^2 \|x - z^i\|^2}{n}$ and $s = (d_{\min})^2$. *dmin* is the minimum distance between cluster centers given by $d_{\min} = \min_{ij} \left\| z_i - z_j \right\|$. n is the number of objects, K is the number of clusters, and Z_i is the cluster centre of cluster C_i. μ_{ij} be the membership value of the objects. In this method every object belongs to exactly one cluster. So the membership is 1. Smaller values of π indicate that the clusters are more compact and larger values of s indicate the clusters are well separated.

3.3.2 Davies Bouldin (DB) Index

Davies-Bouldin index measures the average of similarity between each cluster and its most similar one [28]. It is desirable for the clusters to have the minimum possible similarity to each other [29]. DB index with minimum value shows better cluster configuration. The similarity measure of clusters R_{ij} can be defined as $R_{ij} = \frac{s_i + s_j}{d_{ij}}$ where $d_{ij} = d(v_i, v_j)$; is the cluster dissimilarity measure and $S_i = \frac{1}{\|C_i\|} \sum_{x \in C_i} d(x, v_i)$ is the dispersion measure. Then the Davies-Bouldin index is defined as shown in Eq. 5.

$$f(x) = DB = \frac{1}{K} \sum_{i=1}^{K} R_i \quad (5)$$

where $R_i = max_j = 1 \ldots K,\ i \neq j\ (R_{ij}),\ i = 1 \ldots K$. K is the number of clusters.

3.4 Cluster Validations

Cluster validation determines the clustering tendency of a set of data and used to measure the goodness of clustering structure. It is also used to estimate the optimal number of clusters.

3.4.1 Mean Absolute Error (MAE)

We have used MAE [30] to measure the predictive accuracy of clustering results. It can range from 0 to ∞.

$$MAE = \sum \frac{|x - x'|}{N} \quad (6)$$

In Eq. 6, x is the object belongs to the cluster, x' is the centroid of the cluster N is the total number of objects.

3.4.2 Dunn Index (DI)

Dunn index [31] is used to find the compactness of objects in the cluster. Maximum Dunn index provides better cluster. Minimal distance between points of different clusters is denoted by d_{min} and the largest within cluster distance is denoted by d_{max}. Distance between clusters C_k and $C_{k'}$ is measured by the distance between their closest points. $d_{kk'} = \text{min} i \in I_k, j \in I_{k'} \left\| M_i^{\{k\}} - M_i^{\{k'\}} \right\|$. d_{min} is the smallest of these distances $d_{kk'}$. $d_{\min} = \min_{k \neq k'}$. For each cluster C_k, the diameter of the cluster D_k is given as $D_k = \max_{i,j \in I_k; i \neq j} \left\| M_i^{\{k\}} - M_i^{\{k'\}} \right\|$. Then d_{max} is the largest of these distances D_k. $d_{\max} = \max_{1 \leq k \leq K} D_k$. The Dunn index is defined as the quotient of d_{min} and d_{max} and it is shown in Eq. 7.

$$C = \frac{d_{\min}}{d_{\max}} \quad (7)$$

4 Experimental Analysis

This section discusses the results obtained and the effectiveness of the fitness functions for swarm clustering applied to brain tumor gene expression data. Brain Tumor gene expression dataset is taken from the Broad Institute website (http://www.broadinstitute.org/cgi-bin/cancer/datasets.cgi). It consists of 7,129 genes with 40 samples.

4.1 Performance Analysis of Swarm Clustering with Fitness Functions

In this paper, acceleration coefficients C_1 and C_2 is set to 0.2, which controls personal best position p_i and global best position g in search process. Random variables r_1 and r_2 should be between 0 and 1, so r_1 is set to 0.1 and r_2 is set to 0.5. Inertial constant w is set to 0.7 and maximum number of iterations, $tmax$ is set to 40. Performance of swarm clustering is analyzed by using XB index and DB index as fitness functions and validated using MAE and DI. The results are shown in Table 1. To find the optimal number of clusters the algorithm is executed for 2 clusters to 15 clusters for 40 iterations. Cluster that produce minimum MAE and maximum Dunn Index is considered as the optimal number of clusters.

Based on the performance of fitness functions, optimal number of clusters is selected as 3 and 5. Genes placed in the optimal clusters is shown in Table 2.

Figure 2 depicts the performance of fitness functions for swarm clustering applied to gene expression dataset discussed in this paper based on MAE and Dunn Index. It is clearly shown in the Figs. 2 and 3, when DB index is used as fitness function variation between the clusters is very less for both MAE and DI.

4.2 Interpretation of Results

Swarm clustering is used to group genes that are similar to each other by covering the entire search space. To demonstrate the results obtained, we have taken 50 genes with 40 samples from brain tumor gene expression dataset and executed

Table 1 Performance analysis of swarm clustering with XB and DB as fitness functions

No. of clusters	Mean absolute error (rule: min)		Dunn index (rule: max)	
	XB	DB	XB	DB
2	25.1576	20.2003	1.0000	1.0000
3	20.6668	**20.1860**	0.5684	**0.9024**
4	15.1094	18.4353	0.1877	0.8917
5	**11.7353**	19.1524	**0.9418**	0.8749
6	21.9017	18.8998	0.6809	0.8641
7	20.7539	19.9827	0.8032	0.7476
8	20.0859	19.7833	0.2391	0.7055
9	15.9292	19.7034	0.3158	0.7494
10	17.1885	19.7210	0.0826	0.7029
11	11.0618	21.1497	0.3195	0.7235
12	40.2323	19.2362	0.7861	0.8042
13	15.7250	18.8851	0.5758	0.6707
14	14.0064	20.2355	0.2379	0.7388
15	25.3140	19.5223	0.0690	0.7680

Table 2 Number of genes placed in optimal number of clusters

S. no.	No. of clusters	No. of genes in each cluster	
		XB	DB
1	5	108, 1698, 2004, 1666, 1653	1712, 1840, 1342, 982, 1253
2	3	4579, 235, 2315	1712, 1840, 3577

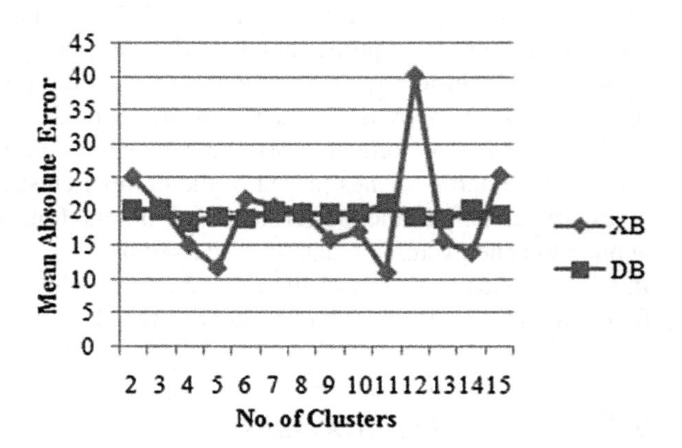

Fig. 2 Performance of swarm clustering based on MAE

for 3 clusters to find highly expressed and suppressed genes. Table 3 Shows the genes placed in 3 optimal clusters by swarm clustering using XB and DB as fitness function. As we use clustering procedure alone, we found gene number and number

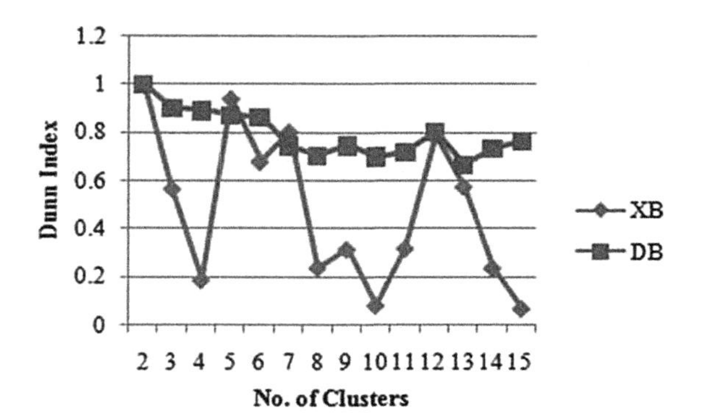

Fig. 3 Performance of swarm clustering based on DI

Table 3 Genes placed in 3 clusters

S. no	No. of clusters	XB		DB	
		No. of genes	Gene no.	No. of genes	Gene no.
1	3	33	1,2,3,4,5,7,8,9,10,11,14, 15,16,17,21,22,23,24,25, 26,27,28,29,30,32,37,38, 39,40,41,48,49,50	35	1,2,3,4,5,6,7,8,9,10,11,12,13,14, 15,16,17,18,20,21,22,23,24,25, 26,27,28,29,30,31,32,33,34,35,36
		7	19,42,43,44,45,46,47	8	19,38,42,43,44,45,46,47
		10	6,12,13,18,20,31,33,34, 35,36	7	37,39,40,41,48,49,50

of genes placed in different clusters. Relationship among the genes can be found using gene annotation tools.

Results obtained for brain tumor gene expression dataset assures that swarm clustering with DB index can also be applied for lung cancer and leukemia cancer gene expressions datasets to find highly suppressed and highly expressed genes.

5 Conclusion

In this paper PSO based clustering is applied to brain tumor gene expression dataset with Xie-Beni index and Davies-Bouldin index as fitness function. MAE and DI are used to measure the quality of clusters obtained by PSO clustering using both the fitness functions. Results shown in this paper proves that PSO clustering with DB index as fitness function works well for clustering brain tumor gene expression data with respect to Dunn index. Fitness functions discussed here provides a way to find optimal number of clusters. Genes in the clusters are either highly suppressed or highly expressed. Highly expressed genes are the cause for malignant tumors. Removal of such tumors may affect other organs of the body. Genes grouped in every cluster helps to gain knowledge and suggest clinicians for suitable treatment to reduce tumor growth. Relationship of genes within the cluster alone is not sufficient for medical analysis; relationship of genes between the clusters should also be known. Group of genes produced by swarm clustering procedure are a key point to find its functionalities. Functionality of genes and its relationship can be found using gene annotation tools in future.

References

1. Bryan, K.: P′adraig cunningham, nadia bolshakova: application of simulated annealing to the biclustering of gene expression data. IEEE Trans. Inf Technol. Biomed. **10**, 519–525 (2006)
2. Eisen, M.B., et al.: Cluster analysis and display of genome-wide expression patterns. Proc. Natl. Acad. Sci. **95,** 14863–14868 (U.S.A. 1998)
3. Tamayo, P., Slonim, D., Mesirov, J., Zhu, Q., Kitareewan, S., Dmitrovsky, E., Lander, E.S., Golub, T.R.: Interpreting patterns of gene expression with self organizing maps: methods and application to hematopoietic differentiation. Proc. Natl. Acad. Sci. **96,** 2907–2912 (U.S.A. 1999)
4. Tavazoie, S., Hughes, D., Campbell, M.J., Cho, R.J., Church, G.M.: Systematic determination of genetic network architecture. Nat. Genet. **22**, 281–285 (1999)
5. Senthilnath, J., Das, V., Omkar, S.N., Mani, V.: Clustering using levy flight cuckoo search. In: Proceedings of Seventh International Conference on Bio-Inspired Computing: Theories and Applications *(BIC-TA 2012).* Advances in Intelligent Systems and Computing, vol. 202, pp. 66–75 (2013)
6. Rashedi, E., Nezamabadi-pour, H., Saryazdi, S.: GSA: A gravitational search algorithm. Inf. Sci. **179**, 2232–2248 (2009)
7. Kennedy, J., Eberhart, R.: Swarm Intelligence, 1st edn. Morgan Kaufmann, San Francisco (2001)
8. Izakian, H., Abraham, A., Snášel, V.: Metaheuristic based scheduling meta-tasks in distributed heterogeneous computing systems. Sensors (Basel) **9**, 5339–5350 (2009)
9. Samadzadegan, F., Abootalebi, A., Alizadeh, A.: Evaluation of the potential of swarm clustering techniques in hyperspectral data. GIS Ostrava **1**, 24–27 (2010)
10. Abraham, A., Das, S., Roy, S.: Swarm Intelligence Algorithms for Data Clustering. Soft Computing for Knowledge Discovery and Data Mining, pp. 279–313. Springer, Berlin (2008)
11. Das, S., Abraham, A., Sarkar, S.K.: A hybrid rough set–particle swarm algorithm for image pixel classification. In: Proceedings of the Sixth International Conference on Hybrid Intelligent Systems, pp. 26–32 (2006)
12. Golub, T.R., Slonim, D.K., Tamayo, P., Huard, C., Gaasenbeek, M., Mesirov, J.P., Coller, H., Loh, M.L., Downing, J.R., Calgiuri, M.A., Bloomfield, C.D., Lander, E.S.: Molecular classification of cancer: class discovery and class prediction by gene expression monitoring. Science **286**, 531–537 (1991)
13. Alon, U., Barkai, N., Notterman, D.A., Gish, K., Ybarra, S., Mack, D., Levine, A.J.: Broad patterns of gene expression revealed by clustering analysis of tumor and normal colon tissues probed by oligonucleotide arrays. Proc. Natl. Acad. Sci. **96,** 6745–6750 (USA 1999)
14. Spellman, P.T., Sherlock, G., Zhang, M.Q., Iyer, V.R., Anders, K., Eisen, M.B., Brown, P.O., Botstein, D., Futcher, B.: Comprehensive identification of cell cycle-regulated genes of the yeast saccharomyces cerevisiae by microarray hybridization. Mol. Biol. Cell **9**, 3273–3297 (1998)

15. Wen, X., Fuhrman, S., Michaels, G.S., Carr, D.B., Smith, S., Barker, J.L.: Roland Somogyi.: Large-scale temporal gene expression mapping of central nervous system development. Proc. Natl. Acad. Sci. USA Neurobiol. **95**, 334–339 (1998)
16. Alizadeh, A.A., Eisen, M.B., Davis, R.E., Ma, C., Lossos, I.S., Rosenwald, A., Boldrick, J.C., et al.: Distinct types of diffuse large B-cell lymphoma identified by gene expression profiling. Nature **403**, 503–511 (2000)
17. Jiang, D., Pei, J., Zhang, A.: DHC: A density-based hierarchical clustering method for time series gene expression data. In Proceedings of BIBE2003: 3rd IEEE International Symposium on Bioinformatics and Bioengineering, pp. 393–400. Bethesda, Maryland (2003)
18. Nizar Banu, P.K., Andrews, S.: Informative gene selection—an evolutionary approach. In: Proceedings of International Conference on Current Trends in Information Technology (CTIT), pp. 129–134 (2013)
19. Friedman, N., Linial, M., Nachman, I., Pe'er, D.: Using Bayesian networks to analyze expression data. J. Comput. Biol. **7**, 601–620 (2000)
20. Dembele, D., Kastner, P.: Fuzzy *c*-means method for clustering microarray data. Bioinformatics **19**, 973–980 (2003)
21. Kim, S.Y., Choi, T.M., Bae, J.S.: Fuzzy types clustering for microarray data. Int. J. Comput. Intell. **2**, 12–15 (2005)
22. Nizar Banu, P.K., Hannah Inbarani, H.: An analysis of gene expression data using penalized fuzzy c-means approach. Int. J. Commun. Comput. Innov. **1**, 101–113 (2011)
23. Lu, Y., Lu, S., Fotouhi, F., Deng, Y., Brown, S.: Fast genetic K-means algorithm and its application in gene expression data analysis. Technical Report TR-DB-06-2003, http://www.cs.wayne.edu/~luyi/publication/tr0603.pdf (2003)
24. Shu, G., Zeng, B., Chen, Y., Smith, O.: Performance assessment of kernel density clustering for gene expression profile data. Comp. Funct. Genomics **4**, 287–299 (2003)
25. Han, J., Kamber, M.: Data Mining: Concepts and Techniques, 2nd edn. Morgan Kaufmann Publishers, San Francisco (2006) ISBN 1-55860-901-6
26. Raskutti, B., Leckie, C.: An evaluation of criteria for measuring the quality of clusters. In: Proceedings of 16th International Joint Conference on Artificial Intelligence, vol. 2, pp. 905–910 (1999)
27. Pal, N.R., Bezdek, J.C.: On cluster validity for the fuzzy c-means model. IEEE Trans. Fuzzy Syst. **3**, 370–379 (1995)
28. Davies, D.L., Bouldin, D.W.: Cluster separation measure. IEEE Trans. Pattern Anal. Mach. Intell. **1**, 95–104 (1979)
29. Halkidi, M., Batistakis, Y., Vazirgiannis, M.: On clustering validation techniques. J. Intell. Inf. Syst. **17**, 107–145 (2001)
30. De Castro, P.A.D., de França, O.F., Ferreira, H.M., Von Zuben, F.J.: Applying biclustering to perform collaborative filtering. In: Proceedings of the Seventh International Conference on Intelligent Systems Design and Applications, pp. 421–426 (2007)
31. Dunn, J.: Well separated clusters and optimal fuzzy partitions. J. Cybern **4**, 95–104 (1974)

Context Based Retrieval of Scientific Publications via Reader Lens

G.S. Mahalakshmi, R. Siva and S. Sendhilkumar

Abstract Scientific publications grow voluminous and there should be a mechanism to organise the related scientific literature from the perspective of research scholar. Again, a scholar could be well in three classes: a skimmed reader, a vocabulary level reader or a comprehensive level reader. Skimmed reader is the level at which the scholar is new to things discussed in the research publication and is finding a bit tougher to understand the discussions in the relevant literature since it connects to a greater depth of understanding. Therefore, at this stage, the reader would be interested to look more superficially than the other matured level readers. The recommendations made to the skimmed researcher should be more contextually significant, better related to the quest of research, and has to contain originality and novelty instead! This paper discusses a framework for modeling the perspective of various reader categories to aid context based retrieval of scientific publications. We have attempted at proof-of-concept experiments to support and validate our claim for skimmed level categories.

Keywords Bibliometrics · Context based content retrieval · Research articles · Novelty estimation · Citation analysis

G.S. Mahalakshmi (✉)
Department of CSE, Anna University, Guindy, Chennai 600025, Tamil Nadu, India
e-mail: gsmaha@annauniv.edu

R. Siva
Department of CSE, KCG College of Engineering, Karappakam, Chennai 600097, India
e-mail: siva465kcg@gmail.com

S. Sendhilkumar
Department of IST, Anna University, Guindy, Chennai 600025, India
e-mail: thamaraikumar@annauniv.edu

L.C. Jain et al. (eds.), *Computational Intelligence in Data Mining - Volume 3*,
Smart Innovation, Systems and Technologies 33, DOI 10.1007/978-81-322-2202-6_53

1 Introduction

Linking documents seems to be a natural proclivity of scholars. In search for additional information or in search for information presented in simpler terms, whatever the case may be, the user is generally misled with abundant pool of publications not knowing the direction of his/her search in relevance to the research problem and often forgets the source that motivated the user to initiate such navigation over research publications. Often budding researchers find themselves misled amidst conceptually ambiguous references while exploring a particular seed scholarly literature either aiming at a more clear understanding of the seed document or trying to find the crux of the concept behind the journal or conference article. Integration of bibliographical information of various publications available on the Web is an important task for researchers to understand the essence of recent advancements and also to avoid misleading of the user over distributed publication information across various sites.

Automatic reference tracking (ART) [1, 2] appears as a good shepherd for the struggling researchers by listing the relevant reference articles to the user's purview. The relevant articles identified and downloaded are stored for later use. ART-listed references (which are the actual references across various levels from the seed article) are represented year wise, author wise or relevance wise [1] for better browsing of ART output. This paper discusses an improvisation on ART focusing the analysis of publication semantics and granularity to appeal to the reader.

2 Related Works

ART involves recursive tracking of reference articles listed for a particular research paper by extracting the references of the input seed publication and further analyzing the relevance [3] of the referred paper with respect to the seed paper. This recursion continues with every reference paper being assumed as seed paper at every track level until the system finds any irrelevant (or farthest) references deep within the reference tracks which does not help much in the understanding of the input seed publication at hand.

Tracking of references implies locating the references in the Web through commercial search engines and successfully downloading them [4] to be fed for subsequent recursions. Web pages, however, change constantly in relation to their contents and existence, often without notification [5]. Therefore, updating of Web page address for every research reference article is not possible.

ART has a dynamic approach for retrieving the articles across the Web. The title of every reference article is extracted from the reference region after successful reference parsing. Later, the title is submitted as a query to search engines and the article is harvested irrespective of the location of the article in the Web. During the tracking of references, pulling down each reference entry and parsing them results in metadata [1, 6], which is populated into the database to promote further search in a recursive fashion. Depending upon the user's needs the data is projected in a suitable representation. The representation is used to track information on specific attribute specified by the user, which includes the author, title, year of publication etc. thereby providing the complete information about the references for the journal cited.

The citation count and author's work is highly influenced by the number of self citations and recitations of the article [7]. ART has not taken into account the nature of self-citations [8] which have become inherent in any research communication [9]. Tracking self-citations would end up with the conceptual linkage of research ideas of a particular author. Self-citation also contributes [10, 11] to the overall citation count of an article which is indirectly influencing the impact factor magic. Journal impact factors do not measure the quality and necessity of self-citations [10, 11]. JIF [12], h-index [13] etc. are greatly affected by the self citations and the pattern in which they are cited [14]. Google Scholar includes self-citations into the total citations measure and Scopus exclude the self-citations for calculation of h-index of a particular researcher.

Analysing the quality of self-citations would help much in deriving the true research progress of a particular research topic by one or group of researchers and will facilitate ART to a greater extent. Inclusion of research community analysis and expert analysis will improvise the filtration process of reference recommendations [15–17]. However this is possible only if validity of self-citations is explored since every self-citation is an indicative of continuing research [18]. The (self) citations are analysed for citation quality via citation sentiment analysis [19–21] which ranks the self-citations qualitatively.

For analyzing the citations, citation relations were introduced. Citation relations describe the nature of connection between the cited and citing articles [22–28].

In addition to citation analysis, semantic analysis of research articles for identifying novelty and originality of the idea expressed is also of utmost importance [29, 30]. With all the above features analysed for a referenced research article, reference suggestions could be revised using ART. However, ART does not account for suggesting reference papers to varying reader categories. This paper attempts to provide the context based research paper recommendation as an extension to ART.

3 Context Based Research Paper Retrieval

The proposed system considers research article as input data. Consider the following case study (Table 1). The first article is the input article (seed paper). The articles A# and A#B# are first level and second level reference articles of the seed paper. The reference articles are initially compared with the seed paper and articles least relevant to the domain are filtered out. The filtered articles are then compared with the seed article and the similarity measure among them is obtained by using cosine similarity (1).

Table 1 Dataset for case study

Article id	Article name
Seed paper	The evolution of transport protocols: An evolutionary game perspective
A1	A game theoretic analysis of protocols based on fountain codes
A2	A game-theoretic approach towards congestion control in communication networks
A6	An evolutionary game perspective to ALOHA with power control
A7	Analysis of scalable TCP
A29	The case for non-cooperative multihoming of users to access points in IEEE 802.11 WLANs
A2B14	Congestion avoidance and control
A2B15	Bottleneck flow control
A2B26	A binary feedback scheme for congestion avoidance in computer networks
A6B1	A survey on networking games
A6B9	On the randomization of transmitter power levels to increase throughput in multiple access radio systems
A1B1	Selfish behavior and stability of the internet: A gametheoretic analysis of TCP
A1B4	A game-theoretic approach towards congestion control in communication networks
A1B6	A mathematical model for the TCP tragedy of the commons
A1B7	LT codes
A29B3	A comparison of overlay routing and multihoming route control
A29B6	New insights from a fixed point analysis of single cell IEEE 802.11 WLANs
A29B7	Fixed point analysis of single cell IEEE 802.11e WLANs: uniqueness, multistability and throughput differentiation
A29B8	Fairness and load balancing in wireless LANs using association control
A29B13	Overlay TCP for multi-path routing and congestion control
A29B14	The multipath utility maximization problem
A29B16	The price of anarchy is independent of the network topology
A29B19	Competitive routing in multiuser communication networks
A29B24	A mathematical model of the paris metro pricing scheme for charging packet networks
A29B25	Telecommunications network equilibrium with price and quality-of-service characteristics

$$\cos\left(s_i, s_j\right) = \frac{\sum_{k=1}^{l} v_i, k \times v_j, k}{\left|s_i\right| \cdot \left|s_j\right|} \tag{1}$$

where s_i represents Sentence vector$(v_{i1}, v_{i2}, \ldots, v_{il})$, l denotes number of documents retrieved from the reference corpus and s_j is sentence vector.

The average of the similarity measure is taken as a threshold and the reference articles with similarity score above the threshold are retained for further processing.

3.1 Reference Quality Estimation

The articles in Table 1 are further processed to determine the quality and worthiness of references made. The context of reference (sentence) around the reference place holders are identified and the sentiment of the context are estimated. The sentences are then brought into being using part-of-speech tag as an annotation on each word or symbol. The sentiment score for each adjective is found from SentiWordNet Lexical Analyzer. All adjective scores are aggregated to obtain overall sentiment score. Adjectives are considered because mostly adjectives represent the sentiment in a sentence.

The references are then classified using the classification scheme [31] with categories as Compare, basis, support, use, modifies, weak and simple. The methodology of citation classification of cited text follows from Sendhilkumar et al. [29]. Sentences with existing references are used as training data after removing the placeholders. For each paper from the dataset, training set is got from the examples annotated with class values. For each context the appropriate features were extracted and the classifier was constructed using Naïve Bayes algorithm. This classification has non numerical label.

The relevance of the seed article with the next level of article and article at level one and that of level two is identified by cosine similarity and outlier determination [32]. Outlier determination is done so as to identify articles that are not greatly relevant to referred article. The outlier is found by Latent Dirichlet allocation (LDA) that is widely used for identifying the topics in a set of documents. The probability distribution is found based on Gibbs sampling and distribution of content over various topics is identified. Then the similarity between the two topic distributions is computed for outlier determination with 50 % similarity threshold to the seed article. Based on the scores obtained above the reference quality score is computed (2).

$$Reference\ Quality\ Score = sentiment_i + Similarity_i + LDA_i \tag{2}$$

where i = highest importance classification category.

If the reference is an outlier then the aggregation process omits the LDA similarity score (3).

$$Reference\ Quality\ Score = sentiment_i + Similarity_i \tag{3}$$

where i = highest importance classification category.

3.2 Novelty Estimation

From the above procedure we obtain a set of articles that are of high reference quality. The seed article is then evaluated with the retained set to evaluate the novel regions in the article. This is done in order to identify sections/sentences of the article that will be prominently looked upon by a skimmed reader. As the skimmed reader is a primitive level of reader, the intention will be to quickly scan the article, running through important sentence and sentence with certain phrases rather than reading the entire article. To ease the reader the system predicts the parts that when looked upon will give enough information for the skimmed reader.

The seed article is initially segmented into sections. Each section is processed to identify sentence carrying keywords and phrases like 'proposed', 'discussed', 'evaluated', 'studied', 'In this paper', 'The study shows', 'We argue' and other terms that captures the contribution of the author or methodology in the article. These sentences must be read by the skimmed reader as it speaks about the core concept of the article. Such sentences are sent to visualization. But reading these sentences alone will not be sufficient to get the essence of work in the article. Hence the important and novel sentences from each section of the article are identified and such sentences are also given to skimmed readers.

The methodology of novelty estimation follows from Deepika and Mahalakhsmi [10]. Each section of the article is segmented into individual sentence and an importance score is given to each of them. The importance of the sentence is calculated using two features: the similarity of sentence with the title and the importance of sentence with importance of terms [33]. The similarity value between the title T and the sentence S_i in a document d is calculated (4).

$$\mathrm{Sim}\left(S_j, \mathrm{T}\right) = \frac{S_j \cdot \mathrm{T}}{\max_{S_j \in \mathrm{d}}\left(S_j \cdot \mathrm{T}\right)} \tag{4}$$

where S_j denotes a vector of the sentence, and T denotes a vector of the title. The importance value of a sentence S_j in a document d is calculated (5).

$$Cen\left(S_j\right) = \frac{\sum_{\mathrm{t} \in S_j} \mathrm{tf}(\mathrm{t}) \times \mathrm{idf}(\mathrm{t}) \times \aleph^2(\mathrm{t})}{\max_{\mathrm{S_j} \in \mathrm{d}}\left\{\sum_{\mathrm{t} \in S_j} \mathrm{tf}(\mathrm{t}) \times \mathrm{idf}(\mathrm{t}) \times \aleph^2(\mathrm{t})\right\}} \tag{5}$$

where $\mathrm{tf}(\mathrm{t})$ denotes the term frequency of term t, $\mathrm{idf}(\mathrm{t})$ denotes the inverted document frequency, and $\aleph^2(\mathrm{t})$ denotes the $\aleph^2$ statistics values.

The importance of each sentence is then taken as aggregation of both the values (6).

$$Impr\left(S_j\right) = \frac{\sum_{j=0}^{n} Sim\left(\mathrm{S_j}, \mathrm{T}\right) + \mathrm{Cen}\left(\mathrm{S_j}\right)}{n} \tag{6}$$

The sentence with importance score above the threshold are retained and novelty for each of the sentence is calculated (7).

$$Novelty = 1 - (Similarity + Relevance) + Divergence \tag{7}$$

The average of the novelty is taken as threshold and the important sentence with novelty score above the threshold are retained. The sentences obtained by end of these processes will carry the information that is both novel and important to understand the seed paper by a skimmed reader.

4 Results

We implemented citation context extraction using supervised learning Precision, recall and F1 score (Table 2) are found based on the total number of annotated fields, total number of correctly identified field and total number of retrieved fields shown in Fig. 1.

Citation classification is the important part for identifying the quality reference article. In line with the novice reader, our approach also gives highest importance to those articles that fall under category like Compare, Base, and Use.

Figure 1 depicts the sentiment identified using the SentiWordNet. In our experiments 17 type of cite sentences are examined which resulted in identifying 3 articles as negative. In our experiments 88 % of results are obtained as positive sentences. The reason may be that we filtered out the articles with highest similarity.

The seed article on average is found to be 80.58 % novel when compared to the first level of reference articles. This is because the seed paper deals with the same problem domain as that of reference articles and the related work section and

Table 2 Precision, recall, F1 score values

Fields	Precision	Recall	F1
Title	1.00	0.94	0.97
Source	0.81	0.76	0.79
Year	1.00	0.94	0.97
Surname	0.94	0.88	0.91
GivenName	0.94	0.88	0.91
Volume	0.75	0.53	0.62
FirstPage	0.94	0.88	0.91
LastPage	0.94	0.88	0.91
Overall	1.00	0.94	0.97

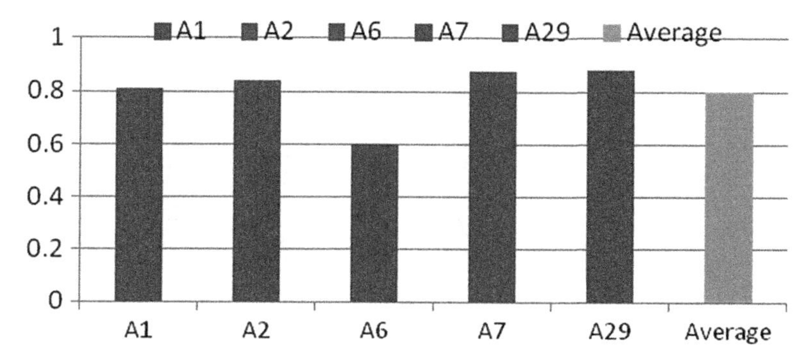

Fig. 1 Novelty of seed paper compared to first level reference articles

introduction part of these articles showed lesser novelty compared to other sections. The distribution is shown in Fig. 1. The seed article on average is found to be 85.8 % novel when compared to the second level of reference articles (refer Fig. 2). This is because the seed paper deals with the similar kind of problem domain as that of reference articles at this level but varies highly in terms of methodology and other sections of the articles. The accuracy and other performance measure of importance estimation and novel sentences identified are shown in Fig. 3. It can be seen that the system showed a high rate of recall, accuracy and specificity. The negative predictive value is also higher as the number of non-important sentences is higher and few of those sentences were falsely predicted as important. This can be overcome by including rule based approaches in the methodology. Section-wise novelty at level 1 is tabulated in Table 3. The inference remains the same as in previous case. Since novelty is estimated in terms of importance of sentence the

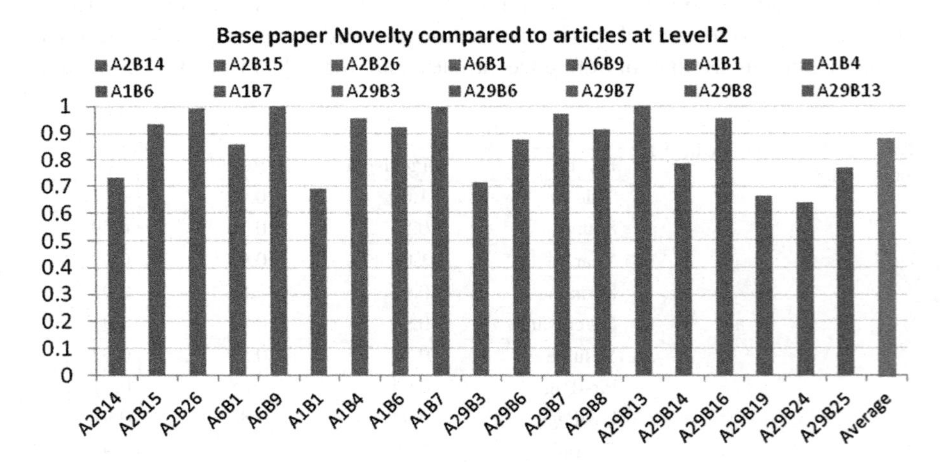

Fig. 2 Novelty of seed paper compared to second level reference articles

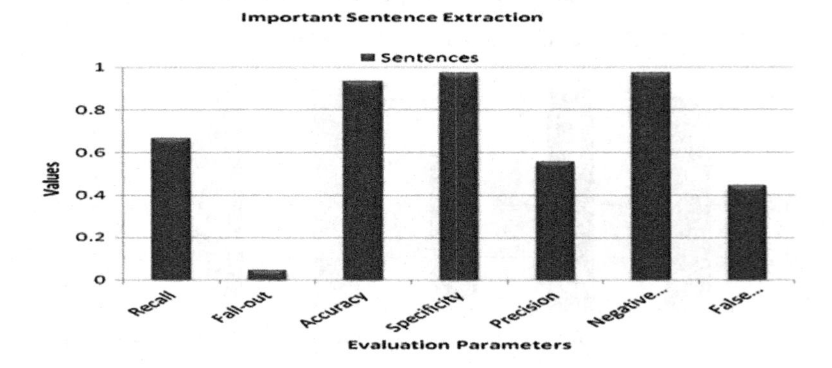

Fig. 3 Performance of important sentence extraction

Table 3 Section-wise novelty at level 1

Section of seed article	A1	A2	A6	A7	A29	Average
Abstract	0.912415	0.901383	0.921414	0.93114	0.90134	0.913538
Introduction	0.456137	0.67312	0.451238	0.413525	0.62451	0.523706
Related work	0.564151	0.765141	0.651523	0.571263	0.741321	0.65868
Remaining part	0.936713	0.8632837	0.981931	0.998771	0.961531	0.948446
Concluding remarks	0.999243	0.992134	0.962145	0.991234	0.978981	0.984747
	Overall document novelty					0.805824

impact of importance estimation is reflected here. Measure to improve the performance of the system will be carried out as an extension of this work. Figure 4 shows the final outcome of the system with all the important and novel sentences highlighted for skimmed readers. The system also recommends a set of reference articles that are to be read by a skimmed reader for better understanding the article as in Table 4.

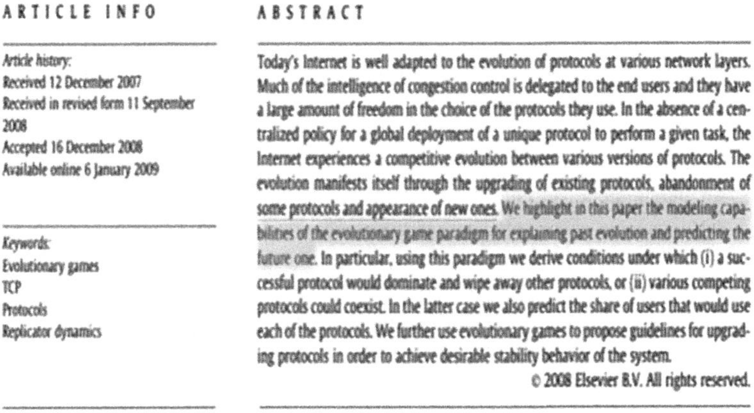

ARTICLE INFO

Article history:
Received 12 December 2007
Received in revised form 11 September 2008
Accepted 16 December 2008
Available online 6 January 2009

Keywords:
Evolutionary games
TCP
Protocols
Replicator dynamics

ABSTRACT

Today's Internet is well adapted to the evolution of protocols at various network layers. Much of the intelligence of congestion control is delegated to the end users and they have a large amount of freedom in the choice of the protocols they use. In the absence of a centralized policy for a global deployment of a unique protocol to perform a given task, the Internet experiences a competitive evolution between various versions of protocols. The evolution manifests itself through the upgrading of existing protocols, abandonment of some protocols and appearance of new ones. We highlight in this paper the modeling capabilities of the evolutionary game paradigm for explaining past evolution and predicting the future one. In particular, using this paradigm we derive conditions under which (i) a successful protocol would dominate and wipe away other protocols, or (ii) various competing protocols could coexist. In the latter case we also predict the share of users that would use each of the protocols. We further use evolutionary games to propose guidelines for upgrading protocols in order to achieve desirable stability behavior of the system.

Fig. 4 Suggestion for skimmed level readers—manual visualisation

Table 4 Recommendation of research articles to skimmed level readers

Article id
A1
A7
A29
A2B15
A6B9
A1B1
A29B3
A29B6
A29B24

Table 5 Sample research articles from vol. 53, computer networks

S. no.	Name of scientific publication
1	Performance analysis of IEEE 802.15.4 wireless sensor networks: An insight into the topology formation process
2	Improving TCP fairness and performance with bulk transmission control over lossy wireless channel
3	Modeling finite buffer effects on TCP traffic over an IEEE 802.11 infrastructure WLANq
4	Efficient and scalable provisioning of always-on multicast streaming services
5	A backward-compatible multiple-round collision avoidance scheme for contention based medium access control
6	A reactive role assignment for data routing in event-based wireless sensor networks
7	VoIP: A comprehensive survey on a promising technology
8	Energy and delay optimized contention for wireless sensor networks
9	On the reliability of large-scale distributed systems—A topological view
10	Mobility label based network: Hierarchical mobility management and packet forwarding architecture
11	Design and analysis of a self-tuning feedback controller for the Internet
12	Joint range assignment and routing to conserve energy in wireless ad hoc networks
13	A per-application mobility management platform for application-specific handover decision in overlay networks
14	A traffic based decomposition of two-class queueing networks with priority service q
15	Minimizing hierarchical routing error
16	An autonomic architecture for optimizing QoE in multimedia access networks
17	Self-managing energy-efficient multicast support in MANETs under end-to-end reliability constraints
18	Minimum disruption service composition and recovery in mobile ad hoc networks
19	On graph-based characteristics of optimal overlay topologies
20	Self-managed topologies in P2P networks
21	A rule-based system for programming self-organized sensor and actor networks
22	The evolution of transport protocols: An evolutionary game perspective
23	An energy-efficient, transport-controlled MAC protocol for wireless sensor networks
24	Trends in the development of communication networks: Cognitive networks
25	A survey on MAC protocols in OSA networks
26	Energy-efficient power/rate control and scheduling in hybrid TDMA/CDMA wireless sensor networks
27	Modeling bufferless packet-switching networks with packet dependencies
28	FAD and SPA: End-to-end link-level loss rate inference without infrastructure
29	Epidemic-based reliable and adaptive multicast for mobile ad hoc networks
30	Stochastic and deterministic performance evaluation of automotive CAN communication

4.1 Empirical Evaluation

The empirical analysis of the proposed approach is experimented over 30 selected research articles as seed (refer Table 5) from Volume 53, Computer Networks of Elsevier. The citation quality and novelty were estimated and tabulated in Table 6. Based on the quality analysis, number of articles were recommended (Table 6) for further reading for skimmed readers. The results did not project a concrete reason

Table 6 Articles recommended for the dataset

S. no.	No. of references per article across two levels	No. of total cites analysed	Citation quality	Novelty	No. of references recommended in level 1 and 2
1	20	29	0.275981	0.8782	12
2	27	40	0.154320	0.9205	14
3	11	30	0.217833	0.9188	5
4	48	51	0.358905	0.8449	19
5	27	35	0.643267	0.8365	12
6	25	39	0.288761	0.6657	13
7	183	216	0.351475	0.8575	56
8	35	50	0.364478	0.7832	20
9	39	30	0.172372	0.8116	23
10	23	48	0.419096	0.8217	5
11	38	83	0.333870	0.9681	19
12	36	49	0.294953	0.9768	21
13	23	36	0.207393	0.893	4
14	22	32	0.268656	0.9649	9
15	23	46	0.106198	0.9037	13
16	38	48	0.371795	0.6501	12
17	46	97	0.500360	0.9979	26
18	38	39	0.287886	0.9287	17
19	29	25	0.411357	0.9716	12
20	23	40	0.449902	0.8321	11
21	31	36	0.438295	0.8571	15
22	29	26	0.453862	0.87	14
23	26	29	0.388270	0.9464	12
24	92	238	0.500359	0.9663	33
25	32	214	0.358576	0.9491	21
26	36	48	0.329431	0.9721	25
27	37	83	0.618448	0.9408	31
28	22	36	0.000000	0.9871	20
29	51	61	0.385005	0.8578	23
30	23	58	0.203269	0.9911	21

behind the variation in recommendation of further reading for the input dataset. The fundamental reason of slow internet connection or research article not being an open source dominated the experiments. Removing the barrier of recommending inclusive research articles would not be a great issue.

5 Conclusion

Through this work we have attempted to model the research articles to a primitive set of reader say skimmed reader. Work on tracking further level of references and recommending a rich set of supplementary materials is on. Further work will also focus on modeling the article for other reader lens categories. In addition, recommending reading research materials and scientific publications outside the reference thread seems to be a promising extension.

Acknowledgments This work was partially funded by the Science and Engineering Research Board (SERB) under sanction number DST sanction No. SR/FTP/ETA-111/2010 for FAST TRACK FOR YOUNG SCIENTIST titled "Automated Tracking of Scholarly Publications to aid Concept Understanding in web based learning".

References

1. Mahalakshmi, G.S., Sendhilkumar, S.: Design and implementation of online reference tracking system. In: Proceedings of First IEEE International Conference on Digital Information Management, Bangalore, India, pp. 418–423 (2006)
2. Mahalakshmi, G.S., Sendhilkumar, S.: Automatic reference tracking. In: Handbook of Research on Text and Web Mining Technologies. Information Science Reference. IGI Global, Hershey (2009)
3. Valerie Cross, V.: Similarity or inference for assessing relevance in information retrieval. In: Proceedings of IEEE International Fuzzy Systems Conference, pp. 1287–1290 (2001)
4. Chen, J., Liu, L., Song, H., Yu, X.: An intelligent information retrieval system model. In: Proceedings of the 4th World Congress on Intelligent Control and Automation, China, pp. 2500–2503 (2001)
5. Gao, X., Murugesan, S., Lo, B.: A dynamic information retrieval system for the web. In: Proceedings of 27th Annual International Computer Software and Applications Conference, IEEE Computer Society, pp. 663–668 (2003)
6. Day, M.Y., Tsai, T.H., Sung, C.L., Lee, C.W., Wu, S.H., Ong, C.S., Hsu, W.L.: A Knowledge-based approach to citation extraction. In: Proceedings of IEEE International Conference on Information Reuse and Integration, pp. 50–55 (2005)
7. Ajiferuke, I., Lu, K., Wolfram, D.: A comparison of citer and citation-based measure outcomes for multiple disciplines. JASIST **61**(10), 2086–2096 (2010)
8. Flower, J.H., Aksnes, D.W.: Does self-citation pay? J. Scientometrics **72**, 427–437 (2007)
9. Costas, R., van Leeuwen, T.N., Bordons, M.: Self-citations at the meso and individual levels: effects of different calculation methods. Scientometrics **82**, 517–537 (2010)
10. Deepika, J., Mahalakhsmi, G.S.: Journal impact factor—a measure of quality or popularity? In: Proceedings of International Conference IICAI Spl Session on Advances in Web Intelligence and Data Mining (2011a)

11. Deepika, J., Mahalakshmi, G.S.: Towards knowledge based impact metrics for open source research publications. Int. J. Internet Distrib. Comput. Syst. **2**(1), 102–108 (2012)
12. Deepika, J., Mahalakhsmi, G.S.: Comparative evaluation of text similarity detection in research publications. In: Proceedings of International Conference IICAI Spl Session on Advances in Natural Language Computing (2011b)
13. Hirsch, J.E.: An index to quantify an individual's scientific research output. Proc. Natl. Acad. Sci. **102**(46), 16569–16572 (2005)
14. Lievers, W.B., Pilkey, A.K.: Characterizing the frequency of repeated citations: the effects of journal, subject area, and self citation. J. Inf. Process. Manage. **48**, 1116–1123 (2012)
15. Mahalakshmi, G.S., Sam, S.D., Sendhilkumar, S.: Establishing knowledge networks via analysis of research abstracts. J. Univ. Comput. Sci. (JUCS) Adv. Soc. Netw. Appl. **18**(8), 993–1021 (2012)
16. Mahalakshmi, G.S., Sam, S.D., Sendhilkumar, S.: Mining research abstracts for exploration of research communities. In: ACM Compute (2012)
17. Sendhilkumar, S., Sam, S.D., Mahalakshmi G.S.: Enhancement of co-authorship networks with content similarity information. In: International Conference on Advances in Computing, Communications and Informatics, ICACCI, ACM Digital Library, pp. 1225–1228 (2012)
18. Mahalakshmi, G.S., Sendhilkumar, S.: Optimising research progress trajectories with semantic power graphs. In: Proceedings of PREMI 2013, Springer LNCS, vol. 8251, pp. 708–713 (2013)
19. Sendhilkumar, S., Mahalakshmi, G.S.: Context based citation retrieval. Int. J. Netw. Virtual Organ. (IJNVO) **8**(1/2), 98–122 (2011)
20. Mahalakshmi, G.S., Sendhilkumar, S., Sam, S.D.: Refining research citations through context analysis. Intell. Inf. Adv. Intell. Syst. Comput. **182**, 65–71 (2013)
21. Sendhilkumar, S., Mahalakshmi, G.S., Harish, S., Karthik, R., Jagadish, M., Sam, S.D.: Assessing novelty of research articles using fuzzy cognitive maps. Intell. Inf. Adv. Intell. Syst. Comput. **182**, 73–79 (2013)
22. Garifield, E.: Can citation indexing be automated? In: Stevens, M.E., Giuliano, V.E., Heilprin, L.B. (eds.) Statistical Association Methods for Mechanized Documentation, Symposium Proceedings, vol. 269, pp. 189–192. National Bureau of Standards Miscellaneous Publication, Washington (1965)
23. Moravcsik, M.J., Murugesan, P.: Some results on the function and quality of citations. Soc. Stud. Sci. **5**, 88–91 (1975)
24. Nanba, H., Kando, N., Okumura, M.: Classification of research papers using citation links and citation types: towards automatic review article generation. In: 11th SIG Classification Research Workshop, Classification for User Support and Learning, pp. 117–134 (2000)
25. Teufel, S., Siddharthan, A., Tidhar, D.: Automatic classification of citation function. In: 2006 Conference on Empirical Methods in Natural Language Processing (EMNLP 2006), Association for Computational Linguistics, Sydney, pp. 103–110 (2006)
26. Pham, S.B., Hoffmann, A.: A new approach for scientific citation classification using cue phrases. In: AI 2003: Advances in Artificial Intelligence. Lecture Notes in Computer Science, vol. 2903, pp. 759–771 (2003)
27. Athar, A.: Sentiment analysis of citations using sentence structure-based features. In: Proceedings of the ACL-HLT 2011 Student Session, Portland, pp. 81–87 (2011)
28. Cavalcanti, D.C., Prudêncio, R.B.C.: Good to be bad? Distinguishing between positive and negative citations in scientific impact. In: 23rd IEEE International Conference on Tools with Artificial Intelligence, pp. 156–162 (2011)
29. Sendhilkumar, S., Elakkiya, E., Mahalakshmi, G.S.: Citation semantic based approaches to identify article quality. In: Wyld, D.C. (ed.) Proceedings of International Conference ICCSEA 2013, ICCSEA, SPPR, CSIA, WimoA, SCAI—2013, vol. 3, pp. 411–420 (2013)

30. Sendhilkumar, S., Nandhini, N.S., Mahalakshmi, G.S.: Novelty detection via topic modeling in research articles. In: Wyld, D.C. (ed.) Proceedings of International Conference ICCSEA 2013, ICCSEA, SPPR, CSIA, WimoA, SCAI—2013, vol. 3, pp. 401–410 (2013)
31. Teufel, S., Siddharthan, A., Tidhar, D.: Automatic classification of citation function. In: EMNLP, pp. 103–110 (2006)
32. Mahalakshmi, G.S., Sendhilkumar, S., Sam, S.D.: Refining research citations through context analysis. In: First International Symposium on Intelligent Informatics ISI 2012, Springer (2012)
33. Ko, Y.J., Park, J., Seo, J.: Improving text categorization using the importance of sentences. Inf. Process. Manage. (IPM) **40**(1), 65–79 (2004)

Numerical Modeling of Cyclone-Generated Wave Around Offshore Objects

Subrata Bose and Gunamani Jena

Abstract Tropical cyclones propagating in the sea hit offshore objects to get diffracted around it. The highly valuable structures such as oil drilling and storage facilities, harbours islands, breakwaters, etc are planned and constructed without resorting to detailed design involving wave diffraction which contributes to 60–70 percent of loading. This paper presents a hybrid numerical method to solve the mathematical wave diffraction problem around offshore objects, large or small. A very large planned island has been investigated by a developed model, SUBRATA/ ISLAND HARBOUR. The results have been validated.

Keywords Wave diffraction · Structures · Numerical model · Impedance · Radiation

1 Introduction

The coastal zone especially deep water of India is vulnerable to the cyclone-waves generated in the Bay of Bengal and the Arabian Sea. Yet the highly valuable structures such as oil drilling and storage facilities, harbours, breakwaters, islands, etc. are ever increasingly being planned and constructed here for benefit of Indian environmental protection and economy but without detailed design. Circular islands in Dubai are also in vogue. The diffraction phenomenon is observed around small or large structures located in the sea. But waves generated by cyclones, play havoc on the offshore objects contributing more than 65 % of wave loading. It requires proper study and computer solution. This paper presents a numerical method to solve the

S. Bose (✉)
Kolkata Port Trust, Kolkata, India
e-mail: subratha_bose@yahoo.com

G. Jena
Roland Institute of Technology, Berhampur, India
e-mail: drgjena@ieee.org

L.C. Jain et al. (eds.), *Computational Intelligence in Data Mining - Volume 3*,
Smart Innovation, Systems and Technologies 33, DOI 10.1007/978-81-322-2202-6_54

mathematical diffraction problem around arbitrary offshore objects of any size in the deep waters where the wave is most severe and structures are vulnerable.

2 Wave Diffraction in Sea

Similarities between sound waves and water waves have been observed (1) Helmholtz equation (derived from the wave equation combining equations of motion and continuity) has been considered as the governing mathematical relationship [2, 3] as below:

$$(\Delta + k^2)\phi_D = 0 \tag{1}$$

in which $\Delta = \partial^2/\partial x^2 + \partial^2/y^2 + \partial^2/z^2$,

$$k = \text{wave no.} = 2\pi/L,$$

and L = design wave length

The boundary conditions are:

$$\begin{aligned}&\text{Sommerfeld radiation condition}\\ &\operatorname{Lt} r^{1/2}[\partial\phi_D - \mathrm{I}\,k\,\phi_D] = 0,\end{aligned} \tag{2}$$

$$r \to \infty$$

The diffracted waves being outgoing waves of decreasing amplitude in cylindrical co-ordinates (r, θ, z).

The independence boundary at fluid–structure interface just outside the boundary layer is

$$\Upsilon\,\partial\,\phi/\partial n - ik(1 - \partial)\phi = 0 \tag{3}$$

in which Υ = reflectivity of the structure $(O < \partial \leq 1)$,

Since no structure in nature fully reflect the water waves.

Equation (3) reduces to

$$\frac{\partial\phi_D}{\delta n} = -\partial\frac{\partial\phi_1}{\delta n} + ikA(1 - \Upsilon)\phi_1 = 0, \tag{3a}$$

neglecting the term with ϕ_D.

3 Numerical Modelling

Numerical solutions can be obtained basically by the following methods:

Finite Difference Method (FDM)
Finite Element Method (FEM)
Finite Integral Technique (FIT)
Transmission Line Matrix (TLM)
Finite Difference Time Domain (FBTD)

Each method is suitable for solution of certain problems. This paper has used Finite Difference Method, which solves with reasonable accuracy the diffraction behaviour of waves around offshore structures.

3.1 Solution of Diffraction Problem

The solution of ϕ_D in Eq. 3a that satisfies the boundary conditions given by Eqs. 2 and 3 is obtained, but using Green's function, $G = i\pi H_0^{(1)}$ which converts 3-D into 2-D along the body contour of the object (Fig. 1):

So,

$$\phi_D^{(P)} = \frac{1}{4\pi} \iint_S \left[\phi_D \frac{\partial G}{\partial n} - G \frac{\partial \phi_D}{\partial n}\right] ds \tag{4}$$

i.e.

$$\phi_D^{(P)} = \frac{i}{4c} \int \left[\phi_D(A) \frac{\partial}{\partial n} H_0^{(1)}(kr) - H_0^{(1)}(kr) \frac{\partial \phi_D^{(A)}}{\partial n} x\right] dA \tag{4a}$$

in which A = arc length and $H_0^{(1)}$ = Hankel's function of first kind and order zero for the argument kr.

Assuming complete reflection at the object, and separating $\phi_D = p + i.q$ after generalizing Eq. 4a (when p lies on the body geometry),

$$\begin{aligned}(p + iq)(1 - \frac{\alpha}{2\pi}) = \int_4^i & [(p + iq)\{-kr_n(J_1 + iY_1)\} \\ & - (u + iv)(J_0 + iY_0)]dA\end{aligned} \tag{5}$$

in which α = the interior angle at corner point, vide Fig. 1

Fig. 1 Definition sketch for finite difference numerical solution

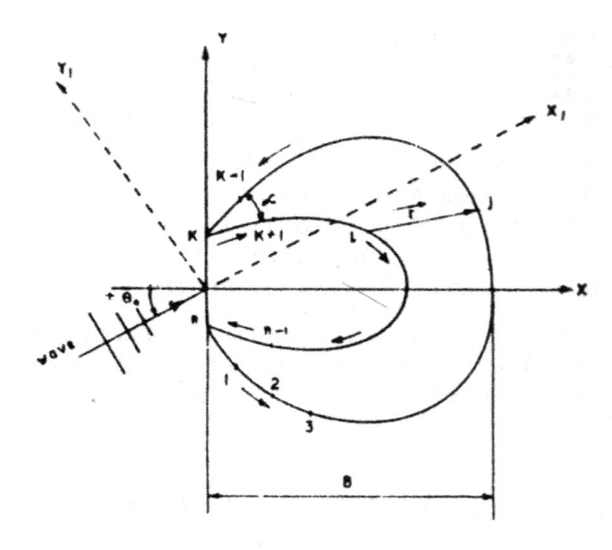

$$r_n = \frac{1}{r}\frac{\partial x}{[(\partial y/\partial A)_j(x_j - x_1) - (\partial x/\partial A)_j(x_j - x_1)]},$$

J_o, J_1; Y_o, Y_1 = Bessel functions of first and second kinds respectively for the argument (kr).
and $u + iv = \partial\phi_D/\partial n = -\partial\phi_1/\partial n$ (complete reflection $\Upsilon = 1$)

$$= ike^{ikx_1}[-(\partial y_1\partial A)\Upsilon + A(1 - \Upsilon)]$$

(reflectivity Υ varying from 0 to 1).

Integral Equation Solution (in FD form).

The real and imaginary parts of Eq. 5, put into FD forms after some approximations are transformed into two sets of simultaneous linear algebraic equations:

$$\sum_{J=1}^{n} p_j\alpha_{j1} + \sum_{j=1}^{n} q_j\beta_{j1} = -0.25\sum_{j=1}^{n} A_j(u_jY_0 + v_jJ_0) \quad (6a)$$

$$-\sum_{J=1}^{n} p_j\beta_{j1} + \sum_{j=1}^{n} q_j\alpha_{j1} = -0.25\sum_{j=1}^{n} A_j(v_jY_0 + v_jJ_0) \quad (6b)$$

in which $\alpha_{j1} = 0.25\,kA_j\,Y_1\,r_n$ and $\beta_{j1} = 0.25\,kA_j\,J_1r_n$;

j and l being source points and observation point where the potential is to be determined (Fig. 1).

The values of p and q are evaluated at n discrete points on the body geometry of the structure solving Eqs. 6a, b.

Diffraction Coefficients, K_D:

K_D is defined as the ratio of diffracted /incident wave heights. Thus

$$K_D = |\phi| = |\phi_I + \phi_D|$$

Since $\phi_I = e^{ikx_1}$ and $\phi_D = p + I\,q$,

$$K_D = \left[1 + p^2 + q^2 + 2(p\cos kx_1 + q\sin kx_1)\right]^{1/2}. \tag{7}$$

4 Wave-Structure Interaction Inputs

Computational needs demand the product of non-dimensionalised, body dimension of the object and wave number to be kept within a definite value. But large objects, wave and fluid mechanic parameters are to be scaled for the modeling in order to arrive at proper result. The dimensionless horizontal lengths are reduced by a scaling ratio. Following Froude scaling law a scale factor for incident wave period, T is derived [4]. h/L ratio for the fictitious model is obtained from the dispersion relations between prototype and the model incorporating model distortion [5]. h around the model is evaluated keeping $(h/g)^{1/2}$. 1/T, a form of Froude no. as constant. L for model is expressed in non-dimensional terms to arrive at proper k for computer input.

Alternatively, Bessel functions involved in computation have to be curtailed at a suitable point to compute diffraction coefficients. These techniques of computation make the numerical model a hybrid one for large objects.

5 Model Validation/Results

A Developed Model, SUBRATA/IS LAND HARBOUR has been used to determine wave-height variation by diffraction around a planned horse-shoe shaped island-harbour in the Bay of Bengal off W. Bengal /Orissa coast (Fig. 2). The major body dimension (12 km) becomes 60 L (Fig. 3) for a design wave parameter of 0.2 km length. The inside harbor is constricted again to obtain results for a 1.5 L diameter circle.

The model is capable to compute large as well as small structures. The computational results are presented in Figs. 4 and 5, representing large and small structures respectively.

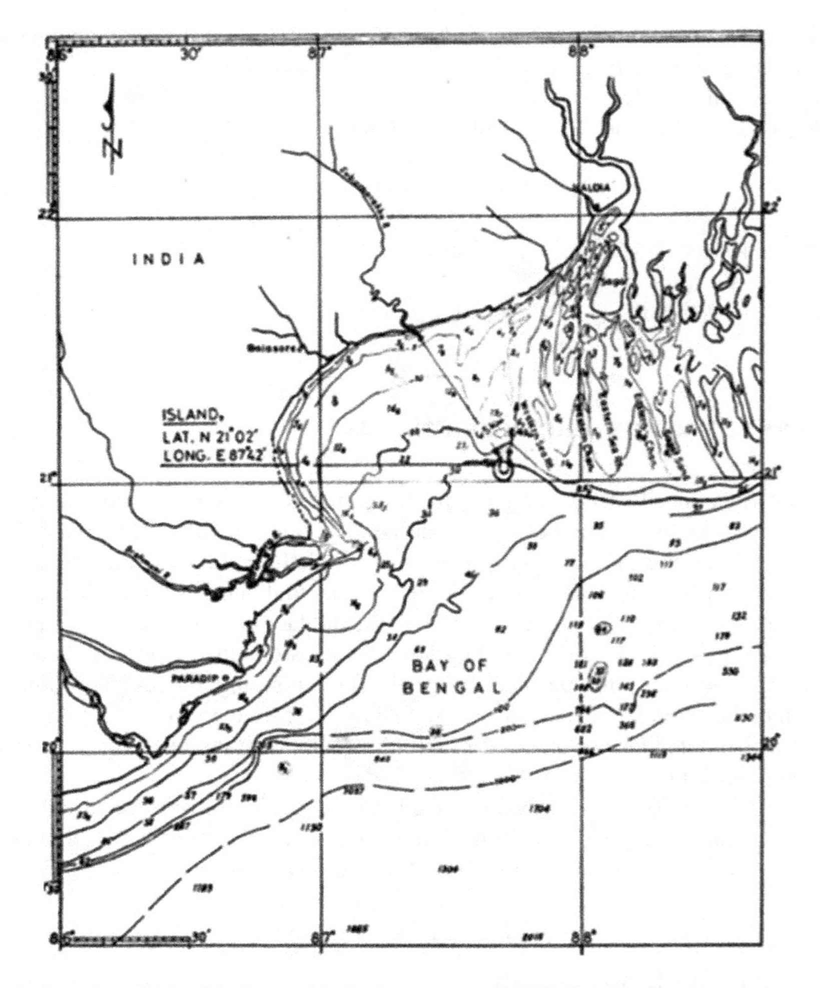

Fig. 2 Location of island harbour with depth contours off W.B./Orissa

The computation was carried out in FORTRAN by Burroughs B6738 system, and was validated against the results available for

(a) 5 L, 4.375 L, and 2.51L breakwater gap problems [6] corresponding to 0°, 30° and 60° wave length penetration in 60 L island-harbour.
(b) Exact eigenfunction expansion around 1.6 dia cylinder of Banaugh [7] corresponding to horizontal wave-hit on a 1.5 L dia. vertical object, Fig. 5.

Mach reflection effects ($K_D \geq$) are observed near the 60 L island-harbour entrance indicating island top here is to be built above twice the design wave height. The systematic rise and fall of the curves in the results correspond to wave height variation guided by smooth body geometry of the island object, resulting in streamlined flow around with little scour.

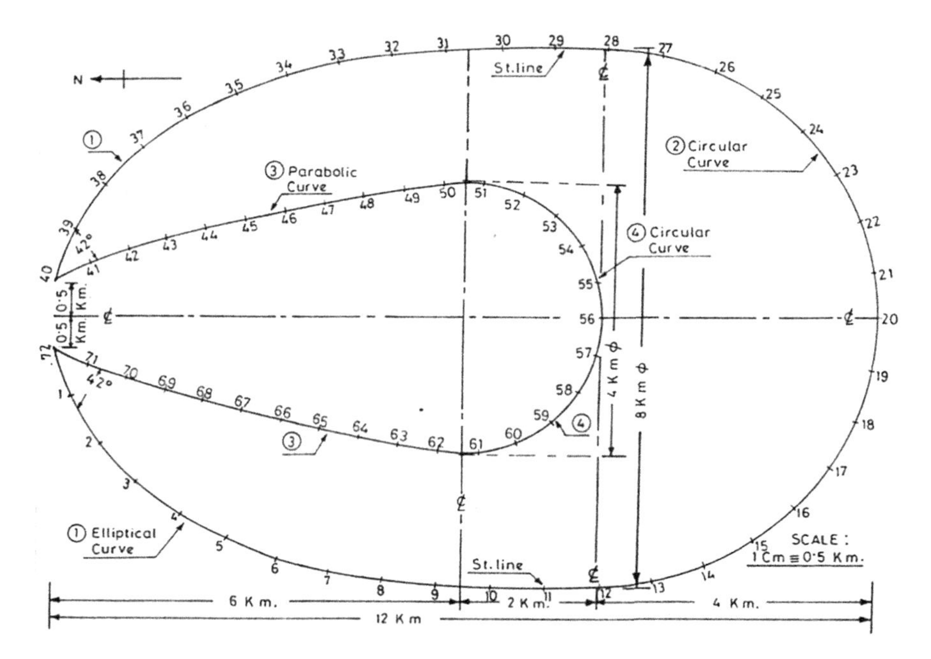

Fig. 3 Layout plan of island harbour

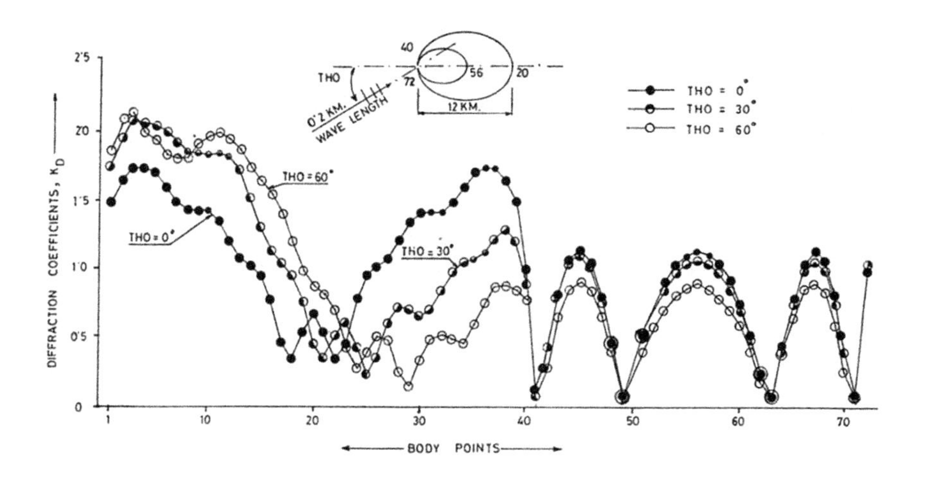

Fig. 4 Diffraction along the island harbour (PERIMETER B = 12 km)

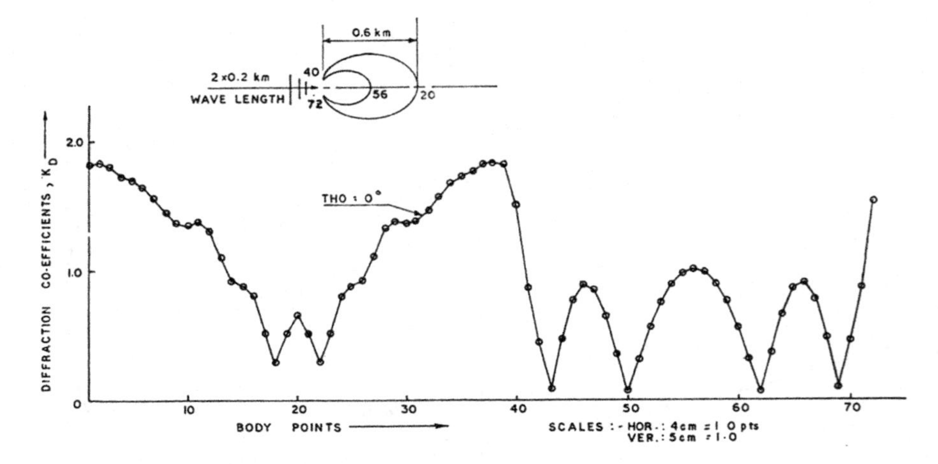

Fig. 5 Wave diffraction along the island harbour, perimeter (B = 0.6 km)

6 Conclusions

- The cyclone-generated wave diffraction around a very large island-harbour (60 L) has been investigated by the developed hybrid model SUBRATA/ISLAND HARBOUR.
- The model requires curtailment of Bessel functions or using a developed fluid-dynamic parameter to arrive at a correct solution, which was validated.
- The numerical modelling technique allows use of the same model for investigating smaller objects, breakwaters or cylindrical piles in the sea.
- The model allows too the study of the reflectively, Υ of a structure like breakwater, using all terms of the impedance boundary condition given by Eq. 3a.

Acknowledgments Acknowledgement is due to Prof. R.L. Wiegel, University of California at Berkley, USA for his comments and discussion with the author at IIT Bombay.

References

1. Wehausen, J.V., Laitone, E.V.: Surface waves. In: Encyclopedia of Physics, vol. IC. pp. 446 −778. Springer, Berlin (1960)
2. Bose, S.: Mathematical modelling of cyclone-wave diffraction by offshore structures. In: Hydrosoft'84, pp. 2/66−2/77. Elsevier Science Publishers, B.V. Amsterdam (1984)
3. Fund, K.Y. : New perspectives on the acoustic impedance. In: CFEMATCON—06. Andhra University, pp. 1−7, (July 24−25, 2006)
4. Bose, S.: Some aspects of cyclone-generated wave and its diffraction around offshore islands in the Bay of Bengal. Ph.D (Engg) Thesis, Jadavpur University (1985)

5. Chatham, C.E.: Lose angles harbor and long beach harbor: design of the hydraulic model. In: Ports'77, Am. Soc. Civ. Eng. pp. 47–64, New York (1977)
6. U.S. Army, Coastal Engg. Research Center: Shore Protection Manual, vol. 1(2), pp. 79–110 (1977)
7. Banaugh, R.P.: The scattering of acoustic and elastic waves by surfaces of arbitrary shape. Ph.D Thesis in Engg. University of California, Berkeley (1962)

Cooperative Resource Provisioning for Futuristic Cloud Markets

Geetika Mudali, Manas Ranjan Patra, K. Hemant K. Reddy and Diptendu S. Roy

Abstract Cost of maintenance for cloud providers and cost of cloud usage for its users need to be kept under check. Cloud market promises as a possible means to deliver this price constraint by offering flexible pricing policies, some pricing models favoring short-term users' requirements, whereas others facilitating long-term requirements. Thus, finding an optimal solution is a challenging proposition. In this paper, an attempt has been made to minimize the service cost for users as well as cloud providers' provisioning cost. We proposed a cooperative resource provisioning algorithm assuming that the cloud providers cooperate among themselves by sharing service demand data from time to time and simulated the cooperative market scenario. Widely accepted demand data and pricing policies have been employed in our simulations. Simulation results show that our proposed algorithm minimizes both types of costs in comparison to non-cooperative counterparts.

Keywords Cloud computing · Cooperative resource provisioning · Futuristic cloud market

---

G. Mudali (✉) · K.H.K. Reddy · D.S. Roy
National Institute of Science and Technology, Berhampur, Odisha
e-mail: geetika_nist@hotmail.com

K.H.K. Reddy
e-mail: khemant.reddy@gmail.com

D.S. Roy
e-mail: diptendu.sr@gmail.com

M.R. Patra
Berhampur University, Berhampur, Odisha
e-mail: mrpatra12@gmail.com

L.C. Jain et al. (eds.), *Computational Intelligence in Data Mining - Volume 3*,
Smart Innovation, Systems and Technologies 33, DOI 10.1007/978-81-322-2202-6_55

1 Introduction

Recent advances in distributed computing have brought a paradigm shift in the computing world. The practices of dedicated access to computers owned by individuals or organizations have been replaced by on-demand accesses to resources shared among many individuals and different organizations. Cloud Computing has largely contributed to this transition by facilitating ever-present, befitting, on-demand network access to a shared pool of configurable computing resources like networks, servers, storage, and services) that can be provisioned fast and incurring minimal overhead, both at cloud providers' premises as well as from the users [1]. Present practice in the Cloud Computing arena is that the Cloud Users (CUs) request for resources and Cloud Providers (CPs) allot the virtualized resources to the CUs according to their requirements. However, while requesting the resources, CUs face a major challenge due to various pricing schemes offered by the CPs. Resources are available on reservation basis, as well as on demand basis. Reservation is to be done for a fixed contract period with a fixed price. However, reservation for a longer period or for larger amount of resources than that required to meet the demand for resources of an application, may lead to over provisioning and higher cost for the CUs. On the other hand, on-demand prices are generally higher than the reserved one and therefore if the resources are allotted only on-demand basis, cost to be paid by the CUs will again be high Therefore, some optimization strategies are required to reduce the resource usage charges from the users' point of view. This paper focuses on the optimizing strategies aimed at reducing the total cost of cloud deployment. Remaining part of the paper is organized as follows. Various pricing schemes offered by different CSPs are discussed in the next section. A brief overview of the related work is given in Sect. 2. An overview of futuristic cloud framework is discussed in Sect. 3. Section 4 presents the proposed heuristics for resource reservation in cooperative cloud market. Results and analysis is discussed in Sect. 5. Section 6 concludes the paper.

2 Related Work

Many researchers in their research work focused on different pricing policies in cloud computing environment. Most of their research work focused on reserved and on-demand pricing policies. Deyuan et al. [1] has proposed a pricing model in which both reserved as well as on-demand instance considered. Mazzucco et al. [2] has proposed a model which focuses on maximizing the profit by dividing the cloud service provider by dividing the cloud users into two categories. Qian et al. [3], proposed a model for cost optimization of SLA-aware scheduling based on a stochastic integer programming. In [4], a cost optimization model is proposed for deadline constrained jobs and budget constrained jobs based on-demand and reserved instances. Chaisiri et al. [5] has proposed an algorithm which provisions of resources in cloud environment is done in optimally by taking resource requirement as input. Mahmoudload et al. [6], developed a demand forecasting model by taking a historical data of 12 months. The research work discussed in [3–6] used integer programming model in the formulation of cost optimization models. The integer programming model comes in the category of NP-hard problems. In [5], Blender decomposition techniques and sample average approximation techniques has been used. With the emergence of services like EC2 and Google Apps, there is a paradigm shift in computing and storage from personal clients to cloud-like datacenters [7, 8]. A study of cost models provided by different vendors of cloud such as Amazon, Google and soon was done in [8–12]. Goiri et al. [13] a virtualization based model is proposed which does consolidates the services to handle situation where there is less utilized servers or where there is situation of operating system heterogeneity. Zaharia et al. [14] proposed a statistical multiplexing technique, for decreasing the server cost by considering the Amazon EC2. Urgaonkar et al. [15] proposed an approach for cost minimization by taking a situation where more resources are booked than the actual requirement.

3 An Overview of Futuristic Cloud Markets

Figure 1 depicts an overview of cooperative cloud market, where set of cloud provider are taking part as market provider. Similarly set of cloud user consume services from different providers. A set of key player (like provider, user, market broker and cloud broker) are involved in resource provisioning in cooperative cloud market. The details of these key players and key features are discussed in subsection.

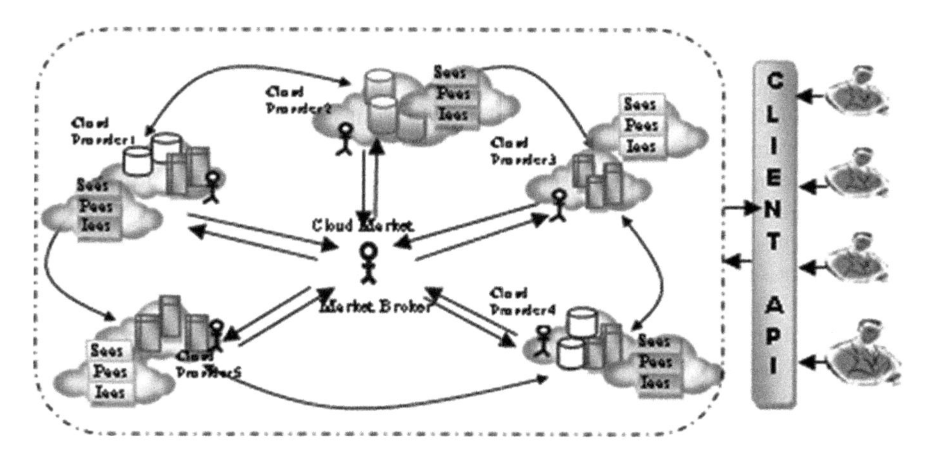

Fig. 1 An overview of CCM

- Any cloud provider can join in the cloud market by providing resource and service details to market broker. Similarly at any point, any cloud provider can leave the cloud market by fulfilling the agreement made to cloud user or migrating user's request SLA to any other provider within the market without effecting the user's task (even during execution) and by taking prior permission from the concerned user for security concerns.
- Market Broker: is a broker, who coordinates among all cloud providers in the market for each and every user request or group of service request.
- User: any user can request for any kind of services by sending request-SLA (complete service request specification) to the market broker.
- Cloud Broker: is a broker, who coordinates the activities within the cloud provider.

3.1 Cooperative Framework

The proposed cooperative cloud marketplace (CCM) has four essential elements, namely buyers, sellers, intermediaries and interaction mechanisms. In the context of the CCM, the distinction between buyers and sellers is blurred, in the sense that by accepting their respective terms and conditions, their individual roles may swing from time to time. An intermediary is referred to as a third party which offers intermediary services like information gathering and sharing, selection, negotiation, payment and so on [10]. A Market Broker (MB) has been proposed that fulfills the role of intermediary in the CCM. A simple schematic representation of the cooperative cloud market is depicted in Fig. 2. The market broker (MB) entity acts as the intermediary. All cloud providers share related information with the MB via a multi-agent system (MAS).

A high-level architecture of the proposed CCM has been depicted in Fig. 2. The cloud service users and the service providers, likened to buyers and sellers for any marketplace, constitute the first two layers in the architecture. The MB entity along with these market actors form a multi agent system, the characteristics of which is presented in the following subsection. Based on information from its agents, the MB takes decisions regarding which cloud service providers' physical machines should be employed for the running VMs. It has to be noted that the bundled physical machines PM_1, PM_2,..., PM_n etc., as depicted in Fig. 2, denote the computing resources within a single cloud provider. The responsibility of resource provisioning for the CCM, as shown in Fig. 2, rests on the MB, which executes appropriate algorithms to achieve its goal. Section 4 presents a cooperative provisioning algorithm for the proposed CCM that takes into account the variety of workload types as well as pricing schemes offered by different provider to achieve cost benefits for the entire market as a whole.

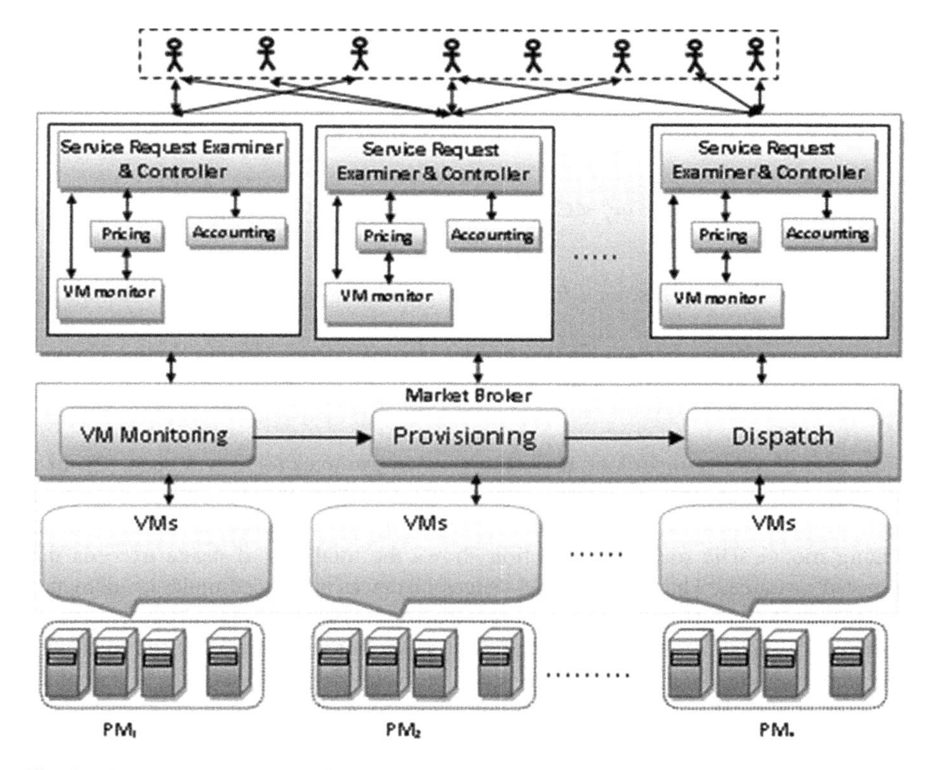

Fig. 2 High level architecture of CCM

4 Cooperative Heuristic Resource Provisioning

This section discusses the problem formulation for finding an optimal resource provisioning with minimum cost of cloud users. User service requests are processed as batch wise. The requested accepted within a fixed interval of time period are clubbed together to form a group and considered for provisioning in cloud market. Suppose 'K' number of user's request collected in a particular time interval for provisioning or for resource reservation. These K = $\{ur_1, ur_2, ur_3, \ldots, ur_k\}$ user requests are with different SLA (i.e. on-demand or reservation), we need to determine the number of VM instances are required for each user request. Total cost of using resource from a single cloud provider and total cost of using resources from more than one cloud provider is: $T_c = \sum_{slot=1}^{T} ur_i^n * C_{(\mathrm{cp_i})}^{(\mathrm{o})}$ $T_c = \sum_{slot=1}^{T} rp_i^n * C_{(\mathrm{cp_i})}^{(\mathrm{o})}$. In case of reservation model: $C_{(\mathrm{cp_i})}^{(f)}$ is the reserved fixed cost provided by cloud provider 'i' for a single VM instance and user request ur_i required ur_i^n number of VM instances. The fixed cost of allocating resources to the user request from a single provider is $ur_i^n * C^{(f)}_{(\mathrm{cp_i})}$ and additional hourly usage cost for the service is $ur_i^n * C^{(\mathrm{r})}_{(\mathrm{cp_i})}$.

Total cost of using resources from a single cloud provider is:

$$ur_i^n * C^{(f)}_{(cp_i)} + \sum_{slot=1}^{T} ur_i^n * C^{(\mathrm{r})}_{(\mathrm{cp_i})}. \tag{1}$$

Total cost of using resource from a multiple cloud providers:

$$T_c = \sum_{i=1}^{S_{rp}} \sum_{k=1}^{K} \left(ur_i^n * C^{(f)}_{(\mathrm{cp_i})} + \sum_{i=1}^{S_{rp}} \sum_{slot=1}^{T} rp_i^n * C^{(\mathrm{r})}_{(\mathrm{cp_i})} \right). \tag{2}$$

In case of reservation, cloud user reserves the resources for 1 or 3 years [2]. In a situation where number of reserve instance is more than the demand of a specific user. In such situation rest of VM instance usage cost calculated on on-demand pricing model. The following equation shows the total cost if usage exceeds the reserved instances. These excess VM usages instances are priced under on-demand. In general the on-demand cost is more than the reserved VM instance cost. A cooperative resource provisioning algorithm is proposed to minimize the total user cost, where excess VM instances are provided from multiple providers to complete the user application. These instances are allocated with minimum cost among the multiple providers. MB allocates the required VM instances to user if lower and upper threshold condition satisfies in order to maintain green computing. The proposed heuristic provisioning algorithm conditions are legibly articulated in the algorithm itself with comments line.

$$T_C(nvm^{(\mathrm{r})}_{(\mathrm{ur})}, nvm^{(\mathrm{D})}_{(ur)}) = \begin{cases} \left[nvm^{(\mathrm{r})} * C^{(f)}_{(\mathrm{cp_i})} + nvm^{(\mathrm{r})} * C^{(\mathrm{r})}\right], & nvm^{(\mathrm{r})} >= nvm^{(\mathrm{D})} \\ \left[nvm^{(\mathrm{r})} * C^{(f)}_{(\mathrm{cp_i})} + nvm^{(\mathrm{r})} * C^{(\mathrm{r})}_{(\mathrm{cp_i})}\right] + \left(nvm^{(\mathrm{D})} - nvm^{(\mathrm{r})}\right) * C^{(\mathrm{o})}_{(\mathrm{cp_i})}, & otherwise \end{cases} \tag{3}$$

4.1 Cooperative Resource Provisioning Algorithm

Algorithm 1 presents the cooperative process of collecting cloud user request and resource provider SLA, which indicate the requirement of resources and available of resources with specific SLA in cooperative market. *rpSLA*[] and *urSLA*[] are become the input for algorithm and heuristic resource allocation *Hra*[] become output of the proposed algorithm. First part of the algorithm check the necessary conditions like available of required resource within provider with min cost and checking of min threshold value in order to maintain the green computing i.e. the minimum number of VM required to start a server. Single cloud user VM request fulfilled within maximum of three resource providers. In second part, it checks *maxcount* i.e. a single user VM request fulfilled within three resource providers or not, if not than the request is queued for next slot. Similarly for *urSLA*, if not satisfied than the user request is queued for next slot.

Algorithm 1: *HeuristicScheduling*(rpSLA[ ], urSLA[ ])

```
Input : rpSLA[ ] is the resouce provider sla vector, urSLA[ ] is the user request sla vector and pricing policy
Output : Hra[ ] is the optimized resouce allocation vector.
Begin :
MaxCount = 3;
rpSLA[ ] = sort(rpSLA[ ], cost, Asc);
for each ur_i : urSLA[ ], do
    Totvm = ur_i SLA[i].num;    rpCount = 0;    j ← 0;
    while ( TotVm > 0  && rpSLA[ j ] != empty) do
        if( rpSLA[ j ].AvlVm > 0 ) thand
            if(TotVm >= rpSLA[ j ].MinVm ) than
                Write("Resources are allocated for provisioning"); status = allocated
            else
                Write("Resources are marked for provisioning"); status = marked
                rpSLA[ j ].MinVm = rpSLA[ j ].MinVm - TotVm;
            endif
            if(TotVm <= rpSLA[ j ].AvlVm) than
                Temp = TotVm; TotVm = 0;
                rpSLA[ j ].AvlVm = rpSLA[ j ].AvlVm - TotVm;
                OnDemand_ra[i].addRecord(ur_i, rpSLA[j], Temp, type);    break
            else
                Temp = pSLA[ j ].AvlVm;        rpSLA[ j ].AvlVm = rpSLA[ j ].AvlVm - TotVm;
                TotVm = TotVm - Temp; TempRec[irpCount].addRecord(ur_i, rpSLA[j], Temp, type); rpCount = rpCount+1;
            endif
        else
            next j;
        endif
        next j;
    endwhile
if(vm = 0) than
    if(rpCount <= MaxCount)
        OnDemand_ra[i].addRecord(TempRec);
    else
        write("ur_i request can't processed within 3 resource providers");
    endif
else
    write("ur_i request can't processed in this slot, to be considered for next slot");
endif
for each rp_i : rpSLA, do:
    if( rp_i .status = marked) than
        cancel allocation requests & insert into queue for next slot
    endif
endfor
end;
```

4.2 Experimental Setup

In an effort to study the performance of cooperative cloud market, the cooperative resource provisioning algorithm has been implemented using Java SDK v6. It has been executed on a sixteen-node cluster of Dell Power Edge R610 blades with Dual Intel Xeon Quad-Core E5620 (2.93 GHz Processor) using standard demand data from SDSC [16].

5 Results and Analysis

Figure 3 depicts the resource provisioning cost with respect to number of VMs used and cooperative heuristic approach show the superior then the other three approaches. Figure 4 depicts the cloud usage cost with respect to number of VMs and the other three approaches used. Heuristic approach shows the superior results than other approaches and more effective as usage VMs instances increases.

Figure 5 depicts percentage of cancelled user request along with total user request and other three cloud provisioning approaches. Where slot1, slot2 are indicate the different time slots. Heuristic approach shows the superior results than other three provisioning approaches.

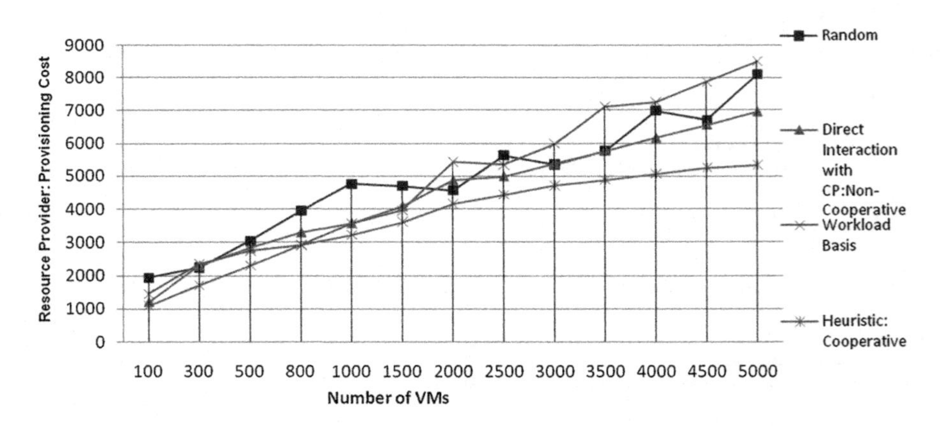

Fig. 3 Cloud provider provisioning cost (CPPC) of cooperative cloud market

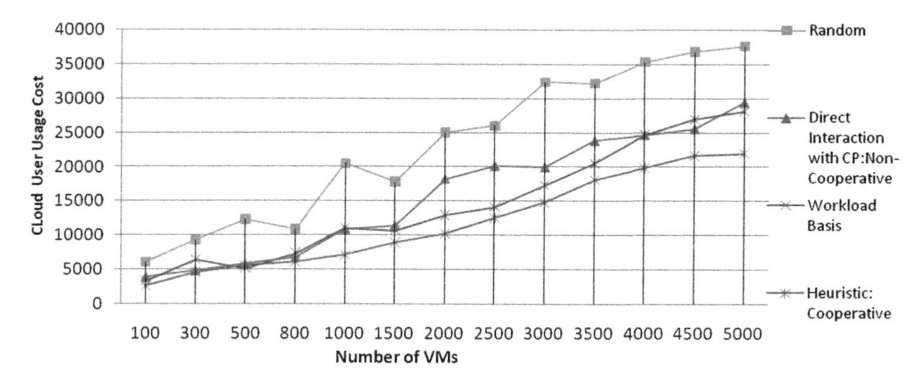

Fig. 4 Cloud user usage cost (CUUC) of cooperative cloud market

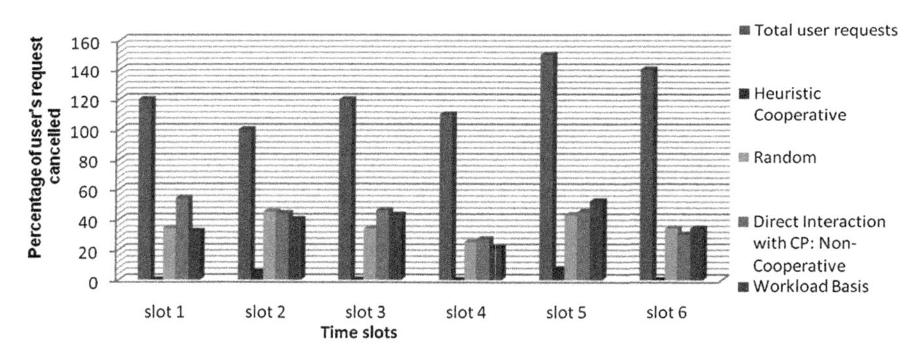

Fig. 5 Percentage of cloud user's request cancelled

6 Conclusion

This paper proposes a heuristic resource provisioning framework for cooperative cloud market that adopts the strategy of provisioning resources from multiple cloud providers in order to serve a common user-base forming a futuristic cloud market. In this context a heuristic algorithm has been proposed. Numerous experiments have been carried out on standard cloud demand data for different provisioning approaches and the results reveal that cost (both providers' and users' viewpoint, abbreviated as CUUC and CPPC respectively) of resource provisioning shows significant reduction for the proposed heuristic provisionSing algorithm. This improvement can be attributed to the cloud market and subsequent resource provisioning among providers in the proposed CCM framework. The emergence of the proposed CCM as a generalized cooperative heuristic provisioning requires further testing it on real test-bed with valid datasets and it is one of the future works of this research.

References

1. Deyuan, W., Ying, W., Jichun, L., Ke, X., Wenjing, L.L., Xuesong, Q.: Pricing reserved and on-demand schemes of cloud computing based on option pricing model. In: 15th Asia-Pacific Network Operations and Management Symposium (APNOMS), pp. 1–3 (2013)
2. Mazzucco, M., Dumas, M.: Reserved or on-demand instances? A revenue maximization model for cloud providers. In: IEEE International Conference on Cloud Computing (CLOUD), pp. 428–435, (2011). doi:10.1109/CLOUD.2011.25
3. Qian, L., Yike, G.: Optimization of resource scheduling in cloud computing. In: 12th International Symposium on Symbolic and Numeric Algorithms for Scientific Computing (SYNASC), pp. 315–320. (2010). doi:10.1109/SYNASC.2010.8
4. Menglan, H., Jun, L., Veeravalli, B.: Optimal provisioning for scheduling divisible loads with reserved cloud resources. In: 18th IEEE International Conference on Networks (ICON), pp. 204–209. (2012). doi:10.1109/ICON.2012.6506559
5. Chaisiri, S., Bu-Sung, L., Niyato, D.: Optimization of resource provisioning cost in cloud computing. IEEE Trans. Serv. Comput. **5**(2), 164–177 (2012). doi:10.1109/TSC.2011.7
6. MahmoudLoad, T.S., Habibi, D., Bass, O., Lachowicz, D.: Demand forecasting: model inputs selection
7. Wang, L.: In cloud, can scientific communities benefit from the economies of scale? IEEE TPDS **23**(20), 296–303 (2012)
8. Hoelzle, U.: The Datacenter as a Computer: An Introduction to the Design of Warehouse-Scale Machines, 1st edn. Morgan and Claypool Publishers, San Rafael
9. Hamilton, J.: Internet-scale service infrastructure efficiency. SIGARCH Comput. Archit. News **37**(3), 232–232 (2009)
10. Moreira, J.: The case for full-throttle computing: an alternative datacenter design strategy. IEEE Micro **30**(4), 25–28 (2010)
11. Iqbal, W.: SLA-driven dynamic resource management for multi-tier web applications in a cloud. In: Proceedings of CCGrid', vol. 10, pp. 832–837 (2010)
12. Vogels, W.: Beyond server consolidation. ACM Queue **6**(1), 20–26 (2008)
13. Goiri, I.: Characterizing cloud federation for enhancing providers' profit. In: Proceedings of cloud'10, pp. 123–130 (2010)
14. Zaharia, M., et al.: Delay scheduling: a simple technique for achieving locality and fairness in cluster scheduling. In: Proceedings of EuroSys'10, pp. 265–278. ACM, New York, NY (2010)
15. Urgaonkar, B., et al.: Resource overbooking and application profiling in shared hosting platforms. In: Proceedings of OSDI'02, Boston, MA (2002)
16. http://www.cs.huji.ac.il/labs/parallel/workload/.

Frequency Based Inverse Damage Assessment Technique Using Novel Hybrid Neuro-particle Swarm Optimization

Bharadwaj Nanda, R. Anand and Dipak K. Maiti

Abstract This study proposes the application of a novel hybrid neuro-particle swarm optimization technique in damage assessment problems. Here, PSO is used to optimally fix various elements of ANN architecture like number of neurons, hidden layer, learning coefficients and momentum coefficient, which by large are decided by trial and error. The pertinency of this method is demonstrated by solving few single and multiple element damage identification cases in a steel bridge truss structure wherein the results are found to be encouraging.

Keywords Damage identification · Hybrid neuro-particle swarm optimization · Radial basis neural network · Natural frequency · Stiffness reduction factor

1 Introduction

Assessment of health of structural systems is an important issue for maintaining the safety and serviceability of the system during their lifetime. Recently, many researches are going on around-the-globe to predict structural health using changes in their vibration response. In this regard, a number of researchers have used modal properties of a structure like natural frequencies or mode shape as damage indicator. The usual approach for vibration based damage identification technique is composed of two steps. In the first step an inverse optimization problem is formulated considering the discrepancies between experimental and numerical vibration data

B. Nanda (✉)
Veer Surendra Sai University of Technology, Burla, India
e-mail: bharadwajnanda@gmail.com

R. Anand · D.K. Maiti
Indian Institute of Technology, Kharagpur, India
e-mail: rananda1990@gmail.com

D.K. Maiti
e-mail: dkmaiti@aero.iitkgp.ernet.in

L.C. Jain et al. (eds.), *Computational Intelligence in Data Mining - Volume 3*,
Smart Innovation, Systems and Technologies 33, DOI 10.1007/978-81-322-2202-6_56

and in the second step this problem is solved by using some optimization technique to get actual damage conditions. Thus, this traditional approach of damage assessment generally involves a combination of heuristic or procedural knowledge and latest soft computing tools like artificial neural networks (ANNs).

ANN is a mathematical model, which is inspired by working process of central nervous system of humans. It can effectively handle qualitative, uncertain and incomplete informations, which it is greatly suitable for damage identification as most of the in situ measured data for large civil, mechanical or aerospace structures. It is observed from literature that several researchers have used ANN for solving such problems [1–10]. But the issues related to basic architecture of ANN like number of hidden layers, number of neurons in the hidden layer, learning rate and momentum coefficients, etc. are still remain unclear and generally chosen by trial and error procedure. In order to get rid of such difficulty, some authors have adopted a hybrid version of ANN and genetic algorithm (GA) for damage assessment purpose [11, 12]. However, slow convergence rate of GA possess few difficulties as, it requires significant computational resources and in many cases it ends with detection of false damages. On contrary, recent particle swarm optimization (PSO) techniques due to their simplicity and convergence speed are getting increasing applications for solving many complex engineering optimization problems including damage identification [13–18].

In this study, an efficient structural damage identification technique is proposed based on a novel hybrid neuro-particle swarm optimization technique wherein PSO algorithm is employed to automate the trial and error procedure of ANN for getting optimized network architecture. The proposed technique is demonstrated using a steel truss-bridge like structure, for single and multiple damage identification cases. The objective function required for the study is constructed using the discrepancies in natural frequencies.

2 Theory and Formulations

2.1 Finite Element Formulation

It is assumed for the present study that, presence of damage cause change in stiffness characteristics of the structural system without any change in its mass. Accordingly, the eigen equation associated with a damaged undamped and freely vibrating system is given by:

$$\left[[\mathrm{K_d}] - \omega_{id}^2[M]\right]\left\{\phi_d^i\right\} = \{0\} \tag{1}$$

where, $[K]$ and $[M]$ indicate the stiffness and mass matrix whereas ω_i and ϕ^i denote the ith natural frequency and corresponding mode shapes respectively. The subscript "d" denotes the damaged values of corresponding parameters. For the present study, damage is represented as "stiffness reduction factor (SRF)" which ranges from [0, 1], where 0 signifies no damage in the element and 1 means the stiffness is completely lost for the element. Mathematically:

$$[\mathrm{K_d}] = \sum_{i=1}^{m} (1 - \beta_i)[\mathrm{k_i}] \tag{2}$$

where, $[k_i]$ indicate the undamaged stiffness matrix for ith element.

The objective function used for the present study is given by Eq. (3), where ω_i^m and ω_i^c denotes the measured and computed frequencies respectively and n denotes the number of natural frequencies considered for the study.

$$F = \sqrt{\frac{1}{n}\sum_{i=1}^{n}\left(\left(\frac{\omega_i^m}{\omega_i^c}\right) - 1\right)^2} \tag{3}$$

2.2 Artificial Neural Network

Artificial neural network is a mathematical model, which is inspired by the biological learning process of the human brain. It consists of a number of processing nodes called neurons which are interconnected to each other by weighted links. Each neuron accepts a set of information from preceding neurons and estimates an output which is propagated to another set of neurons. The relationship between the inputs and output is mapped through an activation function associated to the neuron and the weights associated with the interconnections between the neurons. Initially, these weights are assigned randomly which is modified through an iterative process, called training, so that the behavior of the trained network agrees with actual. The input pattern is propagated forward, and calculated responses are obtained. The error between desired outputs and the calculated outputs are then propagated backward through the network, providing vital information for weight adaptation. The back-propagation algorithm uses this information to adjust the weights such that a "mean-square" error measure is minimized. Mathematically:

$$\Delta w_i = -\eta \frac{\partial E}{\partial w_i} \tag{4}$$

where Δw_i denotes the change in weights and E is the error function, η is a constant called learning coefficient.

$$W(n+1) = W(n) + \eta\delta(n)O(n) + \alpha\nabla E(W(n-1)) \tag{5}$$

Here, α is a constant called momentum coefficient, which forces the search to move in the same direction so as to add the numerical stability and go over the local minima encountered in the search. In contrast to the traditional back propagation neural network (BPNN), radial basis neural network (RBNN) is a type of feed forward neural network that learns using a supervised training technique. Radial functions are a special class of functions, whose response decreases, or increases, monotonically with distance from a center point. Although, RBNN may requires more neuron than a BPNN; they needs significantly lesser time for training the network. Moreover, when the training vectors are not a constraint, it is observed that, RBNN performs better than BPNN.

2.3 Particle Swarm Optimization

Particle swarm optimization (PSO) algorithms are inspired by the collective motion of birds which involves both social interaction and intelligence. Here, each solution of the problem refers to a particle. Initially, the process starts from a swarm of particles placed at random positions in the current search space. During the process of flying each particle adjusts their velocity and positions accordingly to certain velocity and position update rules. Mathematically for a swarm of P-particles in S-dimensional space, if V_i^{t+1} denotes the updated velocity for ith particle for $(t+1)$th iteration as given by:

$$V_i^{(t+1)} = wV_i^t + c_1 r_1 (pbest_i - x_i^t) + c_2 r_2 (gbest - x_i^t) \quad (6)$$

$$x_i^{t+1} = x_i^t + V_i^{t+1} \quad (7)$$

where, V_i^t denotes the velocity of particle i at iteration t, w is a weighting function, all c terms are acceleration coefficients while r terms are random numbers between 0 and 1, x_i^t is the position of particle i at iteration t, $pbest_i$ *and* $gbest$ denotes the personal best solution explored by the particle i and the global best solution explored by any particle in the swarm hood respectively.

2.4 Hybrid Neuro-particle Swarm Optimization

The performance of ANN is significantly depends on various elements of network architecture like number of hidden layers, number of neurons in the hidden layer, learning rate and momentum coefficients, etc. All these parameters are generally selected by trial and error procedure which needs lots of numerical experiments and expertize. As there is no specific relationship between these parameters and

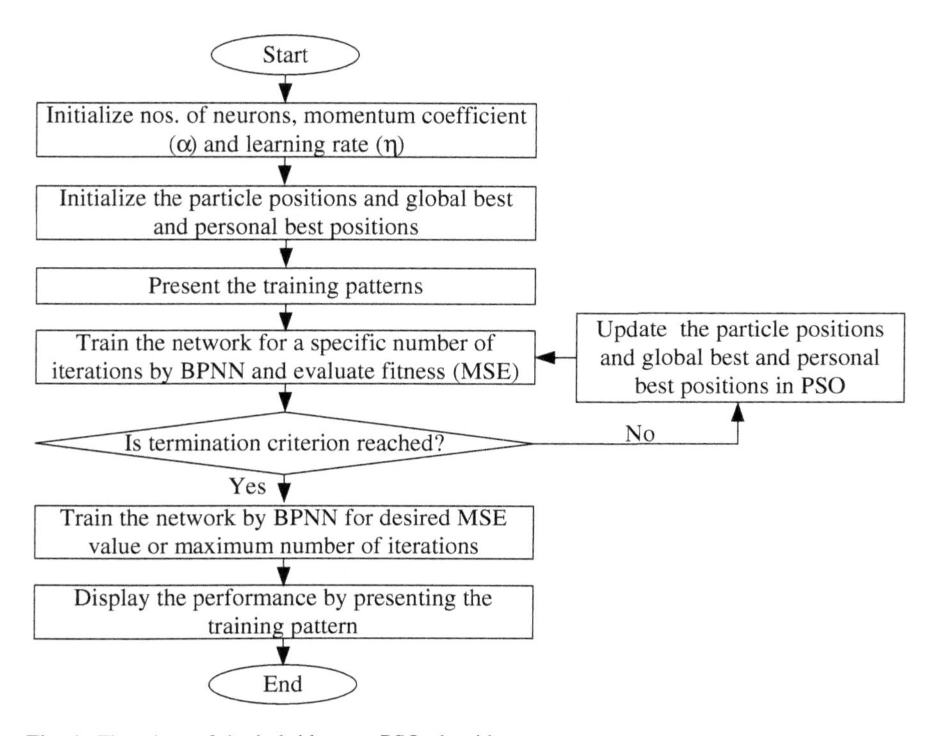

Fig. 1 Flowchart of the hybrid neuro-PSO algorithm

performance of the network; PSO can effectively be used for such purpose. Figure 1 shows the detail procedure in a flowchart form.

3 Results and Discussions

A few numerical simulation studies are carried out to demonstrate the functioning of the proposed damage identification technique. A 21-member steel truss bridge structure as shown in Fig. 2 is selected for this purpose. The Young's modulus and material density is taken as 200 GPa and 7,860 kg/m^3 respectively. Damage is introduced by reducing elemental stiffness matrices in the form of SRF. First four natural frequencies of the structure, estimated from generalized eigen value analysis, is used for construction of the objective function. It is assumed that, damage is located in the left side of the symmetrical structure i.e., within elements 1–11. Accordingly, damage assessment is carried out in these elements only. Two damage cases are considered for the demonstration purpose. First, a single element damage identification problem is discussed wherein damage amounting 30 % is considered to be present at element number 5. Then the performance of current algorithm is

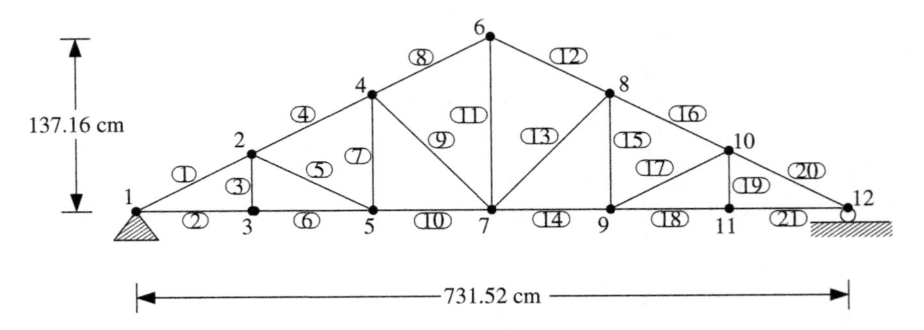

Fig. 2 21-member truss bridge structure

evaluated for problems involving multiple element damage where the damage is considered to be located within elements (4, 5, 6 and 7).

Total 500 different patterns are generated for training and testing of ANN by assigning different SRF values in the range of 0–0.60 to those 11 members located on the left side of the structure. Out of these 500 patterns, 450 are used for training the network whereas the remaining 50 are used for testing purpose. First four frequencies are used as input to the network and the damages in the members are generated as the outputs. The working of hybrid neuro-PSO is described in Fig. 1, where, the elements like number of hidden layers, number of neurons in the hidden layer, learning rate and momentum coefficients, etc., which describes the network architecture, are first optimized using PSO. The swarm size selected for the present study is set to 10. The acceleration coefficients (c_1 and c_2) are selected as 2.05 each and the inertia weight is selected as 0.9. PSO is run up to maximum 50 iterations or till convergence. As the present optimization problem is quite complex in nature, it is very much probable that, the algorithm may stick at some local minima. In order to avoid any such possibility three experiments are conducted in present case. The outputs are presented in Table 1. From these tabulated results, it is observed that both Experiment 1 and 2 provides comparable results in terms of objective function value. However, number of neurons in hidden layer 1 is less in experiment 1, due to which it will require less computational cost for arriving at the result. Therefore, it is selected for construction of hybrid neuro-PSO networks. Further, as discussed earlier, the performances of a BPNN based hybrid network and RBNN based hybrid network are evaluated and compared in damage identification problems. These

Table 1 Network optimization results for different experiments

Experiment no.	Number of neurons		Learning rate (η)	Momentum coefficient (α)	Objective function value
	Layer 1	Layer 2			
1	20	15	0.4036	0.2769	0.0015
2	15	15	0.0777	0.0353	0.0015
3	20	15	0.0237	0.0405	0.0017

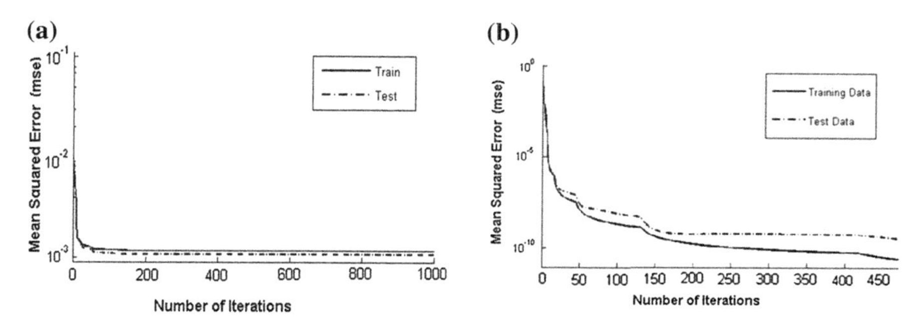

Fig. 3 Variation of training and testing errors for hybrid neuro-PSO algorithms **a** Hybrid BPNN-PSO, **b** Hybrid RBNN-PSO

neural networks are trained and tested by numerically generated data sets. It is observed that, between BPNN and RBNN based hybrid networks; RBNN could achieve an impressive MSE value of 1×10^{-10} within 500 iterations whereas, BPNN could achieve MSE value of 1×10^{-3} even after 1,000 numbers of iterations (Fig. 3).

These trained networks are then employed for damage identification. First, a single element damage case is considered for demonstration, wherein 30 % damage is introduced in the 5th element of the truss. The results are presented in Fig. 4. It is observed from the figure that, both BPNN and RBNN based algorithms are able to provide good quantification of damage. However, the result obtained from RBNN based network is more close to the actual damage. Further, the numbers and amount of false damages are quite less in case of RBNN. Based on this study, it is

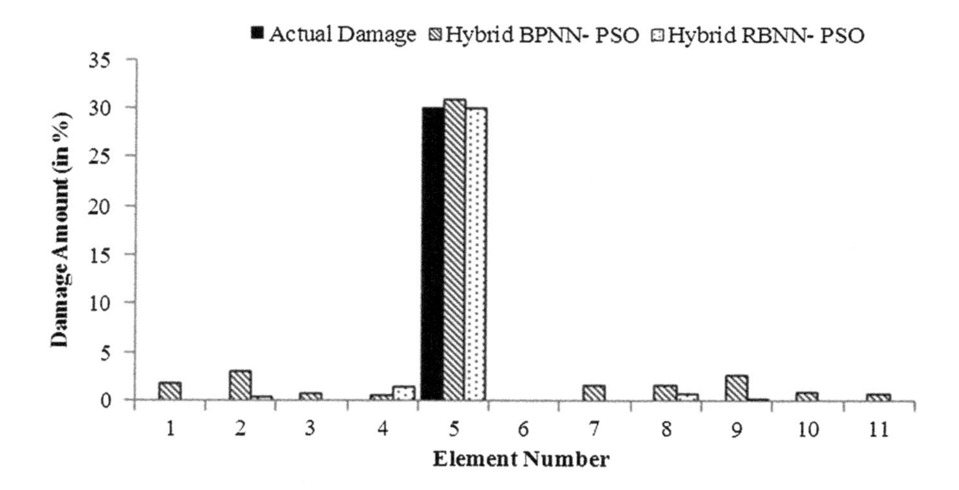

Fig. 4 Single element damage identification result

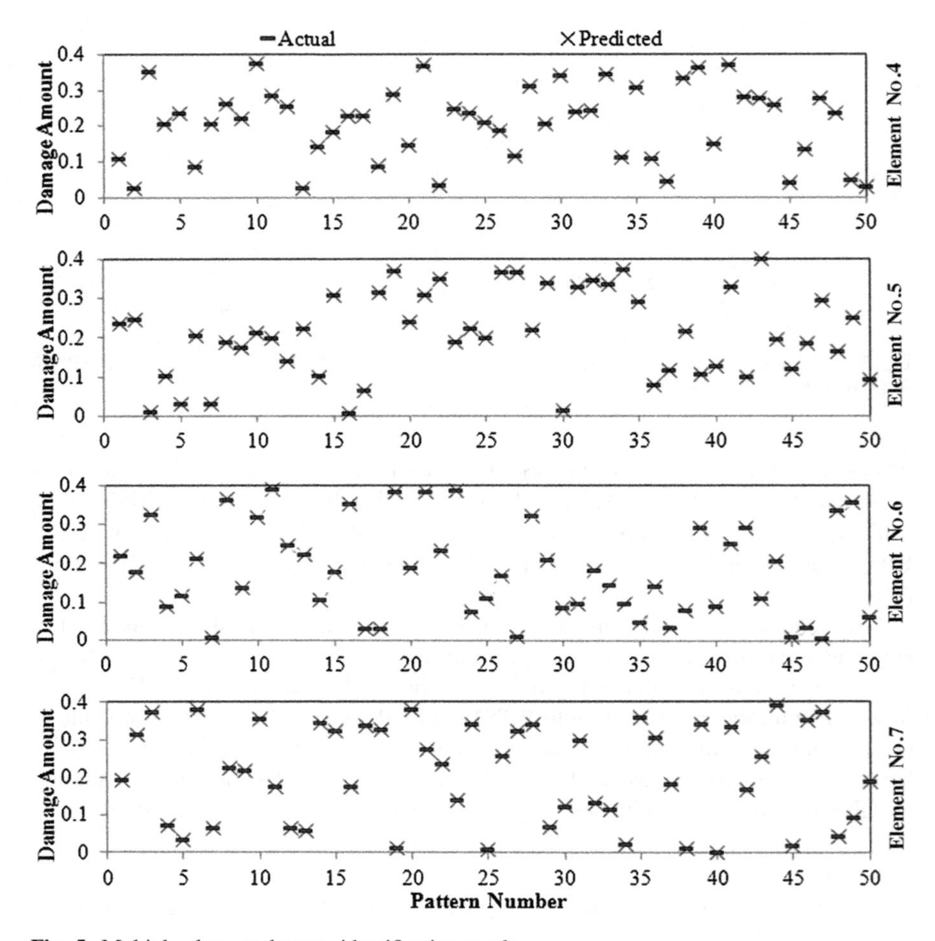

Fig. 5 Multiple element damage identification results

concluded that performance of RBNN based hybrid neuro-PSO algorithm is superior and hence selected for further studies.

Furthermore, the performance of RBNN based hybrid neuro-PSO algorithm is evaluated in assessment of multiple element damage cases. In total 50 numbers of damage patterns are randomly generated considering damages in elements (4, 5, 6 and 7). The graphical results of these multiple damage assessment are shown in Fig. 5, which indicated that the hybrid-neuro PSO algorithm could detect and quantify the damages with a very good accuracy. It is observed that, the maximum error in damage identification is less than 1.0 % for present hybrid neuro-PSO algorithm.

4 Conclusions

In this study, a novel hybrid neuro-particle swarm optimization technique is developed, wherein PSO is used to optimally fix various elements used to define ANN architecture. Thus need of trial and error procedure for fixing the issues like number of hidden layers, number of neurons in the hidden layer, learning rate and momentum coefficients, etc. are eliminated. A comparison study is made among a RBNN based and a BPNN based hybrid neuro-PSO, where it is demonstrated that the performance of the former is superior in comparison to the latter. The efficacy of the developed hybrid algorithm is demonstrated with a damage identification problem in a steel truss-bridge like structure, for single and multiple damage identification cases. It is observed that, the performance of the present algorithm is very encouraging. The developed algorithm is able to detect all the correct damaged members and their quantifications with a very good accuracy.

References

1. Rhim, J., Lee, S.W.: A neural network approach for damage detection and identification of structures. Comput. Mech. **16**(6), 437–443 (1995)
2. Marwala, T.: Damage identification using committee of neural networks. J. Eng. Mech. **126** (1), 43–50 (2000)
3. Yun, C.B., Bahng, E.Y.: Sub-structural identification using neural network. Comput. Struct. **77**(1), 41–52 (2000)
4. Zapico, J.L., González, M.P., Worden, K.: Damage assessment using neural networks. Mech. Syst. Sig. Process **17**(1), 119–125 (2003)
5. Maity, D., Saha, A.: Damage assessment in structure from changes in static parameter using neural networks. Sadhana **29**(3), 315–327 (2004)
6. Tripathy, R.R., Maity, D.: Damage assessment of structures from changes in curvature damage factor using artificial neural network. Ind. J. Eng. Mater. Sci. **11**, 369–377 (2004)
7. Bakhary, N., Hao, H., Deeks, A.J.: Damage detection using artificial neural network with consideration of uncertainties. Eng. Struct. **29**(11), 210–220 (2007)
8. Wang, B.S., Ni, Y.Q., Ko, J.M.: Damage detection utilising the artificial neural network methods to a benchmark structure. Int. J. Struct. Eng. **2**(3), 229–242 (2011)
9. Min, J., Park, S., Yun, C.B., Lee, C.G., Lee, C.: Impedance-based structural health monitoring incorporating neural network technique for identification of damage type and severity. Eng. Struct. **39**, 210–220 (2012)
10. Thatoi, D., Jena, P.K.: Inverse analysis of crack in fixed-fixed structure by neural network with the aid of modal analysis. Adv. Artif. Neural Syst. 150209.1–150209.8 (2013)
11. Suh, M.W., Shim, M.B., Kim, M.Y.: Crack identification using hybrid neuro-genetic technique. J. Sound Vib. **33**(4), 617–635 (2000)
12. Sahoo, B., Maity, D.: Damage assessment of structures using hybrid neuro-genetic algorithm. Appl. Soft Comput. **7**, 89–104 (2007)
13. Yu, L., Wan, Z.Y., Zhu, H.P.: An improved PSO algorithm for structural damage detection based on macro-economic control strategy. Chin. J. Comput. Mech. **25**(1), 14–19 (2008)
14. Seyedpoor, S.M.: Structural damage detection using a multi-stage particle swarm optimization. Adv. Struct. Eng. **14**(3), 533–549 (2011)

15. Nanda, B., Maity, D., Maiti, D.K.: Vibration-based structural damage detection technique using particle swarm optimization with incremental swarm size. Int. J. Aeronaut. Space Sci. **13** (3), 322–329 (2012)
16. Kang, F., Li, J., Liu, S.: Combined data with particle swarm optimization for structural damage detection. Math. Probl. Eng. 416941.1–416941.10 (2013)
17. Nanda, B., Maity, D., Maiti, D.K.: Crack assessment in frame structures using modal data and unified particle swarm optimization technique. Adv. Struct. Eng. **17**(5), 747–766 (2014)
18. Nanda, B., Maity, D., Maiti, D.K.: Modal parameter based inverse approach for structural joint damage assessment using unified particle swarm optimization. Appl. Math. Comput. **242**, 407–422 (2014)

A Hybrid CRO-K-Means Algorithm for Data Clustering

Sibarama Panigrahi, Balaram Rath and P. Santosh Kumar

Abstract Over the past few decades clustering algorithms have been used in diversified fields of engineering and science. Out of various methods, K-Means algorithm is one of the most popular clustering algorithms. However, K-Means algorithm has a major drawback of trapping to local optima. Motivated by this, this paper attempts to hybridize Chemical Reaction Optimization (CRO) algorithm with K-Means algorithm for data clustering. In this method K-Means algorithm is used as an on-wall ineffective collision reaction in the CRO algorithm, thereby enjoying the intensification property of K-Means algorithm and diversification of intermolecular reactions of CRO algorithm. The performance of the proposed methodology is evaluated by comparing the obtained results on four real world datasets with three other algorithms including K-Means algorithm, CRO-based and differential evolution (DE) based clustering algorithm. Experimental result shows that the performance of proposed clustering algorithm is better than K-Means, DE-based, CRO-based clustering algorithm on the datasets considered.

Keywords Data clustering · Chemical reaction optimization · K-Means algorithm

1 Introduction

Clustering is an unsupervised classification technique in which patterns or feature vectors or data items are organized in clusters/groups based on similarity, such that the data within the same cluster have high degree of similarity and patterns

S. Panigrahi (✉)
National Institute of Science and Technology, Palur Hills, Berhampur 761008, Odisha, India
e-mail: panigrahi.sibarama@gmail.com

B. Rath · P. Santosh Kumar
MIRC Lab, MITS Engineering College, Rayagada 765017, Odisha, India
e-mail: balaram.rath@gmail.com

P. Santosh Kumar
e-mail: santosh.mirc@gmail.com

L.C. Jain et al. (eds.), *Computational Intelligence in Data Mining - Volume 3*,
Smart Innovation, Systems and Technologies 33, DOI 10.1007/978-81-322-2202-6_57

belonging to different clusters have high degree of dissimilarity. Jain et al. [1] provides a survey on clustering algorithms and Jain and Gajbhiye [2] provides a comparative performance analysis on different clustering algorithms. However, the clustering techniques can be broadly classified into two types: hierarchical and partitional. Hierarchical clustering proceeds successively by either merging smaller clusters into larger ones, or by splitting larger clusters into smaller ones and generates a hierarchy of partitions in which each partition is nested within the partition at the next level in the hierarchy. On the other hand, partitional clustering attempts to directly decompose the data set into a set of disjoint clusters.

K-Means clustering algorithm [2, 3] is a well-known partitioning based technique which partitions n-patterns into k-clusters with each pattern belongs to the cluster with the nearest mean. For improving the performance and efficiency of K-Means clustering, various and numerous methods have been proposed. Dash et al. [4] proposed a clustering approach by hybridizing K-Means algorithm with principal component analysis (PCA) to reduce dimensionality of data set. Additionally employed a new method obtain the initial centroid values by finding the mean of all the data sets divided into k different sets in ascending order. This approach has the demerit of taking large amount of time and it may eliminate some of the features which are also important. However, Yedla et al. [5] proposed an enhanced K-Means algorithm with improved initial center using the distance from the origin to improve the performance of K-Means algorithm. Although, this approach seems to solve the initialization problem but does not give any guarantee better time complexity. For reducing the computational time Fahim et al. [6] proposed a clustering method by considering initial centroids randomly. However, this method does not provide unique result due to very sensitive to randomly selected initial center points. Zhang and Xia [7] proposed the initial centroids algorithm based on K-Means that have avoided alternate randomness of initial centroids. Despite of several years of research on improving the performance of K-Means algorithm, the major drawback is its chances to trapping to local optima because it is an iterative and hill climbing method. Therefore in order to achieve diversification property hybridization with some global optimization technique is required.

Obtaining the optimal centroids for the clusters can be considered as a multimodal search problem. Therefore, any global optimization like technique differential evolution (DE) [8], genetic algorithm (GA) [9], particle swarm optimization (PSO) [10], ant colony optimization (ACO) [11], a bee colony optimization (BCO) [12], an evolutionary strategy (ES) [13], quantum inspired algorithms (QEA) [14], chemical reaction optimization (CRO) [15–17] etc. can be used for clustering. Motivated by this few authors have applied global optimization techniques like genetic algorithm [18–20], differential evolution [21], Teaching-Learning-Based Optimization [22, 23] algorithms to clustering problems and have shown better performance. However, the major drawback of these approaches is that obtaining a trade-off between diversification and intensification. Motivated by this, a hybridization of K-Means algorithm (used to intensify some of the solutions) and CRO algorithm (diversifies the solution space) is used.

The rest of this paper is organized as follows. Section 2 gives a brief overview of K-Means, DE-based and CRO-based clustering algorithms. Section 3 describes the proposed Hybrid CRO-K-Means Clustering algorithm, Sect. 4 gives the experimental set up and comparison between the proposed algorithm and some of the existing algorithms, and Sect. 5 concludes the paper.

2 Clustering Algorithms

2.1 K-Means Clustering Algorithm

K-Means is one of the simplest centroid based unsupervised learning algorithms that use K number of clusters (fixed a priori) to discriminate the data. Consider a data matrix X with n-data point patterns. The main goal is to find optimal positions of K-centroids (one for each cluster and represented by C_i) so as to partition n-data points into K-disjoint subsets by minimizing the sum of squared error. The sum squared error (E) is calculated as in Eq. 1.

$$E = \sum_{i=1}^{K} \sum_{j=1}^{n} \left\| X_j - C_i \right\|^2 \tag{1}$$

where ||.|| is the Euclidean distance.

Algorithm:

1. Randomly select K centers or centroids.
2. Calculate the distance between each data point and cluster centers and assign the data point to the cluster center whose distance from the cluster center is minimum of all the cluster centers.
3. For each cluster, recompute the cluster centers.
4. Repeat step 2 and 3, until no cluster center changes.

2.2 Differential Evolution Based Clustering Algorithm

The differential evolution (DE) algorithm is a simple and efficient stochastic direct search method for global optimization. It was introduced several years ago (1997) [8]. It has several advantages such as: ability to find the global minimum of a non-differentiable, nonlinear and multimodal function irrespective of initial values of its parameters, parallel implementabilty, ease of use and good convergence properties. Therefore, a wide variants of it has been developed [18]. The major difference between the DE variants lies in the mutation and crossover strategies. The conventions used above is DE/a/b/c, where 'DE' stands for 'differential evolution', 'a' represents the base vector to be perturbed, 'b' represents number of difference vectors used for perturbation of 'a' and 'c' represents the type of crossover used (bin: binary, exp: exponential). Interested reader may go through [8, 18] to have a detail description regarding DE algorithm and its variants.

DE Clustering Algorithm:

1. Initialize each individual solution to contain K randomly chosen points from the dataset.
2. Calculate fitness function for each individual.
3. Perform Crossover and Mutation upon solutions based on the scheme DE/a/b/c.
4. Replace the existing solutions with the new ones if and only if the new solutions have better fitness value.
5. Repeat step 3 and 4, until some termination criteria is satisfied.

Select the best chromosome in the population which will act as the resultant clusters.

2.3 Chemical Reaction Optimization Based Clustering Algorithm

Chemical reaction optimization (CRO) algorithm is a recently established population based metaheuristics optimization technique [15, 16] which is inspired by the nature of chemical reactions. It does not attempt to capture every detail of chemical reaction rather loosely couples chemical reaction with optimization. A chemical reaction starts with some initial species/reactants/molecules in the unstable states and undergoes a sequence of collisions and become the final products in stable states [16]. During the chemical reaction the intra-molecular structure of a reactant changes. Most of the reactions are reversible in nature i.e. a reaction may be turned back due to instability. It enjoys the advantages of both Simulated Annealing (SA) and GA [16]. The major difference between CRO and other evolutionary techniques is that, the population size (that is the number of reactants) may vary from one iteration to the other where as in evolutionary techniques the population size remains fixed. But few authors [25] have proposed fixed population sized CRO algorithms and shown that fixed population sized CRO not only performs better but also easier to implement. The basic unit in CRO is a molecule (chromosome) which is considered as a single possible solution in the population. Each molecule has potential energy (PE) which refers to fitness value, to characterize the molecule. A chemical change of a molecule is triggered by a collision which may be uni-molecular collision (one molecule taking part in chemical reaction) or inter-molecular collision (two or more molecule collide with each other). The corresponding reaction change is termed as Elementary Reaction.

The chemical reaction optimization based Clustering algorithm consists of following steps:

1. Initialize each individual solution to contain K randomly chosen points from the dataset.
2. Evaluation of Potential Energy (PE) of each Reactant representing a set of cluster centers.
3. Applying Chemical reactions to generate new reactants.
4. Reactants update for better potential energy.
5. Check termination criteria. if satisfied go to step-6 otherwise go to step-3
6. Use the reactant having best Potential Energy as the cluster center.

3 Proposed Methodology

In this paper the hybridized CRO-K-Means algorithm uses four kinds of elementary reactions (two uni-molecular and two bi-molecular): on-wall ineffective collision, decomposition, inter-molecular ineffective collision, and synthesis. K-Means algorithm is hybridized as an on-wall ineffective collision of the CRO algorithm. K-Means algorithm can able to find the local minima and the use of other three reactions may assist in trapping from local optima. Therefore chances of obtaining near optimal solution in lesser number of iterations can be guaranteed.

The pseudo-code of clustering using CRO is shown in Algorithm 1 where *ReacNum* is the size of the population 'P' best_mole represents the molecule having the highest potential energy (PE) in current iteration, and the elementary reactions are shown in subsequent algorithms.

Algorithm 1 (CRO Based Clustering Algorithm)

```
Set the iteration-counter i=0
Load the dataset in X.
while Halting criteria not satisfied do
//Initialize the ReacNum of Reactants/Molecules randomly
from the dataset: P^i={R_1^i, R_2^i , R_3^i...., R_ReacNum^i}, with R_j^i ={
W_j,1^i,......,W_j,D^i}   for   j=1,2,3.....   ReacNum,   with   W
representing one dimension of the attribute of cluster
center, D=length of each Reactant (Number of clusters □
Dimension of each cluster centroid), P^i = Population of i^th
iteration.
//Calculate the fitness (PE) of each set of cluster
centers (reactant)
for j=1 to ReacNum
      Calculate the PE of j^th reactant in the population.
end of for
//Apply Chemical Reactions to obtain new cluster centers
While (termination criteria is not satisfied) do
    begin
    for j=1 to ReacNum
       // perform all reaction over the reactants of P^i
       Get rand_1 randomly in an interval [0, 1]
       if rand_1 ≤ 0.5
           Get rand_2 randomly in an interval [0, 1]
           if rand_2 ≤ 0.5
                Decomposition (R_j);
           else
                On-wall ineffective collision(R_j)
           end of if
       else
           Get rand_3 randomly in an interval [0, 1]
           if  rand_3 ≤ 0.5
                Select the best reactant R_k(R_k≠ R_j)
                Synthesis (R_j, R_k)
           else
                Select another reactant R_k(R_k≠ R_j)
                randomly inter-molecular ineffective
        collision(R_j, R_k)
           end of if
       end of if
       Apply strictly greedy Reversible Reaction for
       increased PE to update reactants
    end of for
   Set the iteration counter i=i+1
end of while
best_mole=The Reactant having best PE in the final
iteration.
The resultant clusters centroids will be stored in
best_mole.
```

3.1 On-wall Ineffective Collision Reaction

In this reaction K-Means algorithm is applied on the reactant considering the values of the reactants as the initial cluster centers. The pseudo-code of the decomposition reaction is described in Algorithm 2.

```
Algorithm 2 (On-wall ineffective collision(Rj))
Input: A reactant Rj
Apply the K-Means algorithm (considering the Rj values as
the initial cluster center position) to obtain the new
cluster centers.
Output: A new reactant R1
```

3.2 Decomposition Reaction

In this reaction the value of a randomly selected atom of the reactant is changed in order to perform more local search. Consider a reactant $R_j = \{W_{j,1}, W_{j,2} \ldots, W_{j,D}\}$ with $W_{j,x}$ ($x \in [1, n]$) be an atom of the reactant-j. The rate of change of the atom of the reactant depends on the rate of reaction (λ). The rate of reaction is a random number generated from a Cauchy distribution because it diversifies the solution more as compared to traditional uniform distribution. The pseudo-code is described in Algorithm 3.

```
Algorithm 3 (Decomposition(Rj))
Input: A reactant Rj
Duplicate Rj to produce R1
Select an atom k (k∈ [1:n]) randomly.
W1,k =L+ λ × (U-L)
Where λ= cauchyrnd(0.6,0.1), is a random number generated
from Cauchy distribution with a location parameter 0.6
and scale parameter 0.1. It is regenerated if the random
number falls out of the range [0, 1].
Output: A new reactant R1
```

Here two reactants $R_j = \{W_{j,1}, \ldots, W_{j,D}\}$ and $R_k = \{W_{k,1}, W_{k,2} \ldots, W_{k,D}\}$ will take part in the reaction. These reactions help in diversification of the solution by generating a new solution that is significantly different from the current solution. These reactions occur with a probability of 50 %. Following types of bimolecular reactions are used.

3.3 Synthesis Reaction

In this reaction one reactant is produced due to reaction between a reactant $R_j = \{W_{j,1}, \ldots, W_{j,D}\}$ and another randomly reactant and $R_{rand} = \{W_{k,1}, W_{k,2} \ldots, W_{k,D}\}$; of the iteration. Here, instead of traditional normal or uniform distribution; the rate of reaction (λ) is generated from a Cauchy distribution with a location parameter 0.7 and scale parameter 0.1. The pseudo-code of this reaction is described in Algorithm 4.

Algorithm 4 (Synthesis (R_j, R_k))

Input: Two reactants R_j, R_{rand}

$R_1 = R_J + \lambda \times (R_{rand} - R_J)$

Where λ =cauchyrnd(0.7,0.1), is a random number generated randomly from Cauchy distribution with location parameter 0.7 and scale parameter 0.1. It is regenerated if the random number falls put of the range [0.5, 1].

Output: A new reactant R_1

3.4 Inter-molecular Ineffective Collision Reaction

In this reaction two solutions R_1 and R_2 are obtained from reaction between two reactants R_j and R_k. The pseudo-code of this reaction is described in Algorithm 5.

Algorithm 5 (inter-molecular ineffective collision (R_j, R_k))

Input: Two reactants R_j, R_k

$R_1 = \lambda_t \times R_j + \lambda_t \times (1 - R_k)$

$R_2 = \lambda_t \times R_k + \lambda_t \times (1 - R_j)$

Where λ_t is a random number generated uniformly between[0,1] for t^{th} iteration.

Output: Two reactants R_1 and R_2

3.5 Reactant Update

```
Algorithm 6 (Reversible Reaction())
For Reactions producing single products
     If Rj under goes reaction to produce R1
      If PE (R1) < PE(Rj)
           Replace Rj by R1
      end of if
     end of if

For Reactions producing two products
     If Rj under goes reaction to produce R1 and R2
        If PE(R1)<PE(Rj)
            Replace Rj by R1
        end of if
        If PE(R2)<PE(Rj)
            Replace Rj by R2
        end of if
   end of if
```

Algorithm 6 describes the pseudo-code of the reversible reaction. Every reaction is followed by a reversible reaction to update the reactants (cluster centers) for next iteration i.e. reactants having better fitness (PE) survive for the next iteration. In this reaction, a reaction producing a single product replaces the replaces the used reactant for better PE; and for reactions producing two products, best product replace the used reactant for better PE. Thus the number of reactants of the population remains same throughout the reaction process.

4 Experimental Set up and Simulation Results

For comparative performance analysis of proposed hybrid CRO-K-Means algorithm with K-Means, DE/rand/1/bin, and CRO [26] based clustering algorithms, four real world datasets such as: Lung cancer, Iris, Breast cancer Wisconsin and Haberman's survival from UCI machine learning data repository are considered. *Lung cancer* dataset has 32 instances with predictive attributes which are nominal, and takes integer values from 0 to 3. *Iris* dataset has 150 patterns which are random samples of three species of the iris flower such as: setosa, versicolor, and virginica. *Breast cancer Wisconsin data*set consists of 699 patterns which were collected periodically as Dr. Wolberg reports his clinical cases. *Haberman's survival dataset* contains cases from a study that was conducted between 1958 and 1970 at the University of Chicago's Billings Hospital on the survival of patients who had

Table 1 Properties of the data sets considered

Data set	Number of data points/ instances	Number of attributes	Number of clusters
Lung cancer	32	56	3
Iris	150	4	3
Breast cancer	214	10	3
Haberman's survival	306	3	3

undergone surgery for breast cancer. Number of data points, number of attributes and number of clusters of all four datasets are listed in Table 1.

4.1 Performance Measure

In order to evaluate the effectiveness of proposed methodology with the other algorithms considered following three performance measures are considered.

Clustering metric: The clustering metric 'M' for K clusters $C_1, C_2, \ldots, C_K$ is calculated as in Eq. 2.

$$M(C_1, C_2, \ldots, C_K) = \sum_{i=1}^{K} \sum_{x_j \in C_i} ||x_j - C_i|| \tag{2}$$

The objective is to minimize 'M'.

Note that for the evolutionary algorithms DE, CRO and Proposed methodology clustering metric is used as the fitness function or PE.

Intra-cluster Distance: The intra-cluster distance is mean of maximum distance between two data vectors within a set of clusters and is calculated as in Eq. 3.

$$\frac{1}{K} \sum_{i=1}^{K} \max_{ZpZq \in Ci} d(Zp, Zq) \tag{3}$$

The objective is to minimize the intra cluster distance.

Inter-cluster distance: It is the minimum distance between the centroids of the clusters.

The objective is to maximize the inter-cluster distance.

4.2 Simulation Results

For all the evolutionary algorithms the population size or Reactant number is fixed to 10 and run up to a maximum of 100 iterations or 90 % of chromosomes or reactants converge to same point. As the evolutionary methods are stochastic in nature, ten independent simulations were performed for each data set and for each algorithm. The mean and standard deviation of ten iterations for each algorithm for Lung cancer, Iris, Breast cancer Wisconsin, and Haberman's survival datasets are shown in Tables 2, 3, 4 and 5 respectively. Note that for a particular problem and

Table 2 Comparison of the performance of different clustering algorithms for lung cancer dataset

Clustering algorithms	Clustering metric Mean ± St. Dev.	Intra-cluster distance Mean ± St. Dev.	Inter-cluster distance Mean ± St. Dev.
K-Means	129.48 ± 1.59	7.51 ± 0.29	3.18 ± 0.38
DE-Clustering	127.44 ± 0.67	7.76 ± 0.28	3.27 ± 0.13
CRO-Clustering	126.98 ± 0.22	7.49 ± 0.12	3.38 ± 0.03
Proposed method	126.10 ± 0.15	7.38 ± 0.11	3.46 ± 0.03

Table 3 Comparison of the performance of different clustering algorithms for iris dataset

Clustering algorithms	Clustering metric Mean ± St. Dev.	Intra-cluster distance Mean ± St. Dev.	Inter-cluster distance Mean ± St. Dev.
K-Means	102.42 ± 10.44	2.59 ± 0.17	1.59 ± 0.41
DE-Clustering	97.39 ± 0.17	2.56 ± 0.05	1.79 ± 0.01
CRO-Clustering	97.22 ± 2.92E-14	2.51 ± 4.56E-16	1.80 ± 4.56E-16
Proposed method	97.14 ± 2.90E-14	2.46 ± 4.55E-16	1.84 ± 4.22E-16

Table 4 Comparison of the performance of different clustering algorithms for breast cancer Wisconsin

Clustering algorithms	Clustering metric Mean ± St. Dev.	Intra-cluster distance Mean ± St. Dev.	Inter-cluster distance Mean ± St. Dev.
K-Means	2.88E + 03 ± 18.12	17.24 ± 0.42	3.39 ± 2.14
DE-Clustering	2.86E + 03 ± 4.31	16.64 ± 0.58	4.38 ± 1.84
CRO-Clustering	2.86E + 03 ± 1.56	16.61 ± 0.33	4.37 ± 1.79
Proposed method	2.79E + 03 ± 1.59	15.92 ± 0.23	4.47 ± 1.24

Table 5 Comparison of the performance of different clustering algorithms for Haberman's survival dataset

Clustering algorithms	Clustering metric Mean ± St. Dev.	Intra-cluster distance Mean ± St. Dev.	Inter-cluster distance Mean ± St. Dev.
K-Means	2.29E + 03 ± 12.98	41.78 ± 3.25	13.12 ± 1.24
DE-Clustering	2.28E + 03 ± 18.16	41.76 ± 3.02	12.88 ± 0.47
CRO-Clustering	2.26E + 03 ± 4.12	42.42 ± 3.28	13.22 ± 1.88
Proposed method	2.16E + 03 ± 3.98	40.20 ± 2.08	13.45 ± 1.73

for a particular run the initial value of cluster centers for the three evolutionary methods are kept same. For Differential evolution DE/rand/1/bin strategy with F = 0.5 and Cr = 0.6 is used.

It can be observed from the simulation results that the proposed hybrid clustering method performs better than K-Means, DE-Clustering and CRO-Clustering algorithms for the Lung cancer, Iris, Breast cancer Wisconsin, and Haberman's survival dataset considering Clustering Metric, Intra-Cluster Distance and Inter-Cluster Distance as performance measure.

5 Conclusion

In this paper, a hybrid of chemical reaction optimization and K-Means algorithm for data clustering is proposed. In this hybrid method K-Means algorithm is used as a chemical reaction of the CRO algorithm. The K-Means algorithm intensifies the solution up to local optima meanwhile other chemical reactions assist in overcoming the local optima by exploring the solution space; thereby a trade between intensification and diversification is implicitly maintained and chances of reaching a global optima increases. In order to evaluate the effectiveness of the proposed method four real world datasets have been considered. Results revealed that the proposed method performs better than the K-Means, DE based Clustering and CRO based clustering algorithms for the aforementioned datasets. However, the major drawback of the proposed method is that it takes more amount of time than other algorithms since it uses K-Means algorithm as one of the reactions.

References

1. Jain, A.K., Murty, M.N., Flynn, P.J.: Data clustering: a review. ACM Comput. Surv. **31**, 264–323 (1999)
2. Jain, S., Gajbhiye, S.: A Comparative performance analysis of clustering algorithms. Int. J. Adv. Res. Comput. Sci. Softw. Eng. **2**, 441–445 (2012)
3. Abbas, O.A.: Comparisons between data clustering algorithms. Int. Arab J. Inf. Technol. **5**, 320–325 (2008)
4. Dash, R., Mishra, D., Rath, A.K., Acharya, M.: A hybridized K-means clustering algorithm for high dimensional dataset. Int. J. Comput. Sci. Technol. **2**, 59–66 (2010)
5. Yedla, M., Pathakota, S.R., Srinivasa, T.M.: Enhancing K means algorithm with improved initial center. Int. J. Comput. Sci. Inf. Technol. **1**, 121–125 (2010)
6. Fahim, A.M., Salem, A.M., Torkey, F.A., Ramadan, M.A., Saake, G.: An efficient K-means with good initial starting points. Georgian Electron. Sci. J. Comput. Sci. Telecommun. **2**, 47–57 (2009)
7. Zhang, C., Xia, S.: K-means clustering algorithm with improved initial center. In: Second International Workshop on Knowledge Discovery and Data Mining, WKDD, pp. 790–792 (2009)
8. Storn, R., Price, K.: Differential evolution—a simple and efficient heuristic for global optimization over continuous spaces. J. Global Optim. **11**, 341–359 (1997)

9. Goldberg, D.: Genetic algorithms in search, optimization and machine learning. Addison-Wesley, Reading (1989)
10. Kennedy, J., Eberhart, R.C., Shi, Y.: Swarm Intelligence. Morgan Kaufmann, San Francisco (2001)
11. Socha, K., Doringo, M.: Ant colony optimization for continuous domains. Eur. J. Oper. Res. **185**, 1155–1173 (2008)
12. Pham, D.T., Ghanbarzadeh, A., Koc, E., Otri, S., Rahim, S., Zaidi, M.: The bees algorithm—a novel tool for complex optimization problems. In: Proceedings of 2nd International Virtual Conference on Intelligent Production Machines and Systems, pp. 454–459 (2006)
13. Beyer, H.G., Schwefel, H.P.: Evolutionary strategies: a comprehensive introduction. Nat. Comput. **1**, 3–52 (2002)
14. Han, K.H., Kim, J.H.: Quantum-inspired evolutionary algorithm for a class of combinatorial optimization. IEEE Trans. Evol. Comput. **6**, 580–593 (2002)
15. Lam, A.Y.S., Li, V.O.K.: Chemical-reaction-inspired metaheuristic for optimization. IEEE Trans. Evol. Comput. **14**, 381–399 (2010)
16. Lam, A.Y.S.: Real-coded chemical reaction optimization. IEEE Trans. Evol. Comput. **16**, 339–353 (2012)
17. Alatas, B.: ACROA: artificial chemical reaction optimization algorithm for global optimization. Expert Syst. Appl. **38**, 13170–13180 (2011)
18. Al-Shboul, B., Myaeng, S.: Initializing K-means clustering algorithm by using genetic algorithm. World Acad. Sci. Eng. Technol. **54**, 114–118 (2009)
19. Maulik, U., Bandyopadhyay, S.: Genetic algorithm-based clustering technique. Pattern Recogn. **33**, 1455–1465 (2000)
20. Bandyopadhyay, S., Maulik, U.: An evolutionary technique based on K-means algorithm for optimal clustering in R^N. Inf. Sci. **146**, 221–237 (2002)
21. Paterlini, S., Krink, T.: High performance clustering with differential evolution. evolutionary computation. In: CEC2004, vol. 2, pp. 2004–2011 (2004)
22. Satapathy, S.C., Naik, A.: Data clustering based on teaching-learning-based optimization. In: Swarm, Evolutionary and Memetic Computing Conference Part II, LNCS, vol. 7077, pp. 148–156 (2011)
23. Naik, A., Satapathy, S.C., Parvathi, K.: Improvement of initial cluster center of c-means using teaching learning based optimization. Procedia Technol. **6**, 428–435 (2012)
24. Das, S., Suganthanam, P.N.: Differential evolution: a survey of the state-of-the-art. IEEE Trans. Evol. Comput. **15**, 4–31 (2011)
25. Sahu, K.K., Panigrahi, S., Behera, H.S.: A novel chemical reaction optimization algorithm for higher order neural network training. J. Theoret. Appl. Inf. Technol. **53**, 402–409 (2013)
26. Baral, A., Behera, H.S.: A novel chemical reaction-based clustering and its performance analysis. Int. J. Bus. Intell. Data Min. **8**, 184–198 (2013)

An Approach Towards Most Cancerous Gene Selection from Microarray Data

Sunanda Das and Asit Kumar Das

Abstract Microarray gene dataset is often very high-dimensional which presents complicated problems, like the degradation of data accessing, data manipulating and query processing performance. Dimensionality reduction efficiently tackles this problem and benefited us to visualize the intrinsic properties hidden in the dataset. Therefore, Rough set theory (RST) has been used for selecting only the relevant attributes of the dataset, called reduct, sufficient to characterize the information system. The investigation has been carried out on the publicly available microarray dataset. The analysis revealed that Rough Set using the concepts of dependency among genes is able to extract the various dominant genes in term of reducts which play an important role in causing the disease. Experimental results show the effectiveness of the algorithm.

Keywords Dependency among genes · Similarity measure · Reduct generation

1 Introduction

Generally, extracting useful information from voluminous data has been highly complicating and challenging. Microarray gene dataset often contains high dimensionalities which cause difficulty in clustering or searching for meaningful patterns or important information. Feature selection [1] is an important preprocessing step of data mining that should be applied effectively to select important

S. Das (✉)
Management and Science, Neotia Institute of Technology, Jhinga, Diamond Harbour, South 24-Pargana, Calcutta, West Bengal 743368, India
e-mail: sunanda_srkr@yahoo.co.in

A.K. Das
Department of Computer Science and Technology, Indian Institute of Engineering Science and Technology, Shibpur, Howrah 711103, India
e-mail: akdas@cs.becs.ac.in

L.C. Jain et al. (eds.), *Computational Intelligence in Data Mining - Volume 3*,
Smart Innovation, Systems and Technologies 33, DOI 10.1007/978-81-322-2202-6_58

genes from the gene datasets prior to building efficient, scalable, and robust clustering or classification models for the identification of cancerous samples in the datasets. Important genes characterize the data and can be combined by using appropriate logic to form the rules that represent knowledge hidden within the data set which is very important for cancer detection.

Hence, feature selection [2–4] takes an important role to select only relevant genes by discarding the insignificant one from the data set. Many problems in machine learning involve high dimensional descriptions of input, as a result much research has been carried out on dimensionality reduction [5–8]. So a technique that can reduce dimensionality using information contained within the data set and that preserves the meaning of the features, is highly desirable. In this paper, Rough set theory (RST) [9, 10] is used to discover data dependencies and to reduce the number of features (here, gene) present in a dataset without considering any additional information. Correlations among the genes are computed based on which gene dependency [11] is measured. For each gene g_t a gene dependency $g_t \rightarrow \{g_i\, g_j \ldots g_k\}$ is computed. Then, similarity among any two genes is measured using corresponding dependencies and average similarity of a gene with respect to all other genes is computed and the genes are ranked based on their decreasing similarity factors. Finally, quick reduct algorithm of RST [9, 10] is applied on the genes to generate the reduct based on their rank. The overall work is described below.

The remaining part of this paper is organized as follows: Sect. 2 describes important gene subset selection based on gene dependency and rough set theory. Experimental results and conclusion are demonstrated in Sect. 3. The proposed method and some well known attribute reduction techniques have been applied on several microarray gene datasets for gene selection. The experimental results show the effectiveness of the method.

2 Gene Selection Based on Gene Dependency and Rough Set Theory

Let DS = (U, G, D) be a decision system where U is the finite, non-empty set of samples, known as universe of discourse, $G = \{g_1\, g_2 \ldots g_n\}$ is a finite, non-empty set of genes, and D is the decision class of the samples representing whether the samples are normal or cancerous. Now correlation R_{ij} between two genes g_i and g_j in gene set G is computed using Eq. (1), where, m_i, m_j are mean and σ_i, σ_j are standard deviation of g_i and g_j respectively.

$$R_{\mathrm{ij}} = \frac{\sum(g_{\mathrm{i}} - m_{\mathrm{i}})(g_{\mathrm{j}} - m_{\mathrm{j}})}{n\sigma_{\mathrm{i}}\sigma_{\mathrm{j}}} \tag{1}$$

Using Eq. (1) correlations between all pair of genes are obtained and a correlation matrix (M) is built. A gene dependency set GD is defined where elements in the set are of the form $\{g_i \rightarrow g_{i_1}\ g_{i_2} \cdots g_{i_k}\}$ where, g_i is the gene corresponding to ith row of matrix M and $g_{i_1}, g_{i_2}, \cdots g_{i_k}$ are the genes correlated to g_i. Thus, the set GD contains n gene dependencies each for one gene. From consequent part of gene dependency $\{g_i \rightarrow g_{i_1}\ g_{i_2} \cdots g_{i_k}\}$, g_i is removed as $g_i \rightarrow g_j$ is the trivial dependency, for all i = 1 to n.

2.1 Similarity Measure

Since all genes are not equally important to detect disease so only important genes are kept in the system. The importance of the genes is measured using RST based quick reduct generation algorithm. Here, quick reduct algorithm is modified in the sense that instead of using any arbitrary gene as the initial reduct, most important gene or the core one is selected first. So, before applying rough set based quick reduct algorithm, genes are ranked with respect to their rank, which is measured by similarity value computing for each gene from their gene dependencies available in GD. For any gene dependency $\{g_i \rightarrow g_{i_1}\ g_{i_2} \cdots g_{i_k}\}$ for gene g_i, antecedent g_i is a single gene and consequent $g_{i_1}, g_{i_2}, \cdots g_{i_k}$ is a collection of genes. Similarity factor S_{ij} between two genes g_i and g_j are computed using corresponding gene dependencies $\{g_i \rightarrow g_{i_1}\ g_{i_2} \cdots g_{i_k}\}$ and $\{g_j \rightarrow g_{j_1}\ g_{j_2} \cdots g_{j_l}\}$ using Eq. (2).

$$S_{ij} = \frac{\{g_{i_1}\ g_{i_2} \cdots g_{j_k}\} \cap \{g_{j_1}\ g_{j_2} \cdots g_{j_l}\}}{\{g_{i_1}\ g_{i_2} \cdots g_{i_k}\} \cup \{g_{j_1}\ g_{j_2} \cdots g_{j_l}\}} \tag{2}$$

$S_{ij} = 0$ implies that g_j and g_j are not related to any common gene, hence considered as dissimilar gene and $S_{ij} = 1$ implies that both are related to same set of genes and so they are highly correlated. Obviously, $0 \le S_{ij} \le 1$ and greater its value implies higher the correlation among g_i and g_j. Higher average similarity factor for a gene implies that the gene is related to more number of genes and so it is more important or higher ranked gene.

2.2 Quick Reduct Generation

Genes are ranked in descending order based on their similarity values. Now the highest ranked gene is considered as the most valuable gene or the core gene which is necessary to distinguish the samples in the dataset. To obtain the sufficient set of genes we need to compute the reduct, which is a minimal set of genes for sample identification. In other words, reduct [9, 10] is a sufficient set of genes which by itself fully characterize the knowledge in the dataset. As the decision attribute (i.e., normal and disease samples) is given, attribute dependency value of decision class D on gene set $g = \{g_1\ g_2 \ldots g_r\}$ is computed using Eq. (3), where $POS_g(D)$ is the positive region of D with respect to g.

$$\gamma_g(D) = \frac{|POS_g(D)|}{|U|} \tag{3}$$

Thus, a reduct R can be thought of as a sufficient set of genes if $\gamma_R(D) = \gamma_G(D)$, in other words, R is a reduct if the dependency of decision class D on subset of genes R is exactly equal to that of D on whole gene set G.

The method first considers the most important gene (i.e., core) computed in Sect. 2.1 as the initial reduct R and then in each iteration it adds next most important gene until the condition $\gamma_R(D) = \gamma_G(D)$ is satisfied and finally gives the actual reduct R. If after considering all the genes, the condition is not satisfied then the latest attributes in set R is considered as approximate reduct. The detailed algorithm is described below.

Algorithm: Gene_Subset_Selection(U, G, D)
Input: U, the set of all samples consist of G genes where, genes are already arranged based on their rank.
Input: D, the sample class with values normal and disease
Output: R, a necessary and sufficient gene subset

1. G = $\{g_{i_1} g_{i_2} \cdots g_{i_n}\}$ be the n genes arranged in descending order of their rank obtained using section 2.1
2. R = $\{g_{i_1}\}$ // core consider as initial reduct
3. For all g ∈ (G − R)
4.   If $\gamma_{R\cup\{g\}}(D) > \gamma_R(D)$
5.   R = R ∪ {g}
6.   If $\gamma_R(D) = \gamma_G(D)$
7.   Return *R*
8. Return R

3 Result and Discussions

Experimental studies presented here provide an evidence of effectiveness of proposed gene selection technique. Experiments were carried out on large number of different kinds of microarray data, few of them publicly available as training and test dataset are summarized below:

Colon Cancer dataset: The colon cancer data contains 62 samples collected from colon cancer patients. Among them, 40 tumor biopsies are from tumors (labeled as "negative") and 22 normal (labeled as "positive") biopsies are from healthy parts of the colons of the same patients. 2,000 out of around 6,500 genes were selected based on the confidence in the measured expression levels. The test set consists of 22 tissues of which 8 are normal and 14 tumor samples. The raw data available at http://microarray.princeton.edu/oncology/affydata/index.html.

Leukemia (ALL vs. AML) dataset: Training dataset consists of 38 bone marrow samples (27 ALL and 11 AML), over 7,129 human genes. Whereas the test dataset contains total 34 number of samples (20 ALL and 14 AML). The raw data is available at http://www-genome.wi.mit.edu/cgibin/cancer/datasets.cgi.

Lung Cancer dataset: Training dataset contains 16 samples labeled as "MPM" and 16 samples labeled as "ADCA" with around 12,533 genes. Total 149 samples (15 MPM and 134 ADCA) are available in this dataset. The raw data available at http://www.chestsurg.org/microarray.htm.

Prostate Cancer dataset: Training dataset contains 52 samples labeled as "relapse" and 50 patients having remained relapse free labeled as "non-relapse" prostate samples with around 12,600 genes. The test dataset has 25 tumor samples and 9 non tumor/normal samples. The raw data available at http://wwwgenome.wi.mit.edu/mpr/prostate.

At first, all the numeric attributes of both training and test samples are discretized by ChiMerge [12] discretization algorithm. Then the proposed method is applied on training sets to compute reduct (i.e., important gene subset) for different kinds of microarray gene datasets (cancerous data). For example, in the Leukemia dataset, five genes X95735_at, M55150_at, M31166_at, M27891_at, U46499_at are identified by the proposed method. Now, the test sample Leukemia cancer dataset is reduced by selecting only above five genes for all 34 test samples. Then the original test dataset and reduced test set are applied on various classifiers to measure their accuracy, as described in Fig. 1. From Fig. 1, it is observed that accuracy of classifiers applied on reduced Leukemia test dataset is higher than that on original test dataset, which shows the usefulness of the method.

Also to demonstrate the effectiveness of the proposed method, the training Leukemia dataset is applied on two popular feature selection methods CON [13] and CFS [14] and test set is reduced accordingly before feed it in various classifiers

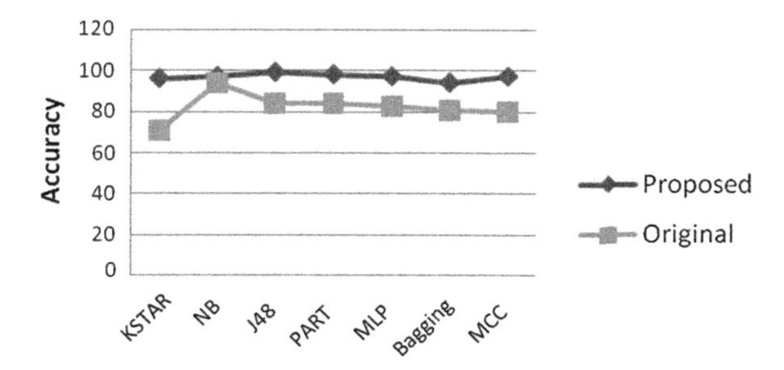

Fig. 1 Classification accuracies of original and reduced dataset for Leukemia data

to measure the accuracies. The accuracy chart for all three methods are plotted in Fig. 2, from where it is noticed that our method provides better accuracies for all classifiers compare to other two methods.

Similar process is followed for all other datasets and the results are listed in Table 1. Thus, Table 1 lists the accuracy of the classifiers on reduced test sample datasets. The methods CFS and CON and the classifiers are run using WEKA tool [15]. Second column of the table contains the method followed by number of gene selected by the method.

In all four datasets, the proposed method selects either minimum or very close to minimum number of genes and at the same time average accuracy of the classifiers are more impressive in case of proposed method than other methods. So we may conclude that final selected genes of the gene dataset are most important than other

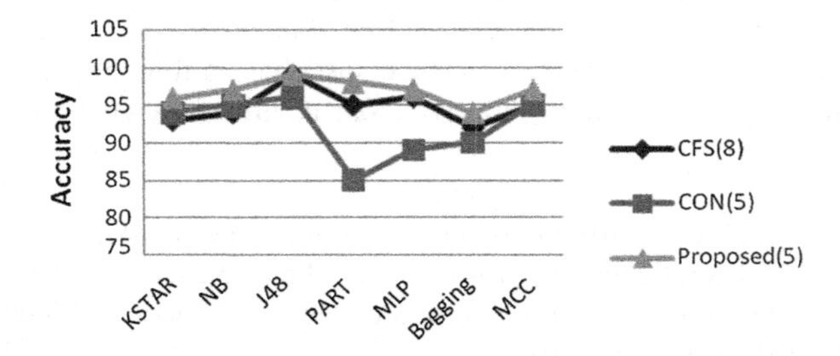

Fig. 2 Classification accuracy between existing and proposed method for Leukemia data

Table 1 Accuracies of various classifiers on reduced cancer data sets

Dataset	Methods (# selected genes)	Classifiers							
		KSTAR	NB	J48	PART	MLP	Bagging	MCC	Average accuracy (%)
Colon cancer	CFS(25)	96	95	77	74	93	85	87	87
	CON(5)	91	91	87	90	90	91	93	91
	Proposed(6)	**92**	**91**	**87**	**91**	**93**	**91**	**90**	**92**
Leukemia	CFS(8)	93	94	99	95	96	92	95	95
	CON(5)	94	95	96	85	89	90	95	92
	Proposed(5)	**96**	**97**	**99**	**98**	**97**	**94**	**97**	**97**
Lung cancer	CFS(5)	80	88	80	56	80	81	80	78
	CON(6)	93	94	94	65	92	91	93	89
	Proposed(4)	**94**	**96**	**94**	**85**	**93**	**92**	**93**	**92**
Prostate cancer	CFS(9)	92	94	94	96	91	95	96	94
	CON(6)	85	89	81	79	77	82	81	82
	Proposed(6)	**94**	**95**	**94**	**96**	**93**	**95**	**97**	**95**

genes with respect to classification accuracy. This implies that the selected genes play an important role in the samples to predict whether it is a normal sample or cancerous one.

4 Conclusion

The work extracts various cancerous genes from the microarray data. It is a novel concept of important gene selection in the field of bioinformatics where the concept of rough set theory is explored. The method is very simple and efficient in the sense that it takes $O(n^2)$ time complexity for finding important gene subset. The only demerit is that the reduct obtains in some cases may not be the exact reduct for which accuracy of the system may degrade. But, it happens as reduct generation itself is a NP-hard problem. So, as a future work, the method may be compared with some evolutionary algorithm based gene selection algorithm with respect to both time and classifier accuracy.

References

1. Chandrashekar, G., Sahin, F.: A survey on feature selection methods. Comput. Electr. Eng. **40** (1), 16–28 (2014)
2. Velayutham, C., Thangavel, K.: Unsupervised quick reduct algorithm using rough set theory. J. Electr. Sci. Technol. **9**(3), 193–201 (2011)
3. Lazar, C., Taminau, J., Meganck, S., Steenhoff, D., Coletta, A., Molter, C., de Schaetzen, V., Duque, R., Bersini, H., Nowe, A.: A survey on filter techniques for feature selection in gene expression microarray analysis. IEEE/ACM Trans. Comput. Biol. Bioinform. **9**(4), 1106–1119 (2012)
4. Dash, M., Liu, H.: Feature selection for classification. Intell. Data Anal. **1**(3), 131–156 (1997)
5. Kira, K., Rendell, L.A.: The feature selection problem: traditional methods and a new algorithm. In: Proceedings of Ninth National Conference on Artificial Intelligence, pp. 129–134 (1992)
6. Langley, P.: Selection of relevant features in machine learning. In: Proceedings on AAAI Fall Symposium Relevance, pp. 1–5 (1994)
7. Liu, H., Motoda, H.: Feature Extraction, Construction and Selection: A Data Mining Perspective (Kluwer International Series in Engineering & Computer Science). Academic Publishers, New York (1998)
8. Miller A.J., Hall, C.: Subset Selection in Regression (1990)
9. Pawlak, Z.: Rough Sets: Theoretical Aspects of Reasoning About Data. Kluwer Academic Publishing, Norwell (1991)
10. Polkowski, L.: Rough Sets: Mathematical Foundations. Advances in Soft Computing. Physica Verlag, Heidelberg (2002)
11. Baixeries, J.: A formal concept analysis framework to mine functional dependencies. In: Proceeding of the Workshop on Mathematical Methods for Learning (2004)
12. Kerber, R., ChiMerge.: Discretization of Numeric Attributes. In: Proceedings of AAAI-92, Ninth International Conference on Artificial Intelligence, AAAI-Press, pp. 123–128 (1992)

13. Yu, L., Liu, H.: Efficient feature selection via analysis of relevance and redundancy. J. Mach. Learn. Res. **5**, 1205–1224 (2004)
14. Hall, M.A.: Correlation-based feature selection for machine learning. The University of Waikato, New Zealand (1999)
15. Hall, M., Frank, E., Holmes, G., Pfahringer, B., Reutemann, P., Witten, I.H.: The WEKA data mining software: an update. SIGKDD Explor. **11**(1), 10–18 (2009)

An Ordering Policy with Time-Proportional Deterioration, Linear Demand and Permissible Delay in Payment

Trailokyanath Singh and Hadibandhu Pattanayak

Abstract In classical inventory models, demand is considered as constant, while in the practical cases the demand changes with time. This paper focuses an economic order quantity (EOQ) model for a deteriorating item with time-proportional deterioration rate and time-dependent linear demand rate under permissible delay in payment. The purpose of this study is to find the EOQ for minimizing the total variable cost. Finally, numerical examples are cited to illustrate the model.

Keywords Deteriorating item · EOQ · Linear demand · Permissible delay in payment · Time-proportional deterioration

1 Introduction

The deterioration of items is a natural phenomenon in any inventory system in our day to day life. Generally, deterioration is defined as decay or damage, spoilage and the loss of the utility of the product. In the past decades, much more researches on inventory models on deteriorating items were carried out by several researchers. In formulating the inventory models on deteriorating items, the following two facts are considered for growing interest, one is deterioration rate of the items and other part is the variation in demand. There are many examples like electronic goods, blood, food items, pharmaceutical, alcohol, volatile liquids which decrease during their storage period because of deterioration. Whitin [23] first considered the fashion items deteriorating after a prescribed storage period. Ghare and Schrader [5] were

T. Singh (✉)
Department of Mathematics, C. V. Raman College of Engineering, Bhubaneswar 752054, Odisha, India
e-mail: trailokyanaths108@gmail.com

H. Pattanayak
Department of Mathematics, KIIT University, Bhubaneswar 751024, Odisha, India
e-mail: h.pattnayak@gmail.com

L.C. Jain et al. (eds.), *Computational Intelligence in Data Mining - Volume 3*,
Smart Innovation, Systems and Technologies 33, DOI 10.1007/978-81-322-2202-6_59

the earliest researchers who developed an inventory model on deteriorating items for exponentially decay in which inventory is not only depleted by the demand rate, but also by direct spoilage, physical depletion or deterioration. Later, Ghare and Schrader's [5] work was extended by Covert and Philip [3] by considering Weibull distribution deterioration. Philip [11] developed the model by considering a three parameter Weibull distribution deterioration to represent the time of deterioration. The literature surveys by Raafat [14], Shah and Shah [16], Goyal and Giri [7] cite up-to-date reviews of deteriorating inventory models.

In the real life situation, it is observed that the demand rate of items may vary with time or with price or with the instantaneous level of inventory displayed in various market and super markets. Silver and Meal [18], the first researchers, who studied a heuristic approach to determine EOQ in case of a time varying demand rate. A full analytic solution of the inventory replenishment problem with a linearly time-dependent demand over a finite-time horizon was studied by Donaldson [4]. After that, the models with time-dependent demand rates have been studied by several researchers like Ritchie [12, 13], Mitra et al. [9] etc. Singh and Pattnayak [21] developed a two-warehouse inventory model with linear demand under conditionally permissible delay in payment. An inventory model with Weibull distribution deterioration and trapezoidal type demand was studied by Singh and Pattnayak [20].

In the business scenario, it is seen that payment is made to the supplier for items just after getting the consignment. However, it is frequently observed that a customer is allowed some grace period before settling the account with the supplier or the producer. During this grace period, no interest is charged, but beyond this period, the supplier will charge interest for the further period. Before settling the account with the supplier, the customer may sell the goods, accumulate revenues on the sales and earn interest on that revenue. In an economic sense, this makes the customer delay payment of the replenishment account up to the last day of the settlement period allowed by the supplier or the producer. This concept of payment is known as permissible delay in payments. In a competitive market, the supplier offers different delay periods to encourage the retailer to order more quantities. The permissible delay in payment produces the advantages to attract more customers, to sale more items and to reduce the holding costs to the suppliers. In this respect, Goyal [6], the first researcher, who established an EOQ model when a supplier offers its retailer a permissible delay in payments. Later, Aggarwal and Jaggi [2], Shinn et al. [17] etc. studied EOQ models where delay in payment is permissible. A generalized deterministic EOQ model with delays in payments and price-discount offers was developed by Sana and Chaudhuri [15]. Khanra et al. [8] studied an EOQ model for a deteriorating item with time dependent quadratic demand under permissible delay in payment. An inventory model of decaying-lot depleted by declining market demand and extended with discretely variable holding costs by Tyagi [22]. Wu and Zhao [24] studied the supplier-retailer inventory system with inventory-dependent and linear-trend demand. Pal et al. [10] studied a production inventory model with three stage trade credit policy. Singh and Pattnayak [19] studied the EOQ model with Weibull distribution deterioration under the permissible delay in payment.

The proposed inventory model is based on deteriorating items like fruits and vegetables whose deterioration rate increases with time. Among the various time-varying demands in EOQ models, the more realistic demand approach is to consider linear demand with time-proportional deterioration. For settling the account, mathematical models are developed by considering the credit period is either less than or equal to or greater than the cycle time. Examples are provided which stands in support of the developed model. In this study, an EOQ model for a deteriorating item in which time-dependent linear demand and time-proportional deterioration rate under permissible delay in payment is considered in order to make the model relevant and more realistic. An algorithm is provided to derive the optimal replenishment policy when the total variable cost is minimized. Finally, the results have been validated by numerical examples.

2 Assumptions

The model is developed under the following assumptions.

(i) The inventory system involves only one item.
(ii) The rate of replenishment is assumed to be infinite and the lead time is zero.
(iii) The demand rate is deterministic and is a linear function of time.
(iv) The deterioration rate of the item is considered as time-proportional.
(v) Shortage is not permitted.
(vi) There is no repair or replenishment of deteriorating items during the period under consideration.
(vii) The inventory system is considered over an infinite time horizon.

3 Notations

The following notations are used throughout the paper.

(i) $I(t)$: the inventory level at time t.
(ii) $D(t)$: the time-dependent demand rate at time t and is given by $D(t) = a + bt$ where a is the demand at time zero and b is the rate with which the demand rate increases.
(iii) $\theta(t)$: the time-proportional deterioration rate and is given by $\theta(t) = \theta t$, $0 < \theta \ll 1$.
(iv) P: the purchase cost per unit item.
(v) C_O: the ordering cost of the inventory (Rs./order).
(vi) C_H: the inventory holding cost (Rs./unit/time) excluding interest charges.

(vii) I_C: the interest charged (Rs./year) in stock by the supplier.
(viii) I_E: the interest earned (Rs./year) by the supplier.
(ix) t_1: the allowable permissible delay period (in year) in settling the accounts.
(x) Q: the optimum order quantity.
(xi) T: the fixed length of each ordering cycle.

4 Mathematical Formulation and Solution of the Model

Consider an inventory system in which the depletion of the inventory occurs due to variable rate of demand and time-proportional deterioration rate only. During the interval [0, T], the instantaneous inventory level $I(t)$ at any time t will satisfy the following differential equation

$$\frac{dI(t)}{dt} + \theta(t)I(t) = -D(t), \quad 0 \le t \le T \tag{1}$$

with the boundary conditions $I(0) = Q$ and $I(T) = 0$, $\theta(t) = \theta t, 0 < \theta \ll 1$ and $D(t) = a + bt, a, b > 0$.

The solution of Eq. (1) is given by

$$I(t) = \left[a(T-t) + \frac{b(T^2-t^2)}{2} + \frac{\theta}{2}\left(\frac{a(T^3-t^3)}{3} + \frac{b(T^4-t^4)}{4}\right)\right]e^{\frac{-\theta t^2}{2}}, \quad 0 \le t \le T, \tag{2}$$

(neglecting the higher power of θ as $0 < \theta \ll 1$)
and the order quantity is given by

$$I(0) = Q = aT + \frac{b}{2}T^2 + \frac{\theta}{2}\left(\frac{a}{3}T^3 + \frac{b}{4}T^4\right). \tag{3}$$

The total variable cost consists of the following components:
(i) The ordering cost (OC):

$$OC = C_O. \tag{4}$$

(ii) The deterioration cost (DC):
The total number of deteriorated items (DI) during the time [0, T] is

$$DI = \int_0^T D(t)dt = \frac{\theta}{2}\left(\frac{a}{3}T^3 + \frac{b}{4}T^4\right).$$

Thus, the deterioration cost (DC) during the time [0, *T*] is

$$DC = \frac{P\theta}{2}\left(\frac{a}{3}T^3 + \frac{b}{4}T^4\right). \tag{5}$$

(iii) The inventory holding cost (IHC):

The total inventory holding cost (IHC) during the time [0, *T*] is given by

$$IHC = h\int_0^T I(t)dt = h\left[\frac{a}{2}T^2 + \frac{b}{3}T^3 + \frac{\theta}{6}\left(\frac{a}{2}T^4 + \frac{2b}{5}T^5\right)\right], \tag{6}$$

where $h = PC_H$, (neglecting the higher power of θ as $0 < \theta \ll 1$).

(iv) The interest charged (IP_1) and earned (IE_1):

Case (I): ($t_1 < T$).

The total variable cost per cycle is comprised of the sum of the ordering cost, deterioration cost, inventory holding cost and interest charged minus the interest earned.

The interest charged (IP_1) during the time [0, *T*] for the inventory not being sold after due time ($T - t_1$) is

$$\begin{aligned} IP_1 &= PI_C\int_{t_1}^T I(t)dt \\ &= PI_C\left[\begin{array}{l} \frac{a}{2}\left(T^2 - 2Tt_1 + t_1^2\right) + \frac{b}{6}\left(2T^3 - 3T^2t_1 + t_1^3\right) \\ + \frac{\theta}{120}\left[\begin{array}{l} 5a\left(2T^4 - 4T^3t_1 + t_1^4 + 4t_1^3T - 3t_1^4\right) \\ + b\left(8T^5 - 15T^4t_1 + 3t_1^5 + 10T^2t_1^3 - 6t_1^5\right) \end{array}\right] \end{array}\right], \end{aligned} \tag{7}$$

(neglecting the higher power of θ as $0 < \theta \ll 1$).

The interest earned (IE_1) during the time [0, *T*] is

$$IE_1 = PI_E\int_0^T tD(t)dt = PI_E\left(\frac{a}{2}T^2 + \frac{b}{3}T^3\right). \tag{8}$$

From Eqs. (4)–(8), the total variable cost (TVC_1) per cycle per unit time is defined as

$$TVC_1(T) = \frac{1}{T}[OC + DC + IHC + IC_1 - IE_1]$$
$$= \frac{1}{T}\left[C_0 + \frac{P\theta}{2}\left(\frac{a}{3}T^3 + \frac{b}{4}T^4\right) + h\left[\frac{a}{2}T^2 + \frac{b}{3}T^3 + \frac{\theta}{6}\left(\frac{a}{2}T^4 + \frac{2b}{5}T^5\right)\right]\right]$$
$$+ \frac{PI_C}{T}\left[\begin{array}{l}\frac{a}{2}(T^2 - 2Tt_1 + t_1^2) + \frac{b}{6}(2T^3 - 3T^2t_1 + t_1^3) \\ + \frac{\theta}{120}\left[\begin{array}{l}5a(2T^4 - 4T^3t_1 + t_1^4 + 4t_1^3T - 3t_1^4) \\ + b(8T^5 - 15T^4t_1 + 3t_1^5 + 10T^2t_1^3 - 6t_1^5)\end{array}\right]\end{array}\right] - PI_E\left(\frac{aT}{2} + \frac{b}{3}T^2\right). \quad (9)$$

Now for minimizing ($TVC_1(T)$) per unit time, the optimal value of t_1 can be obtained by solving

$$\frac{d(TVC_1)}{dT} = (a + bT)\left[h\left(1 + \frac{\theta T}{3}\right) + \frac{\theta PT}{2}\right]$$
$$+ \frac{(a + bT)}{T}\left[PI_C\left[\begin{array}{l}T - t_1 + \\ \frac{\theta}{6}(2T^4 - 4T^3t_1 + t_1^4 + 4t_1^3T - 3t_1^4)\end{array}\right] - PI_ET\right] - \frac{(TVC_1)}{T} = 0, \quad (10)$$

provided it satisfies the sufficient condition $\frac{d^2(TVC_1)}{dT^2} > 0, \forall T$.

Case (II): ($t_1 > T$).
The total variable cost per cycle is comprised of the sum of the ordering cost, deterioration cost and inventory holding cost minus the interest earned.

The interest earned (IE_2) during the time [0, T] is

$$IE_2 = PI_E\left[\int_0^T tD(t)dt + \int_0^T (t_1 - T)D(t)dt\right]$$
$$= PI_E\left[\frac{a}{2T^2} + \frac{b}{3T^3} + (t_1 - T)\left(aT + \frac{b}{2T^2}\right)\right]. \quad (11)$$

From Eqs. (4), (5), (6) and (11), the total variable cost (TVC_2) per cycle per unit time is defined as

$$TVC_2(T) = \frac{1}{T}[OC + DC + IHC - IE_2]$$
$$= \frac{1}{T}\left[\begin{array}{l}C_0 + \frac{P\theta}{2}\left(\frac{a}{3}T^3 + \frac{b}{4}T^4\right) + h\left[\frac{a}{2}T^2 + \frac{b}{3}T^3 + \frac{\theta}{6}\left(\frac{a}{2}T^4 + \frac{2b}{5}T^5\right)\right] \\ - PI_E\left[\frac{a}{2}T^2 + \frac{b}{3}T^3 + (t_1 - T)\left(aT + \frac{b}{2}T^2\right)\right]\end{array}\right]. \quad (12)$$

Similarly, for minimizing $(TVC_1(T))$ for a given value of t_1 can be obtained by solving

$$\frac{d(TVC_2)}{dT} = \frac{1}{T}\left[\begin{array}{l} T(a+bT)\left[h\left(1+\frac{\theta}{3}T^2\right)+\frac{\theta PT}{2}\right] \\ -PI_E\left[\frac{b}{2}T^2+(t_1-T)(a+bT)\right]-(TVC_2) \end{array}\right] = 0. \quad (13)$$

provided it satisfies the sufficient condition $\frac{d^2(TVC_2)}{dT^2} > 0, \forall T$.

Case (III): $(t_1 = T)$
In this case, the total variable cost $TVC_3(t_1)$ is obtained from either Eq. (9) or (12) by substituting $(t_1 = T)$ and is given by

$$TVC_3(t_1) = \frac{1}{t_1}\left[\begin{array}{l} C_0+\frac{P\theta}{2}\left(\frac{a}{3}t_1^3T^3+\frac{b}{4}t_1^4\right) \\ +h\left[\frac{a}{2}t_1^2+\frac{b}{3}t_1^3+\frac{\theta}{6}\left(\frac{a}{2}t_1^4+\frac{2b}{5}t_1^5\right)\right]-PI_E\left(\frac{a}{2}t_1^2+\frac{b}{3}t_1^3\right) \end{array}\right]. \quad (14)$$

The following algorithm is proposed to determine the optimal cost and EOQ unless $(T = t_1)$.

Algorithm

Step I: Find T_1^* from Eq. (10). If $t_1 < T_1^*$, then determine $TVC_1(T_1^*)$ from Eq. (9).

Step II: Find T_2^* from Eq. (13). If $t_1 > T_2^*$, then determine $TVC_1(T_2^*)$ from Eq. (12).

Step III: If $T_1^* > t_1 > T_2^*$ is satisfied, then go to step IV; otherwise, go to step V.

Step IV: Compare both the total variable costs $TVC_1(T_1^*)$ and $TVC_1(T_2^*)$ and write the respective minimum total variable cost.

Step V: If $t_1 < T_1^*$ is satisfied, but the condition $t_1 < T_2^*$ holds, then $TVC_1(T_1^*)$ is the required minimum total variable cost; else if, $t_1 < T_1^*$ is satisfied, but the condition $t_1 < T_2^*$ holds, then $TVC_2(T_2^*)$ is the required minimum total variable cost.

Step VI: Finally, determine the respective EOQ $Q(T_1^*)$ or $Q(T_2^*)$.

5 Numerical Examples

In this section, we provide several numerical examples to illustrate the above theory.

Example 1 Let us consider a system with the following parameter values: D_0 = Rs. 220/order a = 1,200 units/year, b = 120 units/year, P = Rs. 20/unit, h = Rs. 0.12/year, θ = 0.4, I_p = 0.16/year, I_E = 0.13/year and t_1 = 0.3/year.

According to the algorithm, the optimal solutions of the system are $T_1^* = 0.45142$ year and $TVC_1(T_1^*) = \text{Rs.}\,235.16$ respectively which are obtained from Case I. Similarly, the optimal solutions of the system are $T_2^* = 0.291787$ year and $TVC_2(T_2^*) = \text{Rs.}\,424.68$ respectively which are obtained from Case II. Here both the Cases I and II hold together as $T_1^* > t_1 > T_2^*$. Now comparing both costs $TVC_1(T_1^*)$ and $TVC_1(T_2^*)$, we obtained $TVC_1(T_1^*) < TVC_2(T_2^*)$. Hence the optimum length of cycle period and the optimum cost are $T_1^* = 0.45142$ year and $TVC_1(T_1^*) = \text{Rs.}\,235.16$ respectively. The EOQ in this case is given by $Q(T_1^*) = Rs.\,561.54$.

Example 2 Let us consider a system with the following parameter values: $D_0 =$ Rs. 220/order a = 1,200 units/year, b = 120 units/year, P = Rs. 20/unit, h = Rs. 0.12/year, $\theta = 0.4$, $I_p = 0.16$/year, $I_E = 0.13$/year and $t_1 = 0.15$/year.

According to the algorithm, the optimal solutions of the system are $T_1^* = 0.387732$ year and $TVC_1(T_1^*) = \text{Rs.}\,514.32$ respectively which are obtained from Case I. Similarly, the optimal solutions of the system are $T_2^* = 0.290691$ year and $TVC_2(T_2^*) = \text{Rs.}\,899.49$ respectively which are obtained from Case II. Here the Case I holds only as $T_1^* > T_2^* > t_1$. Thus, the optimum length of cycle period and the optimum cost are $T_1^* = 0.387732$ year and $TVC_1(T_1^*) = \text{Rs.}\,899.49$ respectively. The EOQ in this case is given by $Q(T_1^*) = \text{Rs.}\,479.1$.

6 Conclusions

To be precise, an EOQ model for the deteriorating item has been illustrated for the determination of optimal order quantity and the total minimum cost with time-dependent linear demand and the time-proportional deterioration. The proposed inventory model is based on deteriorating items like fruits and vegetables whose deterioration rate increases with time. The model considered above is suited for items having time-proportional deterioration rate, earlier models have considered items having constant deterioration rate. The practical aspects of inventory management like opportunity cost and the effect of permissible delay in payment are also considered. The total cost obtained then can be used to obtain an average inventory variable cost, which can be optimized using calculus techniques.

There is an ample scope for further extension for research. Here, the suggested model can be extended for items with stock dependent demand, price dependent demand or power demand. Also, we could extend the model to incorporate some more features like two-ware houses with multi items in the system, quantity discount and permissible delay in payment with cash discount.

Acknowledgments The authors are most grateful to two anonymous referees for their constructive comments.

References

1. Aggarwal, S.P.: A note on an order-level model for a system with constant rate of deterioration. Opsearch **15**, 184–187 (1978)
2. Aggarwal, S.P., Jaggi, C.K.: Ordering policies of deteriorating items under permissible delay in payments. J. Oper. Res. Soci. **36**, 658–662 (1995)
3. Covert, R., Philip, G.C.: An EOQ model for items with Weibull distribution. AIIE Trans. **5**, 323–326 (1973)
4. Donaldson, W.A.: Inventory replenishment policy for a linear trend in demand-an analytical solution. Oper. Res. Q. **28**, 663–670 (1977)
5. Ghare, P.M., Schrader, G.H.: A model for exponentially decaying inventory system. Int. J. Prod. Res. **21**, 449–460 (1963)
6. Goyal, S.K.: EOQ under conditions of permissible delay in Payments. J. Oper. Res. Soci. **36**, 335–338 (1985)
7. Goyal, S.K., Giri, B.C.: Recent trends in modeling of deteriorating inventory. Eur. J. Oper. Res. **134**, 1–16 (2001)
8. Khanra, S., Ghosh, S.K., Chaudhuri, K.S.: An EOQ model for a deteriorating item with time dependent quadratic demand under permissible delay in payment. Appl. Math. Comput. **218**, 1–9 (2011)
9. Mitra, A., Fox, J.F., Jessejr, R.R.: A note on determining order quantities with a linear trend in demand. J. Oper. Res. Soci. **35**, 141–144 (1984)
10. Pal, B., Sana, S.S., Chaudhuri, K.: Three stage trade credit policy in a three layer supply chain-a production-inventory model. Int. J. Syst. Sci. **45**(9), 1844–1868 (2014)
11. Philip, G.C.: A generalized EOQ model for items with Weibull distribution. AIIE Trans. **6**, 159–162 (1974)
12. Ritchie, E.: Practical inventory replenishment policies for a linear trend in demand followed by a period of steady demand. J. Oper. Res. Soci. **31**, 605–613 (1980)
13. Ritchie, E.: Stock replenishment quantities for unbounded linear increasing demand: an interesting consequence of the optimal policy. J. Oper. Res. Soc. **36**, 737–739 (1985)
14. Raafat, F.: Survey of literature on continuously deteriorating inventory models. J. Oper. Res. Soc. **42**(1), 27–37 (1991)
15. Sana, S.S., Chaudhuri, K.S.: A deterministic EOQ model with delays in payments and price discount offers. Eur. J. Oper. Res. **184**, 509–533 (2008)
16. Shah, N.H., Shah, Y.K.: Literature survey on inventory model for deteriorating items. Econom. Ann. (Yugoslavia) XLIV, 221–237 (2000)
17. Shinn, S.W., Hwang, H.P., Sung, S.: Joint price and lot size determination under condition of permissible delay in payments quantity discounts for freight cost. Eur. J. Oper. Res. Soc. **91**, 528–542 (1996)
18. Silver, E.A., Meal, H.C.: A simple modification of the EOQ for the case of a varying demand rate. Prod. Inventory Manag. **10**(4), 52–65 (1969)
19. Singh, T., Pattnayak, H.: An EOQ model for a deteriorating item with varying time-dependent constant demand and Weibull distribution deterioration under permissible delay in payment. J. Orissa Math. Soc. **31**(2), 61–76 (2012)
20. Singh, T., Pattnayak, H.: An EOQ inventory model for deteriorating items with varying trapezoidal type demand rate and Weibull distribution deterioration. J. Inf. Optim. Sci. **34**(6), 341–360 (2013)
21. Singh, T., Pattnayak, H.: A two-warehouse inventory model for deteriorating items with linear demand under conditionally permissible delay in payment. Int. J. Manag. Sci. Eng. Manag. **9** (2), 104–113 (2014)
22. Tyagi, A.: An optimization of an inventory model of decaying-lot depleted by declining market demand and extended with discretely variable holding costs. Int. J. Ind. Eng. Comput. **5**(1), 71–86 (2014)

23. Whitin, T.M.: The Theory of Inventory Management. Princeton University Press, Princeton (1953)
24. Wu, C., Zhao, Q.: Supplier-retailer inventory coordination with credit term for inventory-dependent and linear-trend demand. Int. Trans. Oper. Res. **21**, 797–818 (2014)

Computation of Compactly Supported Biorthogonal Riesz Basis of Wavelets

Mahendra Kumar Jena and Manas Ranjan Mishra

Abstract In this paper, we compute compactly supported biorthogonal Riesz basis of wavelets. We solve the Bezout equation resulting from biorthogonality of the scaling function with its dual in a simple and algebraic way. We provide some examples of the biorthogonal wavelets showing their detail construction. Two algorithms for their construction are also given.

Keywords Wavelet · Biorthogonal · Subdivision algorithm

1 Introduction

The term wavelet was introduced by Grossmann and Morlet in their work on constant shapes [11]. It denotes a univariate function ψ, defined on R, which when subjected to the fundamental operations of dyadic dilations and integer translations $f_{j,k}$, yields an orthogonal basis of $L_2(R)$. That is, the functions $\psi_{j,k} := 2^{\frac{j}{2}}\psi(2^j \cdot -k),\ j,k \in Z$, form a complete orthonormal basis for $L_2(R)$. One of the pioneer work in this field is on orthogonal wavelets by Daubechies [8] (see also [10]). For more details on wavelets see [2, 4, 9, 10].

In many numerical applications, instead of orthogonality it is desired that the translated dilates $\psi_{j,k}$ $j,k \in Z$ form a Riesz basis for $L_2(R)$. Note that, an orthogonal wavelet basis is itself a Riesz basis. Some other important Riesz basis of $L_2(R)$ are semiorthogonal wavelet bases of Chui and Wang [3] (*CW-wavelet*), biorthogonal

M.K. Jena (✉)
Department of Mathematics, Veer Surendra Sai University of Technology, Burla, Sambalpur 768018, Odisha, India
e-mail: maheny1@yahoo.com

M.R. Mishra
Department of Mathematics, O.P. Jindal Institute of Technology, Raigarh, Chhattisgarh, India
e-mail: manas.mishra@opjit.edu.in

L.C. Jain et al. (eds.), *Computational Intelligence in Data Mining - Volume 3*,
Smart Innovation, Systems and Technologies 33, DOI 10.1007/978-81-322-2202-6_60

wavelet bases of Cohen et al. [5] (*CDF-wavelet*), and compactly supported wavelet bases of Jia et al. [12] (*JWZ-wavelet*).

For given function *f*, let $f_{j,k}$ denotes the function $2^{\frac{j}{2}}f(2^j \cdot -k)$. Let $\phi, \tilde{\phi}$, ψ *and* $\tilde{\psi}$ be such that

$V_j := span\left\langle \phi_{j,k}; j,k \in Z\right\rangle$, $\tilde{V}_j := span\left\langle \tilde{\phi}_{j,k}; j,k \in Z\right\rangle$,

$W_j := span\left\langle \psi_{j,k}; j,k \in Z\right\rangle$ and $\tilde{W}_j := span\left\langle \tilde{\psi}_{j,k}; j,k \in Z\right\rangle$.

The functions ϕ and $\tilde{\phi}$ are called the scaling function and the dual scaling function, respectively. Similarly, ψ and $\tilde{\psi}$ are called the wavelet and the dual wavelet, respectively. The biorthogonality condition is

$$\left\langle \psi_{j,k}, \tilde{\psi}_{j',k'} \right\rangle = \delta_{j,j'}\delta_{k.k'} \tag{1}$$

In this case, the pair $(\psi, \tilde{\psi})$ is called a biorthogonal wavelet pair.

We give here a brief discussion on *CW, JWZ* and *CDF* wavelets. In all these cases the scaling functions as well as the wavelets are compactly supported, and are symmetric or antisymmetric. In *CDF* and *CW* cases we have the dual scaling functions and the dual wavelets. While they are compact in *CDF* cases, they are noncompactly supported in *CW* cases. The biorthogonality property (1) is clear in case of *CDF* and *CW*-wavelets while this property is not established in case of *JWZ*-wavelets which are recently introduced. A characterization of the wavelet Riesz basis is given recently by Han and Jia [13]. Generalization of biorthogonal wavelets to local fields is given recently by Behera and Jahan [1].

The pair $\psi, \tilde{\psi}$ satisfy the two scale relations

$$\hat{\phi}(\xi) = m_0\left(\frac{\xi}{2}\right)\hat{\phi}\left(\frac{\xi}{2}\right) \text{ and } \hat{\tilde{\phi}}(\xi) = \tilde{m}_0\left(\frac{\xi}{2}\right)\hat{\tilde{\phi}}\left(\frac{\xi}{2}\right) \tag{2}$$

for some trigonometric polynomials $m_0(\xi)$ and $\tilde{m}_0(\xi)$ which are called as transfer functions of ϕ and $\tilde{\phi}$, respectively. The necessary and sufficient condition [5] for exact reconstruction is

$$m_0(\xi)\tilde{m}_0(\xi) + m_0(\xi + \pi)\tilde{m}_0(\xi + \pi) = 1 \tag{3}$$

This is also one of the necessary conditions [5] for biorthogonality of the pair $(\psi, \tilde{\psi})$. It is also shown [5] that for suitable choice of x, the above equation reduces to *Bezout equation*

$$(1 - x)^k P(x) + x^k P(1 - x) = 1 \tag{4}$$

where *P(x)* is a polynomial of degree $k - 1$ (this is briefly outlined in Sect. 2). The paper consists of four sections. In Sect. 1, we give an introduction to *CDF* wavelets. In Sect. 2, we briefly outline the construction of *CDF* wavelets. Moreover, a new procedure to compute these wavelets is also given in Sect. 2. Two algorithms for this construction and a case study on these wavelets are given in Sect. 3. Conclusion is given in Sect. 4.

2 Biorthogonal Riesz Bases of Wavelets

In this section, we briefly outline the construction of *CDF* wavelets. These wavelets form Riesz bases of $L_2(R)$. We also provide a simple procedure for the computation of these wavelets.

The following definition of Riesz basis is taken from [4].

Definition 1 The following two conditions define a Riesz basis.

- A sequence $(u_{j,k})_{j,k \in Z}$ is said to be a Riesz sequence if and only if for all $\xi \in R$ there exists constants $0 < A < B < \infty$, such that

$$A < \sum_k |\hat{u}(\xi + 2k\pi)|^2 < B \tag{5}$$

- The Riesz sequence is a Riesz basis if and only if the elements of the sequence are linearly independent.

In our construction, we are looking for compactly supported wavelets. Therefore, our transfer functions m_o and $\tilde{m}_o$ are polynomials of exponential type. That is,

$$m_0(\xi) = \frac{1}{2}\sum_{n=N_1}^{N_2} h_n e^{-in\xi} \quad \text{and} \quad \tilde{m}_0(\xi) = \frac{1}{2}\sum_{n=\tilde{N}_1}^{\tilde{N}_2} \tilde{h}_n e^{-in\xi}$$

where $m_0(0) = \tilde{m}_0(0) = 1$. The two scale equations are

$$\phi(x) = \sum_k h_k \phi(2x - k) \quad \text{and} \quad \tilde{\phi}(x) = \sum_k \tilde{h}_k \tilde{\phi}(2x - k)$$

which in Fourier domain are equivalent to (2). Parallel to orthonormal case ((3.47) of [8] or (6.2.6) of [9]), we define

$$\psi(x) := \sum_k (-1)^k \tilde{h}_{-k-1} \phi(2x - k), \tag{6}$$

and

$$\tilde{\psi}(x) := \sum_k (-1)^k h_{-k-1} \tilde{\phi}(2x - k). \tag{7}$$

Equivalently

$$\hat{\psi}(\xi) = e^{i\frac{\xi}{2}} \overline{\tilde{m}_0\left(\frac{\xi}{2} + \pi\right)}\, \hat{\phi}\left(\frac{\xi}{2}\right) \tag{8}$$

and

$$\hat{\tilde{\psi}}(\xi) = e^{i\frac{\xi}{2}} \overline{m_0\left(\frac{\xi}{2} + \pi\right)}\, \hat{\tilde{\phi}}\left(\frac{\xi}{2}\right) \tag{9}$$

The function ψ and $\tilde{\psi}$ can only be candidates for Riesz bases of wavelets if they satisfy $\hat{\psi}(0) = 0$ and $\hat{\tilde{\psi}}(0) = 0$. In terms of m_0 and $\tilde{m}_0$ these are equivalent to

$$\tilde{m}_0(\pi) = 0 = m_0(\pi) \tag{10}$$

The following Lemma [5] gives a necessary and sufficient condition for linear independence of $\psi_{j,k}$ and $\tilde{\psi}_{j,k}$.

Lemma 1 *The sequences* $(\psi_{j,k})$ *and* $(\tilde{\psi}_{j,k})$ *are linearly independent if and only if* $\left\langle \psi_{j,k}, \tilde{\psi}_{j',k'} \right\rangle = \delta_{j,j'}\delta_{k.k'}$

The above biorthogonality of $\psi_{j,k}$ and $\tilde{\psi}_{j',k'}$ is equivalent to a biorthogonality condition between $\phi_{0,k}$ and $\tilde{\phi}_{0,l}$.

Lemma 2 [5] $\left\langle \psi_{j,k}, \tilde{\psi}_{j',k'} \right\rangle = \delta_{j,j'}\delta_{k.k'}$ *if and only if* $\left\langle \phi_{0,k}, \tilde{\phi}_{0,l} \right\rangle = \delta_{k.l}$

The stability of the biorthogonal wavelets and the condition $\left\langle \phi_{0,k}, \tilde{\phi}_{0,l} \right\rangle = \delta_{k.l}$ are shown with the help of two operators T_0 and $\tilde{T}_0$ called as *transition operators* which act upon 2π-periodic functions. They are defined by

$$(T_0 f)(\xi) = \left| m_0\left(\frac{\xi}{2}\right)\right|^2 f\left(\frac{\xi}{2}\right) + \left| m_0\left(\frac{\xi}{2} + \pi\right)\right|^2 f\left(\frac{\xi}{2} + \pi\right) \tag{11}$$

and

$$(\tilde{T}_0 \tilde{f})(\xi) = \left| \tilde{m}_0\left(\frac{\xi}{2}\right)\right|^2 \tilde{f}\left(\frac{\xi}{2}\right) + \left| \tilde{m}_0\left(\frac{\xi}{2} + \pi\right)\right|^2 \tilde{f}\left(\frac{\xi}{2} + \pi\right) \tag{12}$$

Theorem 1 *Let* m_0 *and* $\tilde{m}_0$ *satisfy* (3). *Then the following two conditions are equivalent*

(a) $\phi, \tilde{\phi} \in R$ *and* $\left\langle \phi_{0,k}, \tilde{\phi}_{0,l} \right\rangle = \delta_{k.l}$

(b) *There exist strictly positive trigonometric polynomials* f_0 *and* $\tilde{f}_0$ *so that* $T_0 f_0 = f_0$ *and* $\tilde{T}_0 \tilde{f}_0 = \tilde{f}_0$ *and they are the only trigonometric polynomials (up to normalization) invariant under* T_0 *and* $\tilde{T}_0$*, respectively.*

Let us begin designing linear phase filters associated with the *transfer functions* m_0 and $\tilde{m}_0$ that lead to symmetric scaling functions ϕ and $\tilde{m}_0$. The transfer functions are symmetric in nature and satisfy (3). Moreover, they are polynomials. They are either in form

(a) $$m_0(\xi) = \cos\left(\frac{\xi}{2}\right)^{2l} p_0(\cos\xi), \ \tilde{m}_0(\xi) = \cos\left(\frac{\xi}{2}\right)^{2\tilde{l}} \tilde{p}_0(\cos\xi)$$

or

(b) $$m_0(\xi) = e^{-i\frac{\xi}{2}}\cos\left(\frac{\xi}{2}\right)^{2l+1} p_0(\cos\xi), \ \tilde{m}_0(\xi) = e^{-i\frac{\xi}{2}}\cos\left(\frac{\xi}{2}\right)^{2\tilde{l}+1} \tilde{p}_0(\cos\xi)$$

where, $l, \tilde{l} \in N$, p_0 and $\tilde{p}_0$ are polynomials with $p_0(-1) \neq 0$ and $\tilde{p}_0(-1) \neq 0$. Substituting m_0 and $\tilde{m}_0$ in (3)

$$\left(\cos\frac{\xi}{2}\right)^{2K} p_0(\cos\xi)\,\tilde{p}_0(\cos\xi) + \left(\sin\frac{\xi}{2}\right)^{2K} p_0(-\cos\xi)\,\tilde{p}_0(-\cos\xi) = 1 \quad (13)$$

with $K = l + \tilde{l}$ in case (a) and $K = l + \tilde{l} + 1$ in case (b).

Let us take $\sin^2\frac{\xi}{2} = \frac{1-\cos\xi}{2}$, then (13) reduces to *Bezout equation*

$$(1 - x)^K P(x) + x^K P(1 - x) = 1 \quad (14)$$

where $P(x) = p_0(\cos\xi)\,\tilde{p}_0(\cos\xi)$ is a polynomial in x. This equation has a polynomial solution *P(x)* of degree $K-1$ which is guaranteed by Theorem 6.3 of [5]. Let

$$P(x) = a_0 + a_1 x + a_2 x^2 + \cdots + a_{K-1} x^{K-1} \quad (15)$$

It is easy to check that for $m = 1, 2, 3, \ldots K$

$$a_{K-m} = \sum_{l=1}^{K-m} (-1)^{l+1} \binom{K}{l} a_{K-m-l}$$

Checking condition (b) of Theorem 1 needs two transition operators T_0 and $\tilde{T}_0$, which are defined by (11) and (12) respectively. Let M_T and $M_{\tilde{T}}$ are the Lawton matrices [5, 7, 14],

$$M_T = \left(2H_{i-2j}\right)_{i,j=-N}^{N} \quad \text{and} \quad M_{\tilde{T}} = \left(2\tilde{H}_{i-2j}\right)_{i,j=-\tilde{N}}^{\tilde{N}}$$

where $H_K = \sum_r h_r h_{r+k}$ and $\tilde{H}_K = \sum_r \tilde{h}_r \tilde{h}_{r+k}$

The matrices M_T and $M_{\tilde{T}}$ are well studied in [6, 7]. It is easily checked that both T_0 and M_T have same eigenvalues. Now checking condition (b) of Theorem 1 is much easier. This means, now we only have to check whether 1 is the non-degenerate eigenvalue of both M_T and $M_{\tilde{T}}$ and whether their corresponding eigenvectors give strictly positive trigonometric polynomials.

3 Case Studies

In this section, we construct biorthogonal Riesz bases of wavelets for the case $K = 2$. The construction for other cases follows similarly. For general K, the construction follow from two different algorithms which are described at the end of this section. When $K = 2$, the possible cases for the pair $(l.\tilde{l}), l + \tilde{l} = K$ are (2,0), (1,1) and (0,2). Moreover, in this case $P(x) = 1 + 2x$. Therefore, $P(\sin^2 \xi) = 2 - \cos \xi = -0.5e^{-i\xi} + 2 - 0.5e^{i\xi}$. It has a factorization $P(\sin^2 \xi) = p_0(\cos \xi)\,\tilde{p}_0(\cos \xi)$ where $p_0(\cos \xi) = -0.5e^{-i\xi} + 2 - 0.5e^{i\xi}$ and $\tilde{p}_0(\cos \xi) = 1$

Note that $\cos^2 \frac{\xi}{2}$ in exponential form is equals to $0.25e^{-i\xi} + 0.5 + 0.25e^{i\xi}$

(A1) $l = 2,\ \tilde{l} = 0$
Case(a) does not leads to a Riesz basis. In case(b), both the transition operator T_0 and $\tilde{T}_0$ have eigenvalue 1. They have unique eigenfunctions f_0 and $\tilde{f}_0$ respectively and both correspond to eigenvalue 1.

It is easy to observe that f_0 and $\tilde{f}_0$ are strictly positive trigonometric polynomials. This leads to a biorthogonal Riesz wavelet basis pair $(\psi, \tilde{\psi})$ of $L_2(R)$. More precisely, we denote the pair by $\left(\psi^{2,0,b}, \tilde{\psi}^{2,0,b}\right)$. They are shown in the Figs. 1b and 2b respectively.

Some more wavelet Riesz bases are shown in Figs. 3, 4, 5 and 6 respectively.

In case of general K, we can construct the Riesz wavelet pair and study their qualitative properties once we find the transfer functions $m_0(\xi)$ and $\tilde{m}_0(\xi)$ The

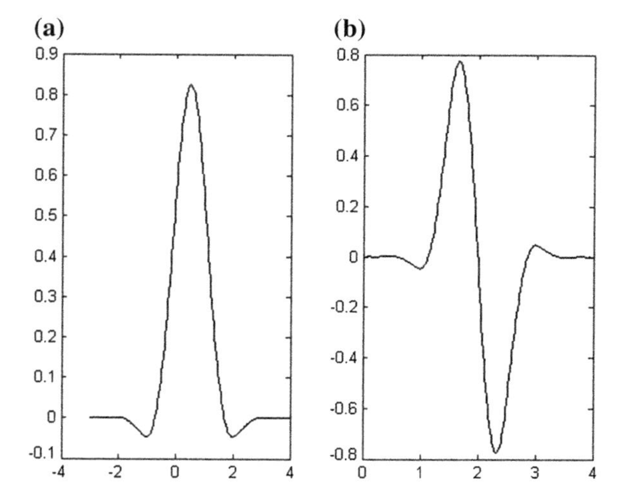

Fig. 1 **a** The scaling function $\phi^{2,0,b}$. **b** The wavelet $\psi^{2,0,b}$

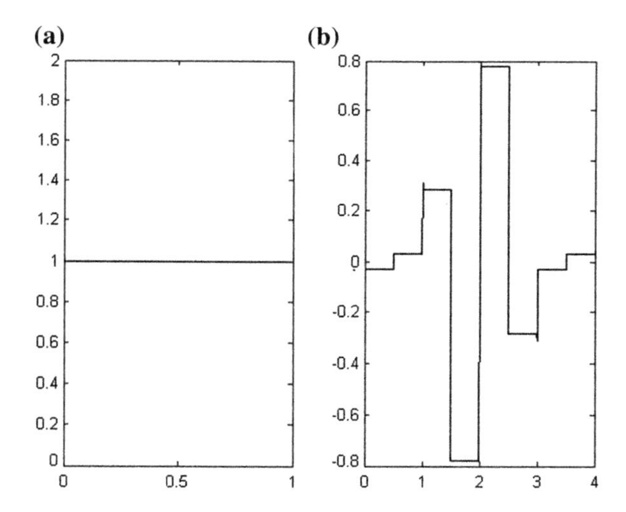

Fig. 2 **a** The scaling function $\tilde{\phi}^{2,0,b}$. **b** The wavelet $\tilde{\psi}^{2,0,b}$

construction is done by two different *subdivision algorithms*. In Algorithm 1 corresponds to case(a) where we assume that the transfer function $m_0(\xi)$ has $2M + 1$ no. of filters and the transfer function $\tilde{m}_0(\xi)$ has $\tilde{M}$ no of filters. Applying Algorithm 1, we obtain the Riesz wavelet pair $\left(\phi^{l,\tilde{l},a}, \psi^{l,\tilde{l},a}\right)$. Algorithm 2 corresponds to case(b)

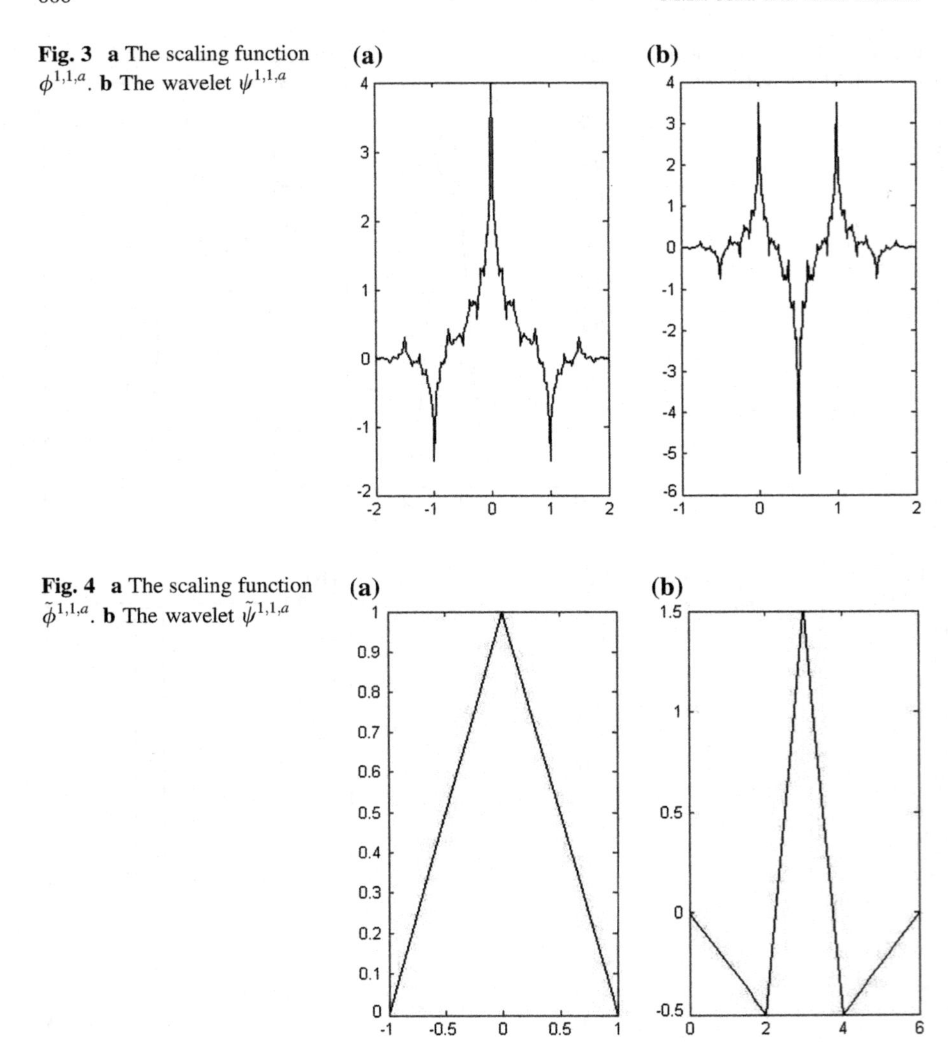

Fig. 3 **a** The scaling function $\phi^{1,1,a}$. **b** The wavelet $\psi^{1,1,a}$

Fig. 4 **a** The scaling function $\tilde{\phi}^{1,1,a}$. **b** The wavelet $\tilde{\psi}^{1,1,a}$

where we assume that the transfer function $m_0(\xi)$ has *2M* no. of filters and the transfer function $\tilde{m}_0(\xi)$ has $\tilde{M}$ no. of filters. Applying Algorithm 2, we obtain the Riesz wavelet pair $\left(\phi^{l,\tilde{l},b}, \psi^{l,\tilde{l},b}\right)$. The dual scaling functions and dual wavelets are obtained by interchanging the role of $m_0(\xi)$ and $\tilde{m}_0(\xi)$ in Algorithms 1 and 2.

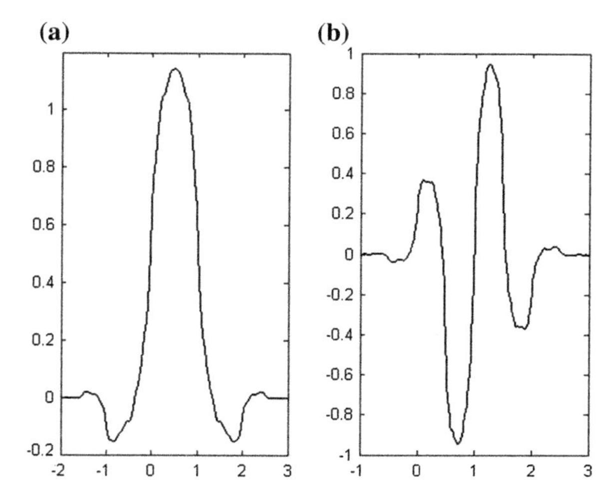

Fig. 5 **a** The scaling function $\phi^{1,1,b}$. **b** The wavelet $\psi^{1,1,b}$

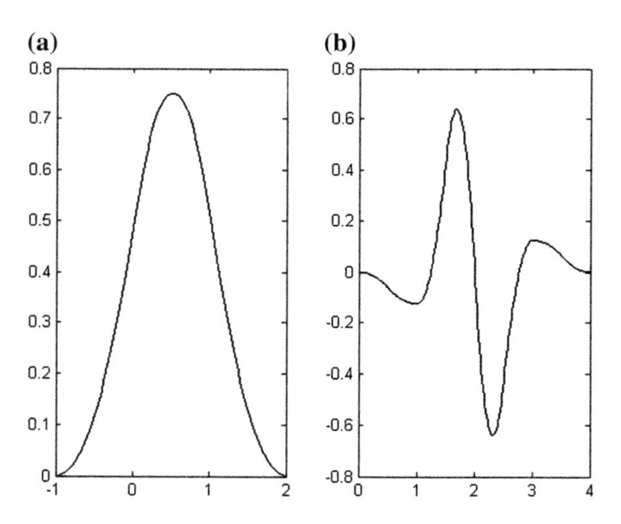

Fig. 6 **a** The scaling function $\tilde{\phi}^{1,1,b}$. **b** The wavelet $\tilde{\psi}^{1,1,b}$

Algorithm 1: Case(a)

```
/* Filters of the transfer function */
Input (b(1),b(2),b(3)·····,b(2M+1));
/* Initialization of starting point and  end point*/
xs(1)=1+M; xe(1)=3M+1;
/* Initialization of data */
for i=1:xe(1)
P_i^1 = 0;
end
P_{2M+1}^1 = 1;
/* Starting point and end point for k-th iteration*/
xs(k)=2 xs(k-1)+M+1;
xe(k)=2 xe(k-1)+M+1;
/* Data after k-th iteration */
       /* Even points */
for i=xs(k-1):xe(k-1)+M
P_{2i}^{K+1} = 2(b(2)P_{i-1}^k + b(4)P_{i-2}^k ····· +b(2M)P_{i-M}^k);
end
       /* Odd points */
for i=xs(k-1):xe(k-1)+M
P_{2i+1}^{K+1} = 2(b(1)P_i^k + b(3)P_{i-1}^k ····· +b(2M+1)P_{i-M}^k)
end
/* Parametrization after  ns iteration */
 mid=(xs(ns)+xe(ns))/2;
for i=xs(ns)  : xe(ns)
 x(i)=(i-mid)/power(2,ns-1);
end
/* The scaling function and its plot */
for i=1 : (xe(ns)-xs(ns)+1)
F(i) = P_{i+xs(ns)-1}^{ns};
X(i)=x(i+xs(ns)-1);
end
plot(X,F);
/* Dilation of the scaling function */
np=xs(ns)-xs(ns);
for i=1:np+1
F2(i)=F(i);
X2(i)=X(i)/2;
end
```

```
/* r-th translation of the dilated scaling function,
r=0,1,......, $\tilde{N}-1$ */
npi=np/(2M);
for i=(1+r*npi) : (np+r*npi+1)
$F_{2,r}$(i)=F2(i-r*npi);
end
/* Filter of the dual transfer function */
input(bt(1), bt(2), …., bt( $\tilde{N}$ );
/* The Wavelet and its plot */
for i=1: (( $\tilde{N}$ -1)npi+np+1)
psi(i)=2( bt(1)*$F_{2,0}$(i)-bt(2)*$F_{2,1}$(i)}+bt(3)*$F_{2,2}$(i)-......
              +$(-1)^{\tilde{N}-1}$ *bt( $\tilde{N}$ ) $F_{2,\tilde{N}-1}(i)$ ;
Xpsi(i)=(i-1)/power(2,ns);
end
plot(Xpsi,psi);
```

Algorithm 2: Case(b)

```
/* Filters of the transfer function */
Input$(b(1),b(2),b(3)\cdots\cdots,b(2M))$;
/* Initialization of starting point and  end point*/
xs(1)=1+M; xe(1)=3M;
/* Initialization of data */
for i=1:xe(1)
$P_i^1 = 0$;
end
$P_{2M}^1 = 1$;
/* Starting point and end point for k-th iteration*/
xs(k)=2 xs(k-1)+M;
xe(k)=2 xe(k-1)+M;
/* Data after k-th iteration */
       /* Even points */
for i=xs(k-1):xe(k-1)+M
$P_{2i}^{K+1} = 2\left(b(2)P_{i-1}^k + b(4)P_{i-2}^k \cdots\cdots + b(2M)P_{i-M}^k\right)$;
end
       /* Odd points */
for i=xs(k-1):xe(k-1)+M
$P_{2i+1}^{K+1} = 2\left(b(1)P_i^k + b(3)P_{i-1}^k \cdots\cdots + b(2M-1)P_{i-M+1}^k\right)$
end
/* Parametrization after  ns iteration */
mid=(xs(ns)+xe(ns))/2;
for i=xs(ns) : xe(ns)
x(i)=(i-mid)/power(2,ns-1);
end
```

```
/* The scaling function and its plot */
for i=1 : (xe(ns)-xs(ns)+1)
F(i) = P^{ns}_{i+xs(ns)-1};
X(i)=x(i+xs(ns)-1);
end
plot(X+0.5F);
/* Dilation of the scaling function */
np=xs(ns)-xs(ns);
for i=1:np+1
F2(i)=F(i);
X2(i)=X(i)/2;
end
/* r-th translation of the dilated scaling function,
r=0,1,......, Ñ-1 */
npi=np/(2M-1);
for i=(1+r*npi) : (np+r*npi+1)
F_{2,r}(i)=F2 (i-r*npi);
end
/* Filter of the dual transfer function */
input(bt(1), bt(2), ...., bt(Ñ);
/* The Wavelet and its plot */
for i=1: ((Ñ-1)npi+np+1)
psi(i)=2( bt(1)*F_{2,0}(i)-bt(2)*F_{2,1}(i)}+bt(3)*F_{2,2}(i)-......
          +(-1)^{Ñ-1} *bt(Ñ) F_{2,Ñ-1}(i);
Xpsi(i)=(i-1)/power(2,ns);
end
plot(Xpsi,psi);
```

4 Conclusion

In this paper, we compute the transfer functions of the wavelets algebraically. Using these transfer functions biorthogonal wavelets are computed with the help of Algorithms 1 and 2.

References

1. Behera, B., Jahan, Q.: Biorthogonal wavelets on local fields of positive characteristic. Commun. Math. Anal. **15**(2), 52–75 (2013)
2. Christian, B.: Wavelets a Primer. Universities Press, Hyderabad (1998)

3. Chui, C.K., Wang, J.Z.: On compactly supported spline wavelets and a duality principle. Trans. Am. Math. Soc. **330**, 903–915 (1992)
4. Chui, C.K.: An Introduction to Wavelets. Academic Press, San Diego (1992)
5. Cohen, A., Daubechies, I., Feauveau, J.C.: Biorthogonal bases of compactly supported wavelets. Commun. Pure Appl. Math. **XLV**, 485–560 (1992)
6. Cohen, A., Daubechies, I.: A stability criterion for biorthogonal wavelet bases and their related subband coding scheme. Duke Math. J. **68**, 313–335 (1992)
7. Curran, P.F., McDarby, G.: Stability of biorthogonal wavelet bases. Int. J. Wavelet Multiresolut. Inf. Process. **1**(1), 1–18 (2003)
8. Daubechies, I.: Orthonormal bases of compactly supported wavelets. Commun. Pure Appl. Math. **41**, 909–996 (1988)
9. Daubechies, I.: Ten Lectures on Wavelets. CBMS-NSF Series of Applied Mathematics, vol. 61. SIAM, Philadelphia (1992)
10. Goswami, J.C., Chan, A.K.: Fundamentals of Wavelets, Theory, Algorithms, and Applications. Wiley, New York (1999)
11. Grossmann, A., Morlet, J.: Decomposition of Hardy functions into square integrable wavelets of constant shape. SIAM J. Math. Anal. **15**, 723–736 (1984)
12. Jia, R.Q., Wang, J., Zhou, D.X.: Compactly supported wavelet bases for Sobolev spaces. Appl. Comput. Harmonic Anal. **15**, 224–241 (2003)
13. Han, B., Jia, R.Q.: Characterization of Riesz bases of wavelets generated from multiresolution analysis. Appl. Comput. Harmonic Anal. **23**, 321–345 (2007)
14. Lawton, W.: Necessary and sufficient conditions for constructing orthonormal wavelet bases. J. Math. Phys. **32**, 57–61 (1991)

Data Mining Approach for Modeling Risk Assessment in Computational Grid

Sara Abdelwahab and Ajith Abraham

Abstract Assessing Risk in a computational grid environment is an essential need for a user who runs applications from a remote machine on the grid, where resource sharing is the main concern. As Grid computing is the ultimate solution believed to meet the ever-expanding computational needs of organizations, analysis of the various possible risks to evaluate and develop solutions to resolve these risks is needed. For correctly predicting the risk environment, we made a comparative analysis of various machine learning modeling methods on a dataset of risk factors. First we conducted an online survey with international experts about the various risk factors associated with grid computing. Second we assigned numerical ranges to each risk factor based on a generic grid environment. We utilized data mining tools to pick the contributing attributes that improve the quality of the risk assessment prediction process. The empirical results illustrate that the proposed framework is able to provide risk assessment with a good accuracy.

Keywords Grid computing · Risk assessment · Feature selection · Data mining

S. Abdelwahab (✉)
Faculty of Computer Science and Information Technology, Sudan University of Science and Technology, Khartoum, Sudan
e-mail: saabdelghani@pnu.edu.sa

A. Abraham
Machine Intelligence Research Labs (MIR Labs), Washington, USA
e-mail: ajith.abraham@ieee.org

A. Abraham
IT4Innovations—Center of Excellence, VSB—Technical University of Ostrava, Ostrava, Czech Republic

L.C. Jain et al. (eds.), *Computational Intelligence in Data Mining - Volume 3*,
Smart Innovation, Systems and Technologies 33, DOI 10.1007/978-81-322-2202-6_61

1 Introduction

A Computational Grid is a collection of diverse computers and resources spread across several administrative domains with the purpose of resource sharing. Currently, grid has been applied in many applications to solve large-scale scientific and e-commerce problems [1]. Therefore, risk reduction is needed to avoid security breaches. In order to offer reliable grid computing services, a mechanism is needed to assess the risks and take necessary precautions to avoid them. As reported in the literature, there are many risk factors associated with grid computing. In this work we utilize a data mining approach for feature selection to determine the most significant risk factors that threaten the computational grid. Feature selection is a method of data processing that aims to remove the redundant or irrelevant features that do not affect the performance of the prediction algorithms by selecting important features from a given pool of input features. Feature selection algorithms are classified into two methods. The first one is the Filter Method in which the goodness of feature subset is evaluated, according to basic properties of the data; and the second is Wrapper methods that evaluate the feature subsets by using induction algorithms [2]. The rest of the paper is organized into sections as follows. The related work is presented in Sect. 2 while Sect. 3 highlights the risk factors associated with computational grid. Proposed method is presented in Sect. 4, followed by Dataset and Feature Selection Method in Sect. 4.2. Experimental results and discussion are depicted in Sect. 5. Conclusion is drawn out in Sect. 6.

2 Related Works

Over the past years, there are many researchers focusing on the security issues in grid computing [3, 4, 5]. Chakrabarti et al. [5] addressed the security issues by classifying them into three levels. The first level is host level issue, which includes data protection issue and job starvation. In data protection issue, the main concern is protecting the pre-existing data of the host that is associated with the grid system. While job starvation happens when the resources used by local job are taken away by stranger job scheduled on the host. In the second level (architecture level), the following issues are addressed: policy mapping, denial of services (DOS), resource hacking, information security, authentication, Integrity and confidentiality. The third level is the credential level that is categorized into: credential repositories which are responsible for user's credential storage, and the second is a credential federation system that supports managing credentials among multiple systems and regality. Butt et al. [3] mentioned that sharing of resource in grid environment involve execution of unreliable code from arbitrary users, which cause security risks such as securing access to shared resources. In addition, the authors addressed two possible situations that can occur in grid environment and has the power to affect both the program executing in shared resource and the integrity of the resource. In the first scenario, the resource may be malicious that can affect the programs using these resource, or the grid program may be malicious which affect the integrity of resource. Chakrabarti [4], investigated a taxonomy of grid security issues that is composed of three categories. The first class addressed the architectural issues, which include data confidentiality, integrity and authentication. While the second category is security issues coupled with infrastructure such as data protection, job starvation, and host availability. The most critical security threats found in the grid infrastructure is malicious service disruption. The third class addressed the security issues that are related to management which include credentials management, and trust management. Some researchers [6, 7] analyzed the security threats in on-demand grid computing. Authors analysis is built on trust relationship between grid actors. They provided three different levels that address the security issues in on-demand computing and each level has possible forms of misuse. Level 1 addresses the threats that arise between users and solution producer. Whereas, level 2 provides the threats made by solution producer to use the resource provider in a bad manner, resulting in possible type of misuse. At the third level resource provider posed threats to users and solution producer caused different type of misuse. Smith et al. [7] addressed the security issues related to on demand grid computing by categorizing it to three types: internal versus external attacks, software attacks, and privilege threats and shared use threats, these threats threaten traditional grid as well. Cody et al. [8] reported the most common vulnerabilities in the three different types of grid system. Each of the three identified grid systems has vulnerabilities common to them. The first type is computational grid, where the grid architecture is responsible for resource allocation to gain computing power to solve complex problems on high performance servers. The most popular vulnerability is node downfall that diminishes the functionality of the system. This could happen when the program contains infinite loops. In the second type of grid system called data grid, the main focus of grid architecture here is on storage and offering access to large amount of data across multiple organizations. The possible risk that is associated with this type is overwrite or data corruption that occurs when user override their obtainable space. Denial of Service attack (DoS) is the most widespread attack that threatens the service grid that is considered as the third type of grid system. Kar et al. [9] provided vulnerabilities of grid computing in context of Distributed Denial of Service (DDoS) attack. Using spoofed IP address by attackers make DoS attacks so hard to detect, especially in large distributed system like grid, where it becomes more complicated. He presented four types of grid intrusions: unauthorized access, misuse, grid exploit, and host or network specific attacks. Kussul [10] addressed the most significant security threats for a utility based reputation model in grid. Based on the resource behavior observed in the past, the reputation can be seen as quality and reliability expected from that resource. The authors mentioned nine types of attack according the reputation model: Individual malicious peers, Malicious collectives, Malicious collectives with camouflage, Malicious spies, Sybil attack, Man in the middle attack, Driving down the reputation of a reliable peer, Partially malicious collectives, and Malicious pre-trusted peers. Hassan et al. [11] proposed the problem of Cross Domain Attack (CDA). Authors viewed that when a grid node is compromised it is so difficult to determine it, due to the existence of different administrative domains collaborating with each other, each with multiple nodes. In this case, the attack is likely propagated to another organization's network that is part of the grid network, resulting in cross-domain attack. Carlsson [12] addressed the problem of Services Level Agreements (SLAs) in Grid. The authors mentioned that a resource provider [RP] in grid computing offers resources and services to other Grid users based on agreed service level agreements [SLAs]. The research problem that they have addressed is formulated as follows: the RP is running a risk of violating SLA if one or more of the resources offered to prospective customers will fail when carrying out the tasks.

Three types of failures were addressed by Lee [13], the process failure, which is expanded into two types that are Process stop failure and a starvation of process failure. Processor failure is the second type of failure which is further categories into a processor crash (Processor stop failure) and a decrease of processor throughput due to burst job (Processor QoS failure) while the third type of failure is a network failure that is classified into a network disconnection and partition (Network disconnection failure) and a decrease of network bandwidth due to communication traffic (Network QoS failure).

3 Risk Factors Associated with Grid Computing

Many risk factors were reported in the literature [9, 14, 15]. These factors affect the grid computing in many ways, such as:

(a) **Availability**: Threats to Availability indicates the percentage of time, usually on a monthly basis, in which the Grid service supplied by the provider will be available [16]. The risk factors that threaten the availability are: Service Level Agreement Violation: SLAs are contracts between service providers and users, specifying acceptable QoS levels. Cross-Domain Attack: Cross-Domain Attack in which the attacker compromises one site and can then spread his attack easily to the other federated sites. Job Starvation: Job starvation happens when the resources used by local job are taken away by stranger job scheduled on the host. Resource Failure: It is a failure if and only if one of the following two conditions is satisfied. a. Resource stops due to resource crash; and b. Availability of resources does not meet the minimum levels of QoS. Distributed Denial of Service Attack, in a DDoS attack the computing power of thousands of compromised machines known as "zombies" are used for a target a victim. Zombies are gathered to send useless service requests, packets at the same time. Resource Attacks: Illegal use of software or physical resources.

(b) **Integrity**: refers to the trustworthiness of data or resources, and it is usually phrased in terms of preventing improper or unauthorized change [17], and gives the assurance that the data received are not altered [18]. The Integrity related risk factors are: Integrity Violation, in computational grid, the grid program may be malicious and threatens the integrity of the resource. Privilege Attack: Many grids are designed to give remote users interactive access to a command shell so they can run their own application. However, giving the user access to a command shell within a predefined script or application is extremely dangerous, because when the users access the command shell they can obtain an extra privilege. Hosting Illegal content: By exploiting the leased nodes to send junk mail and host illegal content for others. The last factor is stealing the input or output or modifies the result of computation.

(c) **Confidentiality**: refers to the protection of information from unauthorized disclosure. The risk factors related to Confidentiality are: Confidentiality Breaches: Indicates that all data sent by users should be accessible to only legitimate users. Data Exposure: The data protection issue is concerned about protecting the pre-existing data on the host that is associated with the grid system. Data Attacks: Illegal access to or modification of data. Man in the Middle Attack: When a message between peers is intercepted and modified. Sybil Attack: When a large number of malicious peers in the system are launched by an enemy, the peers in the system exchange the role of a resource provider and at each time one of them is scheduled, and then provide malicious service before it is replaced by another peer and be disconnected. And Privacy Violation: Compromising the passwords and security system, by exploiting the large computation power that the grid provides.

(d) **Access Control**: The overall security of the grid is at stake if the authentication and authorization mechanisms are not strong enough to handle the access control properly. This will result in an unauthorized access to the resources. The risk factors related to Access Control are: Credential Violation: Credentials are tickets or tokens used to identify, authorize, or authenticate a user. Policy Mapping: Due to the spread of VO across multiple administrative domains with multiple policies, users might be concerned with how to map different policies across the grid. As a result of the grid's heterogeneous nature and its promise of virtualization at the user level, such mapping policies are a very important issue. Shared Use Threats: These issues are caused due to incompatibility between the attributes of grid users and conventional users of the computing resources that form the basis of the grid. The last factor is stealing software or the information contained in the database.

4 Proposed Method

4.1 Experimental Design and Survey

We conducted an online survey with international experts to evaluate the risk factors associated with grid computing. We asked the experts to determine the influence of these factors by categorizing them under three levels: severe, moderate, and marginal. We received responses from 21 experts from nine different countries: France, Czech Republic, Romania, Canada, China, Malaysia, Brazil, Sudan and Ethiopia. All respondents agreed that all the predefined risk factors affect the grid in

Table 1 Risk factors (attributes)

Risk Factor	Abbreviation	Range
Service level agreement violation	SLAV	[0–1]
Cross domain attacks	CDA	[1–3]
Job starvation	JS	[0–1]
Resource failure	RF	[0–1]
Resource attacks	RA	[0–1]
Privilege attack	PA	[0–1]
Confidentiality breaches	CB	[0–2]
Integrity violation	IV	[0–2]
DDoS attacks	DDoS	[1–3]
Data attack	DA	[0–2]
Data exposure	DE	[1–3]
Credential violation	CV	[0–1]
Man in the middle attack	MMA	[0–1]
Privacy violation	PV	[0–2]
Sybil attack	SA	[1–3]
Hosting illegal content	HIC	[0–1]
Stealing the input or output	SIO	[0–1]
Shared use threats	ShUTh	[1–3]
Stealing or altering the software	SS	[0–1]
Policy mapping	PM	[1–3]

a major way. At the next step we assigned a numeric range to each factor depending on its concept and chance of occurrence. Based on expert knowledge and some statistical approaches we then simulated 1950 instances based on a generic grid environment. The original data set consists of 20 input attributes (risk factors), and one output (risk value). The attributes are summarized in Table 1.

4.2 Feature Selection for Prediction

Feature selection is a preprocessing step that reduces dimensionality from a dataset, in order to have better prediction performance [19]. Feature Selection can be viewed as a search problem, searching of a subset from the search space in which each state represents a subset of the possible features. To avoid high computational cost and enhance the prediction accuracy, irrelevant input features are reduced from the dataset before constructing the prediction model. To find the feature subset we used WEKA platform [20]. We used Correlation based Feature Selection (CFS Sub Eval), that evaluates the worth of a subset of attributes by considering the individual predictive ability of each feature along with the degree of redundancy between them. ReliefF Attribute Evaluator (ReliefF Attribute Eval), evaluates the worth of an attribute by repeatedly sampling an instance and considering the value of the given attribute for the nearest instance of the same and different class. Different search methods are applied such as Ranker, Evolutionary Search, Best First Search and Exhaustive Search for feature selection. The Filter methods apply feature-ranking function to select the best feature. A relevance score on the basis of a sequence of examples is assigned to the input feature by the ranking function. Then features with the highest rank are selected [21]. Filter-based feature ranking techniques (rankers) arrange the features, without relying on any prediction algorithm and the best features are chosen from the rank list [19]. Evolutionary Search algorithms rely on aggregated learning process within a population of individuals; each individual denotes a search point in the space of possible solutions to a given problem [22]. The main characteristic of evolutionary search is the fitness function, which has a value that expresses the performance of an individual so the individual can be evaluated and compared with other individuals. Best first search is a method that saves all attribute subsets that were evaluated before, and terminates when the performance starts to drop. The attribute subsets are arranged according to the performance measure; therefore an earlier configuration can be reviewed. Exhaustive search is a complete search that leads it to the optimal solution based on the identified evaluation criteria. Exhaustive search grantees that all reachable nodes are visited in the same level and then proceeding to the next level of the tree, so the possible moves in the search space are examined regularly [23].

After implementing the different attribute selection algorithms described above, we obtained 8 different sub datasets. Table 2 illustrates the number of attributes in each dataset and summarizes the search method. We divided the datasets into training and testing data with different percentages to investigate the effectiveness of data splitting.

- A: Split 60 % training, 40 % testing
- B: Split 70 % training, 30 % testing
- C: Split 80 % training, 20 % testing
- D: Split 90 % training, 10 % testing

4.3 Risk Assessment (Prediction) Algorithms Used

Isotonic Regression algorithm (Isoreg)
Isotonic regression is a regression method that uses the weighted least squares to evaluates linear regression models [20].

Instance Based Knowledge (IBk) Algorithm
Instance Based Knowledge uses the instances themselves from the training set to represent what are learned, and be kept. When an unseen instance is provided the memory is searched for the training instance.

Randomizable Filter Classifier (RFC) Algorithm
This method used an arbitrary classifier on data that has been passed through an arbitrary filter. Like the classifier, the structure of the filter is based exclusively on the training data and test instances will be processed by the filter without changing their structure {Hall, 2009 #48}.

Table 2 Attributes selection methods

Dataset	Evaluator	Search method	Selected attributes	Attributes
Original dataset	–	–	SLAV, CDA, JS, RF, RA, PA, CB, IV, DDoS, DA, DE, CV, MMA, PV, SA, HIC, SIO, ShUTh, SS, PM	20
1	RelifF attribute evaluation	Ranker	DDoS, PM, DE, SA, ShUTh, HIC, CV, RA, SIO, CDA, RF, SLAV, JS, MMA, SS, PA, PV, IV, CB,DA	20
2	Reliff attribute evaluation	Ranker	DDoS, PM, DE, SA, ShUTh, HIC, CV, RA, SIO, CDA, RF, SLAV, JS, MMA, SS, PA, PV, IV	18
3	Reliff attribute evaluation	Ranker	DDoS, PM, DE, SA, ShUTh, HIC, CV, RA, SIO, CDA, RF, SLAV, JS, MMA, SS	15
4	Reliff attribute evaluation	Ranker	DDoS, PM, DE, SA, ShUTh, HIC, CV, RA, SIO, CDA, RF, SLAV	12
5	Reliff attribute evaluation	Ranker	DDoS, PM, DE, SA, ShUTh, HIC, CV, RA, SIO	9
6	CFS subset eval	Evolutionary search	SLAV, JS, RA, CV, HIC, SIO	6
7	CFS subset eval	Best first search backward	CV, HIC, SIO	3
8	CFS subset eval	Exhaustive search	RA, CV, HIC	3

Extra Tree Algorithm (Etree)

This method is an extremely randomized decision tree that uses another randomization process. At each node of an extra tree, partitioned rules are depicted randomly, then on the basis of a computational score the rule that proceed well is selected to be linked with that node [24].

Bagging Algorithm

In bagging, a classifier model is learned with every training set, which has been classified as tuples. Bagging is the most common method that synchronously processes samples. It merges the various outputs of learned predictors into a single computational model, that results in improved accuracy [25].

Ensemble Selection (EnsmS) Algorithm

The principle of ensemble methodology is to combine a set of models that solve the same original mission, with the aim of achieving more accurate and reliable estimates than that achieved from a single model [25].

Random Subspace (RsubS) Algorithm
Random subspace method utilizes random subsets of the available features to train the individual classifiers in an ensemble. Random Subspace is a random combination of models [26].

Random Forest (Rforest) Algorithm
Random forests create a lot of classification and regression trees, by recursively using partitioning, and then combining the results. Utilizing a bootstrap sample of the training data each tree can be created [26].

5 Experimental Results and Discussions

In the preprocessing phase, the data is filtered to remove irrelevant and redundant features and to improve the quality. Table 3 reports the empirical results (for test data) illustrating the root mean squared error (RMSE) for the four datasets. As illustrated in Table 3, Isotonic Regression algorithm, IBK algorithm, Randomizable Filter Classifier algorithm and the Extra tree algorithm performed well for all the training and testing combinations and for the 4 different datasets. All these algorithms exhibited the best performance in the case of all 4 datasets. It is noticed that the higher the percentage of training data (Dataset D) the better for achieving good results. However the empirical result shows that, the prediction algorithm required the least number of attributes (3 attributes only out of 20 attributes) to achieve high performance.

Table 3 Evaluation of different algorithms with different data sets

	Data split	9 attributes	6 attributes	3 attributes	3 attributes
		RMSE			
IsoReg	A	0.0023	0.0024	0.0023	0.0023
	B	0.002	0.002	0.002	0.002
	C	0.0018	0.0018	0.0018	0.0018
	D	**0.0017**	**0.0017**	**0.0017**	**0.0017**
IBk	A	0.0023	0.0022	0.0021	0.0021
	B	0.0019	0.0019	0.0018	0.0018
	C	0.0018	0.0018	0.0016	0.0016
	D	0.0016	0.0016	**0.0015**	**0.0015**
RFC	A	0.0023	0.0023	0.0022	0.0022
	B	0.002	0.002	0.0019	0.0019
	C	0.0018	0.0019	0.0017	0.0017
	D	0.0017	0.0018	**0.0016**	**0.0016**
Etree	A	0.0046	0.3677	0.0045	0.0045
	B	0.0038	0.0039	0.0038	0.0038
	C	0.0033	0.0034	0.0033	0.0033
	D	0.0032	0.0031	**0.0029**	**0.0029**

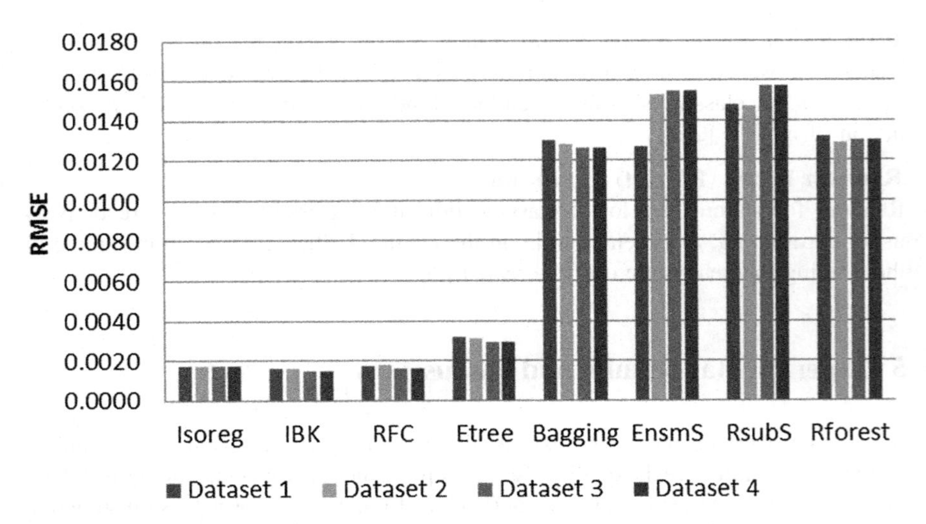

Fig. 1 Performance comparisons of 4 datasets for the test data

The best result is accomplished with the Correlation Coefficient (CC) equal to 1 and the root mean squared error (RMSE) equal to 0.0015 for datasets 3 and 4. Bagging, Ensemble Selection, Random subspace, and Random forest algorithms performed slightly well with the CC equal to 0.9999 and the RMSE varied according to the used algorithm and splitting of data. With Bagging and Random subspace algorithms the higher performance is achieved with 70 % training and 30 % testing, while Random forest gives the best performance with 60 % training and 40 % testing.

Figure 1 illustrates that Isotonic Regression, IBK, Randomizable Filter Classifier and extra tree performed well for datasets 3 and 4 (with 4 features), which indicates that feature reduction can improve the predictor performance.

6 Conclusions

In this paper, we defined risk factors associated with grid computing, and categorized these factors to show how they affect the performance of a grid in a major way. Then we investigated the performance of several machine-learning methods for the risk assessment prediction process for the identified risk factors. We used four different algorithms for feature selection namely Ranker, Evolutionary Search, Best first search, and Exhaustive search. We considered 8 algorithms for Grid Risk factor prediction and then a comparison was made among different subset of features.

We also illustrated the effect of different subsets of training and testing data by randomly splitting them into four different groups. Empirical results illustrate that applying feature reduction can enhance the classifier performance. In our future research, we will investigate an ensemble model in order to demonstrate that, the main issues in constructing a powerful ensemble include training a set of diverse and accurate base predictors, and effectively combining them.

References

1. Foster, I.K., Nick, C., Jeffrey, M., Tuecke, S.: The physiology of the grid. In: Grid Computing: Making the Global Infrastructure a Reality, pp. 217–249 (2003)
2. Zhu, Z., Ond, Y.-S., Dash, M.: Wrapper–filter feature selection algorithm using a memetic framework. IEEE Trans. Syst. Man Cybern. Part B Cybern. **37**(1), 70–76 (2007)
3. Butt, A.R.A., Kapadia, S., Figueiredo, N.H., Fortes, R.J., José, A.B.: Grid-computing portals and security issues. J. Parallel Distrib. Comput. **63**(10), 1006–1014 (2003)
4. Chakrabarti, A.: Taxonomy of grid security issues. In: Grid Computing Security, pp. 33–47. Springer, Berlin (2007)
5. Chakrabarti, A., Damodaran, A., Sengupta, S.: Grid computing security: a taxonomy. IEEE Secur. Priv. **6**(1), 44–51 (2008)
6. Smith, M., Engel, M., Friese, T., Bernd, F.: Security issues in on-demand grid and cluster computing. In: Sixth IEEE International Symposium on Cluster Computing and the Grid, CCGRID 06, pp. 14–24 (2006a)
7. Smith, M., Friese, T., Engel, M., Freisleben, B.: Countering security threats in service-oriented on-demand grid computing using sandboxing and trusted computing techniques. J. Parallel Distrib. Comput. **66**(9), 1189–1204 (2006b)
8. Cody ES, Raj Rao, Raghav H Upadhyaya, Shambhu: Security in grid computing: a review and synthesis. Decis Support Syst 44 (4) (2008) 749-764
9. Kar, S., Sahoo, B.: An Anamaly detection system for DDOS attack in grid computing. Int. J. Comput. Appl. Eng. Technol. Sci. (IJ-CA-ETS) **1**(2), 553–557 (2009)
10. Kussul, O., Kussul, N., Skakun, S.: Assessing security threat scenarios for utility-based reputation model in grids. Comput. Secur. **34**, 1–15 (2013)
11. Syed, R.S., Syrame, M., Bourgeois, J.: Protecting grids from cross-domain attacks using security alert sharing mechanisms. Future Gener. Comput. Syst. **29**, 536–547 (2012)
12. Carlsson, C., Robert, F.: Risk assessment in grid computing. In: Possibility for Decision, pp. 145–165. Springer, Berlin (2011)
13. Lee, H.M.C., Jin, K.S., Lee, S.H., Lee, D.-W., Jung, W.G., Yu, S.Y., Chang, H.: A fault tolerance service for QoS in grid computing. In: Computational Science—ICCS 2003, pp. 286–296. Springer, Berlin (2003)
14. Rana, O.F., Martijn, W., Quillinan, T.B., Brazier, F., Cojocarasu, D.: Managing violations in service level agreements. In: Grid Middleware and Services, pp. 349–358. Springer, Berlin (2008)
15. Syed, R.H., Syrame, M., Julien, B.: Protecting grids from cross-domain attacks using security alert sharing mechanisms. Future Gener. Comput. Syst. **29**(2), 536–547 (2013)
16. Parrilli, D.: Legal issues in grid and cloud computing. In: Stanoevska-Slabeva, K., Wozniak, T., Ristol, S. (ed) Grid and Cloud Computing, pp. 97–118. Springer, Berlin (2010). doi:10.1007/978-3-642-05193-7_7
17. Bishop, M.: What is computer security? IEEE Secur. Priv. **1**(1), 67–69 (2003)
18. Selvi, R.K., Kavitha, V.: Authentication in grid security infrastructure-survey. Procedia Eng. **38**, 4030–4036 (2012)
19. Wang, H., Khoshgoftaar, T.M., Seliya, N.: How many software metrics should be selected for defect prediction? In: FLAIRS Conference (2011)

20. Wu, C.-H.S., Su, W.-H., Ho, Y.-W.: A study on GPS GDOP approximation using support-vector machines. IEEE Trans. Instrum. Meas. **60**(1), 137–145 (2011)
21. Bellotti, T., Ilia, N., Yang, M., Gammerman, A.: Feature selection. In: Conformal Prediction for Reliable Machine Learning, pp. 115–130. Morgan Kaufmann, Boston (2014). doi:http://dx.doi.org/10.1016/B978-0-12-398537-8.00006-7
22. Bäck, T., Schwefel, H.-P.: An overview of evolutionary algorithms for parameter optimization. Evol. Comput. **1**(1), 1–23 (1993)
23. http://www-rci.rutgers.edu/~cfs/472_html/AI_SEARCH/ExhaustiveSearch.html (2014)
24. Désir, C., Petitjean, C., Heutte, L., Salaün, M., Thiberville, L.: Classification of endomicroscopic images of the lung based on random subwindows and extra-trees. IEEE Trans. Biomed. Eng. **59**(9), 2677–2683 (2012)
25. Rokach, L.: Ensemble methods in supervised learning. In: Maimon, O., Rokach, L. (eds.) Data Mining and Knowledge Discovery Handbook, pp. 959–979. Springer, Berlin (2010)
26. Zhang, L.S., Nagaratnam, P.: Random forests with ensemble of feature spaces. Pattern Recognit. **47**, 3429–3437 (2014)

Degree of Approximation of Conjugate Series of a Fourier Series by Hausdroff and Norlund Product Summability

Sunita Sarangi, S.K. Paikray, M. Dash, M. Misra and U.K. Misra

Abstract Lipchitz class of function had been introduced by McFadden. Recently dealing with degree of approximation of conjugate series of a Fourier series of a function of Lipchitz class Misra et al. and Paikray et al. have established certain theorems. Extending their results, in this paper a theorem on degree of approximation of *a* function *f* belongs to a class of $Lip(\alpha, r)$ by using product summability $(E, q)(N, p_n)$ has been established.

Keywords Degree of approximation · $Lip(\alpha, r)$ class of function · (E, q)-mean · (N, p_n)-mean · $(E, q)(N, p_n)$-mean · Fourier series · Conjugate series · Lebesgue integral

S. Sarangi (✉) · M. Dash
Department of Mathematics, Ravenshaw University, Cuttack, Odisha, India
e-mail: sunitas1970@gmail.com

M. Dash
e-mail: dashminakshi@gmail.com

S.K. Paikray
Department of Mathematics, VSSUT, Burla, Odisha, India
e-mail: spaikray2001@yahoo.com

M. Misra
Department of Mathematics, B.A. College, Berhampur, Odisha, India
e-mail: mehendramisra2007@gmail.com

U.K. Misra
Department of Mathematics, NIST, Berhampur, Odisha, India
e-mail: umakanta_misra@yahoo.com

L.C. Jain et al. (eds.), *Computational Intelligence in Data Mining - Volume 3*,
Smart Innovation, Systems and Technologies 33, DOI 10.1007/978-81-322-2202-6_62

1 Introduction

Suppose $\sum a_n$ be a given infinite series with the sequence of partial sums $\{s_n\}$ [5]. Let $\{p_n\}$ be a sequence of positive real numbers such that

$$P_n = \sum_{\upsilon=0}^{n} p_\upsilon \to \infty, \quad \text{as}\, n \to \infty, \; (P_{-i} = p_{-i} = 0\;, i \geq 0) \tag{1}$$

The sequence-to-sequence transformation [4]

$$t_n = \frac{1}{P_n} \sum_{\upsilon=0}^{n} p_{n-\upsilon}\, s_\upsilon \tag{2}$$

defines the sequence $\{t_n\}$ of the (N, p_n)-mean of the sequence $\{s_n\}$ generated by the sequence of coefficient $\{p_n\}$. If

$$t_n \longrightarrow s \quad \text{as} \quad n \longrightarrow \infty, \tag{3}$$

then, the series $\sum a_n$ is said to be (N, p_n) summable to s.

The conditions for regularity of (N, p_n)-summability are easily seen to be [1]

$$\begin{cases} (i) & \frac{p_n}{P_n} \longrightarrow 0, \text{ as } n \longrightarrow \infty \\ (ii) & \sum\limits_{k=0}^{n} p_k = O(P_n), \text{ as } n \longrightarrow \infty \end{cases} \tag{4}$$

The sequence-to-sequence transformation, [1]

$$T_n = \frac{1}{(1+q)^n} \sum_{\upsilon=0}^{n} \binom{n}{\upsilon} q^{n-\upsilon} s_\upsilon, \tag{5}$$

defines the sequence $\{T_n\}$ of the (E, q) mean of the sequence $\{s_n\}$. If

$$T_n \longrightarrow s, \quad \text{as} \quad n \longrightarrow \infty, \tag{6}$$

then the series $\sum a_n$ is said to be (E, q) summable to s.

Clearly (E, q) method is regular. Further, the (E, q) transform of the (N, p_n) transform of $\{s_n\}$ is defined by

$$\begin{aligned} \tau_n &= \frac{1}{(1+q)^n} \sum_{k=0}^{n} \binom{n}{k} q^{n-k}\, T_k \\ &= \frac{1}{(1+q)^n} \sum_{k=0}^{n} \binom{n}{k} q^{n-k} \left\{ \frac{1}{P_k} \sum_{\upsilon=0}^{k} p_\upsilon s_\upsilon \right\} \end{aligned} \tag{7}$$

If

$$\tau_n \to s, \quad \text{as} \quad n \to \infty, \tag{8}$$

then $\sum a_n$ is said to be $(E,q)(N,p_n)$-summable to s.

Let $f(t)$ be a periodic function with period 2π and L-integrable over $(-\pi, \pi)$. The Fourier series associated with f at any point x is defined by

$$f(x) \sim \frac{a_0}{2} + \sum_{n=1}^{\infty} (a_n \cos nx + b_n \sin nx) \equiv \sum_{n=0}^{\infty} A_n(x) \tag{9}$$

and the conjugate series of the Fourier series (9) is

$$\sum_{n=1}^{\infty} (b_n \cos nx - a_n \sin nx) \equiv \sum_{n=1}^{\infty} B_n(x). \tag{10}$$

Let $\bar{s}_n\,(f;x)$ be the nth partial sum of (10). The L_∞-norm [4] of a function $f : R \to R$ is defined by

$$\|f\|_\infty = \sup\{|f(x)| : x \in R\} \tag{11}$$

and the L_υ-norm is defined by

$$\|f\|_\upsilon = \left(\int_0^{2\pi} |f(x)|^\upsilon \right)^{\frac{1}{\upsilon}}, \quad \upsilon \geq 1. \tag{12}$$

The degree of approximation of a function $f : R \to R$ by a trigonometric polynomial $P_n(x)$ of degree n under norm $\| \, . \, \|_\infty$ is defined by Zygmund [5].

$$\|P_n - f\|_\infty = \sup\{|p_n(x) - f(x)| : x \in R\} \tag{13}$$

and the degree of approximation $E_n(f)$ of a function$f \in L_\upsilon$ is given by

$$E_n(f) = \min_{P_n} \|P_n - f\|_\upsilon \tag{14}$$

A function f is said to satisfy Lipschitz condition (here after we write $f \in Lip\,\alpha$) if

$$|f(x+t) - f(x)| = O(|t|^\alpha), \quad 0 < \alpha \leq 1 \tag{15}$$

and $f(x) \in Lip(\alpha, r)$, for $0 \leq x \leq 2\pi$, if

$$\left(\int_0^{2\pi} |f(x+t) - f(x)|^r dx\right)^{\frac{1}{r}} = O(|t|^\alpha), \quad 0<\alpha\leq 1,\ r\geq 1,\ t>0 \tag{16}$$

We use the following notation throughout this paper:

$$\psi(t) = \frac{1}{2}\{f(x+t) - f(x-t)\} \tag{17}$$

and

$$\bar{K}_n(t) = \frac{1}{\pi(1+q)^n}\sum_{k=0}^{n}\binom{n}{k}q^{n-k}\left\{\frac{1}{P_k}\sum_{\upsilon=0}^{k}p_\upsilon\frac{\cos\frac{t}{2} - \cos\left(\upsilon+\frac{1}{2}\right)t}{\sin\frac{t}{2}}\right\} \tag{18}$$

Further, the method $(E,q)(N,p_n)$ is assumed to be regular.

2 Known Theorems

Dealing with degree of approximation by the product mean Misra et al. [2] proved the following theorem using $(E,q)\left(\overline{N},p_n\right)$ mean of conjugate series of Fourier series. The result is as below.

Theorem 2.1 *If f is a 2π—periodic function of class Lip α, then degree of approximation by the product $(E,q)\left(\overline{N},p_n\right)$ summability means of the conjugate series* (10) *of the Fourier series* (9) *is given by* $\|\tau_n - f\|_\infty = O\left(\frac{1}{(n+1)^\alpha}\right), 0<\alpha<1$ *where τ_n is as defined in* (7).

Extending the above result Paikray et al. [3] established a theorem on degree of approximation by the product mean $(E,q)\left(\overline{N},p_n\right)$ of the conjugate series of Fourier series of a function under certain weaker condition of class $Lip(\alpha,r)$. The result is as follows:

Theorem 2.2 *If f is a 2π —Periodic function of class $Lip(\alpha,r)$, then degree of approximation by the product $(E,q)\left(\overline{N},p_n\right)$ summability means of on the conjugate series* (10) *of the Fourier series* (9) *is given by* $\|\tau_n - f\|_\infty = O\left(\frac{1}{(n+1)^{\alpha-\frac{1}{r}}}\right)$, $0<\alpha<1,\ r\geq 1$, *where τ_n is as defined in* (7).

3 Proposed Theorem

In this paper, replacing $(\overline{N}, p_n)$ by (N, p_n), we introduced a new theorem on degree of approximation by the product mean $(E, q)(N, p_n)$ of the conjugate series of the Fourier series of a function of class $Lip(\alpha, r)$. The proposed theorem is as below:

Theorem 3.1 *If* f *be* 2π—Periodic function of class $Lip(\alpha, r)$, $0<\alpha<1$, $r\geq 1$, $t>0$. *Then degree of approximation by the product* $(E, q)(N, p_n)$ *summability means on the conjugate series* (10) *of the Fourier series* (9) *is given by* $\|\tau_n - f\|_\infty = O\left(\frac{1}{(n+1)^{\alpha+1-\frac{1}{r}}}\right), 0<\alpha<1,\ r\geq 1,$ *where* τ_n *is as defined in* (7).

4 Required Lemmas

We require the following Lemmas to prove the theorem.

Lemma 4.1: [3]

$$|\bar{K}_n(t)| = O(n), \quad 0\leq t\leq \frac{1}{n+1}.$$

Lemma 4.2: [3]

$$|\bar{K}_n(t)| = O\left(\frac{1}{t}\right), \quad for \ \frac{1}{n+1}\leq t\leq \pi.$$

5 Proof of Theorem 3.1

Using Riemann—Lévesque theorem, for the nth partial sum $\bar{s}_n\,(f; x)$ of the conjugate series of Fourier series (10) of $f(x)$ and then following Titchmarch [6], we have

$$\overline{s_n}(f; x) - f(x) = \frac{2}{\pi}\int_0^\pi \psi(t)\ \overline{K_n}\ dt,$$

the (N, p_n) transform of $\overline{s_n}\,(f; x)$ using (2) is given by

$$t_n - f(x) = \frac{2}{\pi P_n}\int_0^\pi \psi(t)\sum_{k=0}^{n} p_{n-k}\ \frac{\cos\frac{t}{2} - \sin\left(n+\frac{1}{2}\right)t}{2\sin\left(\frac{t}{2}\right)}dt,$$

denoting the $(E,q)(N,p_n)$ transform of $\overline{s_n}(f;x)$ by τ_n, we have

$$\begin{aligned}
\|\tau_n - f\| &= \frac{2}{\pi(1+q)^n}\int_0^{\pi}\psi(t)\sum_{k=0}^{n}\binom{n}{k}q^{n-k}\left\{\frac{1}{P_k}\sum_{\upsilon=0}^{k}p_{k-\upsilon}\frac{\cos\frac{t}{2}-\sin\left(\upsilon+\frac{1}{2}\right)t}{2\sin\left(\frac{t}{2}\right)}\right\}dt\\
&= \int_0^{\pi}\psi(t)\overline{K_n}(t)\,dt\\
&= \left\{\int_0^{\frac{1}{n+1}} + \int_{\frac{1}{n+1}}^{\pi}\right\}\psi(t)\overline{K_n}(t)dt\\
&= I_1 + I_2,\ say
\end{aligned} \tag{19}$$

Now

$$\begin{aligned}
|I_1| &= \frac{2}{\pi(1+q)^n}\left|\int_0^{1/n+1}\psi(t)\sum_{k=0}^{n}\binom{n}{k}q^{n-k}\left\{\frac{1}{P_k}\sum_{\upsilon=0}^{k}p_{k-\upsilon}\frac{\cos\frac{t}{2}-\cos\left(\upsilon+\frac{1}{2}\right)t}{2\sin\frac{t}{2}}\right\}dt\right|\\
&\leq \left|\int_0^{\frac{1}{n+1}}\psi(t)\overline{K_n}(t)dt\right|\\
&= \left(\int_0^{\frac{1}{n+1}}(\psi(t))^r dt\right)^{\frac{1}{r}}\left(\int_0^{\frac{1}{n+1}}(\bar{K}_n(t))^s dt\right)^{\frac{1}{s}},\ \text{using Holder's inequality}\\
&\leq O\left(\frac{1}{(n+1)^{\alpha+1}}\right)\left(\int_0^{\frac{1}{n+1}} n^s\,dt\right)^{\frac{1}{s}},\ \text{using Lemma 4.1}\\
&= O\left(\frac{1}{(n+1)^{\alpha+1}}\right)\left(\frac{n^s}{n+1}\right)^{\frac{1}{s}} = O\left(\frac{n}{(n+1)^{\alpha+\frac{1}{s}+1}}\right)\\
&= O\left(\frac{1}{(n+1)^{\alpha+\frac{1}{s}}}\right) = O\left(\frac{1}{(n+1)^{\alpha-\frac{1}{r}+1}}\right)
\end{aligned} \tag{20}$$

Next

$$|I_2| \leq \left(\int_{\frac{1}{n+1}}^{\pi} (\psi(t))^r dt \right)^{\frac{1}{r}} \left(\int_{\frac{1}{n+1}}^{\pi} (\bar{K}_n(t))^s dt \right)^{\frac{1}{s}} \text{ using Holder's inequality}$$

$$\leq O\left(\frac{1}{(n+1)^{\alpha+1}}\right) \left(\int_{\frac{1}{n+1}}^{\pi} \left(\frac{1}{t}\right)^s dt \right)^{\frac{1}{s}}, \text{ using Lemma 4.2} \tag{21}$$

$$= O\left(\frac{1}{(n+1)^{\alpha+1}}\right) \left(\frac{1}{n+1}\right)^{\frac{1-s}{s}} = O\left(\frac{1}{(n+1)^{\alpha-\frac{1}{r}+1}}\right)$$

Then from (20) and (21), we have

$$|\tau_n - f(x)| = O\left(\frac{1}{(n+1)^{\alpha-\frac{1}{r}+1}}\right), \text{ for } \quad 0<\alpha<1.$$

Hence,

$$\|\tau_n - f(x)\|_{\infty} = \sup_{-\pi<x<\pi} |\tau_n - f(x)| = O\left(\frac{1}{(n+1)^{\alpha-\frac{1}{r}+1}}\right), 0<\alpha<1,\ r\geq 1$$

This completes the proof of the theorem.

6 Conclusion

The paper provides an analytic idea in the field of summability theory using product mean under certain weaker conditions. In the future, the present work can be extended to establish some theorems on different indexed summability factors of Fourier series as well as conjugate series of Fourier series under certain order of Lipschitz condition.

References

1. Hardy, G.H.: Divergent Series, vol. 70, no. 19, 1st edn. Oxford University press, Oxford
2. Misra, U.K., Misra, M., Padhy, B.P., Buxi, S.K.: On degree of approximation by product means of conjugate series of Fourier series. Int. J. Math. Sci. Eng. Appl. **6**(122), 363–370 (2012)

3. Paikray, S.K., Misra, U.K., Jati, R.K., Sahoo, N.C.: On degree of approximation of Fourier series by product means. Bull. Soc. Math. Serv. Stand. **1**(4), 12–20 (2012)
4. Titchmarch, E.C.: The Theory of Functions, pp. 402–403. Oxford University Press, Oxford (1939)
5. Zygmund, A.: Trigonometric Series, vol. I, 2nd edn, Cambridge University press, Cambridge (1959)

A Novel Approach for Intellectual Image Retrieval Based on Image Content Using ANN

Anuja Khodaskar and Sidharth Ladhake

Abstract This paper deals with a novel approach for intellectual Image Retrieval (IIR) based on image content analysis using Artificial Neural Network. The Multilayer Back Propagation Feed Forward algorithm is proposed for interactive image retrieval, which takes query by image as an input and retrieves the most relevant images from the image dataset. The content based semantic features are extracted for around 500 images from image corpus and applied for training of the Neural Network. The outcome of the rigorous experimentations reveals that application of ANN for IIR enhances the effectiveness of the performance of image retrieval.

Keywords Artificial neural network · Image content analysis · Intellectual image retrieval · Multilayer back propagation feed forward algorithm · Semantic features

1 Introduction

Content based image retrieval (CBIR) has become one of the most active research areas in the past few years. Image retrieval from internet or web is getting overwhelming response from the users. Image retrieval is the processing of searching and retrieving images from a huge dataset. As the images grow complex and diverse, retrieval the right images becomes a difficult challenge. For centuries, most of the images retrieval is text-based which means searching is based on those keyword and text generated by human's creation [1]. The text-based image retrieval systems only concern about the text described by humans, instead of looking into the content of images. Images become a mere replica of what human has seen since

A. Khodaskar (✉) · S. Ladhake
SIPNA COET, Amravati 444606, India
e-mail: annujha51@gmail.com

S. Ladhake
e-mail: sladhake@yahoo.com

L.C. Jain et al. (eds.), *Computational Intelligence in Data Mining - Volume 3*,
Smart Innovation, Systems and Technologies 33, DOI 10.1007/978-81-322-2202-6_63

birth, and this limits the images retrieval. This may leads to many drawbacks which will be state in related works.

Content-based image retrieval [1] has become a prominent research topic because of the proliferation of video and image data in digital form. The increased bandwidth availability to access the internet in the near future will allow the users to search for and browse through video and image databases located at remote sites. Therefore, fast retrieval of images from large databases is an important problem that needs to be addressed. High retrieval efficiency and less computational complexity are the desired characteristics of CBIR systems. In conventional image databases, images are text-annotated and image retrieval is based on keyword searching. Some of the disadvantages of this approach are: 1. Keyword based image retrieval is not appropriate because there is no fixed set of words that describes the image content; 2. keyword annotation is very subjective. To avoid manual annotation, an alternative approach is content-based image retrieval (CBIR), by which images would be indexed by their visual content such as color, texture, shape etc. And the desired images are retrieved from a large collection, on the basis of features that can be automatically extracted from the images themselves [2]. Considerable research work has been done to extract these low level image features [3, 4], evaluate distance metrics, and look for efficient searching schemes [5, 6]. Basically, most CBIR systems work in the same way: A feature vector is extracted from each image in the database and the set of all feature vectors is organized as a database index. At query time, a feature vector is extracted from the query image and it is matched against the feature vectors in the index. The crucial difference between the various systems lies in the features that they extract and in the algorithms that are used to compare feature vectors.

1.1 Image Retrieval Systems

The content-based image retrieval (CBIR) is an image retrieval system based automatic retrieval of images from a database by its content such as color, texture and shape. The typical CBIR system performs two major tasks. The first one is feature extraction, where a set of features is extracted to describe the content of each image in the database. The second task is the similarity measurement between the query image and each image in the database, using the feature extraction. The feature extraction values for a given image are stored in a descriptor that can be used for retrieving similar images. Image descriptors are descriptions of the visual features of the contents in images that produce such descriptions. They describe elementary characteristics, such as color, texture, shape or motion, among others. The key to a successful retrieval system is choosing the right features to accurately represent the images and the size of the feature vector. The features are either global, for the entire image, or local, for a small group of pixels. According to the methods used in CBIR, features can be classified into low-level and high-level features. Color features are the most widely used low-level features for image retrieval because it is one of the most straightforward features utilized by humans for visual recognition. However, image retrieval using color features often gives disappointing results because, in many cases, images with similar colors do not have similar content.

1.2 Image Feature Extraction

Image Feature Extraction is an important process involved in image retrieval. Image features are extracted in spatial-domain and frequency domain. For extracting features in spatial domain we have used the Principal Component Analysis (PCA) algorithm. The advantage of PCA algorithm is that it reduces the dimensions of the feature space. The output of PCA algorithm is Eigen values and Eigen vectors which are representing spatial-domain features of the image. Frequency domain feature extraction part involves computation of DCT and DWT of the image. The DCT and DWT coefficients serve as frequency-domain features. The feature extraction process is depicted in Fig. 1.

1.3 Artificial Neural Network

Artificial neural networks are computational models based on concept of human central nervous systems, designed interconnected neurons which computed values from input, used in machine learning and pattern recognition. Pattern recognition, prediction, optimization, control etc. are application area of artificial neural. Neural networks are emerging successful classifier which search for a functional relationship between the group membership and the attributes of the

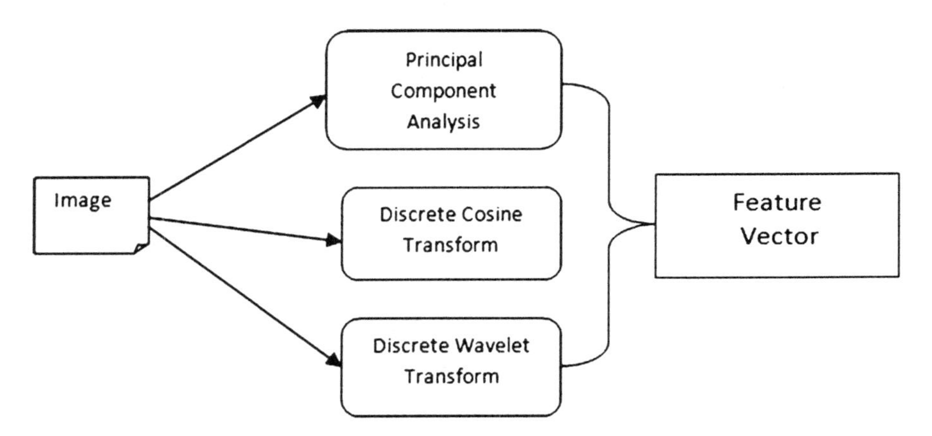

Fig. 1 Feature extraction process

object. Neural networks have been successfully applied to a variety of real world classification applications include bankruptcy prediction, handwriting recognition, speech recognition, product inspection, fault detection, medical diagnosis. Neural networks is used which improve classifier performance. Neural networks work as competitive alternative to traditional classifiers for many practical classification problems. Neural classification includes network training, model design and selection sample size issues.

2 Related Work

Sadek et al. presented Image Retrieval using Cubic Splines Neural Networks [7]. They proposed a new architecture for a CBIR system; the Splines Neural Network-based Image Retrieval (SNNIR) system. SNNIR utilized a rapid and precise network model that employs a cubic-splines activation function. By using the cubic-splines network, the proposed system could determine nonlinear relationship between images features, which gives more accurate similarity comparison between images.

Olkiewicz and Markowska-Kaczmar proposed Emotion-based Image Retrieval using artificial Neural Network Approach [8]. They have presented the approach for content based image retrieval systems based on its emotional content. They examine possibilities of use of an artificial neural network for labelling images with emotional keywords based on visual features only and examine an influence of used emotion filter on process of similar images retrieval.

Venkatraman and Kulkarni designed MapReduce neural network framework for efficient content based image retrieval from large datasets in the cloud computing background, which make CBIR system effective and scalable for real-time processing of very large image collections [9]. Neural network in combination with other techniques like fuzzy logic are used for images retrieval to improve performance of image retrieval based on their content. Verma and Kulkarni proposed methodology by combining neural networks and fuzzy logic for interpretation of queries, feature extraction and classification of features in CBIR [10], which improve the overall performance of the CBIR systems. Chowdhry proposed artificial neural network in combination with support vector machine and single value decomposition for image retrieval which improve system performance as well as bridge semantic gap [11]. They use back propagation algorithm for training and testing data.

2.1 Advanced Development in ANN and Image Retrieval

Images retrieval system worked based on two phases, enrolment phase and retrieval phase for retrieval performance improvement, enrolment phase consist of feature extraction and second, retrieving phase use the artificial neural network and similarity measurement [2]. The features of multilevel Discrete Wavelet Transform and Feed Forward Artificial Neural Network are combined to denoising image [12]. Low-level attributes presented in form of high-level semantic concepts by using a high- level characteristics vector, which is formed by using the artificial neural network Intelligence [8].

Human emotions is one of the important factor of searching images in an image database through content based image retrieval systems in which use an artificial neural network for labeling images with emotional keywords based on visual features [7]. The artificial intelligence explosive ordnance disposal system is a neural network AI-based multiple-incident identification, recording, and tracking system, featuring state-of-the-art search, retrieval, and image and text management [12]. Artificial neural network based model is designed to retrieval of the direct normal, diffuse horizontal and global horizontal irradiances using SEVIRI images [13]. A new matching strategy for content based image retrieval system is based on artificial neural network which selected features of query image are the input and its output is one of the multi classes that have the largest similarity to the query image [14]. Feature space of a content-based retrieval system is nonlinearly transformed into a new space, where the distance between the feature vectors is adjusted by learning and, transformed by an artificial neural network architecture [15]. In the proposed system, artificial neural network is used for interpretation of semantic concepts extraction. Multi-layer feed forward back propagation is trained for 500 sample images. The colour, texture and shape features are used for training.

3 Proposed System

Let $f(x, y)$ be an input image with dimension $[\mathrm{m} \times \mathrm{n}]$.

Let $S = \{s_i,\ i = 1, \ldots, N\}$ denote the set of regions produced for an image by segmentation,

$O = \{o_j,\ j = 1, \ldots, M\}$ denote the set of objects extracted from segments.

$C = \{c_k,\ k = 1, \ldots, M^0\}$ denote the semantic concepts.

$F_s = \{\ f_1, \ldots, M + M^0\}$ is the set of semantic features.

Let $\lambda : f \to F_s$ is a transformation function which maps from input image to set of semantic features.

$R = \{r_k,\ k = 1, \ldots, K\}$ denote the set of supported spatial relations. Then, the degree to which *si* satisfies relation r_k with respect to s_j can be denoted as *Irk* (*si, sj*), where the values of function *Irk*.

More specifically, the mean values, I_{rkmean}, of I_{rk} are estimated, for every k over all region pairs of segments assigned to objects (o_p, o_q), $p \neq q$. Additionally, the variance values σ^2_{rk} are obtained for each of the relations.

$$\sigma_{rk}^2 = \frac{\sum_{i=1}^{N} (I_{r_k i} - I_{r_k mean})^2}{N} \tag{1}$$

where, N denotes the plurality of object pairs (op, oq) for which rk is satisfied.

Feed forward back propagation algorithm of ANN is used to classify images. Ability of neural network is to provide machine intelligence to system and actively support pattern recognition. Backpropagation algorithm is used to handle errors in multilayer. Multi-layer means multiple layers of weights. The back propagation algorithm uses the Delta Rule. A multi-layer, feedforward, backpropagation neural network consist of three layers nodes, input layer, one or more intermediate layer and output layer. Depending on the problem, the output layer can consist of one or more nodes. In classification applications, number of output nodes depends on classes.

3.1 Proposed Framework

A framework for Interactive Image Retrieval (IIR) based on image content analysis using Artificial Neural Network is shown in Fig. 2. The Multilayer Back Propagation Feed Forward algorithm is proposed for interactive image retrieval, which takes query by image as an input and retrieves the most relevant images from the image dataset.

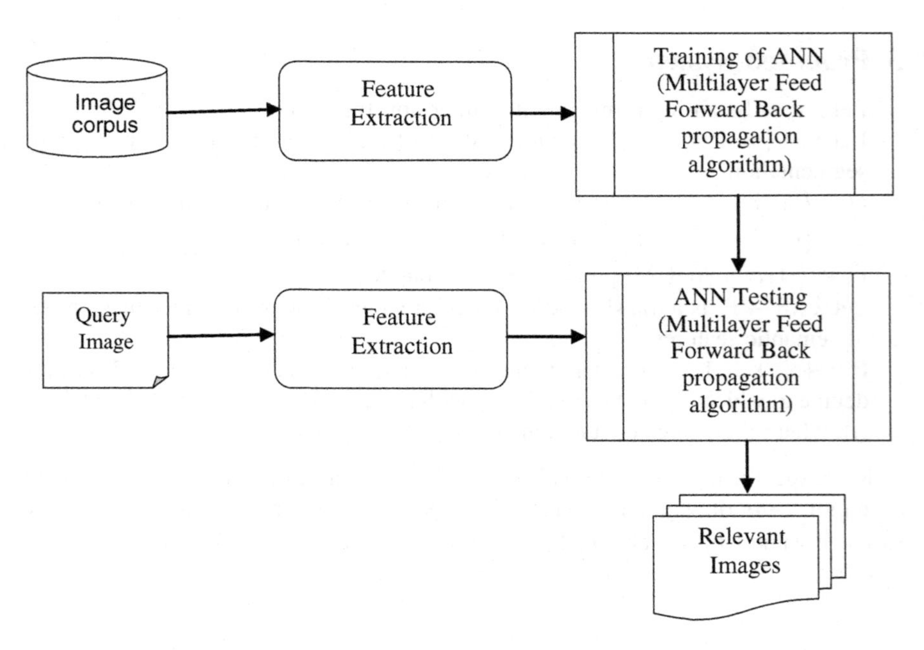

Fig. 2 Proposed framework

3.2 Algorithm

- **Input**

  f(x,y) → Input image having dimension [M X N]

  Num → Number of images used for training

- **Output**

  F(x,y) → Output images with size [M X N]

- **Method :**

Step 1) *for* i = 1 : Num

```
        Select input image fi(x,y)
        Img[M x N, i] = reshape(fi(x,y), M x N], 1);
        X(i)=PCA(Img);
        Y(i)=DCT(Img);
        Z(i)=DWT(Img);
        P = [X Y Z];
        i=i+1;
    end
```

Step 2) net = newff(P,T,[10 10 10], {'tansig', 'tansig'

```
        'tansig'},'learndgm','mse'} /* training ANN */
          /* P → training matrix & T→ target vector
```

Step 3) net=train(P,T);

```
        Y=sim(P,Inp);
```

In the first part, Semantic features are extracted from the image database. This extracted feature space is provided to ANN's Multilayer back-propagation network as training matrix. In the second part, the query image is given for feature extraction and applied to Multilayer back-propagation network for testing. The result of ANN will be used for accomplishing the image content analysis.

4 Result

Multi-Layer Feed Forward Back-Propagation Network gives human intelligence classification and learning ability in relevance feedback. We analyzed 500 images data set in training phase by changing threshold value as per requirement. Experimental result shows improves classification performance. Network weights are randomly varies in training phase for every execution by keeping other factors like training data, learning rate, are kept constant. The algorithm gives deviation in results when work with small data set. When we increase data set on the basis of ground truth, result is variations are minimized and in order optimise result combine the results of multiple neural network classifications.

4.1 Image Database

In this research work, we have used the standard image database comprising of 500 images. It contains images from different categories such as buses, flowers, nature, river etc. The subset of this database is shown in Fig. 3.

A given query image is feature extracted and searched for similar images. For each query image, relevant images (Fig. 5) are considered to be those and only those which belong to the same category as the query image as shown in Fig. 4.

The performance of ANN training and testing is shown in Fig. 6. It shows number epochs required for training, Mean Square Error (MSE) for Train, Validation, Test processes.

In this section, we present experimental results from testing the proposed approach in the domain of beach vacation images. A set of 50 randomly selected images belonging to the beach vacation domain were used to assemble a training set for the low-level implicit knowledge acquisition. In Table 1, quantitative performance measures are given in terms of precision and recall.

Fig. 3 Sub-set of images used for training

Fig. 4 Query image

Fig. 5 Relevant images retrieved

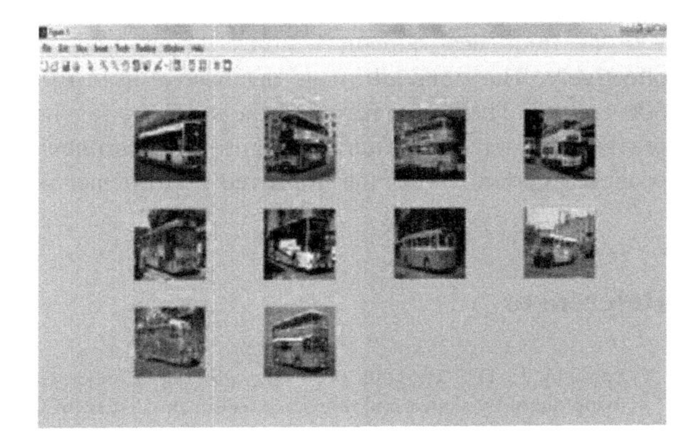

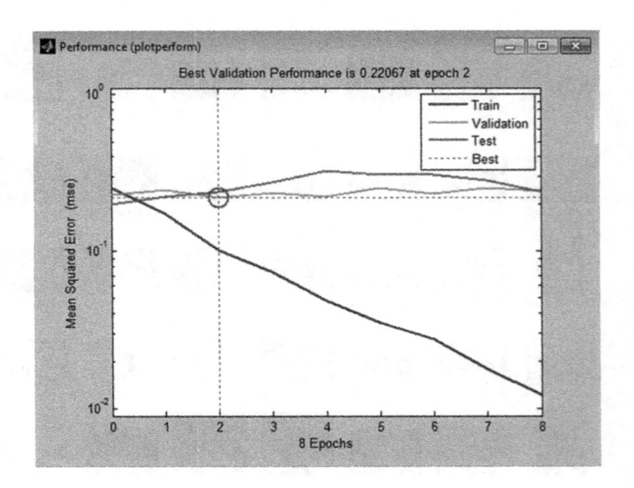

Fig. 6 ANN performance

Table 1 Numerical evaluation for the beach vacation domain

Object	Precision	Recall
Sky	57.31	95.11
Sea	93.68	72.48
Sand	87.15	90.24
Person	80.00	76.28
Accuracy	79.53	83.52

5 Conclusion

In this paper, a novel approach for intellectual Image Retrieval (IIR) based on image content analysis using Artificial Neural Network is presented. The Multilayer Back Propagation Feed Forward algorithm is proposed for interactive image retrieval. The feature vector used for ANN training comprises of spatial-domain as well as frequency domain features which helps ANN to classify the images effectively. The standard image database is used for ANN training with a size of 500 images. The performance of the system is evaluated on the basic parameters, precision and recall. After the rigorous experimentation it is revealed that, the proposed system shows the improved performance of image retrieval.

References

1. Sciascio, E. Di., Mingolla G., Mangiallo, M.: Content-based image retrieval over the web using query by sketch and relevance feedback. Visual Inf. Inf. Syst. Lect. Notes Comput. Sci. **1614**, 123–130 (1999)
2. Bhagat, A.P., Atique, M.: Web based image retrieval system using color, texture and shape analysis: comparative analysis. Int. J. Adv. Comput. Res. **3** (2013)

3. Philippe, H.G., Matthieu, C.: Active learning methods for interactive image retrieval. IEEE Trans. Image Process. (2008)
4. Jian, M., Dong, J., Tang, R.: Combining color, texture and region with objects of user's interest for CBIR. IEEE (2007)
5. Marco, A., Silvana, G.D., Marcello, G.: Design and implementation of web-based systems for image segmentation and CBIR. IEEE Trans. Instrum. Meas. **55** (2006)
6. Bispo dos Santos, J., de Almeida, J.R., Silva, L.A.: Pattern recognition in mammographic images used by the residents in mammography. In: IEEE Conference on Computer Medical Applications, pp. 1–6 (2013)
7. Sadek, S., Al-Hamadi, A., Michaelis, B., Sayed, U.: Image retrieval using cubic splies neural networks. IJVIPNS-IJENS **9**, 5–10 (2009)
8. Olkiewicz, K.A., Markowska-Kaczmar, U.: Emotion-based image retrieval—an artificial neural network approach. In: IEEE Proceedings, pp. 89–96 (2010)
9. Venkatraman, S., Kulkarni, S.: Hybrid intelligent systems. In: IEEE International Conference pp. 63–68 (2012)
10. Verma, V., Kulkarni, S.: Neural network for content based image retrieval. IGI pub (2007)
11. Chowdhry, B.S.: Image Retrieval Based on Color and Texture Feature using Artificial Neural Network. Springer IMTIC, pp. 501–512 (2012)
12. Saikia, T., Sarma, K.K.: Multilevel-DWT based image de-noising using feed forward artificial neural network. In: IEEE Conference on Signal Processing and Integrated Networks, pp. 791–794 (2014)
13. Holmström, L., Koistinen, P.: Using additive noise in back-propagation training. IEEE Trans. Neural Netw. 24–38 (1992)
14. Leverington, D.W.: Discriminating lithology in arctic environments from earth orbit: an evaluation of satellite imagery and classification algorithms. Ph.D. Thesis (2001)
15. Awad, Mohamad: Sea water chlorophyll-a estimation using hyperspectral images and supervised artificial neural network. Elsevier Ecol. Inf. **24**, 60–68 (2014)
16. George, L.E., Mohammed, E.Z.: Tissues image retrieval system based on co-occuerrence, run length and roughness features. In: IEEE Conference on Computer Medical Applications, pp. 1–6 (2013)
17. Houmao, W., Jiakui T.: Aerosol retrieval from remote sensing image using artificial neural network. In: IEEE International Conference on Computer Application and System Modeling, vol. 5, pp. V5-551–V5-555 (2010)
18. ORibeiro, E.F., Barcelos, C.A.Z., Batista, M.A.: High-level semantic based image characterization using artificial neural networks. In: IEEE Conference on Intelligent Systems Design and Applications, pp. 357–362 (2007)
19. Madrid, R., Williams, B., Holland, J.: Artificial intelligence for explosive ordnance disposal system. IEEE Conf. Neural Netw. **1**, 378–383 (1992)
20. Murari, A. (ed.): New techniques and technologies for information retrieval and knowledge extraction from nuclear fusion massive databases. In: IEEE Symposium on Intelligent Signal Processing, pp. 1–6 (2007)
21. Hongwei, Z., Xiong, X., Hasegawa, O. A.: Load-balancing self-organizing incremental neural network. IEEE Trans. Neural Netw. Learn. Syst. **25**, 1096–1105 (2014)
22. David. L.: A Basic Introduction to Feedforward Backpropagation Neural Networks. Geosciences Department, Texas Tech University, Lubbock (2009)
23. Zhang, G.P.: Neural networks for classification: a survey. IEEE Trans. Syst. Man Cybern—Part c: Appl. Rev. **30**. 451–455 (2000)
24. Yehia, E. (ed.): Artificial neural network based model for retrieval of the direct normal, diffuse horizontal and global horizontal irradiances using SEVIRI images. Elsevier Solar Energy **89** 1–16 (2013)
25. Elalami, M.E.: A new matching strategy for content based image retrieval system. Elsevier Appl. Soft Comput. **14**, 407–418 (2014)

26. Pulvirenti, L.E.: Empirical algorithms to retrieve surface rain-rate from special sensor Microwave Imager over a mid-latitude basin. IEEE Geosci. Remote Sens. Sympos. **3**, 1872–1874 (2002)
27. Thilagavathy, A., Aarthi, K., Chilambuchelvan, A.: A hybrid approach to extract scene text from videos. IEEE Comput. 1017–1022 (2012)
28. Carkacioglu, A., Vural, F.-Y.: Learning similarity space. IEEE Conf. Image Process. **1**, I-405–I-408 (2002)
29. Wuryandari, A.I., Wijaya, R.: Gathering information realtime and anywhere (GIRA) using an ANN algorithm. In: IEEE Conference on System Engineering and Technology, pp. 1–6 (2012)
30. Nilpanich, S., Hua, K.A., Petkova, A., Ho, Y.H.: A Lazy Processing approach to user relevance feedback for content-based image retrieval. In: IEEE Symposium on Multimedia, pp. 342–346 (2010)

Hierarchical Agents Based Fault-Tolerant and Congestion-Aware Routing for NoC

Chinmaya Kumar Nayak, Satyabrata Das and Himnsu Sekhar Behera

Abstract In communication medium a single fault will affect the complete system. So in the designing of Network-on-Chip (NoC) based systems the reliability is an important aspect. Also we have to concentrate on the performance improvement in the fault tolerant NoC architectures. In this paper, we are going to achieve high level performance by using hierarchical agents by proposing Fault-tolerant NoC architecture. The fault information will be collected and will be distributed after processing from these agents which are placed everywhere in the network. Along with that the permanent Faults information that occur in the interfaces of network, links and in different parts of the routers will be exploited from the enhanced fault tolerant and congestion aware routing method.

Keywords NoC · Congestion aware routing · DTM

1 Introduction

The faults will incurs in more numbers in the NoC based systems on chip as the technologies scale down, mainly while running at high clock frequencies. In previous work such as paper [1, 2], it is stated that Noc is the favorable interconnection infrastructure for inter-core communication due to its scalability, higher throughput and reusability. The overall performance, reliability and power consumption are some of the important issues in NoC-based many-core systems, which will be more

C.K. Nayak (✉)
Department of CSE, GITA, BBSR, Burla, Odisha, India
e-mail: cknayak85@gmail.com

S. Das · H.S. Behera
Department of CSE & IT, VSSUT, Burla, Odisha, India
e-mail: sb_das@hotmail.com

H.S. Behera
e-mail: hsbehera_india@yahoo.com

L.C. Jain et al. (eds.), *Computational Intelligence in Data Mining - Volume 3*,
Smart Innovation, Systems and Technologies 33, DOI 10.1007/978-81-322-2202-6_64

important when there is increase in number of cores and the number of network nodes.

In NoC-based many-core systems, fault tolerance against permanent faults is necessary since the current and future process technologies have a considerable amount of device failures that may occur in both manufacturing and operational phases. One of the reasons for the performance degradation in NoCs tolerating permanent faults is the lack of non-local fault awareness. In this paper, we propose a very low cost architecture to acquire regional and global fault information through hierarchical and distributed agents [1, 3]. This architecture includes an agent-based management structure and a new routing method capable to exploit nonlocal fault information to tolerate permanent faults in the links, network interfaces and different parts of the routers. So far, permanent fault tolerance has been considered in many fault tolerant routing algorithms. However, due to the size of complex NoC-based systems which may include tens or hundreds of cores, as the methods introduced in paper [3] through paper [4], we only consider distributed and scalable routing algorithms that are capable to utilize the proposed architecture.

The previous works related to the hierarchical agents can be found in paper [5] through paper [6]. A NoC monitoring scheme based on hierarchical agents is addressed mainly to minimize network power consumption is given in paper [5]. A Dynamic Thermal Management (DTM) solution is proposed that uses a system-level approach featuring agent-based power distribution to balance the power consumption of multi- and many-core architectures is given in paper [7]. It uses software-based agents and they should be run on the processing elements of the system. System level design principles and the basic concepts of the general approach for the hierarchical agent monitoring is proposed for parallel and distributed systems in the form of high level abstraction given in paper [8]. In addition, it includes an approach for dynamic voltage and frequency scaling used in power monitoring. However, it does not present any detailed design or physical platform especially for fault-tolerance. Had proposed a preliminary agent based management structure, the basic agent tasks and a simple fault information classification is given in paper [6].

In this paper, we address the design and usage of hardware agents inside the network nodes in a distributed and hierarchical management structure [4, 9]. The proposed architecture optimally utilizes local and non-local fault information in addition to the congestion information. For this purpose, a detailed fault information classification is provided for the routing process. The proposed routing method decreases the overall packet latencies and handles the fault information and congestion information, simultaneously. Using the hierarchical agents, the appropriate portions of the local fault information in each node will be sent to the direct and indirect neighbour nodes to be used as regional or global fault information. The routing algorithms should be able to utilize the new fault information provided during the life time of the system, in their structures. This architecture can be applied to any router incorporating a distributed and scalable fault-tolerant routing algorithm.

Further, In Sect. 2 the fault information classification needed for the routing process is presented, in Sect. 3 the proposed agent-based management architecture is introduced, and in Sect. 4 the proposed fault-tolerant routing method is explained. The experimental results are presented in Sect. 5 and finally, conclusion is given in Sect. 6.

2 Fault Information Classifications

In this section, we use the basic information of routing process to obtain a comprehensive classification of the fault information beneficial for distributed and fault-aware routing algorithms. A typical NoC router architecture is depicted in Fig. 1. This router consists of a controller including routing unit, switch allocator and virtual channel (VC) allocator. In addition, it comprises a crossbar switch, input ports and output ports. The crossbar switch includes a multiplexer for each output port. Each input port includes the buffers for virtual channels, a multiplexer and a demultiplexer, and each output port is directly connected to an outgoing link. Based on paper [10] some test and fault detection circuits can be incorporated in the NoC routers and links to detect the permanent faults in each sub-block with acceptable overheads. Therefore, it is assumed that we are aware about the faultiness of five input ports, four unidirectional outgoing links, routing unit, switch allocator, VC allocator, and at most five multiplexers in the crossbar switch, in a five-port router in addition to the faultiness of the Network Interface (NI) and the local core or Processing Element (PE).

If we assume X can be N, S, E or W which mean north, south, east or west directions, respectively, then we can say the X output direction of a router is unusable for the routing process if the X outgoing link or the crossbar multiplexer

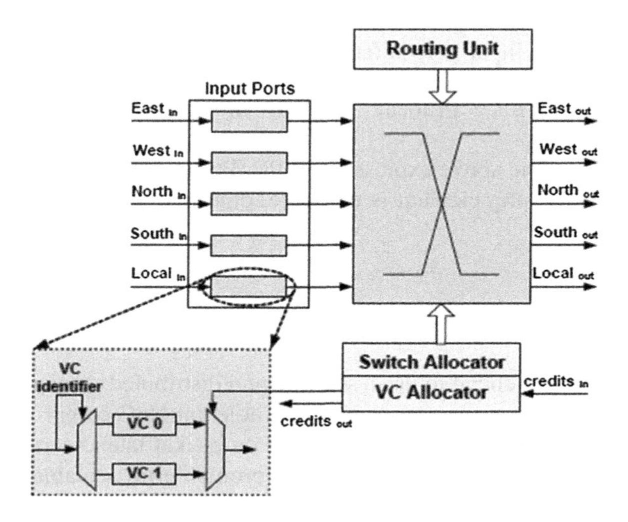

Fig. 1 A typical NoC router architecture with two virtual channels

in the X output direction or the input port in the X neighbour router is faulty. Logical OR can be used to state this condition as (1) by using the appropriate signals from the fault detection circuits:

$$Dir_X^{out} = Link_X^{out} \quad \text{or} \quad MUX_X \quad \text{or} \quad In\_Port_{1-X}^{X\_router} \tag{1}$$

In (1) all terms are one-bit status data showing that if any term equals '1' its corresponding component is faulty, otherwise it is healthy. In this equation, LinkX out and MUXX mean the status of unidirectional outgoing link and the crossbar multiplexer, respectively, in the X output direction of the current router, and In_Port(1-X) X_router stands for the status of the input port in the X neighbor router. In addition, (1-X) stands for the opposite direction of X, which means S, N, W and E for N, S, E and W directions, respectively. This equation is based on the fact that we can model a faulty crossbar multiplexer by assuming its corresponding unidirectional outgoing link to be faulty, and a faulty input port by assuming its unidirectional incoming link to be faulty.

The whole node should be considered as faulty due to the impact of some faulty components inside a router because of that the router is unable to perform its main task. This case occurs when the controller is faulty based on the expression (Exp. 1)

$$\text{Ctrl} = \text{routing\_unit or switch\_allocator or VC\_allocator} \tag{Exp.1}$$

As per exp. (Exp. 1) we should consider the control part, as a result the whole node is faulty if the routing unit, switch allocator or VC allocator is faulty. For this case, we dedicate a bit called Node as a part of fault information regarding to this situation.

In NoC-based many-core systems, the local cores or the processing elements are connected to the routers via the network interfaces [7, 8]. If a processing element is unusable, the high level system manager should either migrate its task to other processing elements or perform a remapping process. In a NoC router we assume the processing element is unusable if it is faulty, its network interface is faulty, its related input port is faulty or output crossbar multiplexer is faulty based on (2):

$$\text{PE} = \text{PElocal} \quad \text{or} \quad \text{NI} \quad \text{or} \quad \text{In\_Portlocal} \quad \text{or} \quad \text{MUXlocal} \tag{Exp.2}$$

In the above expression, if PE (Processing elements) equals '1', the local core or processing element is unusable, otherwise it is usable.

3 Proposed Agent-Based Management Architecture

It is beneficial to use a scalable and distributed management method to decrease the overall packet latency in a fault-tolerant NoC comprising a large number of nodes. Here, agent-based management on packet latencies architecture is proposed which is also hierarchical to be more profitable for scalable NoCs.

3.1 Background

There are two types of agents in the proposed management structure:

- **Node agent (NA)**: Each node includes an agent called the node agent which collects, combines and distributes the fault information related to the components of its own node in addition to the local congestion information. Besides, it updates the LFR, RFR1 and RFR2.
- **Cluster agent (CA)**: Each cluster that includes a number of nodes is controlled by a cluster agent. A CA configures the NAs inside the cluster by sending the new fault information obtained from the other NAs inside the cluster or from the other CAs.

We need a cluster-based NoC if we want to obtain an efficient management approach when running different tasks. The operating system should map different tasks onto the network clusters. The incorporated agent hierarchy is shown in Fig. 2. This agent hierarchy differs from that of proposed in the previous works paper [5], paper [8]. This is due to the fact that in the proposed structure, for faster reconfiguration the NAs communicate with their neighbour NAs even if they are located inside different clusters. This is advantageous because in general, a task running in a system with many processing elements may require more than a cluster. On the other hand, clusters running the same task are not necessarily neighbour clusters. However, the routers should be aware about their neighbours to select the best path for sending the packets to their destinations, and to expedite this awareness their NAs should exchange the required fault and congestion information.

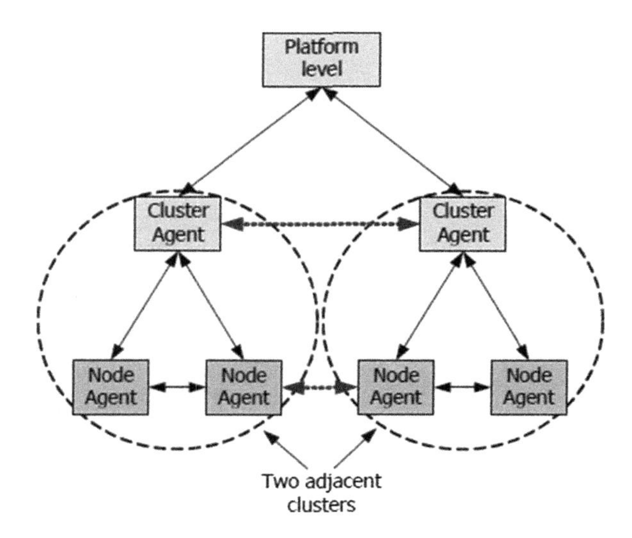

Fig. 2 Hierarchical agents in two neighbour clusters

3.2 Interconnections for Hierarchical Agents

Figure 3 shows a small 3 × 3 network with an agent in each node. In this figure, R, NI and PE correspond to the router, network interface and processing element, respectively. In addition, the agents can be node or cluster agents but the number of cluster agents is much less than the number of node agents. For example, in a regular mesh network each 3 × 3 sub network can be a cluster with a cluster agent in the centre. The proposed agent-based management architecture uses two types of communications: a physically separate network for only peer to peer communication between the NAs, and the baseline data network for control packet communication with a higher priority compared to data packets. The former is shown in Fig. 3 between the agents and the latter is the network connecting the NoC routers. The control packets including the reconfiguration and fault information are exchanged between a NA and its related CA in addition to the communication between the CAs. The manner in which these communication structures are utilized is discussed in the next subsection.

3.3 Agent Tasks

After detecting faults, they should be processed to obtain the useful and classified fault information. Moreover, the fault and congestion information should be distributed among the appropriate nodes. The proposed agent-based management architecture performs these tasks by utilizing different agents and interconnections.

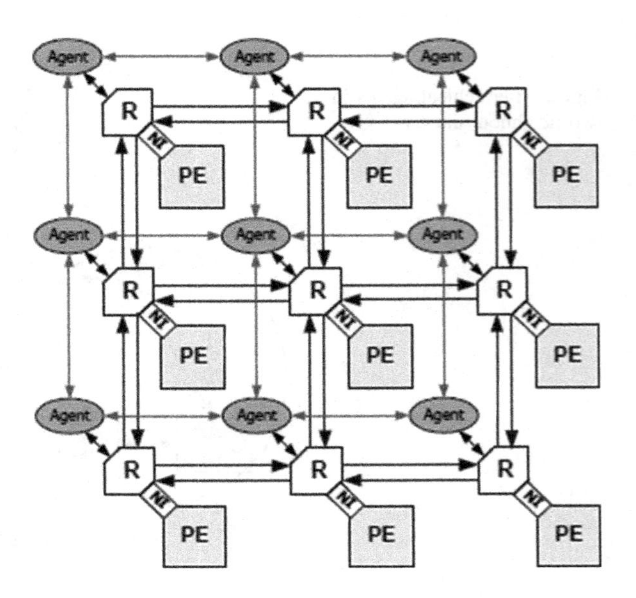

Fig. 3 A 3 × 3 agent-based NoC with two types of interconnections

3.3.1 Node Agent

In the proposed agent hierarchy shown in Fig. 3, the node agents distribute the essential fault information needed for the routing process to their direct neighbours in addition to the local congestion information.

3.3.2 Cluster Agent

A cluster agent informs the platform level manager about the critical failures (PE or node failures) occurred inside the cluster and receives reconfiguration commands from it after remapping the tasks on the nodes or performing task migration. However, a CA itself is informed by the NAs about the critical failures.

4 Proposed Fault-Tolerant Routing Algorithms

To utilize the agent-based architecture and different types of fault and congestion information, a fault-tolerant routing method is required. In addition, as the proposed architecture uses the same wires to transmit fault and congestion information, the wires utilized in the congestion-aware routing methods can be used to convey the fault information, too. The convenient routing methods for the proposed architecture are the modified methods incorporating one of two neighbouring regions in Fig. 3 with unidirectional or bidirectional links. In this paper, a routing method is formed based on the principles introduced in paper [4]. Similar to paper [4] it is congestion-aware, contour based, and it uses only two VCs. This method uses 5 bit wires to send and receive congestion information and in order to transmit the fault information, too, it uses an extra one bit wire in addition to the same 5-bit wires. Moreover, it utilizes the LFR, RFR1 and RFR2. The RFR2 is formed based on the unknown destination area in regular 3 × 3 clusters. Based on the path selection strategy used in paper [4] and according to Fig. 4a, the source node S1 only needs the status of the output directions 1, 2, 3, 4 (obtainable from the cluster agent) when it wants to send the packets to a destination outside its neighbouring region (D1, D1′, D1″). However, S2 only needs the status of the output directions 3, 5 when it wants to send the packets to D1 and D2, respectively. There is not any need to other directions in the unknown destination area when the source node is S1 or S2. Thus, the size of RFR2 will be at most four bits. As a result, by using the new fault information (RFR2) the designed routing algorithm definitely selects the shortest paths in clusters (P1 instead of P2 and P3 instead of P4 in Fig. 4b) unlike the method in paper [4].

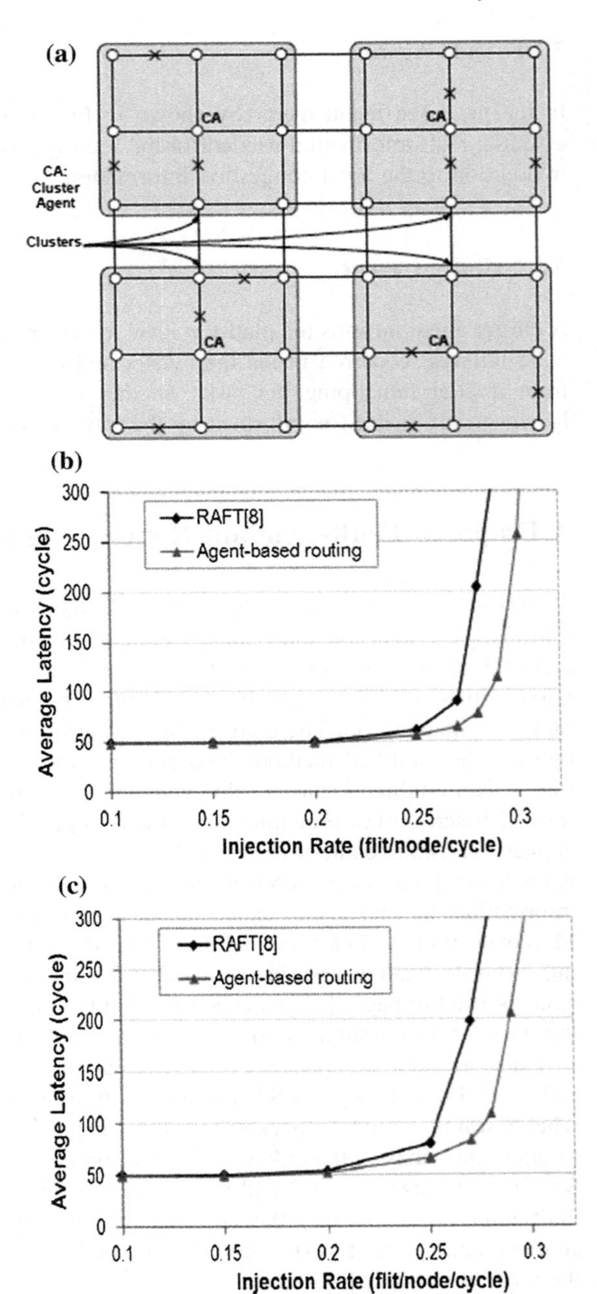

Fig. 4 **a** A clustered 6 × 6 NoC with 20 % faulty links. **b** Average packet latencies under the local traffic pattern. **c** Under the hot-spot traffic pattern

5 Experiment Results

To analyze the effect of the proposed agent-based architecture on packet latencies, a 6 × 6 NoC modeled by VHDL is simulated with the input buffer size of four flits in each of two virtual channels and the packet length of 16 flits under the local uniform and hot-spot traffic patterns. Two different routing algorithms are used: The RAFT method paper [4] not using any agents, and the proposed method that uses hierarchical agents. In Fig. 4a, each cluster is a 3 × 3 sub-network in which the cluster agent is located in the central node. The network shown in Fig. 4a includes 12 scattered faulty links that is rather a high fault percentage (20 %). The load-latency diagrams for this network are shown in Fig. 4b, c, under the local uniform and hot-spot traffic patterns, respectively. The local uniform traffic includes random local traffic in each cluster, and the hot-spot traffic also includes four hot-spot central nodes that each one receives 5 % more traffic compared to other nodes in each cluster. According to these figures, the proposed agent-based architecture leads to higher performance and saturation points compared to the method introduced in paper [4].

6 Conclusions

In this paper, an enhanced fault-tolerant NoC architecture is proposed based on low cost hierarchical agents beneficial for large NoC-based systems. Besides, different types of fault information are classified in addition to the congestion information for distribution in the network through the hierarchical agents. Each level of agent hierarchy collects, combines and distributes a specific type of fault or congestion information. Lower packet latencies are achieved by utilizing intra cluster information. Based on the simulation and synthesis results, the proposed architecture along with the appropriate routing method enhances the network performance with a small hardware overhead. In future, the utilization of inter clusters information will be investigated as global fault or congestion information.

References

1. Tsai, W., Zheng, D., Chen, S., Hu, H.Y.: A fault-tolerant NoC scheme using bidirectional channel. In: Proceedings of 48th Design Automation Conference (DAC), pp. 918–923 (2011)
2. Jantsch, A., Tenhunen, H.: Network on Chip. Kluwer Academic Publishers, Boston (2003)
3. Xu, C.T., Liljeberg, P., Tenhunen, H.: An optimized network-onchip design for data parallel FFT. Proc. Eng. **30**, 313–318 (2012)
4. Valinataj, M., Mohammadi, S., Plosila, J., Lijeberg, P.: A fault tolerant and congestion-aware routing algorithm for networks-on-chip : In: Proceedings of 13th IEEE Symposium on Design and Diagnostics of Electronic Circuits and Systems (DDECS), pp. 139–144 (2010)

5. Valinataj, M., Mohammadi, S., Plosila, J., Liljeberg, P., Tenhunen, H.: A reconfigurable and adaptive routing method for fault-tolerant mesh based networks-on-chip. Elsevier, Int. J. Electron. Commun. (AEÜ) **65**(7), 630–640 (2011)
6. Guang, L., Nigussie, E., Rantala, P., Isoaho, J., Tenhunen, H.: Hierarchical agent monitoring design approach towards self-aware parallel systems-on-chip. ACM Trans. Embed. Comput. Syst. **9**(3), article 25 (2010)
7. Guang, L., Yang, B., Plosila, J., Latif, K., Tenhunen, H.: Hierarchical power monitoring on NoC—a case study for hierarchical agent monitoring design approach. In: Proceedings of 28th NORCHIP Conference (2010)
8. Faruque, A.M., Jahn, J., Ebi, T., Henkel, J.: Runtime thermal management using software agents for multi- and many-core architectures. IEEE Des. Test Comput. **27**(6), 58–68 (2010)
9. Feng, C., Lu, Z., Jantsch, A., Li, J., Zhang, M.: FoN: fault-on neighbor aware routing algorithm for networks-on-chip. In: Proceedings of 23th IEEE International System-on-Chip Conference (SOCC), pp. 441–446 (2010)
10. Valinataj, M., Liljeberg, P., Plosilo, J.: Reliable on-chip network design using an agent-based management method. In: Proceedings of 19th International Conference on Mixed Design of Integrated Circuits and Systems (MIXDES), pp. 447–451 (2012)
11. Saokar, S.S., Banakar, M.R., Siddamal, S.: High speed signed multiplier for digital signal processing applications. IEEE International Conference on Signal Processing, Computing and Control (ISPCC), pp. 1–6 (2012)
12. Kang, H.Y., Kwon, J.T., Draper, J.: Fault-tolerant flow control in on-chip networks. In: Proceedings of 4th ACM/IEEE International Symposium on Networks-on-Chip (NOCS), pp. 79–86 (2010)
13. Zou, Y., Pasricha, S.: NARCO: neighbor aware turn model-based fault tolerant routing for NoCs. IEEE Embed. Syst. Lett. **2**(3), 85–89 (2010)

Author Index

L.C. Jain et al. (eds.), *Computational Intelligence in Data Mining - Volume 3*,
Smart Innovation, Systems and Technologies 33, DOI 10.1007/978-81-322-2202-6

K

L

M

N

O

P

R

S

Printed in the United States
By Bookmasters